CRITERIA AND METHODS OF STRUCTURAL OPTIMIZATION

DEVELOPMENTS IN CIVIL AND FOUNDATION ENGINEERING

Also in this series:

In preparation:

Brandt AM ed: Foundations of Optimum Design in Civil Engineering

CRITERIA AND METHODS OF STRUCTURAL OPTIMIZATION

Andrzej M. BRANDT, Wojciech DZIENISZEWSKI,
Stefan JENDO, Wojciech MARKS,
Stefan OWCZAREK, Zbigniew WASIUTYŃSKI

edited by
Andrzej M. BRANDT

translated from Polish by
Antoni POL

1986 **MARTINUS NIJHOFF PUBLISHERS**
a member of the KLUWER ACADEMIC PUBLISHERS GROUP
DORDRECHT / BOSTON / LANCASTER

PWN—POLISH SCIENTIFIC PUBLISHERS
WARSZAWA

Distributors:
for the United States and Canada

Kluwer Boston, Inc.
190 Old Derby Street
Hingham, MA 02043
USA

for all other countries

Kluwer Academic Publishers Group
Distribution Centre
P.O. Box 322
3300 AH Dordrecht
The Netherlands

for Albania, Bulgaria, Chinese People's Republic, Cuba, Czechoslovakia, German Democratic Republic, Hungary, Korean People's Democratic Republic, Mongolia, Poland, Rumania, the U.S.S.R., Vietnam and Yugoslavia

Ars Polona
Krakowskie Przedmieście 7
00-068 Warszawa
Poland

Library of Congress Cataloging in Publication Data CIP

Kryteria i metody optymalizacji konstrukcji.
English.
Criteria and methods of structural optimization.
(Developments in civil and foundation engineering; v. 1)
Translation of: Kryteria i metody optymalizacji konstrukcji.
Bibliography: p.
Includes index.
1. Structural design—Mathematical models.
2. Structural design—Data processing. I. Brandt, Andrzej Marek. II. Title. III. Series.
TA658.2.K7913 624.1′7 81-14046
AACR2

ISBN-13: 978-94-010-7015-7 e-ISBN-13: 978-94-009-1159-8
DOI: 10.1007/978-94-009-1159-8

Contents

Preface

This book is intended to serve all those who are interested in structural optimization, whether they work in this field or study it for other purposes. Rapid growth of interest in the cognitive aspects of optimization and the increasing demands that the present day engineer has to meet in modern design have created the need of a monographic treatment of the subject. The vast number and wide range of structural optimization problems formulated and investigated in the last twenty years call for an attempt to sum up the present state of knowledge in this domain and to outline the directions of its further development. The present authors undertook this task, hoping that the result would stimulate further work towards finding new methods and solutions and increasing the range of applications of the optimization methods to structural design. The immediate aim of the book is to present the basic criteria and methods of optimization and to provide a reference guide to the most important publications in the field.

The book consists of fourteen chapters.

Chapter 1 introduces the basic concepts, definitions and assumptions relating to structural optimization.

Chapter 2 gives the foundations of optimization for minimum elastic strain potential or maximum rigidity, and sets a basis for optimization of bar, plate and lattice structures.

Chapter 3 presents criteria of strength design and their applications to plane structures.

Chapter 4 deals with the minimum volume, weight and cost designs of elastic structures and structures in limit states; a probabilistic approach to the reliability-based optimum design problem is also discussed. This closes Part I of the book, devoted to the optimality criteria.

Part II is concerned with the methods of optimization. Chapter 5 introduces the background of mathematical programming and sets the stage for the following four chapters, 6, 7, 8 and 9, in which the methods of linear, non-linear, dynamic, and stochastic programming are discussed in some detail. The prominent part played by these methods in optimization makes them a necessary element of the structural engineer's education.

The next two chapters, 10 and 11, deal with the classical and non-classical variational methods from the viewpoint of their applications to optimum structural design. Chapter 12 presents the fundamentals of the extremum problem theory. This and the preceding chapter show some of the modern trends in the theory of optimization, which hold out prospects of fruitful application to structural engineering design. Both chapters are somewhat more mathematical in form than the other chapters of the book.

Chapter 13 gives a survey of iterative, numerical and experimental methods used in the shape-optimization of structures and their components.

The account of the mathematical methods given in Part II cannot be regarded as a self-sufficient and comprehensive treatment of the subject; it should rather be viewed as an encyclopaedic guide, which, by providing the basic information on the most widely used or the most effective methods and referring the reader to more specialized books or articles, facilitates further study.

Part III of the book contains a thematic review of the literature on structural optimization (Chapter 14) and a chronological bibliography of the subject, comprising the most important works in optimization, either books or articles in generally available journals. The mathematical and mechanical references are given in page footnotes.

The choice of material and scope of the book has resulted from various circumstances. In the first place, it expresses the personal bias of the authors, whose training, research experience and interests relate most closely to building and civil engineering design. The authors' attempt to give an account as complete as possible of the criteria and methods of structural optimization has been prompted by the lack of an adequate presentation of the subject in the literature. In order to keep the size of the book within reasonable limits, however, it has been necessary to restrict the number of particular solutions and examples. Many mathematical methods relating to optimization in other fields of engineering than those mentioned above are not considered. Problems which at the present time can only be of historical interest have also been omitted. Information about the early solutions in structural optimization can be found in well-known books on the history of mechanics and in the original papers listed in the bibliography. On the other hand, some of the methods presented have not been used in structural optimization so far but are included because of their theoretical importance and forseeable applicability in the near future.

The book includes many of the authors' original results, partly published in professional journals. Some fragments of the book, therefore, may appear to the reader to be treated in more detail than the rest. Topics lying outside the personal research experience of the authors are presented on the basis of fundamental publications and the latest available results. This duality of approach has sometimes made it difficult to expound the material in a unified and coherent manner. Some approaches and examples were omitted because of the lack of space. The self-taught student of structural optimization should bear in mind that the present book can only serve as an introduction to the original work.

Seeking an optimum design for a structure does not consist merely in applying a set of mathematical criteria and methods relating, to a lesser or greater extent, to the economics of the project, for example to material saving or the cost of construction and use of the product. Cost-effectiveness is merely one of the motives of optimization. In reality, optimization embraces a much wider range of problems and aims, which follow from the necessity to satisfy the material needs of society, in the broad sense, and the desire to achieve aesthetic values. These two tendencies, wrote Professor Zbigniew Wasiutyński, are natural for all human activity. Optimization of the shape of a structure, for example, yields a greater load capacity for a given amount of material or a given cost; if need be, concepts such as minimum deformability or maximum durability can be introduced. The problem can be reversed, and one can seek the least-cost forms of the structure for a given load capacity, deformability or durability. The system of internal forces acting in a structure satisfying the optimality conditions is ordered. By avoiding or reducing concentrations of forces, local stress points and unnecessary shape discontinuities, by an appropriate choice of the internal force systems for the best transmission of the applied loads to the bearings and foundations, and by designing the shape of structural members for the best transmission of forces, the engineer obtains structures answering the natural aesthetic requirements. Undertaking and solving the optimum design problems so understood is thus motivated, in part at least, by the psychological and the philosophical outlook of the designer.

In spite of the intense development and the increasing scope of optimization methods, their use in the practice of structural design is not yet sufficiently widespread. Although examples of their successful application can be shown in every field of engineering design and planning, the shapes of most structures continue to be chosen intuitively. One of the ways of propagating the optimum-seeking approach is to present the methods of optimization in an intelligible form, convincing to the practising designer. This thought has also been one of the guiding principles which the authors adopted in writing this book.

The present book has grown out of the many years' work of Professor Zbigniew Wasiutyński and his pupils and collaborators. Professor Wasiutyński's interest in the maximum-strength structural design dated back to 1936. He formulated the fundamental theorems of minimum elastic strain potential design and strength design, and solved many special problems. His illness and sudden death in January 1974 foiled his taking a greater part in the preparation of the first Polish edition of the book in 1977.

The idea of writing the present book first arose at the Summer School on Structural Optimization organized by the Polish Academy of Sciences in 1968 in Jabłonna near Warsaw, for which a number of papers had been prepared. The next stage was a project sponsored by the Electronic Computation Centre of Building Industry. The goal of the project was to produce a uniform and systematic account of the foundations of structural optimization, in a form suitable for implementation in building design. The project was completed in 1972. The typescript of the first Polish edition of the book was ready in 1974.

The present, English edition differs from the original one in having many additions and modifications. Additions have been made to all chapters, introducing the most important results in structural optimization obtained within the last 6 years. Chapters 5, 8 and 9 are new. The bibliography has also been considerably extended. However, the general idea and the character of the book have remained unchanged.

The division of work among the authors reflects, to some extent, their individual interests. Professor A. M. Brandt prepared Chapter 1, Sections 4.1, 4.2. 4.6 13.7, 13.8, Chapter 14 and Bibliography, coordinated the work of all the authors and edited the whole of the book. Dr W. Dzieniszewski wrote Sections 2.1, 2.3–2.7 and Chapter 11, Dr S. Jendo Section 4.3 and Chapter 7, Dr W. Marks Chapters 3, 6 and 10, Dr S. Owczarek Sections 4.4 and 4.5 and Chapter 12. Chapters 5, 8 and 9 were prepared jointly by Dr S. Jendo and Dr W. Marks. Professor Z. Wasiutyński was the author of Sections 2.2 and 13.1–13.6.

The criteria and methods presented in this book are further developed in another monograph, prepared by an enlarged team of authors, in which applications to optimum structural design are studied in detail. The monograph, laid-out in subject order and furnished with a large number of examples and tables, is accomodated to direct use in the design of real structures.[1]

The authors express their thanks and appreciation to Professor S. Łukasiewicz and Professor M. Życzkowski for their kind and discriminating criticism, helpful suggestions, comments and additions, which in many ways improved the book.

---

[1] *Foundations of optimum design in civil engineering*, ed. by A. M. BRANDT; to appear.

The authors of the book are deeply indebted to the Translator for his personal contribution in improving the original Polish text before and during translation.

Fully aware of the limitations and deficiencies of this book, the authors will gratefully welcome all comments and suggestions from the readers.

ANDRZEJ M. BRANDT

Warsaw, June 1980

PART I

CRITERIA OF STRUCTURAL OPTIMIZATION

1

Aims, Basic Concepts and Assumptions

1.1 The aims and scope of optimization

Structural optimization is a part of mechanics and the earliest ideas of optimum design found in the works of Galileo stemmed from considerations of the bending strength of beams. Bernoulli, Lagrange, Navier, and other great scientists sought the 'best' shapes for structural elements to sastisfy strength requirements. Further development of optimization, concerned with more complex structures such as trusses, frames and plates, was also based on applied mechanics.

These origins have determined the concepts and solutions used in optimum-structural-design problems. They have been formulated in close relation to the ideas and achievements of mechanics—starting from the hypotheses of cross-sectional stress distribution in bent elements, through various strength hypotheses, up to the applications of rheology and material fatigue laws.

Optimization is increasingly applied not only in structural design but also, for example, in the design of electronic equipment and planning of chemical processes. Transport and road network problems are solved by finding extrema of certain functionals. Optimization is applied with success in the planning of technological processes, solving strategic problems of economy and war, in town and country planning, and in many other fields.

The problems of optimization dealt with in this book concern the choice of geometric parameters and strength properties of structural elements and whole structures. The choice aims at reaching design solutions which would be optimal in the sense of criteria set by the requirements of saving material and labour, minimizing maintenance costs, correctly arranging the internal forces system or other factors considered decisive for the value of the structure. Often, several criteria have to be reckoned with simultaneously.

Bringing optimization techniques openly and consistently into design practice, makes it possible for a designer to produce optimal structures even if he does not have an exceptional intuition or extraordinary talent. A design process has always been a process of choice in which the designer's intuition and experience played an important role. Designers and builders are rated not only according to the reliability and performance of the structures they produce but also according to their skill in reaching specific goals by the least means, or—in achieving the 'best' results given limited means.

The number of structures built nowadays, their cost and social importance, have increased so much that it has become impossible to rely exclusively on the intuition and experience of engineers. The need to create objective methods of effective optimum design has been the primary cause of the rapid development of structural optimization observed in the last few decades.

The methods of solving optimum-design problems have changed with the development of mathematical methods. Application of mathematical optimization to structural design has shown new ways of formulating optimum problems and provided new methods of solving them. Indeed, the effectiveness of the optimum-seeking solutions of the increasingly complex design problems is strictly dependent on the use of new mathematical tools. The advent of the digital computer has given rise to a whole new field of mathematical programming, without which modern structural optimization is unthinkable.

The approach to optimization methods adopted in this book differs from that used in mathematical texts. Less emphasis is placed on the strictness and completeness of proofs and more on the effectiveness of methods and their applicability to optimum-structural-design problems. Approximate solutions, easier to derive and cheaper than exact solutions, often prove satisfactory, leading to near-optimum structures.

In spite of the fact that methods and even criteria of optimization are fundamentally interdisciplinary in character and can be used in widely diverse disciplines, they are not indifferent to the context of a particular discipline. Information specific to a given discipline, for example that certain parameters are constant or that other parameters are subject to obvious constraints, often makes it possible to reduce very considerably the analytical or numerical work necessary to obtain the solution. In this sense, the present book, by its choice of optimality oriteria, methods of solutions and numerical examples, leans towards optimization of building and civil engineering structures.

1.2 Basic concepts and definitions

Optimization is a domain of knowledge concerned with methods of choosing optimum solutions of any problems, for example technological or economic. Structural optimization, in particular, involves the choice of shapes and physical properties of structures. The choice consists in finding the optimal solution according to a prescribed criterion.

By the *shape* of a structure we shall understand not only its geometrical characteristics but also all its other properties, for example strength and strain properties. Instead of *optimization*, therefore, we shall sometimes use the term *shaping*.

Shape parameters or *design variables* are the quantities which we use to specify the shape of the structure. As examples, we can mention the following groups of geometrical parameters:

— cross-sectional shape, e.g. of a bar,
— cross-section variation along the bar axis,
— boundary shape of a member, e.g. of a plate or a shell,
— thickness or depth of a member, e.g. of a plate or a shell,
— spatial distribution of bars and joints, e.g. in a truss,
— cross-sectional area and distribution of cross-sectional components, e.g. ordinary or prestressing reinforcement,

and some groups characterizing the material of the structure:

— moduli of deformability,
— plasticity limits and strength,
— mass density,
— coefficients of heat transfer or permeability.

If $x_1, x_2, \ldots, x_n$ are the selected design variables for a given structure, then any design of this structure is represented by a vector $\mathbf{x} = x_i$ and can be regarded as a point in an abstract n-dimensional space. We call this space the *design space*.

In any optimum-design problem, the design variables are subject to various *constraints*, or restrictions which have to be satisfied in order for the design to be acceptable. They follow from *a priori* assumptions concerning the kind and the material of the structure, its geometry, strength, deformability, etc., relating to the function of the structure, its performance and also the method of construction. Common examples of constraints are limitations on displacements or stresses, prescribed variation or constancy of various geometrical parameters, or bounds on these parameters. Most often, constraints take the form of inequalities; more rarely, they are expressed by equalities. In the design space, constraints delimit a subspace of admissible solutions, which is termed the *feasible subspace*.

In the simplest case, an *optimal design* can be defined as follows.

Let x_i, $i = 1, \ldots, n$, be the design variables selected for a given structure and let the relations

$$g_j(\mathbf{x}) = 0, \quad j = 1, 2, \ldots, \bar{j},$$
$$h_k(\mathbf{x}) \leqslant 0, \quad k = 1, 2, \ldots, \bar{k},$$

where $\mathbf{x} = (x_1, x_2, \ldots, x_n)$, be the constraints imposed on these variables.

The feasible design subspace Q consists of all points $\mathbf{x}$ satisfying the above relations.

Now choose a function (or a functional) F defined on Q and expressing a certain property of a design, important from the viewpoint of realization or utilization of the structure to be designed. The function F will be called the *optimality criterion* or the *objective function.* A design $\mathbf{x}_0 \in Q$ is optimal with respect to the criterion F if $F(\mathbf{x}_0)$ is a minimum.

The choice of the optimality criterion determines the logical sense of the optimization problem thus formulated, its applicability and prospects of solution. Let us mention here three general groups of criteria which are considered in subsequent chapters.

The first group are the minimum-cost criteria. The purpose they serve reflects the natural tendency of man to satisfy his needs at a minimum expenditure of resources. The cost of a structure can be defined in various ways. In the simplest approach the cost can be assumed to be proportional to the volume or weight of the material used. Hence follow the minimum-weight and the minimum-volume optimality criteria, often encountered in the design practice. The calculated cost can also be made to include the building and operating costs, which complicates the problem but brings it closer to the real conditions. In any case, however, simplifications are unavoidable and the defined cost has to be regarded only as a substitute related more or less closely to the real cost of the structure expressed in monetary units.

The second group comprises the maximum-rigidity or the minimum-deformability criteria. Structures designed according to these criteria have their geometrical and mechanical properties shaped for minimum work of the loads on the displacements of the structure. Since this work can be expressed in terms of an energy potential, the least-deformability criterion can be reduced to a minimum-potential criterion. First formulated by Wasiutyński in 1939, the minimum-potential criterion has since been applied to many problems of optimum design of various engineering structures.

Criteria of the third group postulate equalization of various kinds of stresses or their functions throughout the structure. Depending on the strength hypothesis adopted, the design may aim at equalization of normal stresses, principal stresses, certain functions of stresses, or unit strain potentials. An optimal design is then one in which the stress of a given kind under a given load is equal at all those points of the structure at which this is possible. This last limitation follows from the conditions of execution and use of the structure, including the obvious requirement that the dimensions of the structure must be finite.

Sometimes optimality criteria belonging to different groups can be shown to be consistent or even to lead to the same results. Iu several structural problems more than one criterion should be considered simultaneously and the vast domain of multidimensional optimization is open.

Optimum design should not be confused with *dimensioning*, in which the objective is to determine the dimensions of a structure in accordance with the design code regulations. The optimum properties sought by optimization should be consistent with the design code but do not necessarily follow from it. Optimization precedes dimensioning in the design process. The properties of the structure determined by optimization can subsequently be rectified by dimensioning to satisfy the requirements not included in the optimization problem either as objective functions or as constraints.

Let us also note that a structure which is optimal in a certain limit state is not, in general, optimal in other limit states or any states prior to the given limit state.

1.3 Preliminary assumptions

The problems of structural optimization discussed in subsequent chapters are formulated and solved under various simplifying and limiting assumptions. These are presented in detail in the appropriate contexts; here let us outline their general classification.

Material assumptions

The basic assumptions concerning the properties of the materials to be used in a structure relate to the admissible ranges of stresses and deformations in that structure. In the optimization of elastic structures, deformations are assumed to be reversible and proportional to stresses. Plasticity conditions and flow laws determine the materials of structures in which the optimal shape is associated with plastic effects and limit states.

Optimization problems can be formulated and solved under different assumptions concerning the internal structure of the material used. Most solutions obtained so far apply to continuous, homogeneous and isotropic media, but optimum-design studies of structures made of non-homogeneous and anisotropic materials have also been pursued in the recent years. In fact, it is inhomogeneity in the internal structure which is the sought-for function in composite material optimization problems.

Loading assumptions

In seeking the optimum shape of a structure, the designer has to make various assumptions regarding the applied loads. Some of them specify the order or simultaneity of application of different loads, their magnitudes, directions, points of application, etc. Thus, a given structure can be considered in several different loading states.

Another group of assumptions define the conditions of load application. In particular, they are concerned with the regions of transmission of concentrated loads and with continuous load distributions. If local effects are outside the scope of a given optimization problem, the loads are usually assumed not to be concentrated in very small regions. In problems in which the distribution of continuous loads is of secondary importance, it is convenient to assume that they are distributed uniformly.

Problems involving dynamic, moving or repeated loads require accurate assumptions specifying the parameters of the loads.

Assumptions of the kinds mentioned above aim at simplifying the real loading conditions so as to facilitate the solution or to make it actually possible.

Dimensional assumptions

In order to simplify both the analysis and the optimization of a structure, it is usually assumed that its three-dimensional strain and stress fields can be reduced to two- or one-dimensional fields, or that the structure itself can be reduced to a two- or one-dimensional system.

Reducing the number of dimensions of the strain and stress fields means neglecting their variation in one or two coordinates, and therefore replacing the real fields by approximate ones. Reduction in the number of dimensions of the structure itself has the same significance.

For exact solutions, three-dimensional structures and fields have to be considered without such simplifying assumptions.

Probabilistic assumptions

All variables involved in an optimization problem, whether they characterize the mechanical properties of the materials, applied loads, sizes, support conditions or internal joints, are random variables. Therefore constraints, objective functions and unknown parameters are also random functions and should be given or determined in terms of probability distributions. In most considerations, however, all these variables are treated as deterministic quantities, the simplification resulting from such an approach being considered admissible and having no essential effect on the solution. But even in those rare cases where a probabilistic approach is followed throughout and the random nature of some variables is taken into account, various simplifying assumptions regarding probability distributions have to be made if the solution is to be reached without excessive complications.

2

Optimization for Minimum Potential Energy, Maximum Rigidity and Minimum Deformability

2.1 The optimality criterion

The criterion of maximum rigidity against deformations, or of minimum deformability, is a mechanical criterion of optimization of elastic structures designed to carry specific loads. A structure whose mechanical and geometrical properties are shaped to satisfy this criterion is the least deformable among admissible solutions, i.e. the work performed by the applied loads on the displacements of the structure has a minimum value. Since the work of loads on the displacements (translations or rotations) of an equilibrium configuration is equal to the elastic strain energy (potential), this energy is also minimum in such structures. Hence the term the *minimum-potential criterion* is often used instead of the *least-deformability criterion*. The criterion was first formulated by Wasiutyński in 1939 with reference to the problem of shape of free unloaded surfaces of homogeneous isotropic bodies. Since then, the minimum-potential criterion has proved widely applicable in the design of engineering structures of various kinds (cf. Chapter 14 and Bibliography).

Formulation and solution of a minimum-potential design problem for deformable structures comprises a number of operations. They involve

- formulation of preliminary assumptions appropriate to the given problem and determination of the set of admissible structures,
- analysis of the influence of different design variables on the deformability of the structure,
- derivation of the necessary and sufficient conditions for the least deformability of the structure from the relations provided by that analysis,
- determination of the unknown optimum variables and the stress and strain fields of the optimum structure from the least-deformability conditions and deformation equations.

The preliminary assumptions mentioned above concern the kind of the applied loads, dead-weight distribution, support and boundary conditions, structural joints and other data regarded as given in the optimization process.

The set of admissible structures in determined by these assumptions and by constraints. The latter may concern the configuration of structural members, their mechanical and geometric properties and the properties of the materials used.

The necessity of defining the set of admissible structures follows from two premises of different nature. First of all, real structures with optimum strain fields are always deformable and cannot have an infinite rigidity against deformation; their dimensions are finite and their material is deformable. Furthermore, some of the quantities describing the properties of the structure, or the structure itself, should satisfy various conditions implied by technological, executive, functional, economic or other requirements. Not always can they be simply unknowns in the optimization problem.

Thus the sense in which an optimum-strain-field structure is least deformable is limited by the conditions determining the feasible subspace from which this structure is chosen.

Determination of the least-deformability parameters, the stress and strain states and the displacements of optimum structures falls within the domain of the mechanics of structures. Problems of this kind can be solved by analytical or computer-based numerical methods.

The present chapter illustrates the application of the minimum-potential criterion as an optimality criterion for structural design. Section 2.2 presents the original formulations of Wasiutyński's theorems on shaping for minimum potential as given in his papers of 1939. Sections 2.3 and 2.4 follow Wasiutyński's approach in a more precise manner, in dealing with the problem of determining the optimum shape of free surfaces of spatial solid structures. Sections 2.5 and 2.6 give a very general formulation of the optimum-design problems for plane bar structures (prestressed or not) and for plate structures. The formulae and relations presented there provide a basis for solving optimum-design problems in specific cases. Finally, Section 2.7 gives a general formulation and solution of the problem of optimization of material distribution in systems with regular internal structures, such as frames and lattices.

2.2 Wasiutyński's theorems[1]

Reduction of potential through reinforcement of the structure

Consider an arbitrary structure with material volume V in elastic equilibrium. Let the components of the external forces per unit area of the surface $\mathring{A}$ they act on be $\mathring{X}_i$ $(i = 1, 2, 3)$. Neglect the body forces.

---

[1] In this chapter an original formulation of Wasiutyński's theorems is given. Because of its historical value the way of presentation of proofs and the symbols used are presented in their primary form.

Consider three states of the system:

state I —the original one,
state II —in which the structure is reinforced on a surface ΔA and subjected to the same external forces as originally,
state III—in which the original structure is subject to the forces equivalent to the reaction of the reinforcing element.

Forces, displacements, strains and potentials occurring in these three states are set together in Table 2.1.

Table 2.1

	Original structure with volume V		New element with volume ΔV
	Surface $\mathring{A}$	Surface ΔA	Surface ΔA
	I. Original structure		
Surface forces	$\mathring{X}$		
Displacements	$\mathring{u}_i$	u_i	
Strains	ε_{ij}		
Potential	$U' = \frac{1}{2}\int_{\mathring{A}} \mathring{X}_i \mathring{u}_i d\mathring{A}$		
	II. Reinforced structure		
Surface forces	$\mathring{X}$	$-X_i$	X_i
Displacements	$\mathring{u}'_i$	u'_i	u'_i
Strains	ε'_{ij}		ε'_{ij}
Potentials	$U' = \frac{1}{2}\int_{\mathring{A}} \mathring{X}_i \mathring{u}'_i d\mathring{A}$		$\Delta U_{\Delta V} = \frac{1}{2}\int_{\Delta A} X_i u'_i dA$
	III. Original structure under the action of the new element		
Surface forces		$-X_i$	
Displacements	$\mathring{u}'_i - \mathring{u}_i$	$u'_i - u_i$	
Strains	$\varepsilon'_{ij} - \varepsilon_{ij}$		

In the first state, the original structure is subject to the external forces $\mathring{X}_i$ acting on the surface $\mathring{A}$. Under this action, points of $\mathring{A}$ undergo displacements $\mathring{u}_i$ and those of ΔA displacements u_i. The strain tensor in this state will be denoted by ε_{ij} and the potential stored in the structure by U.

To reach the second state, we first unload the system, then reinforce it by introducing a new element with arbitrarily small volume ΔV, coupled with the original structure through the surface ΔA, and load it again with the same external forces $\mathring{X}_i$, producing displacements $\mathring{u}_i'$ and u_i' in the surfaces $\mathring{A}$ and ΔA, respectively. The potential U' accumulated in the structure in this case will in general be different from the original potential U. Every point of ΔA carries forces of interaction between the original structure and the new element. Denote the forces exerted by the original structure on the new element by X_i and the reaction of the new element by $-X_i$. The forces X_i and the displacements u_i' have the same directions; $-X_i$ have opposite directions. The strain tensor of the system in this state will be denoted by ε_{ij}'.

In the third state, the original structure is assumed to be acted upon only by the forces $-X_i$ across the surface ΔA. According to the previous notation, the surface $\mathring{A}$ now undergoes the displacements $\mathring{u}_i'-\mathring{u}_i$ and the surface ΔA the displacements $u_i'-u_i$. The strain tensor in this state is $\varepsilon_{ij}'-\varepsilon_{ij}$.

The reciprocity of the deformations in states I and III implies the relation

$$\int_{\mathring{A}} \mathring{X}_i(\mathring{u}_i'-\mathring{u}_i)\mathrm{d}\mathring{A} = -\int_{\Delta A} X_i u_i \mathrm{d}A.$$

Hence

$$\Delta U = U'-U = \tfrac{1}{2}\int_{\mathring{A}} \mathring{X}_i\mathring{u}_i'\mathrm{d}\mathring{A}-\tfrac{1}{2}\int_{\mathring{A}} \mathring{X}_i\mathring{u}_i\mathrm{d}\mathring{A} = -\tfrac{1}{2}\int_{\Delta A} X_i u_i \mathrm{d}A. \tag{2.1}$$

It can be seen that the increment ΔU is negative, and so we come to the following

THEOREM 1. *Addition of a new, active* (i.e. $u_i \neq 0 \neq X_i$) *element reduces the elastic strain potential of the system. The greater are the displacements u_i of the original structure on the surface contiguous to the new element and the stronger are the reactions X_i of the structure to the new element, the greater is the reduction of the potential.*

Let us compare the potential change due to reinforcement with the potential of the new element, $\Delta U_{\Delta V}$. If the volume ΔV of the new element is arbitrarily small, the displacements of the points of the surface ΔA after the reinforcement u_i' will differ arbitrarily little from the displacements of the same points before the reinforcement u_i. Indeed, the displacements u_i' are continuous functions of ΔV and are equal to u_i when $\Delta V = 0$. Hence

$$-\Delta U = \tfrac{1}{2}\int_{\Delta A} X_i u_i \Delta A = \tfrac{1}{2}\int_{\Delta A} X_i u_i' \Delta A + \eta \Delta V, \tag{2.2}$$

with $\eta \to 0$ as $\Delta V \to 0$. The integral on the right-hand side represents the principal part of the potential decrease due to the reinforcement of the structure. This leads us to the following theorem:

THEOREM 2. *The elastic strain potential of a structure reinforced by an additional, arbitrarily small element is reduced by a value equal to the poten-*

tial contained in the new element; the potential of the original part of the structure is decreased by the double value of the potential of the new element.

Example. A bar with cross-section A_1, reinforced by a cross-section A_2 is in tension under a force P. The potentials before and after the reinforcement are

$$U = \frac{1}{2}\frac{P^2 l}{A_1 E}, \qquad U' = \frac{1}{2}\frac{P^2 l}{(A_1+A_2)E},$$

and hence the potential increment is

$$\Delta U = U' - U = -\frac{1}{2}\frac{P^2 A_2 l}{A_1(A_1+A_2)E}.$$

The force acting in the new element is

$$P\frac{A_2}{A_1+A_2},$$

and the potential of the new element is

$$\Delta U_{\Delta V} = \frac{P^2 A_2 l}{2(A_1+A_2)^2 E}.$$

If the area A_2 approaches zero, the potential of the new element and the absolute value of the potential increment tend to the same limit. Indeed,

$$\lim_{\Delta V\to 0}\frac{\Delta U}{\Delta V} = -\lim_{\Delta V\to 0}\frac{\Delta U_{\Delta V}}{\Delta V},$$

$$\lim_{A_2\to 0} -\frac{1}{2}\frac{P^2 A_2 l}{A_1(A_1+A_2)E} = -\lim_{A_2\to 0}\frac{1}{2}\frac{P^2 A_2 l}{(A_1+A_2)^2 E} = -\frac{1}{2}\frac{P^2}{A_1^2 E}.$$

Dependence of the potential change on strains and stresses

The surface integrals representing the potential increments ΔU and $\Delta U_{\Delta V}$ can be transformed into volume integrals. To begin with, let us transform the integral representing the increment $\Delta U_{\Delta V}$. Applying a transformation known from the theory of elasticity, we find

$$\int_{\Delta A} X_i u_i' \mathrm{d}A = \int_{\Delta V} \varepsilon_{ij}' \sigma_{ij}' \mathrm{d}V = \int_{\Delta V} (\lambda\theta'^2 + 2\mu\varepsilon_{ij}'\varepsilon_{ij}')\mathrm{d}V,$$

and so

$$\Delta U_{\Delta V} = \tfrac{1}{2}\int_{\Delta A} X_i u_i' \mathrm{d}A = \tfrac{1}{2}\int_{\Delta V} (\lambda\theta'^2 + 2\mu\varepsilon_{ij}'\varepsilon_{ij}')\mathrm{d}V.$$

Similar transformations lead to the equations

$$\Delta U = -\tfrac{1}{2}\int_{\Delta A} X_i u_i \mathrm{d}A = -\tfrac{1}{2}\int_{\Delta V} (\lambda\theta\theta' + 2\mu\varepsilon_{ij}'\varepsilon_{ij})\mathrm{d}V$$

and

$$\Delta U = -\tfrac{1}{2}\int\limits_{\Delta A} X_i u_i \mathrm{d}A = -\tfrac{1}{2}\int\limits_{\Delta V} [\lambda\theta(\theta'-\theta)+2\mu\varepsilon_{ij}(\varepsilon'_{ij}-\varepsilon_{ij})]\mathrm{d}V.$$

Consequently, the equation

$$\Delta U = -\tfrac{1}{2}\int\limits_{\Delta A} X_i u_i \mathrm{d}A = -\tfrac{1}{2}\int\limits_{\Delta A} X_i u'_i \mathrm{d}A - \eta\Delta V$$

will take the form

$$\Delta U = -\tfrac{1}{2}\int\limits_{\Delta V} (\lambda\theta'\theta+2\mu\varepsilon'_{ij}\varepsilon_{ij})\mathrm{d}V$$

$$= -\tfrac{1}{2}\int\limits_{\Delta V} (\lambda\theta'^2+2\mu\varepsilon'_{ij}\varepsilon'_{ij})\mathrm{d}V - \eta\Delta V = -\tfrac{1}{2}\int\limits_{\Delta V} \Psi'\mathrm{d}V - \eta\Delta V,$$

where $\eta \to 0$ as $\Delta V \to 0$.

Denoting by $\Psi' = \lambda\theta'^2+2\mu\varepsilon'_{ij}\varepsilon'_{ii}$ the unit potential of the reinforced system at the point of reinforcement, we find

$$\frac{\mathrm{d}U}{\mathrm{d}V} = -\frac{1}{2}\Psi'$$

and

$$\Delta U = -\tfrac{1}{2}\Psi'\Delta V - \eta\Delta V. \tag{2.3}$$

The above equations permit us to state the following theorems:

THEOREM 3'. *The reinforcement of a structure by addition of an arbitrarily small element reduces the potential of the structure the more, the greater is the unit potential of the new element after the reinforcement.*

THEOREM 3''. *The reinforcement of a structure by addition of a new element with volume ΔV reduces the potential of the structure the more, the greater is the integral over ΔV of the product of the unit deformations before and after the reinforcement.*

If the volume ΔV of the new element decreases to zero, the unit deformations ε'_{ij} tend to the original deformations ε_{ij}, because the continuity of the function expressing the changes of the strain due to the addition of a portion of material in terms of the added volume implies that these changes are arbitrarily small if the added volume is arbitrarily small. Hence

$$\Delta U = -\tfrac{1}{2}\int\limits_{\Delta V} (\lambda\theta'\theta+2\mu\varepsilon'_{ij}\varepsilon_{ij})\mathrm{d}V$$

$$= -\tfrac{1}{2}\int\limits_{\Delta V} (\lambda\theta^2+2\mu\varepsilon_{ij}\varepsilon_{ij})\mathrm{d}V - \eta\Delta V = -\tfrac{1}{2}\int\limits_{\Delta V} \Psi\mathrm{d}V - \eta\Delta V. \tag{2.4}$$

Approximately, this can be written as

$$\Delta U = -\tfrac{1}{2}\Psi\Delta V - \eta\Delta V,$$

where the unit potential $\Psi = \lambda\theta^2 + 2\mu\varepsilon_{ij}\varepsilon_{ij}$ is taken at the point of reinforcement and η tends to zero as ΔV tends to zero. The last formula proves

THEOREM 3. *The reinforcement of a structure by addition of an arbitrarily small volume of material reduces the potential of the structure the more, the greater is the pre-reinforcement unit potential at the point of reinforcement.*

Minimum potential

If we reverse the transformation considered so far, i.e. if instead of adding a new element to the structure we remove a quantity of material from it, the resulting potential change will be given by the same formulae as before except that now it will be positive.

Subtraction of a volume ΔV of the material increases the potential of the structure the more, the greater is the unit potential Ψ at the point from which the material is removed. The formulae for the potential increment are

$$\Delta U = \tfrac{1}{2}\Psi'\Delta V + \eta'\Delta V, \qquad \Delta U = \tfrac{1}{2}\Psi\Delta V + \eta\Delta V.$$

However, if we adopt the convention that ΔV is negative when the material is removed from the structure and positive when it is added to it, the formulae for the potential increment will in either case read

$$\Delta U = \tfrac{1}{2}\Psi'\Delta V - \eta'\Delta V, \qquad \Delta U = -\tfrac{1}{2}\Psi\Delta V - \eta\Delta V.$$

THEOREM 4′. *If there are two points in the structure,* 1 *and* 1′, *such that the unit potential at the point* 1, Ψ_1, *is greater than the unit potential* $\Psi_{1'}$ *at the point* 1′, *then a transfer of a sufficiently small volume of material* $\Delta_1 V$ *from point* 1′ *to* 1 *reduces the potential of the structure.*

Indeed, if $\Psi_1 > \Psi_{1'}$, then

$$\Delta_{1'}U = \tfrac{1}{2}\Psi_{1'}\Delta_1 V + \eta_{1'}\Delta_1 V, \qquad \Delta_1 U = -\tfrac{1}{2}\Psi_1\Delta_1 V - \eta_1\Delta_1 V,$$

$$\Delta_{1'}U + \Delta_1 U = -\tfrac{1}{2}(\Psi_1 - \Psi_{1'})\Delta_1 V - (\eta_1 - \eta_{1'})\Delta_1 V.$$

If $\Delta_1 V$ is sufficiently small, the term $(\eta_1 - \eta_{1'})\Delta_1 V$ has no effect on the sign of the increment $\tfrac{1}{2}(\Psi_1 - \Psi_{1'})\Delta_1 V$. The sum of the increments $\Delta_{1'}U + \Delta_1 U$ is then negative, because $\Psi_1 - \Psi_{1'} > 0$, and so the total potential of the system is reduced.

The transformation can be applied repeatedly as long as it is possible to transfer material within the structure from points where the function Ψ takes larger values to those where it is smaller. If the indices of the former are $k = 1, 2, \ldots$, and the indices of the latter are $k' = 1', 2', \ldots$, such a volume-conserving (V = const) series of transformations will reduce the potential of the structure by the sum

$$\sum_k (\Delta_{k'} U + \Delta_k U) = -\tfrac{1}{2} \sum_k (\Psi_k - \Psi_{k'}) \Delta_k V - \sum_k (\eta_k - \eta_{k'}) \Delta_k V. \tag{2.5}$$

The process ends when the potential difference $\Psi_k - \Psi_{k'}$ between any two points which admit a material transfer from one to the other within the limits set by the conditions restricting the shape of the structure is zero. Thus, we can state the following

THEOREM 4. *Among all acceptable forms of a structure with a given volume, that has the least potential in which the unit potential takes equal values at all points between which it is possible to transfer material.*

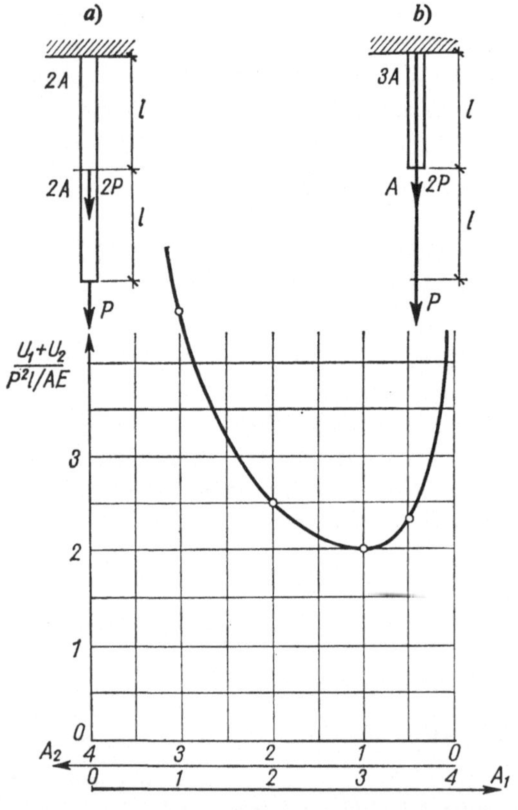

Fig. 2.1. Potential vs. cross-sectional area variation for a bar in tension

Example. Two steel bars with cross-section A each, suspended vertically, are in tension under two concentrated forces, $2P$ and P, applied at the distances l and $2l$, respectively, from the mounting of the bars (Fig. 2.1a). Let the modulus of elasticity of the steel be E.

The elongations of the upper and the lower segments of the bars are

$$\Delta_1 l = \frac{3Pl}{2AE}, \quad \Delta_2 l = \frac{Pl}{2AE}, \quad \Delta_1 l + \Delta_2 l = 2\frac{Pl}{AE}$$

and the strain potentials of these segments are

$$U_1 = \frac{9}{4}\frac{P^2 l}{AE}, \quad U_2 = \frac{1}{4}\frac{P^2 l}{AE}, \quad U_1 + U_2 = \frac{5}{2}\frac{P^2 l}{AE}.$$

The unit potentials are constant along each segment and are given by

$$\Psi_1 = \frac{9}{8}\frac{P^2}{A^2 E}, \quad \Psi_2 = \frac{1}{8}\frac{P^2}{A^2 E}.$$

The unit potential in the upper segment is 9 times that in the lower segment.

Now transform the structure by moving one of the bars of the lower segment to the upper segment as shown in Fig. 2.1b. The cross-sectional area of the upper segment is now $3A$ and that of the lower one is A. The elongation of either segment in this configuration is

$$\Delta_1' l = \Delta_2' l = \frac{Pl}{AE}.$$

The new strain potentials are given by

$$U_1' = \frac{3}{2}\frac{P^2 l}{AE}, \quad U_2' = \frac{1}{2}\frac{P^2 l}{AE}, \quad U_1' + U_2' = 2\frac{P^2 l}{AE}$$

and the unit potentials are equal and constant:

$$\Psi_1' = \Psi_2' = \frac{1}{2}\frac{P^2 l}{AE},$$

According to Theorem 4, this distribution of the unit potentials yields the minimum of the total potential of the structure for the given volume of the material.

The ratios of the quantities involved before and after the transformation are

$$\begin{gathered} \frac{\Delta_1 l}{\Delta_1' l} = \frac{3}{2}, \quad \frac{\Delta_2 l}{\Delta_2' l} = \frac{1}{2}, \quad \frac{U_1}{U_1'} = \frac{3}{2}, \quad \frac{U_2}{U_2'} = \frac{1}{2}, \\ \frac{U_1 + U_2}{U_1' + U_2'} = \frac{5}{4}, \quad \frac{\Psi_1}{\Psi_{1'}} = \frac{9}{4}, \quad \frac{\Psi_2}{\Psi_{2'}} = \frac{1}{4}. \end{gathered} \tag{2.6}$$

The transformation of the structure in the above example can also be performed in a continuous manner, by carrying arbitrarily small portions of material from one of the segments to the other, with the uniform distribution of the material along each of them preserved throughout the process. The variation of the potential of the whole structure under such a transformation is shown in Fig. 2.1. At the ends of the domain, where the cross-sectional area of the lower or the upper segment falls to zero, the potential increases unboundedly.

Equivalence of optimization for minimum potential with optimization for minimum weight

THEOREM 5. *Design for* $U = \min$ *at* $V = \text{const}$ *is equivalent to design for* $V = \min$ *at* $U = \text{const}$.

Represent the potential and the volume of the structure in the form

$$U = \int_{a_k}^{b_k} F(x_k)\mathrm{d}x_k \quad \text{and} \quad V = \int_{a_k}^{b_k} G(x_k)\mathrm{d}x_k,$$

where x_k are design variables and a_k, b_k their respective ranges, $k = 1, 2, \ldots, n$.

Euler's conditions for extremum subject to isoperimetric constraints corresponding to the minimum-potential and the minimum-weight criteria take the form

for $U = \min$, $V = \text{const}$: for $U = \text{const}$, $V = \min$:

$$\frac{\partial}{\partial x_k}[(x_k) - KG(x_k)] = 0, \qquad \frac{\partial}{\partial x_k}[G(x_k) - LF(x_k)] = 0,$$

where K and L are constants. Hence follow the systems of equations

$$\begin{array}{ll}
\dfrac{\partial F}{\partial x_1} - K\dfrac{\partial G}{\partial x_1} = 0, & \dfrac{\partial G}{\partial x_1} - L\dfrac{\partial F}{\partial x_1} = 0, \\
\cdots\cdots\cdots & \cdots\cdots\cdots \\
\dfrac{\partial K}{\partial x_k} - K\dfrac{\partial G}{\partial x_k} = 0, & \dfrac{\partial G}{\partial x_k} - L\dfrac{\partial F}{\partial x_k} = 0, \\
\cdots\cdots\cdots & \cdots\cdots\cdots \\
\dfrac{\partial F}{\partial x_n} - K\dfrac{\partial G}{\partial x_n} = 0, & \dfrac{\partial G}{\partial x_n} - L\dfrac{\partial F}{\partial x_n} = 0.
\end{array} \tag{2.7}$$

If there is an extremum of U, the first system of equations is satisfied for some values of $\partial F/\partial x_k$ and $\partial G/\partial x_k$. The second system is then satisfied by the same values of these derivatives, provided that $KL = 1$, and so both criteria lead to identical shapes of the structure.

Design for least deformability with allowance for dead weight

Theorem 6 given below is a generalization of the theorems formulated previously in that it takes into account the influence of the dead weight of a structure on its elastic strain energy. This influence becomes important when the displacements of the points of the structure under the loads for which it is designed are sufficiently large, as it happens, for example, in large-span bridges.

If we remove a volume ΔV of the material from a point k of an elastic structure subjected to a useful load, the resulting increase of the elastic strain potential of the structure will consist of two components: one—due to the disappearance of forces between the removed element and the rest of the structure, and the other—due to the disappearance of the element itself, which otherwise would undergo a displacement v_k under the applied loads and the weight of the structure; thus

$$\Delta_k U = \Psi_k \Delta V - \gamma v_k \Delta V.$$

Injection of a volume ΔV at a point k', on the other hand, reduces the potential by

$$\Delta_{k'} U = -\Psi_{k'} V + \gamma v_{k'} \Delta V.$$

A transfer of the volume ΔV from point k to k' will therefore cause a potential change of

$$\Delta U = \Delta_k U + \Delta_{k'} U = [(\Psi_k - \Psi_{k'}) + \gamma(v_{k'} - v_k)]\Delta V.$$

Note that the unit potentials Ψ_k and $\Psi_{k'}$ are produced by the useful load while the displacements v_k and $v_{k'}$ by both the useful load and the dead weight.

For the least-potential structure $\Delta U = 0$, and so

$$\Psi_k = \Psi_{k'} - \gamma(v_{k'} - v_k).$$

If the largest displacement occurs at the point k', i.e. if $v_{k'} = \bar{v}$, then

$$\Psi_k = \Psi_{k'} - \gamma(\bar{v} - v_k),$$

and $\Psi_{k'}$ is the greatest value of the unit potential: $\Psi_{k'} = \overline{\Psi}$. Hence

$$\Psi_k = \overline{\Psi} - \gamma(\bar{v} - v_k) \tag{2.8}$$

and this equality applies to all points of the structure. Relation (2.8) serves as a criterion of optimization for maximum rigidity under a useful load.

If $v_k = \bar{v}$, then $\Psi_k = \overline{\Psi}$, i.e. the greatest unit potential occurs at points which suffer the largest displacement $\bar{v}$ under the useful load. If v_k has the least possible value $\underline{v}$, then $\Psi_k = \overline{\Psi} - \gamma(\bar{v} - \underline{v})$ is minimum.

The foregoing considerations can be summed up in the following

THEOREM 6. *An elastic structure with a given volume V accumulates the least potential and has the greatest rigidity under a given useful load if equation* (2.8) *holds at all points of the structure.*

Local effect of a local reinforcement

According to Theorem 2, addition of an element with an arbitrarily small volume ΔV to the neighbourhood of a point at which the unit potential is Ψ reduces the potential of the whole system by

$$\Delta U = -\Psi \Delta V - \eta \Delta V,$$

and that of the original part of the structure by twice as much.

Divide the original structure by means of a surface S into two parts I and II in such a way that the area A, which will be in contact with the element ΔV after the reinforcement, lies entirely within part I and the distances of all its points from S are finite.

Let ε_{ijS} be the strain and σ_{ijS} the stress present on the surface S in the original structure under the given load. They can be regarded as acting in two ways: on the surface S as the boundary of part I or on the surface S as the boundary of part II.

Separate part I from part II by applying surface forces X_{iS} to the first and the forces $-X_{iS}$ to the second so as to counterbalance the stress σ_{ijS} on the surface S in each part. As a result, each of the two parts of the original structure is in equilibrium and the sum of the potentials they contain, $U_{\mathrm{I}}+$ $+U_{\mathrm{II}}$, is equal to the potential of the original structure.

Now reinforce part I, which is maintained in equilibrium by forces X_{iS}, by joining to it the element ΔV in exactly the same way as it would be added to the whole structure. According to Theorem 2, the potential of the reinforced part I will be reduced by

$$\Delta_{\mathrm{I}} U = -\Psi \Delta V - \eta_{\mathrm{I}} \Delta V,$$

The principal part of this decrement is the same as that in the potential drop which occurs in the whole structure as a result of reinforcement:

$$\Delta U = -\Psi \Delta V - \eta \Delta V.$$

The difference between the two decrements,

$$\Delta U - \Delta_{\mathrm{I}} U = -(\eta - \eta_{\mathrm{I}}) \Delta V, \tag{2.9}$$

is small of order higher than ΔV, because both η and η_1 tend to zero with ΔV. We can thus state

THEOREM 7. *Reinforcement of a structure with a new element ΔV produces changes of order ΔV in deformations only in the immediate neighbourhood of this element. The increase of deformations at finite distances from the reinforcing element is small of higher order.*

Similarly, the potential loss ΔU caused by the addition of a new element ΔV is concentrated in a differential neighbourhood of this element. The in-

crements of the potential which occur at finite distances from the new element inside the original structure, are small of higher order.

The above theorem explains the distribution of deformations in the neighbourhood of notches. If the unit potential Ψ_1 in the neighbourhood of a notch with volume ΔV is greater than the values of Ψ at finite distances from the notch, then by filling the notch with an element of volume ΔV one can reduce Ψ_1 to match those values. According to Theorem 7, the region of the original structure in which the changes in the unit potential are of order $\Psi_1 - \Psi$ has a volume of order ΔV. Hence follows the conclusion:

The region about a notch in which the deformation is increased has a volume of the order of the notch volume.

This conclusion can be extended to structures where notches are filled with materials less or more deformable than the material of the structure. The concentration of strain in the immediate neighbourhood of a notch can then be evened out by replacing the foreign material by the same material as that of the region surrounding the notch.

Besides increased deformations, also reduced deformations can occur in the neighbourhood of a notch. Filling the notch compensates both increased and reduced deformations. This equalization can matter a great deal when the notch divides the structure into two parts, as in a two-span beam or plate, or when it occurs between a beam or plate and a support.

2.3 Deformability of a structure and the free surface

Assumptions and subject of optimization

We shall consider a spatial structure of volume $\overset{\bullet}{V}$, made of arbitrary material which can be anisotropic and non-homogeneous but which obeys Hooke's law. The structure occupies a region $\mathring{\Omega}$ bounded by a closed surface $\mathring{S}$ with an external normal $\mathbf{n}$ (Fig. 2.2). Part S_p of the surface $\mathring{S}$ is acted upon

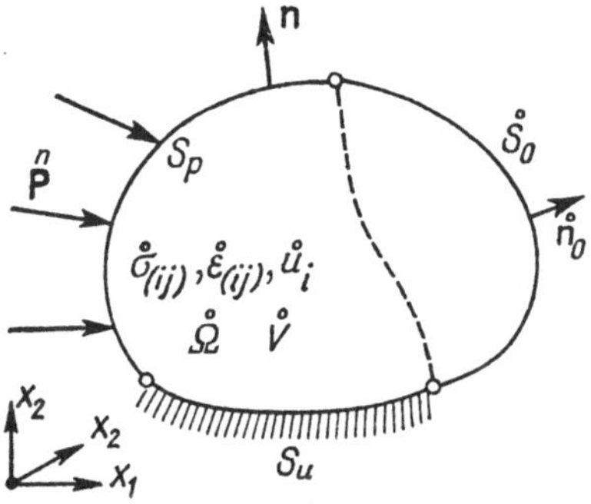

Fig. 2.2. Support and loading scheme

by surface forces having intensity $\overset{n}{\mathbf{P}}$ per unit area of S_p. Part S_u of $\mathring{S}$ is supported rigidly, i.e. the displacement vector $\mathring{\mathbf{u}}$ is zero on S_u. The remaining part of the surface of the structure, $\mathring{S}_0 = \mathring{S} - S_u - S_p$, whose normal vector will be denoted by $\mathbf{n}_0$, is free from external loads and kinematic constraints.

The forces $\overset{n}{\mathbf{P}}$ generate in the structure a stress field $\mathring{\sigma}_{(ij)}$ $(i, j = 1, 2, 3)$ which satisfies the equilibrium conditions

$$\begin{aligned} &\mathring{\sigma}_{ij,j} = 0, \quad \mathring{\sigma}_{ij} = \mathring{\sigma}_{ji} = \mathring{\sigma}_{(ij)} \text{ in } \mathring{\Omega}, \\ &\mathring{\sigma}_{(ij)} n_j = \overset{n}{P}_i \text{ on } S_p, \quad \sigma_{(ij)} \mathring{n}_{j0} = 0 \text{ on } \mathring{S}_0. \end{aligned} \tag{2.10}$$

The strain produced in the structure by these stresses is described by a strain tensor field $\mathring{\varepsilon}_{(ij)}$. The tensor $\mathring{\varepsilon}_{(ij)}$ is connected with the displacement vector $\mathring{\mathbf{u}}_i$ by the geometrical relation

$$\mathring{\varepsilon}_{(ij)} = \tfrac{1}{2}(\mathring{u}_{i,j} + \mathring{u}_{j,i}) = \mathring{u}_{(i,j)}.$$

The relationship between the stress $\mathring{\sigma}_{(ij)}$ and the strain $\mathring{\varepsilon}_{(i))}$ constitutes Hooke's law, which in the general case of an inhomogeneous anisotropic material takes the form

$$\mathring{\sigma}_{(ij)} = A_{(ij)(kl)} \mathring{\varepsilon}_{(kl)}. \tag{2.11}$$

When the material of the structure is isotropic and homogeneous, Hooke's law reads

$$\mathring{\sigma}_{(ij)} = \lambda \mathring{\varepsilon}_{kk} \delta_{ij} + 2\mu \mathring{\varepsilon}_{(ij)}.$$

Here λ and μ are Lamé's constants. The work done by the static load $\overset{n}{\mathbf{P}}$ on the displacement $\mathring{\mathbf{u}}$ is $\mathring{L} = \frac{1}{2} \oint_{S_p} \overset{n}{\mathbf{P}} \cdot \mathring{\mathbf{u}} ds_p$. According to the Clapeyron theorem, it is equal to the elastic strain energy $\mathring{U} = \frac{1}{2} \oint_{\mathring{\Omega}} \mathring{\sigma}_{(ij)} \mathring{\varepsilon}_{(ij)} d\mathring{\Omega}$ stored in the structure:

$$\int_{\mathring{\Omega}} \mathring{\sigma}_{(ij)} \mathring{\varepsilon}_{(ij)} d\mathring{\Omega} = \oint_{S_p} \overset{n}{P}_i \mathring{u}_i dS_p.$$

The unknown of our optimization problem is the shape of the free surface $\mathring{S}_0$. The optimum solution, i.e. the least deformable structure, will be sought in the set of all structures with the same volume $\mathring{V}$, made of the same materials, loaded with the same forces $\overset{n}{\mathbf{P}}$ on S_p and identically supported on S_u.

Effect of a change of the free surface S_0 on the elastic strain potential

To find out how a change in the shape of the surface affects the deformability of the structure, let us modify the surface $\mathring{S}$ by replacing its free part $\mathring{S}_0$ by a different surface S_0 with an external normal $\mathbf{n}_0$. As a result, a new structure is formed, occupying a region Ω bounded by the surface $S = S_p + S_u + S_0$ and having a volume V, in general different from $\mathring{V}$ (Fig. 2.3).

Under the same loads $\overset{n}{\mathbf{P}}$, the modified structure will undergo different deformations than the original structure. There will occur stresses $\sigma_{(ij)} \neq \mathring{\sigma}_{(ij)}$ satisfying the equilibrium equations

$$\sigma_{(ij),j} = 0, \quad \sigma_{(ij)} = \sigma_{(ji)} \text{ in } \Omega, \qquad \sigma_{(ij)} n_j = \overset{n}{P}_i \text{ on } S_p, \quad \sigma_{(ij)} n_{j0} = 0 \text{ on } S_0, \tag{2.12}$$

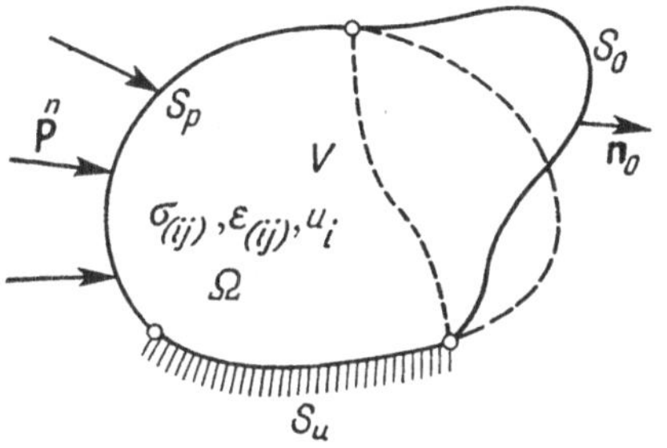

Fig. 2.3. Transformation scheme

and producing strain $\varepsilon_{(ij)}$ and displacement $\mathbf{u}_i$, which is subject to the kinematic constraints $\mathbf{u}_i = 0$ on S_u. The strain tensor $\varepsilon_{(ij)}$ and the displacement vector $\mathbf{u}_i$ are related by

$$\varepsilon_{(ij)} = \tfrac{1}{2}(u_{i,j} + u_{j,i}) = u_{(i,j)}.$$

The work done by the forces $\overset{n}{\mathbf{P}}$ on the modified structure is $L = \frac{1}{2}\oint_{S_p} \overset{n}{\mathbf{P}} \cdot \mathbf{u}\, dS_p$, which, according to the Clapeyron theorem, equals the elastic strain energy $U = \frac{1}{2}\int_{\Omega} \sigma_{(ij)} \varepsilon_{(ij)} d\Omega$.

Now consider again the original structure. Its stress and strain fields can be brought to exactly the same state as that of the modified structure by applying additional surface forces to the free surface $\mathring{S}_0$. Indeed, if we load the surface $\mathring{S}_0$ with the force $\mathring{X}_i = [\sigma_{(ij)}]_{\mathring{S}_0}^{\mathring{}} \mathring{n}_{j0}$ per unit surface area (Fig. 2.4.)

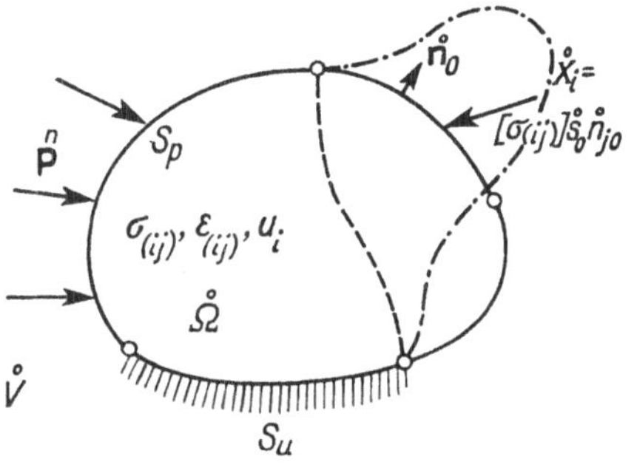

Fig. 2.4. Stress, strain and displacement in the original structure

the stress field that will appear in the structure under the loads $\overset{n}{\mathbf{P}}$ will be $\sigma_{(ij)}$, and the resulting strain tensor and displacement vector in the region $\mathring{\Omega}$ will be $\varepsilon_{(ij)}$ and u_i.

Similarly, if we load the free component $\mathring{S}_0$ of the surface S of the modified structure with the forces $X_i = [\mathring{\sigma}_{(ij)}]_{S_0} n_{j0}$ per unit surface area, then the stress, strain and displacement fields in the region Ω will be the same as in the original structure, i.e. $\mathring{\sigma}_{(ij)}$, $\mathring{\varepsilon}_{(ij)}$ and $\mathring{u}_i$, respectively.

The stress $\sigma_{(ij)}$, strain $\varepsilon_{(ij)}$ and displacement u_i in the region $\mathring{\Omega}$ follow from the solution of the boundary problem (2.12), and $\mathring{\sigma}_{(ij)}$, $\mathring{\varepsilon}_{(ij)}$ and $\mathring{u}_i$—from the solution of the boundary problem (2.10).

The potential of the original structure and that of the modified structure are given by

$$\begin{aligned} 2\mathring{U} &= \int_{\mathring{\Omega}} \mathring{\sigma}_{(ij)} \mathring{\varepsilon}_{(ij)} \mathrm{d}\mathring{\Omega} = \oint_{S_p} \overset{n}{P}_i \mathring{u}_i \mathrm{d}S_p, \\ 2U &= \int_{\Omega} \sigma_{(ij)} \varepsilon_{(ij)} \mathrm{d}\Omega = \oint_{S_p} \overset{n}{P}_i u_i \mathrm{d}S_p. \end{aligned} \tag{2.13}$$

In addition to these expressions we shall also use other equivalent forms. By the Green–Gauss theorem,

$$\int_{\Omega} A_{i,i} \mathrm{d}\Omega = \oint_{S} A_i n_i \mathrm{d}S;$$

if we use the equilibrium equations (2.10) and (2.12), we find

$$\begin{aligned} \int_{\mathring{\Omega}} \mathring{\sigma}_{(ij)} \varepsilon_{(ij)} \mathrm{d}\mathring{\Omega} &= \int_{\mathring{\sigma}} \mathring{\sigma}_{(ij)} \varepsilon_{ij} \mathrm{d}\mathring{\Omega} = \tfrac{1}{2} \int_{\mathring{\Omega}} \mathring{\sigma}_{(ij)} (u_{i,j} + u_{j,i}) \mathrm{d}\mathring{\Omega} \\ &= \int_{\mathring{\Omega}} \mathring{\sigma}_{(ij)} u_{i,j} \mathrm{d}\mathring{\Omega} = \int_{\mathring{\Omega}} [(\mathring{\sigma}_{(ij)} u_i)_{,j} - \mathring{\sigma}_{(ij),j} u_i] \mathrm{d}\mathring{\Omega} \\ &= \oint_{\mathring{S}} \mathring{\sigma}_{(ij)} n_j u_i \mathrm{d}\mathring{S} = \oint_{S_p} \overset{n}{P}_i u_i \mathrm{d}S_p, \\ \int_{\Omega} \sigma_{(ij)} \mathring{\varepsilon}_{(ij)} \mathrm{d}\mathring{\Omega} &= \int_{\Omega} \sigma_{(ij)} \mathring{\varepsilon}_{(ij)} \mathrm{d}\Omega = \tfrac{1}{2} \int_{\Omega} \sigma_{(ij)} (\mathring{u}_{i,j} + \mathring{u}_{j,i}) \mathrm{d}\Omega \\ &= \int_{\Omega} \sigma_{(ij)} \mathring{u}_{i,j} \mathrm{d}\Omega = \int_{\Omega} [(\sigma_{(ij)} \mathring{u}_i)_{,j} - \sigma_{ij,j} \mathring{u}_i] \mathrm{d}\Omega \\ &= \oint_{S} \sigma_{(ij)} n_j \mathring{u}_i \mathrm{d}S = \oint_{S_p} \overset{n}{P}_i \mathring{u}_i \mathrm{d}S_p. \end{aligned} \tag{2.14}$$

On the other hand, the physical relations (2.11) imply the equalities

$$\begin{aligned} \int_{\mathring{\Omega}} \mathring{\sigma}_{(ij)} \varepsilon_{(ij)} \mathrm{d}\mathring{\Omega} &= \int_{\mathring{\Omega}} \sigma_{(ij)} \mathring{\varepsilon}_{(ij)} \mathrm{d}\mathring{\Omega}, \\ \int_{\Omega} \sigma_{(ij)} \mathring{\varepsilon}_{(ij)} \mathrm{d}\Omega &= \int_{\Omega} \mathring{\sigma}_{(ij)} \varepsilon_{(ij)} \mathrm{d}\Omega. \end{aligned} \tag{2.15}$$

Comparing formulae (2.13), (2.14) and (2.15), we come to the following alternative expressions for the potentials of the original and the modified structure:

$$
\begin{aligned}
2\mathring{U} &= \int_{\mathring{\Omega}} \mathring{\sigma}_{(ij)}\mathring{\varepsilon}_{(ij)}\mathrm{d}\mathring{\Omega} = \int_{\Omega} \sigma_{(ij)}\mathring{\varepsilon}_{(ij)}\mathrm{d}\Omega = \int_{\Omega} \mathring{\sigma}_{(ij)}\varepsilon_{(ij)}\mathrm{d}\Omega = \oint_{S_p} \overset{n}{P}_i \mathring{u}_i \mathrm{d}S_p, \\
2U &= \int_{\Omega} \sigma_{(ij)}\varepsilon_{(ij)}\mathrm{d}\Omega = \int_{\mathring{\Omega}} \mathring{\sigma}_{(ij)}\varepsilon_{(ij)}\mathrm{d}\mathring{\Omega} = \int_{\mathring{\Omega}} \sigma_{(ij)}\mathring{\varepsilon}_{(ij)}\mathrm{d}\mathring{\Omega} = \oint_{S_p} \overset{n}{P}_i u_i \mathrm{d}S_p.
\end{aligned}
\tag{2.16}
$$

These formulae provide a basis for the analysis of the effect that a change in the shape of a spatial structure has on the deformability of that structure. Wasiutyński's theorems about the reduction or rise of the elastic strain potential due to a reinforcement or weakening of the structure (Section 2.2) formulate the principles of this analysis. We shall now revise this approach, making use of formulae (2.16).

Suppose that the region Ω occupied by the structure after modification of the free surface $\mathring{S}_0$ contains the original region $\mathring{\Omega}$, i.e. $\mathring{\Omega} \subset \Omega$. The volume of the modified structure, V, is therefore greater than the original volume $\mathring{V}$. The volume increment $\Delta V = V - \mathring{V} > 0$ characterizes the reinforcement of the structure.

Using formulae (2.16), we can represent the change $U–\mathring{U}$ in the elastic strain potential due to reinforcement in the form

$$
\begin{aligned}
U - \mathring{U} &= \tfrac{1}{2}\int_{\Omega} \sigma_{(ij)}\varepsilon_{(ij)}\mathrm{d}\Omega - \tfrac{1}{2}\int_{\mathring{\Omega}} \mathring{\sigma}_{(ij)}\mathring{\varepsilon}_{(ij)}\mathrm{d}\mathring{\Omega} = \tfrac{1}{2}\int_{\mathring{\Omega}} (\sigma_{(ij)} - \mathring{\sigma}_{(ij)})\mathring{\varepsilon}_{(ij)}\mathrm{d}\mathring{\Omega} \\
&= \tfrac{1}{2}\int_{\mathring{\Omega}} (\sigma_{(ij)} - \mathring{\sigma}_{(ij)})\varepsilon_{(ij)}\mathrm{d}\mathring{\Omega} - \tfrac{1}{2}\int_{\mathring{\Omega}} (\sigma_{(ij)} - \mathring{\sigma}_{(ij)})(\varepsilon_{(ij)} - \mathring{\varepsilon}_{(ij)})\mathrm{d}\mathring{\Omega} \\
&= \tfrac{1}{2}\int_{\mathring{\Omega}} \sigma_{(ij)}\varepsilon_{(ij)}\mathrm{d}\mathring{\Omega} - \tfrac{1}{2}\int_{\Omega} \sigma_{(ij)}\varepsilon_{(ij)}\mathrm{d}\mathring{\Omega} - \tfrac{1}{2}\int_{\mathring{\Omega}} \Delta\sigma_{(ij)}\Delta\varepsilon_{(ij)}\mathrm{d}\mathring{\Omega} \\
&= -\int_{\Omega - \mathring{\Omega}} \Psi \mathrm{d}\mathring{\Omega} - \tfrac{1}{2}\int_{\mathring{\Omega}} \Delta\sigma_{(ij)}\Delta\varepsilon_{(ij)}\mathrm{d}\mathring{\Omega} \leqslant 0,
\end{aligned}
\tag{2.17}
$$

where $\Psi = \frac{1}{2}\sigma_{(ij)}\varepsilon_{(ij)}$ is the unit elastic strain potential, $\Delta\sigma_{(ij)} = \sigma_{(ij)} - \mathring{\sigma}_{(ij)}$ and $\Delta\varepsilon_{(ij)} = \varepsilon_{(ij)} - \mathring{\varepsilon}_{(ij)}$. Both integrals in formula (2.18) are positive, because Ψ and $\Delta\sigma_{(ij)}\,\Delta\varepsilon_{(ij)}$ are positive and the region $\mathring{\Omega}$ of the original structure is a proper part of the region Ω of the modified structure by the definition of reinforcement. Hence Theorem 1 of Section 2.2 can be reformulated as follows:

If we reinforce a structure by increasing its volume and modifying its free surface so that the region occupied by the original structure is contained in the region occupied by the reinforced structure, the elastic strain potential of the system is reduced; the greater is the potential $\int_{\Omega-\mathring{\Omega}} \Psi \mathrm{d}\mathring{\Omega}$ *of the reinforcing part, the greater is the reduction of the total potential.*

Similarly, Theorem 2 can be restated as follows:

The elastic strain potential of the reinforced structure is less than the potential of the original structure by more than the potential of the reinforcing part; hence the potential $\frac{1}{2}\int_{\mathring{\Omega}} \mathring{\sigma}_{(ij)}\mathring{\varepsilon}_{(ij)} \mathrm{d}\mathring{\Omega}$ *contained in the part* $\mathring{\Omega}$ *of the region* Ω *is reduced by more than double the potential* $\int_{\Omega-\mathring{\Omega}} \Psi \mathrm{d}\mathring{\Omega}$ *of the reinforcing part.*

Subtracting equality $(2.16)_2$

$$\int_{\mathring{\Omega}} \sigma_{(ij)}\mathring{\varepsilon}_{(ij)} \mathrm{d}\mathring{\Omega} = \oint_{S_p} \overset{n}{P}_i u_i \mathrm{d}S_p$$

from the equality

$$\begin{aligned}\int_{\mathring{\Omega}} \mathring{\sigma}_{(ij)}\mathring{\varepsilon}_{(ij)} \mathrm{d}\mathring{\Omega} &= \int_{\mathring{\Omega}} \sigma_{(ij)}\mathring{\varepsilon}_{(ij)} \mathrm{d}\mathring{\Omega} = \tfrac{1}{2}\int_{\mathring{\Omega}} \sigma_{(ij)}(\mathring{u}_{i,j}+\mathring{u}_{j,i}) \mathrm{d}\mathring{\Omega} \\ &= \int_{\mathring{\Omega}} \sigma_{(ij)}\mathring{u}_{i,j} \mathrm{d}\mathring{\Omega} = \int_{\mathring{\Omega}} [(\sigma_{(ij)}\mathring{u}_i)_{,j} - \sigma_{(ij),j}\mathring{u}_i] \mathrm{d}\mathring{\Omega} \\ &= \oint_{S_p} \overset{n}{P}_i u_i \mathrm{d}S_p + \int_{\mathring{S}_0} \sigma_{(ij)}\mathring{n}_{j0}\mathring{u}_i \mathrm{d}\mathring{S} = \oint_{S_p} \overset{n}{P}_i \mathring{u}_i \mathrm{d}S_p + \oint_{\mathring{S}_0} \mathring{X}_i \mathring{u}_i \mathrm{d}\mathring{S}_0,\end{aligned}$$

which holds for the original structure loaded by the forces $\overset{n}{\mathbf{P}}$ and $\mathring{X}_i$ as shown in Fig. 2.4, we obtain the relation

$$\oint_{S_p} \overset{n}{P}_i u_i \mathrm{d}S_p = \oint_{S_p} \overset{n}{P}_i \mathring{u}_i \mathrm{d}S_p + \oint_{\mathring{S}_0} \sigma_{ij}\mathring{n}_{j0}\mathring{u}_i \mathrm{d}\mathring{S}_0, \tag{2.18}$$

which expresses the Betti law. It follows that the potential change $U-\mathring{U}$ due to the reinforcement can be written in terms of the force $X_i = [\sigma_{(ij)}]_{\mathring{S}_0}\mathring{n}_{j0}$, exerted by the reinforcing part on the original structure, and the displacement $\mathring{u}_i$ at the surface $\mathring{S}_0$ before the reinforcement as

$$U-\mathring{U} = \tfrac{1}{2}\oint_{S_p} \overset{n}{P}_i(u_i-\mathring{u}_i) \mathrm{d}S_p = \tfrac{1}{2}\oint_{\mathring{S}_0} \mathring{X}_i \mathring{u}_i \mathrm{d}S_0. \tag{2.19}$$

We see that the reinforcement will reduce the elastic strain potential of the structure if the surface contiguous with the reinforcing element undergoes displacement before the reinforcement (cf. Theorem 3, Sec. 2.2).

The effect of a weakening of the structure can be determined by considering a transformation inverse to the one discussed so far. We assume that the modification of the free surface $\mathring{S}_0$ weakens the structure, i.e. the new region Ω is now contained in the original region $\mathring{\Omega}$ and so the volume V of the modified structure is less than $\mathring{V}$. The volume decrement $V-\mathring{V} < 0$ characterizes the weakening of the structure. Using relations (2.16), we can represent the potential change $\mathring{U}-U$ due to the weakening as

$$U-\mathring{U} = \tfrac{1}{2}\int_{\Omega}\sigma_{(ij)}\varepsilon_{(ij)}\mathrm{d}\Omega-\tfrac{1}{2}\int_{\mathring{\Omega}}\mathring{\sigma}_{(ij)}\mathring{\varepsilon}_{(ij)}\mathrm{d}\mathring{\Omega} = \tfrac{1}{2}\int_{\Omega}(\sigma_{(ij)}-\mathring{\sigma}_{(ij)})\varepsilon_{ij}\mathrm{d}\Omega$$

$$= \tfrac{1}{2}\int_{\Omega}(\sigma_{(ij)}-\mathring{\sigma}_{(ij)})\mathring{\varepsilon}_{(ij)}\mathrm{d}\Omega+\tfrac{1}{2}\int_{\Omega}(\sigma_{(ij)}-\mathring{\sigma}_{(ij)})(\varepsilon_{(ij)}-\mathring{\varepsilon}_{(ij)})\mathrm{d}\Omega$$

$$= -\tfrac{1}{2}\int_{\Omega}\mathring{\sigma}_{(ij)}\mathring{\varepsilon}_{(ij)}\mathrm{d}\Omega+\tfrac{1}{2}\int_{\mathring{\Omega}}\mathring{\sigma}_{(ij)}\mathring{\varepsilon}_{(ij)}\mathrm{d}\mathring{\Omega}+\tfrac{1}{2}\int_{\Omega}\Delta\sigma_{(ij)}\,\Delta\varepsilon_{(ij)}\mathrm{d}\Omega$$

$$= \tfrac{1}{2}\int_{\mathring{\Omega}-\Omega}\mathring{\sigma}_{(ij)}\mathring{\varepsilon}_{(ij)}\mathrm{d}\mathring{\Omega}+\tfrac{1}{2}\int_{\Omega}\Delta\sigma_{(ij)}\Delta\varepsilon_{(ij)}\mathrm{d}\Omega$$

$$= \int_{\mathring{\Omega}-\Omega}\mathring{\Psi}\mathrm{d}\mathring{\Omega}+\tfrac{1}{2}\int_{\Omega}\Delta\sigma_{(ij)}\Delta\varepsilon_{(ij)}\mathrm{d}\Omega > 0. \tag{2.20}$$

Since $\mathring{\Psi}$ and $\Delta\sigma_{(ij)}\,\Delta\varepsilon_{(ij)}$ are positive and the region Ω of the weakened structure is a proper part of the region occupied by the original structure, the difference $U-\mathring{U}$ is positive. Hence follow the statements:

If we weaken a structure by modifying its free surface and reducing its volume so that the region occupied by the new structure is contained in the region occupied by the original structure, the elastic strain potential of the system will increase; it will increase the more the greater is the potential $\int_{\mathring{\Omega}-\Omega}\mathring{\psi}\mathrm{d}\mathring{\Omega}$ *of the removed element.*

The elastic strain potential of the weakened structure is greater than the potential of the original structure by more than the potential of the removed part; hence the potential $\tfrac{1}{2}\int_{\Omega}\mathring{\sigma}_{(ij)}\mathring{\varepsilon}_{(ij)}\mathrm{d}\mathring{\Omega}$ *contained in the part* Ω *of the region* $\mathring{\Omega}$ *is increased by more than double the potential of the removed part.*

Subtracting equality $(2.16)_1$,

$$\int_{\Omega}\mathring{\sigma}_{(ij)}\varepsilon_{(ij)}\mathrm{d}\Omega = \oint_{S_p}\overset{n}{P}_i\mathring{u}_i\mathrm{d}S_p,$$

which holds in the modified structure, from the equality

$$\int_{\Omega}\mathring{\sigma}_{(ij)}\varepsilon_{(ij)}\mathrm{d}\Omega = \oint_{S_p}\overset{n}{P}_i u_i\mathrm{d}S_p+\int_{S_0}\mathring{\sigma}_{(ij)}n_{j0}\,u_i\mathrm{d}S = \oint_{S_p}\overset{n}{P}_i u_i\mathrm{d}S_p+\oint_{S_0}X_i u_i\mathrm{d}S_0,$$

which holds in the modified structure loaded by the two systems of forces as shown in Fig. 2.5, we obtain the relation

$$\oint_{S_p}\overset{n}{P}_i\mathring{u}_i\mathrm{d}S_p = \oint_{S_p}\overset{n}{P}_i u_i\mathrm{d}S_p+\oint_{S_p}X_i u_i\mathrm{d}S_0 \tag{2.21}$$

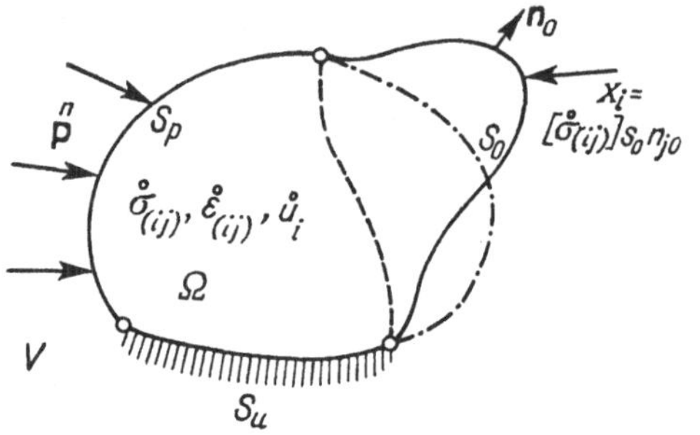

Fig. 2.5. Stress, strain and displacement in the modified structure

expressing the Betti law. The change $U-\mathring{U}$ of the potential caused by the weakening can therefore be written in the form

$$U-\mathring{U} = \tfrac{1}{2}\oint_{S_p} \overset{n}{P}_i(u_i-\mathring{u}_i)\mathrm{d}S_p = -\tfrac{1}{2}\oint_{S_0} \mathring{\sigma}_{(ij)} n_{j0} u_i \mathrm{d}S_0 = -\oint_{S_0} u_i X_i \mathrm{d}S_0, \quad (2.22)$$

where X_i are the forces exerted by the part to be removed on the rest, i.e. on the modified structure, and u_i are the displacements at the surface S_0 after the weakening. This permits us to state

THEOREM 8. *Weakening a structure increases the elastic strain potential if the surface between the part to be removed and the rest of the original structure undergoes displacement after the weakening.*

2.4 The necessary and sufficient conditions of minimum deformability

We shall derive conditions which determine the shape of the free surface for which the potential U of the structure is absolute minimum for a given volume V. A structure whose free surface satisfies those conditions is the least deformable among structures having the same volume V and differing only in the shape of free surface.

The transformations which we considered so far consisted in modifying the free surface so as to reinforce the structure and increase its volume or to weaken it and reduce the volume. In order to derive the conditions for the least deformability at a given volume V, let us consider a transformation in which the material is carried over from some points of the neighbourhood of the free surface $\mathring{S}_0$ to other points in this neighbourhood without breaking the connectivity of the region. The transformation can be interpreted geometrically as shown in Fig. 2.6. The function $\mathring{z}(x^\alpha)$ represents the original surface $\mathring{S}_0$ and the function $z(x^\alpha)$ the modified surface S_0.

If, as before, $\mathring{\Omega}$ is the region of the original structure and Ω the region of the modified structure, the potential change ΔU caused by the modification of the free surface can be represented by means of formula (2.17) as

$$\Delta U = \mathring{U} - U = \int_{\Omega} \Psi \mathrm{d}\Omega - \int_{\mathring{\Omega}} \Psi \mathrm{d}\mathring{\Omega} + \tfrac{1}{2} \int_{\mathring{\Omega}} \Delta\sigma_{(ij)} \Delta\varepsilon_{(ij)} \mathrm{d}\mathring{\Omega}$$

or, by means of formula (2.20), as

$$\Delta U = \mathring{U} - U = \int_{\Omega} \mathring{\Psi} \mathrm{d}\Omega - \int_{\mathring{\Omega}} \mathring{\Psi} \mathrm{d}\mathring{\Omega} - \tfrac{1}{2} \int_{\Omega} \Delta\sigma_{(ij)} \Delta\varepsilon_{(ij)} \mathrm{d}\Omega .$$

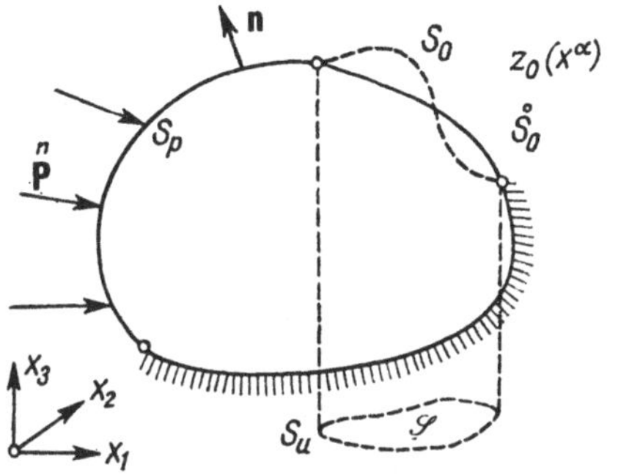

Fig. 2.6. Modification of the free boundary at constant volume

Since, as a result of the transformation, some parts of the structure are reinforced and some are weakened, the change ΔU can in general be positive as well as negative.

The differences of the volume integrals,

$$\int_{\Omega} \Psi \mathrm{d}\Omega - \int_{\mathring{\Omega}} \Psi \mathrm{d}\mathring{\Omega}, \qquad \int_{\Omega} \mathring{\Psi} \mathrm{d}\Omega - \int_{\mathring{\Omega}} \mathring{\Psi} \mathrm{d}\mathring{\Omega},$$

occurring in the expressions for ΔU, can be written in terms of single and double integrals.

As a reference frame in the space we choose a Cartesian coordinate system x^i. To simplify the calculations, let us assume that the parts $\mathring{S}_0$ and S_0 of the surfaces $\mathring{S}$ and S bounding the regions $\mathring{\Omega}$ and Ω do not intersect any straight line parallel to the axis x^3 at more than one point. Projecting the surfaces $\mathring{S}_0$ and S_0 on the plane x^1x^2, we obtain a two-dimensional region $\mathscr{S}$, the same for each surface. If we express the volume integrals by single and double integrals, the formulae for ΔU take the form

$$\Delta U = - \oint_{\mathscr{S}} \mathrm{d}\mathscr{S} \int_{z}^{\mathring{z}} \Psi(x^i) \mathrm{d}x^3 + \tfrac{1}{2} \int_{\mathring{\Omega}} \Delta\sigma_{(ij)} \Delta\varepsilon_{(ij)} \mathrm{d}\mathring{\Omega}$$

or

$$\Delta U = - \oint_{\mathscr{S}} \mathrm{d}\mathscr{S} \int_{z}^{\mathring{z}} \mathring{\Psi}(x^i) \mathrm{d}x^3 - \tfrac{1}{2} \int_{\Omega} \Delta\sigma_{(ij)} \Delta\varepsilon_{(ij)} \mathrm{d}\Omega ,$$

Since by assumption the volume of the structure does not change under the transformation, $\mathring{V} = V$ and

$$\Delta V = \mathring{V} - V = \int_{\mathring{\Omega}} d\mathring{\Omega} - \int_{\Omega} d\Omega = \oint_{\mathscr{S}} d\mathscr{S} \int_{z}^{\mathring{z}} dx^3 = \oint_{\mathscr{S}} d\mathscr{S}(\mathring{z}-z) = 0. \quad (2.23)$$

The potential U of the modified structure is absolute minimum if for any modification $\Delta z = \mathring{z}-z$ such that $\Delta V = 0$ the increment $\Delta U = \mathring{U}-U$ is positive.

To find the necessary and the sufficient conditions for the absolute minimum of the potential, let us represent the potential $\Psi(x^i)$ as

$$\Psi(x^i) = \Psi(x^\alpha)|_{x^3=z} + \Delta\Psi(x^i),$$

where $\Psi(x^\alpha)|_{x^3=z}$ is the unit potential on the surface z and $\Delta\Psi(x^i)$ is the increment of the unit potential between any point x^i in the region undergoing modification and its projection along the x^3-axis on the surface $\mathring{z}$ (Fig. 2.7). The formula for ΔU takes the form

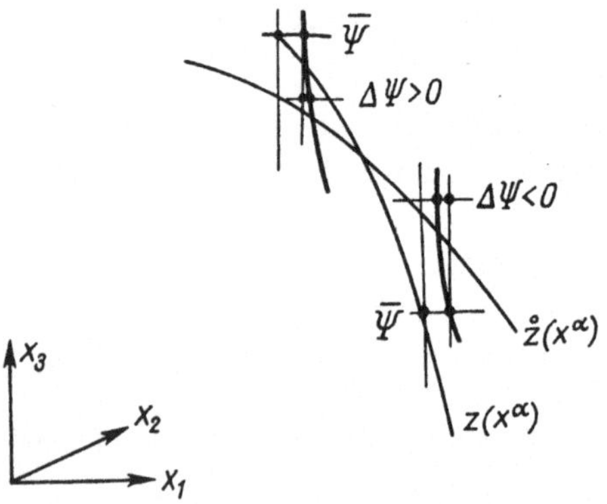

Fig. 2.7. Signs of the unit potential increments at the modified surface

$$\Delta U = -\oint_{\mathscr{S}} d\mathscr{S} \int_{z}^{\mathring{z}} (\Psi(x^\alpha)|_{x^3=z} + \Delta\Psi(x^i))dx^3 + \tfrac{1}{2}\int_{\mathring{\Omega}} \Delta\sigma_{(ij)}\Delta\varepsilon_{(ij)}d\mathring{\Omega}. \quad (2.24)$$

Hence, for U to be absolute minimum, i.e. for $\Delta U = \mathring{U} - U > 0$, the necessary condition is

$$\Psi(x^\alpha)|_{x^3=z} = \overline{\Psi} = \text{const} \quad (2.25)_1$$

and the sufficient condition for the existence of an absolute minimum of the potential in the class of admissible structures is

$$\begin{aligned} &\Delta\Psi(x^i) \geqslant 0 \quad \text{when } \mathring{z} \leqslant z, \\ &\Delta\Psi(x^i) \leqslant 0 \quad \text{when } \mathring{z} \geqslant z. \end{aligned} \quad (2.25)_2$$

Indeed, using conditions (2.25) and the constant-volume condition (2.23), we obtain from (2.24)

$$\Delta U = U - \mathring{U} = -\oint_{\mathscr{S}} \mathrm{d}\mathscr{S} \int_{z}^{\mathring{z}} \Delta\Psi(x^i)\mathrm{d}x^3 + \tfrac{1}{2}\int_{\mathring{\Omega}} \Delta\sigma_{(ij)}\Delta\varepsilon_{(ij)}\mathrm{d}\mathring{\Omega} > 0 \qquad (2.26)$$

for arbitrary changes Δz with $\Delta V = 0$, because $\frac{1}{2}\int_{\Omega} \Delta\sigma_{(ij)}\Delta\varepsilon_{(ij)}\mathrm{d}\mathring{\Omega} > 0$.

When the sufficient conditions, concerning the sign of the increment $\Delta\Psi$, are not satisfied, an absolute minimum of the potential may or may not exist. The conditions show that in a least-deformability structure the unit potential is a function increasing or constant along the inward-directed normal to the optimized surface.

The necessary conditions for the minimum deformability of elastic structures were first derived by Wasiutyński in 1939, becoming the basis for the formulation of many optimum-design problems. The sufficient conditions were given by Mróz (1961).

In some kinds of structures, the sufficient conditions for the existence of the absolute-minimum potential are satisfied identically. Shells in membrane state and plates in plane stress with variable thickness are examples of this situation. In either case $\Delta\Psi(x^i) = 0$ across the thickness, which can be taken as a parameter of optimization.

A different situation arises in structures which are subject to bending, such as beams, plates or solid shells made of isotropic and homogeneous materials. If the shape in the plane of a structure of this kind is assumed to be given, the variable thickness in the case of plates or shells, or the variable height in the case of beams, is left as the only design variable. Now, granted that the Love–Kirchhoff geometrical hypothesis about the normals to the middle surface before and after deformation is correct, we observe that the function Ψ in such a structure is decreasing away from the optimized limit surface towards the middle surface. Therefore the sufficient conditions (2.25) cannot be satisfied and, the signs of the two terms in (2.26) being different, there is no absolute minimum of the potential.

Satisfying the necessary condition $\Psi = \text{const}$ on the outer surfaces, we can then obtain only a local minimum. To derive the necessary and sufficient conditions for a local minimum of the potential, we restrict the class of the surfaces S_0 to those for which the functions $z(x^\alpha)$ and $\mathring{z}(x^\alpha)$ are close in order zero, i.e. satisfy the inequality $|\mathring{z} - z| < \varepsilon$ with ε sufficiently small.

Into the potential difference

$$\Delta U = -\oint_{\mathscr{S}} \mathrm{d}\mathscr{S} \int_{z}^{\mathring{z}} \Psi(x^i)\mathrm{d}x^3 + \tfrac{1}{2}\int_{\mathring{\Omega}} \Delta\sigma_{(ij)}\Delta\varepsilon_{(ij)}\mathrm{d}\mathring{\Omega}$$

we insert the following expansion of the integral $\int_{z}^{\overset{\circ}{z}} \Psi(x^i)\mathrm{d}x^3$ into a series with respect to $\Delta z = \overset{\circ}{z} - z$:

$$\int_{z}^{z+\Delta z} \Psi(x^i)\mathrm{d}x^3 = [\Psi]_{x^3=z}\Delta z + \tfrac{1}{2}[\Psi_{,z}]_{x^3=z}(\Delta z)^2 + O(\Delta z)^3.$$

Neglecting terms with $(\Delta z)^3$, we get

$$\Delta U = -\oint_{\mathscr{S}} \mathrm{d}\mathscr{S}[\Psi]_{x^3=z}\Delta z - \tfrac{1}{2}\oint_{\mathscr{S}} \mathrm{d}\mathscr{S}[\Psi_{,z}]_{x^3=z}(\Delta z)^2 + \tfrac{1}{2}\int_{\overset{\circ}{\Omega}} \Delta\sigma_{(ij)}\Delta\varepsilon_{(ij)}\mathrm{d}\overset{\circ}{\Omega}. \tag{2.27}$$

We can see from this expression that a necessary condition for U to be minimum is

$$\delta U \overset{\mathrm{df}}{=} \oint_{\mathscr{S}} \mathrm{d}\mathscr{S}[\Psi]_{x^3=z}\Delta z = 0,$$

with

$$\Delta V = \oint_{\mathscr{S}} \Delta z \mathrm{d}\mathscr{S} = 0,$$

and hence $[\Psi]_{x^3=z} = \text{const}$.

The sufficient condition for a local minimum of the potential is the inequality $\delta^2 U > 0$ for arbitrary Δz. The symbol $\delta^2 U$ denotes the second variation, i.e. the part of the increment ΔU produced by $(\Delta z)^2$. The inequality is satisfied if

$$[\Psi_{,z}]_{z=x^3} \leqslant 0 \quad \text{on } S_0.$$

When $\Psi_{,z} > 0$ on S_0, the situation is uncertain, because the expression for ΔU then contains two terms of which one is negative and the other is positive. In this case, there may be no least-deformability solution in the set of admissible structures.

The least-deformability conditions obtained above permit us to formulate the following statement: a structure of volume V is the least-deformable—in the sense of local minimum—among other structures with the same volume but with different free surfaces if and only if the unit potential Ψ is constant on the surface S_0 and grad $\Psi \leqslant 0$ on S_0.

The second condition can also be derived in a different way. To this end, consider three orthogonal directions at an arbitrary point of the optimum surface: $\mathbf{n}_0$—normal to the surface S_0, and $\mathbf{t}_1$ and $\mathbf{t}_2$—both lying in the tangent plane. The vectors $\mathbf{t}_1$ and $\mathbf{t}_2$ are tangent to certain curves l_1 and l_2 in S_0. Since the function Ψ is constant on S_0, we have

$$\Psi_{,l_1} = \Psi_{,l_2} = 0.$$

Hence

$$[\Psi_{,z}]_{S_0} = \Psi_{,t_1}\cos(\mathbf{i}_3, \mathbf{t}_1)+\Psi_{,t_2}\cos(\mathbf{i}_3, \mathbf{t}_2)+\Psi_{,n}\cos(\mathbf{i}_3, \mathbf{n}) = \Psi_{,n}\cos(\mathbf{i}_3, \mathbf{n}),$$

and since $\cos(\mathbf{i}_3, \mathbf{n}) \geqslant 0$, we get grad $\Psi \leqslant 0$ on S_0.

Equivalence of optimization for minimum potential at constant volume with minimum-weight design

The minimum-potential-at-constant-volume (or weight) criterion for optimization of structures made of homogeneous materials is equivalent to the criterion of minimum volume at constant potential. To prove this equivalence, consider the potential difference

$$\Delta U = -\oint_{\mathscr{S}} \mathrm{d}\mathscr{S} \int_{z}^{\overset{\circ}{z}} \Psi|_{x^3=z}\mathrm{d}x^3 - \oint_{\mathscr{S}} \mathrm{d}\mathscr{S} \int_{z}^{\overset{\circ}{z}} \Delta\Psi(x^i)\mathrm{d}x^3 + \tfrac{1}{2}\int_{\overset{\circ}{\Omega}} \Delta\sigma_{(ij)}\Delta\varepsilon_{(ij)}\mathrm{d}\overset{\circ}{\Omega}$$

between an original structure occupying a region $\overset{\circ}{\Omega}$ and a modified structure occupying a region Ω. The modified structure differs from the original one in the shape of the free part of its boundary surface. The free surfaces of the two structures are given by the functions

$$\overset{\circ}{z}(x^\alpha) \quad \text{and} \quad z(x^\alpha).$$

Assuming that $\Delta U = 0$ and, in keeping with the minimum-potential criterion, that $[\Psi(x^3)]_{x^3=z} = \overline{\Psi} = \text{const}$, we find

$$\Delta V = \frac{1}{\overline{\Psi}}\left[-\int_{\mathscr{S}} \mathrm{d}\mathscr{S} \int_{z}^{\overset{\circ}{z}} \Delta\Psi(x^i)\mathrm{d}x^3 + \frac{1}{2}\int_{\overset{\circ}{\Omega}} \Delta\sigma_{(ij)}\Delta\varepsilon_{(ij)}\mathrm{d}\overset{\circ}{\Omega}\right].$$

Hence the condition $\Delta V > 0$ for an absolute minimum of the volume is satisfied if

$$\Delta\Psi(x^i) \leqslant 0 \quad \text{when} \quad z < \overset{\circ}{z},$$
$$\Delta\Psi(x^i) \geqslant 0 \quad \text{when} \quad z > \overset{\circ}{z}.$$

2.5 Optimization of plane bar structures

Plane bar structure

Plane structures composed of straight prismatic bars are widely applied in building and civil engineering as elements carrying loads in specified planes.

According to the kinds of internal forces which are transmitted by the bars of a given structure, three kinds of bar structures are customarily distinguished:

1. *Frames* (Fig. 2.8), in which bars are interconnected by rigid joints, which means that a rotation of a joint produces equal rotations of the ends of

all the bars converging at that joint. The bars of a frame transmit axial forces N, shearing forces T and bending moments M.

2. *Trusses* (Fig. 2.9), in which bars are connected by hinged joints, thereby enjoying a freedom of rotation. The bars of a truss transmit only axial forces N.
3. *Beam structures* (Fig. 2.10), shaped along a single straight axis and sustaining bending moments M and shearing forces T under transverse loads

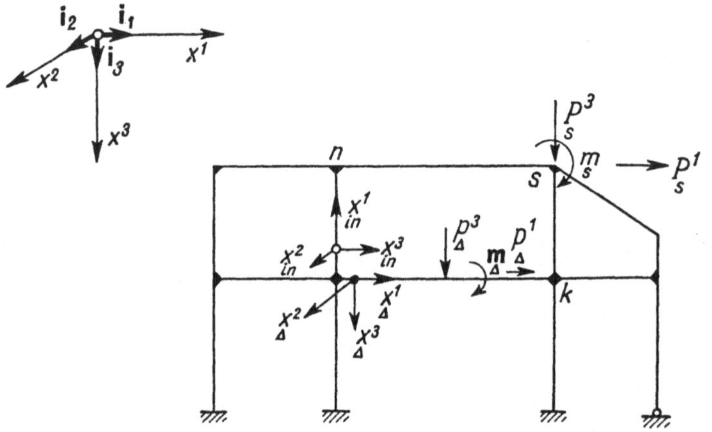

Fig. 2.8. A frame scheme

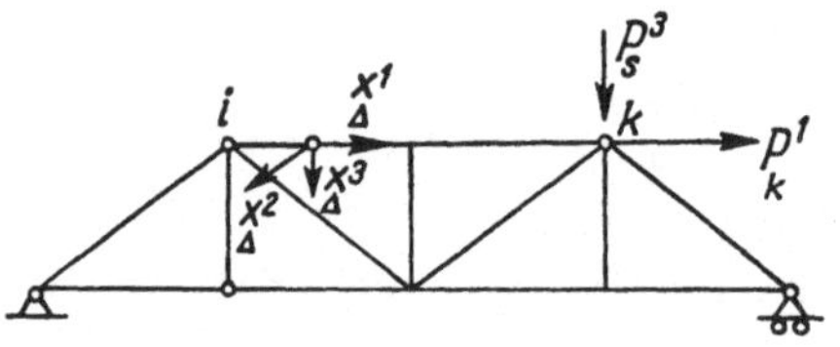

Fig. 2.9. A truss scheme

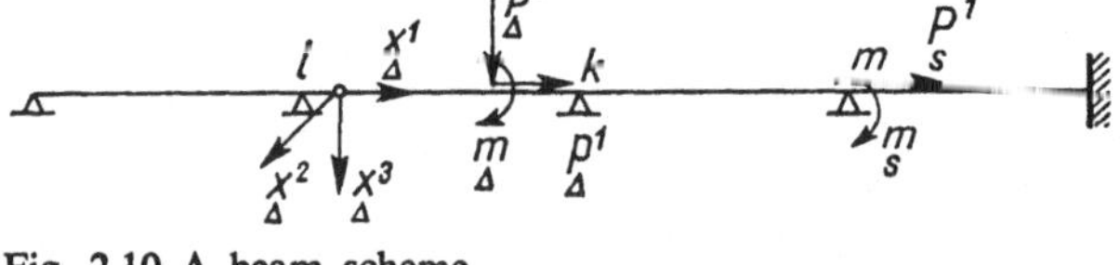

Fig. 2.10 A beam scheme

Among bar structures of the above kinds one can also distinguish isostatic structures, in which the internal forces N, T and M are independent of deformations, and hyperstatic structures, in which these forces depend on the deformations and mechanical properties of the structure.

In the optimization problems which we shall consider in this section it is assumed that the geometry of the structure to be optimized is given, being determined, for example, by functional reasons or by the type of a larger system of which the given structure is an element. It is further assumed that the kind of the elastic or near-elastic material used for construction is also given, its choice being determined by technological, economic or other requirements. The structure subject to optimization can be made of metal, reinforced or prestressed concrete, or a combination of these materials. The methods of construction and assembly are also assumed to be given.

Under these conditions, optimization of a structure will be understood as seeking an optimum distribution of given volumes of structural materials over a given geometrical form of the structure. It is also possible to formulate an optimization problem in which an optimum configuration of bars would be sought as well. Correct distribution of structural materials, taking full advantage of their mechanical properties, is a guiding principle in the design and dimensioning of building and civil engineering structures. Among the design conditions determining the usefulness of a structure in serving the purpose it is designed for one can distinguish: strength conditions, which define the degree of safety of the structure against destruction, and deformational constraints, which are decisive for the functional quality and reliability of the structure.

As the technology of structural materials provides the engineer with stronger and stronger materials, deformational constraints come into greater and greater prominence. A structure with a sufficient margin of safety against destruction may be useless if its deformations under the loads it is supposed to carry are too large. Bar structures used in building and civil engineering appear mostly as skeleton elements, to which other, non-structural elements are attached, serving their specific purposes according to the general function of the building, such as, for example, walls protecting from weather effects, partitions and fills. Deformations of the skeleton bring about deformations of the connected filling elements and can cause their cracking and failure. Thus the strain properties of a bar structure can be crucial for the functional quality of the system of which it is an element.

As a measure of deformability of a bar structure we shall use the elastic strain energy which is equal to the work done by the applied loads on the displacements of the members they act upon.

By an *optimum structure* we shall understand a bar structure showing the greatest resistance to deformations (the least deformability, the greatest rigidity against deformations) of all structures with the same geometry, made of materials of the same kind and volume but with different material distributions.

In the following considerations we shall seek just such optimum or least-deformability bar structures. The design variables will be the distributions of structural materials, given in terms of cross-sectional areas, variable along the bars.

Basic assumptions and definitions

We consider a plane structure composed of bars $\triangle$ and having N nodes. As a reference frame we take a right-handed Cartesian coordinate system $\underset{\triangle}{x}^i$ $(i = 1, 2, 3)$ and assume that the $\underset{\triangle}{x}^1\underset{\triangle}{x}^3$ plane is the plane of the structure.

The geometry of the structure is described by the coordinates $\underset{\triangle}{x}^i$ of the nodes and by the directions $\underset{\triangle}{\mathbf{T}}^1 = \cos\underset{\triangle}{\alpha}\,\mathbf{i}_1 + \sin\underset{\triangle}{\alpha}\,\mathbf{i}_3$ of the bars $\triangle$, where $\underset{\triangle}{\alpha}$ is the angle of inclination of a bar $\triangle$ to the $\underset{\triangle}{x}^1$ axis. The length of a bar $\triangle$ will be denoted by $\underset{\triangle}{l}$.

The stress and strain states and the geometrical properties of a bar $\triangle$ will be described in a local right-handed coordinate system $\underset{\triangle}{x}^i$ $(i = 1, 2, 3)$ whose axes have the directions $\underset{\triangle}{\mathbf{T}}^i$ $(i = 1, 2, 3)$ of the principal axes of inertia of the cross-section of the bar (the axis $\underset{\triangle}{\mathbf{T}}^3$ being also the symmetry axis of the cross-section) and whose origin is placed at one of the ends of the bar (Fig. 2.11).

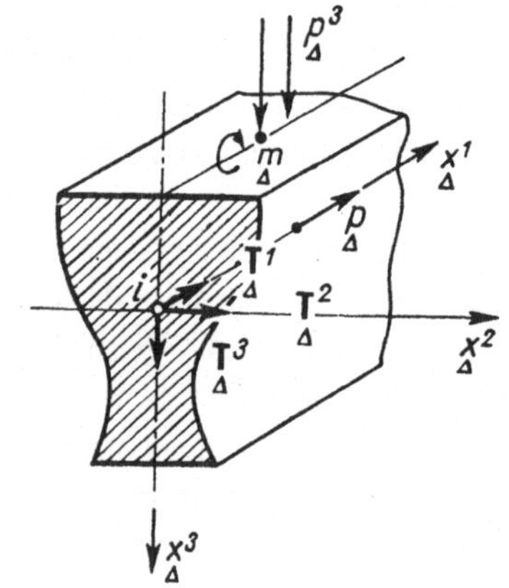

Fig. 2.11. Coordinate system in bar $\triangle$

The bar is acted upon by the loading forces $\underset{\triangle}{p}^3(\underset{\triangle}{x}^1)$ and $\underset{\triangle}{p}^1(\underset{\triangle}{x}^1)$ in the directions $\underset{\triangle}{\mathbf{T}}^1$ and $\underset{\triangle}{\mathbf{T}}^3$, and moment $\underset{\triangle}{\mathbf{m}}(\underset{\triangle}{x}^1)$ perpendicular to the plane x^1x^3 of the structure. They are all functions of $\underset{\triangle}{x}^1$. In addition, any joint s $(s = 1, 2, \ldots, N)$ of the structure may carry forces $\underset{s}{P}^1$ and $\underset{s}{P}^3$ in the directions $\mathbf{i}_1$ and $\mathbf{i}_3$, and a moment $\underset{s}{\mathbf{m}}$ parallel to $\mathbf{i}_2$.

The action of the loads induces internal forces in the bar, namely, an axial force $\underset{\triangle}{N}(\underset{\triangle}{x}^1)$, a shear force $\underset{\triangle}{T}(\underset{\triangle}{x}^1)$ and a bending moment $\underset{\triangle}{M}(\underset{\triangle}{x}^1)$. In the interval $\underset{\triangle}{x}^1(0, \underset{\triangle}{l})$ they satisfy the equilibrium conditions

$$\underset{\triangle}{N}_{,1}+\underset{\triangle}{p}^1 = 0, \quad \underset{\triangle}{T}_{,1}+\underset{\triangle}{p}^3 = 0, \quad \underset{\triangle}{M}_{,1}+\underset{\triangle}{m} = \underset{\triangle}{T} \tag{2.28}$$

and at the ends $\underset{\triangle}{x}^1 = 0$ and $\underset{\triangle}{x}^1 = \underset{\triangle}{l}$ of this interval the conditions

$$\begin{aligned}
&[\underset{\triangle}{M}]_{\underset{\triangle}{x}^1=0} = \underset{\triangle}{M}_{KL}, \quad && [\underset{\triangle}{M}]_{\underset{\triangle}{x}^1=\underset{\triangle}{l}} = \underset{\triangle}{M}_{LK},\\
&[\underset{\triangle}{T}]_{\underset{\triangle}{x}^1=0} = \underset{\triangle}{T}_{KL}, \quad && [\underset{\triangle}{T}]_{\underset{\triangle}{x}^1=\underset{\triangle}{l}} = \underset{\triangle}{T}_{LK},\\
&[\underset{\triangle}{N}]_{\underset{\triangle}{x}^1=0} = \underset{\triangle}{N}_{KL}, \quad && [\underset{\triangle}{N}]_{\underset{\triangle}{x}^1=\underset{\triangle}{l}} = \underset{\triangle}{N}_{LK},
\end{aligned} \tag{2.29}$$

The forces $\underset{\triangle}{N}_{KL}$, $\underset{\triangle}{T}_{KL}$, $\underset{\triangle}{M}_{KL}$ and $\underset{\triangle}{N}_{LK}$, $\underset{\triangle}{T}_{LK}$, $\underset{\triangle}{M}_{LK}$ are the interactions at the ends K and L of the bar $\triangle$, respectively (Fig. 2.12).

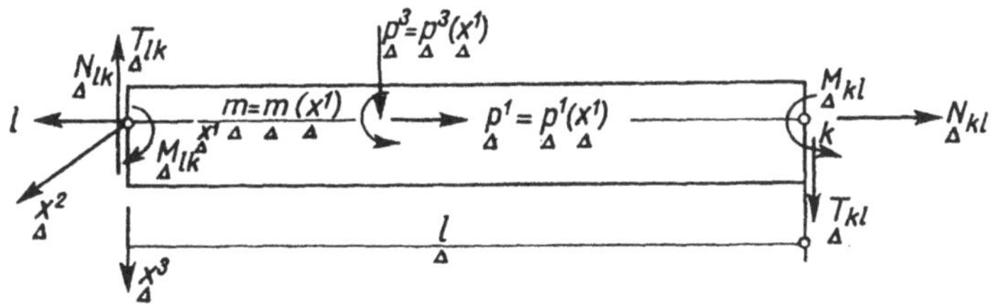

Fig. 2.12. Forces acting on bar $\triangle$

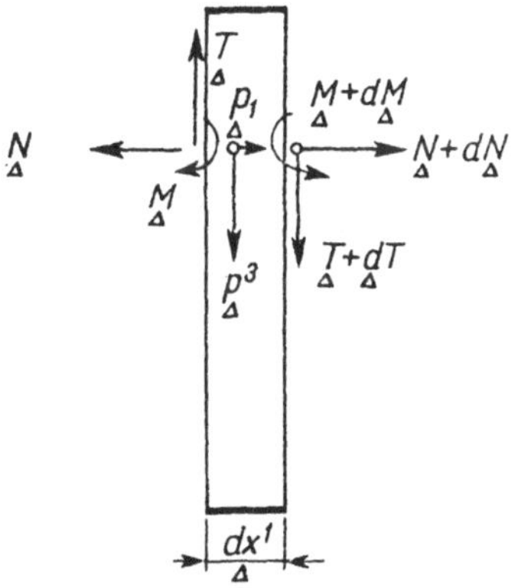

Fig. 2.13. Internal forces in the bar cross-section

Before it is fully loaded, the structure may carry initial internal forces $\underset{\triangle}{N}^s$, $\underset{\triangle}{T}^s$ and $\underset{\triangle}{M}^s$, produced by prestressing, dead weight or other factors. The prestressing can be done by means of tension inserts in the form of cables or rods anchored in the cross-sections of the bars (Fig. 2.13).

In the general case, the prestressing parameters of a bar $\triangle$ are the resultant force $\underset{\triangle}{S}(\underset{\triangle}{x}^1)$ of the tension inserts anchored in the bar and the eccentricity $\underset{\triangle}{e}(\underset{\triangle}{x}^1)$ of this force relative to the principal axis of inertia $\underset{\triangle}{T}^2$.

The initial internal forces $\underset{\triangle}{N^s}$, $\underset{\triangle}{T^s}$ and $\underset{\triangle}{M^s}$ satisfy the following equilibrium conditions:

$$
\begin{aligned}
&\underset{\triangle}{N^s}_{,1}+\underset{\triangle}{S}_{,1}+\underset{\triangle}{\tilde{p}^1} = 0,\\
&\underset{\triangle}{T^s}_{,1}+(\underset{\triangle}{S}\underset{\triangle}{e}_{,1})_{,1}+\underset{\triangle}{\tilde{p}^3} = 0,\\
&\underset{\triangle}{M^s}_{,1}+\underset{\triangle}{S}_{,1}\underset{\triangle}{e}+\underset{\triangle}{\tilde{m}} = \underset{\triangle}{T^s}
\end{aligned}
\tag{2.30}
$$

in the interval $\underset{\triangle}{x^1}(0, \underset{\triangle}{1})$ and

$$
\begin{aligned}
&[\underset{\triangle}{N^s}+\underset{\triangle}{S}]_{\underset{\triangle}{x^1}=0} = \underset{\triangle}{N^s_{KL}}, \quad && [\underset{\triangle}{N^s}+\underset{\triangle}{S}]_{\underset{\triangle}{x^1}=\underset{\triangle}{l}} = \underset{\triangle}{N^s_{LK}},\\
&[\underset{\triangle}{T^s}+\underset{\triangle}{S}\underset{\triangle}{e}_{,1}]_{\underset{\triangle}{x^1}=0} = \underset{\triangle}{T^s_{KL}} \quad && [\underset{\triangle}{T^s}+\underset{\triangle}{S}\underset{\triangle}{e}_{,1}]_{\underset{\triangle}{x^1}=\underset{\triangle}{l}} = \underset{\triangle}{T^s_{LK}},\\
&[\underset{\triangle}{M^s}+\underset{\triangle}{S}\underset{\triangle}{e}]_{\underset{\triangle}{x^1}=0} = \underset{\triangle}{M^s_{KL}}, \quad && [\underset{\triangle}{M^s}+\underset{\triangle}{S}\underset{\triangle}{e}]_{\underset{\triangle}{x^1}=\underset{\triangle}{l}} = \underset{\triangle}{M^s_{LK}}
\end{aligned}
\tag{2.31}
$$

at the ends K and L of the bar. The forces $\underset{\triangle}{N^s_{KL}}, \underset{\triangle}{T^s_{KL}}, \underset{\triangle}{M^s_{KL}}$ and $\underset{\triangle}{N^s_{LK}}, \underset{\triangle}{T^s_{LK}}, \underset{\triangle}{M^s_{LK}}$ are the reactions of the structure on the ends K and L of the bar, induced by prestressing.

The internal forces present in the prestressed structure after the external loads are applied are sums of the initial forces $\underset{\triangle}{N^s}$, $\underset{\triangle}{T^s}$ and $\underset{\triangle}{M^s}$ and the forces induced by the loads.

The initial prestressing and the external loads deform the structure. The nodes k undergo displacements $\underset{k}{u}$ with the components $\underset{k}{u^1}$ and $\underset{k}{u^3}$ and rotations φ_k in the $\underset{\triangle}{x^1}\underset{\triangle}{x^3}$ plane. The position of a bar $\triangle$ in the deformed state is described by the displacements $\underset{\triangle}{u^1}(\underset{\triangle}{x^1}) \overset{\mathrm{df}}{=} \underset{\triangle}{u}(\underset{\triangle}{x^1})$ and $\underset{\triangle}{u^3}(\underset{\triangle}{x^1}) \overset{\mathrm{df}}{=} \underset{\triangle}{w}(\underset{\triangle}{x^1})$ of its normal cross-section in the directions $\underset{\triangle}{\mathbf{T}^1}$ and $\underset{\triangle}{\mathbf{T}^3}$, respectively, and by the rotation $\underset{\triangle}{\varphi}(\underset{\triangle}{x^1})$ of this cross-section in the plane $\underset{\triangle}{x^1}\underset{\triangle}{x^3}$.

If the joints of the structure are perfectly rigid, the displacements and rotations of the ends $\underset{\triangle}{x^1} = 0$ and $\underset{\triangle}{x^1} = \underset{\triangle}{l}$ of the bar $\triangle$ are equal to the displacements and rotations of the corresponding nodes K and L:

$$
\begin{aligned}
&\cos\underset{\triangle}{\alpha}[\underset{\triangle}{u}]_{\underset{\triangle}{x^1}=0}-\sin\underset{\triangle}{\alpha}[\underset{\triangle}{w}]_{\underset{\triangle}{x^1}=0} = \underset{K}{u^1}, \quad && \cos\underset{\triangle}{\alpha}[\underset{\triangle}{u}]_{\underset{\triangle}{x^1}=\underset{\triangle}{l}}-\sin\underset{\triangle}{\alpha}[\underset{\triangle}{w}]_{\underset{\triangle}{x^1}=\underset{\triangle}{l}} = \underset{L}{u^1},\\
&\sin\underset{\triangle}{\alpha}[\underset{\triangle}{u}]_{\underset{\triangle}{x^1}=0}+\cos\underset{\triangle}{\alpha}[\underset{\triangle}{w}]_{\underset{\triangle}{x^1}=0} = \underset{K}{u^3}, \quad && \sin\underset{\triangle}{\alpha}[\underset{\triangle}{u}]_{\underset{\triangle}{x^1}=\underset{\triangle}{l}}+\cos\underset{\triangle}{\alpha}[\underset{\triangle}{w}]_{\underset{\triangle}{x^1}=\underset{\triangle}{l}} = \underset{L}{u^3},\\
&[\underset{\triangle}{\varphi}]_{\underset{\triangle}{x^1}=0} = \underset{K}{\varphi}, \quad && [\underset{\triangle}{\varphi}]_{\underset{\triangle}{x^1}=\underset{\triangle}{l}} = \underset{L}{\varphi}.
\end{aligned}
\tag{2.32}
$$

The state of deformation of the bar $\triangle$ is described by the unit elongation $\underset{\triangle}{\varepsilon}$ of the bar axis, the angle $\underset{\triangle}{\vartheta}$ between the normals to the bar axis before and after deformation, and the curvature $\underset{\triangle}{\chi}$ of the deformed axis. The quantities $\underset{\triangle}{\varepsilon}$, $\underset{\triangle}{\vartheta}$ and $\underset{\triangle}{\chi}$ are connected with the displacements $\underset{\triangle}{u}$, $\underset{\triangle}{w}$ and the rotation $\underset{\triangle}{\varphi}$ by the geometrical relations

$$\underset{\triangle}{\varepsilon} = \underset{\triangle}{u}_{,1}, \qquad \underset{\triangle}{\varphi} = -\underset{\triangle}{w}_{,1} + \underset{\triangle}{\vartheta}, \qquad \underset{\triangle}{\chi} = \underset{\triangle}{\varphi}_{,1}.$$

For a bar element with volume $\underset{\triangle}{A}\,\underset{\triangle}{dx^1}$ and length $\underset{\triangle}{dx^1}$, bounded by two normal sections $\underset{\triangle}{x^1} = \text{const}$ and $\underset{\triangle}{x^1} + \underset{\triangle}{dx^1} = \text{const}$, the product $\underset{\triangle}{\varepsilon}\,\underset{\triangle}{dx^1}$ gives the increment of the length of this element due to the action of the force $\underset{\triangle}{N}$. The products $\underset{\triangle}{\vartheta}\,\underset{\triangle}{dx^1}$ and $\underset{\triangle}{\chi}\,\underset{\triangle}{dx^1}$ represent the relative displacement and rotation of the sections $\underset{\triangle}{x^1} = \text{const}$ and $\underset{\triangle}{x^1} + \underset{\triangle}{dx^1} = \text{const}$, caused by the force $\underset{\triangle}{T}$ and moment $\underset{\triangle}{M}$.

In keeping with Hooke's law, the internal forces $\underset{\triangle}{N}$, $\underset{\triangle}{T}$ and $\underset{\triangle}{M}$ are proportional to the deformations they produce:

$$\begin{aligned} \underset{\triangle}{N} &= \underset{\triangle}{E}\underset{\triangle}{A}\underset{\triangle}{\varepsilon}, \\ \underset{\triangle}{T} &= \underset{\triangle}{G}\underset{\triangle}{A}\underset{\triangle}{\vartheta}/\underset{\triangle}{\varkappa}, \\ \underset{\triangle}{M} &= \underset{\triangle}{E}\underset{\triangle}{J}\underset{\triangle}{\chi}. \end{aligned}$$

The proportionality factors $\underset{\triangle}{E}\underset{\triangle}{A}$, $\underset{\triangle}{G}\underset{\triangle}{A}/\underset{\triangle}{\varkappa}$ and $\underset{\triangle}{E}\underset{\triangle}{J}$ characterize the stiffness of the bar against linear, shearing and flexural strains, respectively. If the cross-section of the bar is variable, they are all functions of $\underset{\triangle}{x^1}$.

Influence of geometrical and mechanical properties of bars on deformability of the structure

As a measure of deformability of a bar structure under given loads we take the double work L of these loads on the displacements produced by the strain. The work L is the sum of the works $\underset{\triangle}{L}$ done on displacements of different bars by the loads acting on those bars. For a bar $\triangle$, the work $\underset{\triangle}{L}$ is

$$\begin{aligned}\underset{\triangle}{L} = {} & \int_{\underset{\triangle}{l}} \gamma \underset{\triangle}{A}(\underset{\triangle}{u}\sin\underset{\triangle}{\alpha}+\underset{\triangle}{w}\cos\underset{\triangle}{\alpha})\,d\underset{\triangle}{x}^1+\int_{\underset{\triangle}{l}}(\underset{\triangle}{p}^1\underset{\triangle}{u}+\underset{\triangle}{p}^3\underset{\triangle}{w}+\underset{\triangle}{m}\underset{\triangle}{\varphi})\,d\underset{\triangle}{x}^1+\\ & +\underset{\triangle}{N}_{LK}[\underset{\triangle}{u}]_{\underset{\triangle}{x}^1=\underset{\triangle}{l}}-\underset{\triangle}{N}_{KL}[\underset{\triangle}{u}]_{\underset{\triangle}{x}^1=0}+\underset{\triangle}{T}_{LK}[\underset{\triangle}{w}]_{\underset{\triangle}{x}^1=\underset{\triangle}{l}}-\underset{\triangle}{T}_{KL}[\underset{\triangle}{w}]_{\underset{\triangle}{x}^1=0}+\\ & +\underset{\triangle}{M}_{LK}[\underset{\triangle}{\varphi}]_{\underset{\triangle}{x}^1=\underset{\triangle}{l}}-\underset{\triangle}{M}_{KL}[\underset{\triangle}{\varphi}]_{\underset{\triangle}{x}^1=0}.\end{aligned} \tag{2.33}$$

The first integral in this expression is equal to the work of the dead weight of the bar, the second integral gives the work of the external loads applied along the bar, and the remaining terms represent the work of the reactions of the structure on the ends K and L of the bar.

Each bar of the structure, as well as the whole structure, is in equilibrium while it carries the loads. Therefore, by Clapeyron theorem, the work $\underset{\triangle}{L}$ is double the elastic strain energy of the bar $\triangle$, i.e.

$$\underset{\triangle}{L} = 2\underset{\triangle}{U}.$$

Denoting by $\underset{\triangle}{\Psi} = \frac{1}{2}(\underset{\triangle}{N}\underset{\triangle}{\varepsilon}+\underset{\triangle}{T}\underset{\triangle}{\vartheta}+\underset{\triangle}{M}\underset{\triangle}{\chi})$ the elastic strain energy stored between the sections $\underset{\triangle}{x}^1 = \text{const}$ and $\underset{\triangle}{x}^1+d\underset{\triangle}{x}^1 = \text{const}$ of the bar $\triangle$, we can write

$$\underset{\triangle}{U} = \int_{\underset{\triangle}{l}} \underset{\triangle}{\Psi}\,d\underset{\triangle}{x}^1 = \frac{1}{2}\int_{\underset{\triangle}{l}}(\underset{\triangle}{N}\underset{\triangle}{\varepsilon}+\underset{\triangle}{T}\underset{\triangle}{\vartheta}+\underset{\triangle}{M}\underset{\triangle}{\chi})\,d\underset{\triangle}{x}^1. \tag{2.34}$$

In order to examine the influence of the properties of the bars on deformability of the structure, let us compare the works L in two structures: the original one and one with the same geometry and identically supported but with bars having different geometrical and mechanical properties. We assume that both structures are in equilibrium under the same external loads.

The quantities relating to the original structure will be marked with a circle on top; those without a circle will refer to the modified structure, or to both—if there is no difference. The algebraic difference between a quantity in the original structure and its counterpart in the modified structure will be denoted with a 'Δ'.

We begin by considering the difference $\Delta\underset{\triangle}{L} = \underset{\triangle}{\mathring{L}}-\underset{\triangle}{L}$ for an arbitrary bar $\triangle$. Since the bar $\triangle$ is in equilibrium both before and after modification, $\underset{\triangle}{\mathring{L}} = 2\underset{\triangle}{\mathring{U}}$ and $\underset{\triangle}{L} = 2\underset{\triangle}{U}$; hence

$$\Delta\underset{\triangle}{U} = \Delta\underset{\triangle}{L}-\Delta\underset{\triangle}{U}.$$

Substituting expressions (2.33) and (2.34) (with circles on top of the symbols where necessary) for $\underset{\triangle}{\mathring{L}}$, $\underset{\triangle}{L}$, $\underset{\triangle}{\mathring{U}}$ and $\underset{\triangle}{U}$, we get

$$
\begin{aligned}
\underset{\triangle}{\Delta U} = & \int\limits_{\underset{\triangle}{l}} [\underset{\triangle}{\mathring{\gamma}}\underset{\triangle}{\mathring{A}}(\Delta \underset{\triangle}{u} \sin \underset{\triangle}{\alpha} + \Delta \underset{\triangle}{w} \cos \underset{\triangle}{\alpha}) + \Delta(\underset{\triangle}{\gamma}\underset{\triangle}{A})(\underset{\triangle}{u} \sin \underset{\triangle}{\alpha} + \underset{\triangle}{w} \cos \underset{\triangle}{\alpha})] \mathrm{d}\underset{\triangle}{x^1} + \\
& + \int\limits_{\underset{\triangle}{l}} [\underset{\triangle}{p^1} \Delta \underset{\triangle}{u} + \underset{\triangle}{p^3} \Delta \underset{\triangle}{w} + \underset{\triangle}{m} \Delta \underset{\triangle}{\varphi}] \mathrm{d}\underset{\triangle}{x^1} + \underset{\triangle}{\mathring{N}_{LK}} [\Delta \underset{\triangle}{u}]_{\underset{\triangle}{x^1} = \underset{\triangle}{l}} + \\
& + \Delta \underset{\triangle}{N_{LK}} [\underset{\triangle}{u}]_{\underset{\triangle}{x^1} = \underset{\triangle}{l}} - \underset{\triangle}{\mathring{N}_{KL}} [\Delta \underset{\triangle}{u}]_{\underset{\triangle}{x^1} = 0} - \Delta \underset{\triangle}{N_{KL}} [\underset{\triangle}{u}]_{\underset{\triangle}{x^1} = 0} + \\
& + \underset{\triangle}{\mathring{T}_{LK}} [\Delta \underset{\triangle}{w}]_{\underset{\triangle}{x^1} = \underset{\triangle}{l}} + \Delta \underset{\triangle}{T_{LK}} [\underset{\triangle}{w}]_{\underset{\triangle}{x^1} = \underset{\triangle}{l}} - \underset{\triangle}{\mathring{T}_{KL}} [\Delta \underset{\triangle}{w}]_{\underset{\triangle}{x^1} = 0} - \\
& - \Delta \underset{\triangle}{T_{KL}} [\underset{\triangle}{w}]_{\underset{\triangle}{x^1} = 0} + \underset{\triangle}{\mathring{M}_{LK}} [\Delta \underset{\triangle}{\varphi}]_{\underset{\triangle}{x^1} = \underset{\triangle}{l}} + \Delta \underset{\triangle}{M_{LK}} [\underset{\triangle}{\varphi}]_{\underset{\triangle}{x^1} = \underset{\triangle}{l}} - \underset{\triangle}{\mathring{M}_{KL}} [\Delta \underset{\triangle}{\varphi}]_{\underset{\triangle}{x^1} = 0} - \\
& - \Delta \underset{\triangle}{M_{KL}} [\underset{\triangle}{\varphi}]_{\underset{\triangle}{x^1} = 0} - \tfrac{1}{2} \int\limits_{\underset{\triangle}{l}} (\underset{\triangle}{\mathring{N}} \Delta \underset{\triangle}{\varepsilon} + \Delta \underset{\triangle}{N} \underset{\triangle}{\varepsilon} + \underset{\triangle}{\mathring{T}} \Delta \underset{\triangle}{\vartheta} + \Delta \underset{\triangle}{T} \underset{\triangle}{\vartheta} + \underset{\triangle}{\mathring{M}} \Delta \underset{\triangle}{\chi} + \Delta \underset{\triangle}{M} \underset{\triangle}{\chi}) \mathrm{d}\underset{\triangle}{x^1}.
\end{aligned}
$$

Using the equality

$$
\begin{aligned}
& \int\limits_{\underset{\triangle}{l}} \underset{\triangle}{\mathring{\gamma}}\underset{\triangle}{\mathring{A}}(\Delta \underset{\triangle}{u} \sin \underset{\triangle}{\alpha} + \Delta \underset{\triangle}{w} \cos \underset{\triangle}{\alpha}) \mathrm{d}\underset{\triangle}{x^1} + \int\limits_{\underset{\triangle}{l}} (\underset{\triangle}{p^1} \Delta \underset{\triangle}{u} + \underset{\triangle}{p^3} \Delta \underset{\triangle}{w} + \underset{\triangle}{m} \Delta \underset{\triangle}{\varphi}) \mathrm{d}\underset{\triangle}{x^1} + \\
& + \underset{\triangle}{\mathring{N}_{LK}} [\Delta \underset{\triangle}{u}]_{\underset{\triangle}{x^1} = \underset{\triangle}{l}} - \underset{\triangle}{\mathring{N}_{KL}} [\Delta \underset{\triangle}{u}]_{\underset{\triangle}{x^1} = 0} + \underset{\triangle}{\mathring{T}_{LK}} [\Delta \underset{\triangle}{w}]_{\underset{\triangle}{x^1} = \underset{\triangle}{l}} - \underset{\triangle}{\mathring{T}_{KL}} [\Delta \underset{\triangle}{w}]_{\underset{\triangle}{x^1} = 0} + \\
& + \underset{\triangle}{\mathring{M}_{LK}} [\Delta \underset{\triangle}{\varphi}]_{\underset{\triangle}{x^1} = \underset{\triangle}{l}} - M_{KL} [\Delta \underset{\triangle}{\varphi}]_{\underset{\triangle}{x^1} = 0} = \int\limits_{\underset{\triangle}{l}} (\underset{\triangle}{\mathring{N}} \Delta \underset{\triangle}{\varepsilon} + \underset{\triangle}{\mathring{T}} \Delta \underset{\triangle}{\vartheta} + \underset{\triangle}{\mathring{M}} \Delta \underset{\triangle}{\chi}) \mathrm{d}\underset{\triangle}{x^1},
\end{aligned}
$$

which expresses the principle of virtual work in application to the bar $\triangle$ in the original structure, and the equality

$$
\begin{aligned}
& \Delta \underset{\triangle}{N} \underset{\triangle}{\varepsilon} + \Delta \underset{\triangle}{T} \underset{\triangle}{\vartheta} + \Delta \underset{\triangle}{M} \underset{\triangle}{x} = \underset{\triangle}{\mathring{N}} \Delta \underset{\triangle}{\varepsilon} + \underset{\triangle}{\mathring{T}} \Delta \underset{\triangle}{\vartheta} + \underset{\triangle}{\mathring{M}} \Delta \underset{\triangle}{x} + \\
& \quad + (\underset{\triangle}{\varepsilon})^2 \Delta (\underset{\triangle}{E}\underset{\triangle}{A}) + (\underset{\triangle}{\vartheta})^2 \Delta (\underset{\triangle}{G}\underset{\triangle}{A}/\underset{\triangle}{\varkappa}) + (\underset{\triangle}{\chi})^2 \Delta (\underset{\triangle}{E}\underset{\triangle}{J}) - \\
& \quad - [\underset{\triangle}{\mathring{E}}\underset{\triangle}{\mathring{A}}(\Delta \underset{\triangle}{\varepsilon})^2 + \underset{\triangle}{\mathring{G}}\underset{\triangle}{\mathring{A}}/\underset{\triangle}{\mathring{\varkappa}}(\Delta \underset{\triangle}{\vartheta})^2 + \underset{\triangle}{\mathring{E}}\underset{\triangle}{\mathring{J}}(\Delta \underset{\triangle}{\chi})^2],
\end{aligned}
$$

which follows from the physical relations, we can represent $\frac{1}{2} \Delta \underset{\triangle}{L} = \Delta \underset{\triangle}{U}$ in the form

$$
\begin{aligned}
\tfrac{1}{2} \Delta \underset{\triangle}{L} = \Delta \underset{\triangle}{U} = & \int\limits_{\underset{\triangle}{l}} (\underset{\triangle}{u} \sin \underset{\triangle}{\alpha} + \underset{\triangle}{w} \cos \underset{\triangle}{\alpha}) \Delta (\underset{\triangle}{\gamma}\underset{\triangle}{A}) \mathrm{d}\underset{\triangle}{x^1} + \\
& + \Delta \underset{\triangle}{N_{LK}} [\underset{\triangle}{u}]_{\underset{\triangle}{x^1} = \underset{\triangle}{l}} - \Delta \underset{\triangle}{N_{KL}} [\underset{\triangle}{u}]_{\underset{\triangle}{x^1} = 0} + \Delta \underset{\triangle}{T_{KL}} [\underset{\triangle}{w}]_{\underset{\triangle}{x^1} = \underset{\triangle}{l}} - \\
& - \Delta \underset{\triangle}{T_{KL}} [\underset{\triangle}{w}]_{\underset{\triangle}{x^1} = 0} + \Delta \underset{\triangle}{M_{KL}} [\underset{\triangle}{\varphi}]_{\underset{\triangle}{x^1} = \underset{\triangle}{l}} - \Delta \underset{\triangle}{M_{KL}} [\underset{\triangle}{\varphi}]_{\underset{\triangle}{x^1} = 0} - \\
& - \tfrac{1}{2} \int\limits_{\underset{\triangle}{l}} [(\underset{\triangle}{\varepsilon})^2 \Delta (\underset{\triangle}{E}\underset{\triangle}{A}) + (\underset{\triangle}{\vartheta})^2 \Delta (\underset{\triangle}{G}\underset{\triangle}{A}/\underset{\triangle}{\varkappa}) + (\underset{\triangle}{\chi})^2 \Delta (\underset{\triangle}{E}\underset{\triangle}{J})] \mathrm{d}\underset{\triangle}{x^1} + \\
& + \tfrac{1}{2} \int\limits_{\underset{\triangle}{l}} [\underset{\triangle}{\mathring{E}}\underset{\triangle}{\mathring{A}}(\Delta \underset{\triangle}{\varepsilon})^2 + \underset{\triangle}{\mathring{G}}\underset{\triangle}{\mathring{A}}/\underset{\triangle}{\mathring{\varkappa}}(\Delta \underset{\triangle}{\vartheta})^2 + \underset{\triangle}{\mathring{E}}\underset{\triangle}{\mathring{J}}(\Delta \underset{\triangle}{\chi})^2] \mathrm{d}\underset{\triangle}{x^1}.
\end{aligned}
\tag{2.35}
$$

Different terms occurring in the above formula reflect the influence of various factors on deformability of the structure. The first term expresses the influence of the distribution of the dead weight $\gamma \underset{\triangle}{A}$ of the bar. The group of the terms proportional to the differences $\underset{\triangle}{\Delta N_{LK}}-\underset{\triangle}{\Delta N_{KL}}$, $\underset{\triangle}{\Delta T_{LK}}-\underset{\triangle}{\Delta T_{KL}}$, $\underset{\triangle}{\Delta M_{LK}}$, $-\underset{\triangle}{\Delta M_{KL}}$ represents the effect of the changes in the reactions of the structure at the ends of the bar due to the changes in the geometrical and mechanical properties of this bar and all other bars of the structure. The next term results from changes in the longitudinal, shearing and flexural rigidities of the bar. The last term of the formula is always positive.

In the particular case where the bar $\triangle$ is left unchanged, so that

$$\underset{\triangle}{\Delta}(\underset{\triangle}{\gamma A}) = \underset{\triangle}{\Delta}(\underset{\triangle}{EA}) = \underset{\triangle}{\Delta}(\underset{\triangle}{GA}/\underset{\triangle}{\varkappa}) = \underset{\triangle}{\Delta}(\underset{\triangle}{EJ}) = 0,$$

the difference $\underset{\triangle}{\Delta U}$ depends on changes in the properties of other bars, and hence in the reactions of the structure on the bar $\triangle$:

$$\begin{aligned}\tfrac{1}{2}\underset{\triangle}{\Delta L} = \underset{\triangle}{\Delta U} = {} & \underset{\triangle}{\Delta N_{LK}}[\underset{\triangle}{u}]_{\underset{\triangle}{x^1}=\underset{\triangle}{l}}-\underset{\triangle}{\Delta N_{KL}}[\underset{\triangle}{u}]_{\underset{\triangle}{x^1}=0}+\underset{\triangle}{\Delta T_{LK}}[\underset{\triangle}{w}]_{\underset{\triangle}{x^1}=\underset{\triangle}{l}}-\\ & -\underset{\triangle}{\Delta T_{KL}}[\underset{\triangle}{w}]_{\underset{\triangle}{x^1}=0}+\underset{\triangle}{\Delta M_{LK}}[\underset{\triangle}{\varphi}]_{\underset{\triangle}{x^1}=\underset{\triangle}{l}}-\underset{\triangle}{\Delta M_{KL}}[\underset{\triangle}{\varphi}]_{\underset{\triangle}{x^1}=0}+\\ & +\tfrac{1}{2}\int_{\underset{\triangle}{l}}[\underset{\triangle}{EA}(\Delta\underset{\triangle}{\varepsilon})^2+\underset{\triangle}{GA}/\underset{\triangle}{\varkappa}(\Delta\underset{\triangle}{\vartheta})^2+\underset{\triangle}{EJ}(\Delta\underset{\triangle}{\chi})^2]\,d\underset{\triangle}{x^1}.\end{aligned}$$

The total difference ΔU can be obtained by adding up the differences $\underset{\triangle}{\Delta U}$ for all bars $\triangle$ of the structure: $\Delta U = \sum_{\triangle}\underset{\triangle}{\Delta U}$. In the state of equilibrium, both in the original structure and in the modified structure, the work done by the reaction forces on the displacements of the ends of the bars is equal to the work of the nodal forces on the displacements of the nodes. Hence the equalities

$$\begin{aligned}&\sum_{\triangle}[\underset{\triangle}{\mathring{N}_{LK}}[\underset{\triangle}{\mathring{u}}]_{\underset{\triangle}{x^1}=\underset{\triangle}{l}}-\underset{\triangle}{\mathring{N}_{KL}}[\underset{\triangle}{\mathring{u}}]_{\underset{\triangle}{x^1}=0}+\underset{\triangle}{\mathring{T}_{LK}}[\underset{\triangle}{\mathring{w}}]_{\underset{\triangle}{x^1}=\underset{\triangle}{l}}-\underset{\triangle}{\mathring{T}_{KL}}[\underset{\triangle}{\mathring{w}}]_{\underset{\triangle}{x^1}=0}+\\ &\qquad+\underset{\triangle}{\mathring{M}_{LK}}[\underset{\triangle}{\mathring{\varphi}}]_{\underset{\triangle}{x^1}=\underset{\triangle}{l}}-\underset{\triangle}{\mathring{M}_{KL}}[\underset{\triangle}{\mathring{\varphi}}]_{\underset{\triangle}{x^1}=\underset{\triangle}{l}}-\underset{\triangle}{\mathring{M}_{KL}}[\underset{\triangle}{\mathring{\varphi}}]_{\underset{\triangle}{x^1}=0}] = \sum_{s=1}^{n}\underset{s}{p^1}\underset{s}{\mathring{u}^1}+\underset{s}{P^3}\underset{s}{\mathring{u}^3}+\underset{s}{m}\underset{s}{\mathring{\varphi}},\\ &\sum_{\triangle}\{\underset{\triangle}{N_{LK}}[\underset{\triangle}{u}]_{\underset{\triangle}{x^1}=\underset{\triangle}{l}}-\underset{\triangle}{N_{KL}}[\underset{\triangle}{u}]_{\underset{\triangle}{x^1}=0}+\underset{\triangle}{T_{LK}}[\underset{\triangle}{w}]_{\underset{\triangle}{x^1}=\underset{\triangle}{l}}-\underset{\triangle}{T_{KL}}[\underset{\triangle}{w}]_{\underset{\triangle}{x^1}=0}+\\ &\qquad+\underset{\triangle}{M_{LK}}[\underset{\triangle}{\varphi}]_{\underset{\triangle}{x^1}=\underset{\triangle}{l}}-\underset{\triangle}{M_{KL}}[\underset{\triangle}{\varphi}]_{\underset{\triangle}{x^1}=0}\} = \sum_{s=1}^{n}\underset{s}{P^1}\underset{s}{u^1}+\underset{s}{P^3}\underset{s}{u^3}+\underset{s}{m}\underset{s}{\varphi}.\end{aligned}\tag{2.36}$$

By the principle of virtual work,

$$\sum_{\triangle}\{\overset{\circ}{\underset{\triangle}{N}}_{LK}[\underset{\triangle}{u}]_{\underset{\triangle}{x^1}=\underset{\triangle}{l}}-\overset{\circ}{\underset{\triangle}{N}}_{KL}[\underset{\triangle}{u}]_{\underset{\triangle}{x^1}=0}+\overset{\circ}{\underset{\triangle}{T}}_{LK}[\underset{\triangle}{w}]_{\underset{\triangle}{x^1}=\underset{\triangle}{l}}-\overset{\circ}{\underset{\triangle}{T}}_{KL}[\underset{\triangle}{w}]_{\underset{\triangle}{x^1}=0}+$$

$$+\overset{\circ}{\underset{\triangle}{M}}_{LK}[\underset{\triangle}{\varphi}]_{\underset{\triangle}{x^1}=\underset{\triangle}{l}}-\overset{\circ}{\underset{\triangle}{M}}_{KL}[\underset{\triangle}{\varphi}]_{\underset{\triangle}{x^1}=0}\}=\sum_{s=1}^{n}\underset{s}{P^1}\underset{s}{u^1}+\underset{s}{P^3}\underset{s}{u^3}+\underset{s}{m}\underset{s}{\varphi}. \tag{2.37}$$

Subtracting this equality from $(2.36)_2$ gives

$$\sum_{\triangle}[\underset{\triangle}{\Delta N}_{LK}[\underset{\triangle}{u}]_{\underset{\triangle}{x^1}=\underset{\triangle}{l}}-\underset{\triangle}{\Delta N}_{KL}[\underset{\triangle}{u}]_{\underset{\triangle}{x^1}=0}+\underset{\triangle}{\Delta T}_{LK}[\underset{\triangle}{w}]_{\underset{\triangle}{x^1}=\underset{\triangle}{l}}-$$

$$-\underset{\triangle}{\Delta T}_{KL}[\underset{\triangle}{w}]_{\underset{\triangle}{x^1}=0}+\underset{\triangle}{\Delta M}_{LK}[\underset{\triangle}{\varphi}]_{\underset{\triangle}{x^1}=\underset{\triangle}{l}}-\underset{\triangle}{\Delta M}_{KL}[\underset{\triangle}{\varphi}]_{\underset{\triangle}{x^1}=0}]=0. \tag{2.38}$$

Accordingly, the formula for the elastic strain energy difference between the original and the modified structure, calculated from (2.35), takes the form

$$\Delta U=\tfrac{1}{2}\Delta L=\sum_{\triangle}\int_{\underset{\triangle}{l}}(\underset{\triangle}{u}\sin\underset{\triangle}{\alpha}+\underset{\triangle}{w}\cos\underset{\triangle}{\alpha})\Delta(\underset{\triangle}{\gamma}\underset{\triangle}{A})\,d\underset{\triangle}{x^1}-$$

$$-\tfrac{1}{2}\sum_{\triangle}\int_{\underset{\triangle}{l}}[(\underset{\triangle}{\varepsilon})^2\Delta(\underset{\triangle}{E}\underset{\triangle}{A})+(\underset{\triangle}{\vartheta})^2\Delta(\underset{\triangle}{G}\underset{\triangle}{A}/\underset{\triangle}{\varkappa})+(\underset{\triangle}{\chi})^2\Delta(\underset{\triangle}{E}\underset{\triangle}{J})]\,d\underset{\triangle}{x^1}+Q^2, \tag{2.39}$$

where

$$Q^2=\tfrac{1}{2}\sum_{\triangle}\int_{\underset{\triangle}{l}}[\overset{\circ}{\underset{\triangle}{E}}\overset{\circ}{\underset{\triangle}{A}}(\Delta\underset{\triangle}{\varepsilon})^2+\overset{\circ}{\underset{\triangle}{G}}\overset{\circ}{\underset{\triangle}{A}}/\overset{\circ}{\underset{\triangle}{\varkappa}}(\Delta\underset{\triangle}{\vartheta})^2+\overset{\circ}{\underset{\triangle}{E}}\overset{\circ}{\underset{\triangle}{J}}(\Delta\underset{\triangle}{\chi})^2]\,d\underset{\triangle}{x^1}.$$

Formula (2.39), expressing the difference ΔU in terms of the differences $\Delta(\underset{\triangle}{\gamma}\underset{\triangle}{A})$, $\Delta(\underset{\triangle}{E}\underset{\triangle}{A})$, $\Delta(\underset{\triangle}{G}\underset{\triangle}{A}/\underset{\triangle}{\varkappa})$ and $\Delta(\underset{\triangle}{E}\underset{\triangle}{J})$ of the quantities determining the geometrical and the mechanical properties of the structures considered, constitutes a basis for formulating and solving the least-deformability design problem for bar structures.

Let us prove that $\Delta U=0$ if

$$\Delta(\underset{\triangle}{\gamma}\underset{\triangle}{A})=\Delta(\underset{\triangle}{E}\underset{\triangle}{A})=\Delta(\underset{\triangle}{G}\underset{\triangle}{A}/\underset{\triangle}{\varkappa})=\Delta(\underset{\triangle}{E}\underset{\triangle}{J})=0$$

for all bars $\triangle$. By the principle of virtual work, the following equalities hold for the original structure and for the modified structure:

$$\sum_{\triangle}\int_{\underset{\triangle}{l}}\overset{\circ}{\underset{\triangle}{\gamma}}\overset{\circ}{\underset{\triangle}{A}}(\Delta\underset{\triangle}{u}\sin\underset{\triangle}{\alpha}+\Delta\underset{\triangle}{w}\cos\underset{\triangle}{\alpha})\,d\underset{\triangle}{x^1}+\sum_{\triangle}\int_{\underset{\triangle}{l}}(\underset{\triangle}{p^1}\Delta\underset{\triangle}{u^1}+\underset{\triangle}{p^3}\Delta\underset{\triangle}{w}+\underset{\triangle}{m}\Delta\underset{\triangle}{\varphi})\,d\underset{\triangle}{x^1}+$$

$$+\sum_{s=1}^{n}(\underset{s}{P^1}\Delta\underset{s}{u^1}+\underset{s}{P^3}\Delta\underset{s}{u^3}+\underset{s}{m}\Delta\underset{s}{\varphi})=\sum_{\triangle}\int_{\underset{\triangle}{l}}(\overset{\circ}{\underset{\triangle}{N}}\Delta\underset{\triangle}{\varepsilon}+\overset{\circ}{\underset{\triangle}{T}}\Delta\underset{\triangle}{\vartheta}+\overset{\circ}{\underset{\triangle}{M}}\Delta\underset{\triangle}{\chi})\,d\underset{\triangle}{x^1}, \tag{2.40$_1$}$$

$$\sum_{\triangle}\int_{\underset{\triangle}{l}}\underset{\triangle}{\gamma}\underset{\triangle}{A}(\Delta\underset{\triangle}{u}\sin\underset{\triangle}{\alpha}+\Delta\underset{\triangle}{w}\cos\underset{\triangle}{\alpha})d\underset{\triangle}{x}^1+\sum_{\triangle}\int_{\underset{\triangle}{l}}(\underset{\triangle}{p}^1\Delta\underset{\triangle}{u}+\underset{\triangle}{p}^3\Delta\underset{\triangle}{w}+\underset{\triangle}{m}\Delta\underset{\triangle}{\varphi})d\underset{\triangle}{x}^1+$$

$$+\sum_{s=1}^{n}(\underset{s}{P}^1\Delta\underset{s}{u}^1+\underset{s}{P}^3\Delta\underset{s}{u}^3+\underset{s}{m}\Delta\underset{s}{\varphi})=\sum_{\triangle}\int_{\underset{\triangle}{l}}(\underset{\triangle}{N}\Delta\underset{\triangle}{\varepsilon}+\underset{\triangle}{T}\Delta\underset{\triangle}{\vartheta}+\underset{\triangle}{M}\Delta\underset{\triangle}{\chi})d\underset{\triangle}{x}^1. \qquad (2.40)_2$$

Subtracting the second equality from the first, we obtain

$$\sum_{\triangle}\int_{\underset{\triangle}{l}}\Delta(\underset{\triangle}{\gamma}\underset{\triangle}{A})(\Delta\underset{\triangle}{u}\sin\underset{\triangle}{\alpha}+\Delta\underset{\triangle}{w}\cos\underset{\triangle}{\alpha})d\underset{\triangle}{x}^1$$

$$=\sum_{\triangle}\int_{\underset{\triangle}{l}}(\Delta\underset{\triangle}{N}\Delta\underset{\triangle}{\varepsilon}+\Delta\underset{\triangle}{T}\Delta\varrho\underset{\triangle}{\vartheta}+\Delta\underset{\triangle}{M}\Delta\underset{\triangle}{\chi})d\underset{\triangle}{x}^1, \qquad (2.41)$$

and so

$$Q^2=\tfrac{1}{2}\sum_{\triangle}\int_{\underset{\triangle}{l}}[\underset{\triangle}{\mathring{E}}\underset{\triangle}{\mathring{A}}(\Delta\underset{\triangle}{\varepsilon})^2+\underset{\triangle}{\mathring{G}}\underset{\triangle}{\mathring{A}}/\underset{\triangle}{\mathring{\varkappa}}(\Delta\underset{\triangle}{\vartheta})^2+\underset{\triangle}{\mathring{E}}\underset{\triangle}{\mathring{J}}(\Delta\underset{\triangle}{\varkappa})^2]d\underset{\triangle}{x}^1$$

$$=\sum_{\triangle}\int_{\underset{\triangle}{l}}\Delta(\underset{\triangle}{\gamma}\underset{\triangle}{A})(\Delta\underset{\triangle}{u}\sin\underset{\triangle}{\alpha}+\Delta\underset{\triangle}{w}\cos\underset{\triangle}{\alpha})d\underset{\triangle}{x}^1-$$

$$-\tfrac{1}{2}\sum_{\triangle}\int_{\underset{\triangle}{l}}[\underset{\triangle}{\varepsilon}\Delta(\underset{\triangle}{E}\underset{\triangle}{A})\Delta\underset{\triangle}{\varepsilon}+\underset{\triangle}{\vartheta}\Delta\underset{\triangle}{\vartheta}\Delta(\underset{\triangle}{G}\underset{\triangle}{A}/\underset{\triangle}{\varkappa})+\underset{\triangle}{\chi}\Delta\underset{\triangle}{\chi}\Delta(\underset{\triangle}{E}\underset{\triangle}{J})]d\underset{\triangle}{x}^1. \qquad (2.42)$$

It follows that if the bars of the two structures do not differ in geometrical and mechanical properties, then $Q^2=0$.

Least deformable isostatic bar structures

For an isostatic structure, in which the internal forces are independent of deformations and deformational rigidities of the bars, we have $\Delta\underset{\triangle}{M}=\Delta\underset{\triangle}{T}=\Delta\underset{\triangle}{N}=0$, if also the dead weight distribution is assumed to have no effect on deformability of the structure.

Under the isostatic conditions $\Delta\underset{\triangle}{M}=\Delta\underset{\triangle}{T}=\Delta\underset{\triangle}{N}=0$, the physical relations give

$$\Delta\underset{\triangle}{\chi}=-\underset{\triangle}{\chi}\frac{\Delta\underset{\triangle}{J}}{\underset{\triangle}{J}+\Delta\underset{\triangle}{J}},\qquad \Delta\underset{\triangle}{\vartheta}=-\underset{\triangle}{\vartheta}\frac{\Delta(\underset{\triangle}{A}/\underset{\triangle}{\varkappa})}{\underset{\triangle}{A}/\underset{\triangle}{\varkappa}+\Delta(\underset{\triangle}{A}/\underset{\triangle}{\varkappa})},\qquad \Delta\underset{\triangle}{\varepsilon}=-\underset{\triangle}{\varepsilon}\frac{\Delta\underset{\triangle}{A}}{\underset{\triangle}{A}+\Delta\underset{\triangle}{A}},$$

and so formula (2.39) takes the form

$$\Delta U=\frac{1}{2}\sum_{\triangle}\int_{\underset{\triangle}{l}}\left[M^2\frac{1}{E}\Delta\left(\frac{1}{J}\right)+\frac{\underset{\triangle}{N}^2}{E}\Delta\left(\frac{1}{\underset{\triangle}{A}}\right)+\frac{1}{G}\underset{\triangle}{T}^2\Delta\left(\frac{\underset{\triangle}{\varkappa}}{\underset{\triangle}{A}}\right)\right]d\underset{\triangle}{x}^1$$

$$= -\frac{1}{2}\sum_{\triangle}\int_{\underset{\triangle}{l}}\left[\frac{1}{\underset{\triangle}{EJ}}M^2\frac{\underset{\triangle}{\Delta J}}{\underset{\triangle}{\Delta J}+\underset{\triangle}{J}}+\frac{\underset{\triangle}{N^2}}{\underset{\triangle}{EA}}\,\frac{\underset{\triangle}{\Delta A}}{\underset{\triangle}{\Delta A}+\underset{\triangle}{A}}+\right.$$

$$\left.+\frac{\underset{\triangle}{T^2}}{\underset{\triangle}{GA}}\underset{\triangle}{\varkappa}\,\frac{\Delta(\underset{\triangle}{A}/\underset{\triangle}{\varkappa})}{\underset{\triangle}{A}/\underset{\triangle}{\varkappa}+\Delta(\underset{\triangle}{A}/\underset{\triangle}{\varkappa})}\right]d\underset{\triangle}{x}^1.$$

Using the expansions

$$\Delta\left(\frac{1}{\underset{\triangle}{J}}\right) = -\frac{\underset{\triangle}{\Delta J}}{\underset{\triangle}{J}(\underset{\triangle}{J}+\underset{\triangle}{\Delta J})} = -\frac{1}{\underset{\triangle}{J^2}}\underset{\triangle}{\Delta J}+\frac{2}{\underset{\triangle}{J^3}}(\underset{\triangle}{\Delta J})^2+\;\dots\;+0(\underset{\triangle}{\Delta J})^3,$$

$$\Delta\left(\frac{\underset{\triangle}{\varkappa}}{\underset{\triangle}{A}}\right) = -\frac{\Delta(\underset{\triangle}{A}/\underset{\triangle}{\varkappa})}{\underset{\triangle}{A}/\underset{\triangle}{\varkappa}[\underset{\triangle}{A}/\underset{\triangle}{\varkappa}+\Delta(\underset{\triangle}{A}/\underset{\triangle}{\varkappa})]}$$

$$= -\frac{1}{(\underset{\triangle}{A}/\underset{\triangle}{\varkappa})^2}\Delta(\underset{\triangle}{A}/\underset{\triangle}{\varkappa})+\frac{2}{(\underset{\triangle}{A}/\underset{\triangle}{\varkappa})^3}[\Delta(\underset{\triangle}{A}/\underset{\triangle}{\varkappa})]^2+\;\dots$$

and neglecting the terms of order higher than 2 in the differences, we obtain the basic formula

$$\Delta U \approx \frac{1}{2}\sum_{\triangle}\int_{\underset{\triangle}{l}}\left[\frac{\underset{\triangle}{M^2}}{\underset{\triangle}{EJ^2}}\underset{\triangle}{\Delta J}+\frac{\underset{\triangle}{N^2}}{\underset{\triangle}{EA^2}}\underset{\triangle}{\Delta A}+\frac{\underset{\triangle}{T^2}}{\underset{\triangle}{GA^2}}\underset{\triangle}{\varkappa}^2\Delta(\underset{\triangle}{A}/\underset{\triangle}{\varkappa})\right]d\underset{\triangle}{x}^1-$$

$$-\sum_{\triangle}\int_{\underset{\triangle}{l}}\left[\frac{\underset{\triangle}{M^2}}{\underset{\triangle}{EJ^3}}(\underset{\triangle}{\Delta J})^2+\frac{\underset{\triangle}{N^2}}{\underset{\triangle}{EA^3}}(\underset{\triangle}{\Delta A})^2+\frac{\underset{\triangle}{T^2}}{\underset{\triangle}{GA^3}}\underset{\triangle}{\varkappa}^3(\Delta(\underset{\triangle}{A}/\underset{\triangle}{\varkappa}))^2\right]d\underset{\triangle}{x}^1, \tag{2.43}$$

a starting point for optimum design of isostatic bar structures.

The influence of prestressing on deformability of a bar structure

To measure the effect that an initial prestressing of the bars of a structure has on the elastic strain energy of this structure, let us calculate the difference $\Delta U = \overset{\circ}{U}-\overset{s}{U}$ between the energies before and after prestressing.

All quantities relating to the original structure are marked, as before, with a circle on top, and those referring to the prestressed structure—with an 's'.

First, let us calculate the energy difference $\underset{\triangle}{\Delta U} = \underset{\triangle}{\overset{\circ}{U}}-\underset{\triangle}{\overset{s}{U}}$ for an arbitrary bar $\triangle$ with ends K and L.

The works $\underset{\triangle}{\overset{\circ}{L}}$ and $\underset{\triangle}{\overset{s}{L}}$ and the energies $\underset{\triangle}{\overset{\circ}{U}}$ and $\underset{\triangle}{\overset{s}{U}}$ of the bar $\triangle$ before and after prestressing are given by the formulae

$$\begin{aligned}
\underset{\triangle}{\overset{\circ}{L}} = {} & \int_{\underset{\triangle}{l}} \underset{\triangle}{\gamma}\underset{\triangle}{A}(\underset{\triangle}{\mathring{u}}\sin\underset{\triangle}{\alpha}+\underset{\triangle}{\mathring{w}}\cos\underset{\triangle}{\alpha})\mathrm{d}x^1+\int_{\underset{\triangle}{l}}(\underset{\triangle}{p^1}\underset{\triangle}{\mathring{u}}+\underset{\triangle}{p^3}\underset{\triangle}{\mathring{w}}+\underset{\triangle}{m}\underset{\triangle}{\mathring{\varphi}})\mathrm{d}x^1+\underset{\triangle}{\mathring{N}_{LK}}[\underset{\triangle}{\mathring{u}}]_{x^1=\underset{\triangle}{l}}- \\
& -\underset{\triangle}{\mathring{N}_{KL}}[\underset{\triangle}{\mathring{u}}]_{x^1=0}+\underset{\triangle}{\mathring{T}_{LK}}[\underset{\triangle}{\mathring{w}}]_{x^1=\underset{\triangle}{l}}-\underset{\triangle}{\mathring{T}_{KL}}[\underset{\triangle}{\mathring{w}}]_{x^1=0}+\underset{\triangle}{\mathring{M}_{LK}}[\underset{\triangle}{\mathring{\varphi}}]_{x^1=\underset{\triangle}{l}}-\underset{\triangle}{\mathring{M}_{LK}}[\underset{\triangle}{\mathring{\varphi}}]_{x^1=0},
\end{aligned}$$

$$\underset{\triangle}{\overset{\circ}{U}} = \int_{\underset{\triangle}{l}}\underset{\triangle}{\mathring{\Psi}}\mathrm{d}x^1 = \frac{1}{2}\int_{\underset{\triangle}{l}}(\underset{\triangle}{\mathring{N}}\underset{\triangle}{\mathring{\varepsilon}}+\underset{\triangle}{\mathring{T}}\underset{\triangle}{\mathring{\vartheta}}+\underset{\triangle}{\mathring{M}}\underset{\triangle}{\mathring{\chi}})\mathrm{d}x^1,$$

$$\begin{aligned}
\underset{\triangle}{\overset{s}{L}} = {} & \int_{\underset{\triangle}{l}} \underset{\triangle}{\gamma}\underset{\triangle}{A}(\underset{\triangle}{\overset{s}{u}}\sin\underset{\triangle}{\alpha}+\underset{\triangle}{\overset{s}{w}}\cos\underset{\triangle}{\alpha})\mathrm{d}x^1+\int_{\underset{\triangle}{l}}(\underset{\triangle}{p^1}\underset{\triangle}{\overset{s}{u}}+\underset{\triangle}{p^3}\underset{\triangle}{\overset{s}{w}}+\underset{\triangle}{m}\underset{\triangle}{\overset{s}{\varphi}})\mathrm{d}x^1+\underset{\triangle}{\overset{s}{N}_{LK}}[\underset{\triangle}{\overset{s}{u}}]_{x^1=\underset{\triangle}{l}}- \\
& -\underset{\triangle}{\overset{s}{N}_{KL}}[\underset{\triangle}{\overset{s}{u}}]_{x^1=0}+\underset{\triangle}{\overset{s}{T}_{LK}}[\underset{\triangle}{\overset{s}{w}}]_{x^1=\underset{\triangle}{l}}-\underset{\triangle}{\overset{s}{T}_{KL}}[\underset{\triangle}{\overset{s}{w}}]_{x^1=0}+\underset{\triangle}{\overset{s}{M}_{KL}}[\underset{\triangle}{\overset{s}{\varphi}}]_{x^1=\underset{\triangle}{l}}- \\
& -\underset{\triangle}{\overset{s}{M}_{LK}}[\underset{\triangle}{\overset{s}{\varphi}}]_{x^1=0}-\int_{\underset{\triangle}{l}}\underset{\triangle}{S}(\underset{\triangle}{\overset{s}{\varepsilon}}+\underset{\triangle}{e}\underset{\triangle}{\overset{s}{\chi}}+\underset{\triangle}{e_{,1}}\underset{\triangle}{\overset{s}{\vartheta}})\mathrm{d}x^1,
\end{aligned}$$

$$\underset{\triangle}{\overset{s}{U}} = \int_{\underset{\triangle}{l}}\underset{\triangle}{\overset{s}{\Psi}}\mathrm{d}x^1 = \frac{1}{2}\int_{\underset{\triangle}{l}}(\underset{\triangle}{\overset{s}{N}}\underset{\triangle}{\overset{s}{\varepsilon}}+\underset{\triangle}{\overset{s}{T}}\underset{\triangle}{\overset{s}{\vartheta}}+\underset{\triangle}{\overset{s}{M}}\underset{\triangle}{\overset{s}{\chi}})\mathrm{d}x^1.$$

By the Clapeyron theorem, $\underset{\triangle}{\overset{\circ}{L}} = 2\underset{\triangle}{\overset{\circ}{U}}$. A procedure similar to that used in deriving formula (2.35) leads to the following expression for $\underset{\triangle}{\Delta U}$:

$$\begin{aligned}
\underset{\triangle}{\Delta U} = \underset{\triangle}{\overset{\circ}{U}}-\underset{\triangle}{\overset{s}{U}} = {} & \underset{\triangle}{\Delta \mathring{N}_{LK}}[\underset{\triangle}{\overset{s}{u}}]_{x^1=\underset{\triangle}{l}}-\underset{\triangle}{\Delta N_{KL}}[\underset{\triangle}{\overset{s}{u}}]_{x^1=0}+\underset{\triangle}{\Delta T_{LK}}[\underset{\triangle}{\overset{s}{w}}]_{x^1=\underset{\triangle}{l}}- \\
& -\underset{\triangle}{\Delta T_{KL}}[\underset{\triangle}{\overset{s}{w}}]_{x^1=0}+\underset{\triangle}{\Delta M_{LK}}[\underset{\triangle}{\overset{s}{\varphi}}]_{x^1=\underset{\triangle}{l}}-\underset{\triangle}{\Delta M_{KL}}[\underset{\triangle}{\overset{s}{\varphi}}]_{x^1=0}+ \\
& +\int_{\triangle}\underset{\triangle}{S}(\underset{\triangle}{\overset{s}{\varepsilon}}+\underset{\triangle}{e}\,\underset{\triangle}{\overset{s}{\chi}}+\underset{\triangle}{e_{,1}}\underset{\triangle}{\overset{s}{\vartheta}})\mathrm{d}x^1+ \\
& +\tfrac{1}{2}\int_{\underset{\triangle}{l}}[\underset{\triangle}{EA}(\underset{\triangle}{\Delta\varepsilon})^2+\underset{\triangle}{GA/\varkappa}(\underset{\triangle}{\Delta\vartheta})^2+\underset{\triangle}{EJ}(\underset{\triangle}{\Delta\chi})^2]\mathrm{d}x^1.
\end{aligned} \tag{2.44}$$

To find the total difference ΔU of the elastic strain energies of the original structure and the prestressed structure we take the sum of the differences $\underset{\triangle}{\Delta U}$ for all bars $\triangle$. Since the principle of virtual work implies that (cf. (2.38))

$$\sum_{\triangle}\Big[\Delta \underset{\triangle}{N}_{LK}[\overset{s}{\underset{\triangle}{u}}]_{\underset{\triangle}{x^1}=\underset{\triangle}{l}} - \Delta \underset{\triangle}{N}_{KL}[\overset{s}{\underset{\triangle}{u}}]_{\underset{\triangle}{x^1}=0} + \Delta \underset{\triangle}{T}_{LK}[\overset{s}{\underset{\triangle}{w}}]_{\underset{\triangle}{x^1}=\underset{\triangle}{l}} - \Delta \underset{\triangle}{T}_{KL}[\overset{s}{\underset{\triangle}{w}}]_{\underset{\triangle}{x^1}=0} +$$

$$+\Delta \underset{\triangle}{M}_{LK}[\overset{s}{\underset{\triangle}{\varphi}}]_{\underset{\triangle}{x^1}=\underset{\triangle}{l}} - \Delta \underset{\triangle}{M}_{KL}[\overset{s}{\underset{\triangle}{\varphi}}]_{\underset{\triangle}{x^1}=0}\Big] = 0,$$

the result can be written as

$$\Delta U = \sum_{\triangle} \int_{\underset{\triangle}{l}} \underset{\triangle}{S}(\overset{s}{\underset{\triangle}{\varepsilon}} + \underset{\triangle}{e}\,\overset{s}{\underset{\triangle}{\chi}} + \underset{\triangle}{e}_{,1}\overset{s}{\underset{\triangle}{\vartheta}})\mathrm{d}\underset{\triangle}{x^1} +$$

$$+\tfrac{1}{2}\sum_{\triangle}\int_{\underset{\triangle}{l}} [\underset{\triangle}{E}\underset{\triangle}{A}(\Delta \underset{\triangle}{\varepsilon})^2 + \underset{\triangle}{G}\underset{\triangle}{A}/\underset{\triangle}{\varkappa}(\Delta \underset{\triangle}{\vartheta})^2 + \underset{\triangle}{E}\underset{\triangle}{J}(\Delta \underset{\triangle}{\chi})^2]\mathrm{d}\underset{\triangle}{x^1}, \tag{2.45}$$

with

$$\sum_{\triangle}\int_{\underset{\triangle}{l}} [\underset{\triangle}{E}\underset{\triangle}{A}(\Delta \underset{\triangle}{\varepsilon})^2 + \underset{\triangle}{G}\underset{\triangle}{A}/\underset{\triangle}{\varkappa}(\Delta \underset{\triangle}{\vartheta})^2 + \underset{\triangle}{E}\underset{\triangle}{J}(\Delta \underset{\triangle}{\chi})^2]\mathrm{d}\underset{\triangle}{x^1}$$

$$= \sum_{\triangle}\int_{\underset{\triangle}{l}} \underset{\triangle}{S}(\Delta \overset{s}{\underset{\triangle}{\varepsilon}} + \underset{\triangle}{e}\Delta \overset{s}{\underset{\triangle}{\chi}} + \underset{\triangle}{e}_{,1}\Delta \overset{s}{\underset{\triangle}{\vartheta}})\mathrm{d}\underset{\triangle}{x^1}.$$

Expression (2.45) permits us to conclude that prestressing will reduce deformability of the structure if

$$\sum_{\triangle}\int_{\underset{\triangle}{l}} \underset{\triangle}{S}(\overset{s}{\underset{\triangle}{\varepsilon}} + \underset{\triangle}{e}\,\overset{s}{\underset{\triangle}{\chi}} + \underset{\triangle}{e}_{,1}\overset{s}{\underset{\triangle}{\vartheta}})\mathrm{d}\underset{\triangle}{x^1} \geqslant 0. \tag{2.46}$$

A building structure may carry several different loads during its normal use. In general, deformability of a structure in different loading states will be different. However, by a suitable prestressing, it is possible to obtain a structure which will have the same deformability in two independent loading states.

Consider a prestressed bar structure in two different loading states $\Lambda = \mathrm{I}, \mathrm{II}$, for example in two extreme (minimum and maximum) states. We shall find the conditions under which the elastic strain energy of the structure is the same in each state.

The works $\overset{\mathrm{I}}{L}$ and $\overset{\mathrm{II}}{L}$ and the energies $\overset{\mathrm{I}}{U}$ and $\overset{\mathrm{II}}{U}$ corresponding to the two states are given by

$$\overset{\mathrm{I}}{L} = \sum_{\triangle}\Big[\int_{\underset{\triangle}{l}} \underset{\triangle}{\gamma}\underset{\triangle}{A}(\overset{\mathrm{I}}{\underset{\triangle}{u}}\sin\underset{\triangle}{\alpha} + \overset{\mathrm{I}}{\underset{\triangle}{w}}\cos\underset{\triangle}{\alpha})\mathrm{d}\underset{\triangle}{x^1} - \int_{\underset{\triangle}{l}} \underset{\triangle}{S}(\overset{\mathrm{I}}{\underset{\triangle}{\varepsilon}} + \underset{\triangle}{e}\,\overset{\mathrm{I}}{\underset{\triangle}{\chi}} + \underset{\triangle}{e}_{,1}\overset{\mathrm{I}}{\underset{\triangle}{\vartheta}})\mathrm{d}\underset{\triangle}{x^1} +$$

$$+ \int_{\underset{\triangle}{l}} (\overset{\mathrm{I}}{\underset{\triangle}{p}}{}^1\overset{\mathrm{I}}{\underset{\triangle}{u}} + \overset{\mathrm{I}}{\underset{\triangle}{p}}{}^3\overset{\mathrm{I}}{\underset{\triangle}{w}} + \overset{\mathrm{I}}{\underset{\triangle}{m}}\overset{\mathrm{I}}{\underset{\triangle}{\varphi}})\mathrm{d}\underset{\triangle}{x^1}\Big],$$

$$\overset{\mathrm{I}}{U} = \tfrac{1}{2}\sum_{\triangle}\int_{\underset{\triangle}{l}} (\overset{\mathrm{I}}{\underset{\triangle}{N}}\overset{\mathrm{I}}{\underset{\triangle}{\varepsilon}} + \overset{\mathrm{I}}{\underset{\triangle}{T}}\overset{\mathrm{I}}{\underset{\triangle}{\vartheta}} + \overset{\mathrm{I}}{\underset{\triangle}{M}}\overset{\mathrm{I}}{\underset{\triangle}{\chi}})\mathrm{d}\underset{\triangle}{x^1},$$

$$\begin{aligned}\overset{\mathrm{II}}{L} = \sum_{\triangle}\Big[&\int_{\underset{\triangle}{l}} \underset{\triangle}{\gamma}\underset{\triangle}{A}(\overset{\mathrm{II}}{\underset{\triangle}{u}}\sin\underset{\triangle}{\alpha}+\overset{\mathrm{II}}{\underset{\triangle}{w}}\cos\underset{\triangle}{\alpha})\,\mathrm{d}\underset{\triangle}{x}^1-\int_{\underset{\triangle}{l}}\underset{\triangle}{S}(\overset{\mathrm{II}}{\underset{\triangle}{\varepsilon}}-\underset{\triangle}{e}\,\overset{\mathrm{II}}{\underset{\triangle}{\chi}}+\underset{\triangle}{e}_{,1}\overset{\mathrm{II}}{\underset{\triangle}{\vartheta}})\,\mathrm{d}\underset{\triangle}{x}^1+\\ &+\int_{\underset{\triangle}{l}}(\overset{\mathrm{II}}{\underset{\triangle}{p}}{}^1\overset{\mathrm{II}}{\underset{\triangle}{u}}+\overset{\mathrm{II}}{\underset{\triangle}{p}}{}^3\overset{\mathrm{II}}{\underset{\triangle}{w}}+\overset{\mathrm{II}}{\underset{\triangle}{m}}\overset{\mathrm{II}}{\underset{\triangle}{\varphi}})\,\mathrm{d}\underset{\triangle}{x}^1\Big],\\ \overset{\mathrm{II}}{U} = \tfrac{1}{2}\sum_{\triangle}&\int_{\underset{\triangle}{l}}(\overset{\mathrm{II}}{\underset{\triangle}{N}}\overset{\mathrm{II}}{\underset{\triangle}{\varepsilon}}+\overset{\mathrm{II}}{\underset{\triangle}{T}}\overset{\mathrm{II}}{\underset{\triangle}{\vartheta}}+\overset{\mathrm{II}}{\underset{\triangle}{M}}\overset{\mathrm{II}}{\underset{\triangle}{\chi}})\,\mathrm{d}\underset{\triangle}{x}^1.\end{aligned}$$

The difference of the energies $\overset{\mathrm{I}}{U}$ and $\overset{\mathrm{II}}{U}$, $\Delta U = \Delta L - \Delta U$, can be written

$$\begin{aligned}\Delta U = \sum_{\triangle}\Big[&\int_{\underset{\triangle}{l}} \underset{\triangle}{\gamma}\underset{\triangle}{A}(\Delta\underset{\triangle}{u}\sin\underset{\triangle}{\alpha}+\Delta\underset{\triangle}{w}\cos\underset{\triangle}{\alpha})\,\mathrm{d}\underset{\triangle}{x}^1+\int_{\underset{\triangle}{l}}[\Delta(\underset{\triangle}{p}{}^1\underset{\triangle}{u})+\Delta(\underset{\triangle}{p}{}^3\underset{\triangle}{w})+\\ &+\Delta(\underset{\triangle}{m}\underset{\triangle}{\varphi})]\,\mathrm{d}\underset{\triangle}{x}^1\Big]-\sum_{\triangle}\Big[\int_{\underset{\triangle}{l}}\underset{\triangle}{S}(\Delta\underset{\triangle}{\varepsilon}+\underset{\triangle}{e}\Delta\underset{\triangle}{\chi}+\underset{\triangle}{e}_{,1}\Delta\underset{\triangle}{\vartheta})\,\mathrm{d}\underset{\triangle}{x}^1+\\ &+\int_{\underset{\triangle}{l}}[\Delta(\underset{\triangle}{N}\underset{\triangle}{\varepsilon})+\Delta(\underset{\triangle}{T}\underset{\triangle}{\vartheta})+\Delta(\underset{\triangle}{M}\underset{\triangle}{\chi})]\,d\underset{\triangle}{x}^1\Big].\end{aligned}\tag{2.47}$$

Adding and subtracting the equalities

$$\begin{aligned}&\sum_{\triangle}\Big[\int_{\underset{\triangle}{l}} \underset{\triangle}{\gamma}\underset{\triangle}{A}(\Delta\underset{\triangle}{u}\sin\underset{\triangle}{\alpha}+\Delta\underset{\triangle}{w}\cos\underset{\triangle}{\alpha})\,\mathrm{d}\underset{\triangle}{x}^1+\int_{\underset{\triangle}{l}}(\overset{\mathrm{II}}{\underset{\triangle}{p}}{}^1\Delta\underset{\triangle}{u}+\overset{\mathrm{II}}{\underset{\triangle}{p}}{}^3\Delta\underset{\triangle}{w}+\overset{\mathrm{II}}{\underset{\triangle}{m}}\Delta\underset{\triangle}{\varphi})\,\mathrm{d}\underset{\triangle}{x}^1\Big]-\\ &\quad-\sum_{\triangle}\int_{\underset{\triangle}{l}}\underset{\triangle}{S}(\Delta\underset{\triangle}{\varepsilon}+\underset{\triangle}{e}\Delta\underset{\triangle}{\chi}+\underset{\triangle}{e}_{,1}\Delta\underset{\triangle}{\vartheta})\,\mathrm{d}\underset{\triangle}{x}^1=\sum_{\triangle}\int_{\underset{\triangle}{l}}(\overset{\mathrm{II}}{\underset{\triangle}{N}}\Delta\underset{\triangle}{\varepsilon}+\overset{\mathrm{II}}{\underset{\triangle}{T}}\Delta\underset{\triangle}{\vartheta}+\overset{\mathrm{II}}{\underset{\triangle}{M}}\Delta\underset{\triangle}{\chi})\,\mathrm{d}\underset{\triangle}{x}^1,\\ &\sum_{\triangle}\Big[\int_{\underset{\triangle}{l}} \underset{\triangle}{\gamma}\underset{\triangle}{A}(\Delta\underset{\triangle}{u}\sin\underset{\triangle}{\alpha}+\Delta\underset{\triangle}{w}\cos\underset{\triangle}{\alpha})\,\mathrm{d}\underset{\triangle}{x}^1+\int_{\underset{\triangle}{l}}(\overset{\mathrm{I}}{\underset{\triangle}{p}}{}^1\Delta\underset{\triangle}{u}+\overset{\mathrm{I}}{\underset{\triangle}{p}}{}^3\Delta\underset{\triangle}{w}+\overset{\mathrm{I}}{\underset{\triangle}{m}}\Delta\underset{\triangle}{\varphi})\,\mathrm{d}\underset{\triangle}{x}^1-\\ &\quad-\sum_{\triangle}\int_{\underset{\triangle}{l}}\underset{\triangle}{S}(\Delta\underset{\triangle}{\varepsilon}+\underset{\triangle}{e}\Delta\underset{\triangle}{\chi}+\underset{\triangle}{e}_{,1}\Delta\underset{\triangle}{\vartheta})\,\mathrm{d}\underset{\triangle}{x}^1\Big]=\sum_{\triangle}\int_{\underset{\triangle}{l}}(\overset{\mathrm{I}}{\underset{\triangle}{N}}\Delta\underset{\triangle}{e}+\overset{\mathrm{I}}{\underset{\triangle}{T}}\Delta\underset{\triangle}{\vartheta}+\overset{\mathrm{I}}{\underset{\triangle}{M}}\Delta\underset{\triangle}{\chi})\,\mathrm{d}\underset{\triangle}{x}^1,\end{aligned}\tag{2.48}$$

which express the principle of virtual work in each state, we obtain the relations

$$\begin{aligned}&\sum_{\triangle}\int_{\underset{\triangle}{l}} \underset{\triangle}{\gamma}\underset{\triangle}{A}(\Delta\underset{\triangle}{u}\sin\underset{\triangle}{\alpha}+\Delta\underset{\triangle}{w}\cos\underset{\triangle}{\alpha})\,\mathrm{d}\underset{\triangle}{x}^1+\tfrac{1}{2}\sum_{\triangle}\int_{\underset{\triangle}{l}}(\overset{\mathrm{II}}{\underset{\triangle}{p}}{}^1\Delta\underset{\triangle}{u}+\overset{\mathrm{II}}{\underset{\triangle}{p}}{}^3\Delta\underset{\triangle}{w}+\overset{\mathrm{II}}{\underset{\triangle}{m}}\Delta\underset{\triangle}{\varphi})\,\mathrm{d}\underset{\triangle}{x}^1+\\ &\quad+\tfrac{1}{2}\sum_{\triangle}\int_{\underset{\triangle}{l}}(\overset{\mathrm{I}}{\underset{\triangle}{p}}{}^1\Delta\underset{\triangle}{u}+\overset{\mathrm{I}}{\underset{\triangle}{p}}{}^3\Delta\underset{\triangle}{w}+\overset{\mathrm{I}}{\underset{\triangle}{m}}\Delta\underset{\triangle}{\varphi})\,\mathrm{d}\underset{\triangle}{x}^1-\sum_{\triangle}\int_{\underset{\triangle}{l}}\underset{\triangle}{S}(\Delta\underset{\triangle}{\varepsilon}+\underset{\triangle}{e}\Delta\underset{\triangle}{\chi}+\underset{\triangle}{e}_{,1}\Delta\underset{\triangle}{\vartheta})\,\mathrm{d}\underset{\triangle}{x}^1\\ &=\tfrac{1}{2}\sum_{\triangle}\int_{\underset{\triangle}{l}}[(\overset{\mathrm{I}}{\underset{\triangle}{N}}+\overset{\mathrm{II}}{\underset{\triangle}{N}})\Delta\underset{\triangle}{\varepsilon}+(\overset{\mathrm{I}}{\underset{\triangle}{T}}+\overset{\mathrm{II}}{\underset{\triangle}{T}})\Delta\underset{\triangle}{\vartheta}+(\overset{\mathrm{I}}{\underset{\triangle}{M}}+\overset{\mathrm{II}}{\underset{\triangle}{M}})\Delta\underset{\triangle}{\chi}]\,\mathrm{d}\underset{\triangle}{x}^1,\end{aligned}\tag{2.49$_1$}$$

$$\sum_{\triangle}\int_{\underset{\triangle}{l}}(\underset{\triangle}{\Delta p^1}\,\underset{\triangle}{\Delta u}+\underset{\triangle}{\Delta p^3}\,\underset{\triangle}{\Delta w}+\underset{\triangle}{\Delta m}\,\underset{\triangle}{\Delta \varphi})\underset{\triangle}{\mathrm{d}x^1}=\sum_{\triangle}\int_{\underset{\triangle}{l}}(\underset{\triangle}{\Delta N}\,\underset{\triangle}{\Delta \varepsilon}+\underset{\triangle}{\Delta T}\,\underset{\triangle}{\Delta \vartheta}+\underset{\triangle}{\Delta M}\,\underset{\triangle}{\Delta \chi})\underset{\triangle}{\mathrm{d}x^1}. \tag{2.49$_2$}$$

Using these relations, we transform (2.47) to

$$\Delta U=\sum_{\triangle}\int_{\underset{\triangle}{l}}(\underset{\triangle}{\Delta p^1}\,\underset{\triangle}{\overset{\mathrm{I}}{u}}+\underset{\triangle}{\Delta p^3}\,\underset{\triangle}{\overset{\mathrm{I}}{w}}+\underset{\triangle}{\Delta m}\,\underset{\triangle}{\overset{\mathrm{I}}{\varphi}})\underset{\triangle}{\mathrm{d}x^1}+$$
$$+\tfrac{1}{2}\sum_{\triangle}\int_{\underset{\triangle}{l}}(\underset{\triangle}{\overset{\mathrm{II}}{N}}\,\underset{\triangle}{\Delta \varepsilon}+\underset{\triangle}{\overset{\mathrm{II}}{T}}\,\underset{\triangle}{\Delta \vartheta}+\underset{\triangle}{\overset{\mathrm{II}}{M}}\,\underset{\triangle}{\Delta \chi})\underset{\triangle}{\mathrm{d}x^1}-\tfrac{1}{2}\sum_{\triangle}\int_{\underset{\triangle}{l}}(\underset{\triangle}{\Delta N}\,\underset{\triangle}{\overset{\mathrm{I}}{\varepsilon}}+\underset{\triangle}{\Delta T}\,\underset{\triangle}{\overset{\mathrm{I}}{\vartheta}}+\underset{\triangle}{\Delta M}\,\underset{\triangle}{\overset{\mathrm{I}}{\chi}})\underset{\triangle}{\mathrm{d}x^1};$$

but since

$$\underset{\triangle}{\overset{\mathrm{I}}{N}}\,\underset{\triangle}{\Delta \varepsilon}+\underset{\triangle}{\overset{\mathrm{I}}{T}}\,\underset{\triangle}{\Delta \vartheta}+\underset{\triangle}{\overset{\mathrm{I}}{M}}\,\underset{\triangle}{\Delta \chi}=\underset{\triangle}{\Delta N}\,\underset{\triangle}{\overset{\mathrm{I}}{\varepsilon}}+\underset{\triangle}{\Delta T}\,\underset{\triangle}{\overset{\mathrm{I}}{\vartheta}}+\underset{\triangle}{\Delta M}\,\underset{\triangle}{\overset{\mathrm{I}}{\chi}},$$

we get

$$\Delta U=\sum_{\triangle}\int_{\underset{\triangle}{l}}(\underset{\triangle}{\Delta p^1}\,\underset{\triangle}{\overset{\mathrm{I}}{u}}+\underset{\triangle}{\Delta p^3}\,\underset{\triangle}{\overset{\mathrm{I}}{w}}+\underset{\triangle}{\Delta m}\,\underset{\triangle}{\overset{\mathrm{I}}{\varphi}})\underset{\triangle}{\mathrm{d}x^1}+$$
$$+\tfrac{1}{2}\sum_{\triangle}\int_{\underset{\triangle}{l}}(\underset{\triangle}{\Delta N}\,\underset{\triangle}{\Delta \varepsilon}+\underset{\triangle}{\Delta T}\,\underset{\triangle}{\Delta \vartheta}+\underset{\triangle}{\Delta M}\,\underset{\triangle}{\Delta \chi})\underset{\triangle}{\mathrm{d}x^1}.$$

Finally, if we apply equality $(2.49)_2$, we obtain the following expression for the energy difference:

$$\Delta U=\tfrac{1}{2}\sum_{\triangle}\int_{\underset{\triangle}{l}}[\underset{\triangle}{\Delta p^1}(\underset{\triangle}{\overset{\mathrm{I}}{u}}+\underset{\triangle}{\overset{\mathrm{II}}{u}})+\underset{\triangle}{\Delta p^3}(\underset{\triangle}{\overset{\mathrm{I}}{w}}+\underset{\triangle}{\overset{\mathrm{II}}{w}})+\underset{\triangle}{\Delta m}(\underset{\triangle}{\overset{\mathrm{I}}{\varphi}}+\underset{\triangle}{\overset{\mathrm{II}}{\varphi}})]\underset{\triangle}{\mathrm{d}x^1}. \tag{2.50}$$

The prestressing should be chosen so as to make ΔU equal to zero. If we assume that all bars of the structure are prestressed, we can write

$$\underset{\triangle}{\Delta U}=\tfrac{1}{2}\int_{\underset{\triangle}{l}}[\underset{\triangle}{\Delta p^1}(\underset{\triangle}{\overset{\mathrm{I}}{u}}+\underset{\triangle}{\overset{\mathrm{II}}{u}})+\underset{\triangle}{\Delta p^3}(\underset{\triangle}{\overset{\mathrm{I}}{w}}+\underset{\triangle}{\overset{\mathrm{II}}{w}})+\underset{\triangle}{\Delta m}(\underset{\triangle}{\overset{\mathrm{I}}{\varphi}}+\underset{\triangle}{\overset{\mathrm{II}}{\varphi}})]\underset{\triangle}{\mathrm{d}x^1}=0. \tag{2.51}$$

If the prestressing forces and the eccentricities are variable along the bars, we can require that

$$\underset{\triangle}{\Delta \Psi}=\tfrac{1}{2}[\underset{\triangle}{\Delta p^1}(\underset{\triangle}{\overset{\mathrm{I}}{u}}+\underset{\triangle}{\overset{\mathrm{II}}{u}})+\underset{\triangle}{\Delta p^3}(\underset{\triangle}{\overset{\mathrm{I}}{w}}+\underset{\triangle}{\overset{\mathrm{II}}{w}})+\underset{\triangle}{\Delta m}(\underset{\triangle}{\overset{\mathrm{I}}{\varphi}}+\underset{\triangle}{\overset{\mathrm{II}}{\varphi}})]=0. \tag{2.52}$$

In the case of a beam loaded transversely, with $\underset{\triangle}{\Delta m}=0$, the condition $\underset{\triangle}{\Delta \Psi}=0$ is reduced to the condition $\underset{\triangle}{\overset{\mathrm{I}}{w}}+\underset{\triangle}{\overset{\mathrm{II}}{w}}=0$, which means that the prestressing must be chosen in such a way that no deflection appears in the beam under the load equal to $\tfrac{1}{2}(\overset{\mathrm{I}}{p}+\overset{\mathrm{II}}{p})$.

Formulation of optimum design problems for bar structures

The knowledge of how to evaluate the effect of a change in the properties of bars on the deformability of a structure permits the formulation of specific optimum-design problems. Formula (2.39),

$$\Delta U = \sum_{\triangle} \int_{\underset{\triangle}{l}} (\underset{\triangle}{u}\sin\underset{\triangle}{\alpha}+\underset{\triangle}{w}\cos\underset{\triangle}{\alpha})\Delta(\underset{\triangle}{\gamma}\underset{\triangle}{A})\,\mathrm{d}\underset{\triangle}{x}^1 -$$

$$-\tfrac{1}{2}\sum_{\triangle} \int_{\underset{\triangle}{l}} [(\underset{\triangle}{\varepsilon})^2\Delta(\underset{\triangle}{E}\underset{\triangle}{A})+(\underset{\triangle}{\vartheta})^2\Delta(\underset{\triangle}{G}\underset{\triangle}{A}/\underset{\triangle}{\varkappa})+(\underset{\triangle}{\chi})^2\Delta(\underset{\triangle}{E}\underset{\triangle}{J})]\,\mathrm{d}\underset{\triangle}{x}^1+Q^2,$$

relates the difference ΔU to the differences $\Delta(\underset{\triangle}{\gamma}\underset{\triangle}{A})$, $\Delta(\underset{\triangle}{G}\underset{\triangle}{A}/\underset{\triangle}{\varkappa})$ and $\Delta(\underset{\triangle}{E}\underset{\triangle}{J})$ representing the changes, due to modification, in weight distributions and deformational rigidities of the bars.

The modified structure will be a least-deformability structure if its energy U is at an absolute minimum. For this to occur, the inequality $\Delta U > 0$ must hold for arbitrary differences $\Delta(\underset{\triangle}{\gamma}\underset{\triangle}{A})$, $\Delta(\underset{\triangle}{G}\underset{\triangle}{A}/\underset{\triangle}{\varkappa})$ and $\Delta(\underset{\triangle}{E}\underset{\triangle}{J})$, regarded as independent variables.

Without any additional assumptions on the set of bar structures among which we seek the solution, we can see that the condition $U = \min$ will be satisfied if

$$\underset{\triangle}{u}\sin\underset{\triangle}{\alpha}+\underset{\triangle}{w}\cos\underset{\triangle}{\alpha} = (\underset{\triangle}{\varepsilon})^2 = (\underset{\triangle}{\vartheta})^2 = (\underset{\triangle}{\varkappa})^2 = 0,$$

which is possible only when

$$\underset{\triangle}{\gamma A} = 0, \quad \underset{\triangle}{E}\underset{\triangle}{A} = \underset{\triangle}{G}\underset{\triangle}{A}/\underset{\triangle}{\varkappa} = \underset{\triangle}{E}\underset{\triangle}{J} = \infty.$$

The zeroing of $\underset{\triangle}{\gamma}\underset{\triangle}{A}$ and the infinite rigidities imply that the optimum structure would have to be weightless and undeformable. In such a structure the energy U would be zero, and hence $\Delta U = \mathring{U}$. This unrealistic result shows that we must restrict the set of admissible structures.

Let us assume, for example, that all the bars of the original and the modified structures are made of the same elastic materials having the same coefficients $\mathring{\underset{\triangle}{E}} = \underset{\triangle}{E} = E$ and $\mathring{\underset{\triangle}{G}} = \underset{\triangle}{G} = G$ and the same bulk density $\mathring{\underset{\triangle}{\gamma}} = \underset{\triangle}{\gamma}$, so that $\Delta E = \Delta G = \Delta\gamma = 0$.

Furthermore, suppose that each structure has the same volume of material: $\mathring{V} = V$, and so $\Delta V = 0$. Since the volumes $\mathring{V}$ and V are the sums of the volumes $\mathring{\underset{\triangle}{V}} = \int_{\underset{\triangle}{l}} \mathring{\underset{\triangle}{A}}\, d\underset{\triangle}{x}^1$ and $\underset{\triangle}{V} = \int_{\underset{\triangle}{l}} \underset{\triangle}{A}\, d\underset{\triangle}{x}^1$ of the bars of the respective structures, the condition $\Delta V = 0$ can be written

$$\Delta V = \sum_{\triangle} (\underset{\triangle}{\mathring{V}} - \underset{\triangle}{V}) = \sum_{\triangle} \int_{\underset{\triangle}{l}} (\underset{\triangle}{\mathring{A}} - \underset{\triangle}{A})\,\mathrm{d}\underset{\triangle}{x}^1 = \sum_{\triangle} \int_{\underset{\triangle}{l}} \Delta \underset{\triangle}{A}\,\mathrm{d}\underset{\triangle}{x}^1 = 0.$$

If we denote by

$$\underset{\triangle}{\Psi} = \frac{1}{2}\left(\frac{\underset{\triangle}{N^2}}{\underset{\triangle}{EA}} + \underset{\triangle}{\varkappa}\frac{\underset{\triangle}{T^2}}{\underset{\triangle}{GA}} + \frac{\underset{\triangle}{M^2}}{\underset{\triangle}{EJ}}\right) = \frac{1}{2}\left(E\underset{\triangle}{\varepsilon}^2 + \frac{1}{\underset{\triangle}{\varkappa}}G\underset{\triangle}{\vartheta}^2 + E\underset{\triangle}{i}^2\underset{\triangle}{\varkappa}^2\right)\underset{\triangle}{A}$$

the elastic strain energy of a bar element bounded by the sections $\underset{\triangle}{x}^1 = \text{const}$ and $\underset{\triangle}{x}^1 + d\underset{\triangle}{x}^1 = \text{const}$, and take into account the above restrictions, we obtain the following expression for ΔU:

$$\Delta U = \sum_{\triangle} \int_{\underset{\triangle}{l}} [\gamma(\underset{\triangle}{u}\sin\underset{\triangle}{\alpha} + \underset{\triangle}{w}\cos\underset{\triangle}{\alpha}) - \underset{\triangle}{\Psi}/\underset{\triangle}{A} + C]\Delta\underset{\triangle}{A}\,\mathrm{d}\underset{\triangle}{x}^1 -$$

$$-\frac{1}{2}\sum_{\triangle} \int_{\underset{\triangle}{l}} \left[G\underset{\triangle}{\vartheta}^2\Delta\underset{\triangle}{\left(\frac{1}{\varkappa}\right)} + E(\chi)^2\Delta\underset{\triangle}{i}^2\right](\Delta\underset{\triangle}{A} + \underset{\triangle}{A})\,\mathrm{d}\underset{\triangle}{x}^1 + Q^2. \qquad (2.53)$$

For U to attain an absolute minimum in the defined set of admissible structures the inequality $\Delta U > 0$ must hold for arbitrary differences $\Delta\underset{\triangle}{A}$, $\Delta\underset{\triangle}{(1/\varkappa)}$ and $\Delta\underset{\triangle}{i}^2$. Since the bars of the optimized structures can differ in the shape of their cross-sections, these differences should be regarded as independent variables. It is readily seen from formula (2.53) that the absolute minimum condition will be satisfied if

$$\gamma(\underset{\triangle}{u}\sin\underset{\triangle}{\alpha} + \underset{\triangle}{w}\cos\underset{\triangle}{\alpha}) - \underset{\triangle}{\Psi}/\underset{\triangle}{A} + C = 0$$

and

$$\underset{\triangle}{\vartheta} = \underset{\triangle}{\chi} = 0.$$

The vanishing of $\underset{\triangle}{\vartheta}$ and $\underset{\triangle}{\chi}$ is possible in weightless bar trusses with hinged joints, in which $\underset{\triangle}{M} = \underset{\triangle}{T} = 0$, or in other bar structures whose bars have radii of gyration $\underset{\triangle}{i} \to \infty$. However, in dealing with deformable bar structures transmitting bending and shearing forces, it is necessary to restrict the set of admissible structures still further.

Let us assume that the slenderness ratio $\underset{\triangle}{\lambda}(\underset{\triangle}{x}^1) = \underset{\triangle}{l}/\underset{\triangle}{i}$ and the coefficient $\underset{\triangle}{\varkappa}(\underset{\triangle}{x}^1)$ of any bar $\triangle$ are the same in each of the structures considered. Then

$$\Delta\underset{\triangle}{\lambda} = \underset{\triangle}{l}\Delta\underset{\triangle}{\left(\frac{1}{i}\right)} = \underset{\triangle}{l}\frac{\Delta\underset{\triangle}{i}}{\underset{\triangle}{i}(\underset{\triangle}{i} + \Delta\underset{\triangle}{i})} = 0, \qquad \Delta\underset{\triangle}{\varkappa} = 0,$$

and so

$$\underset{\triangle}{\Delta i} = 0, \quad \underset{\triangle}{\Delta\left(\frac{l}{\varkappa}\right)} = \frac{\underset{\triangle}{\Delta\varkappa}}{\underset{\triangle}{\varkappa}(\underset{\triangle}{\varkappa}+\underset{\triangle}{\Delta\varkappa})} = 0.$$

As a result of the vanishing of the differences $\underset{\triangle}{\Delta i}$ and $\Delta(1/\underset{\triangle}{\varkappa})$, formula (2.53) takes the form

$$\Delta U = \sum_{\triangle} \int_{\underset{\triangle}{l}} [\gamma(\underset{\triangle}{u}\sin\underset{\triangle}{\alpha}+\underset{\triangle}{w}\cos\underset{\triangle}{\alpha})-\underset{\triangle}{\Psi}/\underset{\triangle}{A}+C]\Delta\underset{\triangle}{A}\,\mathrm{d}\underset{\triangle}{x}^1+Q^2.$$

It follows that the necessary and sufficient condition for U to be absolute minimum in the set considered is the equality

$$\gamma(\underset{\triangle}{u}\sin\underset{\triangle}{\alpha}+\underset{\triangle}{w}\cos\underset{\triangle}{\alpha})-\underset{\triangle}{\Psi}/\underset{\triangle}{A}+C = 0. \tag{2.54}$$

The constant C can be found from the constant-volume condition

$$\sum_{\triangle}\int_{\underset{\triangle}{l}} \underset{\triangle}{A}\,\mathrm{d}\underset{\triangle}{x}^1 = \sum_{\triangle}\int_{\underset{\triangle}{l}} \underset{\triangle}{\mathring{A}}\,\mathrm{d}\underset{\triangle}{x}^1 = V.$$

Multiplying equality (2.54) by $\underset{\triangle}{A}$ and integrating from 0 to $\underset{\triangle}{l}$, we get

$$-\int_{\underset{\triangle}{l}} \gamma\underset{\triangle}{A}(\underset{\triangle}{u}\sin\underset{\triangle}{\alpha}+\underset{\triangle}{w}\cos\underset{\triangle}{\alpha})\mathrm{d}\underset{\triangle}{x}^1+\underset{\triangle}{U} = C\underset{\triangle}{V}.$$

Summation over all bars gives

$$-\sum_{\triangle}\int_{\underset{\triangle}{l}} \gamma\underset{\triangle}{A}(\underset{\triangle}{u}\sin\underset{\triangle}{\alpha}+\underset{\triangle}{w}\cos\underset{\triangle}{\alpha})\,\mathrm{d}\underset{\triangle}{x}^1+U = CV,$$

whence

$$C = \frac{1}{V}\left[-\sum_{\triangle}\int_{\underset{\triangle}{l}} \gamma\underset{\triangle}{A}(\underset{\triangle}{u}\sin\underset{\triangle}{\alpha}+\underset{\triangle}{w}\cos\underset{\triangle}{\alpha})d\underset{\triangle}{x}^1+U\right]. \tag{2.55}$$

Substituting this expression in (2.54), we obtain

$$\underset{\triangle}{\Psi}/\underset{\triangle}{A}-\gamma(\underset{\triangle}{u}\sin\underset{\triangle}{\alpha}+\underset{\triangle}{w}\cos\underset{\triangle}{\alpha}) = \frac{1}{V}\left[U-\sum_{\triangle}\int_{\underset{\triangle}{l}} \gamma\underset{\triangle}{A}(\underset{\triangle}{u}\sin\underset{\triangle}{\alpha}+\underset{\triangle}{w}\cos\underset{\triangle}{\alpha})\mathrm{d}\underset{\triangle}{x}^1\right].$$

Thus, a bar structure with a given volume V, formed from bars $\triangle$ with given slenderness ratios $\underset{\triangle}{\lambda}(\underset{\triangle}{x}^1)$ and cross-sectional shape coefficients $\underset{\triangle}{\varkappa}(\underset{\triangle}{x}^1)$, is a least-deformability structure if and only if the material distributions $\underset{\triangle}{A}$ of the bars satisfy the condition

$$\underset{\triangle}{A} = V \frac{\underset{\triangle}{\Psi}}{U - \sum\limits_{\triangle} \int\limits_{\underset{\triangle}{l}} \gamma \underset{\triangle}{A}(\underset{\triangle}{u}\sin \underset{\triangle}{\alpha} + \underset{\triangle}{w}\cos \underset{\triangle}{\alpha})\mathrm{d}x^1 - \gamma V(\underset{\triangle}{u}\sin \underset{\triangle}{\alpha} + \underset{\triangle}{w}\cos \underset{\triangle}{\alpha})}.$$

In the particular case where the bars of the structure are assumed to have constant cross-sections, the material-distribution condition takes the form

$$\underset{\triangle}{A} = V \frac{\int\limits_{\underset{\triangle}{l}} \underset{\triangle}{\Psi}\mathrm{d}\underset{\triangle}{x}^1}{\underset{\triangle}{l}\left[U - \sum\limits_{\triangle} \underset{\triangle}{A} \int\limits_{\underset{\triangle}{l}} \gamma(\underset{\triangle}{u}\sin \underset{\triangle}{\alpha} + \underset{\triangle}{w}\cos \underset{\triangle}{\alpha})\mathrm{d}x^1\right] - \gamma V \int\limits_{\underset{\triangle}{l}} (\underset{\triangle}{u}\sin \underset{\triangle}{\alpha} + \underset{\triangle}{w}\cos \underset{\triangle}{\alpha})\mathrm{d}x^1}.$$

The above formulae can serve as a basis for numerical construction of optimum distribution of materials in bar structures.

In many cases, the distribution of the dead weight of the bars can be considered insignificant for the deformability of the structure and the corresponding dead loads can be assumed to be distributed uniformly along the bars. Under this assumption, the formula for the cross-sectional areas of the bars of a least-deformability structure becomes

$$\underset{\triangle}{A} = V \frac{\underset{\triangle}{\Psi}(\underset{\triangle}{x}^1)}{U}, \tag{2.56}$$

implying the equality of the mean cross-sectional potentials of all the bars of the structure.

To explain the mechanical sense of the constraints $\Delta \underset{\triangle}{i^2} = 0$ and $\Delta \underset{\triangle}{\varkappa} = 0$ restricting the set of admissible structures, let us first note that the vanishing of these differences implies that the differences $\Delta(\underset{\triangle}{E}\underset{\triangle}{A})$, $\Delta(\underset{\triangle}{G}\underset{\triangle}{A}/\underset{\triangle}{\varkappa})$ and $\Delta(\underset{\triangle}{E}\underset{\triangle}{J})$ are not independent. Indeed, since

$$\underset{\triangle}{i^2} = \frac{\underset{\triangle}{J}}{\underset{\triangle}{A}}, \qquad \underset{\triangle}{\varkappa} = \frac{\underset{\triangle}{A}}{\underset{\triangle}{J^2}} \int\limits_{\underset{\triangle}{l}} \underset{\triangle}{S^2}/\underset{\triangle}{b^2}\mathrm{d}\underset{\triangle}{A},$$

the equalities $\Delta \underset{\triangle}{i^2} = \Delta \underset{\triangle}{\varkappa} = 0$ give the relationships

$$\underset{\triangle}{A}\Delta \underset{\triangle}{J} = \underset{\triangle}{J}\Delta \underset{\triangle}{A}, \qquad \underset{\triangle}{A} \int\limits_{\underset{\triangle}{\mathring{A}}} \underset{\triangle}{\mathring{S}^2}/\underset{\triangle}{\mathring{b}^2}\mathrm{d}\underset{\triangle}{\mathring{A}} = (\underset{\triangle}{A} + \Delta \underset{\triangle}{A}) \int\limits_{\underset{\triangle}{A}} \underset{\triangle}{S^2}/\underset{\triangle}{b^2}\mathrm{d}\underset{\triangle}{A}, \tag{2.57}$$

which constrain the differences between the cross-sectional parameters of different structures.

In some optimum-design problems, the cross-sectional shapes of bars are given, and the unknowns of optimization are parameters of these sections variable along the bars, for example the thickness of a web in a flanged beam. Conditions (2.57) permit us to derive formulae for such parameters.

To see how this is done, consider a normal section $\underset{\triangle}{x}^1 = \text{const}$ of an arbitrary bar of a structure to be optimized, with height $\underset{\triangle}{\mathring{h}}$ and an arbitrary shape described by a width-function $[\underset{\triangle}{\mathring{b}}(\underset{\triangle}{x}^3, \underset{\triangle}{x}^1)]_{x^1=\text{const}} = \underset{\triangle}{\mathring{\varphi}}\underset{\triangle}{\xi}(\underset{\triangle}{x}^3/\underset{\triangle}{\mathring{h}})$. In general, the cross-section is not symmetrical about the principal axis $\underset{\triangle}{x}^2$, whose position is defined by the parameters $\underset{\triangle}{\mathring{a}}_1$ and $\underset{\triangle}{\mathring{a}}_2$ satisfying the equality $\underset{\triangle}{\mathring{a}}_1 + \underset{\triangle}{\mathring{a}}_2 = \underset{\triangle}{\mathring{h}}$ (Fig. 2.14).

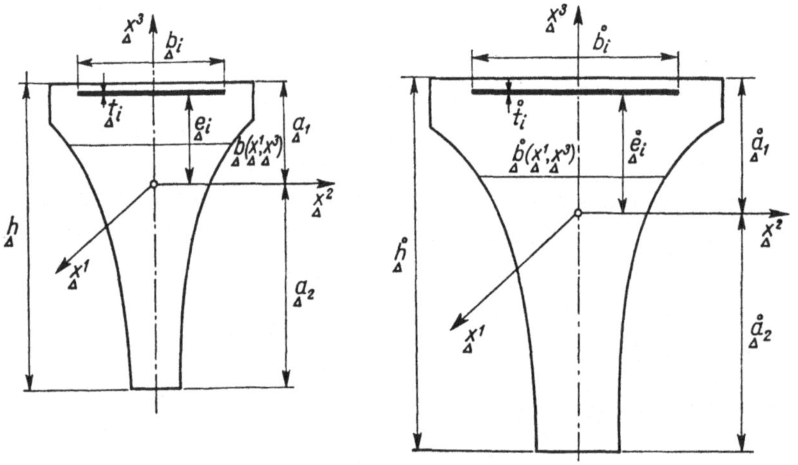

Fig. 2.14. Optimization parameters in the cross-section of a reinforced element

Suppose that the cross-section of the bar is reinforced with a system of n thin rectangular inserts with dimensions $\underset{\triangle}{\mathring{t}_i}$ and $\underset{\triangle}{\mathring{b}_i}$, placed as shown in Fig. 2.14 at distances $\underset{\triangle}{\mathring{e}_i}$ from the axis $\underset{\triangle}{x}^2$. The materials of the inserts have moduli $\underset{\triangle}{E_i} = \underset{\triangle}{n_i}E$.

We seek an optimum solution in the class of cross-sections whose parameters $\mathring{t}_i = \eta_i\mathring{h}(x^1)$, $\mathring{e}_i = \zeta_i\mathring{h}(x^1)$ vary along the bar in proportion to the height $\underset{\triangle}{\mathring{h}}$ and whose width $\underset{\triangle}{\mathring{b}}$ and insert widths $\underset{\triangle}{\mathring{b}_i}$ are functions given as

$$\underset{\triangle}{\mathring{b}}(\underset{\triangle}{x}^1, \underset{\triangle}{x}^3) = \underset{\triangle}{\mathring{\varphi}}(\underset{\triangle}{x}^1)\underset{\triangle}{\xi}(\underset{\triangle}{x}^3/\underset{\triangle}{\mathring{h}}); \qquad \underset{\triangle}{\mathring{b}_i} = \underset{\triangle}{\mathring{\varphi}}(\underset{\triangle}{x}^1)\underset{\triangle}{\varrho_i} < \underset{\triangle}{\mathring{b}}(\underset{\triangle}{x}^1\underset{\triangle}{\mathring{e}_i}).$$

Let us transform the original cross-section, preserving its shape but changing uniformly the dimensions of its elements in the directions of the axes x^1 and x^3; the parameters of the modified cross-section will be

$$\begin{gathered} \underset{\triangle}{b}(\underset{\triangle}{x}^1, \underset{\triangle}{x}^3) = \underset{\triangle}{\varphi}(\underset{\triangle}{x}^1)\underset{\triangle}{\xi}(\underset{\triangle}{x}^3/\underset{\triangle}{h}), \\ \underset{\triangle}{t_i} = \underset{\triangle}{\eta_i}\underset{\triangle}{h}(\underset{\triangle}{x}^1), \quad \underset{\triangle}{b_i} = \underset{\triangle}{\varphi}(\underset{\triangle}{x}^1)\underset{\triangle}{\varrho_i}, \quad e_i = \underset{\triangle}{\zeta_i}\underset{\triangle}{h}(\underset{\triangle}{x}^1). \end{gathered} \tag{2.58}$$

Hence the widths $\underset{\triangle}{\mathring{b}}{}^*$ and $\underset{\triangle}{b}{}^*$ of the equivalent cross-sections before and after modification are

$$\begin{aligned}
\underset{\triangle}{\mathring{b}}{}^* &= \underset{\triangle}{\mathring{\varphi}}(\underset{\triangle}{x}{}^1)\underset{\triangle}{\xi}(\underset{\triangle}{x}{}^3/\underset{\triangle}{\mathring{h}}) + \sum_{i=1}^{n} (\underset{\triangle}{n}_i - 1)\underset{\triangle}{\mathring{t}}_i \underset{\triangle}{\mathring{b}}_i \,\delta(\underset{\triangle}{x}{}^3 - \underset{\triangle}{\mathring{e}}_i),\\
\underset{\triangle}{b}{}^* &= \underset{\triangle}{\varphi}(\underset{\triangle}{x}{}^1)\underset{\triangle}{\xi}(\underset{\triangle}{x}{}^3/\underset{\triangle}{h}) + \sum_{i=1}^{n} (\underset{\triangle}{n}_i - 1)\underset{\triangle}{t}_i \underset{\triangle}{b}_i \,\delta(\underset{\triangle}{x}{}^3 - \underset{\triangle}{e}_i).
\end{aligned} \tag{2.59}$$

The vanishing of the static moments of the equivalent cross-sections, i.e. $\int_{-\underset{\triangle}{\mathring{a}}_2}^{\underset{\triangle}{\mathring{a}}_1} \underset{\triangle}{\mathring{b}}{}^* \underset{\triangle}{x}{}^3 \mathrm{d}\underset{\triangle}{x}{}^3 = \int_{-\underset{\triangle}{a}_2}^{\underset{\triangle}{a}_1} \underset{\triangle}{b}{}^* \underset{\triangle}{x}{}^3 \mathrm{d}\underset{\triangle}{x}{}^3 = 0$, implies the equalities

$$\underset{\triangle}{\mathring{a}}_1/\underset{\triangle}{\mathring{h}} = \underset{\triangle}{a}_1/\underset{\triangle}{h} \quad \text{and} \quad \underset{\triangle}{\mathring{a}}_2/\underset{\triangle}{\mathring{h}} = \underset{\triangle}{a}_2/\underset{\triangle}{h}.$$

The cross-sectional areas $\underset{\triangle}{\mathring{A}}$ and $\underset{\triangle}{A}$ can now be calculated as

$$\begin{aligned}
\underset{\triangle}{\mathring{A}} &= \int_{-\underset{\triangle}{a}_2}^{\underset{\triangle}{\mathring{a}}_1} \underset{\triangle}{\mathring{b}}{}^* \mathrm{d}\underset{\triangle}{x}{}^3 = \underset{\triangle}{\mathring{h}} \int_{-\underset{\triangle}{a}_1/\underset{\triangle}{\mathring{h}}}^{\underset{\triangle}{\mathring{a}}_2/\underset{\triangle}{\mathring{h}}} \underset{\triangle}{\mathring{b}}{}^* \mathrm{d}(\underset{\triangle}{x}{}^3/\underset{\triangle}{\mathring{h}})\\
&= \underset{\triangle}{\mathring{\varphi}}(\underset{\triangle}{x}{}^1)\underset{\triangle}{\mathring{h}}(\underset{\triangle}{x}{}^1)\Big[\int_{-\underset{\triangle}{a}_2/\underset{\triangle}{h}}^{\underset{\triangle}{a}_1/\underset{\triangle}{h}} \xi(\underset{\triangle}{x}{}^3/\underset{\triangle}{\mathring{h}})\mathrm{d}(\underset{\triangle}{x}{}^3/\underset{\triangle}{\mathring{h}}) + \sum_{i=1}^{n}(\underset{\triangle}{n}_i - 1)\underset{\triangle}{\varrho}_i \underset{\triangle}{\eta}_i\Big] = \underset{\triangle}{\mathring{\varphi}}(\underset{\triangle}{x}{}^1)\underset{\triangle}{\mathring{h}}(\underset{\triangle}{x}{}^1)\underset{\triangle}{\lambda}_A,\\
\underset{\triangle}{A} &= \int_{-\underset{\triangle}{a}_1}^{\underset{\triangle}{a}_2} \underset{\triangle}{b}{}^* \mathrm{d}\underset{\triangle}{x}{}^3 = \underset{\triangle}{h} \int_{-\underset{\triangle}{a}_1/\underset{\triangle}{h}}^{\underset{\triangle}{a}_2/\underset{\triangle}{h}} \underset{\triangle}{b}{}^* \mathrm{d}(\underset{\triangle}{x}{}^3/\underset{\triangle}{h})\\
&= \underset{\triangle}{\varphi}(\underset{\triangle}{x}{}^1)\underset{\triangle}{h}(\underset{\triangle}{x}{}^1)\Big[\int_{-\underset{\triangle}{a}_2/\underset{\triangle}{h}}^{\underset{\triangle}{a}_1/\underset{\triangle}{h}} \xi(\underset{\triangle}{x}{}^3/\underset{\triangle}{h})\mathrm{d}(\underset{\triangle}{x}{}^3/\underset{\triangle}{h}) + \sum_{i=1}^{n}(\underset{\triangle}{n}_i - 1)\underset{\triangle}{\varrho}_i \underset{\triangle}{\eta}_i\Big] = \underset{\triangle}{\varphi}(\underset{\triangle}{x}{}^1)\underset{\triangle}{h}(\underset{\triangle}{x}{}^1)\underset{\triangle}{\lambda}_A.
\end{aligned} \tag{2.60}$$

Hence we find the difference $\underset{\triangle}{\Delta A} = \underset{\triangle}{\mathring{A}} - \underset{\triangle}{A} = \underset{\triangle}{\lambda}_A \Delta[\underset{\triangle}{\varphi}(\underset{\triangle}{x}{}^1)\underset{\triangle}{h}]$. Similarly, we find the moments of inertia

$$\begin{aligned}
\underset{\triangle}{\mathring{J}} &= \int_{-\underset{\triangle}{\mathring{a}}_1}^{\underset{\triangle}{\mathring{a}}_2} \underset{\triangle}{\mathring{b}}{}^* \underset{\triangle}{x}{}^2 \mathrm{d}\underset{\triangle}{x}{}^3 = \underset{\triangle}{\mathring{\varphi}}(\underset{\triangle}{x}{}^1)\underset{\triangle}{\mathring{h}}{}^3(\underset{\triangle}{x}{}^1)\Big[\int_{-\underset{\triangle}{\mathring{a}}_2/\underset{\triangle}{h}}^{\underset{\triangle}{\mathring{a}}_1/\underset{\triangle}{h}} \xi^2(\underset{\triangle}{x}{}^3/\underset{\triangle}{\mathring{h}})\mathrm{d}(\underset{\triangle}{x}{}^3/\underset{\triangle}{h}) +\\
&\quad + \sum_{i=1}^{n}(n_i - 1)\underset{\triangle}{\varrho}_i \underset{\triangle}{\eta}_i(\underset{\triangle}{\zeta}_i)^2\Big] = \underset{\triangle}{\mathring{\varphi}}(\underset{\triangle}{x}_1)\underset{\triangle}{\mathring{h}}{}^3(\underset{\triangle}{x}{}^1)\underset{\triangle}{\lambda}_J = \underset{\triangle}{\mathring{A}}\underset{\triangle}{\mathring{h}}{}^2\underset{\triangle}{\lambda}_J/\underset{\triangle}{\lambda}_A,\\
\underset{\triangle}{J} &= \underset{\triangle}{\varphi}(\underset{\triangle}{x}{}^1)\underset{\triangle}{h}{}^3(\underset{\triangle}{x}{}^1)\underset{\triangle}{\lambda}_J = \underset{\triangle}{A}\underset{\triangle}{h}{}^2\underset{\triangle}{\lambda}_J/\underset{\triangle}{\lambda}_A
\end{aligned} \tag{2.61}$$

and the difference $\underset{\triangle}{\Delta J} = \Delta(\underset{\triangle}{A}\underset{\triangle}{h}{}^2)\underset{\triangle}{\lambda}_J/\underset{\triangle}{\lambda}_A$.

Comparing the differences $\Delta \underset{\triangle}{A}$ and $\Delta \underset{\triangle}{J}$, we come to the conclusion that the condition $\underset{\triangle}{A}\Delta \underset{\triangle}{J} = \underset{\triangle}{J}\Delta \underset{\triangle}{A}$ can be satisfied only if $\underset{\triangle}{\mathring{h}} = \underset{\triangle}{h}$, i.e. when the heights of the original and the modified cross-sections are the same. The second of the restricting conditions (2.57) holds identically under the transformation considered. Indeed, expressing the static moments as

$$\underset{\triangle}{S} = \int_{\underset{\triangle}{x_3}}^{\underset{\triangle}{a_2}} \underset{\triangle}{b}^* \underset{\triangle}{x}^3 \mathrm{d}\underset{\triangle}{x}^3 = \underset{\triangle}{\varphi}(\underset{\triangle}{x}^1)\underset{\triangle}{h}^2 \left[\int_{\underset{\triangle\triangle}{a_3/h}}^{\underset{\triangle\triangle}{a_2/h}} \xi(\underset{\triangle}{x}^3/\underset{\triangle}{h})\mathrm{d}(\underset{\triangle}{x}^3/\underset{\triangle}{h}) + \sum_{i=1}^{n} (n_i - 1)\underset{\triangle}{\varrho_i}\underset{\triangle}{\eta_i}\underset{\triangle}{e_i} \right]$$

$$= \underset{\triangle}{\varphi}(\underset{\triangle}{x}^1)\underset{\triangle}{h}^2 \lambda_s(\underset{\triangle}{x}^3/\underset{\triangle}{h}),$$

$$\underset{\triangle}{\mathring{S}} = \underset{\triangle}{\mathring{\varphi}}(\underset{\triangle}{x}^1)\underset{\triangle}{h}^2 \underset{\triangle}{\lambda_s}(\underset{\triangle}{x}^3/\underset{\triangle}{h}),$$

we find

$$\int_{\underset{\triangle}{A}} \underset{\triangle}{S}^2/\underset{\triangle}{b}^2\mathrm{d}\underset{\triangle}{A} = \int_{-\underset{\triangle}{a_2}}^{\underset{\triangle}{a_1}} \underset{\triangle}{\varphi}(\underset{\triangle}{x}^1)\underset{\triangle}{h}^4 \underset{\triangle}{\lambda_s^2}(\underset{\triangle}{x}^3/\underset{\triangle}{h}) \frac{1}{\xi^2(\underset{\triangle}{x}^3/\underset{\triangle}{h})} \mathrm{d}\underset{\triangle}{x}^3$$

$$= \underset{\triangle}{\varphi}(\underset{\triangle}{x}^1)\underset{\triangle}{h}^5 \int_{-\underset{\triangle\triangle}{a_2/h}}^{\underset{\triangle\triangle}{a_1/h}} \left(\frac{\underset{\triangle}{\lambda_s}(\underset{\triangle}{x}^3/\underset{\triangle}{h})}{\underset{\triangle}{\xi}(\underset{\triangle}{x}^3/\underset{\triangle}{h})} \right)^2 \mathrm{d}(\underset{\triangle}{x}^3/\underset{\triangle}{h}) = \underset{\triangle}{\varphi}(\underset{\triangle}{x}^1)\underset{\triangle}{h}^5 \lambda_{\varkappa} = \underset{\triangle}{A}\underset{\triangle}{h}^4 \underset{\triangle}{\lambda_{\varkappa}}/\underset{\triangle}{\lambda_A},$$

$$\int_{\underset{\triangle}{\mathring{A}}} \underset{\triangle}{\mathring{S}}^2/\underset{\triangle}{\mathring{b}}^2\mathrm{d}\underset{\triangle}{\mathring{A}} = \underset{\triangle}{\mathring{\varphi}}(\underset{\triangle}{x}^1)\underset{\triangle}{h}^5 \underset{\triangle}{\lambda_{\varkappa}} = \underset{\triangle}{\mathring{A}}\underset{\triangle}{h}^4 \underset{\triangle}{\lambda_{\varkappa}}/\underset{\triangle}{\lambda_A},$$

and so the two integrals are equal, as required.

In this way, the optimum cross-section problem has been reduced to determining the functions $\underset{\triangle}{\varphi}(\underset{\triangle}{x}^1)$ on the basis of condition (2.56). Knowing $\underset{\triangle}{\varphi}(\underset{\triangle}{x}^1)$, one can determine the parameters $\underset{\triangle}{A}$, $\underset{\triangle}{J}$ and $\underset{\triangle}{\varkappa}$ of an arbitrary normal section of any bar in the least-deformability structure from the following formulae:

$$\underset{\triangle}{A} = \underset{\triangle}{\varphi}(\underset{\triangle}{x}^1)\underset{\triangle}{h}(\underset{\triangle}{x}^1)\underset{\triangle}{\lambda_A}, \quad \underset{\triangle}{\lambda_A} = \int_{-\underset{\triangle\triangle}{a_2/h}}^{\underset{\triangle\triangle}{a_1/h}} \underset{\triangle}{\xi}(\underset{\triangle}{x}^3/\underset{\triangle}{h})\mathrm{d}(\underset{\triangle}{x}^3/\underset{\triangle}{h}) + \sum_{i=1}^{n} (n_i - 1)\underset{\triangle}{\varrho_i}\underset{\triangle}{\eta_i},$$

$$\underset{\triangle}{J} = \underset{\triangle}{\varphi}(\underset{\triangle}{x}^1)\underset{\triangle}{h}^3(\underset{\triangle}{x}^1)\underset{\triangle}{\lambda_J}, \quad \underset{\triangle}{\lambda_J} = \int_{-\underset{\triangle\triangle}{a_2/h}}^{\underset{\triangle\triangle}{a_1/h}} \underset{\triangle}{\xi}^2(\underset{\triangle}{x}^3/\underset{\triangle}{h})\mathrm{d}(\underset{\triangle}{x}^3/\underset{\triangle}{h}) + \sum_{i=1}^{n} (n_i - 1)\underset{\triangle}{\varrho_i}\underset{\triangle}{\eta_i}\underset{\triangle}{\zeta_i^2},$$

$$\underset{\triangle}{\varkappa} = \varphi(\underset{\triangle}{x}^1)\frac{\underset{\triangle}{h}^6 \underset{\triangle}{\lambda}_A \underset{\triangle}{\lambda}_\varkappa}{\underset{\triangle}{h}^6 \underset{\triangle}{\lambda}_J^2} = \varphi(\underset{\triangle}{x}^1)\underset{\triangle}{\lambda}_A/\underset{\triangle}{\lambda}_J^2 \underset{\triangle}{\lambda}_\varkappa,$$

$$\underset{\triangle}{\lambda}_\varkappa = \int\limits_{-\underset{\triangle}{a}_2/\underset{\triangle}{h}}^{\underset{\triangle}{a}_1/\underset{\triangle}{h}} \left(\frac{\underset{\triangle}{\lambda}_s(\underset{\triangle}{x}^3/\underset{\triangle}{h})}{\underset{\triangle}{\xi}(\underset{\triangle}{x}^3/\underset{\triangle}{h})} \right)^2 \mathrm{d}(\underset{\triangle}{x}^3/\underset{\triangle}{h}),$$

$$\underset{\triangle}{\lambda}_s = \int\limits_{\underset{\triangle}{x}^3/\underset{\triangle}{h}}^{\underset{\triangle}{a}_2/\underset{\triangle}{h}} \underset{\triangle}{\xi}(\underset{\triangle}{x}^3/\underset{\triangle}{h})\mathrm{d}(\underset{\triangle}{x}^3/\underset{\triangle}{h}) + \sum_{i=1}^{n} (n_i - 1)\underset{\triangle}{\varrho}_i \underset{\triangle}{\eta}_i \underset{\triangle}{e}_i .$$

For bars with I- or box-shaped cross-section with constant height $\underset{\triangle}{h}$, the optimum solution on the basis of condition (2.56), dead weight being neglected, becomes particularly simple if we assume that

$$\underset{\triangle}{J} = \underset{\triangle}{A}_f \frac{h^2}{4}, \qquad \underset{\triangle}{\varkappa} = \frac{\underset{\triangle}{A}_f + \underset{\triangle}{A}_w}{\underset{\triangle}{A}}, \qquad \underset{\triangle}{A} = \underset{\triangle}{A}_f + \underset{\triangle}{A}_w$$

and, from the condition of constant volumes V_f and V_w (the subscripts f and w refer to the flange and to the web, respectively)

$$\sum_{\triangle} \int_{\underset{\triangle}{l}} \Delta(C_f \underset{\triangle}{A}_f + C_w \underset{\triangle}{A}_w)\mathrm{d}\underset{\triangle}{x}^1 = 0.$$

The general solution of the optimization problem then takes the form

$$\frac{1}{2}\left(\frac{\underset{\triangle}{N}^2}{\underset{\triangle}{A}^2 E} + \frac{\underset{\triangle}{T}^2}{\underset{\triangle}{A}_w^2 G} \right) = C_w, \tag{2.62}$$

$$\frac{1}{2}\left(\frac{\underset{\triangle}{N}^2}{\underset{\triangle}{A}^2 E} + \frac{\underset{\triangle}{M}^2}{E\underset{\triangle}{J}^2}\frac{\underset{\triangle}{h}^2}{4} \right) = C_f. \tag{2.63}$$

2.6 Optimization of plates

Assumptions and definitions

The subject of optimization in this section are non-homogeneous anisotropic medium-thickness plates with arbitrary contours $\Gamma(s)$, made of Hookean materials (Fig. 2.15).

The formulation of the optimization problem will be based on the infinitesimal theory of medium-thickness plates. Underlying this theory are the kinematic assumptions concerning the displacements $\mathbf{v}(x^j) = v_i(x^i)\mathbf{i}_i$ of the points of plate.

We shall assume that the points lying on a normal to the middle surface x^1x^2 are displaced, while the plate is deformed, in such a way that they remain on a straight line, but at an angle $\boldsymbol{\vartheta} = \vartheta_\alpha \mathbf{i}_\alpha$ to the normal to the bent surface of the plate. This is a generalization of the Love–Kirchhoff assumption, in which $\vartheta_\alpha \equiv 0$.

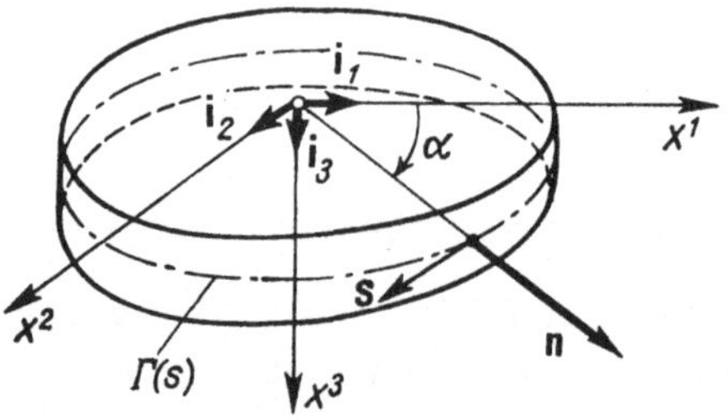

Fig. 2.15. Coordinates in an element of the plate

Denoting by $u_\alpha(x^\beta) = v_\alpha(x^\beta, 0)$ and $w(x^\beta) = v_3(x^\beta, 0)$ the components $[v_i]_{x^3=0}$ of the displacements of the points of the middle surface, we can represent the displacement v_i of an arbitrary point of the plate as

$$v_\alpha(x^i) = u_\alpha(x^\beta)+x^3(-w(x^\beta)_{,\alpha}+\vartheta_\alpha(x^\beta)), \tag{2.64}$$

$$v_3(x^i) = w(x^\beta), \quad \alpha, \beta = 1, 2, \quad i = 1, 2, 3. \tag{2.64}$$

By the geometrical equations of the theory of elasticity $\gamma_{(ij)} = v_{(ij)} = \frac{1}{2}(v_{i,j}+v_{j,i})$, we obtain the equalities

$$\gamma_{(\alpha,\beta)} = u_{(\alpha,\beta)}+x^3(-w_{,\alpha\beta}+\vartheta_{(\alpha,\beta)}) = \varepsilon_{(\alpha\beta)}+x^3\varphi_{(\alpha,\beta)} = \varepsilon_{(\alpha\beta)}+x^3\varkappa_{(\alpha\beta)},$$

$$2\gamma_{(\alpha,\beta)} = \vartheta_\alpha, \quad \gamma_{33} = 0, \tag{2.65}$$

which express the state of deformation of an arbitrary point of the plate in terms of the components $\varepsilon_{(\alpha\beta)}$, $\varkappa_{(\alpha\beta)}$ and ϑ_α. The symbols φ_α denote the components of the rotation of the normal to the middle surface.

The state of stress of the plate is defined by the following internal forces: normal forces $N_{\alpha\beta} = N_{\beta\alpha} = \int\limits_{-h/2}^{h/2} \sigma_{(\alpha\beta)}\,dx^3$ acting in the middle plane x^1x^2, bending and torsional moments

$$M_{\alpha\beta} = M_{\beta\alpha} = \int\limits_{-h/2}^{h/2} x^3\sigma_{(\alpha\beta)}\,dx^3,$$

whose vectors are parallel to the x^1x^2 plane, and transverse forces $Q_\alpha = \int_{-h/2}^{h/2} \sigma_{(\alpha 3)} dx^3$ perpendicular to the middle plane.

The plate is loaded by the surface force and moment fields p_i and m_α and—on the free part of the edge—by the forces $\tilde{N}_n$ and $\tilde{N}_s$ in the directions of the normal **n** and tangent **s** to the edge, the transverse forces $\tilde{Q}_n$ and the normal and torsional moments $\tilde{M}_n$ and $\tilde{M}_{ns}$.

The internal forces $N_{(\alpha\beta)}$, $M_{(\alpha\beta)}$ and Q_α satisfy the equilibrium equations

$$\begin{aligned} N_{(\alpha\beta),\beta}+p_\alpha &= 0, \\ M_{(\alpha\beta),\beta}+m_\alpha &= Q_a, \\ Q_{\alpha,\alpha}+p_3 &= 0 \end{aligned} \tag{2.66}$$

on the middle surface of the plate, and the boundary conditions

$$\begin{aligned} N_{(\alpha\beta)}n_\alpha n_\beta &= \tilde{N}_n \qquad N_{(\alpha\beta)}n_\alpha s_\beta = \tilde{N}_s, \\ M_{(\alpha\beta)}n_\alpha n_\beta &= \tilde{M}_n, \quad M_{(\alpha\beta)}n_\alpha s_\beta = \tilde{M}_{sn}, \\ Q_\alpha n_\alpha &= \tilde{Q}_n \end{aligned} \tag{2.67}$$

on the contour $\Gamma(s)$, whose normal and tangent vectors are $\mathbf{n} = \cos\alpha \mathbf{i}_1 + \sin\alpha \mathbf{i}_2$ and $\mathbf{s} = -\sin\alpha \mathbf{i}_1 + \cos\alpha \mathbf{i}_2$, respectively.

The physical relations for the plate can be derived from generalized Hooke's law

$$\sigma_{(ij)} = A_{ijkl}\,\varepsilon_{(kl)},$$

which involves 21 coefficients $A_{ijkl} = A_{jikl} = A_{ijlk} = A_{klij}$.

We restrict our considerations to the case where the anisotropy is linear, with one plane of symmetry parallel to the middle plane of the plate. Under this assumption we have

$$A_{\alpha\beta\gamma 3} = A_{\alpha 3\beta\nu} = A_{\alpha 333} = 0$$

and so the number of the elastic coefficients is reduced to 13 and the physical relations of the material take the form

$$\begin{aligned} \sigma_{\alpha\beta} &= A_{\alpha\beta\gamma\nu}\gamma_{\gamma\nu}+A_{\alpha\beta 33}\gamma_{33}, \\ \sigma_{\alpha 3} &= A_{\alpha 3\beta 3}\gamma_{\beta 3}, \\ \sigma_{33} &= A_{33\alpha\beta}\gamma_{\alpha\beta}+A_{3333}\gamma_{33}. \end{aligned} \tag{2.68}$$

Expressing the deformations $\gamma_{\gamma\nu}$ and $\gamma_{\beta 3}$ in terms of $\varepsilon_{\alpha\beta}$, $\varkappa_{\alpha\beta}$ and ϑ_α as in (2.65) and substituting formulae (2.68) into the expressions for the internal forces, we obtain the following physical relations for the plate:

$$\begin{aligned} N_{(\alpha\beta)} &= S_{\alpha\beta\gamma\nu}\,\varepsilon_{(\gamma\nu)}+C_{\alpha\beta\gamma\nu}\varkappa_{(\gamma\nu)}, \\ M_{(\alpha\beta)} &= C_{\alpha\beta\gamma\nu}\,\varepsilon_{(\gamma\nu)}+D_{\alpha\beta\gamma\nu}\varkappa_{(\gamma\nu)}, \\ Q_\alpha &= S_{\alpha 3\beta 3}\vartheta_\beta; \end{aligned} \tag{2.69}$$

here $S_{\alpha i\beta j}$, $C_{\alpha\beta\gamma\nu}$ and $D_{\alpha\beta\gamma\nu}$ represent the deformational rigidities of the plate:

$$
\begin{aligned}
S_{\alpha i\beta j} &= \int_{-h/2}^{h/2} A_{\alpha i\beta j}\mathrm{d}x^3, \\
C_{\alpha\beta\gamma\nu} &= \int_{-h/2}^{h/2} A_{\alpha\beta\gamma\nu}x^3\mathrm{d}x^3, \\
D_{\alpha\beta\gamma\nu} &= \int_{-h/2}^{h/2} A_{\alpha\beta\gamma\nu}(x^3)^2\mathrm{d}x^3.
\end{aligned}
\tag{2.70}
$$

In practice one often deals with surface structures in the form of orthotropic plates, whose orthotropy results from purposefully designed regular inhomogeneities. In this case, the elastic symmetry of the body with respect to three orthogonal planes implies the physical relations

$$
\sigma_{\alpha\alpha} = \sum_{\beta=1}^{2} A_{\alpha\alpha\beta\beta}\,\varepsilon_{(\beta\beta)}, \qquad \sigma_{\alpha 3} = A_{\alpha 3\alpha 3}\,\varepsilon_{(\alpha 3)}, \qquad \sigma_{12} = A_{1212}\,\varepsilon_{(12)},
$$

and the deformational rigidities are

$$
\begin{aligned}
&S_{1111} = S_{11}, \quad S_{2222} = S_{22}, \quad S_{1122} = S_{2211} = S_{12} = S_{21}, \\
&S_{1212} = S_{66}, \quad S_{1313} = S_{44}, \quad S_{2323} = S_{55}, \\
&D_{1111} = D_{11}, \quad D_{2222} = D_{22}, \quad D_{1122} = D_{2211} = D_{12} = D_{21}, \\
&D_{1212} = D_{66}.
\end{aligned}
\tag{2.71}
$$

Influence of geometrical and mechanical properties of the plate on elastic strain energy

The effect of a change in the mechanical and geometrical properties of a medium-thickness plate on its deformability can be evaluated by considering the energy difference $\Delta U = \mathring{U} - U$ between the given plate and one with changed properties. The properties concerned are the deformational rigidities and the dead weight distribution g of the plate. The boundary contour $\Gamma(s)$, the way the plate is supported, the surface forces p_i and m_α and the boundary forces $\tilde{N}_n$, $\tilde{M}_n$, $\tilde{N}_s$, $\tilde{M}_s$ and $\tilde{Q}_n$ remain the same.

Denoting the elastic strain energy stored in an infinitesimal element with the surface area dx^1dx^2 and height $\mathring{h}$ of the original plate and that in a similar element but with height h of the modified structure by $\mathring{\Psi}$ and Ψ,

$$
\begin{aligned}
\mathring{\Psi} &= \tfrac{1}{2}[\mathring{N}_{(\alpha\beta)}\mathring{\varepsilon}_{(\alpha\beta)} + \mathring{M}_{(\alpha\beta)}\mathring{\varkappa}_{(\alpha\beta)} + \mathring{Q}_\alpha\mathring{\vartheta}_\alpha], \\
\Psi &= \tfrac{1}{2}[N_{(\alpha\beta)}\varepsilon_{(\alpha\beta)} + M_{(\alpha\beta)}\varkappa_{(\alpha\beta)} + Q_\alpha\vartheta_\alpha],
\end{aligned}
\tag{2.72}
$$

we can write

$$
\Delta U = \mathring{U} - U = \int_B (\mathring{\Psi} - \Psi)\mathrm{d}x^1\mathrm{d}x^2 = \int_B \Delta\Psi \mathrm{d}x^1\mathrm{d}x^2.
$$

In general, ΔU depends on the differences $\Delta S_{\alpha i\beta j}$, $\Delta C_{\alpha\beta\gamma\nu}$ and $\Delta D_{\alpha\beta\gamma\nu}$ of deformational rigidities and on the difference $\Delta g = \mathring{g} - g$ in the dead weight distribution. Finding the relationship between ΔU and these differences is the objective of the present analysis of the influence of the properties of the plate on its deformability.

Since both the original and the modified plate are in equilibrium, by the Clapeyron theorem $\mathring{U} = \mathring{L}$ and $U = L$, where $\mathring{L}$ and L are the works of the statically applied loads on the displacements of the original and the modified plate, respectively:

$$\begin{aligned}\mathring{L} &= \tfrac{1}{2}\Big\{\int_B [(p_3+\mathring{g})\mathring{w}+p_\alpha \mathring{u}_\alpha + m_\alpha \mathring{\varphi}_\alpha]\mathrm{d}x^1\mathrm{d}x^2 + \\ &\quad + \oint_S (\tilde{N}_s\mathring{u}_s+\tilde{N}_n\mathring{u}_n+\tilde{Q}_n\mathring{w}+\tilde{M}_n\mathring{\varphi}_{,n}+\tilde{M}_{sn}\mathring{\varphi}_{,s})\mathrm{d}s\Big\},\\ L &= \tfrac{1}{2}\Big\{\int_B [(p_3+g)w+p_\alpha u_\alpha+m_\alpha\varphi_\alpha]\mathrm{d}x^1\mathrm{d}x^2+\\ &\quad + \oint_S (\tilde{N}_s u_s+\tilde{N}_n u_n+\tilde{Q}_n w+\tilde{M}_n\varphi_n+\tilde{M}_{sn}\varphi_s)\mathrm{d}s\Big\}.\end{aligned} \tag{2.73}$$

Hence ΔU can be represented as

$$\begin{aligned}\Delta U = 2\Delta L - U &= \int_B (p_3\Delta w+p_\alpha\Delta u_\alpha+m_\alpha\Delta\varphi_\alpha+\mathring{g}\Delta w+\Delta g w)\mathrm{d}x^1\mathrm{d}x^2+\\ &\quad+\oint_S (\tilde{N}_s\Delta u_s+\tilde{N}_n\Delta u_n+\tilde{Q}_n\Delta w+\tilde{M}_n\Delta\varphi_n+\tilde{M}_{sn}\Delta\varphi_s)\mathrm{d}s-\\ &\quad-\tfrac{1}{2}\int_B (\mathring{N}_{(\alpha\beta)}\Delta\varepsilon_{(\alpha\beta)}+\Delta N_{(\alpha\beta)}\varepsilon_{(\alpha\beta)}+\mathring{M}_{(\alpha\beta)}\Delta\varkappa_{(\alpha\beta)}+\Delta M_{(\alpha\beta)}\varkappa_{(\alpha\beta)}+\\ &\quad\quad+\mathring{Q}_\alpha\Delta\vartheta_\alpha+\Delta Q_\alpha\vartheta_\alpha)\mathrm{d}x^1\mathrm{d}x^2.\end{aligned}$$

We shall transform this expression, using the equalities

$$\begin{aligned}&\int_B [(p_3+\mathring{g})\Delta w+p_\alpha\Delta u_\alpha+m_\alpha\Delta\varphi_\alpha]\mathrm{d}x^1\mathrm{d}x^2+\\ &\quad+\oint_S (\tilde{N}_s\Delta u_s+\tilde{N}_n\Delta u_n+\tilde{Q}_n\Delta w+\tilde{M}_n\Delta\varphi_n+\tilde{M}_{sn}\Delta\varphi_s)\mathrm{d}s\\ &\quad=\int_B (\mathring{N}_{(\alpha\beta)}\Delta\varepsilon_{(\alpha\beta)}+\mathring{M}_{(\alpha\beta)}\Delta\varkappa_{(\alpha\beta)}+\mathring{Q}_\alpha\Delta\vartheta_\alpha)\mathrm{d}x^1\mathrm{d}x^2,\\ &\int_B [(p_3+g)\Delta w+p_\alpha\Delta u_\alpha+m_\alpha\Delta\varphi_\alpha]\mathrm{d}x^1\mathrm{d}x^2+\\ &\quad+\oint_S (\tilde{N}_s\Delta u_s+\tilde{N}_n\Delta u_n+\tilde{Q}_n\Delta w+\tilde{M}_n\Delta\varphi_n+M_{sn}\Delta\varphi_s)\mathrm{d}s\\ &\quad=\int_B (N_{(\alpha\beta)}\Delta\varepsilon_{(\alpha\beta)}+M_{(\alpha\beta)}\Delta\varkappa_{(\alpha\beta)}+Q_\alpha\Delta\vartheta_\alpha)\mathrm{d}x^1\mathrm{d}x^2\end{aligned} \tag{2.74}$$

which express the principle of virtual work for the two plates. They follow from the plane Green theorem

$$\int_B uv_{,\alpha}\mathrm{d}x^1\mathrm{d}x^2 = \oint_B uvn_\alpha \mathrm{d}s - \int_B vu_{,\alpha}\mathrm{d}x^1\mathrm{d}x^2,$$

geometrical relations (2.65) and equilibrium equations (2.67).

Using $(2.74)_1$, we can write the expression for ΔU in the form

$$\Delta U = \int_B \Delta g w \mathrm{d}x^1\mathrm{d}x^2 - \\ -\frac{1}{2}\int_B [-\mathring{N}_{(\alpha\beta)}\Delta\varepsilon_{(\alpha\beta)} + \Delta N_{(\alpha\beta)}\varepsilon_{(\alpha\beta)} - \mathring{M}_{(\alpha\beta)}\Delta\varkappa_{(\alpha\beta)} + \Delta M_{(\alpha\beta)}\varkappa_{(\alpha\beta)} - \\ -\mathring{Q}_\alpha\Delta\vartheta_\alpha + \Delta Q_\alpha\vartheta_\alpha]\mathrm{d}x^1\mathrm{d}x^2. \tag{2.75}$$

The sum $\Delta N_{(\alpha\beta)}\varepsilon_{(\alpha\beta)} + \Delta M_{(\alpha\beta)}\varkappa_{(\alpha\beta)} + \Delta Q_\alpha\vartheta_\alpha$, which occurs under the second integral, can be represented as

$$\Delta N_{(\alpha\beta)}\varepsilon_{(\alpha\beta)} + \Delta M_{(\alpha\beta)}\varkappa_{(\alpha\beta)} + \Delta Q_\alpha\vartheta_\alpha = \mathring{N}_{(\alpha\beta)}\Delta\varepsilon_{(\alpha\beta)} + \mathring{M}_{(\alpha\beta)}\Delta\varkappa_{(\alpha\beta)} + \mathring{Q}_\alpha\Delta\vartheta_\alpha \\ = \varepsilon_{(\alpha\beta)}\varepsilon_{(\gamma\nu)}\Delta S_{\alpha\beta\gamma\nu} + 2\varkappa_{(\alpha\beta)}\varepsilon_{(\gamma\nu)}\Delta C_{\alpha\beta\gamma\nu} + \varkappa_{(\alpha\beta)}\varkappa_{(\gamma\nu)}\Delta D_{\alpha\beta\gamma\nu} + \vartheta_\beta\vartheta_\alpha\Delta S_{\alpha 3\beta 3} - \\ -(\Delta\varepsilon_{(\alpha\beta)}\Delta\varepsilon_{(\gamma\nu)}\mathring{S}_{\alpha\beta\gamma\nu} + 2\Delta\varkappa_{(\alpha\beta)}\Delta\varepsilon_{(\gamma\nu)}\mathring{C}_{\alpha\beta\gamma\nu} + \Delta\varkappa_{(\alpha\beta)}\Delta\varkappa_{(\gamma\nu)}\mathring{D}_{\alpha\beta\gamma\nu} + \\ +\Delta\vartheta_\alpha\Delta\vartheta_\beta\mathring{S}_{\alpha 3\beta 3}).$$

This follows from the physical relations (2.69) and from the symmetry of the deformational rigidities, which is implied by the symmetry of the elastic coefficients. Inserting this expression into (2.75), we obtain the following relationship between ΔU and $\Delta S_{\alpha i\beta j}$, $\Delta C_{\alpha\beta\gamma\nu}$ and $\Delta D_{\alpha\beta\gamma\nu}$:

$$\Delta U = \int_B \Delta g w \mathrm{d}x^1\mathrm{d}x^2 - \tfrac{1}{2}\int_B (\varepsilon_{(\alpha\beta)}\varepsilon_{(\gamma\nu)}\Delta S_{\alpha\beta\gamma\nu} + 2\varkappa_{(\alpha\beta)}\varepsilon_{(\gamma\nu)}\Delta C_{\alpha\beta\gamma\nu} + \\ + \varkappa_{(\alpha\beta)}\varkappa_{(\gamma\nu)}\Delta D_{\alpha\beta\gamma\nu} + \vartheta_\beta\vartheta_\alpha\Delta S_{\alpha 3\beta 3})\mathrm{d}x^1\mathrm{d}x^2 + \mathscr{P}^2, \tag{2.76}$$

where

$$\mathscr{P}^2 = \tfrac{1}{2}\int_B [\Delta\varepsilon_{(\alpha\beta)}\Delta\varepsilon_{(\gamma\nu)}\mathring{S}_{\alpha\beta\gamma\nu} + 2\Delta\varkappa_{(\alpha\beta)}\Delta\varepsilon_{(\gamma\nu)}\mathring{C}_{\alpha\beta\gamma\nu} + \\ + \Delta\varkappa_{(\alpha\beta)}\Delta\varkappa_{(\gamma\nu)}\mathring{D}_{\alpha\beta\gamma\nu} + \Delta\vartheta_\alpha\Delta\vartheta_\beta\mathring{S}_{\alpha 3\beta 3}]\mathrm{d}x^1\mathrm{d}x^2 \geqslant 0.$$

That $\mathscr{P}^2$ is non-negative follows from the definition of $\mathring{\Psi}$. Formula (2.76) will serve us as a basis for formulating minimum-deformability design problems for plates.

Let us show that if the differences $\Delta S_{\alpha\beta\gamma\nu}$, $\Delta C_{\alpha\beta\gamma\nu}$, $\Delta D_{\alpha\beta\gamma\nu}$, Δg and $\Delta S_{\alpha 3\beta 3}$ all vanish, then $\Delta U = 0$. Subtracting equation $(2.74)_2$ from $(2.74)_1$ gives

$$\int_B \Delta g \Delta w \mathrm{d}x^1\mathrm{d}x^2 = \int_B (\Delta N_{(\alpha\beta)}\Delta\varepsilon_{(\alpha\beta)} + \Delta M_{(\alpha\beta)}\Delta\varkappa_{(\alpha\beta)} + \Delta Q_\alpha\Delta\vartheta_\alpha)\mathrm{d}x^1\mathrm{d}x^2.$$

Taking into account that

$$\Delta N_{(\alpha\beta)} = \mathring{S}_{\alpha\beta\gamma\nu}\Delta\varepsilon_{(\gamma\nu)}+\Delta S_{\alpha\beta\gamma\nu}\varepsilon_{(\gamma\nu)}+\mathring{C}_{\alpha\beta\gamma\nu}\varkappa_{(\gamma\nu)}+\Delta C_{\alpha\beta\gamma\nu}\varkappa_{(\gamma\nu)},$$
$$\Delta M_{(\alpha\beta)} = \mathring{D}_{\alpha\beta\gamma\nu}\Delta\varkappa_{(\gamma\nu)}+\Delta D_{\alpha\beta\gamma\nu}\varkappa_{(\gamma\nu)}+\mathring{C}_{\alpha\beta\gamma\nu}\varepsilon_{(\gamma\nu)}+\Delta C_{\alpha\beta\gamma\nu}\varepsilon_{(\gamma\nu)},$$
$$\Delta Q_{\alpha} = \mathring{S}_{\alpha3\beta3}\Delta\vartheta_{\beta}+\Delta S_{\alpha3\beta3}\vartheta_{\beta}$$

and combining these equalities and the preceding one with the expression for $\mathscr{P}^2$, we obtain the relation

$$2\mathscr{P}^2 = \int_B \Delta g\Delta w\,\mathrm{d}x^1\mathrm{d}x^2 - \int_B (\Delta S_{\alpha\beta\gamma\nu}\Delta\varepsilon_{(\alpha\beta)}\varepsilon_{(\gamma\nu)}+\Delta C_{\alpha\beta\gamma\nu}(\Delta\varepsilon_{(\alpha\beta)}\varkappa_{(\gamma\nu)}+$$
$$+\Delta\varkappa_{(\alpha\beta)}\varepsilon_{(\gamma\nu)})+\Delta D_{\alpha\beta\gamma\nu}\Delta\chi_{(\alpha\beta)}\chi_{(\gamma\nu)}+\Delta S_{\alpha3\beta3}\Delta\vartheta_{\alpha}\vartheta_{\beta})\,\mathrm{d}x_1\mathrm{d}x^2,$$

which, together with formula (2.76), implies the vanishing of ΔU if all the differences mentioned above are zero.

Solid plates of variable thickness

Deformability of a solid plate with a given contour Γ and made of a homogeneous and isotropic material depends on the variable thickness h of the plate and on its dead weight $g = \gamma h$, proportional to the thickness.

Since the elastic coefficients of a homogeneous isotropic material are given by

$$A_{\alpha\beta\gamma\nu} = \frac{E}{1-\nu^2}[\tfrac{1}{2}(1-\nu)(\delta_{\alpha\gamma}\delta_{\beta\nu}+\delta_{\alpha\nu}\delta_{\beta\gamma})+\nu\delta_{\alpha\beta}\delta_{\gamma\nu}],$$
$$A_{\alpha3\beta3} = \delta_{\alpha\beta}\frac{E}{2(1+\nu)} = G\delta_{\alpha\beta}, \tag{2.77}$$

we obtain the following formulae for the rigidities of the plate

$$S_{\alpha\beta\gamma\nu} = \frac{hE}{1-\nu^2}[\tfrac{1}{2}(1-\nu)(\delta_{\alpha\gamma}\delta_{\beta\nu}+\delta_{\alpha\nu}\delta_{\beta\gamma})+\nu\delta_{\alpha\beta}\delta_{\gamma\nu}],$$
$$S_{\alpha3\beta3} = \delta_{\alpha\beta}Gh = \delta_{\alpha\beta}\frac{Eh}{2(1+\nu)}, \tag{2.78}$$
$$D_{\alpha\beta\gamma\nu} = \frac{h^3E}{12(1-\nu^2)}[\tfrac{1}{2}(1-\nu)(\delta_{\alpha\gamma}\delta_{\beta\nu}+\delta_{\alpha\nu}\delta_{\beta\gamma})+\nu\delta_{\alpha\beta}\delta_{\gamma\nu}],$$
$$C_{\alpha\beta\gamma\nu} = 0;$$

the function $h(x^1, x^2)$ is assumed to be continuous.

The optimum-design problem we are considering is that of finding the least deformable plate in the set of plates with the same contour $\Gamma(s)$, made of the same homogeneous and isotropic material and having the same volume V as the given plate, but differing in thickness. If $h(x^1, x^2)$ is the variable thickness of a plate from this class and $\mathring{h}(x^1, x^2)$ is the thickness of the original plate, the constant-volume condition implies that

$$V = \int_B h\, dx^1 dx^2 = \int_B \mathring{h}\, dx^1 dx^2.$$

We further note that the rigidity differences $\Delta S_{\alpha\beta\gamma\nu}$, $\Delta D_{\alpha\beta\gamma\nu}$ and $\Delta S_{\alpha 3\beta 3}$ between the original plate and one with thickness h are related to the difference $\Delta h = \mathring{h} - h$ by the formulae

$$\Delta S_{\alpha\beta\gamma\nu} = \frac{\Delta h}{h} S_{\alpha\beta\gamma\nu}, \qquad \Delta C_{\alpha\beta\gamma\nu} = 0,$$
$$\Delta D_{\alpha\beta\gamma\nu} = \frac{\Delta h^3}{h^3} D_{\alpha\beta\gamma\nu}, \qquad \Delta S_{\alpha 3\beta 3} = \frac{\Delta h}{h} S_{\alpha 3\beta 3}. \tag{2.79}$$

Using these expressions in formula (2.76), we obtain

$$\Delta U = \gamma \int_B w \Delta h\, dx^1 dx^2 - \\ - \frac{1}{2} \int_B \left[\frac{1}{h} (N_{(\alpha\beta)} \varepsilon_{(\alpha\beta)} + Q_\alpha \vartheta_\alpha) \Delta h + \frac{1}{h^3} M_{(\alpha\beta)} \varkappa_{(\alpha\beta)} \Delta h^3 \right] dx^1 dx^2 + \mathscr{P}^2 \tag{2.80}$$

where $\mathscr{P}^2$ is given by

$$\mathscr{P}^2 = \frac{1}{2} \int_B \frac{\mathring{h} E}{1-\nu^2} \{(\Delta \varepsilon_{11} + \Delta \varepsilon_{22})^2 + \\ + 2(1-\nu)[(\Delta \varepsilon_{(12)})^2 - \Delta \varepsilon_{11} \Delta \varepsilon_{22}]\} dx^1 dx^2 + \\ + \frac{1}{2} \int_B \frac{\mathring{h}^3 E}{12(1-\nu^2)} \{(\Delta \varkappa_{11} + \Delta \varkappa_{22})^2 + \\ + 2(1-\nu)[(\Delta \varkappa_{12})^2 - \Delta \varkappa_{11} \Delta \varkappa_{22})\} dx^1 dx^2 + \\ + \frac{1}{2} \int_B G \mathring{h} (\Delta \vartheta)^2 dx^1 dx^2.$$

We shall consider separately the cases where the plate is in a plane-stress state and in a bending-stress state.

In the first case, assuming that the x^2-axis is directed as forces of gravity and that the loading forces act in the plane of the plate, we have $w = \mathring{w} = 0$, $\vartheta = \mathring{\vartheta} = 0$, and hence the expression for ΔU takes the form

$$\Delta U = \int_B (\gamma u_2 - \tfrac{1}{2} N_{(\alpha\beta)} \varepsilon_{(\alpha\beta)} / h) \Delta h\, dx^1 dx^2 + \mathscr{P}^2, \tag{2.81}$$

with

$$\mathscr{P}^2 = \frac{1}{2} \int_B \frac{\mathring{h} E}{1-\nu^2} \{(\Delta \varepsilon_{11} + \Delta \varepsilon_{22})^2 + 2(1-\nu)[(\Delta \varepsilon_{(12)})^2 - \Delta \varepsilon_{11} \Delta \varepsilon_{22}]\} dx^1 dx^2.$$

The condition for U to be absolute minimum is the inequality $\Delta U > 0$, which must hold for arbitrary differences Δh with $\Delta V = \int_B \Delta h \, dx^1 = 0$.

It follows that a plate with variable thickness under plane stress is a least-deformability structure if

$$\gamma u_2 - \Psi/h = \text{const},$$

because then $\Delta U = \mathscr{P}^2 > 0$.

If we neglect the distribution of the dead weight, then γ can be set equal to zero and the least-deformability condition for a plate in plane stress reduces to

$$\Psi/h = \text{const}, \tag{2.82}$$

i.e. the mean potential must be constant over the whole plate.

In a plate loaded transversely the displacement components u_α vanish and therefore the expression for ΔU takes the form

$$\Delta U = \gamma \int_B w \Delta h \, dx^1 dx^2 - \frac{1}{2} \int_B \left[\frac{1}{h} Q_\alpha \vartheta_\alpha \Delta h + \frac{1}{h^3} M_{(\alpha\beta)} \varkappa_{(\alpha\beta)} \Delta h^3 \right] dx^1 dx^2 + \mathscr{P}^2,$$

where

$$\mathscr{P}^2 = \frac{1}{2} \frac{\mathring{h}^3 E}{12(1-\nu^2)} \{(\Delta \varkappa_{11} + \Delta \varkappa_{22})^2 +$$

$$+ 2(1-\nu)[(\Delta \varkappa_{12})^2 - \Delta \varkappa_{11} \Delta \varkappa_{22}]\} dx^1 dx^2 + \frac{1}{2} \int_B G \mathring{h} (\Delta \vartheta)^2 dx^1 dx^2.$$

Taking into account the constant-volume assumption $\Delta V = \int_B \Delta h \, dx^1 dx^2 = 0$ and the formula

$$\Delta(h^3) = \Delta h [3h^2 + 3h \Delta h + (\Delta h)^2],$$

we can represent the difference ΔU as

$$\Delta U = \int_B [\gamma w h - \tfrac{1}{2}(Q_\alpha \vartheta_\alpha + 3 M_{(\alpha\beta)} \varkappa_{(\alpha\beta)}) - hC] \frac{\Delta h}{h} \, dx^1 dx^2 -$$

$$- \frac{1}{2} \int_B \frac{3}{h^2} M_{(\alpha\beta)} \varkappa_{(\alpha\beta)} (\Delta h)^2 dx^1 dx^2 + \mathscr{P}^2 + O(\Delta h^3), \tag{2.83}$$

where C is an arbitrary constant, $C \neq 0$.

The necessary condition for U to be minimum is

$$\gamma w - \tfrac{1}{2}(Q_\alpha \vartheta_\alpha + 3 M_{(\alpha\beta)} \varkappa_{(\alpha\beta)})/h = G.$$

Even if this is satisfied, however, there is no minimum, because the remaining expression

$$-\frac{1}{2}\int\limits_{B}\frac{3}{h^2}M_{(\alpha\beta)}\varkappa_{(\alpha\beta)}(\Delta h)^2\mathrm{d}x^1\mathrm{d}x^2+\mathscr{P}^2$$

can be negative for some Δh. It follows that in the general class of isotropic homogeneous solid plates with smooth external surfaces (the smoothness is implied by the assumed continuity of the function $h(x^1, x^2)$) a minimum-deformability plate does not exist.

However, if we restrict the class of admissible structures to some special kind of plates, the minimum-deformability condition, becoming a local-minimum condition in a sense, can be satisfied.

To illustrate this point, let us consider the problem of optimization of a plate band with width l subject to cylindrical bending, assuming that its supports are externally statically determinate. If the x^2-axis is parallel to the length of the plate, then

$$\mathring{M}_{22} = M_{22} = \mathring{Q}_2 = Q_2 = \mathring{M}_{(12)} = M_{(12)} = 0$$

and since $\mathring{M}_{11} = M_{11} = M$ and $\mathring{Q}_{11} = Q_{11} = Q$, we have the restrictions $\Delta M = \Delta Q = Q$, which express the independence of the internal forces of the deformations $\mathring{\varkappa}_{11} = \varkappa$ and $\vartheta_1 = \vartheta$ and the variable thickness h. The restrictions imply that

$$\Delta\varkappa = -\varkappa\frac{\Delta h^3}{h^3+\Delta h^3}, \qquad \Delta\vartheta = -\vartheta\frac{\Delta h}{h+\Delta h}.$$

Substituting these equalities in formula (2.83) and taking into account that $\varkappa = 12(1-\nu^2)M/Eh^3$ and $\vartheta = Q/Gh$, we obtain the following expression for ΔU:

$$\Delta U = -\frac{1}{2}\int\limits_{l}\left(\frac{1}{h}Q\vartheta\Delta h+\frac{1}{h^3}M\varkappa\Delta h^3\right)\mathrm{d}x^1+\gamma\int\limits_{l}w\Delta h\,\mathrm{d}x^1+$$

$$+\frac{1}{2}\int\limits_{l}\left[\frac{12(1-\nu^2)}{E}\frac{M^2(\Delta h^3)^2}{(h^3+\Delta h^3)}+\frac{1}{G}Q^2\frac{(\Delta h)^2}{h+\Delta h}\right]\mathrm{d}x^1$$

$$= -\frac{1}{2}\int\limits_{l}\left(\frac{Q^2}{Gh}\frac{\Delta h}{h+\Delta h}+\frac{12M^2(1-\nu)^2}{Eh^3}\frac{\Delta h^3}{h^3+\Delta h^3}\right)\mathrm{d}x^1+\gamma\int\limits_{l}w\Delta h\,\mathrm{d}x^1.$$

Using the series expansions

$$\frac{\Delta h}{h+\Delta h} = \frac{1}{h}\Delta h-\frac{2}{h^2}(\Delta h)^2+\ldots+O[(\Delta h)^3],$$

$$\frac{\Delta h^3}{h^3+\Delta h^3} = \frac{1}{h^3}\Delta h^3-\frac{2}{h^6}(\Delta h^3)^2+\ldots = \frac{3}{h}\Delta h-\frac{6}{h^2}(\Delta h)^2+\ldots+O(\Delta h)^3,$$

in which we shall neglect terms of order higher than 2 in Δh, we can write ΔU in the form

$$\Delta U = \delta U + \delta^2 U + \ldots,$$

where

$$\delta U = \int_l \left[\gamma w - \frac{1}{2}\left(\frac{Q^2}{Gh^2} + \frac{36M^2(1-\nu)^2}{Eh^4}\right)\right]\Delta h \, dx^1,$$

$$\delta^2 U = \frac{1}{2}\int_l \left(\frac{2Q^2}{Gh^3} + \frac{72M^2(1-\nu^2)}{Eh^5}\right)(\Delta h)^2 dx^1.$$

For ΔU to be positive for arbitrary Δh such that $\int_l \Delta h \, dx^1 = 0$ it is necessary and sufficient that $\delta U = 0$ and $\delta^2 U > 0$. These conditions will be satisfied if

$$\frac{1}{2}\left(\frac{Q^2}{Gh^2} + \frac{36M^2(1-\nu)^2}{Eh^4}\right) - \gamma w = C. \tag{2.84}$$

Relation (2.84) determines the variable thickness h of the least deformable structure in the set of plates under consideration. The constant C is determined by the given volume $V = \int_l h \, dx^1$ of the unit element of the band.

In the design practice we often deal with the problem of improving the shape of a plate with given volume V, contour Γ and thickness $\mathring{h}$ (e.g. constant thickness) so as to reduce its deformability. Let us show that by a suitable choice of the variable thickness h we can always obtain a plate with lower deformability.

To this end, let us write the expression for ΔU in the form

$$\Delta U = \gamma \int_B w \Delta h \, dx^1 dx^2 - \frac{1}{2}\int_B \left[\frac{1}{h} Q_\alpha \vartheta_\alpha \Delta h + \frac{1}{h^3} M_{(\alpha\beta)} \varkappa_{(\alpha\beta)} \Delta h^3\right] dx^1 dx^2 +$$

$$+ C \int_B \Delta h \, dx^1 dx^2 + \mathscr{P}^2,$$

where we have used the fact that the volume of the plate is unchanged, i.e. $\int_B \Delta h \, dx^1 dx^2 = 0$. Expressing Δh^3 as

$$\Delta h^3 = \Delta h[3h^2 + 3h\Delta h + (\Delta h)^2] = \Delta h[3\mathring{h}^2 + 3\mathring{h}\Delta h + (\Delta h)^2]$$

and rearranging the whole formula, we get

$$\Delta U = \int_B \left[\gamma w - \frac{1}{2}\left(\frac{1}{\mathring{h}} Q_\alpha \vartheta_\alpha + \frac{1}{\mathring{h}^3} 3\mathring{h}^2 M_{(\alpha\beta)} \varkappa_{(\alpha\beta)}\right) + C\right] \Delta h \, dx^1 dx^2 +$$

$$+ \frac{1}{2}\int_B 3\mathring{h}(\Delta h)^2 \frac{1}{\mathring{h}^3} M_{(\alpha\beta)} \varkappa_{(\alpha\beta)} dx^1 dx^2 + \mathscr{P}^2.$$

It follows that for ΔU to be positive it is enough that

$$\frac{1}{2}\left[\frac{1}{h}Q_\alpha \vartheta_\alpha + \frac{3\mathring{h}^2}{h^3}M_{(\alpha\beta)}\varkappa_{(\alpha\beta)}\right] - \gamma w = C. \tag{2.85}$$

This relation can serve as a basis for finding the variable thickness of a plate of reduced deformability as compared with a given plate having the same volume and known thickness.

If the distribution of the dead weight is considered insignificant for the deformability of the structure and is assumed, for example, to be uniform, then a similar reasoning leads to the relation

$$\frac{1}{2}\left[\frac{1}{h}Q_\alpha \vartheta_\alpha + \frac{3\mathring{h}^2}{h^3}M_{(\alpha\beta)}\varkappa_{(\alpha\beta)}\right] = C,$$

which can be obtained by setting $\gamma = 0$ in (2.85).

If we further neglect the influence of the transverse forces on the deform ability of the plate, assuming, after the classical plate theory, that $G \to \infty$ we get

$$\frac{3}{2}\frac{\mathring{h}^2}{h^3}M_{(\alpha\beta)}\varkappa_{(\alpha\beta)} = C$$

or

$$\frac{36\mathring{h}^2}{2E(1-\nu^2)}[(\varkappa_{11}+\varkappa_{22})^2 + 2(1-\nu)(\varkappa_{12}^2 - \varkappa_{11}\varkappa_{22})] = C.$$

In the case where the original plate has a constant thickness h_0, the above condition becomes the following differential equation with respect to the deflection function w:

$$\frac{\mathring{h}^2}{8E(1-\nu^2)}[(w_{,11}+w_{,22})^2 + 2(1-\nu)(w^2{}_{,12} - w_{,11}w_{,22})] = C. \tag{2.86}$$

It has the same form as the deflection equation for a three-sandwich uniform-strength plate (Dzieniszewski 1969_2).

Plates which we have considered so far were assumed to have smooth external surfaces. If we turn our attention to plates with irregular faces, e.g. plates with ribs as shown in Fig. 2.16, homogeneity and isotropy assumptions can no longer be used and optimality considerations have to be based on the anisotropic plate theory.

We shall show that the potential of a smooth plate can be reduced by changing the faces of the plate into ones with ribs.

To this end, consider two plates: one with a smooth surface $\mathring{h}(x^1, x^2)$ (Fig. 2.17a) (for example, it may be a plate satisfying the equal-potential condition over the faces) and one with the same volume but shaped as shown in Fig. 2.17b.

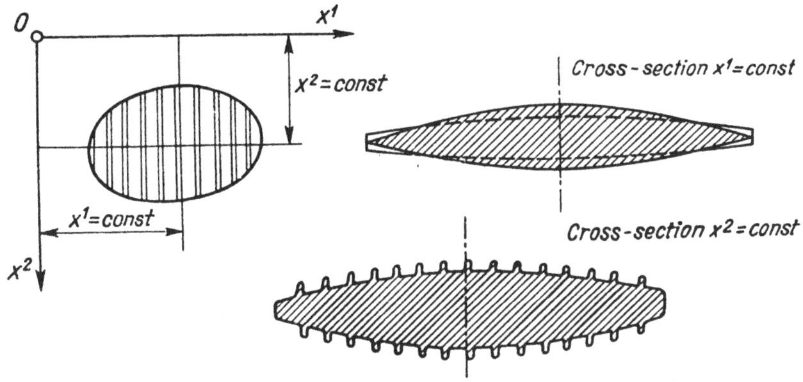

Fig. 2.16. A plate scheme with ribbed faces

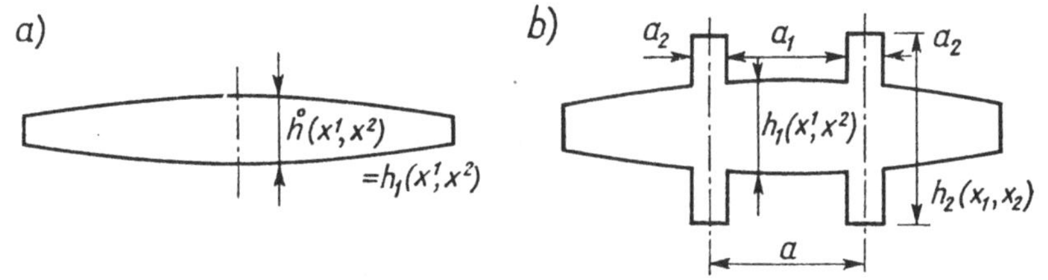

Fig. 2.17. A plate with two ribs

The equality of the volumes

$$\mathring{V} = \int_B h_1 \mathrm{d}x^1 \mathrm{d}x^2 \quad \text{and} \quad V = \int_B \frac{1}{a}(a_1 h_1 + a_2 h_2) dx^1 \mathrm{d}x^2$$

implies that

$$\begin{aligned} \Delta V &= \int_B \left[h_1 - \frac{1}{a}(a_1 h_1 + a_2 h_2) \right] \mathrm{d}x^1 \mathrm{d}x^2 \\ &= \frac{a_2}{a} \int_B (h_1 - h_2) \mathrm{d}x^1 dx^2 = -\frac{a_2}{a} \int_B \Delta h \mathrm{d}x^1 dx^2 = 0. \end{aligned} \tag{2.87}$$

The potential difference between the two plates can be represented in the form

$$\Delta U = -\tfrac{1}{2} \int_B [\Delta D_{11} \varkappa_{11}^2 + 2\Delta D_{(12)} \varkappa_{11} \varkappa_{22} + \Delta D_{22} \varkappa_{22}^2 + 2\Delta D_{66} \varkappa_{(12)}^2] \mathrm{d}x^1 \mathrm{d}x^2,$$

derived on the assumption that the influence of dead weight distribution and transverse forces Q_α is negligible. In fact, taking these two factors into account would not change our final conclusion regarding the inequality $\Delta U > 0$.

The rigidities of the plates are given by

$$\mathring{D}_{11} = \mathring{D}_{22} = \frac{E\mathring{h}_1^3}{12(1-\nu^2)}, \qquad \mathring{D}_{(12)} = \nu\mathring{D}_{11}, \qquad \mathring{D}_{66} = (1-\nu)\mathring{D}_{11},$$

$$D_{11} = \mathring{D}_{11}, \qquad D_{(12)} = \mathring{D}_{(12)}, \qquad D_{66} = \mathring{D}_{66},$$

$$D_{22} = \frac{a_1h_1^3+a_2h_2^3}{a}\,\frac{E}{12(1-\nu^2)}$$

$$= \frac{E}{12(1-\nu^2)}\left[h_1^3+\frac{a_2}{a}(3h_1\Delta h^2+3h_1\Delta h+(\Delta h)^3)\right],$$

and hence the differences

$$\Delta D_{11} = \Delta D_{(12)} = \Delta D_{66} = 0,$$

$$\Delta D_{22} = -\frac{E}{12(1-\nu^2)}\,\frac{a_2}{a}(3h_1\Delta h^2+3h_1^2\Delta h)+O(\Delta h^3).$$

Using these formulae and the constant-volume condition (2.87), we can write the potential difference as

$$\Delta U = \int_B\left[\frac{E}{8(1-\nu^2)}\varkappa_{22}^2h_1^2-C\right]\frac{a_2}{a}\,\Delta h\,\mathrm{d}x^1\mathrm{d}x^2+$$

$$+\frac{1}{2}\int_B\frac{E}{4(1-\nu^2)}\varkappa_{22}^2h_1(\Delta h)^2\frac{a_2}{a}\,\mathrm{d}x^1dx^2+\mathscr{P}^2,$$

from which we conclude that $\Delta U > 0$ if

$$\frac{E}{8(1-\nu^2)}(\varkappa_{22})^2h_1^2 = C.$$

This condition means that the strain $\varepsilon_{22} = \frac{h_1}{2}\varkappa_{22}$ is equalized over the surface h_1 of the plate.

Thus, a ribbed plate shows a greater stiffness than a smooth-faced plate made of the same kind and volume of material. The reason why a solid plate with smooth and regular external surfaces cannot be the least deformable lies in the non-existence of the minimum of the potential. This fact was first pointed out by Mróz (1969_2).

Three-layer plates

A three-layer plate (Fig. 2.18) is an example of a transversely nonhomogeneous plate. Its external layers are usually made of stronger materials and have greater elastic moduli than the internal one.

Quantities relating to the external layers will be marked with a subscript z and those referring to the internal layer with a subscript w.

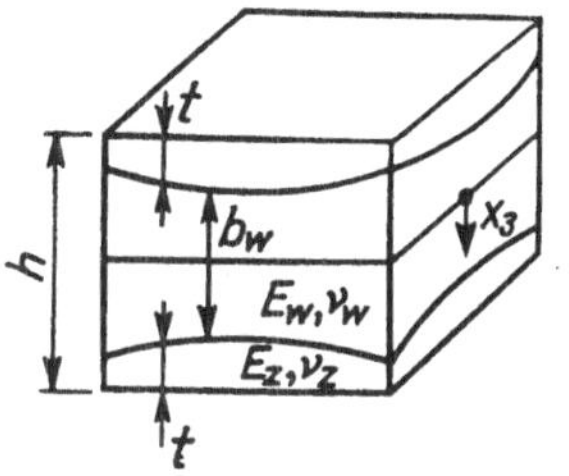

Fig. 2.18. A three-layer plate scheme

To find how the geometrical dimensions bear on the deformability of the plate, which is assumed to be loaded transversely, we first calculate its deformation rigidities, substituting the elastic coefficients

$$A_{\alpha\beta\gamma\nu} = \frac{E(x^3)}{1-\nu^2(x^3)}\left[\frac{1}{2}(1-\nu)(\delta_{\alpha\gamma}\delta_{\beta\nu}+\delta_{\alpha\nu}\delta_{\beta\gamma})+\nu\delta_{\alpha\beta}\delta_{\gamma\nu}\right], \tag{2.88}$$

$$A_{\alpha 3\beta 3} = G\delta_{\alpha\beta},$$

where

$$E(x^3) = E_z[H(z-h_w/2)-H(z-h/2)]+E_w[H(z+h_w/2)-H(z-h_w/2)]+$$
$$+E_z[H(z+h/2)-H(z+h_w/2)],$$
$$\nu(x^3) = \nu_z[H(z-h_w/2)-H(z-h/2)]+\nu_w[H(z+h_w/2)-H(z-h_w(2)]+$$
$$+\nu_z[H(z+h/2)-H(z+h_w/2)]$$

into formulae (2.70). The result is

$$D_{\alpha\beta\gamma\nu} = \frac{1}{12}\left\{\delta_{\alpha\gamma}\delta_{\beta\nu}\left[\frac{E_w h_w^3}{1+\nu_w}+\frac{E_z(h^3-h_w^3)}{1+\nu_z}\right]+\right.$$
$$\left.+\delta_{\alpha\beta}\delta_{\nu\gamma}\left[\frac{E_w\nu_w h_w^3}{1-\nu_w^2}+\frac{\nu_z E_z(h^3-h_w^3)}{1-\nu_z^2}\right]\right\}, \tag{2.89}$$

$$S_{\alpha 3\beta 3} = \frac{1}{2}\delta_{\alpha\beta}\left[\frac{E_w h_w}{1+\nu_w}+\frac{E_z(h-h_w)}{1+\nu_z}\right].$$

If we change the thickness t of each of the external layers, preserving their volume and the height $h \gg 2t$ (not necessarily constant) of the plate, the resulting differences $\Delta D_{\alpha\beta\gamma\nu}$ and $\Delta S_{\alpha 3\beta 3}$ will be

$$\Delta D_{\alpha\beta\gamma\nu} = \frac{1}{12}\left[\frac{1}{2}(\delta_{\alpha\gamma}\delta_{\beta\nu}+\delta_{\alpha\nu}\delta_{\beta\gamma})\frac{E_w(1+\nu_z)-E_z(1+\nu_w)}{(1+\nu_z)(1+\nu_w)}\right]\Delta h_w^3, \tag{2.90}$$

$$\Delta S_{\alpha 3\beta 3} = \frac{1}{2}\delta_{\alpha\beta}\frac{E_w(1+\nu_z)-E_z(1+\nu_w)}{(1+\nu_w)(1+\nu_z)}\Delta h_w, \tag{2.91}$$

where

$$\Delta h_w^3 = -6(h^2-2ht+4t^2)\Delta t+6(h-4t)(\Delta t)^2+ \ldots +O(\Delta t^3)$$
$$\sim -3h^2\Delta 2t+6h(\Delta t)^2.$$

Substituting expressions (2.90) and (2.91) in the potential difference

$$\Delta U = -\tfrac{1}{2}\int_B [\Delta D_{\alpha\beta\gamma\nu}\varkappa_{(\beta\alpha)}\varkappa_{(\gamma\nu)}+\Delta S_{\alpha 3\beta 3}\vartheta_\beta\vartheta_\alpha]\mathrm{d}x^1\mathrm{d}x^2,$$

we obtain

$$\Delta U = \frac{1}{2}\int_B\Bigg\{\frac{1}{12}\Bigg[\frac{-E_w(1+\nu_z)+E_z(1+\nu_w)}{(1+\nu_z)(1+\nu_w)}(\varkappa_{11}^2+2\varkappa_{(12)}^2+\varkappa_{22}^2)+$$
$$+\frac{-E_w\nu_w(1-\nu_z^2)+E_z(1-\nu_w^2)\nu_z}{(1-\nu_w^2)(1-\nu_z^2)}(\varkappa_{11}+\varkappa_{22})^2\Bigg]\times$$
$$\times(-3h^2\Delta 2t+6h(\Delta t)^2)+(G_z-G_w)(\vartheta_1^2+\vartheta_2^2)\Delta 2t\Bigg\}\mathrm{d}x^1\mathrm{d}x^2+\mathscr{P}^2.$$

Since by the constant-volume condition $\int_B 2\Delta t\,\mathrm{d}x^1\mathrm{d}x^2 = 0$, the above formula gives us the following minimum-deformability condition ($\Delta U > 0$ for arbitrary Δt) for the plates considered:

$$\frac{1}{8}\Bigg[\frac{E_w(1+\nu_z)+E_z(1+\nu_w)}{(1+\nu_z)(1+\nu_w)}(\varkappa_{11}^2+2\varkappa_{12}^2+\varkappa_{22}^2)+$$
$$+\frac{E_z(1-\nu_w^2)\nu_z-E_w\nu_w(1-\nu_z^2)}{(1-\nu_w^2)(1-\nu_z^2)}(\varkappa_{11}+\varkappa_{22})^2\Bigg]h^2+ \tag{2.92}$$
$$+(G_w-G_z)(\vartheta_1^2+\vartheta_2^2) = C.$$

The case where $E_w \to 0$ and $G_w \to G_z \to \infty$ is discussed in detail by Dzieniszewski (1969_2).

Box plates

Box plates with one-way ribbing (e.g. along the x^2-axis), as shown in Fig. 2.19, are often used in engineering structures as box slab floors.

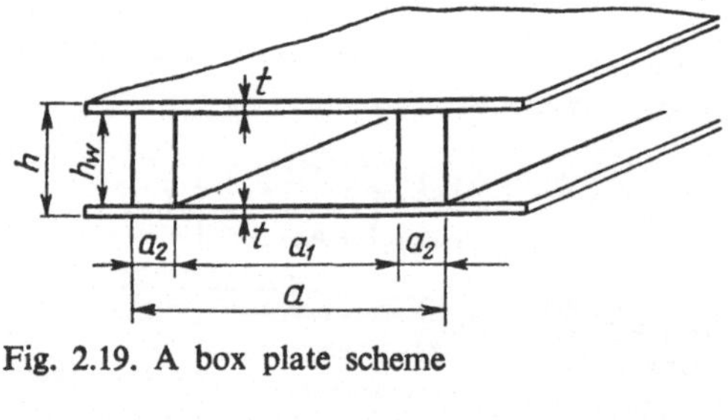

Fig. 2.19. A box plate scheme

Choosing the skin plate thickness t and the web thickness a_2 as the design variables, we seek the least deformable structure in the set of box plates with the same volume V, $h \gg t$ and $a \gg a_2$.

The rigidity differences between two plates in this set are

$$\begin{aligned}
\Delta D_{11} &= \frac{E}{12(1-\nu^2)}[h^3-\mathring{h}_w^3-h^3+h_w^3] = -\frac{E}{12(1-\nu^2)}\Delta h_w^3 \\
&= \frac{E}{12(1-\nu^2)}[(h-2t)^3-(h-2\mathring{t})^3] \\
&= \frac{Eh^2}{2(1-\nu^2)}\Delta t - \frac{E}{(1-\nu^2)}(h-2t)(\Delta t)^2, \\
\Delta D_{22} &= \Delta D_{11}, \quad \Delta D_{12} = \nu\Delta D_{11}, \quad \Delta D_{66} = (1-\nu)\Delta D_{11}, \\
\Delta S_{55} &= Gh\Delta\left(\frac{a_2}{a}\right), \quad \Delta S_{44} \simeq 0;
\end{aligned}$$

the plates are regarded as orthotropic. Inserting these expressions into the elastic strain energy difference

$$\begin{aligned}
\Delta U = -\tfrac{1}{2}\int_B [&\Delta D_{11}\varkappa_{11}^2 + 2\Delta D_{(12)}\varkappa_{11}\varkappa_{22} + \Delta D_{22}\varkappa_{22}^2 + \\
&+ 2\Delta D_{66}\varkappa_{(12)}^2 + \Delta S_{44}(\vartheta_1)^2 + \Delta S_{55}(\vartheta_2)^2]\mathrm{d}x^1\mathrm{d}x^2 + \mathscr{P}^2,
\end{aligned} \tag{2.93}$$

we find

$$\begin{aligned}
\Delta U = &-\frac{1}{2}\frac{E}{2(1-\nu^2)}\int_B h^2(\varkappa_{11}^2 + 2\nu\varkappa_{11}\varkappa_{22} + \varkappa_{22}^2 + 2\varkappa_{(12)}^2)\Delta t\,\mathrm{d}x^1\mathrm{d}x^2 + \\
&+ \frac{1}{2}\frac{E}{(1-\nu^2)}(h-2t)\int_B [\varkappa_{11}^2 + 2\nu\varkappa_{11}\varkappa_{22} + \varkappa_{22}^2 + 2\varkappa_{(12)}^2](\Delta t)^2\mathrm{d}x^1\mathrm{d}x^2 - \\
&- \frac{1}{2}G\int_B h\vartheta_2^2\Delta\left(\frac{a_2}{a}\right)\mathrm{d}x^1\mathrm{d}x^2 + \mathscr{P}^2.
\end{aligned}$$

Remembering that the volume is constant and so

$$\Delta V = -\int_B \left[2\Delta t + h\Delta\left(\frac{a_2}{a}\right)\right]\mathrm{d}x^1\mathrm{d}x^2 = 0,$$

we can see that the minimum-deformability condition $\Delta U > 0$ will be satisfied when

$$\begin{aligned}
&\frac{E}{4(1-\nu^2)}h^2(\varkappa_{11}^2 + 2\nu\varkappa_{11}\varkappa_{22} + \varkappa_{22}^2 + 2\varkappa_{(12)}^2) = C, \\
&\tfrac{1}{2}Gh(\vartheta_2)^2 = C.
\end{aligned} \tag{2.94}$$

Together with the equations of deformation, the above equations provide a basis for determining the dimensions t and a_2 of the optimum box plate.

The case of box plates with two-way ribbing can be treated similarly.

2.7 Optimization of spatial lattice bar structures

Description of lattice structures

Dense lattice structures formed from prismatic bars are inreasingly applied in building and civil engineering. They appear as plane, surface or spatial structures designed to transmit various loads or as skeleton or fibrous elements in structural multi-phase composites. The shapes and dimensions of their networks and bars, the types of their joints and the kinds of their structural materials can vary widely[1]. The foundations of a general theory of lattice structures were only laid a decade ago. The first monographic treatments of the subject were the works of Woźniak, Frąckiewicz and Gutkowski[2].

The problem of choice of shapes and dimensions of bar lattices can be of primary importance in various engineering projects, for example in designing regular skeleton building structures. Indeed, by a suitable choice of shape and dimensions, a lattice structure can be given a rigidity appropriate to its technological function, with a limited expenditure of structural materials. In particular, lattice structures constructed for minimum deformability under given loads, and hence 'better' than other structures (e.g. solid-walled ones) made of the same kind and amount of materials, deserve detailed consideration.

The existing literature on optimization does not comprise many papers dealing with lattice structures. We can mention here four papers by Dzieniszewski (1970_1, 1971_2, 1971_3, 1973), of which the first three concern the optimum design of bar lattices with respect to material distribution and directions of principal axes, and the fourth deals with optimization of spatial lattice structures with uniform deformability. Two papers by Gierliński ($1973_{1,2}$) concern optimum square or annular lattice plates.

In this section we present a general formulation and solution of the minimum-deformability design problem for spatial bar systems with arbitrarily dense lattices. The solution is based on the linear theory of fibrous media developed by Woźniak[3].

---

[1] *Space Structures, A study of methods and developments in three dimensional constructions resulting from The International Conference on Space Structures*, University of Surrey, September 1966, edited by R. M. Davies, Blackwell Scientific Publications, Oxford and Edinburgh.

[2] Woźniak, Cz., *Lattice surface girders*, PWN, Warsaw 1969; Frąckiewicz, H., *Lattice Mechanics*, PWN, Warsaw 1970; Gutkowski, W., *Regular Bar Structures*, PWN, Warsaw 1973 (all in Polish).

[3] Woźniak, Cz., Theory of fibrous media, *Arch. Mech. Stos.*, **5**, 17, 651–669, 1965; **6**, 17, 777–799, 1965; On the equations of the theory of lattice structures, *Arch. Mech. Stos.*, **5**, 21, 539–555, 1969; Continuous models of dense bar lattices (in Polish), *Arch. Inż. Ląd.*, **2**, 11, 175–185, 1965; *Lattice Surface Girders* (in Polish), PWN, Warsaw 1969.

As a measure of deformability of a lattice structure made of a given volume V of materials and subject to a given load system we shall use, as before, the elastic strain energy U stored in the structure. So far, we have used the minimum-energy criterion to solve surface problems, i.e. to determine the optimum shape of free surfaces of structures made of homogeneous and isotropic materials or boundary surfaces between different elements of multiphase structures. By applying this criterion to optimization of spatial lattice structures, and hence to three-dimensional optimum design, we make a significant step forward in the general theory.

A dense lattice is defined by three one-parameter families of curves or broken lines, $\triangle = I, II, III$ (Fig. 2.20). The structure we are considering consists of segments of prismatic bars, tangent to the lines $\triangle$. The lengths $\underset{\triangle}{l}$ of the segments are assumed to vary continuously. The cross-sections of the bundle of all the bars tangent to the lines of a given family $\triangle$ form a family of normal surfaces to $\triangle$. The partial area $\underset{\triangle}{\omega}$ of the bundle cross-section

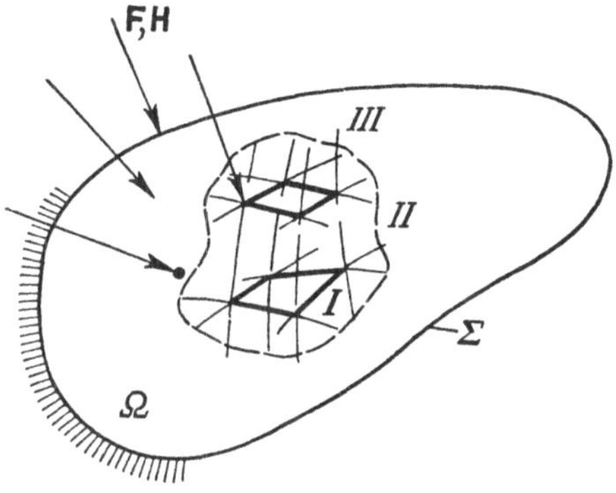

Fig. 2.20. A lattice medium

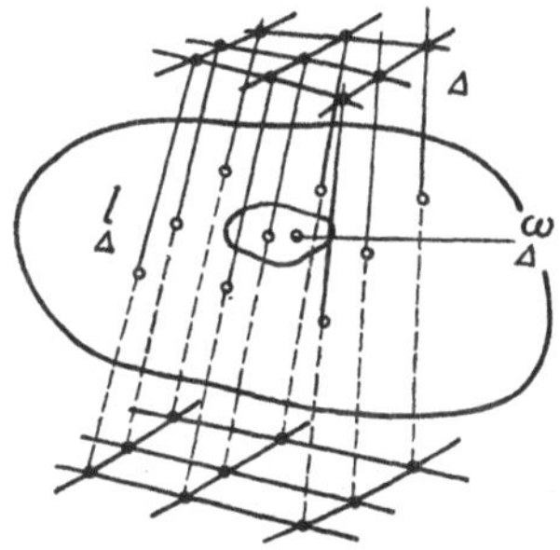

Fig. 2.21. The bundle of bars of a family $\triangle$

which falls to an individual bar is a measure of the density of bars in the bundle (Fig. 2.21). The bar segments belonging to different families $\triangle$ meet only in rigid joints at the nodes of the lattice. Each of the three families of bars can be made of different material, as long as it obeys Hooke's law.

The geometry of a bundle $\triangle$ is described at each point by three orthogonal principal directions: the direction of the longitudinal axis of the bar to which this point belongs (tangent to a $\triangle$-line) and the directions of the principal axes of inertia of the cross-section of the bar at this point. The corresponding unit vectors will be denoted by $\underset{\triangle}{\mathbf{D}_A} = \underset{\triangle}{\mathbf{D}^A}$, $A = 1, 2, 3$, the vector $\underset{\triangle}{\mathbf{D}_1} = \underset{\triangle}{\mathbf{D}^1} = \underset{\triangle}{\mathbf{T}}$ being that of the bar axis. Thus at each point of the bundle $\triangle$ the vectors $\underset{\triangle}{\mathbf{D}_A}$ form a right-handed orthonormal vector basis. Their components, contravariant or covariant, in an arbitrarily chosen curvilinear coordinate system $\theta^i = \theta^i(x^1, x^2, x^3)$, $i = 1, 2, 3$, will be denoted by $\underset{\Delta}{D}{}^i_A = \underset{\Delta}{\mathbf{D}}_A \mathbf{g}^i$ or $\underset{\Delta}{D}{}^A_i = \underset{\Delta}{\mathbf{D}}{}^A \mathbf{g}_i$. In the particular case where θ^i are Cartesian coordinates, $\mathbf{g}^i = \mathbf{g}_i = \mathbf{i}_i = \mathbf{i}^i$. In general, the components $\underset{\Delta}{D}{}^i_A$ are functions of the coordinates θ^i and satisfy at each point of the structure the orthogonality relations

$$\underset{\Delta}{D}{}^i_A \underset{\Delta}{D}{}^A_j = \delta^i_j, \qquad \underset{\Delta}{D}{}^i_A \underset{\Delta}{D}{}^B_i = \delta^B_A. \tag{2.95}$$

The bars and the joints connecting bars of different families can transmit longitudinal and transverse forces and also bending and torsional moments. The mechanical properties of any bar belonging to a family Δ are described by the following matrices of principal rigidities (for a derivation of these formulae see Dzieniszewski 1971):

$$\|\underset{\Delta}{\bar{R}}{}^{AB}\| = \begin{Vmatrix} E\underset{\Delta\Delta}{A} & 0 & 0 \\ 0 & 12E\underset{\Delta\Delta}{J^{33}}\underset{\Delta}{l}^{-2}(1+12\underset{\Delta}{\varrho}^{33})^{-1} & 0 \\ 0 & 0 & 12E\underset{\Delta\Delta}{J^{22}}\underset{\Delta}{l}^{-2}(1+12\underset{\Delta}{\varrho}^{22})^{-1} \end{Vmatrix},$$

$$\|\underset{\Delta}{\bar{S}}{}^{AB}\| = \begin{Vmatrix} G\underset{\Delta\Delta}{J^{11}} & 0 & 0 \\ 0 & E\underset{\Delta\Delta}{J^{22}} & 0 \\ 0 & 0 & E\underset{\Delta\Delta}{J^{33}} \end{Vmatrix}. \tag{2.96}$$

The symbols appearing in the matrices have the following meanings:

$\underset{\Delta}{J}{}^{11}$ —geometrical rigidity of torsion;

$\underset{\Delta}{J}{}^{22} = (\underset{\Delta}{i}{}^{22})^2\underset{\Delta}{A}$ and $\underset{\Delta}{J}{}^{33} = (\underset{\Delta}{i}{}^{33})^2\underset{\Delta}{A}$—the principal moments of inertia of the cross-section of the bar;

$\underset{\Delta}{\varrho}{}^{AA} = 2(1+\nu)(\underset{\Delta}{l}/\underset{\Delta\Delta}{i^{AA}})^{-2}\underset{\Delta}{\bar{\varkappa}}{}^{AA} < 7.5\,\underset{\Delta}{\lambda}^2$—a coefficient representing the influence of transversal forces on deformation of the bar;

$1.2 < \underset{\Delta}{\bar{\varkappa}}{}^{AA} < 2.5$—a coefficient characterizing the form of the cross-section;

E and G—Young's modulus and shear modulus of the material of the bar.

The elements of the matrices (2.96) are the rigidities of the bar in linear strain, shear, torsion and bending. They are all proportional to the cross-sectional area $\underset{\Delta}{A}$ of the bar, as can be seen from the difinitions given above.

By dividing $\underset{\Delta}{\bar{R}}{}^{AB}$ and $\underset{\Delta}{\bar{S}}{}^{AB}$ by $\underset{\Delta}{\omega}$ we define the densities $\underset{\Delta}{R}{}^{AB}$ and $\underset{\Delta}{S}{}^{AB}$ of the different kinds of rigidity. Similarly, the ratio $\underset{\triangle}{\mathscr{A}} = \underset{\triangle}{A}/\underset{\triangle}{\omega}$ describes the distribution of material in the bundle $\triangle$. Since the bars may differ in dimensions and cross-sections, the quantities $\underset{\triangle}{R}^{AB}$, $\underset{\triangle}{S}^{AB}$ and $\underset{\triangle}{\mathscr{A}}$ are functions of the coordinates θ^i.

Formulation of the optimum-design problem

The bar lattice to be optimized is supposed to transmit given force and moment fields, $\mathbf{F}(\theta^1, \theta^2, \theta^3)$ and $\mathbf{H}(\theta^1, \theta^2, \theta^3)$, applied at the nodes of the lattice, both internal and external. The external nodes lie on the surface Γ of the system. The load intensities per unit volume of the structure are $\mathbf{f}$ and $\mathbf{h}$. The form of the surface Γ, the geometry of the structure and the way it is supported are fixed. The structure is in stable equilibrium.

The design variables are the three scalar fields $\underset{\triangle}{\mathscr{A}}(\theta^1, \theta^2, \theta^3)$ ($\triangle = I, II, III$) describing the distributions of structural materials in the different bundles of the lattice. The optimum-design problem we are concerned with consists in determining, for given volumes $\underset{\triangle}{V}$ of the materials, the unknown functions $\underset{\triangle}{\mathscr{A}}$ for which the elastic strain energy U of the system is at an absolute minimum. The structure corresponding to the functions $\underset{\triangle}{\mathscr{A}}$ so chosen is the least deformable in the design space under consideration. Indeed, since the structure is in equilibrium (by assumption for any choice of $\underset{\triangle}{\mathscr{A}}$), the work L performed by the loads in the transition from the natural to the deformed state is equal, by the Clapeyron theorem, to the energy U, and therefore a minimum U implies a minimum L, and hence minimum deformability. The present optimum-design problem is thus a classical problem of variational calculus: determine the functions $\underset{\triangle}{\mathscr{A}}$ which minimize the functional U under the isoperimetric conditions $\underset{\triangle}{V} = \text{const}$.

In order to solve the problem stated above, we have to find how a change in the material distributions $\underset{\triangle}{\mathscr{A}}$ affects the energy U. This will be possible once we know how to calculate the strain of the system for arbitrary $\underset{\triangle}{\mathscr{A}}$. We begin by bringing together some necessary notions, definitions and relations pertaining to the theory of strain of lattice structures.

Fundamental notions and relations

Let us consider an arbitrary bar belonging to a family $\triangle$ (Fig. 2.22). Its length is $\underset{\triangle}{l}$ and the directions of its principal axes are $\underset{\triangle}{\mathbf{D}}_A = \underset{\triangle}{\mathbf{D}}^A$. The ends

$\underline{i}$ and $\underline{k}$ of the bar are loaded by the forces $-\underset{\triangle}{\overline{\mathbf{P}}}_{\underline{ik}}$, $\underset{\triangle}{\overline{\mathbf{P}}}_{\underline{ki}}$ and moments $-\underset{\triangle}{\overline{\mathbf{M}}}_{\underline{ik}}$, $\underset{\triangle}{\overline{\mathbf{M}}}_{\underline{ki}}$ representing the reactions of the system induced by the forces $\mathbf{F}$ and moments $\mathbf{H}$ applied at the nodes of the lattice.

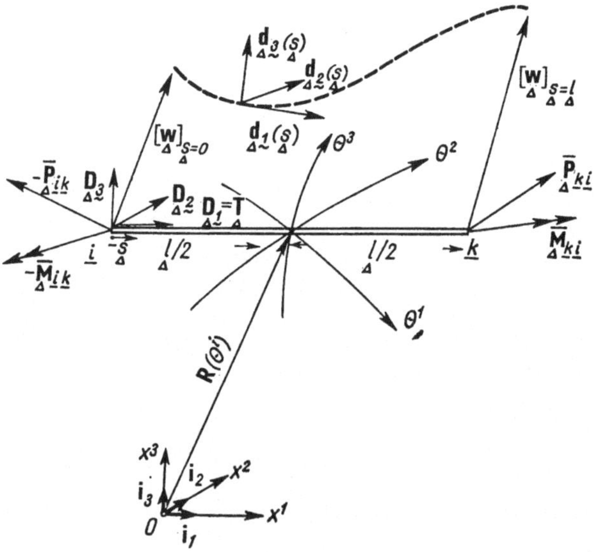

Fig. 2.22. Load scheme for a bar segment in family $\triangle$

The loads produce rotations $\underset{\triangle}{\mathbf{u}}(\underset{\triangle}{s}) = \underset{\triangle}{u}_A \underset{\triangle}{\mathbf{D}}^A$ of the principal axes of the bar, and displacements $\underset{\triangle}{\mathbf{w}}(\underset{\triangle}{s}) = \underset{\triangle}{w}_A \underset{\triangle}{\mathbf{D}}^A$ of their points. The rotations $\underset{\triangle}{\mathbf{u}}$ and displacement $\underset{\triangle}{\mathbf{w}}$ are functions of a variable $\underset{\triangle}{s}$ $[0, \underset{\triangle}{l}\,]$ measured along the bar. We assume that $\underset{\triangle}{w}_A$ are small compared with the dimensions of the bar and that $\underset{\triangle}{u}_A$ are much smaller than unity.

Under these assumptions, the directions $\underset{\triangle}{\mathbf{d}}_A = \underset{\triangle}{\mathbf{d}}^A$ of the principal axes of the bar in the deformed state are related with the unstrained directions $\underset{\triangle}{\mathbf{D}}_A$ by the equation

$$\underset{\triangle}{\mathbf{d}}_A(\underset{\triangle}{s}) = \underset{\triangle}{\mathbf{D}}_A + \underset{\triangle}{\mathbf{u}}(\underset{\triangle}{s}) \times \underset{\triangle}{\mathbf{D}}_A . \tag{2.97}$$

The vector $\underset{\triangle}{\mathbf{l}} = \underset{\triangle}{l}\underset{\triangle}{\mathbf{T}}$ becomes

$$\underset{\triangle}{\mathbf{l}} + [\underset{\triangle}{\mathbf{w}}]_{\underset{\triangle}{s} = \underset{\triangle}{l}} - [\underset{\triangle}{\mathbf{w}}]_{s=0} .$$

The state of strain of the bar can be described by the formulae

$$\begin{aligned}\frac{1}{2}(\underset{\triangle}{\bar{\mathbf{P}}_{ik}}+\underset{\triangle}{\bar{\mathbf{P}}_{ki}}) &= \underset{\triangle}{\mathbf{R}}^{AB}\Big[\frac{1}{l}([\underset{\triangle}{w_B}]_{\underset{\triangle}{s}=\underset{\triangle}{l}}-[\underset{\triangle}{w_B}]_{\underset{\triangle}{s}=0})+ \\ &\qquad + \frac{1}{2}\in_{B\overset{C}{l}.}([\underset{\triangle}{u_C}]_{\underset{\triangle}{s}=\underset{\triangle}{l}}+[\underset{\triangle}{u_C}]_{\underset{\triangle}{s}=0})\Big]\underset{\triangle}{\mathbf{D}}_A, \\ \frac{1}{2}(\underset{\triangle}{\bar{\mathbf{M}}_{ik}}+\underset{\triangle}{\bar{\mathbf{M}}_{ki}}) &= \underset{\triangle}{S}^{AB}\frac{1}{\underset{\triangle}{l}}([\underset{\triangle}{u_B}]_{\underset{\triangle}{s}=\underset{\triangle}{l}}-[\underset{\triangle}{u_B}]_{\underset{\triangle}{s}=0})\underset{\triangle}{\mathbf{D}}_A,\end{aligned} \tag{2.98}$$

which follow from the equations of statics of prismatic bars and are well known in structural mechanics as the transformation equations in the method of displacements (for details see e.g. Dzieniszewski 1971). The symbol $\underset{\triangle}{\in_{AB.}^{\;\;\;C}} = \underset{\triangle}{D^i_A}\underset{\triangle}{D^j_B}\underset{\triangle}{D^C_k}\underset{\triangle}{\in_{ij.}^{\;\;\;k}}$ denotes a Ricci three-vector.

Denoting by $\underset{\triangle}{\mathbf{P}} = \frac{1}{2}(\underset{\triangle}{\bar{\mathbf{P}}_{ik}}+\underset{\triangle}{\bar{\mathbf{P}}_{ki}})\underset{\triangle}{\omega}^{-1}$ and $\underset{\triangle}{\mathbf{M}} = \frac{1}{2}(\underset{\triangle}{\bar{\mathbf{M}}_{ik}}+\underset{\triangle}{\bar{\mathbf{M}}_{ki}})\underset{\triangle}{\omega}^{-1}$ the intensity vectors of the internal forces in the middle section $\underset{\triangle}{s} = \underset{\triangle}{l}/2$ of the bar, and setting

$$\begin{aligned}\underset{\triangle}{\mathbf{e}} &= \underset{\triangle}{e_B}\underset{\triangle}{\mathbf{D}}^B = \Big[\frac{1}{\underset{\triangle}{l}}([\underset{\triangle}{w_B}]_{\underset{\triangle}{s}=\underset{\triangle}{l}}-[\underset{\triangle}{w_B}]_{\underset{\triangle}{s}=0})+\frac{1}{2}\in_{B\overset{C}{l}.}([\underset{\triangle}{u_C}]_{\underset{\triangle}{s}=\underset{\triangle}{l}}+[\underset{\triangle}{u_C}]_{\underset{\triangle}{s}=0})\Big]\underset{\triangle}{\mathbf{D}}^B, \\ \underset{\triangle}{\boldsymbol{\varphi}} &= \underset{\triangle}{\varphi_B}\underset{\triangle}{\mathbf{D}}^B = \frac{1}{\underset{\triangle}{l}}([\underset{\triangle}{u_B}]_{\underset{\triangle}{s}=\underset{\triangle}{l}}-[\underset{\triangle}{u_B}]_{\underset{\triangle}{s}=0})\underset{\triangle}{\mathbf{D}}^B,\end{aligned} \tag{2.99}$$

we can write the physical relations (2.98) in the form

$$\underset{\triangle}{P}^A = \underset{\triangle}{R}^{AB}\underset{\triangle}{e_B}, \qquad \underset{\triangle}{M}^A = \underset{\triangle}{S}^{AB}\underset{\triangle}{\varphi_B}. \tag{2.100}$$

The components $\underset{\triangle}{P}^A = \underset{\triangle}{\mathbf{P}}\cdot\underset{\triangle}{\mathbf{D}}^A = \underset{\triangle}{P^i}\underset{\triangle}{D^A_i}$ represent the intensities of the longitudinal and the transverse forces in the two principal planes of inertia of the bar. The components $\underset{\triangle}{M}^A = \underset{\triangle}{\mathbf{M}}\cdot\underset{\triangle}{\mathbf{D}}^A = \underset{\triangle}{M^i}\underset{\triangle}{D^A_i}$ are densities of the torsional and the bending moments in these planes.

The component $\underset{\triangle}{e_1} = \underset{\triangle}{\mathbf{T}}\cdot\underset{\triangle}{\mathbf{e}}$ defines the unit elongation of the bar axis; $\underset{\triangle}{e_2}$ and $\underset{\triangle}{e_3}$ are equal to the arithmetic means of the angles between the projections of the bar chord on the planes $\underset{\triangle}{\mathbf{D}_1}$, $\underset{\triangle}{\mathbf{D}_2}$ and $\underset{\triangle}{\mathbf{D}_1}$, $\underset{\triangle}{\mathbf{D}_3}$, respectively, and the direction $\mathbf{d}_1$ of the bar axis in the deformed state at the ends of the bar.

The components $\underset{\triangle}{\varphi_A} = \underset{\triangle}{\boldsymbol{\varphi}}\cdot\underset{\triangle}{\mathbf{D}_A}$ represent the unit increments of the angle of twist and of the bending angles along the axis $\underset{\triangle}{\mathbf{T}}$ of the bar in the principal planes of inertia.

Since the joints of the lattice are assumed to be perfectly rigid, the displacements and rotations of the ends $\underset{\triangle}{s} = 0$ and $\underset{\triangle}{s} = \underset{\triangle}{l}$ of the bar under consideration are equal to the displacements $\mathbf{w}$ and rotations $\mathbf{u}$ of the nodes connecting this bar with the bars of the other two families $\Sigma \neq \triangle$. Hence

$$
\begin{aligned}
[w_{\underset{\triangle}{A}}]_{\underset{\triangle}{s}=0,\,\underset{\triangle}{s}=\underset{\triangle}{l}} &= D^i_{\underset{\triangle}{A}}[w_i]_{\underset{\triangle}{s}=0,\,\underset{\triangle}{s}=\underset{\triangle}{l}},\\
[u_{\underset{\triangle}{A}}]_{\underset{\triangle}{s}=0,\,\underset{\triangle}{s}=\underset{\triangle}{l}} &= D^i_{\underset{\triangle}{A}}[u_i]_{\underset{\triangle}{s}=0,\,\underset{\triangle}{s}=\underset{\triangle}{l}}.
\end{aligned}
\tag{2.101}
$$

Assuming that the functions w_i and u_i are continuous and have total derivatives at $\underset{\triangle}{s} = \underset{\triangle}{l}/2$, we can resolve them into Taylor series in the interval $[0, \underset{\triangle}{l}]$:

$$
w_i(\underset{\triangle}{s}) = [w_i]_{\underset{\triangle}{s}=\underset{\triangle}{l}/2} + \frac{\underset{\triangle}{s}-\underset{\triangle}{l}/2}{1!}\left[\frac{dw_i}{d\underset{\triangle}{s}}\right]_{\underset{\triangle}{s}=\underset{\triangle}{l}/2} + \dots,
$$

$$
u_i(\underset{\triangle}{s}) = [u_i]_{\underset{\triangle}{s}=\underset{\triangle}{l}/2} + \frac{\underset{\triangle}{s}-\underset{\triangle}{l}/2}{1!}\left[\frac{du_i}{d\underset{\triangle}{s}}\right]_{\underset{\triangle}{s}=\underset{\triangle}{l}/2} + \dots
$$

Substituting the linear terms of these series for w_i and u_i in (2.101) and using the result in (2.99), we get

$$
e_{\underset{\triangle}{A}} = D^i_{\underset{\triangle}{A}}\left[\frac{dw_i}{d\underset{\triangle}{s}}\right]_{\underset{\triangle}{s}=\underset{\triangle}{l}/2} + \in_{\underset{\triangle}{A}i.}^{\;\;B} D^i_B[u_i]_{\underset{\triangle}{s}=\underset{\triangle}{l}/2},
$$

$$
\varphi_{\underset{\triangle}{A}} = D^i_{\underset{\triangle}{A}}\left[\frac{du_i}{d\underset{\triangle}{s}}\right]_{\underset{\triangle}{s}=\underset{\triangle}{l}/2}.
$$

Setting

$$
[w_i]_{\underset{\triangle}{s}=\underset{\triangle}{l}/2} = w_i(\theta^1, \theta^2, \theta^3), \qquad [u_i]_{\underset{\triangle}{s}=\underset{\triangle}{l}/2} = u_i(\theta^1, \theta^2, \theta^3)
$$

and expressing the total derivatives in terms of the covariant derivatives

$$
\left[\frac{dw_i}{d\underset{\triangle}{s}}\right]_{\underset{\triangle}{s}=\underset{\triangle}{l}/2} = \underset{\triangle}{T^j} w_{i|j}, \qquad \left[\frac{du_i}{d\underset{\triangle}{s}}\right]_{\underset{\triangle}{s}=\underset{\triangle}{l}/2} = \underset{\triangle}{T^j} u_{i|j},
$$

we can write the above equations as

$$
e_{\underset{\triangle}{A}} = D^i_{\underset{\triangle}{A}}\underset{\triangle}{T^j}(w_{i|j}+\varepsilon_{ij.}^{\;\;k}u_k), \qquad \varphi_{\underset{\triangle}{A}} = D^i_{\underset{\triangle}{A}}\underset{\triangle}{T^j}u_{i|j}. \tag{2.102}
$$

Formulae (2.102) represent the geometrical relations of the lattice structure under consideration. If we compare them with the geometrical equations

$$
\gamma_{ij} = w_{j|i}+\in_{ij.}^{\;\;k}u_k, \qquad \varkappa_{ij} = u_{j|i} \tag{2.103}
$$

of a Cosserat medium, which relate the displacements w_i and rotations u_i of material points with the asymmetric strain tensor γ_{ij} and torsion tensor $\varkappa_{ij}$, we find that

$$\underset{\triangle}{e}_A = \underset{\triangle}{D}{}^i_A \underset{\triangle}{T}{}^j \gamma_{ji}, \quad \underset{\triangle}{\varphi}_A = \underset{\triangle}{D}{}^i_A \underset{\triangle}{T}{}^j \varkappa_{ji}, \tag{2.104}$$

and so

$$\underset{\triangle}{e}_i = \underset{\triangle}{T}{}^j \gamma_{ji}, \quad \underset{\triangle}{\varphi}_i = \underset{\triangle}{T}{}^j \varkappa_{ji}. \tag{2.105}$$

Formulae (2.104) and (2.105) show that the state of strain of the lattice can be described by means of the strain tensors γ_{ij} and $\varkappa_{ij}$ of Cosserat media.

The state of stress of the lattices can be determined by considering the conditions of equilibrium for a single bundle $\triangle$. Suppose that the bundle occupies a region $\hat{\Omega}$ bounded by a closed surface $\hat{\Gamma}$ with an external normal $\underset{\triangle}{\hat{\mathbf{n}}}$. The dimensions of the bundle and the number of bars in it are arbitrary.

The separated part of the structure is acted upon by the following loads:

$\underset{\triangle}{\mathbf{f}}, \underset{\triangle}{\mathbf{h}}$ —the densities of the external forces $\underset{\triangle}{\mathbf{F}}$ and moments $\underset{\triangle}{\mathbf{H}}$ (parts of the external loads $\mathbf{F}$ and $\mathbf{H}$),

$\sum_{\Sigma} \underset{\triangle\Sigma}{\mathbf{f}} = -\sum_{\Sigma} \underset{\Sigma\triangle}{\mathbf{f}}, \ \sum_{\Sigma} \underset{\triangle\Sigma}{\mathbf{h}} = -\sum_{\Sigma} \underset{\Sigma\triangle}{\mathbf{h}}, \ \triangle \neq \Sigma$ —the densities of forces and moments representing the reactions of the bars of the other two families,

$\underset{\triangle}{\mathbf{P}}(\hat{n}) = \underset{\triangle}{\mathbf{P}}(\underset{\triangle}{\mathbf{T}} \cdot \underset{\triangle}{\hat{\mathbf{n}}}) = \underset{\triangle}{\mathbf{P}} \underset{\triangle}{T}{}^k \underset{\triangle}{\hat{n}}_k, \ \underset{\triangle}{\mathbf{M}}(\hat{n}) = \underset{\triangle}{\mathbf{M}}(\underset{\triangle}{\mathbf{T}} \cdot \underset{\triangle}{\hat{\mathbf{n}}}) = \underset{\triangle}{\mathbf{M}} \underset{\triangle}{T}{}^k \underset{\triangle}{\hat{n}}_k$ —the densities of the internal forces and moments acting in the bundle across the surface $\hat{\omega}$ with the external normal $\hat{n}$.

The equations of equilibrium for the above loads are

$$\oint_{\hat{\omega}} \underset{\triangle}{\mathbf{P}}(\underset{\triangle}{\hat{n}}) \mathrm{d}\hat{\omega} + \int_{\hat{\Omega}} \Big(\underset{\triangle}{\mathbf{f}} + \sum_{\Sigma} \underset{\triangle\Sigma}{\mathbf{f}}\Big) \mathrm{d}\hat{\Omega} = 0,$$

$$\oint_{\hat{\omega}} (\underset{\triangle}{\mathbf{M}}(\hat{n}) + \mathbf{R} \times \underset{\triangle}{\mathbf{P}}(\hat{n})) \mathrm{d}\hat{\omega} + \int_{\hat{\Omega}} \Big[\underset{\triangle}{\mathbf{h}} + \sum_{\Sigma} \underset{\triangle\Sigma}{\mathbf{h}} + \mathbf{R} \times \Big(\underset{\triangle}{\mathbf{f}} + \sum_{\Sigma} \underset{\triangle\Sigma}{\mathbf{f}}\Big)\Big] \mathrm{d}\hat{\Omega} = 0.$$

Expressing the force and moment densities $\underset{\triangle}{\mathbf{P}}(\underset{\triangle}{\hat{\mathbf{n}}})$ and $\underset{\triangle}{\mathbf{M}}(\underset{\triangle}{\hat{\mathbf{n}}})$ in terms of the middle-section densities $\underset{\triangle}{\mathbf{P}}$ and $\underset{\triangle}{\mathbf{M}}$ and using the equalities

$$\oint_{\hat{\omega}} \underset{\triangle}{\mathbf{P}} \underset{\triangle}{T}{}^k \underset{\triangle}{\hat{n}}_k \mathrm{d}\hat{\omega} = \int_{\hat{\Omega}} (\underset{\triangle}{\mathbf{P}} \underset{\triangle}{T}{}^k)|_k \mathrm{d}\hat{\Omega},$$

$$\oint_{\hat{\omega}} (\underset{\triangle}{\mathbf{M}} \underset{\triangle}{T}{}^k + \mathbf{R} \times \underset{\triangle}{\mathbf{P}} \underset{\triangle}{T}{}^k) \underset{\triangle}{\hat{n}}_k \mathrm{d}\hat{\omega} = \int_{\hat{\Omega}} [\underset{\triangle}{T}{}^k (\underset{\triangle}{\mathbf{M}} + \mathbf{R} \times \underset{\triangle}{\mathbf{P}})]|_k \mathrm{d}\hat{\Omega},$$

which follow from the Gauss–Green theorem, we obtain

$$\int_{\hat{\Omega}} \Big[(\underset{\triangle}{\mathbf{P}} T^k)|_k + \underset{\triangle}{\mathbf{f}} + \sum_{\Sigma} \underset{\triangle\Sigma}{\mathbf{f}}\Big] \mathrm{d}\hat{\Omega} = 0,$$

$$\int_{\hat{\Omega}}\Big\{[\underset{\triangle}{T^k}(\underset{\triangle}{\mathbf{M}}+\mathbf{R}\times\underset{\triangle}{\mathbf{P}})]|_k+\underset{\triangle}{\mathbf{h}}+\sum_{\Sigma}\underset{\triangle\Sigma}{\mathbf{h}}+\mathbf{R}\times\Big(\underset{\triangle}{\mathbf{f}}+\sum_{\Sigma}\underset{\triangle\Sigma}{\mathbf{f}}\Big)\Big\}\mathrm{d}\hat{\Omega}=0.$$

As these equations apply to any region $\hat{\Omega}$, we can conclude that

$$(\underset{\triangle}{\mathbf{P}}\underset{\triangle}{T^k})|_k+\underset{\triangle}{\mathbf{f}}+\sum_{\Sigma}\underset{\triangle\Sigma}{\mathbf{f}}=0,$$

$$[T^k(\underset{\triangle}{\mathbf{M}}+\mathbf{R}\times\underset{\triangle}{\mathbf{P}})]|_k+\underset{\triangle}{\mathbf{h}}+\sum_{\Sigma}\underset{\triangle\Sigma}{\mathbf{h}}+\mathbf{R}\times\Big(\underset{\triangle}{\mathbf{f}}+\sum_{\Sigma}\underset{\triangle\Sigma}{\mathbf{f}}\Big)=0.$$

Combining the second equation with the first and differentiating it, we get

$$(\underset{\triangle}{\mathbf{P}}\underset{\triangle}{T^k})|_k+\underset{\triangle}{\mathbf{f}}+\sum_{\Sigma}\underset{\triangle\Sigma}{\mathbf{f}}=0,$$

$$(\underset{\triangle}{\mathbf{M}}\underset{\triangle}{T^k})|_k+\mathbf{g}_k\times\underset{\triangle}{\mathbf{P}}\underset{\triangle}{T^k}+\underset{\triangle}{\mathbf{h}}+\sum_{\Sigma}\underset{\triangle\Sigma}{\mathbf{h}}=0,$$

where $\mathbf{g}_k=\mathbf{R}_{,k}$. The above equations hold for each of the families $\triangle = I, II, III$. In view of the equalities $\underset{\triangle\Sigma}{\mathbf{f}}=-\underset{\triangle\Sigma}{\mathbf{f}}$ and $\underset{\triangle\Sigma}{\mathbf{h}}=-\underset{\triangle\Sigma}{\mathbf{h}}$, summation over $\triangle$ gives

$$\sum_{\triangle}(\underset{\triangle}{T^k}\underset{\triangle}{\mathbf{P}})|_k+\mathbf{f}=0,$$
$$\sum_{\triangle}(\underset{\triangle}{T^k}\underset{\triangle}{\mathbf{M}})|_k+\sum_{\triangle}\underset{\triangle}{\mathbf{T}}\times\underset{\triangle}{\mathbf{P}}+\mathbf{h}=0, \tag{2.106}$$

where $\mathbf{f}=\sum_{\triangle}\underset{\triangle}{\mathbf{f}}$ and $\mathbf{h}=\sum_{\triangle}\underset{\triangle}{\mathbf{h}}$.

Equations (2.106) and the boundary equilibrium conditions on the surface Γ of the structure,

$$\sum_{\triangle}\underset{\triangle}{\mathbf{P}}_{(n)}=\sum_{\triangle}\underset{\triangle}{\mathbf{P}}(\underset{\triangle}{\mathbf{T}}\cdot\mathbf{n})=\hat{\mathbf{p}},\qquad \sum_{\triangle}\underset{\triangle}{\mathbf{M}}_{(n)}=\sum_{\triangle}\mathbf{M}(\mathbf{T}\cdot\mathbf{n})=\hat{\mathbf{m}}, \tag{2.107}$$

where $\hat{\mathbf{p}}$ and $\hat{\mathbf{m}}$ denote the surface densities of the loads **F** and **H**, respectively, determine the state of stress of the structure.

Introducing the tensor notation

$$\underset{\triangle}{p^{kl}}=\underset{\triangle}{T^k}\underset{\triangle}{p^l},\qquad p^{kl}=\sum_{\triangle}\underset{\triangle}{p^{kl}},$$
$$\underset{\triangle}{m^{kl}}=\underset{\triangle}{T^k}\underset{\triangle}{M^L},\qquad m^{kl}=\sum_{\triangle}\underset{\triangle}{m^{kl}}, \tag{2.108}$$

we can write equations (2.106) and boundary conditions (2.107) as

$$p^{kl}|_k+f^l=0,\qquad m^{kl}|_k+\varepsilon_{km\cdot}^{\;\;\;\;l}p^{km}+h^l=0\quad\text{in }\Omega,$$
$$p^{kl}n_k=\hat{p}^l,\qquad m^{kl}n_k=\hat{m}^l\quad\text{on }\Sigma. \tag{2.109}$$

Equations (2.109) have exactly the same form as equilibrium equations for Cosserat media. In the theory of those media, the symbols p^{kl} and m^{kl} represent the force-stresses and couple-stresses, respectively.

The physical relations for the lattice can be obtained by substituting formulae (2.100) and (2.104) in (2.108):

$$\begin{aligned} p^{kl} &= \sum_{\triangle} \underset{\triangle}{p}^{kl} = \sum_{\triangle} \underset{\triangle}{T}^k \underset{\triangle}{P}^l = \sum_{\triangle} \underset{\triangle}{T}^k \underset{\triangle}{D}{}^l_A \underset{\triangle}{P}^A = \sum_{\triangle} \underset{\triangle}{T}^k \underset{\triangle}{D}{}^l_A \underset{\triangle}{R}^{AB} \underset{\triangle}{e}_B \\ &= \sum_{\triangle} \underset{\triangle}{T}^k \underset{\triangle}{D}{}^l_A \underset{\triangle}{R}^{AB} \underset{\triangle}{T}^i \underset{\triangle}{D}{}^j_B \gamma_{ij} = \sum_{\triangle} \underset{\triangle}{A}^{ijkl} \gamma_{ij} = A^{ijkl}\gamma_{ij}, \\ m^{kl} &= \sum_{\triangle} \underset{\triangle}{m}^{kl} = \sum_{\triangle} \underset{\triangle}{T}^k \underset{\triangle}{M}^l = \sum_{\triangle} \underset{\triangle}{T}^k \underset{\triangle}{D}{}^l_A \underset{\triangle}{M}^A = \sum_{\triangle} \underset{\triangle}{T}^k \underset{\triangle}{D}{}^l_A \underset{\triangle}{S}^{AB} \underset{\triangle}{\varphi}_B \\ &= \sum_{\triangle} \underset{\triangle}{T}^k \underset{\triangle}{D}{}^l_A \underset{\triangle}{S}^{AB} \underset{\triangle}{T}^i \underset{\triangle}{D}{}^j_B \varkappa_{ij} = \sum_{\triangle} \underset{\triangle}{C}^{ijkl} \varkappa_{ij} = C^{ijkl}\varkappa_{ij}. \end{aligned} \tag{2.110}$$

The tensorial quantities A^{ijkl} and C^{ijkl} introduced above as

$$\begin{aligned} A^{ijkl} &= \sum_{\triangle} \underset{\triangle}{A}^{ijkl}, \quad \underset{\triangle}{A}^{ijkl} = \underset{\triangle}{A}^{klij} = \underset{\triangle}{T}^i \underset{\triangle}{D}{}^j_A \underset{\triangle}{T}^k \underset{\triangle}{D}{}^l_B \underset{\triangle}{R}^{AB}, \\ C^{ijkl} &= \sum_{\triangle} \underset{\triangle}{C}^{ijkl}, \quad \underset{\triangle}{C}^{ijkl} = \underset{\triangle}{C}^{klij} = \underset{\triangle}{T}^i \underset{\triangle}{D}{}^j_A \underset{\triangle}{T}^k \underset{\triangle}{D}{}^l_B \underset{\triangle}{S}^{AB}, \end{aligned} \tag{2.111}$$

determine the geometrical and mechanical properties of the bar structure of the lattice. Formulae (2.103), (2.109) and (2.110) are the basic relations from which to derive equations describing the state of strain of the lattice. Inserting p^{kl} and m^{kl} given by (2.110) into the equilibrium equations $(2.109)_1$ and then, using (2.103), we obtain the following system of differential equations:

$$\begin{aligned} &(A^{ijkl} w_l|_k)|_i + \varepsilon_{ik\cdot}^{\ \ m}(A^{ijkl} u_m)|_i + f^j = 0, \\ &(C^{ijkl} u_l|_k)|_i + \varepsilon_{il\cdot}^{\ \ j} A^{ilkm}(w_m|_k + \varepsilon_{mk\cdot}^{\ \ \ n} u_n) + h^j = 0. \end{aligned} \tag{2.112}$$

Together with the boundary conditions $(2.109)_2$, equations (2.112) describe the fields $\mathbf{u}$ and $\mathbf{w}$ of displacements of the nodes of a lattice with arbitrary internal structures A^{ijkl} and C^{ijkl}.

Using formulae (2.100), (2.104), (2.95), (2.108), and (2.110), we can set up formulae for the strain energies $\underset{\triangle}{\Psi}$, $\underset{\triangle}{U}$ and U of an arbitrary bar, an entire family of bars $\triangle$ and the whole lattice, respectively:

$$\begin{aligned} \underset{\triangle}{\Psi}(\theta^1, \theta^2, \theta^3) &= \tfrac{1}{2}(\underset{\triangle}{\mathbf{P}} \cdot \underset{\triangle}{\mathbf{e}} + \underset{\triangle}{\mathbf{M}} \cdot \underset{\triangle}{\boldsymbol{\varphi}}) \underset{\triangle}{\omega} \underset{\triangle}{l} = \tfrac{1}{2}(\underset{\triangle}{P}^A \underset{\triangle}{e}_A + \underset{\triangle}{M}^A \underset{\triangle}{\omega}_A) \underset{\triangle}{\omega} \underset{\triangle}{l} \\ &= \tfrac{1}{2}(\underset{\triangle}{p}^{ij}\gamma_{ij} + \underset{\triangle}{m}^{ij}\varkappa_{ij}) \underset{\triangle}{\omega} \underset{\triangle}{l} = \tfrac{1}{2}(\underset{\triangle}{A}^{ijkl}\gamma_{ij}\gamma_{kl} + \underset{\triangle}{C}^{ijkl}\varkappa_{ij}\varkappa_{kl}) \underset{\triangle}{\omega} \underset{\triangle}{l}, \end{aligned} \tag{2.113}$$

$$\underset{\triangle}{U} = \int_{\Omega} \underset{\triangle}{\psi}(\underset{\triangle}{\omega} \underset{\triangle}{l})^{-1} \mathrm{d}\Omega, \qquad U = \sum_{\triangle} \underset{\triangle}{U}.$$

The work done by the loads **F** and **H** on the displacements of the nodes of the lattice is given by

$$L = \oint_{\Sigma} (\hat{p}^i w_i + \hat{m}^i u_i)\mathrm{d}\Sigma + \int_{\Omega} (f^i w_i + h^i u_i)\mathrm{d}\Omega \,. \tag{2.114}$$

The fundamental relations presented in this section provide a starting point for investigating the influence of the internal structure of a bar lattice on the elastic strain energy under given loads.

Influence of internal structure on strain energy

In order to evaluate the effect of a change in the internal structure on the deformability of the lattice, let us consider two bar lattices $\mathscr{P}$ and $\mathscr{O}$ identical in external form but having different internal structures $\underset{\triangle}{\overset{\circ}{A}{}^{ijkl}} \neq \underset{\triangle}{A^{ijkl}}$ and $\underset{\triangle}{\overset{\circ}{C}{}^{ijkl}} \neq \underset{\triangle}{C^{ijkl}}$ (Figs. 2.23 and 2.24). Each system occupies the same region Ω with boundary Γ, is identically supported and is loaded with the same body forces **f**, body couples **h** and surface loads $\hat{\mathbf{p}}$ and $\hat{\mathbf{m}}$ on Γ.

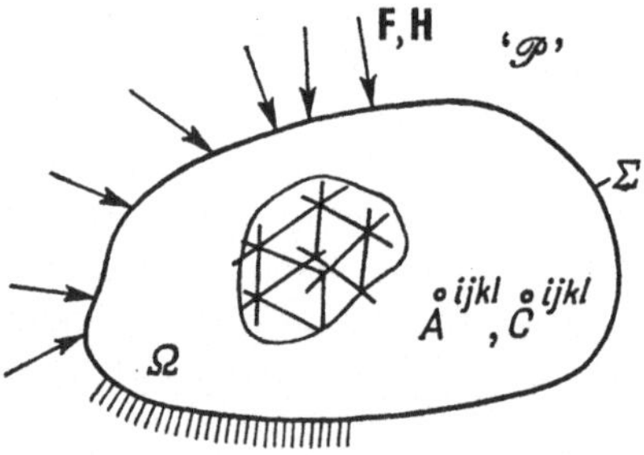

Fig. 2.23. The original system

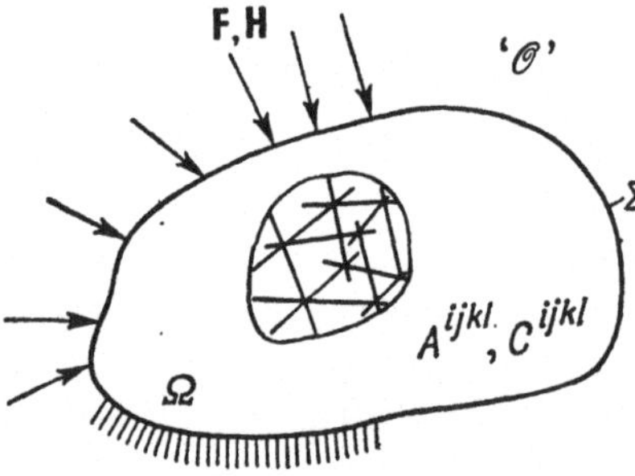

Fig. 2.24. The modified system

Since the internal structures of the two systems differ, the strains they suffer under the loads **f**, **h**, $\hat{\mathbf{p}}$, and $\hat{\mathbf{m}}$ will also differ. Denote by $\overset{\circ}{m}{}^{ij}$, $\overset{\circ}{p}{}^{ij}$, $\overset{\circ}{\gamma}_{ij}$, $\overset{\circ}{\varkappa}_{ij}$, $\overset{\circ}{w}_i$ and $\overset{\circ}{u}_i$ the quantities describing the states of stress, strain and displacement of the system $\mathscr{P}$ and by m^{ij}, p^{ij}, γ_{ij}, $\varkappa_{ij}$, w_i and u_i the corresponding quantities for the system $\mathscr{O}$.

The elastic strain energies $\overset{\circ}{U}$ and U stored in the systems $\mathscr{P}$ and $\mathscr{O}$ will also be different. The energy difference $\Delta U = \overset{\circ}{U} - U$ reflects the difference between the internal structures of the systems and can be considered a measure of the significance of the internal structure for the deformability of the bar lattice.

Since both systems are in equilibrium, the Clapeyron theorem implies that $2U = L$ and $2\overset{\circ}{U} = \overset{\circ}{L}$, and so

$$\Delta U = \overset{\circ}{L} - \overset{\circ}{U} - (L - U) = \Delta L - \Delta U \,.$$

Using formulae (2.113) and (2.114), we find

$$\Delta U = -\tfrac{1}{2}\int_{\Omega} (\mathring{p}^{ij}\Delta\gamma_{ij}+m^{ij}\Delta\varkappa_{ij})\mathrm{d}\Omega - \tfrac{1}{2}\int_{\Omega} (\Delta p^{ij}\gamma_{ij}+\Delta m^{ij}\varkappa_{ij})\mathrm{d}\Omega + \\ + \oint_{\Sigma} (\hat{p}^{i}\Delta w_{i}+\hat{m}^{i}\Delta u_{i})\mathrm{d}\Sigma + \int_{\Omega} (f^{i}\Delta w_{i}+h^{i}\Delta u_{i})\mathrm{d}\Omega, \tag{2.115}$$

where Δp^{ij}, Δm^{ij}, $\Delta\gamma_{ij}$, $\Delta\varkappa_{ij}$, Δu_i and Δw_i denote the differences of the corresponding strain components of the systems $\mathscr{P}$ and $\mathscr{O}$. By (2.110) the differences Δp^{ij} and Δm^{ij} can be written

$$\Delta p^{ij} = \mathring{A}^{ijkl}\mathring{\gamma}_{kl} - A^{ijkl}\gamma_{kl} = \mathring{A}^{ijkl}\Delta\gamma_{kl} + \Delta A^{ijkl}\gamma_{kl}, \\ \Delta m^{ij} = \mathring{C}^{ijkl}\mathring{\varkappa}_{kl} - C^{ijkl}\varkappa_{kl} = \mathring{C}^{ijkl}\Delta\varkappa_{kl} + \Delta C^{ijkl}\varkappa_{kl}. \tag{2.116}$$

Hence, using the symmetry of the tensors A^{ijkl} and C^{ijkl} (cf. (2.111)), we can represent the second integrand of (2.115) as

$$\Delta p^{ij}\gamma_{ij} + \Delta m^{ij}\varkappa_{ij} = \mathring{p}^{ij}\Delta\gamma_{ij} + \mathring{m}^{ij}\Delta\varkappa_{ij} + \Delta A^{ijkl}\gamma_{ij}\gamma_{kl} + \Delta C^{ijkl}\varkappa_{ij}\varkappa_{kl} - (Q)^2. \tag{2.117}$$

In deriving this expression we have put

$$\mathring{A}^{ijkl} - A^{ijkl} = \Delta A^{ijkl}, \quad \mathring{C}^{ijkl} - C^{ijkl} = \Delta C^{ijkl}, \\ \mathring{A}^{ijkl}\Delta\gamma_{ij}\Delta\gamma_{kl} + \mathring{C}^{ijkl}\Delta\varkappa_{ij}\Delta\varkappa_{kl} \\ = \sum_{\triangle}\sum_{A=1}^{3} \underset{\triangle}{\mathring{T}^{i}}\underset{\triangle}{\mathring{D}^{j}_{A}}\underset{\triangle}{\mathring{T}^{k}}\underset{\triangle}{\mathring{D}^{l}_{A}}(\underset{\triangle}{\mathring{R}^{AA}}\Delta\gamma_{ij}\Delta\gamma_{kl} + \underset{\triangle}{\mathring{S}^{AA}}\Delta\varkappa_{ij}\Delta\varkappa_{kl}) \\ = \sum_{\triangle}\sum_{A=1}^{3} \underset{\triangle}{\mathring{R}^{AA}}(\underset{\triangle}{\mathring{T}^{i}}\underset{\triangle}{\mathring{D}^{j}_{A}}\Delta\gamma_{ij})^2 + \underset{\triangle}{\mathring{S}^{AA}}(\underset{\triangle}{\mathring{T}^{i}}\underset{\triangle}{\mathring{D}^{j}_{A}}\Delta\varkappa_{ij})^2 = (Q)^2 > 0, \tag{2.118}$$

and used the fact that $\underset{\triangle}{\mathring{R}^{AB}} = 0$ and $\underset{\triangle}{\mathring{S}^{AB}} = 0$ for $A \neq B$ according to the orthogonality conditions (2.95).

Inserting (2.117) into formula (2.116), we obtain

$$\Delta U = -\tfrac{1}{2}\int_{\Omega} (\Delta A^{ijkl}\gamma_{ij}\gamma_{kl} + \Delta C^{ijkl}\varkappa_{ij}\varkappa_{kl}\mathrm{d}\Omega - \\ - \int_{\Omega} (\mathring{p}^{ij}\Delta\gamma_{ij}+\mathring{m}^{ij}\Delta\varkappa_{ij})\mathrm{d}\Omega + \oint_{\Sigma} (\hat{p}^{i}\Delta w_{i}+\hat{m}^{i}\Delta u_{i})\mathrm{d}\Sigma + \\ + \int_{\Omega} (f^{i}\Delta w_{i}+h^{i}\Delta u_{i})\mathrm{d}\Omega + \tfrac{1}{2}\int_{\Omega} (Q)^2\mathrm{d}\Omega,$$

and making use of the virtual work equation for the system $\mathscr{P}$

$$\int_{\Omega} (\mathring{p}^{ij}\Delta\gamma_{ij}+\mathring{m}^{ij}\Delta\varkappa_{ij})\mathrm{d}\Omega \\ = \oint_{\Sigma} (\hat{p}^{k}\Delta w_{k}+\hat{m}^{k}\Delta u_{k})\Delta\Sigma + \int_{\Omega} (f^{k}\Delta w_{k}+h^{k}\Delta u_{k})\mathrm{d}\Omega, \tag{2.119}$$

we finally get

$$\Delta U = -\tfrac{1}{2}\int_{\Omega}(\Delta A^{ijkl}\gamma_{ij}\gamma_{kl}+\Delta C^{ijkl}\varkappa_{ij}\varkappa_{kl})\mathrm{d}\Omega+\tfrac{1}{2}\int_{\Omega}(Q)^2\mathrm{d}\Omega. \tag{2.120}$$

The above formula represents the effect of changes in A^{ijkl} and C^{ijkl} on the elastic strain energy of the structure.

Let us show that $\Delta U = 0$ if $\Delta A^{ijkl} = \Delta C^{ijkl} = 0$. To this end, it is enough to express the second integrand in (2.120) in terms of these differences. Subtracting the virtual work equation for the system $\mathcal{O}$,

$$\int_{\Omega}(p^{ij}\Delta\gamma_{ij}+m^{ij}\Delta\varkappa_{ij})\mathrm{d}\Omega = \oint_{\Sigma}(\hat{p}^i\Delta w_i+\hat{m}^i\Delta u_i)\mathrm{d}\Sigma+\int_{\Omega}(f^i\Delta w_i+h^i\Delta u_i)\mathrm{d}\Omega,$$

from the corresponding equation for the system $\mathscr{P}$ (2.119), we get

$$\int_{\Omega}(\Delta p^{ij}\Delta\gamma_{ij}+\Delta m^{ij}\Delta\varkappa_{ij})\mathrm{d}\Omega = 0,$$

Hence, using formulae (2.116) and (2.117), we obtain

$$\int_{\Omega}(Q)^2\mathrm{d}\Omega = -\int_{\Omega}(\Delta A^{ijkl}\Delta\gamma_{kl}\gamma_{ij}+\Delta C^{ijkl}\Delta\varkappa_{kl}\varkappa_{ij})\mathrm{d}\Omega.$$

Thus, if we set $\Delta A^{ijkl} = 0$ and $\Delta C^{ijkl} = 0$ in (2.110), the difference ΔU will vanish.

Influence of material distribution

Formula (2.120) permits us to evaluate the effect of the material distribution $\underset{\triangle}{\mathscr{A}}$ on the deformability of the structure. Let us assume that the differences ΔA^{ijkl} and ΔC^{ijkl} occurring in (2.120) are caused exclusively by the differences $\Delta\underset{\triangle}{\mathscr{A}} = \underset{\triangle}{\mathring{\mathscr{A}}}-\underset{\triangle}{\mathscr{A}}$ between the distributions of structural materials in the systems $\mathscr{P}$ and $\mathcal{O}$. Following this assumption, we shall express the differences

$$\Delta A^{ijkl} = \sum_{\triangle}\Delta \underset{\triangle}{A^{ijkl}}, \qquad \Delta C^{ijkl} = \sum_{\triangle}\Delta \underset{\triangle}{C^{ijkl}}$$

in terms of $\Delta\underset{\triangle}{\mathscr{A}}$. Using (2.111), we can write them in the form

$$\Delta A^{ijkl} = \sum_{\triangle}\Delta \underset{\triangle}{R^{AB}}\underset{\triangle}{T^i}\underset{\triangle}{D^j_A}\underset{\triangle}{T^k}\underset{\triangle}{D^l_B}, \qquad \Delta C^{ijkl} = \sum_{\triangle}\Delta \underset{\triangle}{S^{AB}}\underset{\triangle}{T^i}\underset{\triangle}{D^j_A}\underset{\triangle}{T^k}\underset{\triangle}{D^l_B}, \tag{2.121}$$

where $\Delta\underset{\triangle}{R^{AB}} = \underset{\triangle}{\mathring{R}^{AB}} - \underset{\triangle}{R^{AB}}$ and $\Delta\underset{\triangle}{S^{AB}} = \underset{\triangle}{\mathring{S}^{AB}}-\underset{\triangle}{S^{AB}}$. By rigidity formulae (2.96) these last differences can be represented as

$$\Delta\underset{\triangle}{R^{AB}} = (\underset{\triangle}{\mathscr{A}})^{-1}\underset{\triangle}{R^{AB}}\Delta\underset{\triangle}{\mathscr{A}}, \qquad \Delta\underset{\triangle}{S^{AB}} = (\underset{\triangle}{\mathscr{A}})^{-1}\underset{\triangle}{S^{AB}}\Delta\underset{\triangle}{\mathscr{A}}.$$

Inserting the above expressions into (2.121) gives

$$\begin{aligned} \Delta A^{ijkl} &= \sum_{\triangle} \underset{\triangle}{A^{ijkl}} \Delta \underset{\triangle}{\mathscr{A}} (\underset{\triangle}{\mathscr{A}})^{-1}, \\ \Delta C^{ijkl} &= \sum_{\triangle} \underset{\triangle}{C^{ijkl}} \Delta \underset{\triangle}{\mathscr{A}} (\underset{\triangle}{\mathscr{A}})^{-1}. \end{aligned} \tag{2.122}$$

Now if we substitute these formulae in (2.120) and use (2.113), the formula for the elastic strain energy difference ΔU will take the form

$$\Delta U = -\sum_{\triangle} \int_{\Omega} \underset{\triangle}{\psi}(\underset{\triangle}{V})^{-1} \Delta \underset{\triangle}{\mathscr{A}} \mathrm{d}\Omega + \tfrac{1}{2} \int_{\Omega} (Q)^2 \mathrm{d}\Omega . \tag{2.123}$$

where $\underset{\triangle}{V} = \underset{\triangle}{\mathscr{A}} \underset{\triangle}{\omega} \underset{\triangle}{l}$ denotes the volume of a bar in the bundle $\triangle$.

Solution of the optimum-design problem

We assume that the volumes and materials of the corresponding families $\triangle$ in the structures $\mathscr{P}$ and $\mathscr{O}$ are identical, so that

$$\Delta \underset{\triangle}{V} = \underset{\triangle}{\mathring{V}} - \underset{\triangle}{V} = \int_{\Omega} \Delta \underset{\triangle}{\mathscr{A}} \mathrm{d}\Omega = 0 . \tag{2.124}$$

Furthermore, let us assume that the distribution functions are known *a priori*. Clearly, the system $\mathscr{P}$ will not, in general, be optimum. The distributions $\underset{\triangle}{\mathscr{A}}$ of materials in the system $\mathscr{O}$ are unknown functions.

For the system $\mathscr{O}$ to be optimal, i.e. to have maximum resistance against deformations, its elastic strain energy U under the loads $\mathbf{f}$, $\mathbf{h}$, $\hat{\mathbf{m}}$ and $\hat{\mathbf{p}}$ must be absolute minimum. This will occur if ΔU is positive for arbitrary admissible differences $\Delta \underset{\triangle}{\mathscr{A}}(\theta^1, \theta^2, \theta^3)$. Taking into account the constant-volume condition (2.124), we can conclude from formula (2.123) that the condition. $\Delta U > 0$ will be satisfied for arbitrary admissible $\Delta \underset{\triangle}{\mathscr{A}}$ if

$$\underset{\triangle}{\psi}(\underset{\triangle}{V})^{-1} = \underset{\triangle}{C}, \qquad \triangle = I, II, III . \tag{2.125}$$

The constants $\underset{\triangle}{C}$ can be determined by dividing equation (2.123) by $\underset{\triangle}{\omega} \underset{\triangle}{l}$ and integrating the result over the region Ω. This yields $\underset{\triangle}{C} = \underset{\triangle}{U} / \underset{\triangle}{V}$.

Equations (2.125) constitute the general solution of the optimum-design problem in question. Together with the state-of-strain equations (2.112), they determine the distribution of structural materials in optimal bar lattices, We can see that an optimum distribution is one in which the specific elastic strain energy $\underset{\triangle}{\psi} / \underset{\triangle}{V}$ is constant throughout the structure.

3

Uniform Strength Design

3.1 Strength hypotheses

General remarks

Uniform-strength design aims at producing structures in which the state of plastic yield or fracture is reached simultaneously at as many points as possible.

In this approach, determining the strength of the material of the structure to be optimized is of essential importance. The strength of a material depends on many factors, above all on its physical properties and the state of stress. For simple stress states, e.g. axial tension or compression, the strength of a material can be determined experimentally. In practice, however, the variety of possible stress states is unlimited and it is impossible to determine the strengths of various materials in all the complex states of stress by way of experiment. To obviate this difficulty, the so-called *strength hypotheses* have been introduced, according to which the strength of materials in complex stress states is characterized by certain functions of stresses, strains and material constants. These functions will be referred to as *strength functions.*

A universal strength hypothesis which would apply to all materials under various conditions and loadings has not yet been proposed. Finding such a hypothesis seems hardly probable and, besides, is unnecessary. In general, the state of plastic yield or fracture of any of the structural materials used at present in building and civil engineering under normal conditions and loadings can adequately be described by some of the existing hypotheses.

A rational phenomenological strength hypothesis (SH) should meet the following requirements:

(1) An SH should provide a plastic yield or fracture condition for a given material in any complex stress state.
(2) In its analytical form, besides the components of the stress tensor, an SH should involve scalar or tensorial quantities characterizing the strength properties of the material.
(3) An SH should take account of such properties of the material as differ-

ent tensile and compressive strengths, dependence of shearing strength on the direction of tangential stresses (for anisotropic materials), and the like. In simple cases, SH should lead to the standard formulae of strength of materials.

(4) An SH should be invariant under coordinate transformations.

(5) A necessary condition for an SH to be self-consistent is that the relations it implies between the material constants be independent of the coordinate system.

(6) The coefficients characterizing the strength properties of an anisotropic material take different values depending on the orientation of the coordinate system to which a given volume element is referred. Therefore, a law of change under coordinate rotations should be specified for these coefficients. In particular, an SH for an anisotropic material should give the strength of this material in pure tension, compression, shearing, etc., in an arbitrary coordinate system; for example, for a plastic reinforced with glass fibres, the SH should specify the tensile, compressive and bending strengths for any direction of the fibres.

(7) It is desirable that an SH should explicitly take into account the influence of time, temperature and scale on the condition of failure of a given material under various stress states.

(8) Geometrically, an SH can be interpreted as a hypersurface in the nine-dimensional stress space; it is called the *limit surface*. Drucker[1] postulated that the limit surface of plastic yield should be convex. This postulate can also be extended to limit surfaces for brittle materials.

(9) If the material under consideration has different strength in tension and in compression, the SH should involve odd-degree invariants as well as even-degree ones, so that both kinds of strength can be characterized in a unified manner by a single analytical expression.

(10) When the strength constants of the material are increased, the limit surface in the stress space should expand in such a way as to contain the original limit surface entirely within itself.

There are several tens of strength hypotheses known at the present time. Most of them have simple physical interpretations; others involve certain combinations of the invariants of the stress tensor. The former are usually subdivided into stress, strain and energy hypotheses. A division of strength hypotheses according as they allow or do not allow the tensile and the compressive strengths to be different (one- or two-parameter hypotheses) has

---

[1] DRUCKER, D. C., Stress-strain relations in the plastic range of metal experiments and basic concepts, in the book: EIRICH, F. R., *Rheology, Theory and Applications*, vol. 1, Acad. Press, New York 1956.

also been considered by some authors. Hypotheses relating to the plastic yield or fracture of anisotropic materials form a separate group.

Many of the existing strength hypotheses are only of historical interest, their descriptions of failure having proved not to fit accurately enough any of the known structural materials.

In this section, we briefly present only a few selected strength hypotheses which are commonly applied to ductile (steel) or brittle (concrete) materials. Their usefulness has been confirmed by experiment. Descriptions of other hypotheses, and also more detailed discussions of those presented below, can be found in books by Filonienko–Borodich[1], Hoffman and Sachs[2], Nadai[3], Seely and Smith[4] or Stabilini[5].

The yield condition of Tresca (1868) and Guest (1900)

According to the Tresca–Guest hypothesis, plastic yielding of a material occurs when the largest shear stress $\tau_{max} = \frac{1}{2}(\sigma_1 - \sigma_2)$ reaches a critical value characteristic for that material.

In the space of principal stresses σ_u, σ_v, σ_w, we obtain the following set of inequalities determining the region of safe stress states:

$$\tfrac{1}{2}(\sigma_u - \sigma_v) \leqslant \pm\bar{\sigma} \quad \text{for} \quad \sigma_u > \sigma_w > \sigma_v,$$

$$\tfrac{1}{2}(\sigma_v - \sigma_w) \leqslant \pm\bar{\sigma} \quad \text{for} \quad \sigma_v > \sigma_u > \sigma_w,$$

$$\tfrac{1}{2}(\sigma_w - \sigma_u) \leqslant \pm\bar{\sigma} \quad \text{for} \quad \sigma_w > \sigma_v > \sigma_u.$$

The equalities correspond to six planes, pairwise parallel, inclined at 45° to the coordinate axes and forming a hexagonal prism (Fig. 3.1).

For a plane state of stress, putting $\sigma_w = 0$, we get six straight lines bounding the safety region (Fig. 3.2):

$$\tfrac{1}{2}(\sigma_u - \sigma_v) \leqslant \pm\bar{\sigma},$$

$$\tfrac{1}{2}\sigma_v \leqslant \pm\bar{\sigma},$$

$$\tfrac{1}{2}\sigma_u \leqslant \pm\bar{\sigma}.$$

In the case of uniaxial stresses, the condition reduces to

$$\tfrac{1}{2}\sigma_u \leqslant \pm\bar{\sigma}.$$

---

[1] Filonienko-Borodich, M. M., *Mechanical Theories of Strength*, Publ. Mosk. Uni., 1961 (in Russian).

[2] Hoffman, O., and Sachs, G., *Introduction to the Theory of Plasticity for Engineers*, McGraw-Hill, New York 1953.

[3] Nadai, A., *Theory of Flow and Fracture of Solids*, vol. 1, McGraw-Hill, New York 1950.

[4] Seely, F. B., and Smith, J. O., *Advanced Mechanics of Materials*, Wiley, New York 1952.

[5] Stabilini, L., *La Plasticita*, Libreria Editrice, Polit. Tamburini, Milano 1961.

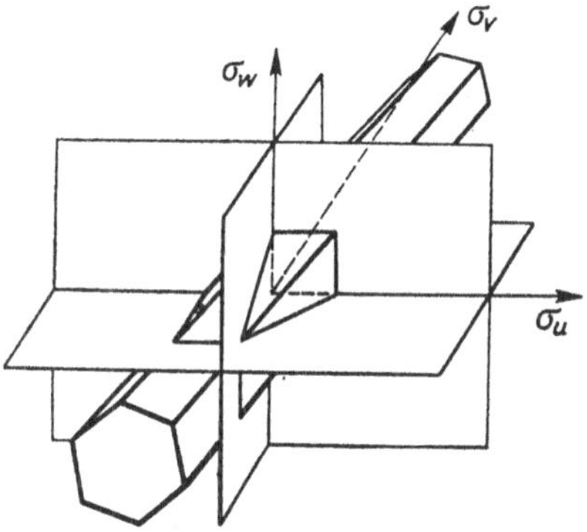

Fig. 3.1. The space of safe stress states according to Tresca–Guest hypothesis

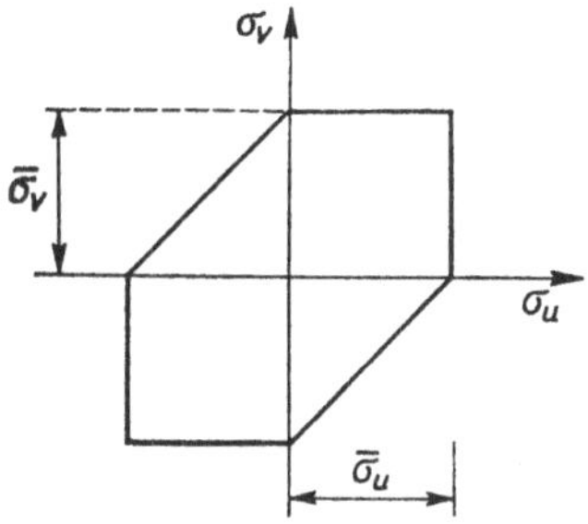

Fig. 3.2. Tresca–Guest yield hexagon

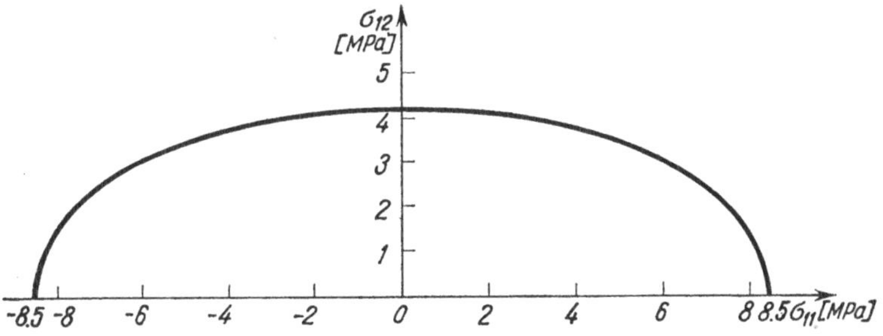

Fig. 3.3. Tresca's region of safe stresses for a beam

For beams, with only σ_{11} and σ_{12} non-zero, the above condition takes the form (Fig. 3.3)

$$\sqrt{\sigma_{11}^2+4\sigma_{12}^2} \leqslant \bar{\sigma}.$$

Some reservations as to the applicability of the Tresca yield condition may be aroused by the fact that it takes no account of the influence of the intermediate principal stress on the strength of the material. Nevertheless, the validity of the hypothesis has been verified experimentally for many ductile materials (e.g. aluminium alloys).

The Beltrami hypothesis (1885)

According to this hypothesis, the onset of fracture or yielding in a material takes place when the strain energy per unit volume reaches a critical value characteristic for this material. The expression for the unit strain energy is

$$\Psi = \frac{1}{E}\left[\frac{1}{2}(\sigma_{11}+\sigma_{12}+\sigma_{33})^2+ \right.$$
$$\left. +(1+\nu)(\sigma_{12}^2+\sigma_{23}^2+\sigma_{31}^2-\sigma_{11}\sigma_{22}-\sigma_{22}\sigma_{33}-\sigma_{33}\sigma_{11})\right].$$

In simple tension, this energy is given by

$$\Psi = \frac{\sigma_{11}^2}{2E} = \frac{\bar{\sigma}^2}{2E}.$$

The region of safe stress states is described by the condition

$$\sqrt{(\sigma_{11}+\sigma_{22}+\sigma_{33})^2+2(1+\nu)(\sigma_{12}^2+\sigma_{23}^2+\sigma_{31}^2-\sigma_{11}\sigma_{22}-\sigma_{22}\sigma_{33}-\sigma_{33}\sigma_{11})} \leqslant \bar{\sigma}$$

in orthogonal coordinates x_1, x_2, x_3, or by

$$\sqrt{(\sigma_u+\sigma_v+\sigma_w)^2-2(1+\nu)(\sigma_u\sigma_v+\sigma_v\sigma_w+\sigma_w\sigma_u)} \leqslant \bar{\sigma}$$

in the space of principal stresses.

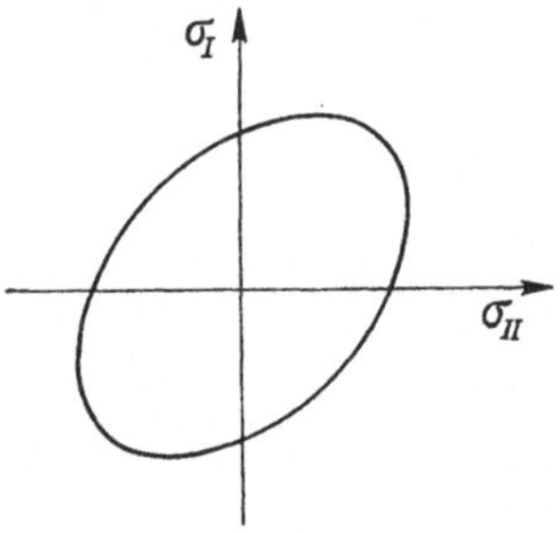

Fig. 3.4. The region of safe states according to Beltrami's hypothesis

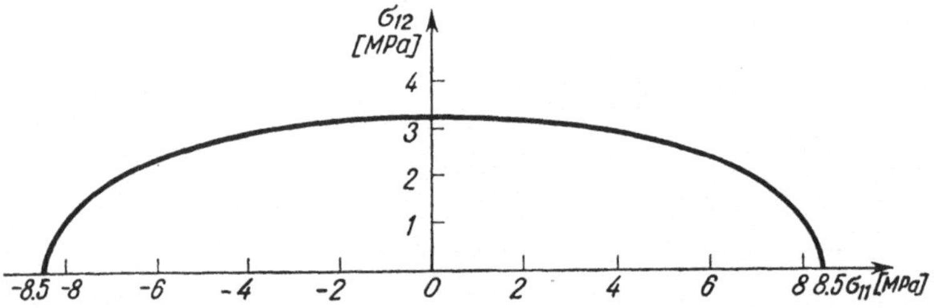

Fig. 3.5. Beltrami's region of safe stresses for a beam

For plane stress states, we obtain the condition (Fig. 3.4)

$$\sqrt{(\sigma_{11}+\sigma_{22})^2+2(1+\nu)(\sigma_{12}^2-\sigma_{11}\sigma_{22})} = \sqrt{(\sigma_u+\sigma_v)^2-2(1+\nu)\sigma_u\sigma_v} \leqslant \bar{\sigma},$$

which for beams takes the form (Fig. 3.5)

$$\sqrt{\sigma_{11}^2+2(1+\nu)\sigma_{12}^2} \leqslant \bar{\sigma}.$$

The yield condition of Huber (1904), Mises (1913) and Hencky (1924)

According to the Huber–Mises–Hencky hypothesis, plastic yielding begins when the strain energy of distortion per unit volume reaches a critical value characteristic for a given material.

The unit distortion energy is given by

$$\Psi_f = \frac{1+\nu}{6E}\Big[(\sigma_{11}-\sigma_{22})^2+(\sigma_{22}-\sigma_{33})^2+(\sigma_{33}-\sigma_{11})^2+ +6(\sigma_{12}^2+\sigma_{23}^2+\sigma_{31}^2)\Big].$$

For simple tension, this expression reduces to

$$\Psi_f = \frac{1+\nu}{3E}\sigma^2 = \frac{1+\nu}{3E}\bar{\sigma}^2.$$

The region of safe stress states is described by the inequality

$$\frac{1}{\sqrt{2}}\sqrt{(\sigma_{11}-\sigma_{22})^2+(\sigma_{22}-\sigma_{33})^2+(\sigma_{33}+\sigma_{11})^2+6(\sigma_{12}^2+\sigma_{23}^2+\sigma_{31}^2)} \leqslant \bar{\sigma}$$

or, in the space of principal stresses, by

$$\frac{1}{\sqrt{2}}\sqrt{(\sigma_u-\sigma_v)^2+(\sigma_v-\sigma_w)^2+(\sigma_w-\sigma_u)^2} \leqslant \bar{\sigma}.$$

The surface of this region, i.e. the yield surface, is an infinitely long, circular cylinder (Fig. 3.6).

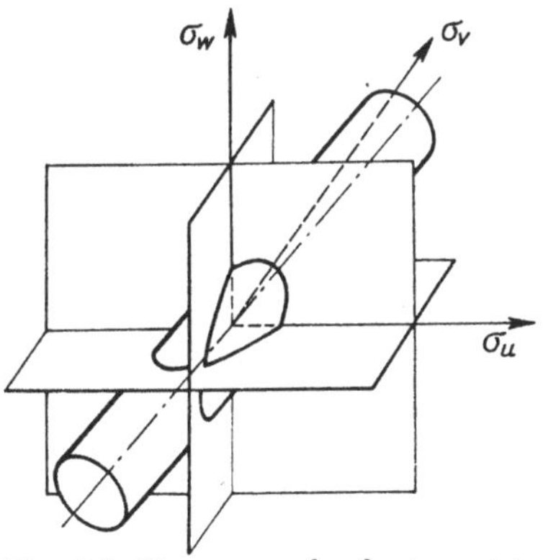

Fig. 3.6. The space of safe stress states according to the Huber–Mises–Hencky hypothesis

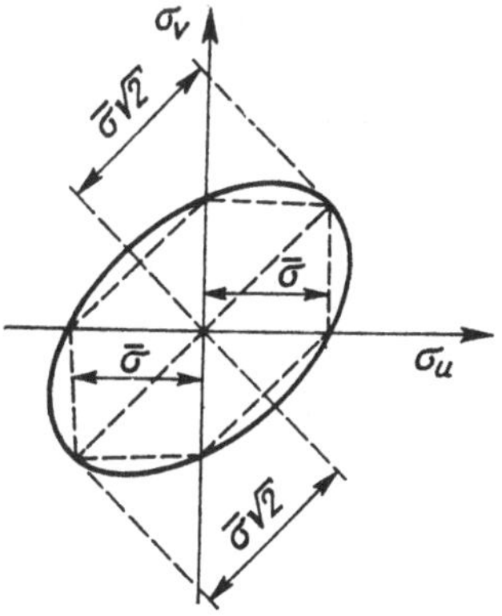

Fig. 3.7. The region of safe stress states according to the Huber–Mises–Hencky hypothesis

For a plane stress distribution, we obtain the condition (Fig. 3.7)

$$\sqrt{\sigma_u^2+\sigma_v^2-\sigma_u\sigma_v} = \sqrt{\sigma_{11}^2+\sigma_{22}^2-\sigma_{11}\sigma_{22}+3\sigma_{12}} \leqslant \bar{\sigma};$$

for beams, it reads (Fig. 3.8)

$$\sqrt{\sigma_{11}^2+3\sigma_{12}^2} \leqslant \bar{\sigma}.$$

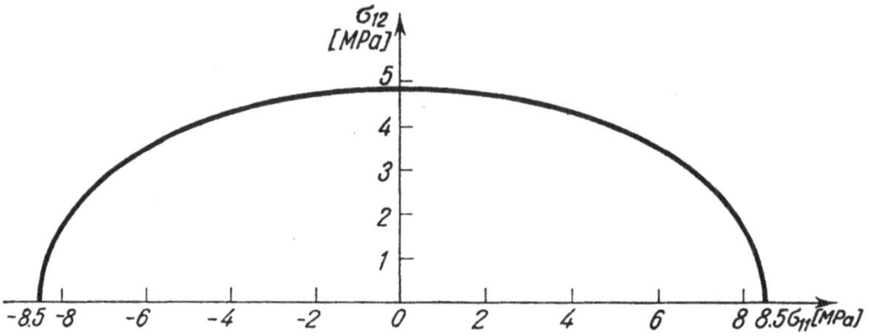

Fig. 3.8. Huber–Mises–Hencky's region of safe stresses for a beam

The Huber–Mises–Hencky hypothesis shows a very good agreement with the experimental data obtained for ductile materials equally resistant to tension and compression.

The Caquot hypothesis

The relationship derived by Caqout between the components of critical deformations applies to homogeneous and isotropic materials in which the first irreversible deformations are due to sliding.[1]

To examine deformations at an arbitrary point O of the material, consider a cross-section through O defined by a normal $\mathbf{n}$. Let dA be a neighbourhood of O in this section, $d\mathbf{P}$ the resultant of the forces acting on dA and $\boldsymbol{\sigma} = d\mathbf{P}/dA$ the corresponding stress (Fig. 3.9). If the first irreversible deform-

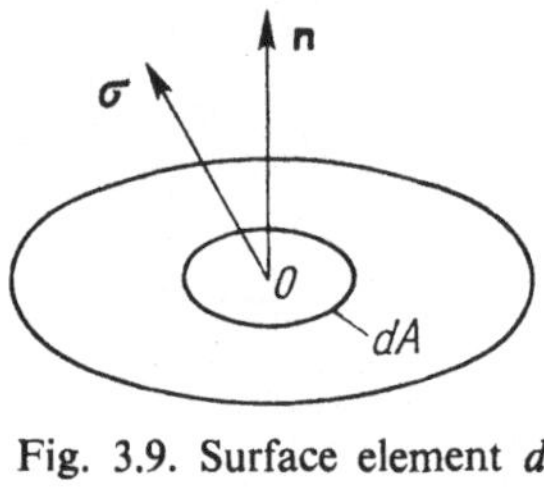

Fig. 3.9. Surface element dA with normal $\mathbf{n}$

ation which occurs while $\boldsymbol{\sigma}$ increases is a slip in the plane dA, then, owing to the isotropy of the material, an identical slip will occur in every plane through O in which the effective stress has the same magnitude and the same inclination to the normal as $\boldsymbol{\sigma}$ in the plane dA.

If the inclination of the effective stress with respect to the normal $\mathbf{n}$ is altered, for a slip to occur in the plane defined by $\mathbf{n}$ the magnitude of $\boldsymbol{\sigma}$ will also have to be altered. The locus of the end of the slip-producing stress vector $\boldsymbol{\sigma}$ is a surface of revolution about the normal $\mathbf{n}$. It is what we call the *limit surface*. Its intersection with an arbitrary plane through the normal $\mathbf{n}$ is a *limit curve*.

[1] Courbon, J., *Résistance des Materiaux*, Vol. 1, Dunod, Paris 1964,

On this basis, Caquot has formulated the following theorem:

The region of elastic deformations at a point of a homogeneous and isotropic material in which the first permanent deformations are due to sliding is delimited by a surface of revolution about the normal to the plane of sliding, formed by the ends of the slip-producing stress vectors acting at this point.

Consider a plane through the normal **n** and the limit curve lying in it (Fig. 3.10). The shear stress axis σ_{12} is drawn along the trace of the area dA; the normal stress axis σ_{11} passes through O in the direction **n**. The segments joining O with the points of the limit curve represent slip-producing stress vectors. To each of these segments there corresponds a Mohr circle tangent to the limit curve. One such circle, with centre at point O_1, tangent to the limit curve at P and P_1, and intersecting the σ_{11}-axis at points σ_1 and σ_2 (the principal stresses), is shown in Fig. 3.10. Circles intersecting the limit curve admit stress vectors going beyond the elastic range, while those lying within the curve and not tangent to it represent stress states at which the material has still a reserve of strength.

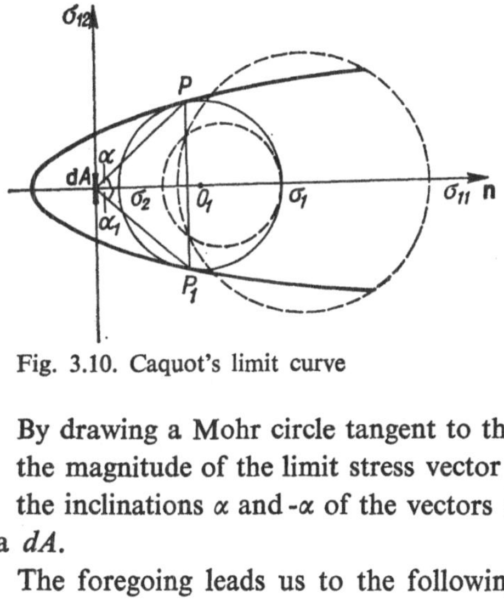

Fig. 3.10. Caquot's limit curve

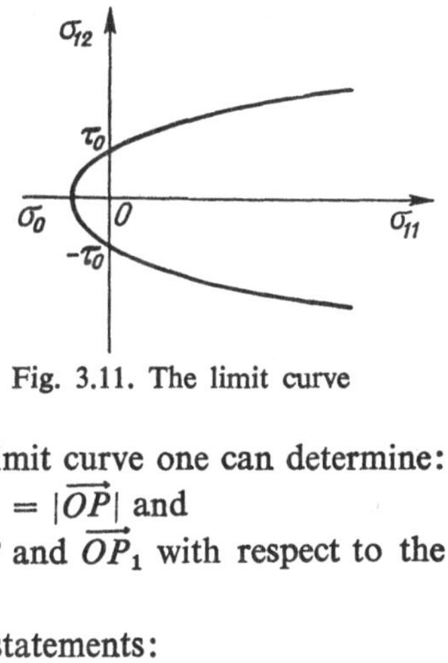

Fig. 3.11. The limit curve

By drawing a Mohr circle tangent to the limit curve one can determine:
the magnitude of the limit stress vector $|\boldsymbol{\sigma}| = |\overrightarrow{OP}|$ and
the inclinations α and $-\alpha$ of the vectors $\overrightarrow{OP}$ and $\overrightarrow{OP_1}$ with respect to the area dA.

The foregoing leads us to the following statements:

(1) *The limit curve is an envelope of the Mohr circles corresponding to the first permanent deformations.*

(2) *To each Mohr circle there correspond two planes of sliding, equally inclined to the normal stress axis.*

In terms of the intercepts of the limit curve on the coordinate axes (Fig. 3.11), the equation of this curve can be written

$$\frac{\sigma_{11}}{\sigma_0} + \left|\frac{\sigma_{12}}{\tau_0}\right|^{3/2} = 1.$$

σ_0, the abscissa of the vertex of the curve, corresponds to the onset of failure in hydrostatic tension, while τ_0, the ordinate of the intercept on the σ_{12} axis, represents the ultimate stress in pure shear.

The Caqout hypothesis has been confirmed by experimental investigations conducted in France. A certain inconvenience in its practical applications is caused by the fact that σ_{11} and σ_{12} represent stresses in the plane of sliding whose inclination to the coordinate plane is unknown. Therefore, in order to verify the safety of a given stress state, one first has to find the plane in which the sum $\sigma_{11}/\sigma_0+|\sigma_{12}/\tau_0|^{3/2}$ is maximum and then to compare this value with unity.

For beams, the equation of the limit curve can be expressed in terms of the normal and shearing stresses in a section normal to the beam axis. In this form it reads

$$\sigma_{12} = \mp\frac{\sigma_0}{2}\sqrt{\frac{1}{3D}(\cosh 2\beta_1-1)\cosh 2\beta_1-\left(\frac{\sigma_{11}}{\sigma_0}\right)^2}, \tag{3.1}$$

where

$$\beta_1 = \frac{1}{3}\operatorname{arcsinh} 3\sqrt{6D}\left(1-\frac{1}{2}\frac{\sigma_{11}}{\sigma_0}\right),$$

$$D = \frac{27}{64}\left(\frac{\tau_0}{\sigma_0}\right)^6.$$

Expression (3.1) can be approximated by

$$\sigma_{12}^* = \left(\frac{\sigma_{11}}{\alpha}+\beta\right)\sqrt{(\sigma_{11}-\underline{\sigma})(\overline{\sigma}-\sigma_{11})}, \tag{3.2}$$

where $\underline{\sigma}$ and $\overline{\sigma}$ are the normal stresses obtained from (3.1) with $\sigma_{12} = 0$ and α and β are constants determined by the least square method for an optimum agreement between curves (3.1) and (3.2) in the interval

$$\underline{\sigma} \leqslant \sigma_{11} \leqslant \overline{\sigma}.$$

Figure 3.12 shows plots of curves (3.1) and (3.2) for concrete with an allowable pure-shear stress of 1.05 MPa and the ratio $\tau_0/\sigma_0 = 3.5$. In this case $\alpha = 92$ MPa and $\beta = 0.19$.

The hypothesis of Balandin (1950) *and Stassi d'Alia* (1952)[1]

As in the Huber–Mises–Hencky hypothesis, here too the strength of a homogeneous isotropic and elastic—up to a limit—material is measured by the strain energy of distortion per unit volume. This time, however, the

[1] This hypothesis is a particular case of Burzynski's hypothesis of invariants; see Burzyński, W., *Selected Works*, Vol. I, PWN, Warsaw 1980 (in Polish).

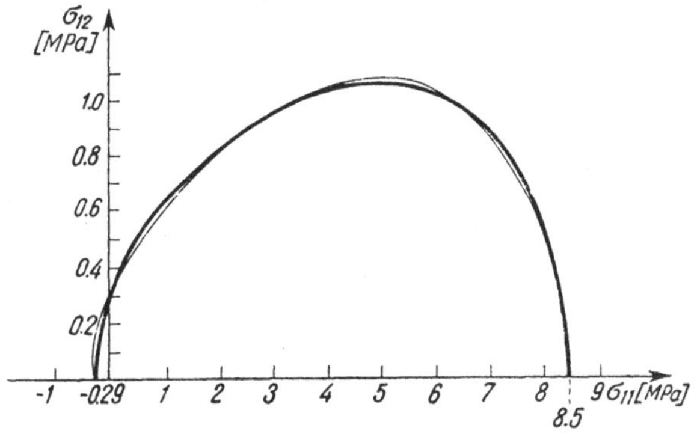

Fig. 3.12. A region of safe states according to Caquot's hypothesis

critical value of this energy is not taken as constant but as dependent on the state of stress; specifically, it is assumed to be a linear function of the mean stress.

The Balandin–Stassi d'Alia condition can be written

$$\Psi_f = \overline{\Psi}_f, \tag{3.3}$$

where

$$\Psi_f = \frac{1+\nu}{3E}\left[\sigma_{11}^2+\sigma_{22}^2+\sigma_{33}^2-\sigma_{11}\sigma_{22}-\sigma_{22}\sigma_{33}-\sigma_{33}\sigma_{11}+\right.$$
$$\left.+3(\sigma_{12}^2+\sigma_{23}^2+\sigma_{31}^2)\right]$$

and

$$\overline{\Psi}_f = \frac{a}{3}(\sigma_{11}+\sigma_{22}+\sigma_{33})+b.$$

For uniaxial tension and compression, we obtain

$$\frac{1+\nu}{3E}\underline{\sigma}^2 = \frac{a}{3}\underline{\sigma}+b$$

and

$$\frac{1+\nu}{3E}\bar{\sigma}^2 = \frac{a}{3}\bar{\sigma}+b.$$

The constants $\bar{\sigma}$ and $\underline{\sigma}$ have to be determined experimentally. Solving these equations for a and b, we find

$$a = \frac{1+\nu}{E}(\bar{\sigma}+\underline{\sigma}), \quad b = -\frac{1+\nu}{3E}\bar{\sigma}\underline{\sigma}.$$

Substituting these expressions in equation (3.3), we obtain

$$\frac{3E}{1+\nu}\Psi_f-(\bar{\sigma}+\underline{\sigma})(\sigma_{11}+\sigma_{22}+\sigma_{33}) = -\bar{\sigma}\underline{\sigma}. \tag{3.4}$$

When the material has the same resistance to tension and to compression, i.e. when $\bar{\sigma} = -\underline{\sigma}$, the Balandin–Stassi d'Alia condition becomes identical with the Huber–Mises–Hencky condition. In the space of principal stresses, equation (3.4) represents a paraboloid of revolution whose axis is a trisectrix between the positive directions of the coordinate axes (Fig. 3.13).

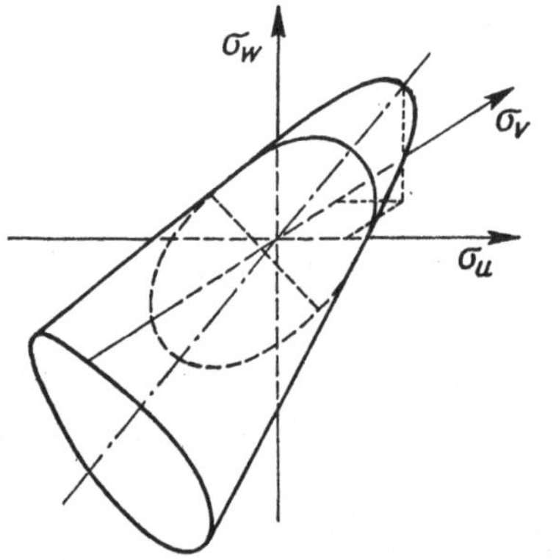

Fig. 3.13. The space of safe stress states according to the Balandin–Stassi d'Alia hypothesis

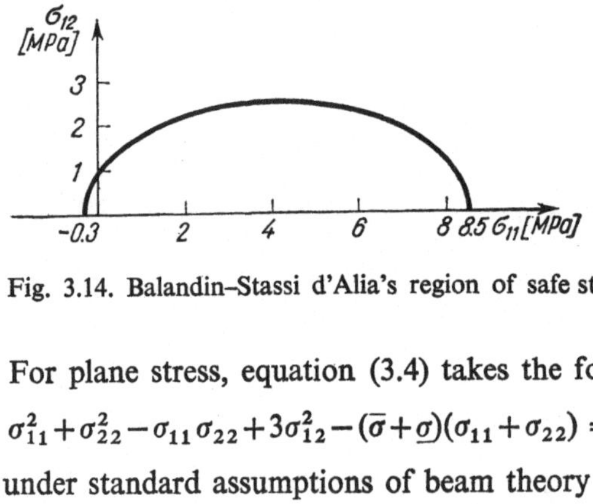

Fig. 3.14. Balandin–Stassi d'Alia's region of safe stresses for a beam

For plane stress, equation (3.4) takes the form

$$\sigma_{11}^2+\sigma_{22}^2-\sigma_{11}\sigma_{22}+3\sigma_{12}^2-(\bar{\sigma}+\underline{\sigma})(\sigma_{11}+\sigma_{22}) = -\bar{\sigma}\underline{\sigma}$$

and under standard assumptions of beam theory it becomes (Fig. 3.14)

$$\sigma_{11}^2+3\sigma_{12}^2-(\bar{\sigma}+\underline{\sigma})\sigma_{11} = -\bar{\sigma}\underline{\sigma}.$$

Experiments have confirmed the validity of these results for both steel and concrete.

The generalized Mohr hypothesis

Mohr's condition is one of the best known strength hypotheses, mostly owing to its simple representation as an envelope of a family of Mohr circles. In its classical form, it relates the strength of materials exclusively to the smallest and the largest principal stresses. Referring the reader to Filonienko-Borodich's book (cf. footnote p. 90) for a comprehensive discussion of the Mohr hypothesis, we only recall here the generalized Mohr condition, which can be written

$$\frac{2E}{1-2\nu}[3-2(1+\varrho)]\Psi_v+\frac{2E(1+\varrho)}{1+\nu}\Psi_f-(\bar{\sigma}+\underline{\sigma})(\sigma_{11}+\sigma_{22}+\sigma_{33})+$$

$$+\bar{\sigma}\underline{\sigma}=0,$$

where Ψ_v denotes the pure dilatational strain energy per unit volume, Ψ_f the distortion energy per unit volume and ϱ a coefficient equal to $\frac{1}{2}$ for plastic materials or ν for brittle materials.

The above condition comprises several of the yield conditions considered so far as particular cases. For example, for $\varrho = 1/2$ we obtain the Balandin–Stassi d'Alia condition, for $\bar{\sigma} = -\underline{\sigma}$ the Huber–Mises–Hencky condition, and for $\varrho = \nu$ and $\bar{\sigma} = -\underline{\sigma}$ the Beltrami condition.

3.2 The criterion of uniform strength

A design is said to be *optimal* with respect to the criterion of uniform strength if the strength function $\Phi(\sigma_{ij})$ defined according to the strength hypothesis used (e.g. τ_{max} for the Tresca–Guest hypothesis, Ψ for the Beltrami hypothesis, Ψ_f for the Huber hypothesis) is constant throughout the structure under the given loading:

$$\Phi(\sigma_{ij}) = \text{const}.$$

In view of the limitations imposed on the design variables by fabrication, construction or serviceability requirements, only certain kinds of structures, e.g. columns in axial compression, tension cables and trusses, admit a uniform-strength design in the strict sense. For other types, the condition of uniform strength may only be satisfied in certain regions or on certain surfaces of the structure. For example, if the cross-section of a beam with a given width b and an unknown depth $h(x)$ is assumed to be rectangular (Fig. 3.15), then only a certain surface of the beam can satisfy the uniform-strength criterion. For an I-shaped cross-section formed of rectangles (Fig. 3.16), several surfaces may satisfy it. If the cross-section of the beam is only required to be inscribed in a rectangle of finite dimensions, the strength can be made uniform almost throughout the beam (Fig. 3.17).

Because of these limitations, by uniform-strength designs we shall also understand designs in which the condition of uniform strength is satisfied at as many points as possible under given conditions.

In some cases, the criterion of uniform strength allows the shape of the structure to be determined uniquely. It permits us, for example, to determine uniquely the cross-sectional areas of columns in axial compression and the

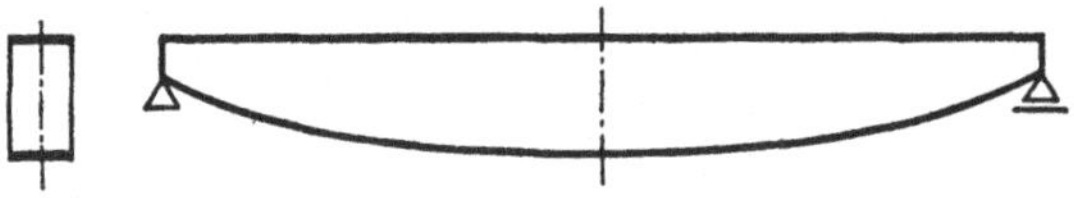

Fig. 3.15. The uniform-strength surface in a rectangular beam

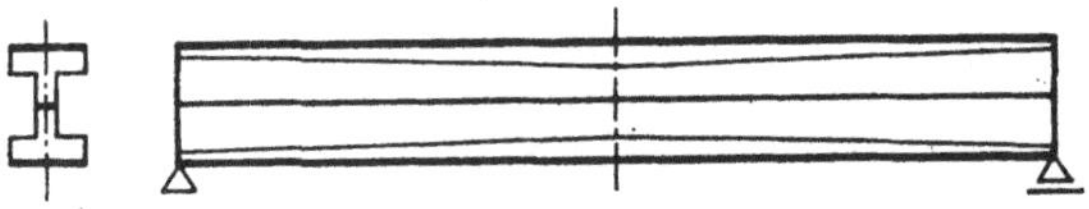

Fig. 3.16. The uniform-strength surfaces in an I-beam

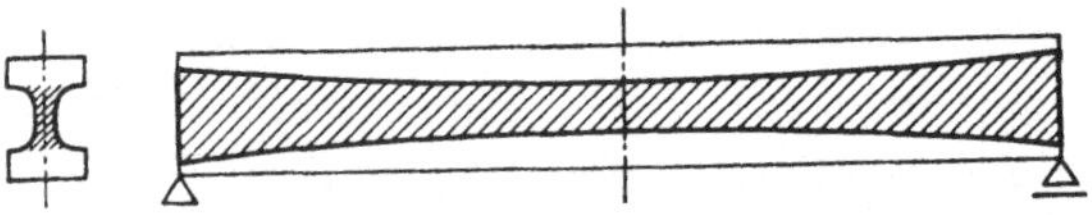

Fig. 3.17. The uniform-strength region in a beam with curved cross-section

depths of rectangular beams in bending. In general, however, a uniform-strength design is not unique. For example, a truss of a given span and under a given loading may be designed for uniform strength in many ways. One of such alternative solutions can then be chosen as optimal with respect to some other criterion, usually that of minimum volume or cost (Chapter 4).

Relations between the uniform-strength criterion and other criteria of structural optimization

A natural generalization of uniform-strength design is minimum-volume design subject to strength constraints. The minimum-volume criterion permits all the parameters of the optimized structure to be determined uniquely. In cases where all the design variables can be calculated from the condition of uniform strength, the uniform-strength design and the minimum-volume design converge. It is so for structures such as isostatic trusses with a given configuration of bars, in which the cross-sectional areas of the bars are the only design variables, for rectangular beams, for columns in compression and for tension cables.

The uniform-strength criterion is also closely related to the criterion of minimum elastic strain energy. According to Wasiutyński's theorems (Sec. 2.1) and Mroz's results (1963), the elastic strain energy accumulated in a structure is minimum when the unit potential is constant throughout the structure or—depending on the conditions—over its free surface, which implies that a minimum-energy design is obtained by evening out the unit potential distribution. And since the potential is used as a measure of the strength of materials in the Beltrami hypothesis, we may conclude that design for minimum energy and design for uniform strength with respect to the Beltrami condition converge.

If the material of a structure is such that the Beltrami hypothesis cannot be used as a condition of failure, the above-mentioned convergence does not occur; however, the methods of solution remain analogous. The same applies to design for minimum energy with allowance for the dead weight of the structure. In this case, the necessary condition for a minimum of the potential is of the form

$$\Psi - \gamma \mathbf{u} = \text{const}.$$

Although this is not a condition of failure, the solving procedure is often similar to that used in uniform-strength design. As examples we may mention minimum-energy design of beams of arbitrary cross-section (Wasiutyński 1939), columns in axial compression with allowance for dead weight (Grycz 1968_1) and moment-free arches (Grycz 1968_3).

3.3 Uniform-strength design of structures under multiple loading conditions

Practically every engineering structure is submitted to many different loading conditions. For some structures we can single out one or two loading states which produce the largest effective stresses in the structure or in a given member. This applies to situations where loads are of variable magnitude but are always applied at the same point.

For a structure subjected to a moving load, such as a bridge, all loading states are equally important, because every position of the load will lead to a maximum effective stress in some part of the structure. It follows that the condition of uniform strength

$$\Phi(\sigma_{ij}) = \text{const}$$

has to be replaced by the condition

$$\max \Phi(\sigma_{ij}) = \text{const},$$

where the maximum is taken with respect to the position of the load. If x_m, $m = 1, 2, \ldots, k$, are the variables defining this position, we obtain a set of k equations

$$\frac{\partial \Phi(\sigma_{ij})}{\partial x_m} = 0, \quad m = 1, 2, \ldots, k, \tag{3.5}$$

which along with the equation

$$\Phi(\sigma_{ij}) = \bar{\Phi}$$

allow us, in principle, to determine the required values of the design variables.

It should be noted that the largest value of $\Phi(\sigma_{ij})$ may occur at a point at which some x_m are at the limits of their ranges, or at a point of discontinuity of Φ or $\partial\Phi/\partial x_m$. Whether the strength function has a proper or an improper maximum must be verified individually in each case.

In practice, the above approach leads to rather complicated equations, from which the determination of the desired quantities may present considerable difficulties. It is therefore advisable to begin by selecting those loading states which are certain to yield a maximum of Φ at a given point of the structure.

In some structures, e.g. in isostatic trusses, it is easy to associate with each member a loading state which produces the largest force in that member. For structures which are hyperstatic, or even isostatic but such that maximum stress components at a given point can be produced by different loads, the problem becomes complicated. But even then it is sometimes possible to separate regions in the structure—still before it is optimized—in which the

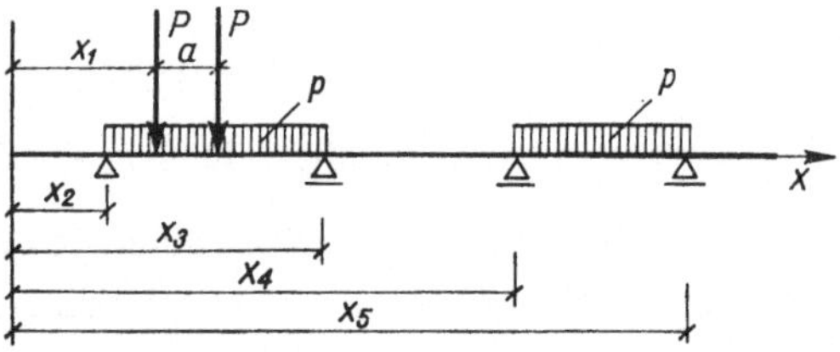

Fig. 3.18. Loading scheme for a steel beam

maximum stresses will be induced by a single specified loading state. This can be done, for example, in the case of isostatic beams. We shall illustrate this procedure on a steel beam of constant depth, loaded in accordance with the standards for road bridges (Fig. 3.18) and optimized for uniform strength on the basis of the Huber–Mises–Hencky hypothesis (Marks 1968_4).

Location of the load for a maximum distortion potential at a given point

For a beam, the unit strain energy of distortion is given by

$$\Psi_f = \frac{1+\nu}{3E}(\sigma_{11}^2+3\sigma_{12}^2). \tag{3.6}$$

The function Ψ_f attains a minimum at $\sigma_{11}=0$, $\sigma_{12}=0$, and increases monotonically with the absolute values of σ_{11} and σ_{12}. It follows that a shift of the load will produce

an increase of Ψ_f if both σ_{12} and σ_{11} are increased,

a decrease of Ψ_f if both σ_{11} and σ_{12} are decreased.

If the shift results in an increase of σ_{11} and a decrease of σ_{12}, or vice versa, it cannot be said in general whether a rise or a drop of Ψ_f will occur.

Next, let us consider conditions (3.5) for the case where $\partial\sigma_{11}/\partial x_m$ is proportional to $\partial\sigma_{12}/\partial x_m$. This relation occurs in isostatic beams loaded as shown in Fig. 3.18, because the influence lines for bending moments and shearing forces consist then exclusively of straight line segments. The proportionality can be written

$$\alpha \frac{\partial \sigma_{11}}{\partial x_m} = \frac{\partial \sigma_{12}}{\partial x_m}; \tag{3.7}$$

the coefficient α depends on the inclination angles of the influence lines and on the cross-sectional shape. Conditions (3.5) take the form

$$\begin{aligned} \frac{\partial \Psi_f}{\partial x_m} &= \frac{2(1+\nu)}{3E}\left(\sigma_{11}\frac{\partial \sigma_{11}}{\partial x_m} + 3\sigma_{12}\frac{\partial \sigma_{12}}{\partial x_m}\right) \\ &= \frac{2(1+\nu)}{3E}(\sigma_{11}+3\alpha\sigma_{12})\frac{\partial \sigma_{11}}{\partial x_m} = 0, \end{aligned} \tag{3.8}$$

and for $\partial \sigma_{11}/\partial x_m \neq 0$ we get

$$\sigma_{11} + 3\alpha\sigma_{12} = 0.$$

If $\partial \sigma_{11}/\partial x_m = 0$, then since by (3.7) also $\partial \sigma_{12}/\partial x_m = 0$, σ_{11} and σ_{12} would be independent of x_m and seeking a maximum of Ψ_f with respect to x_m would not make sense.

The second derivative of Ψ_f with respect to x_m reads

$$\frac{\partial^2 \Psi_f}{\partial x_m^2} = \frac{2(1+\nu)}{3E}\left[(\sigma_{11}+3\alpha\sigma_{12})\frac{\partial^2 \sigma_{11}}{\partial x_m^2} + (1+3\alpha^2)\left(\frac{\partial \sigma_{11}}{\partial x_m}\right)^2\right] < 0. \tag{3.9}$$

At points satisfying condition (3.8), the first component of (3.9) vanishes and the second is always positive, so that $\partial^2 \Psi_f/\partial x_m^2 > 0$. It follows that Ψ_f cannot attain a maximum at any of those points.

Thus, in the case of an isostatic beam loaded as specified, the largest values of Ψ_f occur when the variables x_m^* are at their limits or take values for which Ψ_f is discontinuous or the derivatives of Ψ_f do not exist or are infinite.

The above observation makes it sometimes possible to say at once what distribution of the load will produce the largest potential Ψ_f at a given point of the beam. In other cases, it permits the ranges of the variables defining the position of the load to be narrowed considerably.

The values of the load distribution parameters that yield a maximum Ψ_f in a cross-section $x = u_0$ of the beam (Fig. 3.19) for different kinds of loading are presented in Table 3.1.

Note that to each normal section of the beam there correspond one or two load distributions producing the largest potential at different points of that section. When two such distributions exist, we should determine the

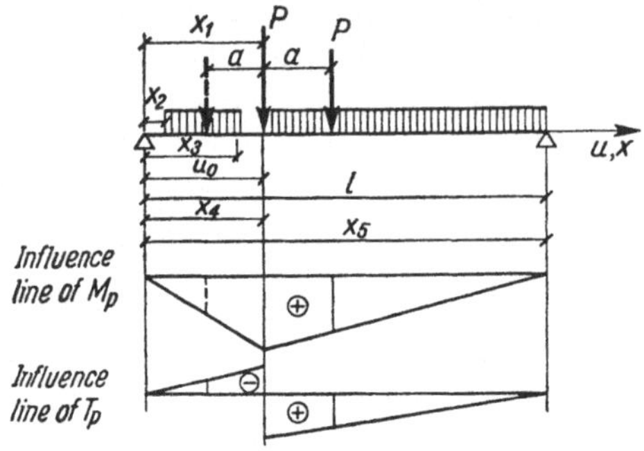

Fig. 3.19. Load distribution which yields a maximum of Ψ_f in the section $x = u_0$

Table 3.1 Load distributions corresponding to a maximum potential Ψ_f in the section $x = u_0$ of a simply supported beam

(Notation as in Fig. 3.19; $0 \leqslant u_0 \leqslant l/2$)

Kind of loading	Load distribution parameters					Number of different load distributions under which Ψ_f can be maximum in the section $x = u_0$
	x_1	x_2	x_3	x_4	x_5	
Concentrated force P	u_0	–	–	–	–	1
Two concentrated forces P at a distance a from each other	u_0	–	–	–	–	1
Continuous load p	—	0	0	u_0	l	2
	—	0	u_0	u_0	l	
Continuous load p +	u_0	0	0	u_0	l	
+ two concentrated forces P	u_0	0	u_0	u_0	l	2

boundary between the region in which the maximum of Ψ_f is produced by one distribution and that in which it is produced by the other.

This can be done by equating the potentials induced by the two loadings:

$$\Psi_f(\sigma_{11}^{(1)}, \sigma_{12}^{(1)}) = \Psi_f(\sigma_{12}^{(2)}, \sigma_{12}^{(2)}).$$

In beams of arbitrary cross-sections, the boundaries depend on the cross-sectional shape. For a beam shaped to satisfy the condition $\Psi_f(\sigma_{11}, \sigma_{12}) = \text{const}$, the boundary can be expressed in terms of the bending moments and shearing forces at each loading state. We now consider these two cases in more detail.

Arbitrary cross-section

Comparing the values of Ψ_f at the two loading states, we get

$$\frac{1+\nu}{3E}(\sigma_{12}^{(1)2}+3\sigma_{12}^{(1)2}) = \frac{1+\nu}{3E}(\sigma_{11}^{(2)2}+3\sigma_{12}^{(2)2}).$$

Replacing $\sigma_{11}^{(1)}$, $\sigma_{12}^{(1)}$, $\sigma_{11}^{(2)}$ and $\sigma_{12}^{(2)}$ by their respective values gives

$$\frac{(M_q+M_1)^2y^2}{J^2}+3\frac{(T_q+T_1)^2S^2}{J^2g^2} = \frac{(M_q+M_2)^2y^2}{J^2}+3\frac{(T_q+T_2)^2S^2}{J^2g^2}.$$

Putting $k^2 = 3\,S^2/g^2y^2$, we find

$$k_{1\text{im}} = \pm\sqrt{\frac{(2M_q+M_1+M_2)(M_1-M_2)}{(2T_q+T_1+T_2)(T_2-T_1)}};$$

y denotes the distance of a given point from the neutral axis of the beam.

Cross-section designed for uniform strength according to the Huber condition $\Psi_f = \overline{\Psi}_f$

In this case we have

$$\left(\frac{S}{g}\right)^2 = \frac{1}{3}\left(\frac{J}{T}\right)^2(\bar{\sigma}-\sigma_{11})^2,$$

where

$$\bar{\sigma} = \frac{3E\overline{\Psi}_f}{1+\nu} = \frac{M_1 a}{J}.$$

Since at the boundary between the maximum-potential regions corresponding to two different load distributions the ratio S/g can be calculated from either loading state, we can write

$$\frac{1}{3}\left(\frac{J}{T_q+T_1}\right)^2\left[\frac{(M_q+M_1)^2a^2}{J^2}-\frac{(M_q+M_1)^2y^2}{J^2}\right]$$
$$= \frac{1}{3}\left(\frac{J}{T_q+T_2}\right)^2\left[\frac{(M_q+M_2)^2a^2}{J^2}-\frac{(M_q+M_2)^2y^2}{J^2}\right].$$

Hence we find

$$\eta = \frac{y}{a} = \pm\sqrt{\frac{\left(\frac{T_q+T_2}{T_q+T_1}\right)^2-1}{\left(\frac{T_q+T_2}{T_q+T_1}\right)^2-\left(\frac{M_q+M_2}{M_q+M_1}\right)^2}}.$$

The maximum-potential-due-to-single-load regions for a 20 m bridge span with a 7 m-wide deck are shown in Fig. 3.20.

A similar reasoning can be used with any other strength hypothesis. For example, for a beam of prestressed concrete loaded as before (Fig. 3.18) and designed for uniform strength according to the Caqout hypothesis, we obtain regions 1 and 2 shown in Fig. 3.21. They correspond to the load dis-

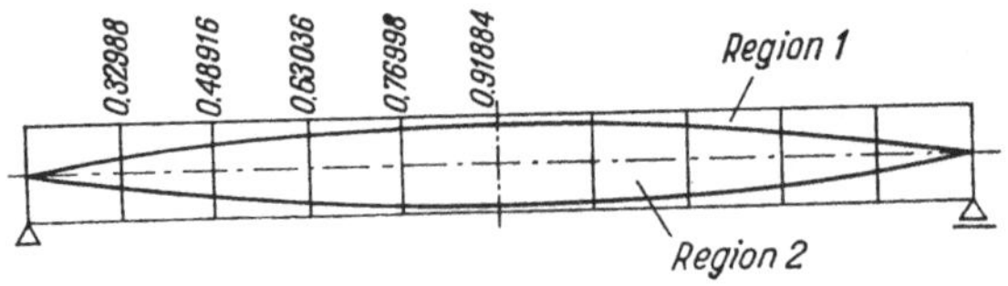

Fig. 3.20. Maximum-potential regions corresponding to two different load distributions over a beam designed from Huber's condition

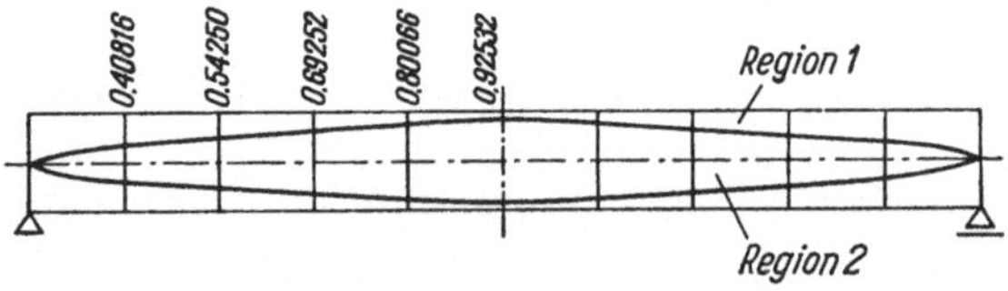

Fig. 3.21. Maximum-potential regions corresponding to two different load distributions over a beam designed from Caquot's condition

tributions producing a maximum bending moment and a maximum shearing force in a given cross-section, respectively. The boundary between the two regions is given by the equation

$$\left(\frac{T_2}{T_1}\right)^2\left(\frac{\bar{\sigma}+\sigma}{2}+\alpha\beta+\frac{\bar{\sigma}-\sigma}{2}\eta_{1\mathrm{lim}}\right)^2(1-\eta_{1\mathrm{lim}}^2)$$

$$=\left(\frac{\bar{\sigma}+\underline{\sigma}}{2}+\alpha\beta+\frac{\bar{\sigma}-\underline{\sigma}}{2}\frac{M_2}{M_1}\eta_{1\mathrm{lim}}\right)^2\left[1-\left(\frac{M_2}{M_1}\eta_{1\mathrm{lim}}\right)^2\right],$$

where M_1 and T_1 are the maximum bending moment and the corresponding shearing force, and T_2 and M_2 are the maximum shearing force and the corresponding bending moment.

3.4 Uniform-strength design of bar structures

In plane bar structures, among the six components of the stress tensor only the normal component σ_{11} and the shearing component σ_{12} are non-zero. Under this assumption, all strength hypotheses take the form $\Phi(\sigma_{11}, \sigma_{12}) = \mathrm{const.}$

In the analysis of bar structures, use is made of the resultant stresses in the cross-sections of the individual bars: the longitudinal force N, the shearing force T and the bending moment M. The most common bar structures in building and civil engineering are: trusses, frames, cables, columns and beams.

Cables, columns in axial compression and bars of trusses can only carry longitudinal forces N and normal stresses σ_{11}. In this case, strength optimization reduces to fully-stressed design, i.e. design such that

$$\frac{N}{A} = \bar{\sigma}.$$

The above condition allows calculation of the cross-sectional areas of the elements involved. Often, the solution so obtained is not unique. For example, a truss with a given configuration of joints and a given load or a cable anchored at a given point and transmitting a given force can be designed for uniform strength in many ways. In these cases, for a unique solution, the criterion of minimum volume, weight or cost has to be used, with an allowable stress imposed as a constraint.

Beams, frames and columns in compression and bending transmit normal and shearing stresses (σ_{11} and σ_{12}). The uniform-strength criterion makes it possible to determine detailed parameters of the cross-sections of such structures. Depending on the restrictions present in a given problem, either functions describing the shape of the cross-section or the dimensioning design for a specified type of shape, e.g. rectangular or I-section, are determined.

Uniform-strength design of beams of arbitrary cross-section

The problem consists in determining the cross-sectional shape of a beam in bending under one or many loading states, so that the strength of the beam is equalized over as many points as possible.

For functional reasons, it is assumed that the cross-section of the beam can be enclosed in a rectangle of depth $2a$ and width b, both given. To make the solution useful in practice, it is also assumed that the section is crowned with flanges of width b (Wasiutyński 1939, Marks 1968_3). The flange thickness c and the web thickness g are the design variables to be determined from the criterion of uniform strength.

As mentioned above, the general form of this criterion in the present case is

$$\Phi(\sigma_{11}, \sigma_{12}) = \bar{\Phi}; \tag{3.10}$$

where Φ is an appropriate strength function and $\bar{\Phi}$ a constant. The stresses σ_{11} and σ_{12} may depend not only on the coordinates of the point they are calculated at, but also on the position x_i of the moving load.

In the form solved for σ_{12}, equation (3.10) can be written

$$\sigma_{12} = \varphi(\sigma_{11}).$$

Substituting an appropriate expression for σ_{12}, we get

$$\frac{S}{g} = \frac{J}{T}\varphi(\sigma_{11}), \tag{3.11}$$

where S denotes the static moment, g the web thickness, J the moment of inertia and T the shearing force.

Since

$$S = bc(a-\tfrac{1}{2}c)+\int_{y}^{a-c} gy\,\mathrm{d}y,$$

equation (3.11) becomes

$$bc(a-\tfrac{1}{2}c)+\int_{y}^{a-c} gy\,\mathrm{d}y = \frac{J}{T}g\varphi(\sigma_{11}).$$

Putting

$$\int_{y}^{a-c} gy\,\mathrm{d}y = F(y), \quad g = -\frac{1}{y}\frac{\mathrm{d}F}{\mathrm{d}y}, \tag{3.12}$$

we obtain a differential equation

$$\frac{\mathrm{d}F}{\mathrm{d}y}+\frac{y}{\dfrac{J}{T}\varphi(\sigma_{11})}F(y) = \frac{-bc(a-\frac{1}{2}c)y}{\dfrac{J}{T}\varphi(\sigma_{11})},$$

whose general solution reads

$$F(y) = D\exp\left(-\int\frac{y}{\dfrac{J}{T}\varphi(\sigma_{11})}\mathrm{d}y\right)-bc(a-\tfrac{1}{2}c). \tag{3.13}$$

Using this result, we find from (3.12)

$$g = \frac{D}{\dfrac{J}{T}\varphi(\sigma_{11})}\exp\left(-\int\frac{y}{\dfrac{J}{T}\varphi(\sigma_{11})}\mathrm{d}y\right). \tag{3.14}$$

The integration constant D can be calculated from the requirement that the web thickness g be equal to the flange width b at $y = a-c$.

The above solution is valid provided that the bending moments and shearing forces can be regarded as given in each cross-section, irrespective of the section shape. This assumption is satisfied in isostatic structures.

For a hyperstatic beam, the problem can be solved by a method of successive approximations-iterations (see Chapter 13). The procedure is then as follows:

— determine the internal forces in a structure with the same statical scheme but a constant moment of inertia,
— design the structure for transmission of these internal forces and calculate the corresponding moments of inertia,
— determine the internal forces in a structure with the moments of inertia calculated in the preceding step,
— design the structure for transmission of these internal forces, etc.

The calculations are continued until the difference between the shapes calculated at two successive steps is sufficiently small. Usually, the process converges very rapidly and a sufficiently good result can be obtained at the third or even second step.

In structures designed in this manner, the strength function Φ is constant on the web but takes larger values on the flanges. This is justified by the small flange thickness as compared with the depth of the cross-section. However, if the allowable value of Φ is not to be exceeded at any point of the structure, either a smaller $\bar{\Phi}$ has to be chosen for the web or a stronger material used the flanges.

An alternative possibility is to level the function Φ up to a value $\bar{\Phi}_1$ over the web and, simultaneously, to another value $\bar{\Phi}_2$ at the external edges of the flanges. In this case the condition

$$g = b \quad \text{at} \quad y = a - c$$

must be replaced by the condition

$$\frac{TS(0)}{Jg(0)} = \bar{\sigma}_{12},$$

where the stress $\bar{\sigma}_{12}$ corresponds to $\bar{\Phi}_1$.

To illustrate the above approach, let us consider a few examples of uniform strength design of beams, based on different strength hypotheses.

Design for uniform unit potential

The condition for the elastic strain energy per unit volume to be constant throughout the beam is (Wasiutyński 1939)

$$\sigma_{12} = \frac{1}{\sqrt{2(1+\nu)}} \sqrt{\bar{\sigma}^2 - \sigma_{11}^2}.$$

Following the procedure presented above (cf. equation (3.14)), we find the web thickness to be given by

$$g = \frac{D}{K\sqrt{L^2 - y^2}} \, \mathrm{e}^{\frac{1}{K}\sqrt{L^2 - y^2}}, \tag{3.15}$$

where

$$K = \frac{M}{T\sqrt{2(1+\nu)}}, \quad L^2 = (a-c)^2 + \left[\frac{c}{K}\left(a-\frac{c}{2}\right)\right]^2. \tag{3.16}$$

The constant D is found from the condition $g(y) = b$ at $y = a-c$ as

$$D = bc\left(a-\frac{c}{2}\right)e^{-\frac{c}{K^2}\left(a-\frac{c}{2}\right)}. \tag{3.17}$$

The function $g(y)$ must be such that the moment of inertia of any cross-section is related to the bending moment M and the limiting stress $\bar{\sigma} = \sqrt{2E\bar{\Phi}}$ by the relation $J = Ma/\bar{\sigma}$:

$$J = \frac{Ma}{\bar{\sigma}} = 4\int_0^{a-c} g(y)y^2 dy + 4\left[bc\left(a-\frac{c}{2}\right)^2 + \frac{1}{12}bc^3\right].$$

Substituting $g(y)$, we obtain a condition relating b and c:

$$\frac{Ma}{4\bar{\sigma}} - bc\left(a-\frac{c}{2}\right)e^{-\frac{c}{K^2}\left(a-\frac{c}{2}\right)}\int_0^{a-c}\frac{y^2 e^{\frac{1}{K}\sqrt{L^2-y^2}}}{K\sqrt{L^2-y^2}}dy - bc\left(a-\frac{c}{2}\right)^2 -$$

$$-\frac{bc^3}{12} = 0.$$

Expanding the integral into series, after simple calculations we get

$$\frac{Ma}{4\bar{\sigma}} - \frac{bc^2}{6}(3a-c) - bc\left(a-\frac{c}{2}\right)(a-c)\times$$

$$\times\left[1 - \frac{(a-c)^2}{6KL} - \frac{(a-c)^4}{40KL^2}\left(\frac{1}{L}-\frac{1}{K}\right) - \dots\right]\exp\left[\frac{1}{LK} - \frac{c}{K^2}\left(a-\frac{c}{2}\right)\right] = 0.$$

It can be seen from this equation that as c decreases to zero, b increases unboundedly. Since the actual flange width is limited, we may conclude that a beam in bending cannot be designed so that the unit potential is constant throughout. However, it is possible to construct beams in which the potential per unit volume will differ very little from point to point. In fact, it can be made constant throughout the web, with only a slight increase towards the external edges of the flanges.

Example 1. The above results are used to calculate the shapes of several cross-sections of a simply supported flanged beam loaded by a concentrated force at mid-span (Wasiutyński 1939).

The depth, the flange width and the section modulus are taken as equal to those of an NP 50 I-bar (Polish standards), viz. $2a = 0.50$ m, $2b = 0.185$ m, $W = 0.00275$ m^3.

The largest bending moment is calculated for the limiting stress $\bar{\sigma} = 140{,}000$ kN/m^2; $M = 140{,}000 \cdot 0.00275 = 385.0$ kNm. The span of the

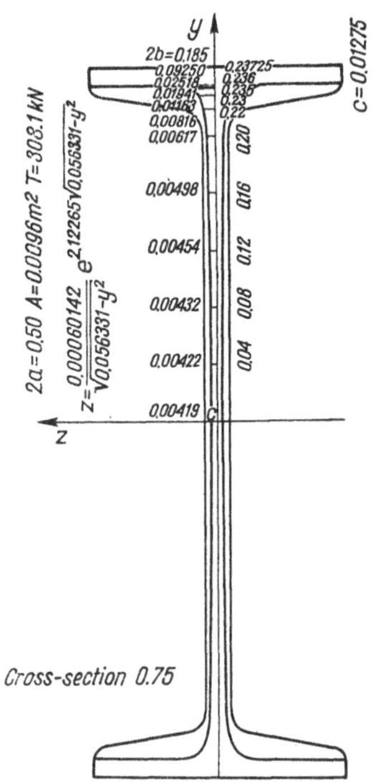

Fig. 3.22. Cross-section $x = 0.75$ m designed for uniform unit potential

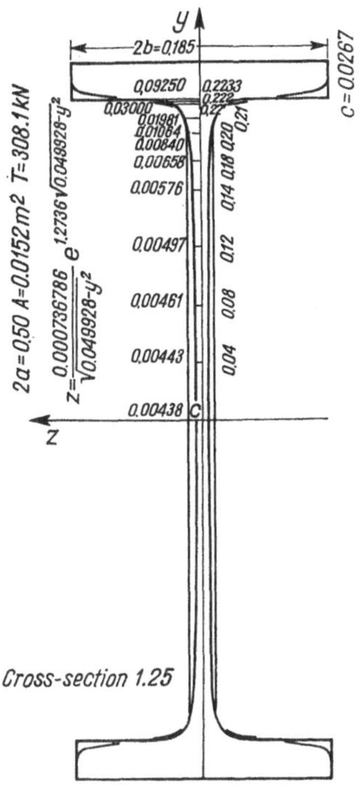

Fig. 3.23. Cross-section $x = 1.25$ m designed for uniform unit potential

beam is $2l = 2.50$ m, and hence the concentrated load $P = 4M/2l = 616.00$ kN. The shearing force $T = 308.00$ kN is constant along the beam.

Given these data, the constants K, c, D and L can be computed from equations (3.16)–(3.18) and then used in formula (3.15) to determine the web thickness in an arbitrary cross-section of the beam.

The calculated shapes of the cross-sections $x = 0.75$ and $x = 1.25$ are shown in Fig. 3.22 and Fig. 3.23.

Design for uniform strength according to the Huber hypothesis

Assuming the Huber hypothesis, the criterion of uniform strength for a beam takes the form (Wasiutyński 1939)

$$\sigma_{12} = \frac{1}{\sqrt{3}} \sqrt{\bar{\sigma}^2 - \sigma_{11}^2}.$$

The formula for the web thickness, (3.14), becomes

$$g = \frac{D}{K\sqrt{L^2 - y^2}} \exp \frac{1}{K} \sqrt{L^2 - y^2}, \tag{3.19}$$

where

$$K = \frac{M}{T\sqrt{6}}, \qquad L^2 = (a-c)^2 + \left[\frac{c(a-c/2)}{K}\right]^2.$$

The integration constant D can be calculated from the requirement that $g(y) = b$ at $y = a-c$:

$$g_{(a-c)} = \frac{D}{c(a-c/2)} \exp\frac{c}{K^2}\left(a-\frac{c}{2}\right),$$

$$D = bc\left(a-\frac{c}{2}\right)\exp\left[-\frac{c}{K^2}\left(a-\frac{c}{2}\right)\right].$$

As before, the moment of inertia of any cross-section must correspond to the bending moment and the limiting normal stress $\bar{\sigma}$ via the relation

$$J = \frac{Ma}{\bar{\sigma}} = 4\int_0^{a-c} g(y)y^2\mathrm{d}y + 4\left[bc\left(a-\frac{c}{2}\right) + \frac{bc^3}{12}\right].$$

Inserting here $g(y)$ given by (3.19) with D found above, we obtain a relationship between b and c:

$$\frac{Ma}{4\bar{\sigma}} - bc\left(a-\frac{c}{2}\right)\exp\left[-\frac{c}{K^2}\left(a-\frac{c}{2}\right)\right]\int_0^{a-c}\frac{y^2}{K\sqrt{L^2-y^2}}\exp\frac{1}{K}\sqrt{L^2-y^2}\,\mathrm{d}y -$$

$$-bc\left(a-\frac{c}{2}\right)^2 - \frac{bc^3}{12} = 0.$$

After computation of the integral and considerable algebraic manipulations, the above condition can be written in the form

$$\frac{Ma}{4\bar{\sigma}} - bc\left(a-\frac{c}{2}\right)^2 - \frac{1}{12}bc^3 - bc\left(a-\frac{c}{2}\right)\exp\left[-\frac{c}{K^2}\left(a-\frac{c}{2}\right)\right]G(y) = 0,$$

where

$$G(y) = L\left[-\beta\exp\alpha\sqrt{1-\beta^2} + \beta^2\left(-\frac{1}{2}+\frac{22}{9}\beta - 4\beta^2 + \frac{32}{15}\beta^3\right) + \right.$$

$$+ e^{\alpha\beta}\left(1 - \frac{25}{6}\beta + \frac{70}{9}\beta^2 - \frac{20}{3}\beta^3 + \frac{32}{15}\beta^4\right) +$$

$$+ e^{0.9682\alpha\beta^2}\left(8 - \frac{208}{9}\beta + 24\beta^2 - \frac{128}{15}\beta^3\right) +$$

$$+ e^{0.8660\alpha\beta^2}\left(-6 + 76\beta - 32\beta^2 + \frac{64}{3}\beta^3\right) +$$

$$\left. + e^{0.6614\alpha\beta^2}\left(\frac{8}{3} - \frac{112}{9}\beta + \frac{56}{3}\beta^2 - \frac{128}{15}\beta^3\right)\right],$$

$$\alpha = \frac{L}{K}, \qquad \beta = \frac{a-c}{L}.$$

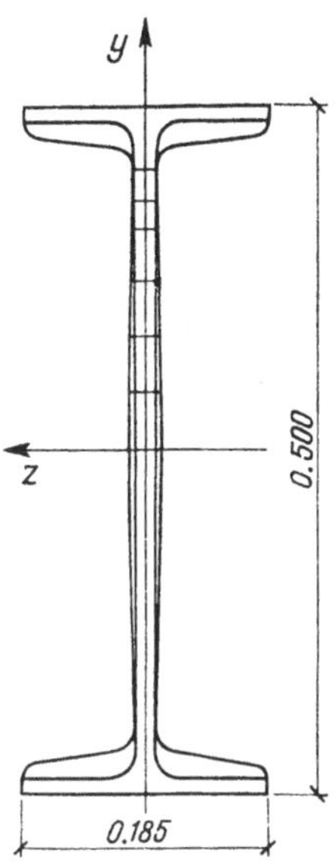

Fig. 3.24. Cross-section $x = 0.75$ m calculated from Huber's condition

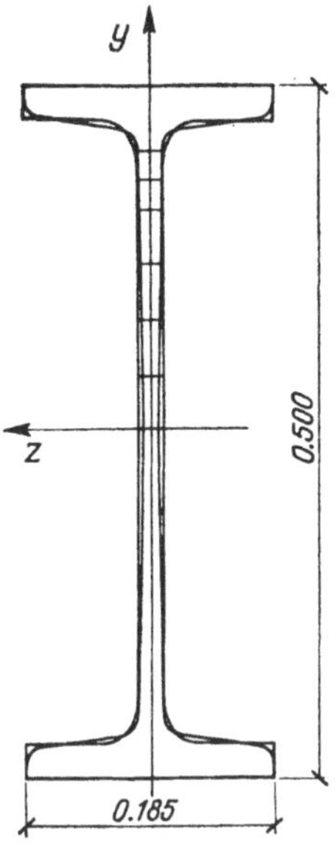

Fig. 3.25. Cross-section $x = 1.25$ m calculated from Huber's condition

Example 2. The web thickness is calculated from formula (3.19) for a beam with the same basic data as in Example 1. The optimized sections $x = 0.75$ m and $x = 1.25$ m are shown in Fig. 3.24 and Fig. 3.25.

Design of a prestressed-concrete beam on the basis of the Caquot criterion

We assume, after Wasiutyński (1939) and Marks (1968_3), that the cross-section of the beam consists of two flanges with a constant width b and a depth c variable along the beam, and a web of continuously variable depth and width. The beam depth a is constant. It is required to determine the web and flange thickness functions g and c so that the Caquot strength function attains a limiting value at each point of the web under at least one of the specified loading conditions. Satisfying a similar requirement in the flanges is impossible because of their constant width.

In the case considered, the design condition takes the form

$$\sigma_{12} = \left(\frac{\sigma_{11}}{\alpha} + \beta\right)\sqrt{(\sigma_{11} - \underline{\sigma})(\bar{\sigma} - \sigma_{11})}.$$

Using the same procedure as before, we obtain the following formula for the web thickness:

$$g = \frac{D}{\dfrac{J}{T_y}\left(\dfrac{P}{A} + \alpha\beta + \dfrac{My}{J}\right)\sqrt{\left(\dfrac{P}{A} - \underline{\sigma} + \dfrac{My}{J}\right)\left(\bar{\sigma} - \dfrac{P}{A} - \dfrac{My}{J}\right)}} \times$$

$$\times \exp \frac{T\alpha J}{M} \left| \arcsin \frac{\bar{\sigma}+\underline{\sigma}-2\frac{P}{A}+\frac{My}{J}}{\bar{\sigma}-\underline{\sigma}} - \frac{\frac{P}{A}+\alpha\beta}{\sqrt{(\alpha\beta+\underline{\sigma})(\alpha\beta+\bar{\sigma})}} \times \right.$$

$$\left. \times \arcsin \frac{(\bar{\sigma}+\underline{\sigma}+2\alpha\beta)\left(\frac{P}{A}+\alpha\beta+\frac{My}{J}\right)-2(\alpha\beta+\underline{\sigma})(\alpha\beta+\bar{\sigma})}{(\bar{\sigma}-\underline{\sigma})\left(\frac{P}{A}+\alpha\beta+\frac{My}{J}\right)} \right].$$

The constant D, determined from the condition $g = b$ at $y = a-c$, is

$$D = \frac{Jb}{\alpha T}\left[\frac{P}{A}+\alpha\beta+\frac{M(a-c)}{J}\right] \times$$

$$\times \sqrt{\left[\frac{P}{A}-\underline{\sigma}-\frac{M(a-c)}{J}\right]\left[\bar{\sigma}-\frac{P}{A}-\frac{M(a-c)}{J}\right]} \times$$

$$\times \exp \frac{-T\alpha J}{M} \left\{ \arcsin \frac{\bar{\sigma}+\underline{\sigma}-2\frac{P}{A}-2\frac{M(a-c)}{J}}{\bar{\sigma}-\underline{\sigma}} - \right.$$

$$- \frac{\frac{P}{A}+\alpha\beta}{\sqrt{(\alpha\beta+\underline{\sigma})(\alpha\beta+\bar{\sigma})}} \times$$

$$\left. \times \arcsin \frac{(\bar{\sigma}+\underline{\sigma}+2\alpha\beta)\left[\frac{P}{A}+\alpha\beta+\frac{M(a-c)}{J}\right]-(\alpha\beta+\underline{\sigma})(\alpha\beta+\bar{\sigma})}{(\bar{\sigma}-\underline{\sigma})\left[\frac{P}{A}+\alpha\beta+\frac{M(a-c)}{J}\right]} \right\}.$$

The function g must satisfy two more conditions: for each section, the bending moment and the cross-sectional area must correspond with the normal stresses σ_u and σ_l at the edges of the section. Assuming optimum prestressing parameters (Marks 1968_3), we obtain

$$J = \frac{2aM}{\sigma_u-\sigma_l} = \frac{bc^3}{6}+2bc\left(a-\frac{c}{2}\right)^2+2\int_0^{a-c} gy^2 dy, \tag{3.20}$$

$$A = \frac{2P}{\sigma_u+\sigma_l} = 2bc+2\int_0^{a-c} g dy. \tag{3.21}$$

These equations allow two of the quantities a, b, c, P, σ_u and σ_l to be expressed in terms of the remaining four. The values of the latter are usually set by the conditions of use or fabrication.

Example 3. The above results are used to calculate the shapes of several cross-sections of a box beam of prestressed concrete under a moving load. The beam constitutes the supporting structure of a three-span railway bridge. The effective spans are 40 m, 8 m and 40 m; their ratio is $l_1:l:l_1 = 5:1:5$. The width of the bridge is 4 m.

The moving load is specified according to the Polish Railway Standards.

The depth and web thicknesses are calculated at 11 points of each of the sections selected every $0.1l_1$ of the shore spans. The cross-section of the centre 8 m-span is assumed to be constant and equal to that at the intermediate supports. The effect of the variation of the moments of inertia and the cross-sectional areas on the internal forces is dealt with by the method of successive approximations.

The depth of the beam is assumed to be constant, $2a = 2.20$ m.

The internal forces in the sections $x = 0.4l_1$ and $x = l_1$ are

x/l_1	M_1	T_1	M_2	T_2
0.4	11640 kNm	−40 kN	9780 kNm	−760 kN
1.0	−12670 kNm	−1460 kN	−8850 kNm	−1960 kN.

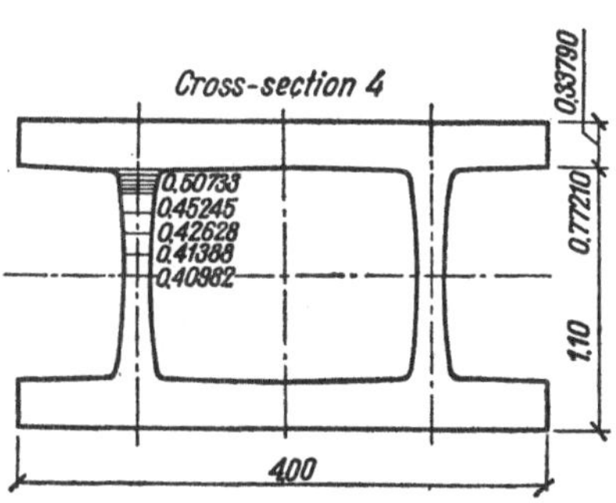

Fig. 3.26. Cross-section $x = 0.4l_1$ designed from Caquot's condition

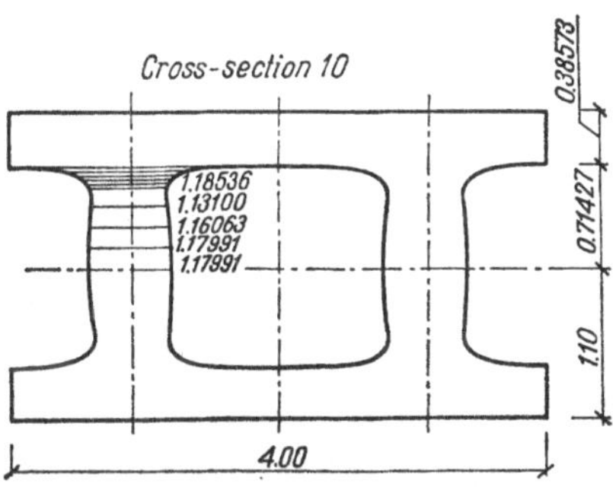

Fig. 3.27. Cross-section $x = l_1$ designed from Caquot's condition

The prestressing forces in these sections are taken to be 17600 kN and 26650 kN, respectively. It is also assumed that $\bar{\sigma} = 8.5$ MPa, $\underline{\sigma} = -0.3$ MPa, $1/\alpha = 0.0109$ and $\beta = 0.19010$. Calculations have been carried out on an Elliott 803-B computer. The optimized sections $x = 0.4l_1$ and $x = l_1$ are shown in Figs. 3.26 and 3.27.

3.5 Uniform-strength design of beams with cross-sections defined up to *m* parameters

When the requirements of execution or use predetermine the cross-sectional shape of a beam to the accuracy of *m* parameters, the design problem is reduced to finding these parameters from the condition of equal strength at *m* points in each cross-section of the beam.

In particular, if the beam section is rectangular with a given width and a variable depth, so that its exact shape depends on only one variable, we can only require that the strength be increased to the limiting value at a single point in each section.

The design procedure in such cases is illustrated below on a rectangular beam and an I-beam.

Huber-hypothesis-based design of a rectangular beam

A rectangular beam of a given width b made of material of equal resistance to tension and compression is loaded by a concentrated force P moving along the beam. It is required to find the beam depth a, variable along the beam (Fig. 3.28).

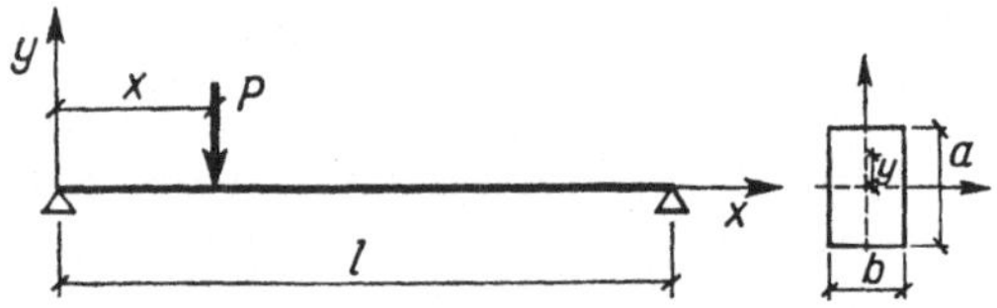

Fig. 3.28. A beam loaded by a concentrated moving force

The Huber–Mises–Hencky condition for a beam takes the form

$$\sigma_{11}^2+3\sigma_{12}^2 = \bar{\sigma}^2.$$

Expressing the normal and shearing stresses in terms of the internal forces and the geometrical dimensions of the beam, we obtain

$$\frac{144M^2\eta^2}{b^2a^4}+\frac{108T^2}{b^2a^2}\left(\frac{1}{4}-\eta^2\right)^2 = \bar{\sigma}^2. \tag{3.22}$$

Equation (3.22) must hold for η for which the expression $\sigma_{11}^2+3\sigma_{12}^2$ is maximum. A maximum of this expression may occur at the edge of the section ($\eta = \pm 1/2$) or for η satisfying the equation

$$\sigma_{11}\frac{\partial\sigma_{11}}{\partial\eta}+3\sigma_{12}\frac{\partial\sigma_{12}}{\partial\eta} = 0.$$

In terms of the internal forces and the geometrical quantities, this can be written

$$\frac{288M^2\eta}{b^2a^4}-\frac{432T^2}{b^2a^2}\left(\frac{1}{4}-\eta^2\right)\eta = 0. \tag{3.23}$$

Hence

$$\eta_1 = 0, \quad \eta_2^2 = \frac{1}{4}-\frac{3}{2}\frac{M^2}{T^2a^2}.$$

η_1 yields a maximum of the sum $\sigma_{11}^2+3\sigma_{12}^2$ if $M^2-\frac{3}{8}T^2a^2<0$; at η_2 no maximum can occur.

Thus, depending on M and T, the largest value of $\sigma_{11}^2+3\sigma_{12}^2$ occurs when $\eta = \pm 1/2$ or $\eta = 0$. In the first case, the depth a is given by

$$a = \sqrt{\frac{6M}{b\bar{\sigma}}} \tag{3.24}$$

and in the second case by

$$a = \frac{3\sqrt{3}}{2}\,\frac{T}{b\bar{\sigma}}. \tag{3.25}$$

Of formulae (3.24) and (3.25), that is used for a given section which yields a greater value.

Example 4. For $P = 200$ kN, $l = 10$ m, $b = 0.12$ m and $\bar{\sigma} = 100$ MPa, the bending moment and the shearing force are

$$M = Pl(1-\xi)\xi = 5000(1-\xi)\xi \text{ kNm},$$
$$T = P(1-\xi) = 500(1-\xi) \text{ kN}.$$

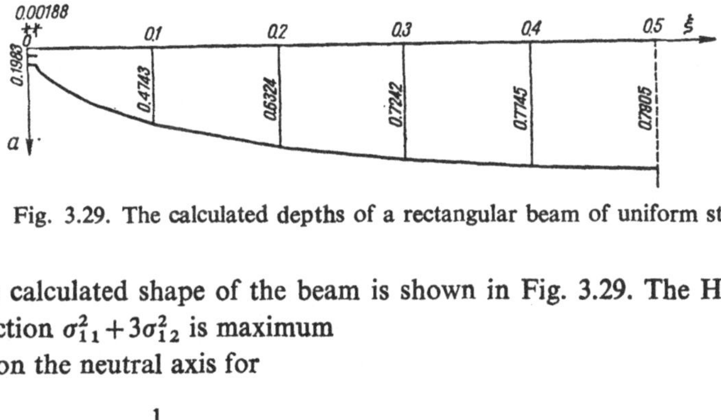

Fig. 3.29. The calculated depths of a rectangular beam of uniform strength

The calculated shape of the beam is shown in Fig. 3.29. The Huber strength function $\sigma_{11}^2+3\sigma_{12}^2$ is maximum

— on the neutral axis for

$$\xi \leqslant \frac{1}{1+\dfrac{8}{9}\dfrac{b\bar{\sigma}l}{P}} = 0.00187 \quad \text{and} \quad \xi \geqslant 0.99813,$$

— at the edge of the section for

$$0.00187 \leqslant \xi \leqslant 0.99813.$$

Huber-hypothesis-based design of an I-*beam*

A simply supported steel I-beam of constant depth $2a$, with both flanges and web rectangular, is subjected to a uniformly distributed constant load p (Fig. 3.30). The flange width b is given and constant. The flange thickness c and the web thickness g are variable along the beam. The design problem consists in determining c and g so that the Huber strength function is equalized over the external edges of the flanges and over the web. It is also required that c and g are not less than certain allowable values

$$c \geqslant \underline{c}, \quad g \geqslant \underline{g}.$$

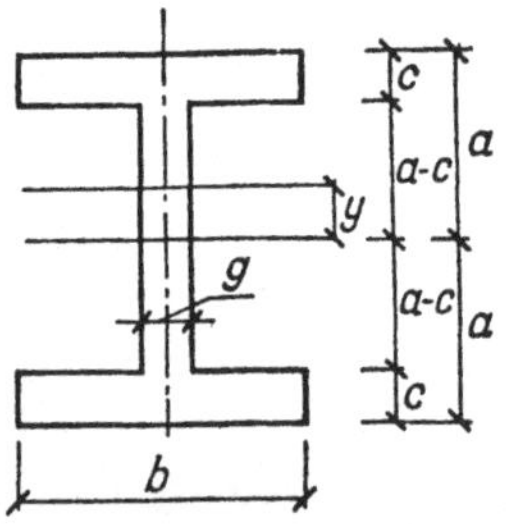

Fig. 3.30. An I-section schematic

With stresses expressed by the internal forces and the geometrical dimensions of the beam, the uniform-strength conditions for the problem read

$$\frac{Ma}{J} = \bar{\sigma},$$

$$\varphi_{(y)} = \frac{M^2y^2}{J^2} + 3\frac{T^2S^2}{J^2g^2} = \bar{\sigma}^2.$$

To determine the values of y that yield a maximum of the function φ in the web, we calculate the first and the second derivatives of this function:

$$\frac{d\varphi}{dy} = \frac{2y}{J}\left(M^2 - 3T^2\frac{S}{g}\right),$$

$$\frac{d^2\varphi}{dy^2} = \frac{2}{J}\left(M^2 - 3T^2\frac{S}{g} + 3T^2y^2\right).$$

The necessary condition for a maximum of φ will be satisfied when

$$y = 0,$$

or

$$M^2 - 3T^2\frac{S}{g} = 0.$$

As for the sufficient condition, when $y = 0$, the inequality $\varphi'' < 0$ implies

$$M^2 - 3T^2\frac{S}{g} < 0,$$

and when $M^2 - 3T^2\frac{S}{g} = 0$, $\varphi'' < 0$ gives

$$3T^2y^2 < 0,$$

which is impossible.

It follows that the largest value of $\varphi(y)$ in the web can be attained either
— on the neutral axis, $y = 0$, or
— at the inner flange edge, $y = \pm(a-c)$.

The sets of equations from which to determine c and g take the form:
— near supports

$$\begin{aligned} &\frac{2}{3}b[a^3-(a-c)^3]+\frac{2}{3}g(a-c)^3 = \frac{Ma}{\bar{\sigma}}, \\ &g\left\{\frac{2}{3}b[a^3-(a-c)^3]+\frac{2}{3}g(a-c)^3\right\} \\ &\qquad = \frac{\sqrt{3}\,T}{\bar{\sigma}}\left[bc\left(a-\frac{c}{2}\right)+\frac{1}{2}g(a-c)^2\right], \end{aligned} \tag{3.26}$$

— in the intermediate region

$$\begin{aligned} &\frac{2}{3}b[a^3-(a-c)^3]+\frac{2}{3}g(a-c)^3 = \frac{Ma}{J}, \\ &3T^2\frac{b^2c^2}{g^2}\left(a-\frac{c}{2}\right)^2 = M^2[a^2-(a-c)^2]. \end{aligned} \tag{3.27}$$

At the boundaries between the two regions we have

$$\begin{aligned} &3\frac{T^2}{g_1^2}\left[bc_1\left(a-\frac{c_1}{2}\right)+\frac{1}{2}g_1(a-c_1)^2\right]^2 \\ &\qquad = 3\frac{T^2b^2c_2^2(a-c_2/2)^2}{g_2^2}+M^2(a-c_2)^2, \end{aligned}$$

where c_1 and g_1 are found from equations (3.26) and c_2 and g_2 from equations (3.27).

In the regions near supports where c obtained from equations (3.26) is less than $\underline{c}$, we take $c = \underline{c}$ and find g from the condition

$$\begin{aligned} &\frac{2}{3}(a-\underline{c})^3g^2+\frac{2}{3}b[a^3-(a-\underline{c})^3]g \\ &\qquad = \frac{\sqrt{3}\,T}{\bar{\sigma}}\left[b\underline{c}\left(a-\frac{c}{2}\right)+\frac{1}{2}g(a-\underline{c})^2\right]. \end{aligned} \tag{3.28}$$

In these regions, the Huber stress will be levelled up to $\bar{\sigma}^2$ on the neutral axis only and its values at the flanges will be smaller.

In the region near the centre of the beam where g determined from (3.27) turns out to be less than $\underline{g}$, we put $g = \underline{g}$; equations (3.26) then give c equal to

$$c = a-\sqrt{\frac{ba^3}{(b-\underline{g})}-\frac{3}{2}\,\frac{Ma}{\bar{\sigma}(b-\underline{g})}}, \tag{3.29}$$

In this case, the Huber stress is equalized to $\bar{\sigma}^2$ on the flange edges and has lower values throughout the web.

A numerical example has been calculated with the following data: beam span $l = 20$ m, depth $2a = 1.60$ m, width $b = 0.40$ m, load $p = 6.4$ kN/m, least allowable flange thickness $\underline{c} = 5$ mm, least allowable web thickness $\underline{g} = 5$ mm.

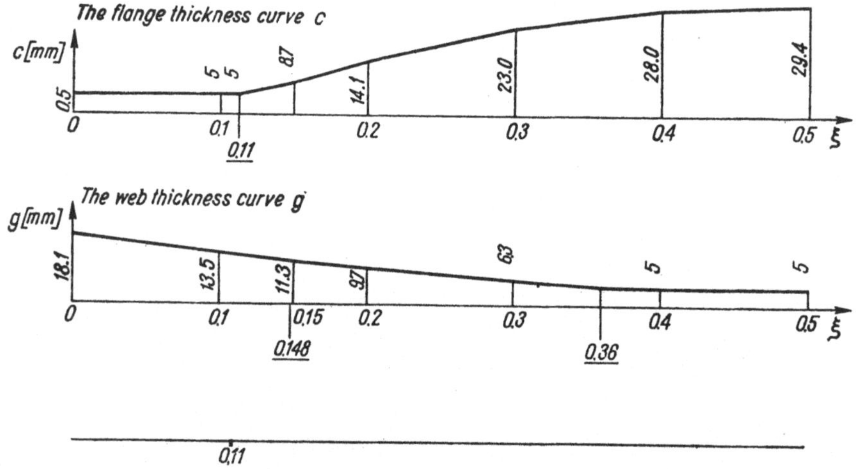

Fig. 3.31. The flange and web thickness curves; 0.11, 0.148 and 0.36 are boundary values between the regions of applicability of equations (3.26)–(3.29)

The flange and web thickness curves $c(x)$ and $g(x)$ are shown in Fig. 3.31. The regions of applicability of equations (3.26), (3.27), (3.28) and (3.29) are also indicated.

3.6 Uniform-strength design of plates

Optimization of prestressed plates under two loading states

The problem we consider in the present section is that of designing a thin prestressed plate of variable thickness $h(x_1, x_2)$ and arbitrary boundary $\Gamma(s)$ (Fig. 3.32) for uniform strength in two loading states (Dzieniszewski 1969_1). The plate is supported in an arbitrary manner but such that its boundary is free to move in the horizontal planes $x_3 = \text{const}$.

The prestressing forces with densities $S_i(x_1, x_2)$, $i = 1, 2$, act on the plate through a dense mesh of thin cables anchored in the material and appropriately curved in the vertical planes $x_i = \text{const}$. The prestressing cables form two sets k_i, $i = 1, 2$, crossing each other at right angles and distributed on surfaces $e_i(x_1, x_2)$ of small rise and small flexure.

For each family k_i, the cross-sectional area density $A_i(x_1, x_2)$, i.e. the cumulative cross-sectional area of the cables k_i per unit length in the direction perpendicular to the cables, is variable.

The plate is considered in two independent loading states $K = \mathrm{I, II}$. The first is a permanent loading state, i.e. one in which the plate sustains its dead weight and a steady useful load. In the second state the plate carries its dead weight and a maximum useful load.

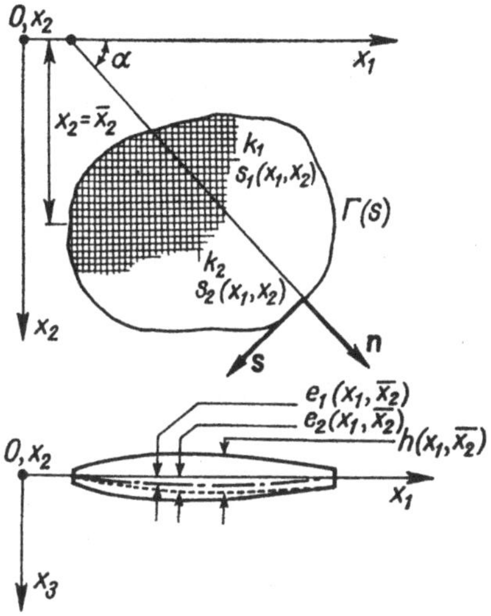

Fig. 3.32. Horizontal projection and cross-section of a prestressed plate

It is assumed that the useful load consists of vertical forces $p(x_1, x_2)$ applied to the surface of the plate, vertical forces $p(s)$ acting on the free boundary, and normal boundary moments $M(s)$.

After initial prestressing, the cable stresses σ_{k_i} are assumed to be the same in the loading states I and II and equal to a limiting allowable value σ_0. The actual changes in the cable stresses due to an increased deformation of the plate in the state II have little effect, in general, on the overall strain state of the plate and can be neglected.

The material of the plate will be treated as elastic but having different allowable compressive and tensile stresses $\bar{\sigma}$ and $\underline{\sigma}$.

The design (optimality) criterion

As a design criterion we choose the generalized Mohr hypothesis (Sec. 3.1). Assuming that the material is brittle (concrete), we put $\varrho = \nu$.

In the present problem, the generalized Mohr condition takes the form

$$F(x_1, x_2, x_3) = 2E\Psi^{(K+S)} - (\bar{\sigma}+\underline{\sigma})(\sigma_{11}+\sigma_{22})^{(K+S)} + \bar{\sigma}\underline{\sigma} = 0, \quad K = \mathrm{I, II}, \quad S = \text{initial prestressing}. \tag{3.30}$$

According to thin plate theory, we have

$$\sigma_{ij}^{(K+S)} = \frac{N_{ij}^{(K+S)}}{h} + \frac{12M_{ij}^{(K+S)}}{h^2} x_3, \quad i, j = 1, 2, \quad K = \text{I}, \text{II},$$

and

$$N_{ij}^{(K+S)} = N_{ij}^{(S)} = N_{ij}.$$

Using these relations in (3.30), we obtain the following expression

$$\begin{aligned}
&F^{(K+S)}(x_1, x_2, x_3) \\
&= \frac{1}{h^2} [(N_{11}+N_{22})^2+2(1+\nu)(N_{12}^2-N_{11}N_{22})]+ \\
&+\frac{144}{h^6} x_3^2\{(M_{11}^{(K+S)}+M_{22}^{(K+S)})^2+ \\
&+2(1+\nu)[(M_{12}^{(K+S)})^2-M_{11}^{(K+S)}M_{22}^{(K+S)}]\}- \\
&-(\bar{\sigma}+\underline{\sigma})\frac{1}{h}(N_{11}+N_{22})-x_3\left\{\frac{24(1+\nu)}{h^4}(N_{11}M_{22}^{(K+S)}+N_{22}M_{11}^{(K+S)})-\right. \\
&-48(1+\nu)\frac{1}{h^4}M_{12}^{(K+S)}N_{12}- \\
&\left.-\frac{12}{h^3}(M_{11}^{(K+S)}+M_{22}^{(K+S)})\left[\frac{2}{h}(N_{11}-N_{22})-(\bar{\sigma}+\underline{\sigma})\right]\right\}+\bar{\sigma}\underline{\sigma}.
\end{aligned} \tag{3.31}$$

For x_i = const ($i = 1, 2$), the above function takes its largest values when $x_3 = \pm h/2$, and hence the uniform-strength condition for our plate is

$$F^{(K+S)}\left(x_1, x_2, \pm\frac{h}{2}\right) = 0, \quad K = \text{I}, \text{II}. \tag{3.32}$$

If the internal forces due to prestressing are chosen to be

$$N_{ij}^{(S)} = \frac{(\bar{\sigma}+\underline{\sigma})}{2(1-\nu)} \delta_{ij} h(x_1, x_2), \quad i, j = 1, 2, \tag{3.33}$$

$$M_{ij}^{(S)} = -\tfrac{1}{2}(M_{ij}^{\text{I}}+M_{ij}^{\text{II}}), \tag{3.34}$$

where δ_{ij} is the Kronecker symbol, the bending moments in the plate will in each loading state be

$$M_{ij}^{(K+S)} = M_{ij}^{(S)}+M_{ij}^{(K)} = \frac{(-1)^K}{2} M_{ij}^{(\text{II}-\text{I})}, \quad K = \text{I}, \text{II}, \tag{3.35}$$

with

$$M_{ij}^{(S+\text{I})}+M_{ij}^{(S+\text{II})} = 0.$$

For the plate so prestressed, the left-hand side of equations (3.32) can be written

$$F^{(K+S)}\left(x_1, x_2, \pm\frac{h}{2}\right) = \frac{9}{h^4}\left[(M_{11}+M_{22})^2+2(1+\nu)(M_{12}^2-M_{11}M_{22})\right] - -\frac{1}{2}\frac{(\bar{\sigma}-\underline{\sigma})^2}{1-\nu}-\bar{\sigma}\underline{\sigma}, \tag{3.36}$$

where M_{ij} denotes $M_{ij}^{(S+I)} = -M_{ij}^{(S+II)}$. Equation (3.36) holds for arbitrary functions $h(x_1, x_2)$ and $\Gamma(s)$.

The choice of the prestressing forces N_{ij}^S in accordance with equation (3.33) also secures the identity

$$\sum_{i,j=1,2}\left\{\frac{1}{h}\left[\delta_{ij}\sum_{k=1,2}N_{kk}^{(S)}-(1-\nu)N_{ij}^{(S)}\right]\right\}_{,ij} = 0,$$

which is the condition of continuity of deformations of the middle surface of the plate (Dzieniszewski 1969_1).

By (3.36), the condition for our plate to be of uniform strength, (3.32), becomes

$$\frac{18(1-\nu)}{h^4}\left[(M_{11}+M_{22})^2+2(1+\nu)(M_{12}^2-M_{11}M_{22})\right] = \underline{\sigma}^2+\bar{\sigma}^2+2\nu\underline{\sigma}\bar{\sigma}. \tag{3.38}$$

Determining the prestressing forces

The prestressing forces $S_i(x_1, x_2)$ $(i = 1, 2)$ can be found from the equations of equilibrium of the forces acting in the middle surface of the plate

$$\sum_{j=1,2} N_{ij,j}^{(S)}+S_{i,i} = 0, \quad i = 1, 2,$$

with the boundary conditions on $\Gamma(s)$

$$(N_{11}^{(S)}+S_1)\cos\alpha+N_{12}^{(S)}\sin\alpha = 0,$$

$$(N_{22}^{(S)}+S_2)\sin\alpha+N_{12}^{(S)}\cos\alpha = 0.$$

Using the equality

$$N_{12}^{(S)} = 0,$$

which follows from (3.34), in the equilibrium equations, we obtain the following solution:

$$S_i = -N_{ii}^{(S)}, \quad i = 1, 2,$$

or

$$S_i = -\frac{\bar{\sigma}+\underline{\sigma}}{2(1-\nu)}h(x_1, x_2), \quad i = 1, 2. \tag{3.39}$$

Determining the shape of the prestressing cables

The cross-sectional area densities $A_i(x_1, x_2)$ of the prestressing cables k_i, $i = 1, 2$, can be calculated as

$$A_i = \frac{1}{\sigma_0} S_i(x_1, x_2) = -\frac{\bar{\sigma}+\underline{\sigma}}{2(1-\nu)\sigma_0} h(x_1, x_2). \tag{3.40}$$

Hence the reinforcement ratio μ is given by

$$\mu = \frac{V_c}{V_p} \frac{\sum_{i=1,2} \iint_{\Gamma} A_i \mathrm{d}x_1 \mathrm{d}x_2}{V_p} = -\frac{\bar{\sigma}+\underline{\sigma}}{(1-\nu)\sigma_0}. \tag{3.41}$$

To derive the equations of the cable axes, we use the conditions of equilibrium for the moments in the prestressed plate

$$\sum_{i,j=1,2} M_{ij.ij}^{(K+S)} + (N_{ij}+\delta_{ij}S_i) w_{.ij}^{(K+S)} = -p^{(K)} - \sum_{i=1,2} (S_i e_i)_{,ii}, \qquad K = \mathrm{I}, \mathrm{II},$$

Adding these equations with $K = \mathrm{I}$ and $K = \mathrm{II}$ and using formulae (3.35) and (3.36), we find

$$\sum_{i=1,2} (S_i e_i)_{,ii} = -\tfrac{1}{2} p^{(K)}, \qquad K = \mathrm{I}, \mathrm{II}.$$

By (3.39) this can be written

$$\sum_{i=1,2} (h_i e_i)_{,ii} = \frac{1-\nu}{\bar{\sigma}+\underline{\sigma}} p^{(K)}, \qquad K = \mathrm{I}, \mathrm{II}, \tag{3.42}$$

which is the required equation of the prestressing cables.

To find the appropriate boundary conditions, note that two of the following must hold on the boundary of the prestressed plate, according to the way the plate is supported (Dzieniszewski 1964):

(a) $w^{(K+S)} = 0$,

(b) $w_{.n}^{(K+S)} = 0$,

(c) $M_n^{(K+S)} + S_1 e_1 \cos^2\alpha + S_2 e_2 \sin^2\alpha = M^{(K)}$,

(d) $Q_n^{(K+S)} + M_{ns.n}^{(K+S)} + S_1 e_{1,1} \cos\alpha + S_2 e_{2,2} \sin\alpha +$
$\quad + \tfrac{1}{2}[\sin 2\alpha (S_2 e_2 - S_1 e_1)]_{,s} = P^K, \qquad K = \mathrm{I}, \mathrm{II};$

conditions (a) and (b) apply to a clamped plate, (a) and (c) to a simply supported plate and (c) and (d) to the case where the boundary Γ is free. $Q_n^{(K+S)}$ denotes the Kirchhoff substitute shear force given by

$$Q_n^{(K+S)} = (S_{1,1} e_1 + M_{11.1}^{(K+S)} + M_{12.2}^{(K+S)}) \cos\alpha + \\ + (S_{2,2} e_2 + M_{22.2}^{(K+S)} + M_{12.1}^{(K+S)}) \sin\alpha.$$

Using equations (3.35), (3.36) and (3.39), we find that the boundary conditions for equation (3.42) are

$$h(e_1\cos^2\alpha+e_2\sin^2\alpha) = -\frac{1-\nu}{\bar{\sigma}+\underline{\sigma}}\sum_{K=\mathrm{I,II}} M^{(K)},$$

$$(he_1)_{,1}\cos\alpha+(he_2)_{,2}\sin\alpha+\tfrac{1}{2}[(e_2-e_1)h\sin 2\alpha]_{,s} = -\frac{1-\nu}{\bar{\sigma}+\underline{\sigma}}\sum_{K=\mathrm{I,II}} P^{(K)}$$

when Γ is free, and

$$h(s)(e_1\cos^2\alpha+e_2\sin^2\alpha) = -\frac{1-\nu}{\bar{\sigma}+\underline{\sigma}}\sum_{K=\mathrm{I,II}} M^{(K)}$$

when it is simply supported. In the case where the boundary of the plate is fixed, the condition

$$\sum_{K=\mathrm{I,II}} w^{(K+S)} = 0$$

implies that the positions e_i of the cables k_i at the boundary can be prescribed arbitrarily.

Formulae (3.40) and (3.41) and equation (3.42) with the appropriate boundary conditions make it possible, in principle, to determine the shape of the prestressing cables in a plate of uniform strength under two loading states, given the relative position e_1-e_2 of the cables. If we assume, for example, that

$$(S_1e_1)_{,22} = (S_2e_2)_{,11}\varkappa,$$

then the prestressing moments S_ie_i $(i = 1,2)$ in the plate will be identical with the bending moments M_{ii} $(i = 1,2)$ in a dense beam-grid subjected to the load that the plate would carry in the state $\frac{1}{2}(\mathrm{I}+\mathrm{II})$ and supported in the same way. Flexural rigidities $B_i = B(1+(\varkappa-1)\delta_{2i})$ of a beam-grid are constant.

The relative position of the cables k_i can also be a subject of optimization; this problem has been investigated, e.g., by Dzieniszewski (1968, 1969_1).

Determining the shape of the plate

To determine the shape of a prestressed plate of uniform strength we use condition (3.38) and the equations of equilibrium

$$\sum_{i,j=1,2} M_{ij,ij} = -(p^{\mathrm{II}}-p^{\mathrm{I}}).$$

Substituting the bending moments given by

$$M_{ij} = -\frac{Eh^3}{12(1-\nu^2)}[\nu\Delta w\delta_{ij}+(1-\nu)w_{,ij}], \quad i,j = 1,2,$$

just as for a non-prestressed plate under the loads $p^{\mathrm{II}}(x_1, x_2)-p^{\mathrm{I}}(x_1, x_2)$, $p^{\mathrm{II}}(s)-p^{\mathrm{I}}(s)$ and $M^{\mathrm{II}}(s)-M^{\mathrm{I}}(s)$, we obtain the set of equations

$$\begin{aligned} &\Delta(h^3\Delta w)+(1+\nu)L(h^3, w) = \frac{12(1-\nu^2)}{E}\, p^{(\mathrm{II}-\mathrm{I})}, \\ &h^2[(\Delta w)^2+(1-\nu)L(w, w)] = \frac{8(1+\nu)}{E^2}\, [(\bar{\sigma}-\underline{\sigma})^2+2(1+\nu)\bar{\sigma}\underline{\sigma}], \end{aligned} \tag{3.43}$$

where the symbols Δ and L denote the operators

$$\Delta(\;) = (\;)_{,11}+(\;)_{,22},$$
$$L(\psi, \Phi) = 2\psi_{,12}\Phi_{,12}-\psi_{,11}\Phi_{,22}-\psi_{,22}\Phi_{,11}.$$

Equations (3.43), supplemented with two boundary conditions to be satisfied on Γ, chosen from the conditions

$$w = 0,$$
$$w_{,n} = 0,$$
$$M_n = -D\Delta w+(1-\nu)D(w_{,22}\cos^2\alpha+w_{,11}\sin^2\alpha-w_{,12}\sin 2\alpha) = M^{(\mathrm{II}-\mathrm{I})},$$
$$Q = Q_n+M_{ns,s} = (1-\nu)[D_{,1}(w_{,2})_{,s}-D_{,2}(w_{,1})]-(D\Delta w)_{,n} = P^{(\mathrm{II})}-P^{(\mathrm{I})},$$

according to the way the plate is supported, provide a basis for determining the function $h(x_1, x_2)$ and the deflection

$$w^{(K+S)} = \frac{(-1)^K}{2}\, w$$

in each of the two specified loading states.

In can be observed that the thickness $h(x_1, x_2)$ of a prestressed concrete plate of uniform strength is proportional to that of a similar plate but non-prestressed and shaped for minimum elastic strain energy with the dead weight γh neglected. This conclusion can be reached directly by comparing equations (3.43) with the corresponding equations for plates of maximum rigidity (Dzieniszewski 1968_1).

Concluding remarks

As we have seen, by prestressing a concrete plate it is possible to ensure that the Mohr strength function will take the same values in two different loading states, irrespective of the shape of the plate and the way it is supported. The strength can be made uniform over the middle surface of the plate but below the allowable value.

Increasing the strength function to a limiting allowable value at each point of the plate and in both loading states is impossible.

The assumption of variable thickness allows a design in which the strength will be equal to the limiting allowable value at each point of the external surfaces of the prestressed plate in two loading states.

Plates made of materials with equal tensile and compressive strength do not require prestressing. In this case, the condition of uniform strength on the faces of the plate in two loading states may only be satisfied when the two loads are antisymmetric with respect to each other.

In the general case, a concrete plate of equal strength in two loading states will be furnished with a uniform ($S_i = S$), two-way prestressing of two sets of curved cables. The prestressing forces and the cross-sectional area density of the cables will be proportional to the plate thickness.

A load equal to the arithmetic mean of the loads acting in thetwo specified states produces a membrane deformation of the plate. In this state, the strength is equalized at all points of the plate below the allowable value.

Uniform-strength design of simply supported three-layer plates

A solution to this problem was given by Dzieniszewski (1969_2); here we present the basic assumptions and results of that work.

The plate under consideration consists of an internal layer—a filler, which transmits vertical shearing forces, and two supporting external layers (Fig. 3.33). The plate is assumed to be in the elastic range and to comply

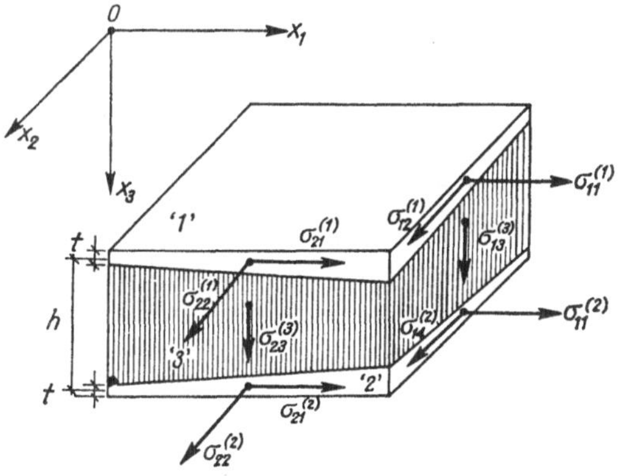

Fig. 3.33. A three-layer plate element

with the theory of thin plates. The design aims at equalizing the strength throughout the external layers in the sense of the Beltrami condition

$$(\sigma_{11}+\sigma_{22})^2+2(1+\nu)(\sigma_{12}^2-\sigma_{11}\sigma_{22}) = \bar{\sigma}^2.$$

With the stresses σ_{11}, σ_{22} and σ_{12} expressed by the deflections of the middle surface, the above condition takes the form

$$\frac{Eh^3}{12(1-\nu^2)}\left\{\frac{1-5\nu}{1-\nu}(w_{,11}+w_{,22})+6(1-\nu)[(w_{,12})^2-w_{,11}w_{,22}]\right\}=\bar{\sigma}^2. \tag{3.44}$$

A particular solution to the problem is obtained in which the rectangular plate region is subdivided into four triangular corner regions and a rhombic central region, with respect to the signs of the principal curvatures of the deflection surface. The deflection function $w(x_1, x_2)$ is found for the middle surface, satisfying equation (3.44) and appropriate boundary conditions. Next, the equations of equilibrium are used to determine the required thickness of the external layers. The considerations are illustrated by an example of design of a uniformly loaded square plate.

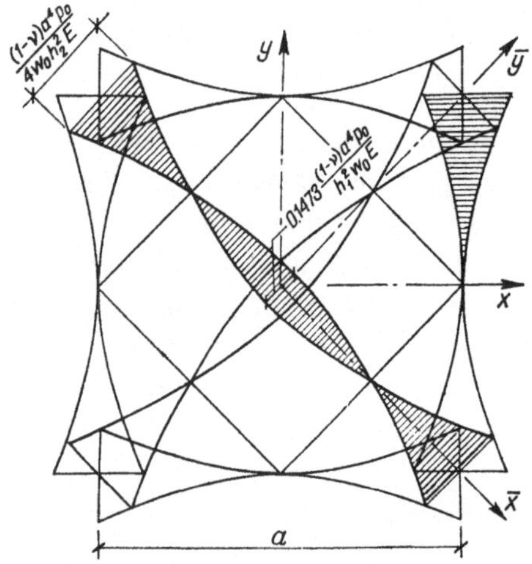

Fig. 3.34. Variation of the external layer thickness

Figure 3.34 shows the calculated variation of the external layer thickness. The indicated division of the plate into regions may be useful, e.g., in precasting of plates for building construction.

4

Optimization for Minimum Volume, Weight or Cost

4.1 The criterion of optimization

Joint treatment of the three criteria named in the title of this chapter is justified by the fact that they are very close and in many problems even identical in meaning.

According to the minimum-volume criterion, of all the structures satisfying given conditions and carrying the same loads the one that is made of the least volume of material is considered optimal. This approach reflects the natural tendency of the designer to save material, both the material directly used in the structure he designs and that required for other structures connected with it, for example the supports of a bridge span or the columns carrying a roof.

Sometimes the task of minimizing the volume of the material to be used in a structure can be reduced to an appropriate choice of the cross-sectional areas of its members.

The criterion of minimum weight is in many cases equivalent to that of minimum volume, but, unlike the latter, it can be used in the design of structures made of non-homogeneous materials, whose specific gravity varies with the coordinates or from member to member.

Underlying these two criteria is the simplified assumption that the cost of the structure is proportional to the volume or weight of the material used. According to this assumption, all the components of the total cost, whether they concern material or labour, should only depend on the quantity of the material used. In reality, this proportionality only holds for some of the components of the cost. For example, an extensive reduction in the volume of a beam may entail an excessive complication of its form, and hence a considerable rise of the cost of fabrication.

The minimum-cost criterion can be understood in various ways. Often optimization for minimum cost does not differ from minimum volume or minimum-weight design, because the cost is treated as proportional to volume or weight. But if the cost of a structure is made to include several different factors, the minimum-cost criterion leads to problems which go beyond the realm of mechanics into the sphere of economics. Two examples of how certain real cost factors are incorporated into an optimization problem are mentioned in Sections 4.2 and 4.6. Apart from that, in the present chapter we adopt the simplest assumption, namely that the volume, weight and cost of the structure to be optimized are proportional.

Besides its independent use as a design criterion, the condition of minimum volume or weight can be used to select the most economical structure of the admissible structures of uniform strength (cf. Chapter 3). The least-volume structure may correspond to a proper minimum of the volume function, but it may also be given by the limit values of the design variables, e.g. when the volume is a monotonic function.

For a fully-stressed truss of constant height and given configuration of nodes and bars, for example, the problem of determining the height so that the truss is of minimum volume has a unique, proper solution, because the volume of the truss increases unboundedly as the height either increases or decreases and the function expressing the volume has exactly one minimum.

As an example of a formulation of the minimum-volume problem for trusses let us present that of Hemp and Chan (1970). Consider a system of concentrated forces in static equilibrium applied at given points of space. Suppose that this system of forces is to be transmitted by a pin-jointed truss. Some of the nodes of the truss will be placed at the given external loading points and some will be prescribed by the conditions of use of the structure. The remaining nodes can be placed arbitrarily; the cross-sections of the bars connecting the nodes are also arbitrary. The problem is to determine the positions of the nodes and the bar cross-sections so that the total weight of the truss is minimum. The quantities involved in the problem are subject to two constraints:

equilibrium equations must hold at each node of the truss,

both compressive and tensile stresses in the bars of the truss must be safe.

Let us adopt the following notation:

t— row matrix of forces at the ends of the bars,
f— row matrix of the external forces acting on the truss (without support reactions),
K— matrix representing the geometry of the structure,
a— row matrix of bar cross-sections,
l— column matrix of bar lengths,
$\pm\sigma$— permissible compressive and tensile stresses.

With the above notation, the condition of equilibrium of forces at the nodes can be written as

$$tK = f \tag{4.1}$$

and the stress constraints as

$$t \leqslant \sigma a, \qquad -t \leqslant \sigma a. \tag{4.2}$$

The problem can now be stated as follows: find l and a statisfying conditions (4.1) and (4.2) and such that

$$V = al \text{ is minimum.} \tag{4.3}$$

Introducing non-negative variables t' and t'' defined by

$$\begin{aligned} t_i' &= t_i, \quad t_i'' = 0 \quad \text{for} \quad t_i \geqslant 0, \\ t'' &= t_i, \quad t_i' = 0 \quad \text{for} \quad t_i \leqslant 0, \end{aligned}$$

so that

$$t_i = t_i' - t_i'',$$

we can write constraints (4.1) and (4.2) in the form

$$(t' - t'')K = f, \quad 0 \leqslant t' \leqslant \sigma a, \quad 0 \leqslant t'' \leqslant \sigma a.$$

We thus arrive at the following mathematical formulation of the problem:

Minimize

$$V = \frac{(t' + t'')l}{\sigma}$$

subject to

$$\begin{aligned} &(t' - t'')K = f, \\ &t', t'' \geqslant 0. \end{aligned}$$

The above set of conditions represents a typical linear-programming problem with non-negative variables. It can be solved by the simplex method or some of its variants.

It will be interesting to recall in this context the optimality criterion for trusses formulated by Michell in 1904:

A truss is optimal if it is fully stressed under given external forces (i.e. all its elements sustain the allowable stresses $\pm\sigma$) *and, furthermore, if it can be subjected to a virtual deformation such that the strains in its tension bars and struts are* $+\varepsilon$ *and* $-\varepsilon$*, respectively, where* ε *is a constant, and no strain numerically larger than* ε *is produced in any direction along which a bar might lie.*

4.2 Optimum elastic design

The basic feature of any optimum design problem for structures in the elastic range, irrespective of the criterion chosen, is the presence of stress, strain or displacement constraints which aim at preserving the elasticity of the material.

There can be various motives for adopting elastic approach in the optimum-design of structures. Until quite recently, one of the most common methods of design was to check the stresses produced in a given structure so as to keep them below certain permissible values. The permissible stresses are chosen differently for different materials but always within the elastic range. Also at present many structures, in view of their complicated form and difficulties arising in their static analysis or on account of the kind of load they are to transmit, are designed by the permissible-stress method.

Another argument for elastic design is the fact that nearly all building structures, and also most mechanical structures, remain within the range of elastic deformations while they last and are utilized. It can be considered natural, therefore, to adapt the design to the conditions and loadings appropriate to the normal modes of exploitation, the more so as imposing limits on stresses safeguards structures, to a certain extent at least, against other undesirable phenomena, such as material fatigue and ageing or strength loss due to wear or damage.

The above remarks apply especially to structures for which the decisive constraints are those concerning stiffness, for example deflection constraints in the case of hanging structures. In general, optimum plastic design of hanging structures is unjustified.

The following example of a formulation of an optimum-elastic-design problem concerns a prestressed-concrete beam structure.

The optimization criterion is that of minimum cost. The cost of the structure is defined as

$$K = \sum_{i=1}^{l} K_i^{(c)} V_i^{(c)} + \sum_{j=1}^{n} K_j^{(s)} B_j^{(s)}; \tag{4.4}$$

where $K_i^{(c)}$ and $K_j^{(s)}$ are the unit costs of the concrete and steel used, and $V_i^{(c)}$ and $B_j^{(s)}$ are the volume of concrete and the weight of steel in the structure. The coefficients $K_i^{(c)}$ are defined according to the cost of concreting under different conditions, e.g. in the precasting plant or on the building site. The coefficients $K_j^{(s)}$ depend on the kind of steel used, which may be, for example, ordinary reinforcing steel or prestressing steel.

The design variables are chosen so as to characterize the cross-sectional shapes of the beams and the arrangement of reinforcement.

The feasible design space is delimited by the constraints which follow from the requirements imposed on the beams; the constraints specify

limit stresses,
permissible displacements,
minimum and maximum sizes,
other structural conditions, e.g. that the upper face of a beam must be flat.

Problems formulated in this manner, with numerous equality or (more often) inequality constraints, lead to designs of direct practical value. In view of their complicated mathematical form, however, a solution can seldom be obtained analytically and an involved numerical treatment is usually necessary.

Examples of solutions of optimum-elastic design problems formulated as shown above can be found in many papers; one of the earliest is that of Haug and Kirmser (1967).

Problems formulated without inequality constraints or with constraints simplified so as to make analytical solution possible generally lead to idealized results, which are of limited use for practical design but can give indications regarding, for example, the effect of changes of different shape parameters on the volume of the structure.

4.3 Examples of optimum elastic design

Optimization of a three-bar truss

Optimization of a statically determinate truss of known configuration reduces to the problem of determining the cross-sectional areas of its bars for given internal forces. When the directions of the bars are unknown, one can also seek an optimum configuration of the truss.

Optimization of statically indeterminate trusses is a more complex problem.[1] In this case, the configuration and the stresses in the bars of the truss are assumed to be given. If we vary the cross-sections of the bars, the minimum volume solution will emerge as one in which a certain number of bars have zero cross-sections. This number is equal to the degree of statical indeterminacy. In other words, among the trusses with a given configuration and under a given loading, that one has the minimum volume which is statically determinate. For fully-stressed trusses this theorem was proved by Lévi as early as in 1873. In this way the problem is reduced to considering a finite number of variants of a given truss, which are obtained by removing redundant bars.

[1] For fundamentals of optimization of such structures see e.g. Rabinovich (1933).

The minimum-weight design problem for a truss can be formulated as follows: Determine the values of the design variables $A_1, A_2, \ldots, A_n$ representing the cross-sectional areas of the members of the truss so as to minimize the weight of the truss

$$W(A_i) = \sum_{i=1}^{n} \varrho_i L_i A_i,$$

where ϱ_i is the weight density and L_i the length of the ith bar. The solution is required to satisfy
geometrical constraints

$$\begin{aligned} A_{i\,\max}-A_i &\geqslant 0, \\ A_i-A_{i\,\min} &\geqslant 0, \end{aligned} \qquad i = 1, 2, \ldots, m, \tag{4.5}$$

stress constraints

$$A_i - \frac{F_{ik}}{\mu\sigma_{ik}} \geqslant 0, \qquad k = 1, 2, \ldots, l, \tag{4.6}$$

strain or displacement constraints

$$\begin{aligned} u_{jk}-u_{jk\,\min} &\geqslant 0, \quad & j = 1, 2, \ldots n, \\ u_{jk\,\max}-u_{jk} &\geqslant 0, \quad & k = 1, 2, \ldots, l. \end{aligned} \tag{4.7}$$

In the above relations
k denotes the loading state index,
i—bar index,
j—node index,
F_{ik}—the force in the ith bar due to the kth loading,
σ_{ik}—the corresponding allowable stress,
μ—the sign of the force F_{ik},
u_{jk}—the displacement of the jth node in the kth loading state,
$u_{jk\,\min}$ and $u_{jk\,\max}$—the allowable displacements of the jth node in the kth loading state.

In addition, the following must be satisfied:
equilibrium and compatibility equations,
buckling constraints.

In practical design, the minimum-weight condition has often been replaced by the condition of equal strength, i.e. by the requirement that each bar of the truss should sustain a limiting allowable stress. This so-called *fully-stressed design* does not necessarily coincide with minimum-weight design. A simple example of optimization of a three-bar truss presented below will demonstrate the difference between the two approaches.

Optimum-design problems for trusses can be formulated in various ways, according to the goals to be accomplished. In the general case, optimization of a truss may aim at determining an optimal configuration of its members, the kind of the material to be used and its distribution. As an illustration of such a general problem let us discuss an example of optimization of a three-bar truss, given by Schmit and Mallet (1963).

We seek a planar, minimum-weight three-bar truss which will transmit a given load P, applied at node S at an angle α to the x-axis, to a support line r–r (Fig. 4.1). It is assumed that the mechanical properties of the material,

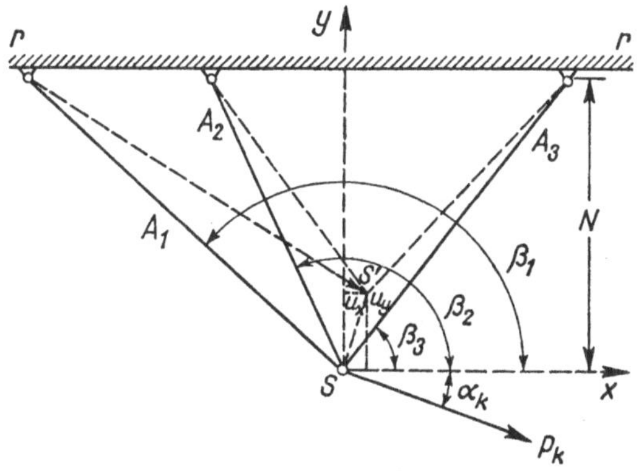

Fig. 4.1. Schematic of a three-bar plane truss

e.g. the modulus of elasticity, limit of plasticity, etc., are continuous functions of its weight density. In order to protect the structure against local buckling, the members of the truss are chosen to be tubular, with an appropriate diameter to wall thickness ratio.

The parameters given in the problem are

N—normal distance from point S to support line r–r,
d/t—mean diameter to wall thickness ratio of the members,
$E(\varrho)$—modulus of elasticity as a function of weight density,
$\sigma(\varrho)$—limit of plasticity,

and the design variables are

ϱ_p—weight density of member p ($p = 1, 2, 3$),
β_p—orientation angle of bar p,
A_p—cross-sectional area of bar p.

The bounds imposed on these variables are as follows:

$$
\begin{aligned}
&1.35 \leqslant \varrho_p \leqslant 8.85; \quad p = 1, 2, 3,\\
&\beta_2 \leqslant \beta_1 < \pi,\\
&\beta_3 \leqslant \beta_2 \leqslant \beta_1,\\
&0 \leqslant \beta_3 \leqslant \beta_2,\\
&0 < A_p \leqslant A_{p\,\max};
\end{aligned}
\tag{4.8}
$$

$A_{p\max}$ is the upper limit of the cross-sectional area of bar p. Constraints imposed on stresses and displacements are also of the form of inequalities. Tensile, compressive and buckling stresses derived from conditions of structural failure under various loadings should not exceed certain allowable values. Displacement of the node S under the load P should not be greater than a certain allowable displacement.

The weight W of the structure is a non-linear function of the nine design variables:

$$W(A_p, \beta_p, \varrho_p) = \sum_{p=1}^{3} \varrho_p \frac{N}{\sin\beta_p} A_p. \tag{4.9}$$

A detailed solution of this problem was given by Schmit and Mallet (1963). The procedure they used allowed the configuration, member sizing and the kind of material to be determined all at the same time. The results of the optimization of the truss are summarized in Table 4.1. Different cases are designated here with two figures, jk, of which the first is the basic case

Table 4.1 Results of optimization of a three-bar truss

Case	Material densities [kN/m³]			Cross-sectional areas [mm²]			Orientation angles [deg.]			Weight W
	ϱ_1	ϱ_2	ϱ_3	A_1	A_2	A_3	β_1	β_2	β_3	[N]
13	28.0	28.0	28.0	725.1	337.4	1038.6	135	90	45	40.23
23	45.7	45.7	45.7	483.8	12.9	761.9	135	90	45	41.17
33	78.3	78.3	78.3	358.7	9.7	580.6	135	90	45	53.25
43	28.0	28.0	28.0	607.0	489.6	1529.5	150	90	30	67.65
53	45.7	45.7	45.7	317.3	298.0	1183.1	150	90	30	76.56
63	78.3	78.3	78.3	247.1	214.8	912.8	150	90	30	100.87
73	28.0	28.0	28.0	1308.3	6.5	1365.0	120	90	60	43.93
83	45.7	45.7	45.7	845.7	6.5	858.0	120	90	60	45.81
93	78.3	78.3	78.3	443.8	6.5	563.1	120	90	60	46.53
16	40.7	29.1	32.1	494.8	136.8	936.0	135	90	45	38.10
	41.0	24.9	31.8	490.3	170.3	936.7	135	90	45	38.10
26	19.9	33.5	19.4	892.8	434.8	1819.2	150	90	30	61.38
	19.9	30.7	19.7	872.8	480.0	1816.0	150	90	30	61.39
36	35.7	15.2	26.0	888.9	6.5	1532.1	120	90	60	41.91
	38.5	15.5	24.1	779.9	7.7	1726.9	120	90	60	42.17
	41.8	25.2	34.3	560.6	9.7	939.9	125	93	51	35.78
09	41.0	26.6	34.9	570.3	6.5	912.8	123	103	29	35.80
	38.5	19.9	33.8	618.0	6.5	962.5	125	79	51	35.94

indicator and the second gives the number of design variables that have not been fixed in advance.

Cases 13 to 93 show the results obtained on the assumption that only the cross-sectional areas are continuous design variables; the configuration and material parameters are varied discretely. The optimum solutions corresponding to these nine combinations of three different pre-determined configurations and three different materials are summarized separately in Table 4.2.

Table 4.2 Optimum designs for nine combinations of material and configuration

Configuration / Material	A	B	C
	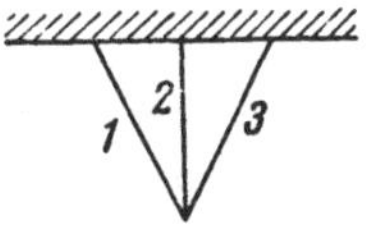	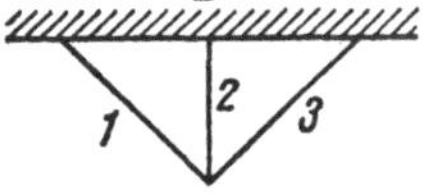	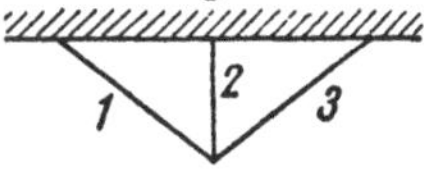
	$\beta_1 = 120°$, $\beta_2 = 90°$, $\beta_3 = 60°$	$\beta_1 = 135°$, $\beta_2 = 90°$, $\beta_3 = 45°$	$\beta_1 = 150°$, $\beta_2 = 90°$, $\beta_3 = 30°$
Aluminium	Case 73	Case 13	Case 43
$\varrho = 28.0$ kN/m^3	$A_1 = 1308.3$ mm^2	$A_1 = 725.1$ mm^2	$A_1 = 607.0$ mm^2
$E = 73.0$ GPa	$A_2 = 6.5$ mm^2	$A_2 = 337.4$ mm^2	$A_2 = 489.6$ mm^2
	$A_3 = 1365.0$ mm^2	$A_3 = 1038.6$ mm^2	$A_3 = 1529.5$ mm^2
$\sigma = 503$ MPa	$W = 43.93$ N	$W = 40.23$ N	$W = 67.65$ N
Titanium	Case 83	Case 23	Case 53
$\varrho = 45.76$ kN/m^3	$A_1 = 845.7$ mm^2	$A_1 = 483.8$ mm^2	$A_1 = 317.3$ mm^2
$E = 119.5$ GPa	$A_2 = 6.5$ mm^2	$A_2 = 12.9$ mm^2	$A_2 = 298.0$ mm^2
	$A_3 = 858.0$ mm^2	$A_3 = 761.9$ mm^2	$A_3 = 1183.1$ mm^2
$\sigma = 908$ MPa	$W = 45.81$ N	$W = 41.17$ N	$W = 76.56$ N
Steel	Case 93	Case 33	Case 63
$\varrho = 78.3$ kN/m^3	$A_1 = 443.8$ mm^2	$A_1 = 358.7$ mm^2	$A_1 = 247.1$ mm^2
$E = 205.0$ GPa	$A_2 = 6.5$ mm^2	$A_2 = 009.7$ mm^2	$A_2 = 214.8$ mm^2
	$A_3 = 563.1$ mm^2	$A_3 = 580.6$ mm^2	$A_3 = 912.8$ mm^2
$\sigma = 1230$ MPa	$W = 46.53$ N	$W = 53.25$ N	$W = 100.87$ N

It can be seen from Table 4.2 that the minimum-weight design is that of case 13, with the external bars inclined at 45° to the level and aluminium as the structural material.

It should be noted that the minimum-weight structure of case 13 is statically indeterminate and is not a fully-stressed structure in the sense that each bar is fully stressed in at least one loading state. That a minimum-weight structure need not be fully stressed had been pointed out earlier by Schmit (1960). The question of when a minimum-weight structure can be expected to be a fully-stressed structure was explored by Mayerjack (1962).

Cases 16, 26 and 36 show the results of optimization for a fixed configuration, with the cross-sectional areas and material densities treated as continuous design variables.

In case 09, configuration, cross-sectional areas and material parameters are all treate as dcontinuous design variables. The optimum structure obtained in this case has the least weight of all admissible structures. The material of the idealized optimum structure can be approximated by closely allied available materials.

Optimization of suspension cables

Suspension cables are main carrying members of hanging structures. Statical analysis of many cable structures can be reduced to analysis of single cables working independently of other cables in a given structure and subjected to identical or similar loads. For this reason, analysis of a single suspension cable, and particularly the question of optimum choice of shape parameters for a suspension cable subject to an arbitrary external loading, is of considerable importance. In considering this problem, we shall adopt the minimum-weight (or volume) criterion. Let us note that the horizontal component of the tensile force (thrust) in an optimal cable is the smallest possible, which is not without importance for the design of the supporting structures, e.g. supporting rings, columns and footings.

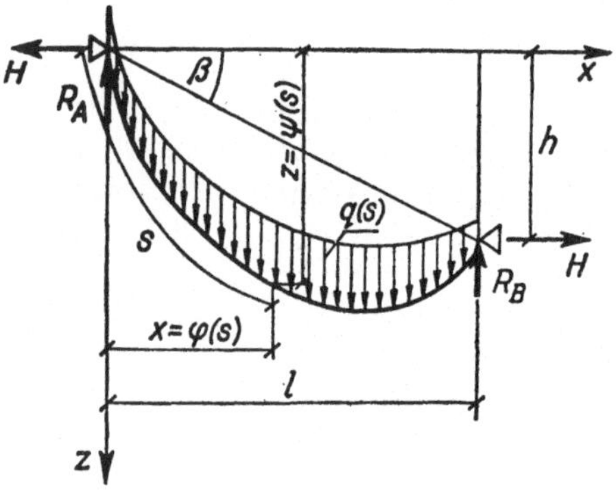

Fig. 4.2. Cable under a distributed load

The shape of a cable under an arbitrary loading can be found from equilibrium conditions. It is convenient to represent it in a parametric form, in terms of the arc length s of the cable (Fig. 4.2). The functions $x = \varphi(s)$ and $z = \psi(s)$ describing the shape of the cable are given by (Mazurkiewicz 1958)

$$x = \varphi(s) = \pm\int \frac{ds}{\sqrt{1+[F(s)]^2}} + D_1,$$
$$z = \psi(s) = \pm\int \frac{F(s)ds}{\sqrt{1+[F(s)]^2}} + D_2. \tag{4.10}$$

where $F(s) = \left(\int q(s)\mathrm{d}s + C\right)/H$; H denotes the thrust of the cable and s_1 its length.

Thus, the shape of the cable is determined by equilibrium conditions up to five arbitrary constants C, D_1, D_2, s_1 and H. The unknowns C, D_1, D_2 and s_1 can be calculated from the boundary conditions

$$\begin{aligned} x &= \varphi(0) = 0, \quad & z &= \psi(0) = 0, \\ x &= \varphi(s_1) = l, \quad & z &= \psi(s_1) = h. \end{aligned} \tag{4.11}$$

To find the remaining unknown, H, we use the criterion of optimization. In minimum-weight design, the condition which serves this purpose is the necessary condition for the existence of a minimum of the weight function W with respect to the variables H or ω:

$$\frac{\mathrm{d}W}{\mathrm{d}H} = 0 \quad \text{or} \quad \frac{\mathrm{d}\varrho}{\mathrm{d}\omega} = 0, \quad \text{where} \quad \omega = \frac{H}{ql}, \quad \varrho = \frac{W\bar{\sigma}}{\gamma q l^2}. \tag{4.12}$$

For cables with ends fixed at different levels and carrying arbitrary loads, it is impossible, in general, to determine optimal sags in a closed form, because the integrals occurring in formulae (4.10) cannot be expressed in terms of elementary functions. Whether a solution can be obtained by means of elementary functions depends on the form of the function $F(s)$. Even for loads varying linearly along the cable the integrals (4.10) are elliptic, while for more complicated loads they can only be calculated numerically. In the case of axisymmetric cable structures, with both shape and loading axially symmetric, integrals (4.10) can be expressed in terms of special functions (error functions). A detailed solution of this problem can be found in a paper by Jendo (1969).

The system of four equations (4.11) representing the boundary conditions can be written more explicitly as

$$\begin{aligned} &\int \frac{\mathrm{d}s}{\sqrt{1+[F(s)]^2}}\bigg|_{s=0} + D_1 = 0, \quad && \int \frac{\mathrm{d}s}{\sqrt{1+[F(s)]^2}}\bigg|_{s=s_1} + D_1 = l, \\ -&\int \frac{F(s)\mathrm{d}s}{\sqrt{1+[F(s)]^2}}\bigg|_{s=0} + D_2 = 0, \quad && -\int \frac{F(s)\mathrm{d}s}{\sqrt{1+[F(s)]^2}}\bigg|_{s=s_1} + D_2 = h. \end{aligned} \tag{4.13}$$

The weight of the cable is $W = \gamma s_1 A$; the cross-sectional area A is given by

$$A = \frac{N_{\max}}{\bar{\sigma}} = \frac{1}{\bar{\sigma}}\frac{H}{\cos\psi_{\max}} = \frac{H}{\bar{\sigma}}\sqrt{1+\tan^2\psi_{\max}} = \frac{H}{\bar{\sigma}}\sqrt{1+[F(s)]^2}, \tag{4.14}$$

and so

$$W = \frac{\gamma}{\bar{\sigma}} s_1 H \sqrt{1+[F(s)]^2}.$$

The system of five equations (4.12) and (4.13) can be reduced to three equations with the unknowns s_1, H and C. For an arbitrary external loading, the problem can be solved numerically.

In the case of a load distributed uniformly along the cable, if we perform the integration in equation (4.10) and make use of boun dary conditions $(4.11)_1$, we obtain the functions $\varphi(s)$ and $\psi(s)$ in the form

$$\begin{aligned} \varphi(s) &= \frac{H}{q}\operatorname{arcsinh}\left(u\sqrt{1+t^2}+t\sqrt{1+u^2}\right), \\ \psi(s) &= \frac{H}{q}\left(\sqrt{1+t^2}-\sqrt{1+u^2}\right), \end{aligned} \tag{4.15}$$

where

$$u = \frac{q}{H}s-t, \quad t = \frac{C}{H}. \tag{4.16}$$

The new unknowns u and t can be determined from the remaining two boundary conditions $(4.11)_2$. H can be found from the minimum-weight condition $(4.12)_2$.

To illustrate the above considerations, we present the solution of the problem for a cable with the ends fixed at the same level. Solving equations $(4.11)_2$ gives

$$u = t = \sinh\frac{1}{2\omega}, \qquad \omega = \frac{H}{ql}.$$

The length of the cable calculated from Eq. (4.16) is

$$s_1 = 2\frac{H}{q}\sinh\frac{1}{2\omega}. \tag{4.17}$$

The cross-sectional area of the cable can be expressed as

$$A = \frac{1}{\bar{\sigma}}\frac{H}{\cos\psi} = \frac{H}{\bar{\sigma}}\sqrt{1+(z')^2} = \frac{H}{\bar{\sigma}}\sqrt{1+t^2} = \frac{H}{\bar{\sigma}}\cosh\frac{1}{2\omega}.$$

Using these results, we find the weight of the cable to be

$$\varrho = \omega^2\sinh\frac{1}{\omega}. \tag{4.18}$$

The necessary condition for the function ϱ to have an extremum leads to a transcendental equation

$$2\omega\sinh\frac{1}{\omega}-\cosh\frac{1}{\omega} = 0. \tag{4.19}$$

The root of this equation is $\omega = 0.522153$. The optimum sag of the cable can be calculated from the formula

$$\varkappa = \omega\left(\cosh\frac{1}{2\omega}-1\right), \qquad \varkappa = \frac{f}{l}. \tag{4.20}$$

Substituting $\omega = 0.522153$, we get $\varkappa = 0.258244$.

The optimum shape of a cable suspended between equal-level supports under a uniformly distributed load is given by parametric equations (4.15). With φ and ψ replaced by dimensionless coordinates ξ and η, they finally take the form

$$\xi = \frac{1}{2}+\omega\operatorname{arcsinh}\left[(2\nu-1)\sinh\frac{1}{2\omega}\right],$$

$$\eta = \omega\left[\cosh\frac{1}{2\omega}-\sqrt{1+(2\nu-1)^2\sinh^2\frac{1}{2\omega}}\right].$$

The dimensionless parameter ν represents the distance measured along the cable axis, so that $s = \nu s_1$.

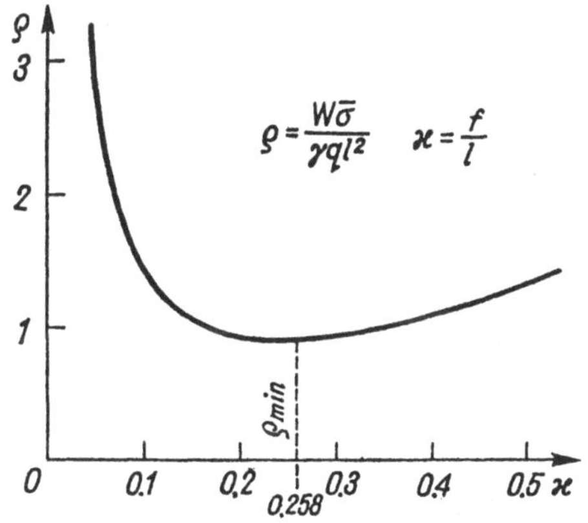

Fig. 4.3. The weight vs. sag curve

Since the minimum weight occurs at large sags, which may be undesirable, it is useful to know how the function ϱ changes with sag. The weight vs. sag curve plotted according to equation (4.18) is shown in Fig. 4.3. It can be seen from the graph that the sag has a considerable effect on the weight of the cable. The minimum weight occurs at $\varkappa = 0.258$.

Optimization of cables with sag, deflection and stress constraints

The foregoing considerations indicate that minimum-volume suspension cables have large sags. The curve shown in Fig. 4.3 illustrates the variation of the cable volume with the value of sag. One can calculate from it how much the volume will increase if the sag is reduced to comply with some additional requirements, e.g. constructional or architectural.

We shall formulate an optimum-design problem for a suspension cable subject to sag, deflection and stress constraints. We shall assume that the load $q(x)$ acts vertically, the material of the cable deforms within the elastic range and the cable is a small-sag cable. In view of this last assumption the cable will be treated as extensible and the load as distributed along the span of the cable rather that the cable itself.

Let $A = A(f, w, \sigma)$ be the cross-sectional area of the cable, expressed as a function of cable sag, deflection and stress. In the language of mathematical programming, the problem can be formulated as follows:

Minimize the objective function

$$A = A(f, w, \sigma), \tag{4.21}$$

subject to the constraints

$$\begin{aligned} &f \leqslant \bar{f}, \\ &w \leqslant \bar{w}, \\ &\sigma \leqslant \bar{\sigma} \end{aligned} \tag{4.22}$$

and the equations of equilibrium and deformation.

The allowable sag $\bar{f}$ and the allowable deflection $\bar{w}$ are determined by architectural, constructional, functional or other requirements.

Thus, for a given external load, span and material, we are to determine the optimal physical and geometrical parameters of the cable, yielding sufficient stiffness and strength at a minimum expenditure of material.

The fundamental equations obtained from the analysis of stress and strain in the cable under the external load $q(x)$ are

$$\begin{aligned} &H^3 + \frac{\mu EA}{l} H^2 = \frac{DEA}{2l}, \\ &H = \frac{M}{z_{10}+w}, \end{aligned} \tag{4.23}$$

where M and T are the bending moment and the shear force calculated as for a simply supported beam of length l, $D = \int T^2 \mathrm{d}x$ and $\mu = s_0 - l$ is the difference between the initial length of the cable and its span.

From equations (4.23), making the natural assumption

$$A = \frac{H}{\bar{\sigma}}, \tag{4.24}$$

we find

$$\begin{aligned} &H = \sqrt{\frac{DE}{2l(\bar{\sigma}+\mu E)}}, \\ &H = \frac{M}{z_{10}+w}. \end{aligned} \tag{4.25}$$

For a cable hanging freely under its own weight, assuming that its equilibrium shape is that of a quadratic parabola, we obtain

$$z_{10} = \frac{4f}{l^2}\, x(l-x), \qquad \mu = \frac{8f^2}{3l}. \tag{4.26}$$

Using relations (4.24), (4.25) and (4.26) in equation $(4.23)_1$, we find[1]

$$\sigma = \frac{DE}{2M^2 l}\left[\frac{4f}{l^2}\, x(l-x)+w\right]^2 - \frac{8f^2}{3l^2}\, E \tag{4.27}$$

or

$$w^2 + w\left[\frac{8fx(l-x)}{l^2}\right] + \left[\frac{16f^2x^2(l-x)^2}{l^4} - \frac{16f^2M^2}{3Dl} - \frac{2\sigma M^2 l}{DE}\right] = 0. \tag{4.28}$$

The optimum-design problem formulated at the beginning can now be stated as follows:

Minimize the objective function

$$A = \frac{M}{\left[\frac{4f}{l^2}\, x(l-x)+w\right]}\cdot\frac{1}{\bar{\sigma}} \tag{4.29}$$

subject to the constraints

$$\begin{aligned} &f \leqslant \bar{f},\\ &w \leqslant \bar{w},\\ &\frac{DE}{2M^2 l}\left[\frac{4f}{l^2}\, x(l-x)+w\right]^2 - \frac{8f^2}{3l^2}\, E \leqslant \bar{\sigma}. \end{aligned} \tag{4.30}$$

Geometrically, equation (4.29) represents a conicoid, a non-central hyperbolic cylinder (Fig. 4.4). The sought-for optimum parameters of the cable, i.e. the solution realizing the minimum of the objective function, lie on the boundary of the region cut out from this cylinder by the planes $(4.30)_{1,2}$ and the cylindrical surface $(4.30)_3$. There are several cases one can consider:

Case 1. Assume $f = \bar{f}$ and $w = \bar{w}$. If the stress σ calculated from (4.27) satisfies the additional condition $\sigma > 0$, the stress constraints become passive and the optimum parameters which minimize the objective function are $\bar{f}$, $\bar{w}$ and σ (Fig. 4.5a).

Case 2. Assume $f = \bar{f}$ and $\sigma = \bar{\sigma}$. If the deflection w calculated from formula (4.28) satisfies the condition $|w| \leqslant \bar{w}$, the minimum of the objective

[1] DMITRIEV, L. G. and KASILOV, K. W., *Cable Roofs*, Budivelnik, Kiev 1968 (in Russian).

function occurs at the point of intersection of the boundary planes $\bar{\sigma}$ and $\bar{f}$. In this case, $\bar{f}$ and $\bar{\sigma}$ are the optimum parameters (Fig. 4.5b). In a special case, the optimal design can be given by $\bar{f}$, $\bar{w}$ and $\bar{\sigma}$.

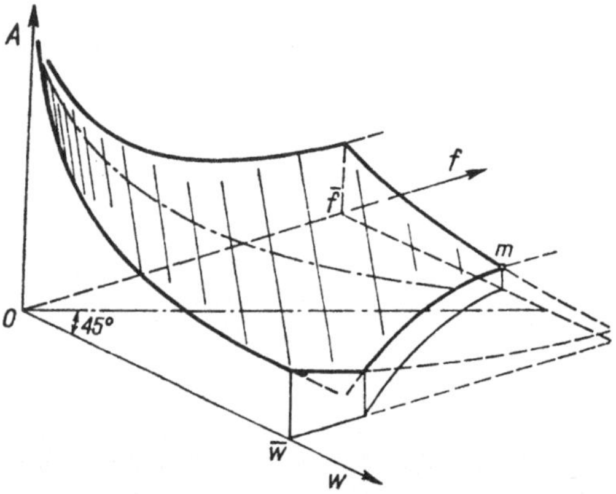

Fig. 4.4. Geometrical representation of equation (4.29) with sag and deflection constraints

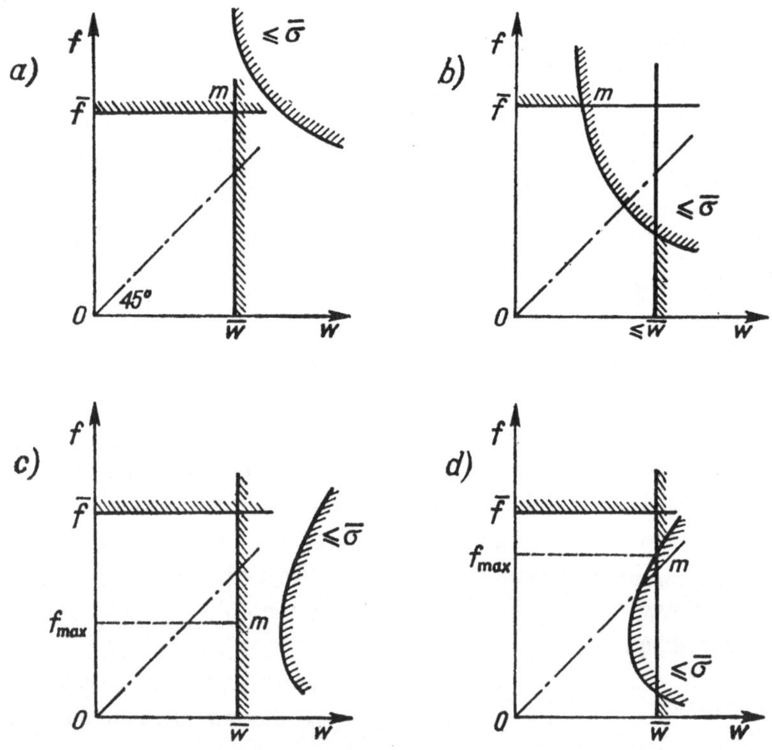

Fig. 4.5. Four cases of the feasible region

Case 3. Assume that the minimum of the objective function lies on the plane $w = \bar{w}$. We seek the value of f corresponding to this minimum. To this end, we differentiate the objective function with respect to f and set the result equal to zero, obtaining an equation for f_{max} at $w = \bar{w}$:

$$f^2\left[\frac{3Dx^2(l-x)^2}{M^2l^3}-1\right]+$$
$$+f\left[\frac{3Dx^2(l-x)^2}{M^2l^3}-\frac{1}{3}\right]\frac{l^2w}{2x(l-x)}+\frac{3Dlw^2}{16M^2}=0. \tag{4.31}$$

Knowing f_{max} and $w = \bar{w}$, we find the stress σ from formula (4.27). The following special cases can occur here:

(a) $f_{max} \leqslant \bar{f}$ and $0 < \sigma < \bar{\sigma}$; in this case the optimal design is given by f_{max}, $\bar{w}$ and σ (Fig. 4.5c).
(b) $f_{max} \leqslant \bar{f}$ but $\sigma < 0$; in this case there is no solution since cables cannot transmit compressive stresses.
(c) $f_{max} \leqslant \bar{f}$, $f_{max} = f_{min}$, $\sigma = \bar{\sigma}$; the optimal parameters are f_{max}, $\bar{w}$, $\bar{\sigma}$.
(d) $f_{max} \leqslant \bar{f}$, $f_{max} \neq f_{min}$, $\sigma = \bar{\sigma}$; as in the preceding case, the optimum design is given by f_{max}, $\bar{w}$ and $\bar{\sigma}$ (Fig. 4.5d).

Optimization of prestressed double-layer cable structures

Prestressed cable structures are principal elements of large-span suspension roofs. Usually, they consist of two systems of cables linked together (Fig. 4.6). The positive-curvature cables, convex downwards, are termed *carrying*, and those with negative curvature—*prestressing*. The links between the carrying and the prestressing cables may be tension rods or struts. Figure 4.6 shows three types of arrangement of carrying and prestressing cables used in two-layer cable structures. A description of the properties, structural

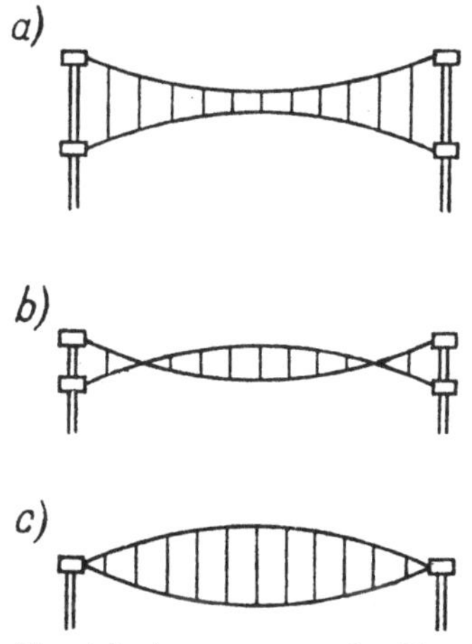

Fig. 4.6. Arrangement of cables in two-layer cable structures

and functional, of such structures can be found in a book by Rainus[1] or that by Jendo and Stachowicz (1974). Here, we are concerned with the problem of optimum choice of the shapes of the cables, their cross-sectional areas and initial prestressing so as to ensure the kinematic invariance of the structure and the least possible expenditure of material.

The following assumptions are made:

— the cables are perfectly flexible, i.e. incapable of transmitting bending moments or compressive and shear forces,
— the cross-sectional areas of the cables are constant along their length,
— the external loads act vertically and are continuously distributed over the span of the structure,
— the effect of the displacements of the supporting rings on cable forces is negligible,
— the stress-strain relationship is linear,
— all the cables have small sagging,
— the links between the two families of cables (tension rods or struts) are undeformable, i.e. displacements of the upper and the lower cables are identical,
— the links are spaced densely enough to be considered continuously distributed and allow a differential treatment of the stress and strain states,
— the links remain vertical after deformations of the structure.

The optimum-design problem for a double-layer cable structure can be stated as follows: Determine the shapes of the carrying and the prestressing cables, i.e. the functions $z_{11}(x)$ and $z_{21}(x)$,[2] the cross-sectional cable areas A_1 and A_2 and the initial cable prestressings H_{10} and H_{20}, for which the cost function defined as

$$V = k_1 V_1 + k_2 V_2 + k_3 V_3 \tag{4.32}$$

is minimum, where V_1, V_2 and V_3 denote the volumes of the carrying cables, the prestressing cables and the links, respectively, and k_1, k_2 and k_3 are coefficients depending on the unit costs of the materials used. In addition, it is required that the maximum stresses be equal to prescribed allowable values:

$$\sigma_{i\,\max} = \bar{\sigma}_i, \qquad i = 1, 2, 3$$

The objective function (4.32) is a functional of the cable shape functions z_{11} and z_{21} and the cable forces H_{ij}:

$$V = k_1 \int A_1 \mathrm{d}s_1 + k_2 \int A_2 \mathrm{d}s_2 + k_3 \int b(z_{11} - z_{21}) \mathrm{d}x .$$

---

[1] RAINUS, G. E., *Statical analysis of cable trusses* (in Russian), Leningrad 1962.

[2] The first subscript of z and H refers to the type of cable: 1—carrying, 2—prestressing. The second subscript refers to the loading state: 0—initial prestressing, 1—the first loading state, 2—the second loading state.

To within the first two terms of their power series expansions, the line elements ds_1 and ds_2 can be written

$$ds_1 = [1+\tfrac{1}{2}(z'_{11})^2]\,dx, \quad ds_2 = [1+\tfrac{1}{2}(z'_{21})^2]\,dx.$$

The cross-sectional areas A_1 and A_2 and the link layer thickness b are related to the internal forces by the formulae

$$A_1 = \frac{H_{11}}{\bar{\sigma}_1}, \quad A_2 = \frac{H_{20}}{\bar{\sigma}_2}, \quad b = \frac{N_r}{\bar{\sigma}_3} = \frac{H_{11}z''_{11}}{\bar{\sigma}_3}.$$

N_r denotes the link force per unit length of the structure. The functions z_{11} and z_{21} and the cable forces H_{ij} must satisfy equilibrium and deformation equations, which in the case of a single loading take the form

$$H_{10}z''_{10}+H_{20}z''_{20} = 0, \quad H_{11}z''_{11}+H_{21}z''_{21} = -p_1(x),$$

$$w = z_{11}-z_{10} = z_{21}-z_{20},$$

$$\frac{(H_{11}-H_{10})l}{E_1A_1} = \frac{1}{2}\int_0^l [(z'_{11})^2-(z'_{10})^2]dx,$$

$$\frac{(H_{21}-H_{20})l}{E_2A_2} = \frac{1}{2}\int_0^l [(z'_{21})^2-(z'_{20})^2]dx.$$

By a suitable transformation of these equations, it is possible to express the functions z_{10}, z_{20} and z_{21} in terms of the function z_{11} and the cable forces H_{ij}. Hence the functional V will involve only one unknown shape function, z_{11}, and the unknown cable forces H_{ij}; the same can be done with the deformation equations. In this way the optimum-design problem is reduced to an isoperimetric variation problem with the deformation equations as the isoperimetric conditions. The necessary conditions for the existence of an extremum lead to a system of differential and algebraic equations from which the desired shape function z_{11} and the cable forces H_{ij} can be found (Jendo, 1972).

For uniformly distributed loads the problem is considerably simplified, because the objective functional becomes a function of several variables subject to side constraints. Extremum of this function can be sought by the Lagrange multipliers method.

In the approach adopted by Jendo (1970) the problem is treated in two stages. In the first, the structure is optimized under a single load acting vertically downwards. In the second, the structure is additionally subjected to a vertical upward loading. This load, unlike the first, need not occur in reality.

In either case, both arbitrary and uniformly distributed loads have been considered. For arbitrary loads the procedure of solution has been shown and a basic system of equations derived, soluble once the load is specified.

For uniformly distributed loads, the solution has been carried through to the end; the optimal sags of the cables and the corresponding internal forces are calculated.

The results of optimization of a prestressed double-layer cable structure under a single loading distributed uniformly along the span are shown in Fig. 4.7. It can be seen that the minimum volume corresponds to the sag $\varkappa = f/l \simeq 0.20$. With respect to the initial prestressing characteristic p_0, the volume is a monotonically increasing function.

To deal with the case where the structure is subject to two loads as defined above, two methods of solution have been worked out: an exact method

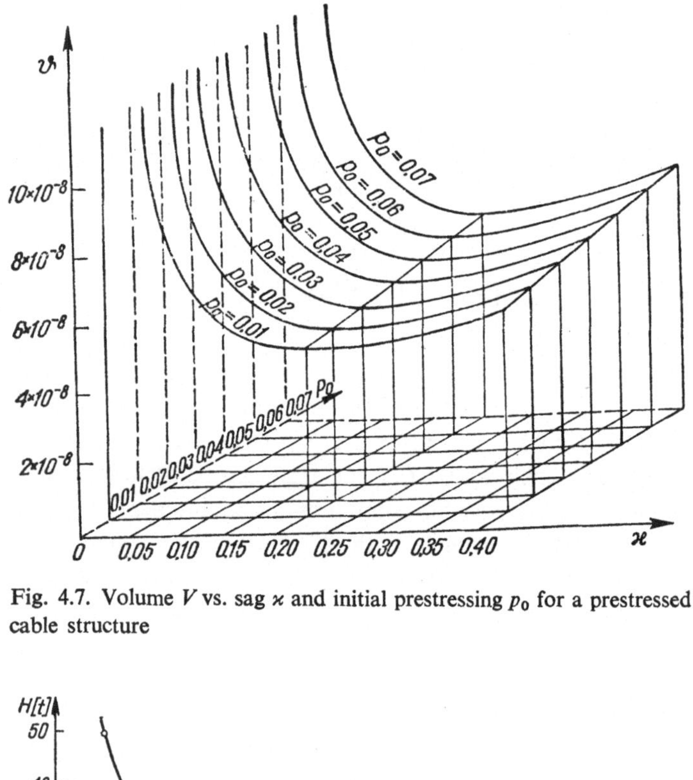

Fig. 4.7. Volume V vs. sag $\varkappa$ and initial prestressing p_0 for a prestressed double-layer cable structure

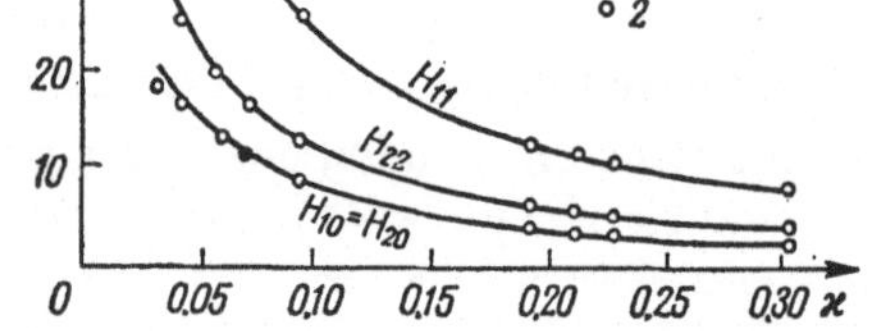

Fig. 4.8. Cable forces H_{ij} vs. sag $\varkappa$ of the carrying cables. 1—solution by the approximate method, 2—solution by the exact method

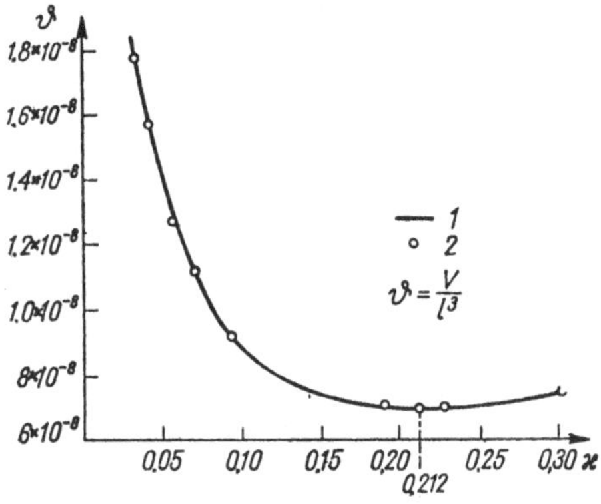

Fig. 4.9. Volume vs. sag curve in the second loading state. 1—solution by the approximate method, 2—solution by the exact method

and an approximate method. In the latter, the squares of the deflection derivatives are neglected in the deformation equations wherever they are compared with the products of these derivatives and the derivatives of the cable shape functions. A comparison of the results obtained by the two methods shows that the approximation is good enough to be applicable in practical design (see Figs 4.8 and 4.9). The following, more detailed discussion of the problem is based on the approximate method.

Equilibrium and deformation equations for different loading states

The deformation and the equilibrium equations for a two-layer cable structure under the different loading states described before are as follows:

In the initial prestressing state:

$$\begin{aligned} &H_{10}z''_{10} = -p_0(x), \qquad H_{20}z''_{20} = p_0(x), \\ &z_{10} = \frac{M_0(x)}{H_{10}}, \qquad z_{20} = -\frac{M_0(x)}{H_{20}}, \\ &z'_{10} = \frac{T_0(x)}{H_{10}}, \qquad z'_{20} = -\frac{T_0(x)}{H_{20}}. \end{aligned} \tag{4.33}$$

In the first loading state $p_1 = p_1(x)$:

$$\begin{aligned} &H_{11}z''_{11}+H_{21}z''_{21} = -p_1(x), \qquad w_1 = z_{11}-z_{10} = z_{21}-z_{20}, \\ &\frac{(H_{11}-H_{10})l}{E_1A_1} = \frac{1}{2}\int_0^l [(z'_{11})^2-(z'_{10})^2]\mathrm{d}x, \\ &\frac{(H_{21}-H_{20})l}{E_2A_2} = \frac{1}{2}\int_0^l [(z'_{21})^2-(z'_{20})^2]\mathrm{d}x. \end{aligned} \tag{4.34}$$

In the second loading state $p_3(x) = -p_1(x)+p_2(x)$:

$$H_{12}z''_{12}+H_{22}z''_{22} = -p_1(x)+p_2(x), \quad w_2 = z_{12}-z_{10} = z_{22}-z_{20},$$

$$\begin{aligned} \frac{(H_{12}-H_{10})l}{E_1A_1} &= \frac{1}{2}\int_0^l [(z'_{12})^2-(z'_{10})^2]\mathrm{d}x, \\ \frac{(H_{22}-H_{20})l}{E_2A_2} &= \frac{1}{2}\int_0^l [(z'_{22})^2-(z'_{20})^2]\mathrm{d}x. \end{aligned} \tag{4.35}$$

In addition, the extremum volume condition gives

$$H_{21} = 0, \quad H_{12} = 0. \tag{4.36}$$

The cross-sectional cable areas can be found from the formulae

$$A_1 = \frac{H_{11}}{\bar{\sigma}_1}, \quad A_2 = \frac{H_{22}}{\bar{\sigma}_2}. \tag{4.37}$$

Solution for arbitrary loads

The equations set out above form a system from which the internal forces, cable shapes, initial prestressing and displacements for the optimum structure may be derived. There are, in all, 14 equations in 15 unknowns: 4 equilibrium equations, 4 inextensibility conditions for links and 6 deformation equations for cables. The necessary condition for the existence of a minimum of the equivalent volume of material provides the 15th equation. The unknows are:

— the shape functions $z_{10}, z_{20}, z_{11}, z_{12}, z_{21}, z_{22}$ for different loading states,
— the cable forces $H_{10}, H_{20}, H_{11}, H_{12}, H_{21}, H_{22}$,
— the vertical deflection of cables w_1, w_2,
— the function $M_0 = \frac{l-x}{l}\int_0^l \xi p(\xi)\,\mathrm{d}\xi - \int_x^l (\xi-x)p(\xi)\mathrm{d}\xi$.

The fourteen equations listed above can be reduced to a system of four equations with the unknowns $M_0(x)$, H_{10}, H_{20}, H_{11} and H_{22}, and the optimization problem comes down to that of determining a minimum of the functional

$$\begin{aligned} V = \int_0^l \Bigg\{ & mH_{11}\left[1+\frac{1}{2}\left(\frac{T_0}{H_{10}}\right)^2\right]+mH_{22}\left[1+\frac{1}{2}\left(\frac{T_0}{H_{20}}\right)^2\right]- \\ & -k_3mn\left(\frac{1}{H_{10}}+\frac{1}{H_{20}}\right)p_3(x)M_0(x)\Bigg\}\mathrm{d}x \end{aligned} \tag{4.38}$$

subject to the four isoperimetric conditions

$$\frac{(H_{11}-H_{10})l}{E_1A_1} = \frac{1}{2}\int_0^l \left[\left(\frac{T_1}{H_{11}}\right)^2-\left(\frac{T_0}{H_{10}}\right)^2\right]\mathrm{d}x, \tag{4.39$_1$}$$

$$-\frac{H_{20}l}{E_2A_2} = \frac{1}{2}\int_0^l\left[\left(\frac{T_1}{H_{11}}\right)^2+\left(\frac{T_0}{H_{10}}\right)^2+\right.$$

$$\left.+\frac{2T_0^2}{H_{10}H_{20}}-2T_1T_0\frac{1}{H_{11}}\left(\frac{1}{H_{10}}+\frac{1}{H_{20}}\right)\right]\mathrm{d}x, \qquad (4.39)_2$$

$$-\frac{H_{10}l}{E_1A_1} = \frac{1}{2}\int_0^l\left[\left(\frac{T_2}{H_{22}}\right)^2+\left(\frac{T_0}{H_{20}}\right)^2+\right.$$

$$\left.+2\frac{T_0^2}{H_{10}H_{20}}-2T_2T_0\frac{1}{H_{22}}\left(\frac{1}{H_{10}}+\frac{1}{H_{20}}\right)\right]\mathrm{d}x, \qquad (4.39)_3$$

$$\frac{(H_{22}-H_{20})l}{E_2A_2} = \frac{1}{2}\int_0^l\left[\left(\frac{T_2}{H_{22}}\right)^2-\left(\frac{T_0}{H_{20}}\right)^2\right]\mathrm{d}x. \qquad (4.39)_4$$

The necessary condition for a constrained minimum of the functional V is the existence of an absolute minimum of the substitute functional

$$V^* = \int_0^l F^*\mathrm{d}x+f^* \qquad (4.40)$$

where

$F^* = F+\lambda_1F_1+\lambda_2F_2+\lambda_3F_3+\lambda_4F_4,$

$f^* = \lambda_1f_1+\lambda_2f_2+\lambda_3f_3+\lambda_4f_4,$

F —the integrand of functional (4.38),

F_1, F_2, F_3, F_4—the integrands in equations (4.39),

f_1, f_2, f_3, f_4 —the left-hand sides of equations (4.39),

$\lambda_1, \lambda_2, \lambda_3, \lambda_4$ —Lagrange multipliers.

The necessary conditions for a minimum of the functional V^* are

$$\frac{\partial F^*}{\partial M_0}-\frac{\mathrm{d}}{\mathrm{d}x}\left[\frac{\partial F^*}{\partial M_0'}\right]=0, \qquad \frac{\partial F^*}{\partial H_{10}}+\frac{\partial f^*}{\partial H_{10}}=0,$$

$$\frac{\partial F^*}{\partial H_{20}}+\frac{\partial f^*}{\partial H_{20}}=0, \qquad \frac{\partial F^*}{\partial H_{11}}+\frac{\partial f^*}{\partial H_{11}}=0, \qquad (4.41)$$

$$\frac{\partial F^*}{\partial H_{22}}+\frac{\partial f^*}{\partial H_{22}}=0.$$

Solving the nine equations (4.39) and (4.41) for $M_0(x)$, H_{10}, H_{20}, H_{11}, H_{22}, λ_1, λ_2, λ_3, and λ_4, we obtain all we need to calculate the desired optimal design.

Solution for uniformly distributed loads

When both the downward load p_1 and the upward load p_2 are constant over the span of the structure, the interaction between the two families of cables is also uniform: p_0 = const. The quantities M_0 and T_0 can now be calculated as

$$M_0(x) = \tfrac{1}{2}p_0(l-x)x, \quad T_0(x) = \tfrac{1}{2}p_0(l-2x),$$

and formulae (4.33) take the form

$$\begin{aligned} z_{10} &= \frac{1}{2}\frac{p_0}{H_{10}}(l-x)x, \quad & z_{20} &= -\frac{1}{2}\frac{p_0}{H_{20}}(l-x)x, \\ z_{10}' &= \frac{1}{2}\frac{p_0}{H_{10}}(l-2x), \quad & z_{20}' &= -\frac{1}{2}\frac{p_0}{H_{20}}(l-2x). \end{aligned} \tag{4.42}$$

The deflection w_1 can be found from equation $(4.34)_1$ after substituting $(4.34)_2$ and (4.36) into it. Double integration of equation $(4.34)_1$ subject to the boundary conditions $w_1(0) = w_1(l) = 0$ and $z_{10}(0) = z_{10}(l) = 0$ gives

$$w_1 = \frac{1}{2}\left(\frac{p_1}{H_{11}} - \frac{p_0}{H_{10}}\right)(l-x)x.$$

Similarly, we find the deflection w_2 from equation $(4.35)_1$, using the boundary conditions $w_2(0) = w_2(l) = 0$ and $z_{20}(0) = z_{20}(l) = 0$. The result is

$$w_2 = \frac{1}{2}\left(\frac{p_3}{H_{22}} + \frac{p_0}{H_{20}}\right)(l-x)x.$$

The volume of the structure is given by (assuming $k_1 = k_2 = 1$)

$$V = A_1\int ds_1 + A_2\int ds_2 + k_3\int b(z_{10}-z_{20})dx.$$

Substituting A_1 and A_2 from (4.37), assuming that

$$b = \frac{N_r}{-\sigma_3} = -\frac{1}{\bar{\sigma}_3}(-H_{22}z_{22}''-p_1) = -\frac{1}{\bar{\sigma}_3}(-p_3),$$

using (4.42) and, finally, integrating, we find

$$\begin{aligned} \vartheta = m\left[H_{11}\left(\delta^2 + \frac{1}{24}\frac{p_0^2}{H_{10}^2}\right) + H_{22}\left(\delta^2 + \frac{1}{24}\frac{p_0^2}{H_{20}^2}\right)\right] + \\ + \frac{1}{12}mnk_3p_3p_0\left(\frac{1}{H_{10}} + \frac{1}{H_{20}}\right). \end{aligned} \tag{4.43}$$

The deformation equations $(4.34)_2$ and $(4.35)_2$ take the form

$$\begin{aligned} &\frac{H_{11}-H_{10}}{H_{11}} - \mu_4K_1\left(\frac{p_0}{H_{10}} + \frac{1}{2}K_1\right) = 0, \\ &\frac{H_{20}}{H_{22}} + \mu_4K_1\left(-\frac{p_0}{H_{10}} + \frac{1}{2}K_1\right) = 0, \\ &\frac{H_{10}}{H_{11}} + \mu_4K_2\left(\frac{p_0}{H_{10}} + \frac{1}{2}K_2\right) = 0, \\ &\frac{H_{22}-H_{20}}{H_{22}} - \mu_4K_2\left(-\frac{p_0}{H_{20}} + \frac{1}{2}K_2\right) = 0, \end{aligned} \tag{4.44}$$

where

$$K_1 = \frac{p_1}{H_{11}} - \frac{p_0}{H_{10}}, \quad K_2 = \frac{p_3}{H_{22}} + \frac{p_0}{H_{20}}, \quad \vartheta = \frac{V}{l^3},$$

$$\mu_1 = \frac{\bar{\sigma}_1 l}{E_1} = \frac{\bar{\sigma}_2 l}{E_2}, \quad \mu_2 = \frac{\bar{\sigma}_1}{E_1} = \frac{\bar{\sigma}_2}{E_2}, \quad \mu_3 = \frac{l^3}{12},$$

$$\mu_4 = \frac{\mu_3}{\mu_2}, \quad m = \frac{1}{\bar{\sigma}_1} = \frac{1}{\bar{\sigma}_2}, \quad n = \frac{\bar{\sigma}_1}{\bar{\sigma}_3}, \quad \delta = \frac{1}{l}.$$

Thus, to determine the optimal structure, we have to find a constrained extremum of the function (4.43) subject to side conditions (4.44) on the variables H_{10}, H_{20}, H_{11}, H_{22} and p_0. By using the Lagrange multipliers method it is possible to reduce the problem to a system of 9 equations with the 9 unknowns H_{10}, H_{20}, H_{11}, H_{22}, p_0, λ_1, λ_2, λ_3 and λ_4.

The problem has been programmed for a GIER digital computer. The computer program includes a solving procedure for system (4.44) at a given p_0 and calculation of the volume function (4.43). A numerical example has been carried out with the following data:

$$\sigma_1 = 500 \text{ MPa}, \quad \sigma_3 = 140 \text{ MPa}, \quad E = 170 \text{ GPa},$$

$$l = 100 \text{ m}, \quad k_1 = k_2 = 1, \quad k_3 = 0.5, \quad p_1 = 2 \text{ kN/m},$$

$$p_2 = 3 \text{ kN/m}.$$

The results are shown in Figs 4.8 and 4.9 as cable force vs. sag and equivalent volume vs. sag curves. Detailed numerical data are given in the said paper by Jendo (1970).

The results of the analysis can be summarized in a number of general conclusions:

- The optimum shape of a two-layer cable structure depends on the distribution of the external loading rather than on its intensity;
- Under maximum external loading, the prestressing cable forces in the optimum structure vanish and the carrying and the prestressing cables follow funicular curves for the loading;
- In the case of large-span roofs, the bulk of the external load is usually uniformly distributed per unit horizontal area. Under such a load, the loads transmitted by individual layers in a parallel-layer arrangement are also uniform. The minimum-weight criterion then leads to square-parabolic shapes of the cables. For radial-layer structures, the loads transmitted by different layers follow a triangular distribution and the cables trace cubic parabolas;
- For a double-layer cable structure under a single uniformly distributed load, the minimum volume of material corresponds to the cable sag $\varkappa$

$\simeq 0.20$, regardless of what is the value of the initial prestressing (cf. Fig. 4.7);

— Double-layer cable structures are geometrically variable and may undergo large displacements. To reduce them, an initial prestressing should be applied— the greater, the greater is the useful load;

— The optimum sag of both carrying and prestressing cables under two vertical opposite loadings is $\varkappa = 0.212$ (cf. Fig. 4.8);

— For the internal forces in a double-layer cable structure under two vertical, opposite, uniform loadings p_1 and p_2 to be real and positive, the initial prestressing p_0 must be less than $(p_1p_2 - p_1^2)/p_2$.

Optimization of thin-walled bars subject to stability constraints

Optimization of annular thin-walled bars in compression and bending subject to local and global stability constraints is one of more interesting examples of minimum-volume design of structures. The problem has been investigated by many authors, among them by Krzyś (1967), Gajewski (1970) and Gajewski and Życzkowski (1969, 1970). Two examples of optimization of bars in bending will be discussed in Section 10.6. Here, let us briefly consider the minimum-volume problem for a bar compressed by an Eulerian force, as presented by Krzyś (1967).

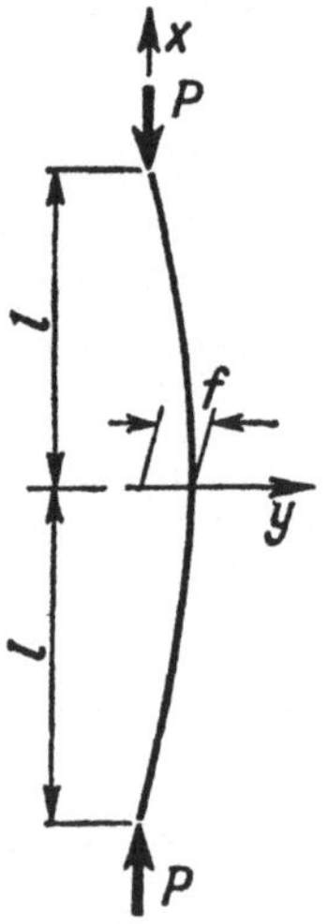

Fig. 4.10. Support and loading conditions for a bar

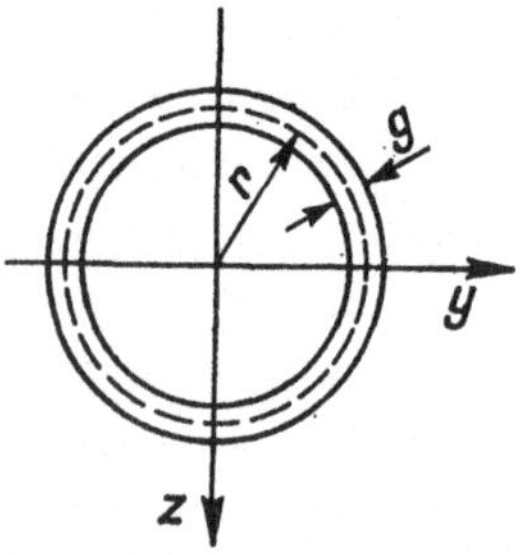

Fig. 4.11. Cross-section of the bar

Consider a bar with length $2l$ and annular cross-section, compressed by an axial force P (Fig. 4.10). The cross-section of the bar is characterized by two independent parameters: the inner radius r and the wall thickness g (Fig. 4.11). We assume that the bar is thin-walled, i.e. $g \ll r$.

The cross-sectional area and the moment of inertia are given by

$$A = 2\pi r g, \quad J = \pi r^3 g. \tag{4.45}$$

The problem is to find functions $r(x)$ and $g(x)$ such that

$$V = 2\int_0^l A\,\mathrm{d}x = \min$$

subject to the local stability constraint, as derived from shell theory,

$$\sigma = \frac{Pj_l}{2\pi r g} \leqslant \sigma_k = \alpha\frac{Eg}{r}, \tag{4.46}$$

where j_l is the local safety factor and $\alpha \sim 0.27$ a critical stress coefficient.

For a given moment of inertia J, the least cross-sectional area A will correspond to the minimum wall thickness g allowed by inequality (4.46). For

$$g = \sqrt{\frac{Pj_l}{2\pi\alpha E}},$$

formulae (4.45) give the following relationship between A and J:

$$A = \sqrt[3]{\frac{4\pi Pj_l}{\alpha E}J}. \tag{4.47}$$

From the deflection equation for the bar

$$EJy'' + Py = 0,$$

we find

$$J = -\frac{Pj_l}{E}\frac{y}{y''}, \tag{4.48}$$

where j_l denotes the safety factor relating to the buckling force. Combining formulae (4.47) and (4.48) gives

$$A = \left(\frac{4\pi P^2 j_l j_l}{\alpha E^2}\right)^{1/3}\left(-\frac{y}{y''}\right)^{1/3} = \beta\left(-\frac{y}{y''}\right)^{1/3}.$$

Thus, the bar volume can be written

$$V = 2\int_0^l A\,\mathrm{d}x = 2\int_0^l \beta\left(-\frac{y}{y''}\right)^{1/3}\mathrm{d}x.$$

Euler's necessary condition for this functional to have a minimum takes the form

$$\frac{\partial A}{\partial(y/y'')}\cdot\frac{1}{y''} - \left[\frac{\partial A}{\partial(y/y'')}\cdot\frac{y}{(y'')^2}\right]'' = 0.$$

Solution of this equation, satisfying the boundary conditions

$$y(0) = f, \quad y(l) = 0, \quad y'(0) = 0,$$

the stability loss being of form shown in Fig. 4.10, is given by

$$\xi = (1+\tfrac{1}{2}\sqrt{\eta})(1-\sqrt{\eta})^{1/2},$$

where

$$\xi = \frac{x}{l}, \quad \eta = \frac{y}{f}.$$

The radius of the minimum-volume bar is

$$r = \left(\frac{81\alpha P j_i^2 l^4}{128\pi E j_l}\right)^{1/6} \eta^{1/2}. \tag{4.49}$$

Note that the volume of the bar so shaped is equal to 0.8924 of the volume of an optimum prismatic bar.

Krzyś (1967) has also solved the minimum-volume bar problem with the constraint

$$\frac{Pn}{A} \leqslant k,$$

where k is the ultimate stress and n the safety factor. This restriction is particularly relevant in the regions near the supports (Fig. 4.10); in the middle part of the bar it will be satisfied with a large reserve.

The problem can be formulated as follows: Find the deflection curve $y = y(x)$ of class C^1 for which the functional

$$V = 2\int_0^l A\,\mathrm{d}x = 2\int_0^l \Phi(y, y'')\mathrm{d}x \tag{4.50}$$

attains a minimum, subject to the constraint

$$A = \Phi(y, y'') \geqslant \frac{Pn}{k} \tag{4.51}$$

and the boundary conditions

$$y(0) = f, \quad y'(0) = 0, \quad y'''(0) = 0, \quad y(l) = 0.$$

A general method for seeking extrema of a functional with constraints in the form of differential inequalities has not been found as yet. Krzyś's solution of the problem stated above is as follows. The extremals of functional (4.50) are given by a fourth-order differential equation, and in order to determine a particular solution we need four boundary conditions. At the point $A(x = 0)$, three boundary conditions are given which reduce the general solution of the equation to a one-parameter family of curves $y_1 = y_1(x, s)$.

It is possible to determine the region R_1 in which these curves satisfy condition (4.51) and the region R_2 in which they do not satisfy it (Fig. 4.12). At the boundary $s(x, y) = 0$ between R_1 and R_2 condition (4.51) holds as an

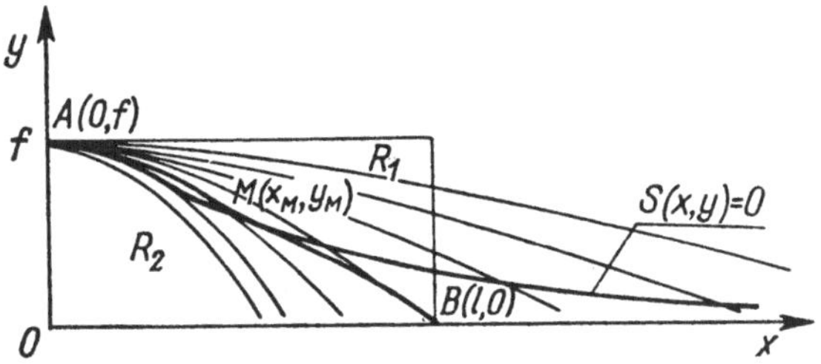

Fig. 4.12. Regions in which condition (4.51) is and is not satisfied

equality. Let $M(x_M, y_M)$ be an arbitrary point on the boundary curve $s(x, y) = 0$ and let us examine functional (4.50) along the path AMB:

$$V = 2\int_0^{x_M} \Phi(y, y'')\mathrm{d}x + 2\int_{x_M}^{l} \Phi(y, y'')\mathrm{d}x.$$

The first component is minimized by a curve of family y_1. The least value of the second component is realized by a curve satisfying the equation

$$\Phi(y, y'') = \frac{Pn}{k}. \tag{4.52}$$

The integral of this second-order differential equation subject to the boundary condition at the point $B(l, 0)$ represents a one-parameter family of curves $y_2 = y_2(x, t)$ passing through B. Thus the extremal AMB consists of two arcs: AM—of family y_1, and MB—of family y_2.

At the point M we have $y_1 = y_2$ and in virtue of equation (4.52) also $y_1'' = y_2''$, but in general the first derivatives will not be equal. However, the equality $y_1' = y_2'$ must hold at M if the extremal is to represent the deflection curve of the bar, and indeed it is possible to find a point M_z on the boundary curve at which this equality is satisfied.

The deflection curve of the bar obtained by this procedure is described by the equations

$$\xi = \frac{2}{3c_1}\eta_z^{-3/4}(2+\sqrt{\eta})(1-\sqrt{\eta})^{1/2} \quad \text{for} \quad \eta \geqslant \eta_z,$$

$$\eta = \eta_z\left(\frac{4}{\sqrt{\eta_z}}-3\right)^{1/2}\sin[c_1(1-\xi)] \quad \text{for} \quad \eta \leqslant \eta_z,$$

where

$$c_1 = l\left(\frac{4\pi j_i j_l k^3}{\alpha E^2 Pn^3}\right)^{1/2},$$

and η_z has to be calculated from

$$\tan\left[c_1 - \frac{2}{3}\left(\frac{2}{\sqrt{\eta_z}}+1\right)\left(\frac{1}{\sqrt{\eta_z}}-1\right)^{1/2}\right] = \frac{1}{2(1/\sqrt{\eta_z}-1)^{1/2}}.$$

The radius r of the least-volume bar is found to vary according to the formulae

$$r = \sqrt{\frac{PE\alpha n^2}{2\pi^2 kj_i}}\sqrt{\frac{\eta}{\eta_z}} \quad \text{for} \quad \eta \geqslant \eta_z,$$

$$r = \sqrt{\frac{PE\alpha n^2}{2k^2 j_i}} \quad \text{for} \quad \eta \leqslant \eta_z.$$

4.4 Optimum limit design

Criteria of optimization and design variables

Structures at limit states can be optimized for

(1) minimum volume of material at a constant load carrying capacity,
(2) maximum load carrying capacity at a constant volume of material,
(3) minimum volume at a constant load carrying capacity, subject to stability constraints,
(4) minimum deformability at constant volume and constant load carrying capacity.

The present discussion of optimum limit design will be based on the minimum-volume criterion.

As design variables one can choose

— geometrical dimensions, e.g. thickness of a plate, width of a beam or a column, cross-sectional area of a rod, etc.,
— material constants,
— distribution of forces.

To be able to formulate optimum-design problems for structures at limit states, it is necessary to know the basic equation governing such states.

The limit state

By a *limit state* we understand here a state corresponding to the beginning of unconstrained plastic flow.

Consider a structure made of a homogeneous and isotropic material and subjected to a load T_i^0. Assume that the structure is at a limit state, i.e. that the increasing forces $T_i = \mu T_i^0$ have already reached a limit value $T_i^c = \mu_c T_i^0$ and any further deformation will be a purely plastic flow under constant T_i^c. Theoretically, to provide a complete description of the limit state one should

(1) find the boundary between the rigid and the plastic regions, i.e. the slide curve or slide envelope;
(2) determine in the plastic region the stress field satisfying the equations of equilibrium, yield condition and boundary conditions;
(3) show that there is a continuation of the plastic stress field into the rigid region, satisfying the equations of equilibrium and boundary conditions and not exceeding the yield point;
(4) determine the displacement rate according to the boundary conditions and to the flow law associated with the yield condition;
(5) verify that the coefficient λ relating the strain rate to the state of stress is positive.

Conditions (1), (2) and (3) amount to satisfying the equilibrium equations (4.53) throughout the volume V of the body, keeping the stress below yield everywhere in V and satisfying the boundary conditions (4.55) on the surface S:

$$\sigma_{ij,j}+X_i = 0 \quad \text{in } V, \tag{4.53}$$

$$\Phi(\sigma_{ij}) \leqslant K \quad \text{in } V, \tag{4.54}$$

$$\sigma_{ij}n_j = p_i \quad \text{on } S. \tag{4.55}$$

Conditions (4) and (5) require the fulfilment of the geometrical relations (4.56), the flow law associated with the yield condition, (4.57), the incompressibility condition (4.58), inequality (4.59)—all these throughout V—and the boundary conditions for the displacement velocity on the surface S of the body (4.60):

$$\left.\begin{aligned} &\varepsilon_{ij} = \tfrac{1}{2}(u_{i,j}+u_{j,i}) && (4.56)\\ &\varepsilon_{ij} = \frac{\partial\Phi}{\partial\sigma_{ij}}\lambda && (4.57)\\ &\varepsilon_{ii} = 0 && (4.58)\\ &\lambda > 0 && (4.59)\end{aligned}\right\} \text{in } V$$

$$u_i = u_i(s) \quad \text{on } S \tag{4.60}$$

In the above, u_i, ε_{ij} and σ_{ij} denote the components of the displacement velocity vector, the strain rate tensor and the stress tensor, respectively, X_i are the body forces and Φ is the yield function.

Solutions satisfying conditions (4.53)–(4.60) are called *complete.*

Solutions satisfying conditions (4.53), (4.54) and (4.55) are called *statically admissible.*

Solutions satisfying conditions (4.54) and (4.56)–(4.60) are called *kinematically admissible.*

The notions of statically admissible and kinematically admissible fields are commonly used both in the limit analysis of structures and in optimum limit design. They permit the designer to estimate the load carrying capacity of a given structure and to set bounds on the minimum volume of that structure in cases where obtaining a complete solution is difficult.

In estimating the load carrying capacity from statically and kinematically admissible fields, the following two principles, known as the first and the second theorem of limit analysis, are observed:

(1) *The limit load calculated from a statically admissible field is smaller than the actual load carrying capacity of the structure.*

(2) *The limit load calculated from a kinematically admissible field is greater than the actual load carrying capacity of the structure.*

In all the above statements, stresses, strain and displacement velocities can be replaced by generalized stresses, generalized strains or generalized velocities.

Formulation of the minimum-volume limit design problem

In most of the papers published so far on the limit design of structures the criterion of optimization used has been that of minimum volume of material at a constant load carrying capacity. The outline given below of a general formulation of the problem is based on a paper by Mróz (1963).

Consider a structure at a limit state, bounded by a surface S. Let S consist of three parts (Fig. 4.13):

S_u—rigidly supported,

S_T—loaded with constant forces T_i,

S_T^0—free from external loads and kinematic constraints.

Let A be the set of all structures obtained from the given one by alterations of the free surface S_T^0 and having the same load carrying capacity as the original structure. The optimum-design problem can be stated as follows:

Find the least-volume structure in the set A.

The conditions for the existence of the least volume structure in A can be derived by the following considerations. Let V_1 and V_2 be two structures chosen from the set A (Fig 4.13a, b). Let the actual states of stress and strain rates in these structures, determined from Eqs. (4.53)–(4.60), be ε_{ij}^1 and σ_{ij}^1

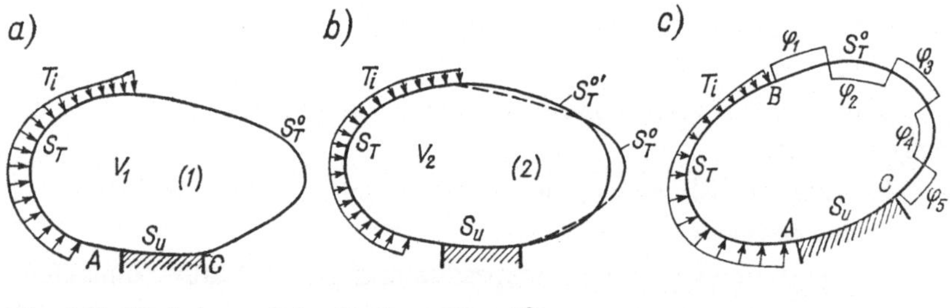

Fig. 4.13. Variations of the free boundary S_T^0

in V_1 and ε_{ij}^2 and σ_{ij}^2 in V_2. In addition, we construct a kinematically admissible field in each of the chosen bodies, namely ε_{ij}^2 in V_1 and ε_{ij}^1 in V_2. The kinematically admissible field ε_{ij}^2 for the body V_1 is obtained by continuing the actual velocity field of V_2 into V_1. In the region $\delta V = V_1 \setminus V_1 \cap V_2$, the field ε_{ij}^2 only satisfies the kinematic compatibility condition (4.56) and the incompressibility condition (4.60). Similarly, the kinematically admissible field ε_{ij}^1 for the body V_2 is constructed by continuing the actual velocity field of V_1 into V_2. And again, in the region $\delta V = V_2 \setminus V_1 \cap V_2$ it only satisfies the kinematic compatibility and the incompressibility conditions.

To sum up, the following fields are considered in the two structures:
in V_1: the actual field σ_{ij}^1, ε_{ij}^1; the kinematically admissible field ε_{ij}^2, u_i^2;
in V_2: the actual field σ_{ij}^2, ε_{ij}^2; the kinematically admissible field ε_{ij}^1, u_i^1.

By the principle of virtual power, for the actual field in V_1 we have

$$\int_{V_1} \sigma_{ij}^1 \varepsilon_{ij}^1 \mathrm{d}V = \int_{S_T} T_i u_i^1 \mathrm{d}S. \tag{4.61}$$

The same principle applied to the real field σ_{ij}^2 and the kinematically admissible field ε_{ij}^1, u_i^1 in V_2 gives

$$\int_{V_2} \sigma_{ij}^2 \varepsilon_{ij}^1 \mathrm{d}V = \int_{S_T} T_i u_i^1 \mathrm{d}S. \tag{4.62}$$

This can be written

$$\int_{V_2} [\sigma_{ij}^2 - \sigma_{ij}^1 + \sigma_{ij}^1]\, \varepsilon_{ij}^1 \mathrm{d}V = \int_{S_T} T_i u_i^1 \mathrm{d}S. \tag{4.63}$$

Subtracting equation (4.61) from equation (4.63), we obtain

$$\int_{\delta V} \varepsilon_{ij}^1 \sigma_{ij}^1 \mathrm{d}V = -\int_{V_2} (\sigma_{ij}^2 - \sigma_{ij}^1)\, \varepsilon_{ij}^1 \mathrm{d}V. \tag{4.64}$$

Similarly, one can derive the relationship

$$\int_{\delta V} \sigma_{ij}^2 \varepsilon_{ij}^2 \mathrm{d}V = -\int_{V_1} (\sigma_{ij}^1 - \sigma_{ij}^2)\, \varepsilon_{ij}^2 \mathrm{d}V.$$

Formula (4.64) leads to the following sufficient condition for the volume of the structure to be an absolute minimum:

If the energy dissipation rate $D(\varepsilon_{ij}^1) = \bar{D} = \text{const}$ *on* S_T^0, $D(\varepsilon_{ij}^1) \geqslant \bar{D}$ *within* V_1 *and* $D(\varepsilon_{ij}^1) \leqslant \bar{D}$ *outside* V_1, *then the volume* V_1 *is absolute minimum.*

Indeed, putting $D(\varepsilon_{ij}^1) = \bar{D} + \Delta D(\varepsilon_{ij}^1)$, we can write equation (4.64) in the form

$$\bar{D}\delta V = -\int_{V_2} (\sigma_{ij}^2 - \sigma_{ij}^1)\, \varepsilon_{ij}^1 \mathrm{d}V - \int_{\delta V} \Delta D \mathrm{d}V. \tag{4.65}$$

Since, owing to the convexity of the yield surface,

$$(\sigma_{ij}^2 - \sigma_{ij}^1)\,\varepsilon_{ij}^1 \leqslant 0,$$

and, by assumption,

$$\Delta D \geqslant 0 \quad \text{when} \quad \Delta V \leqslant 0 \text{ (inside the body)},$$
$$\Delta D \leqslant 0 \quad \text{when} \quad \Delta V \geqslant 0 \text{ (outside the body)},$$

both terms on the right-hand side of equation (4.65) are non-negative. It follows that $\delta V \geqslant 0$ for arbitrary modifications of the free boundary S_T^0, which means that the volume V_1 is absolute minimum.

A different situation arises when $D(\varepsilon_{ij}^1) \leqslant \bar{D}$ inside the structure and $D(\varepsilon_{ij}^1) \geqslant \bar{D}$ outside it. In this case the second term in equation (4.65) is negative and the sum can be positive or negative. If it is positive, the minimum exists, but if it is negative, there exists a local maximum. Whatever the case, however, we should seek a shape of S_T^0 on which the dissipation rate $D(\varepsilon_{ij}^1)$ would be constant. The necessary condition for a local extremum, $D = \text{const}$ on the free boundary, will always be satisfied if both terms in (4.65) are of the second order and, consequently, the first variation vanishes.

In the special case where $D(\varepsilon_{ij}^1) = \bar{D} = \text{const}$ throughout the region V_1, we have $\Delta D = 0$ and the right-hand side of Eq. (4.65) is reduced to the first term, which is always positive. Hence, in such cases, there always exists an absolute-minimum-volume design. It is so, for example, for circular plates, membranes and sandwich plates.

The condition for an extremum volume takes a somewhat different form when the body to be optimized is required to be symmetrical about a plane or a point. A modification of one of the symmetric parts S_{T1}^0 and S_{T2}^0 of the free surface entails a modification of the other part, even if the surface tractions T_i and the kinematically admissible velocity field u_i^1 do not follow the symmetry of the body. In this case the condition that $D(\varepsilon_{ij}^1)$ should be constant on S_T^0 has to be replaced by the condition

$$D^1(\varepsilon_{ij}^1) + D^2(\varepsilon_{ij}^2) = \bar{D} = \text{const}, \tag{4.66}$$

where D^1 and D^2 are the values of $D(\varepsilon_{ij}^1)$ at the symmetrical points of S_{T1}^0 and S_{T2}^0. Condition (4.66) has been applied to the optimum design of shells (see e.g. Freiberger 1956, Shield 1960).

Another case worth noting is that of parametric optimization, in which the free surface S_T^0 is assumed to be given by n known functions $\varphi_i(P)$. The design variables are the distances of the different segments of the boundary from a fixed reference curve (Fig. 4.13c). This approach makes it possible to pre-select the profiles φ_i so as to comply with any additional requirements, e.g. technological or functional. The necessary condition for the volume of the structure to be minimum is the equality of the mean dissipation rates on the subregions S^0_{Tk} of the free surface:

$$D_k = \frac{\int_{S^0_{Tk}} D(\varepsilon_{ij})ds}{S^0_{Tk}} = \text{const}.$$

Solution procedure in optimization for minimum volume at constant limit load

The procedure of minimum-volume limit design is illustrated in Section 4.5 on certain types of plates and frames, for which optimum thicknesses or cross-sectional shapes are sought.

Mathematically, the problem amounts to minimizing the volume V as a functional of a boundary function $\varphi(x, y, z)$ under a constant loading T_i on S_t, subject to the side conditions (4.53)–(4.60):

Minimize $V_m = V[\varphi(x, y, z)]$ subject to the constraints

$$\sigma_{ij,j}+x_i = 0, \quad \Phi(\sigma_{ij}) \leqslant K, \quad \varepsilon_{ij} = \tfrac{1}{2}(u_{i,j}+u_{j,i}),$$

$$\varepsilon_{ij} = \lambda \frac{\partial \Phi}{\partial \sigma_{ij}}, \quad \varepsilon_{ij} = 0, \quad \lambda > 0 \text{ in } V,$$

$$\sigma_{ij}n_j = p_i, \quad u_i = u_i(s) \text{ on } S, \; p_i \text{ fixed}.$$

The problem so formulated is very general and no method of its solution has yet been worked out which would apply to all kinds of structures and to arbitrary loads. Finding such a method is still a live issue.

The existing solutions have been developed on the basis of the sufficient conditions discussed above. The procedure ordinarily used can be summed up in the following five points:

(1) Assume the minimum condition: D = const on the boundary;
(2) Associate stress and strain rate fields with the design according to the loading and the kind of structure, the yield condition and the associated flow law;
(3) Verify the sufficient condition for the existence of a minimum;
(4) Determine the values of the design variables and displacement velocity components from the conditions of equilibrium and the minimum condition D = const;
(5) Check if the solution so obtained satisfies conditions (4.53)–(4.60).

Limit design of bar structures

In the limit analysis of bar structures in bending one usually neglects all forces except the bending moments. The yield condition is taken in the form $\Phi = M - M_0 = 0$ for $M > 0$ and $\Phi = M + \nu M_0 = 0$ for $M < 0$; M_0 and νM_0 are the positive and the negative limit bending moments. Hence a safe internal force field must satisfy at each point of the bar axis the inequality

$$-\nu M_0 \leqslant M \leqslant M_0 .$$

The magnitude of the limiting moments depends on the yield point of the material and on the form of the cross-section. For homogeneous cross-sections, $\nu = 1$; the limiting moments for a few homogeneous cases are written out in Fig. 4.14. In non-homogeneous cross-sections $0 < \nu < \infty$.

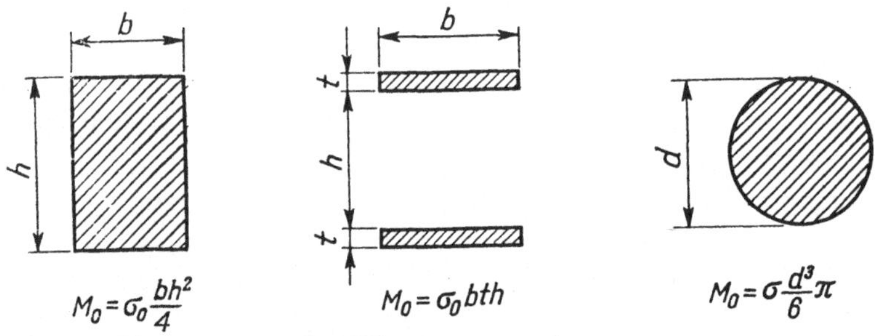

Fig. 4.14. Limit moments for different cross-sections

At the limit state, the bar is required to
(a) satisfy the equations of internal equilibrium,
(b) not to exceed the limit moment condition,
(c) be transformed into a mechanism.

These conditions can be written as

$$\frac{d^2M}{dx^2} = -p, \qquad \frac{dM}{dx} = T,$$

$$-M_0 \leqslant M \leqslant M_0 \quad \text{along the bar,} \tag{4.67}$$

$$u_i \neq 0, \quad L = \int_A \mu p_i u_i dA > 0;$$

here u_i denotes a displacement velocity field satisfying prescribed kinematic constraints and p_i is the external loading.

At a limit state, a bar structure becomes a kinematic chain with at least one degree of freedom. For a structure n-times statically indeterminate to become a kinematic chain, the yield point must be reached in at least $n+1$ cross-sections. Adopting the rigid-plastic model, we assume that dissipation of energy occurs in the plastic hinges only.

In the case of frames, we assume that both beams and columns have constant cross-sections. Let l_i and M_i be the length and the plastic moment of the ith member. Since the cross-section of any member does not vary along its length, the volume of the structure can be minimized by an appropriate choice of the length l_i and the cross-sectional area F_i for each member. Thus, the case we are considering is that of parametric optimization.

In general, the relationship between F_i and M_i will be non-linear, of the type $F_i = cM_i^{\alpha_i}$, where c and $\alpha \leqslant 1$ depend on the shape of the cross-section (Fig. 4.14): for an I-shaped section of constant height $\alpha = 1$, for a rectangular section of constant width $\alpha = 1/2$, and for geometrically similar cross-sections $\alpha = 2/3$. The total volume of the frame is thus given by

$$V = c(l_1 M_1^{\alpha_1} + l_2 M_2^{\alpha_2} + \ldots + l_n M_n^{\alpha_n}). \tag{4.68}$$

Among the frames with a given load carrying capacity and differing only in the cross-sections and lengths of their bars it is required to choose one with the least volume of material. Mathematically, this amounts to minimizing the non-linear function (4.68) subject to constraints (4.67).

According to what has been said so far, an extremum of the volume will occur when the mean powers of plastic dissipation D_i on the surfaces S_{Ti}^0 liable to alterations are identical in all the members. It follows that each member must develop an independent failure mechanism and so the structure will become a mechanism with n degrees of freedom. Indeed, if all the yield hinges of, say, the nth member belonged to the yield mechanism of some other member, the requirement $D_n = D_i$ would not be satisfied. Denoting by $M_i\theta_i$ (no summation over i) the energy dissipation rate in the yield hinges of the ith member, where θ_i is the sum of the rotations of all the yield hinges of this member, we can write

$$D_i = \frac{M_i\theta_i}{cl_i M_i^{\alpha_i}} = \frac{\theta_i}{cl_i(M_i)^{\alpha_i - 1}}.$$

The necessary condition for an extremum of the volume is therefore

$$\frac{\theta_1}{l_1 M_1^{\alpha_1 - 1}} = \frac{\theta_2}{l_2 M_2^{\alpha_2 - 1}} = \ldots = \frac{\theta_n}{l_n M_n^{\alpha_n - 1}} = Dc. \tag{4.69}$$

This condition was first derived by Prager (1956). For an ideal flanged beam ($\alpha = 1$) equation (4.69) takes the simple form

$$\frac{\theta_1}{l_1} = \frac{\theta_2}{l_2} = \ldots = \frac{\theta_n}{l_n}. \tag{4.70}$$

Minimum-volume limit design subject to stability constraints

In most cases, minimum-volume optimization leads to slender thin-walled structures. Forces and stresses in thin-walled elements optimized without stability constraints may exceed the values critical with respect to buckling. In designing such structures, therefore, there arises the necessity of including the conditions of local and global stability.

In their general form, limit design problems with stability constraints do not differ from the problems we have formulated so far except that the side conditions now include in addition local and global stability requirements. Again, the task is reduced to a variation problem of minimizing a volume functional subject to inequality constraints. The typical cases where stability conditions have to be taken into consideration include optimization of thin columns in compression, thin-walled box beams in bending and thin shells.

Bounds on the minimum volume

In general, obtaining an exact solution of a limit-optimum-design problem is a very laborious task. Often, an exact solution is not obtainable at all. In such cases, the designer contents himself with an approximate solution, known to be safe and having a greater volume than the optimal solution: $V_u > V_{\min}$.

In order to find out how far such a design is from the exact solution, we may seek a solution (inadmissible with regard to safety) with a volume $V_l < V_{\min}$. If the upper and the lower estimates of the optimal design, V_u and V_l, are sufficiently close to each other, the upper estimate may, to all practical purposes, be taken as the required solution. An upper estimate can be derived from any statically admissible field assumed for the structure; ower estimates are obtained by considering kinematically admissible fields.

Let V_s be the volume of a design based on a stress field σ_{ij}^s satisfying the equations of internal equilibrium and boundary conditions on the free surfaces and on the surfaces subjected to tractions T_m; T_m is the limit load at which the minimum volume is sought. By the first theorem of limit analysis, the body V_s will not collapse under the load T_m. By definition, the volume V_m of a minimum-volume design is bounded from above by V_s,

$$V_m \leqslant V_s.$$

In order to bound V_m from below, we can reason as follows. Let the actual load carrying capacity of the minimum-volume structure with volume V_m be T_m. Consider a design derived by equating the power dissipated within the body (and calculated from a kinematically admissible velocity field u_k^k [1] defined in the body) with the power developed by the limit load T_m

$$\int_A T_m u_k^k \mathrm{d}A = \int_{V_k} D(u_k^k) \mathrm{d}V. \tag{4.71}$$

Let the volume of this structure be V_k and let its actual load carrying capacity be T_k. The surface A acted upon by the forces T is the same in both de-

[1] The upper index refers to the kind of field: k—kinematically admissible, r—real; the lower index refers to the body: k—V_k, m—$V_{\min}$.

signs. Other surfaces of the two structures differ. By the second theorem of limit analysis the actual carrying capacity of the structure V_k is less than that derived from equation (4.71) with the kinematically admissible field u_k^k; thus, $T_k \leqslant T_m$. Assuming that the load carrying capacity is proportional to the volume, we obtain $V_k \leqslant V_m$, i.e. V_k is a lower bound on the minimum volume.

A lower bound can be obtained in a more formal way by using the following two extremum principles:

1. Among the kinematically admissible displacement velocity fields, that is the actual field at the limit state which corresponds to the minimum power of internal energy dissipation

   $$W^k \geqslant W^r.$$

2. Among the kinematically admissible displacement velocity fields, that is the actual field at the limit state which corresponds to the maximum external force power

   $$\int_A T^r u^k \mathrm{d}A \leqslant \int_A T^r u^r \mathrm{d}A.$$

   Using (4.71) and the first principle, we can write

$$\int_A T_m u_k^k \mathrm{d}A = \int_{V_k} D(u_k^k)\mathrm{d}V \geqslant \int_{V_k} D(u_k^r)\mathrm{d}V. \tag{4.72}$$

By the second principle,

$$\int_A T_m u_k^k \mathrm{d}A \leqslant \int_A T_m u_m^r \mathrm{d}A = \int_{V_m} D(u_m^r)\mathrm{d}V. \tag{4.73}$$

A comparison of inequalities (4.72) and (4.73) gives

$$\int_{V_k} D(u_k^r)\mathrm{d}V \leqslant \int_{V_m} D(u_m^r)\mathrm{d}V.$$

Replacing the energy dissipation rates by their mean values, we get

$$D_k^s \int_{V_k} dV \leqslant D_m^s \int_{V_m} \mathrm{d}V.$$

Hence, assuming that at the limit state the mean dissipation rates per unit volume are identical in the two structures, $D_k^s = D_m^s$, we conclude that

$$V_k \leqslant V_m.$$

Jointly, the bounds on the minimum volume can be written

$$V_k \leqslant V_m \leqslant V_s. \tag{4.74}$$

Inequality (4.74) allows us to conclude that

(1) A design based on a statically admissible field provides a safe estimate of the dimensions of the structure;

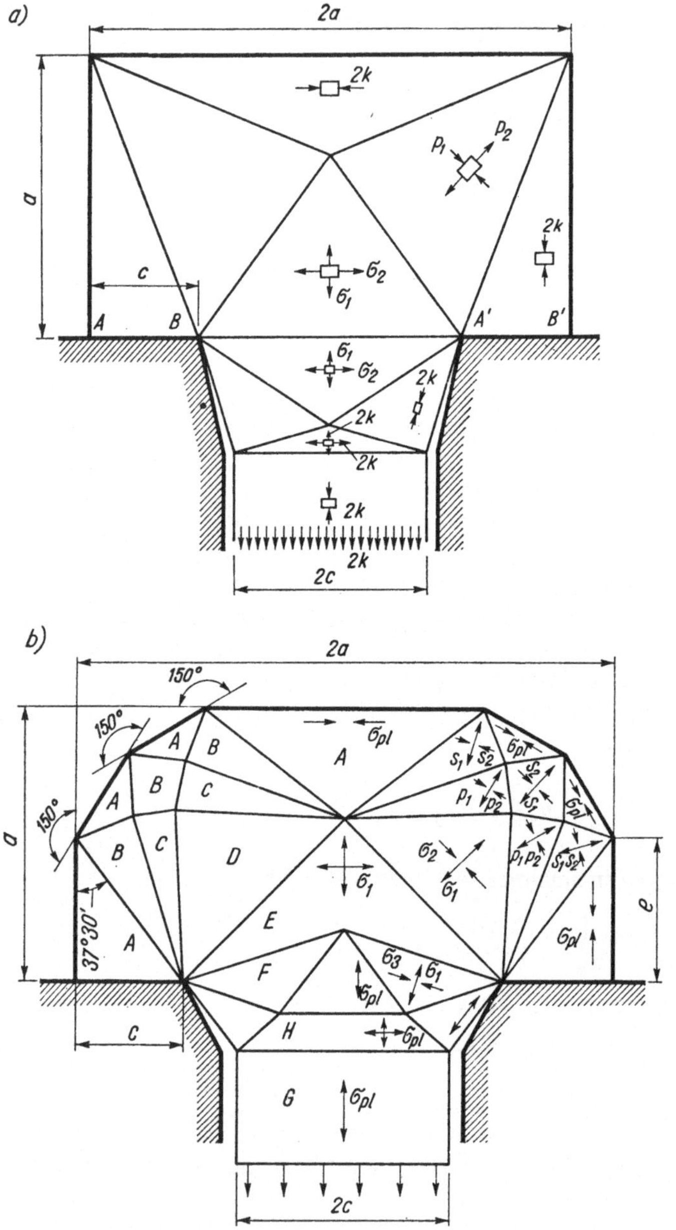

Fig. 4.15. Two statically admissible fields for an anchor element

(2) The volume of a design based on a kinematically admissible field is smaller than that of the minimal design.

The foregoing considerations suggest the following procedure of minimum-volume design from statically and kinematically admissible fields.

Dimensions of the structure are found from a statically admissible stress field. The field must satisfy the static boundary conditions on the free and on the loaded surfaces, and the equations of internal equilibrium; the resultant external forces corresponding to this field must be equal to the required external loading and the field must not be above yield at any point. The borderlines of such a static field delimit the contour of the element to be designed. Considering several different types of such fields, which give different shapes of the structure, we choose the solution having the least volume. The excess in the sizing of a design derived from a statically admissible field can be estimated by means of a kinematically admissible field.

The above method was used by Szczepiński (1968) in optimum design of machine elements such as anchor elements, big ends of connecting rods, etc. Figure 4.15 shows, after Szczepiński, two static fields for an anchor element. The stress field in each subregion is assumed to be constant with the principal directions as indicated. The division lines, called the *discontinuity lines*, follow from the conditions of equilibrium and the requirement that the yield surface be not exceeded in any of the regions. Of the two solutions based on these statically admissible fields, the second is better, because the volume of material it requires is smaller.

4.5 Examples of optimum limit design

Optimization of the thickness of a rotating disc

The present example is based on a paper by Drucker and Shield (1975_1) and one by Chan (1960). A circular annular disc of inner radius r_i and outer radius r_0 is rotated with constant angular velocity about a perpendicular axis through the centre of the disc (Fig. 4.16). The outer edge of the disc is loaded by a uniform tensile force T per unit length and the inner edge is free. The body force (centrifugal force) is given by $\varrho\omega^2 r$ per unit volume, where ϱ is the mass density of the material.

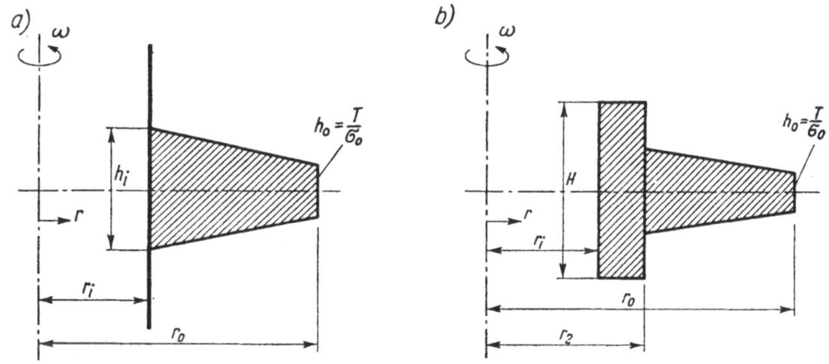

Fig. 4.16. Minimum-volume design of a rotating disc

The material of the disc is assumed to obey the Tresca yield condition (Fig. 4.17). We also assume that the necessary condition for a minimum, D = const, is satisfied on both the upper and the lower faces of the disc. The fact that the disc is in plane stress implies that the energy dissipation rate per unit volume is constant. It follows that in a minimum-volume design D is constant throughout the disc,

$$D = \sigma_r \varepsilon_r + \sigma_\theta \varepsilon_\theta - \varrho \omega^2 r u = \text{const.} \tag{4.75}$$

We have seen before (p. 162) that when D = const throughout the structure there exists an absolute minimum of volume.

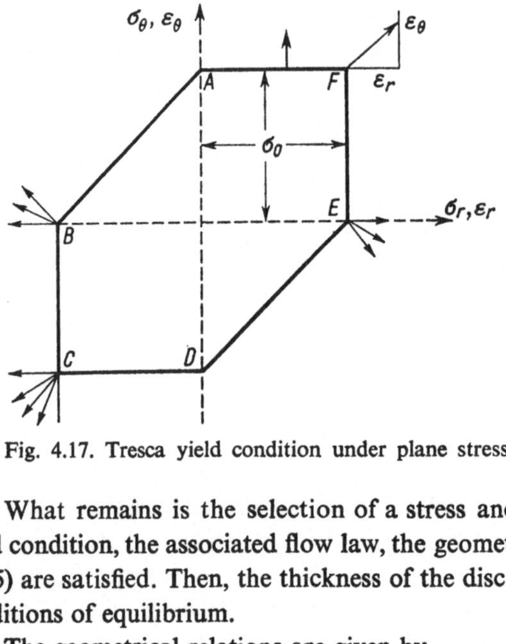

Fig. 4.17. Tresca yield condition under plane stress

What remains is the selection of a stress and a strain field so that the yield condition, the associated flow law, the geometrical relations and equation (4.75) are satisfied. Then, the thickness of the disc will be determined from the conditions of equilibrium.

The geometrical relations are given by

$$\varepsilon_r = \frac{du}{dr}, \qquad \varepsilon_\theta = \frac{u}{r}. \tag{4.76}$$

The stress state is represented by a point of the yield hexagon shown in Fig. 4.17. According to the flow law, the strain vector is normal to the yield surface in the stress space. This implies, for example, that the stress point cannot lie on the side AF, because the normality condition would then require $\varepsilon_r = 0$, u = const, and conditions (4.76) and (4.75) would not hold simultaneously. Similarly, conditions (4.76) and (4.75) and the normality condition cannot all be satisfied if the state of stress is represented by a point of any other side of Tresca's yield polygon.

The only candidates left, therefore, are the vertices A, B, E and F of the polygon (points C and D represent compression states). Points A and B drop out, because in their case the boundary condition at the outer edge $r = r_0$ cannot be satisfied. The stress point E is also inadmissible as it would require inward radial velocity at the inner edge $r = r_i$. There only remains point F, at which $\sigma_r = \sigma_\theta = \sigma_0$. In polar coordinates, the equation of equilibrium for a differential element of the disc takes the form

$$\frac{d}{dr}(h\sigma_r)+h\,\frac{\sigma_r-\sigma_\theta}{r} = -\varrho\omega^2 rh, \tag{4.77}$$

where h is the variable thickness of the disc. Substituting $\sigma_r = \sigma_\theta = \sigma_0$, we get

$$\frac{dh}{dr} = -\frac{\varrho\omega^2}{\sigma_0}rh,$$

from which we find

$$h = \frac{T}{\sigma_0}\exp(x_0^2-x^2), \qquad x_i \leqslant x \leqslant x_0, \tag{4.78}$$

where

$$x = \left[\sqrt{\frac{\varrho\omega^2}{2\sigma_0}}\right]r.$$

In order to satisfy the condition of zero stress at $r = r_i$, a flange must be added at the inner boundary (Fig. 4.16). We assume that the flange has an infinite height but a finite cross-sectional area A. The radial stress at the flange is $\sigma_r = \sigma_0$ and the tangential stress σ_θ is constant. By the equation of equilibrium, we have

$$r_i h_i \sigma_0 + \varrho\omega^2 r_i^2 A = \sigma_0 A, \tag{4.79}$$

and hence the cross-sectional area

$$A = \frac{r_i h_i}{1-2x_i^2}; \tag{4.80}$$

h_i is the value of h given by (4.78) at $x = x_i$. The speed of rotation of the disc is restricted by the condition $x_i^2 \leqslant 1/2$, i.e. $\varrho\omega^2 r_i^2 < \sigma_0$.

Next, we determine the velocity field from eqn. (4.75). For point F, using relations (4.76), we obtain

$$\sigma_0\left(\frac{du}{dr}+\frac{u}{r}\right)-\varrho\omega^2 ru = \text{const.} \tag{4.81}$$

For the associated flow law to be satisfied at F,

$$\varepsilon_r = \frac{du}{dr} \geqslant 0, \qquad \varepsilon_\theta = \frac{u}{r} \geqslant 0, \qquad r \geqslant r_i. \tag{4.82}$$

On the other hand, since the state of stress of the line flange is that of the stress point A on the yield curve, the associated flow law will be satisfied there if

$$\varepsilon_r = \frac{du}{dr} \leqslant 0 \quad \text{for} \quad r \leqslant r_i. \tag{4.83}$$

Inequalities (4.82) and (4.83) imply that

$$\varepsilon_r = \frac{du}{dr} = 0 \quad \text{for} \quad r = r_i. \tag{4.84}$$

With this result, (4.81) gives the displacement velocity field as

$$u = \frac{c}{x}\left[\frac{\exp(x^2 - x_i^2)}{(1-2x_i^2)} - 1\right], \tag{4.85}$$

where c is an arbitrary constant. It can readily be verified that u given by (4.85) satisfies conditions (4.82) and, therefore, that the associated flow law holds. It can also be checked that the solution so obtained satisfies all conditions (4.53)–(4.60) and the sufficient condition for the occurrence of a minimum.

The only weak point of this solution is the unrealistic infinite height of the line flange. To avoid that, we place an upper limit H on the thickness h (Fig. 4.16b), taking the full thickness H for $r_i \leqslant r \leqslant r_2$. In this region, equation (4.75) cannot hold, and the stress state is represented by points on the side AF of the yield hexagon,

$$\sigma_\theta = \sigma_0, \qquad 0 \leqslant \sigma_r \leqslant \sigma_0, \qquad \frac{du}{dr} = 0.$$

It follows that in this region

$$u = \text{const} = B \tag{4.86}$$

and (cf. (4.75))

$$D = B\left(\frac{\sigma_0}{r} - \varrho\omega^2 r\right), \qquad r_i \leqslant r \leqslant r_2.$$

For $r_i \leqslant r \leqslant r_2$, the stress σ_r is found from the equilibrium condition (4.77) with $h = H$. The edge $r = r_2$ can be obtained by equating this stress to the constant σ_0 at $r = r_2$. For $r_2 \leqslant r \leqslant r_0$, the radial velocity is determined by (4.82) and (4.85). With the condition $du/dr = 0$ at $r = r_2$, equation (4.85) gives

$$u = \frac{c_1}{x}\left[\frac{\exp(x^2 - x_2^2)}{(1-2x_2^2)} - 1\right], \qquad x \geqslant x_2. \tag{4.87}$$

The constant c_1 can be found by equating (4.85) and (4.87) at $r = r_2$.

The energy dissipation rate per unit volume, D, is a decreasing function for $r_i \leqslant r \leqslant r_2$ and constant for $r \geqslant r_2$.

Note again that in order to obtain a realistic solution it has been necessary to restrict the height h.

Optimization of a sandwich plate

A simply supported circular sandwich plate of constant core thickness H is loaded by a uniform pressure p over its upper, horizontal surface. The core is assumed not to transmit bending. It is required to find the variable thickness of the two identical face sheets so that they are just at the point of collapse under the pressure p and have minimum volume. The specific weight of the face sheets is denoted by γ and the body forces are taken into account. Since the thickness of the faces is small, we can assume that the state of strain is constant across each of them. The necessary condition for a minimum of volume, i.e. the constancy of the energy dissipation rate on the faces, implies the constancy of this rate throughout the plate. This in turn secures the fulfilment of the sufficient conditions for an absolute minimum of volume. We shall solve the problem using generalized stresses and generalized strains and assuming that the material obeys the Tresca yield condition. The representation of this condition in the moment-curvature space is analogous to that in the stress-strain space (Fig. 4.17), with σ_θ and σ_r replaced by M_θ and M_r, and ε_θ and ε_r by $\varkappa_\theta$ and $\varkappa_r$, respectively.

For the given support and load conditions, the curvatures $\varkappa_\theta$ and $\varkappa_r$ are positive throughout the plate and given by

$$\varkappa_r = -\frac{d^2 w}{dr^2}, \qquad \varkappa_\theta = -\frac{dw}{dr}\frac{1}{r}, \tag{4.88}$$

where w is the deflection rate in the downward direction. Since the curvatures are positive, the state of stress will be represented, as in the previous example, by the point F of the Tresca yield hexagon. Hence

$$M_r = M_\theta = \sigma_0 H t. \tag{4.89}$$

The condition that the energy dissipation rate should be constant across the face sheets takes the form

$$D = \frac{M_r \varkappa_r + M_\theta \varkappa_\theta - \gamma w 2t}{2t} = \text{const.} \tag{4.90}$$

Substitution of (4.88) and (4.89) in (4.90) gives

$$-\frac{\sigma_0 H}{2}\left(\frac{d^2 w}{dr^2} + \frac{1}{r}\frac{dw}{dr}\right) - \gamma w = \text{const.} \tag{4.91}$$

Solving this equation, we find the deflection rate for the plate as

$$w = C[J_0(\alpha r) - J_0(\alpha r_0)], \tag{4.92}$$

where C is a positive constant, $\alpha = (2\gamma/\sigma_0 H)^{1/2}$ and J_0 is the Bessel function of the first kind of zero order. The restriction that $\varkappa_r$ and $\varkappa_\theta$ be positive requires that $\alpha r_0 < 1.84$. The thickness t of the faces can be determined from the equilibrium equation

$$\frac{d^2}{dr^2}(rM_r) - \frac{dM_\theta}{dr} + pr + 2\gamma t r = 0. \tag{4.93}$$

Using (4.89), we get

$$\frac{d^2t}{dr^2}+\frac{1}{r}\frac{dt}{dr}+\alpha^2 t = -\frac{P}{\sigma_0 H}. \tag{4.94}$$

Solving this equation with the boundary condition

$$t = 0, \quad r = r_0, \quad M = 0, \tag{4.95}$$

we obtain the desired thickness

$$t = \frac{P}{\sigma_0 H\alpha^2}\left[\frac{J_0(\alpha r)}{J_0(\alpha r_0)}-1\right]. \tag{4.96}$$

Since M and M_r are positive, αr_0 must be less than the first zero of the Bessel function J_0, i.e. $\alpha r_0 < 2.4$. For $\alpha r_0 < 1.84$, solution (4.96) is a minimum-volume design. Using (4.92) and (4.96), one can easily check that all conditions (4.53)–(4.60) and the sufficient condition for a minimum volume are satisfied. A more detailed discussion of the above example can be found in papers by Drucker and Shield (1957_1) and Prager and Shield (1959).

Optimization of a homogeneous plate

The support conditions and a load perpendicular to the middle surface of the plate are given. It is required to find the variable thickness $2h$, for which, under a constant limit load, the volume of the plate is minimum. The problem has been treated by many authors, among them by Freiberger and Tekinalp (1956), Onat, Schumann and Shield (1957), Mróz (1958_2), Prager (1958_1) and Shield (1960_2).

We presume that in an optimal plate the energy dissipation rate per unit volume is constant on the surfaces of the plate, $D(\varepsilon_{ij}^1) = \overline{D}$. The Love–Kirchhoff hypothesis about normals is adopted, i.e. the strain rate ε_{ij}^1 is assumed to be determined by the curvature $\varkappa_{ij}$ of the middle surface,

$$\varepsilon_{ij}^1 = z\varkappa_{ij} = \frac{z}{h}\,\varepsilon_{ij}^0. \tag{4.97}$$

In an arbitrary layer of the plate, the dissipation rate is given by

$$D = \sigma_0\,\varepsilon_{ij}^1 = \frac{z}{h}\,\varepsilon_{ij}^0\sigma_0 = \frac{z}{h}\overline{D}. \tag{4.98}$$

It can be seen from formula (4.98) that in the present case D cannot be constant throughout the plate. Any continuation of the strain rate (4.97) beyond the plate surface is kinematically admissible. The corresponding function $D(\varepsilon_{ij}^1)$ of (4.98) takes greater values outside the surfaces $z = \pm h$ than between them and so the sufficient condition for an absolute minimum of volume cannot be applied. Therefore, we have to show that the volume increment given by formula (4.65) is positive for arbitrary variations of the plate thickness. Let δh be an infinitesimal variation of the thickness and A the middle surface area. The terms occurring in (4.65) can be calculated as

$$\int_{V_2} (\sigma_{ij}^2-\sigma_{ij}^1)\,\varepsilon_{ij}^1\,dV = 2\int_A dA \int_0^{h+\delta h} (\sigma_{ij}^2-\sigma_{ij}^1)\,\varepsilon_{ij}^0 \frac{z\,dz}{h} = \int_A (\sigma_{ij}^2-\sigma_{ij}^1)\,\varepsilon_{ij}^0 \frac{(h+\delta h)^2}{h}\,dA, \tag{4.99}$$

$$\int_{\delta V} \Delta D\,dV = 2\int_A dA \int_h^{h+\delta h} \left[\frac{|z|}{h}-1\right]\bar{D}\,dz = \bar{D}\int_A \frac{(\delta h)^2}{h}\,dA. \tag{4.100}$$

Substitution in (4.65) gives

$$\delta V = -\int_A \frac{(\delta h)^2}{h}\,dA - \frac{1}{\bar{D}}\int_A \delta M'_{ij}\varkappa_{ij}\left(\frac{h+\delta h}{h}\right)^2 \delta A, \tag{4.101}$$

where $\delta M'_{ij} = (\sigma_{ij}^2-\sigma_{ij}^1)h^2$ is the change in the bending moments resulting from a change in the stress state of the plate.

The first term on the right-hand side of (4.101) is of order $(\delta h)^2$ and is always negative. The second term depends on the yield condition and, in the general case, estimating its value is not easy. Let us assume that the material of the plate obeys the Tresca yield condition. Consider first a design with D = const on the surface of the plate, coresponding to a point on any of the sides of Tresca's hexagon. Then $\delta M'_{ij}\varkappa_{ij} = 0$ and only the first term remains in equation (4.101). Hence $\delta V < 0$, which means that the design considered gives a plate of maximum volume. Next, suppose that the solution corresponds to any of the vertices of the yield polygon, say A (Fig. 4.18).

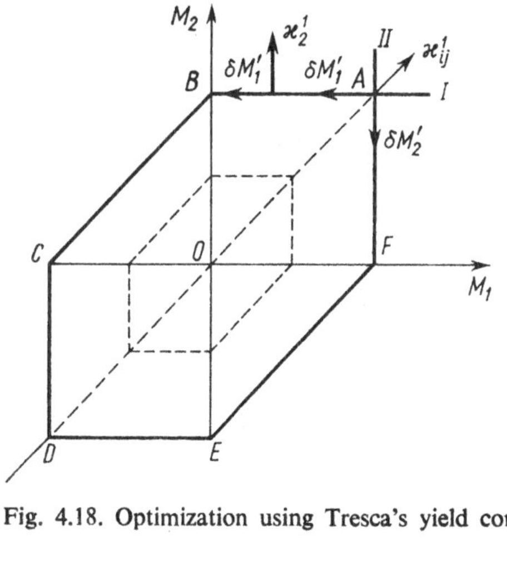

Fig. 4.18. Optimization using Tresca's yield condition

The stress states of the "neighbouring" plates with slightly varied thickness will then lie on the sides converging at A. If $\delta M'_{ij}$ is of the same order of magnitude as δh, the product $\delta M'_{ij}\varkappa_{ij}$ will be negative and also of order δh. This follows from the fact that the vector $\varkappa^1_{ij}$ can only assume a direction between I and II (Fig. 4.18) and so the angle it will form with $\delta M'_{ij}$ will be between $\pi/2$ and π. In this case, therefore, the second positive term in expression (4.101) will determine the sign of δV.

Thus, an optimal solution should be sought on the segments DOA, BOE and COF joining the opposite vertices of Tresca's hexagon.

Optimization of a circular plate

A spherical plate of radius a is simply supported and loaded symmetrically by a uniform load q (Fig. 4.19). Owing to the axial symmetry, the principal moments coincide with the radial and the circumferential directions.

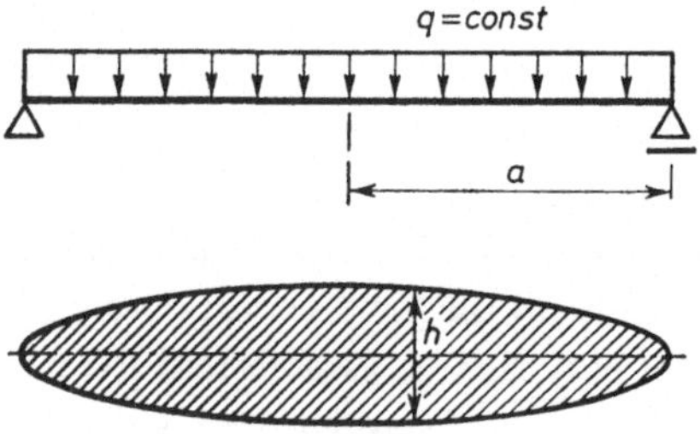

Fig. 4.19. Minimum-volume solution for a circular plate

The state of stress is represented by a point on the line OA of the yield hexagon, i.e. $M^1_r = M^1_\theta = \sigma_0 h^2/4$, where σ_0 is the yield point in simple tension. From the equation of equilibrium

$$\frac{d}{dr}(rM_r) - M_\theta = -\frac{qr^2}{2} \tag{4.102}$$

and the boundary condition $M_r = 0$ at $r = 0$, we find the bending moments as functions of r,

$$M^1_r = M^1_\theta = \frac{q}{4}(a^2-r^2). \tag{4.103}$$

After substituting $M^1_r = M^1_\theta = \sigma_0 h^2/4$ in (4.103), we obtain the thickness of the plate as

$$h = \sqrt{\frac{q}{\sigma_0}(a^2-r^2)}.$$

The deflection rate w can be determined from the condition that the energy dissipation rate $D(\varepsilon^1_{ij})$ is constant:

$$D(\varepsilon^1_{ij}) = \frac{M^1_r \varkappa^1_r + M^1_\theta \varkappa^1}{h} = \text{const}. \tag{4.105}$$

Using relations (4.88), we get

$$\frac{1}{r}\frac{dw}{dr}+\frac{d^2w}{dr^2} = -\frac{\alpha}{\sqrt{a^2-r^2}}, \tag{4.106}$$

where $\alpha_1 = 4\bar{D}/\sqrt{q\sigma_0}$.

Integration of equation (4.106) gives

$$w = \alpha_1\left[-a\ln\left(\frac{r}{a}\right)+\sqrt{a^2-r^2}-a\ln\frac{a+\sqrt{a^2-r^2}}{r}\right]. \tag{4.107}$$

By differentiating this solution, one can verify that $\varkappa_\theta > 0$ and $\varkappa_r > 0$ throughout the plate and therefore the flow law in force at the point A is satisfied.

Optimization of an annular plate

A simply supported annular plate is subjected to a load q as shown in Fig. 4.20. As before, the stress state of the plate is represented by the line OA, i.e. $M_r^1 = M_\theta^1 = \sigma_0 h^2/4$.

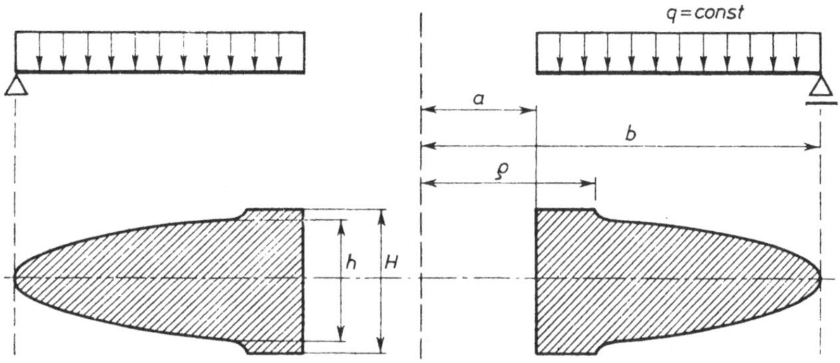

Fig. 4.20. Optimum design of a simply supported annular plate

The equation of equilibrium for the present structure takes the form

$$\frac{d}{dr}(rM_r)-M_\theta = -\frac{q}{2}(r^2-a^2). \tag{4.108}$$

By integrating this equation with the boundary condition $M_r = 0$ at $r = b$, we can find the bending moments as functions of r, and, as in the previous example, the required thickness h of the plate:

$$M_r^1 = M_\theta^1 = \frac{q}{4}\left(b^2-r^2-2a^2\ln\frac{b}{r}\right), \tag{4.109}$$

$$h = \sqrt{\frac{q}{\sigma_0}\left(b^2-r^2-2a^2\ln\frac{b}{r}\right)}. \tag{4.110}$$

As in the case of a rotating disc, there is the problem of satisfying the boundary condition $M_r = 0$ at $r = a$. To ensure that, we have to assume a jump change in the thickness at a radius $r = \varrho$ and to take a suitable constant thickness H for $a \leqslant r \leqslant \varrho$. The stress state in this region will then be represented by the segment AB of the Tresca yield polygon, i.e. $M_\theta^1 = M_0 = \sigma_0 H^2/4$. From the equation of equilibrium M_r is found as

$$rM_r = M_0(r-a) - \frac{q}{6}(r^3 - 3a^2 r + 2a^3). \tag{4.111}$$

Comparing the values of the moments given by (4.109) and (4.111) at $r = \varrho$, we obtain a relationship between H and h:

$$\frac{H^2}{h^2} = \frac{1}{3}\,\frac{3b^2\varrho - \varrho^3 + 4a^3 - 6a^2\varrho(1+\ln(b/\varrho))}{(b^2 - \varrho^2 - 2a^2\ln(b/\varrho))(\varrho - a)}.$$

The deflection rate for $\varrho \leqslant r \leqslant b$ can be found from the condition $D = \text{const}$,

$$\frac{1}{r}\frac{\mathrm{d}w}{\mathrm{d}r} + \frac{\mathrm{d}^2 w}{\mathrm{d}r^2} = -\frac{\alpha_1}{\sqrt{(b^2 - r^2 - 2a^2\ln(b/r))}}. \tag{4.112}$$

Integration gives

$$w = \alpha_1 \int_r^b \frac{1}{\xi}\,\mathrm{d}\xi \int_b^\xi \frac{\eta\,\mathrm{d}\eta}{h(\eta)} - c_1 \ln\frac{b}{r}.$$

The choice of constant thickness H for $a < r < \varrho$ has made it impossible for D to be constant in this interval. The deflection rate there is given by

$$w = C_2 r + C_3.$$

The integration constants C_1, C_2 and C_3 can be calculated from the requirement that w, $\mathrm{d}w/\mathrm{d}r$ and D be continuous at $r = \varrho$. It can be verified that the inequalities $\varkappa_r \geqslant 0$ and $\varkappa_\theta \geqslant 0$ hold throughout the plate and that the flow law associated with the yield condition is also satisfied.

Optimization of a single span frame

A single span frame with members of constant cross-section (Fig. 4.21) is loaded by forces P and Q at the ratio $P{:}Q = 1.8$. The problem is to find optimal cross-sections of the frame, given $h = l$ and $\alpha = 1$. At the limit state, the minimum-volume frame should be a mechanism with three degrees of freedom, satisfying the necessary condition (4.70). Under the loading shown in Fig. 4.21a, plastic hinges may be formed at points A, B, C and D, and at the point of application of the force P. Let M_1, M_2 and M_3 be the limit moments of members 1, 2 and 3, respectively. By considering all possible failure mechanism, one can find that when $M_1 \neq M_2 \neq M_3$, a mechanism with

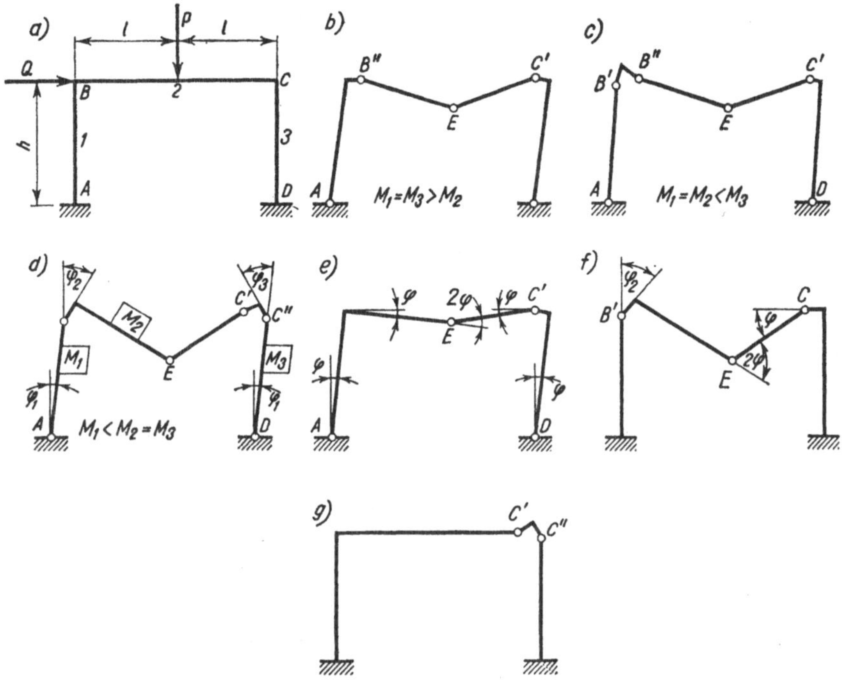

Fig. 4.21. Optimization of a simple span frame

three degress of freedom cannot be formed. The same applies to the case where $M_1 = M_3$ (Fig. 4.21b). The frame can only become a mechanism with three degrees of freedom when yield hinges develop both in the beam and in the column at one of the corners B or C, as in Figs 4.21c, d. Let us analyse the mechanism corresponding to the case of Fig. 4.21d. Since $M_1 < M_2 = M_3$, the first to attain a plastic state will be the sections A and B. This number of hinges is not yet sufficient to destroy a structure three times statically indeterminate. Since $M_2 = M_3$, a further growth of loading will cause simultaneous plastic yield in sections C', C'' and D, whereupon the frame will become a mechanism like the one shown in Fig. 4.21d. We can regard it as the sum of the three independent mechanisms shown in Figs 4.21e, f, g. If φ_1, φ_2 and φ_3 are the rates of rotation at hinges A, B and C'', respectively, the principle of virtual work for the structure can be written

$$(M_1+4M_2+M_3-Qh-Pl)\varphi_1+(M_1+2M_2+M_2-Pl)\varphi_2+$$
$$+(M_3-M_2)\varphi_3 = 0.$$

From the resulting system of three linear algebraic equations we find

$$M_2 = M_3 = \frac{Qh}{2}, \quad M_1 = Pl-\frac{3}{2}Qh. \tag{4.113}$$

Since $M_1 > 0$ and $M_1 < M_2$, the solution is valid for

$$\frac{3}{2} \leqslant \frac{Pl}{Qh} \leqslant 2. \tag{4.114}$$

The sums of the angles of rotation for different members are

$$\theta_1 = \varphi_1 + \varphi_2, \quad \theta_2 = 4\varphi_1 + 3\varphi_2 - \varphi_3, \quad \theta_3 = \varphi_3 + \varphi_1.$$

The necessary condition (4.70) will be satisfied if

$$\frac{\varphi_1 + \varphi_2}{h} = \frac{4\varphi_1 + 3\varphi_2 - \varphi_3}{2l} = \frac{\varphi_3 + \varphi_1}{h} = \bar{D}. \tag{4.115}$$

Solving this equation for φ_1, φ_2 and φ_3 gives

$$\varphi_1 = (l-h)\bar{D}, \quad \varphi_2 = (2h-l)\bar{D}, \quad \varphi_3 = (2h-l)\bar{D}.$$

The angles of rotation are positive and therefore $l-h \geqslant 0$ and $2h-l \geqslant 0$; hence follows the restriction

$$1 \leqslant l/h \leqslant 2. \tag{4.116}$$

For the frame under consideration, restrictions (4.114) and (4.116) are satisfied, and therefore solution (4.113) is an optimal solution. The limit moments are $M_1 = 0.3\,Ql$, $M_2 = M_3 = 0.5\,Pl$. Knowing them, we can calculate the desired cross-sections.

It should be noted that finding an appropriate mechanism for more complex frames is a very difficult problem. No method has yet been proposed which would cover a sufficiently wide range of different cases of minimum-volume limit design of frames. Among the basic publications in this field are the works of Foulkes (1954), Heyman and Prager (1958) and Heyman (1959).

4.6 Probabilistic approach to optimum design

Design and optimization based on probability theory

A moment of reflection on the quantities that occur in structural design and optimization suffices to realize the quite obvious fact that nearly all of them, e.g. loads, actions, geometrical dimensions, material properties, quality of joints or support conditions, are random variables. The first attempts to take into account in an actual design the probabilistic character of the quantities involved date back to the early 1940s, while in optimization probabilistic considerations appeared in the 1960s. Some of the earliest work in this area was done by Hilton and Feigen (1960) and by Turkstra (1962). This is not to say that the fact that almost all design variables are random variables had been completely neglected before. The admissible stresses or loads prescribed by design codes and regulations had always been based, to a greater or lesser extent, on probabilistic estimates derived from observation of real structures and their loadings or from laboratory tests. However, no consistent probabilistic approach was used, and the codified parameters were largely due to chance or intuition, even though the concepts of mean values and dispersion were utilized.

This state of the methods of structural design and optimization resulted above all from the lack of data which would allow a complete statistical analysis. Even at the present time the observational or experimental data are insufficient, which impedes the development of application of probabilistic methods. Also, as has been pointed out by Lind (1969), the traditional views and habits prevailing in building practice have hindered the introduction of probability into design. Unlike other processes, in which mathematical statistics and probability theory can be used throughout, structural analysis must lead to deterministic conclusions defining precise dimensions of structural elements and their configurations.

The present development of the experimental methods of testing and observation of structures and their loadings, new facilities for data collection and data processing, and also the development of suitable mathematical methods, make it possible to replace intuitive handling of non-deterministic inputs in design problems by a rational application of probabilistic methods. The quantitative growth of building and the increasing sizes of individual special-purpose engineering structures further stimulate the search for safe and economical solutions using this approach.

Probabilistic approach to safety and reliability of structures

The safety of a structure can be estimated and represented by the probability of its failure, i.e. the occurrence of an ultimate limit state manifesting itself as, say, the formation of plastic hinges, rupture, fracture, overturning, etc. In addition to ultimate limit states, we also distinguish limit states of serviceability, involving phenomena such as excessive cracking, displacements or vibrations, which do not cause a structural failure but make it impossible to use the structure according to its intended function; states in which a given structure is unserviceable for reasons independent of the design are not included in this category.

The concepts of *safety* and of *serviceability* so understood make up jointly what we call the *reliability* of the structure. Let us note that these terms are not always used in the sense given above and, depending on context, various authors define them in a somewhat different manner.

The probability of the occurrence of a limit state in a given structure can be calculated from the probability distribution of the loading and strength of the structure and all the other relevant parameters. The probability distributions of the random variables characterizing the loading can be determined by observation and measurement, and those of material and structural strength—by laboratory or field tests.

In the present section we are concerned with the ultimate limit states; extending the theory to include the limit states of serviceability does not present any serious conceptual or formal difficulties.

Failure of a structure may concern the structure as a whole or may be attributed to the failure of its individual members. Depending on the kind of the structure, the relationship between the failure of a single member and that of the whole structure may take different forms. In statically determinate structures, for example, a failure of the weakest member causes a failure of the whole structure (weakest-link structures), while hyperstatic structures fail only if several members reach their capacity simultaneously (*fail-safe structures*) (Moses 1969). In this second case, the calculations are more complicated and require electronic computers.

The probability of various failure modes and the influence of the statistical correlation between them on the overall probability of failure P_f were examined theoretically by Stevenson (1967). The relationship between the failure probability and the optimum weight of a structure was also investigated by Lind (1969). Moses and Kinser (1967) showed that the effect of the correlation of the optimal design depends largely on the load to strength variation coefficient ratio and only to a lesser extent on the chosen allowable failure probability P_f.

Let us consider the simplest case of probabilistic safety analysis, in which the structure consists of a single bar of strength R, loaded by a tensile force P, where R and P are random variables of known distributions (Moses 1969). The probability P_f of failure for such a model, i.e. the probability that $P > R$, may be computed from

$$P_f = \int_0^\infty [F_R(t)] f_P(t) \mathrm{d}t = 1 - \int_0^\infty [F_B(t)] f_R(t) \mathrm{d}t, \tag{4.117}$$

where $F(t)$ denotes the probability distribution and $f(t)$ the density or frequency distribution.

If we assume, for simplicity, that both the load and the strength are normally distributed, P_f can be computed as follows.

P and R being normal, the difference $Z = P - R$ is also normal, with mean value

$$\bar{Z} = \bar{P} - \bar{R}$$

and variance

$$\sigma_Z^2 = \sigma_P^2 + \sigma_R^2.$$

By definition, $P_f = P(Z > 0)$, and so

$$P_f = 1-\varphi\left[\frac{\bar{P}-\bar{R}}{\sqrt{\sigma_P^2+\sigma_R^2}}\right], \tag{4.118}$$

where φ is the normal distribution function in standard form. The result may conveniently be expressed in terms of the ratio $n = \bar{R}/\bar{P}$ as

$$n = \bar{R}/\bar{P},$$

$$P_f = 1-\varphi\left[-\frac{n-1}{\sqrt{\dfrac{\sigma_P^2}{\bar{P}^2}+\dfrac{\sigma_R^2 n^2}{\bar{R}^2}}}\right].$$

Introducing the coefficients of variation

$$\gamma_P = \frac{\sigma_P}{\bar{P}}; \quad \gamma_R = \frac{\sigma_R}{\bar{R}},$$

we get

$$P_f = 1-\varphi\left[-\frac{n-1}{\sqrt{\gamma_P^2+n^2\gamma_R^2}}\right].$$

This formula gives the failure probability as a function of the coefficient n and the coefficients of variation of the two random variables involved. Figure 4.22 shows a plot of P_f vs. n for $\gamma_P = 0.20$ and $\gamma_R = 0.10$.

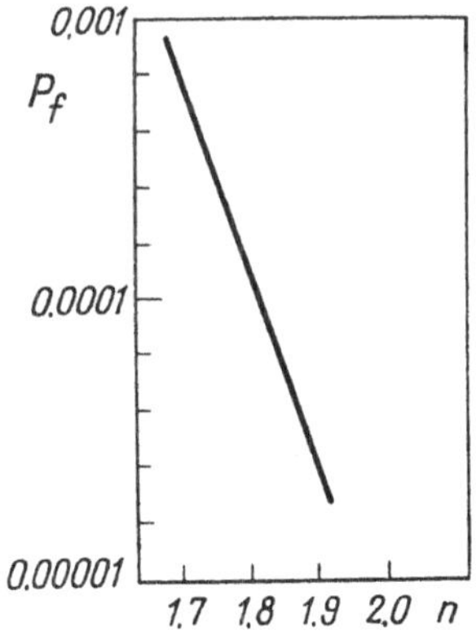

Fig. 4.22. Failure probability vs. safety factor

The above example is useful as an illustration of a manner of reasoning and a way to compute the failure probability. For a real structure, all structural members and their possible failure modes under various loading conditions during the entire lifetime of the overall structure must be taken into consideration.

The coefficient n introduced above corresponds to what in a deterministic approach is termed the *safety factor*. The conventional safety factors in most specifications have been developed in an evolutionary manner according to the experience based on the existing structures. In recent years, a great deal of work has been done on deriving safety factors from probabilistic safety analysis; their definitions vary depending on the parameters they relate to, aiming at computational facility. A set of partial safety factors accomodated to semi-probabilistic limit state design was proposed by the European Committee of Concrete and the International Federation of Prestressing in the form of International Recommendations in 1970.[1] A new, improved version of the Recommendations was published in 1976.[2] For want of sufficient statistical data, only conservative estimates were given for most of the coefficients, but underlying their derivation was the probabilistic concept of reliability; hence the term "semi-probabilistic method". This approach permits the results of new investigations and observations to be introduced gradually as they are acquired and elaborated statistically.

Reliability-based optimization

The classical formulation of an optimum design problem reads (Moses 1969, Brandt 1973):

$$\textit{Minimize } F(x_i) \textit{ subject to } g_j(x_i) \geqslant 0; \tag{4.120}$$

x_i denote design variables, $F(x_i)$ is an objective function and $g_j(x_i)$ are constraints.

The design variables x_i represent the geometrical and the mechanical properties of the structure that must be determined. The objective function $F(x_i)$ measures the volume, weight or cost of the structure, or some other quantity chosen as a criterion of optimization, e.g. elastic strain energy. Constraints $g_j(x_i)$ follow from strength and strain requirements, but may also include fabrication or functional requirements. Those constraints which limit the stresses or strains to some permissible values involve the conventional safety factors, which in the best of situations are determined from probabilistic and statistical analyses. Constraints imposed by the conditions of execution or use of the structure are regarded as deterministic.

In the reliability-based approach the numerous constraints on stresses and strains are replaced by a single condition on the failure probability P_f as a function of the design variables:

---

[1] *Comité Européen du Béton et Fédération Internationale de la Précontrainte—Récommandations Internationales pour le calcul et l'exécution des ouvrages en béton*, VI-ème Congrés de la FIP, Juin 1970, Prague.

[2] Common Unified Rules for Different Types of Construction and Material, *CEB Bulletin d'Information, No.* 116, Paris 1976.

Minimize $F(x_i)$ *subject to* $P_f(x_i) \leqslant P_f$ *permissible.* (4.121)

Fabrication or other side constraints can also be imposed of the form $g_j(x_i) \geqslant 0$.

The limit value of the failure probability should be determined with regard to all possible failure modes, the value of the structure and the cost of its failure and its consequences. The probability P_f is derived from the statistical distributions of the random variables representing the loading and strength of the structure. The optimization problem consists in determining the values of the design variables such that the objective funtion F, e.g. volume or cost, will be minimum and the probability of failure will not exceed the allowable value.

For multi-member structures, the failure probability P_f can be approximated by the sum of the probabilities of failure of the individual members; these can be determined in a manner similar to that discussed above in the fundamental one-member one-load case. The approximation is good if all the probabilities involved are small. The optimal structure has its members proportioned so that the overal objective function F is minimum and the failure probabilities of the individual members add up to an overall probability of failure not exceeding a prescribed permissible value. Problems formulated in this fashion were considered by Hilton and Feigen (1960), Switzky (1964), Khachaturian and Haider (1966) and Moses and Kinser (1967), among others.

If the function $F(x_i)$ represents the cost of the structure, it can—in the simplest case—be taken as proportional to the volume or weight. For more realistic design, however, it is necessary to include in it both economy and safety components. One alternative formulation (Moses, 1969) defines the total cost C as the sum

$$C = C_i + P_f C_f, \tag{4.122}$$

where C_i is the initial cost of erecting the structure and C_f is the cost associated with structural failure; the failure cost consists of the cost of reconstruction assumed to be equal to the initial cost and another term C' expressing the consequences of failure (damage):

$$C_f = C_i + C'. \tag{4.123}$$

Another approach to reliability optimization is to set a permissible value of $F(x_i)$, e.g. the material volume of the structure, and to seek the distribution of material over different parts or members for a minimum of the failure probability P_f. If the optimum P_f is too large, then either the assigned volume or the feasibility of the kind of the structure adopted must be re-evaluated (Hilton and Feigen, 1960; Moses and Kinser, 1967; Moses, 1973).

Both approaches to reliability optimization—design for minimum total cost subject to a failure probability constraint and design for minimum failure probability with a volume constraint—are discussed and illustrated with examples in a recent paper by Frangopol and Rondal (1978).

An instructive example of a minimum-volume reliability-based design of an isostatic truss was given by Khachaturian (1969). The problem is solved for cases where both the concentrated load and strength of each bar of the truss follow the Lognormal or the Gamma probability distributions, with the overall probability of failure prescribed as 10^{-3}, 10^{-4} or 10^{-5} and for several different values of the coefficients of variation of load (v_x) and strength (v_{yk}). The truss is treated as a "weakest-link" structure. The results of calculations are assembled in tables, showing the influence of the distributions of the random variables involved, the coefficients of variation and the value of P_f on the optimal volumes of the individual bars and the overall volume of the truss. The relationship between the optimal volume V and the specified probability of failure is plotted in Fig. 4.23 for three different values of v_x

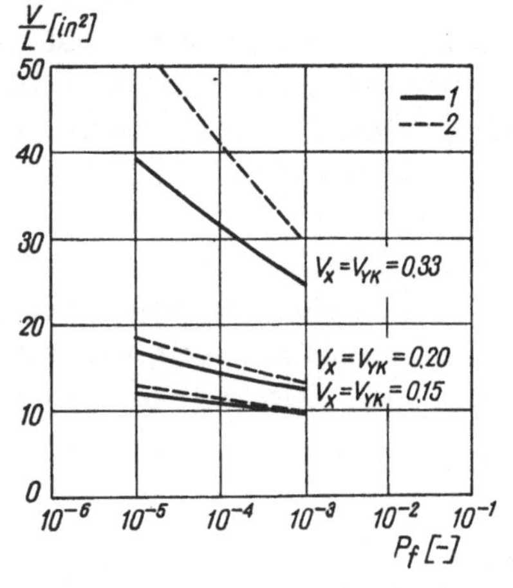

Fig. 4.23. Variation of optimal volume with probability of failure for three different values of $v_x = v_{yk}$. 1—Lognormal distribution, 2—Gamma distribution

$= v_{yk}$. The results show that the validity of the optimal solution depends on the degree of knowledge of the load and strength probability distributions and on the choice of an appropriate level of safety.

Let us note, after Khachaturian, that the design considered in the example ignores certain important factors, e.g. deformability or serviceability requirements, that the type of structure and its geometry are fixed and that the structure is idealized to a pin-jointed truss.

Stevenson (1967) considered reliability-based optimum design of hyperstatic structures. His analysis can be applied to frames, trusses and grids, and even to plates treated by the fracture line method, that is, to all cases where the ultimate limit state function can be represented as a linear combination of the random variables of load and strength.

By analysing optimal solutions obtained under various assumptions it is possible to observe and formulate certain relationships between different parameters of an optimum structure. Such relationships can be useful in simplifying specific optimization problems. As an example, let us mention a result obtained by Switzky (1964); developing the ideas of Hilton and Feigen (1960), he showed that in an optimum structure under a single load the following relationship holds:

$$\frac{\text{weight of member } i}{\text{total weight}} = \frac{P_f \text{ of member } i}{P_f \text{ overall allowable}} .$$

Prospects of reliability-based structural optimization

The goals to be achieved by exercising a probabilistic approach to safety are easy to formulate, although their realization does not seem possible in the near future. The first objective is to base the code safety factors, which relate to strengths, loads and other random variables determining the safety of a structure, on a firm probabilistic foundation. An important issue is also to eliminate the inconsistencies present in deterministic design. The general objective is to reach economic gains by designing structures cheaper than those based on deterministic premises but equally reliable; or, alternatively, to increase the safety of structures without increasing the cost.

Reliability-based optimization leads to designs in which structure and member sizes are optimal with respect to an adopted criterion, say of minimum volume or cost, the overall failure probability being equal to a prescribed allowable value. The safety factors of the individual members in such a design are not equal and depend on the degree of statistical correlation between the failures of those members. Also, to different failure modes of a given structure there correspond different safety factors.

The efforts which are being made to incorporate a probabilistic approach to safety into structural optimization should stimulate studies of random variables encountered in structural engineering. As we have indicated before, the choice of probability distributions and their parameters for a design strongly influences the optimal solution; the lack of sufficient observational and experimental data makes it impossible to choose correctly, and hence to obtain a useful result. Empirical studies of load conditions are particularly desirable as the existing statistical data in this area are much poorer than those relating to the strength of materials and structural elements. As pointed out by Lind (1968), small failure probability limitations and very meagre statistical data on load and strength probability distributions do not allow probability statements of the same confidence levels that statisticians usually have in other fields.

There is also the problem of providing a rational basis for determining the allowable probability of failure for a given structure. Although complete elimination of the subjective element and intuition is hardly conceivable, detailed statistical data are indispensable if serious errors are to be avoided. The data should include detailed information about the safety of the existing structures of the kind considered; information on safety in other areas, e.g. transport, would also be useful for comparative studies.

At the present state of knowledge the value of the existing solutions of probabilistic structural design problems is primarily theoretical. They allow verification of the design procedures currently used and give an insight into the possibilities of their further evolution. Above all, however, they make it possible to re-evaluate deterministic design solutions and to asess the potential advantages of further work in this direction, based on thorough statistical studies.

Thus, although structural design and optimization continue to rely on deterministic methods, which have indeed led to many interesting and important solutions, it seems more rational—even at the present state of knowledge—to treat loads, strengths, deformability and other relevant quantities as random variables. We can thus expect a further development of structural optimization methods based on a probabilistics approach to safety and an analogous evolution of load standards, material specification and methods of observation and testing of structures. New, improved mathematical methods will also be called for, considering that probabilistics design problems formulated in compliance with actual conditions and requirements, i.e. without oversimplification, can be reduced to stochastic programming (cf. Chapter 9).

PART II

METHODS OF STRUCTURAL OPTIMIZATION

5

Introduction to Mathematical Methods of Optimization

5.1 Problem formulation in optimum structural design

The basic concepts involved in the formulation of any optimum-design problem are: the design variables, the objective function and the feasible design space.

Design variables are quantities which describe the given structure and are subject to variation in the optimization process. They may represent the geometrical dimensions of individual members, the configuration or geometrical layout of the structure and the mechanical or physical properties of the material as well as other relevant aspects of the design.

The simplest to handle are "size variables", such as the width or depth of a member section, the cross-sectional area of a bar in a frame, the moment of inertia of a flexural member, or the thickness of a plate.

Configuration variables are often represented by the coordinates of element joints. Finding an optimum configuration for a structure is usually a much more complicated problem than optimizing the section properties for members in a given layout. The mechanical properties of the structural materials, next in order of difficulty, may also be a subject of optimization.

Design variables may be continuous or discrete. Many of the variables that occur in actual design are discrete, e.g. standard structural section sizes or sheet thicknesses. However, when the values of a discrete variable are distributed sufficiently densely over an interval, a continuous variable treatment followed by the selection of an acceptable close-to-optimum discrete value is often a satisfactory expedient.

Mathematically, the full set of design variables for a given structure is represented as a vector $\mathbf{x}^T = (x_1, x_2, \ldots, x_n)$ in an n-dimensional space (hyperspace) which we call the *design space*. A point q in that space corresponds to a design with variables $\mathbf{x}_q$. The search for an optimum design is often realized in a series of design modifications, or moves along a sequence of points in the design space. Transition from a point q to a point $q+1$ can be described by the equation

$$\mathbf{x}_{q+1} = \mathbf{x}_q + \lambda_q \mathbf{s}_q; \tag{5.1}$$

the vector $\mathbf{s}_q$ defines the direction of the move and λ_q gives its length.

The objective function in structural optimization, denoted here by F, is a mathematical expression describing a certain property of the structure, e.g. weight or cost, chosen as a basis or criterion for the selection of one of acceptable alternative designs. In most formulations, the "best" of the acceptable designs is the one for which the objective function takes the smallest value, but even if a maximum is originally required, the problem can easily be converted into one of minimization by changing the sign of the objective function: $\max F(\mathbf{x}) = -\min(-F(\mathbf{x}))$.

The arguments of the objective function are the design variables $\mathbf{x}$. The design variables may be unknown functions or coefficients of known functions or simply numerical variables. When they are functions, the objective function is a functional.

When the objective function is linear, it represents a hyperplane in the n-dimensional design space. When it is non-linear, the locus of all design points with a given value of F is a hypersurface in this space. Valuable information about the behaviour of the objective function in the design space can be derived from its gradient, ∇F. The gradient represents a vector normal to the surface $F = \text{const}$; its components are the derivatives of F with respect to each of the design variables:

$$\nabla F^T = \left[\frac{\partial F}{\partial x_1}, \frac{\partial F}{\partial x_2}, \ldots, \frac{\partial F}{\partial x_n}\right]. \tag{5.2}$$

The symbol T denotes transposition.

The feasible design space is a region in the design space delimited by the constraints, or conditions, which the design is required to satisfy. They may be imposed directly on the design variables or may represent limitations on certain quantities dependent on the design variables. In algebraic terms, constraints take the form of equalities or inequalities:

$$\begin{aligned} g_i(\mathbf{x}) &= 0, \quad i = 1, 2, \ldots, r, \\ g_i(\mathbf{x}) &\leqslant 0, \quad i = r+1, \ldots, m. \end{aligned} \tag{5.3}$$

When an equality constraint is given, we may try to reduce the number of dimensions of the design space by solving the equality and eliminating one of the design variables. In practice, however, this is not always possible.

Two general classes of constraints can be distinguished: side constraints and behaviour constraints. Side constraints limit the maximum or the minimum dimensions of structural elements. In some cases they may specify the commercially available range of a product, e.g. the maximum or minimum thickness of metal sheets. Side constraints are explicit in form and may be imposed on individual design variables, e.g. geometrical dimensions, or on relationships involving several variables, e.g. cross-sectional areas and cross-sectional moments of inertia.

Behaviour constraints are usually limitations on stresses or displacements but may also concern such factors as vibrational frequency or buckling strength. Generally, they occur in an implicit form, i.e. are imposed on quantities not directly expressible in terms of the design variables.

In the design space, each constraint is represented by a hypersurface consisting of all points at which the constraint is satisfied as an equality constraint. The normal vector to a constraint surface g_i is given by the gradient

$$\nabla g_i^T(\mathbf{x}) = \left[\frac{\partial g_i(\mathbf{x})}{\partial x_1}, \ldots, \frac{\partial g_i(\mathbf{x})}{\partial x_n}\right]. \tag{5.4}$$

When the design variables are real numbers, the design space can be interpreted geometrically as a Cartesian n-space. The case $n = 2$ is illustrated in Fig. 5.1. The optimum design problem often amounts to finding the point

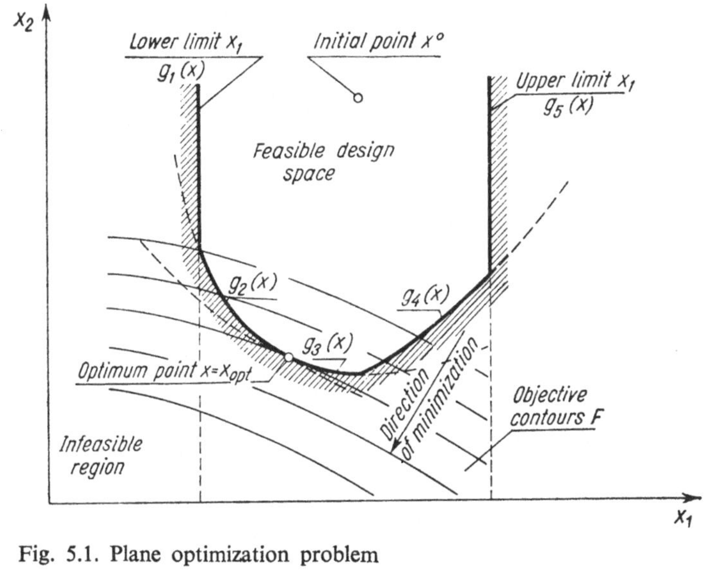

Fig. 5.1. Plane optimization problem

of tangency between a curve (or straight line) representing the objective function and the boundary of the feasible design space.

The general optimization problem can be stated as follows. Let X be a vector space (design space), Q_i $(i = 1, 2, \ldots, m)$ a family of its subsets (constraints) and F: $X \to R$ a functional defined in X (objective function). It is required to find a vector $\hat{\mathbf{x}}$ in X such that

$$\hat{\mathbf{x}} \in Q = \bigcap_{i=1}^{m} Q_i \quad \text{(feasible design space)} \tag{5.5}$$

and

$$F(\hat{\mathbf{x}}) = \inf_{\mathbf{x} \in Q} F(\mathbf{x}). \tag{5.6}$$

The constraint sets Q_i are usually of the form

$$\begin{aligned} Q_i &= \{\mathbf{x}: g_i(\mathbf{x}) = 0\}, \quad i = 1, 2, \dots, k, \\ Q_i &= \{\mathbf{x}; g_i(\mathbf{x}) \leqslant 0\}, \quad i = k+1, \dots, m, \end{aligned} \tag{5.7}$$

where g_i, $i = 1, \dots, m$, are functionals defined in X. The necessary and sufficient condition for a vector $\hat{\mathbf{x}} \in Q$ to be the required solution is that the common part of Q and the set

$$Q_0 = \{\mathbf{x}: F(\mathbf{x}) < F(\hat{\mathbf{x}})\}$$

be void:

$$Q_0 \cap Q = \emptyset. \tag{5.8}$$

When the design variables are numerical variables, the objective function is a function of one or more variables (and not a functional) and the problem is termed a *mathematical programming problem* or a *mathematical program*. In a standard form, a general mathematical program is stated as

Minimize

$$F(\mathbf{x}) \tag{5.9}$$

subject to

$$\begin{aligned} g_j(\mathbf{x}) &= 0, \quad j = 1, 2, \dots, r, \\ g_j(\mathbf{x}) &\leqslant 0, \quad j = r+1, \dots, m. \end{aligned} \tag{5.10}$$

In a more popular formulation, only inequality constraints are included.

According as both the objective function and constraints in a mathematical programming problem are or are not linear, we speak of *linear* or *non-linear programming*. Methods of linear programming are discussed in Chapter 6. Two important particular cases of non-linear programming are *quadratic programming* and *geometric programming*. The first method is used when the constraints are linear while the objective function is the sum of a linear form and a quadratic form. The method of geometric programming has been developed to deal with cases where the objective function and constraints are polynomials of arbitrary degrees (Chapter 7).

When the design variables are functions, $x_i = x_i(\mathbf{t})$, the objective function becomes a functional, which for a large class of problems can be represented as

$$F[x_i(\mathbf{t})] = \int G[x_i(\mathbf{t}), x_i'(\mathbf{t}), \mathbf{t}]\,dt, \tag{5.11}$$

with constraints also in the form of integral conditions:

$$\begin{aligned} &\int g_j[x_i(t), x_i'(t), t]dt = 0, \quad j = 1, 2, \ldots, r, \\ &\int g_j[x_i(t), x_i'(t), t]dt \leqslant 0, \quad j = r+1, \ldots, m. \end{aligned} \tag{5.12}$$

Problems of this kind are solved by the methods of the calculus of variations. When only equality integral constraints are present, the problem becomes a classical isoperimetric variation problem (Chapter 10). Otherwise, non-classical variational methods have to be used (Chapter 11).

According as the design variables are time-independent or time-dependent, we distinguish *static* or *dynamic problems*. The optimum-design problems considered in Chapters 1–4 are examples of static problems. A typical example of a dynamic problem is the optimum control of a process, in which the decision variables depend on time and are corrected as the conditions under which the given system performs change. Many optimum-structural-design processes can also be represented as decision processes in which decisions are made in a sequence of steps, even though the design variables have nothing to do with time. Dynamic programming is a mathematical approach

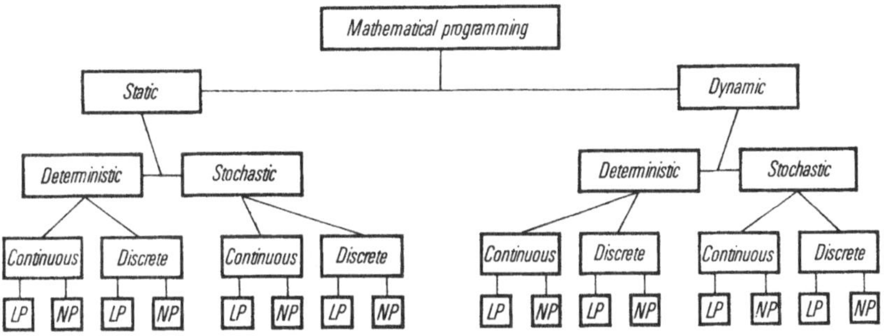

Fig. 5.2. Mathematical programming classification scheme. LP—linear programming, NP—non-linear programming

to solving such multistage decision problems. In Chapter 8 we present methods of dynamic programming in application to the optimum design of multimember structures.

Another division of optimization problems is into deterministic and non-deterministic problems. In deterministic formulations, all the quantities involved are assumed to have precise and unambiguous values. In a probability-based approach (cf. Section 4.6), the quantities describing the state of a given system can be random variables. Non-deterministic problems can be solved with the use of stochastic programming, which reduces them to deterministic non-linear programming problems. Stochastic programming is discussed in Chapter 9.

Finally, depending on whether the design variables are continuous or discrete, i.e. take all or only some values from an interval, we deal with *continuous* or *discrete* programming problems.

A classification scheme of mathematical programming problems is shown in Fig. 5.2.

5.2 Classical extremum theory

The classical methods of optimization are analytical methods based on differential calculus. They are useful in determining the extrema of continuous and differentiable functions. They have also provided a theoretical basis for the development of the numerical techniques of optimization. Below, we bring together the fundamental theorems and formulae of the classical extremum theory for functions of one variable and for functions of several variables with and without constraints.

Necessary and sufficient conditions for an extremum of a function of one variable:

THEOREM 1. *If a function $f(x)$ defined in an interval $a \leqslant x \leqslant b$ has an extremum (a local minimum or maximum) at a point $x = \hat{x}$, $a < \hat{x} < b$, and the derivative f' exists at $\hat{x}$, then $f'(\hat{x}) = 0$.*

THEOREM 2. *Let $f'(\hat{x}) = f''(\hat{x}) = \ldots = f^{(n-1)}(\hat{x}) = 0$ and $f^{(n)}(\hat{x}) \neq 0$. If n is even and $f^{(n)}(\hat{x}) > 0$ then the function f has a minimum at $x = \hat{x}$. If n is even and $f^{(n)}(\hat{x}) < 0$, the function f has a maximum at $x = \hat{x}$. If n is odd then there is neither a maximum nor a minimum at $\hat{x}$.*

Necessary and sufficient conditions for an unconstrained extremum of a function of several variables:

THEOREM 3. *If a function $f(\mathbf{x})$ attains an extremum at a point $\mathbf{x} = \hat{\mathbf{x}}$ and the first partial derivatives of $f(\mathbf{x})$ exist at $\hat{\mathbf{x}}$, then*

$$\frac{\partial f}{\partial x_1}(\hat{\mathbf{x}}) = \frac{\partial f}{\partial x_2}(\hat{\mathbf{x}}) = \ldots = \frac{\partial f}{\partial x_n}(\hat{\mathbf{x}}) = 0. \tag{5.13}$$

Every point $\hat{\mathbf{x}}$ satisfying condition (5.13) is called a *stationary point* of the function f.

THEOREM 4. *A sufficient condition for a stationary point $\hat{\mathbf{x}}$ to be a minimum (maximum) point is that the second differential*

$$\mathrm{d}^2 f = \sum_{i=1}^{n} \sum_{j=1}^{n} \frac{\partial^2 f}{\partial x_i \partial x_j} \mathrm{d}x_i \mathrm{d}x_j \tag{5.14}$$

be positive-definite (negative-definite) at $\hat{\mathbf{x}}$. If differential (5.14) *is indefinite, there is no extremum at $\hat{\mathbf{x}}$; if it is only semi-definite, no conclusion can be drawn.*

Differential (5.14) is a quadratic form of the variables dx_k, $k = 1, 2, \ldots, n$. Let us recall that a quadratic form is said to be *positive-definite* (*negative-definite*) if it takes a positive (negative) value for every set of arguments not all zero.

By matrix algebra, Theorem 4 can be restated as

THEOREM 4′. *A sufficient condition for a function* $f(\mathbf{x})$ *to have a minimum (a maximum) at a stationary point* $\hat{\mathbf{x}}$ *is that the matrix of the second partial derivatives of* f *(the Hessian matrix) be positive-definite (negative-definite) at* $\hat{\mathbf{x}}$.

A matrix $\mathbf{A}$ is *positive-definite* if all its eigenvalues, i.e. all the numbers λ such that $\det(\mathbf{A} - \lambda\mathbf{I}) = \mathbf{0}$, where $\mathbf{I}$ is the unit matrix, are positive.

In order to verify whether a matrix $\mathbf{A} = [a_{ij}]_{i,j \leqslant n}$ is positive or negative-definite we calculate the determinants $A_k = \det[a_{ij}]_{i,j \leqslant k}$, $k = 1, 2, \ldots, n$. The matrix $\mathbf{A}$ is *positive-definite* if and only if $A_k > 0$ for all $k = 1, 2, \ldots, n$; it is *negative-definite* if and only if $(-1)^k A_k > 0$, $k = 1, 2, \ldots, n$. If non-sharp inequalities are allowed, the matrix is only *semi-definite*.

Theorem 4 can be generalized as follows:

THEOREM 5. *Let all the partial derivatives up to an order* $k \geqslant 2$ *of a function* $f(\mathbf{x})$ *be continuous in the neighbourhood of a point* $\hat{\mathbf{x}}$ *and let* $d^k f|_{\hat{\mathbf{x}}}$ *be the lowest-order non-vanishing differential at* $\hat{\mathbf{x}}$, *i.e.* $d^r f|_{\hat{\mathbf{x}}} = 0$, $r = 1, 2, \ldots, k-1$, *and* $d^k f|_{\hat{\mathbf{x}}} \neq 0$. *If* k *is even, then*

(1) f *has a local minimum at* $\hat{\mathbf{x}}$ *if* $d^k f|_{\hat{\mathbf{x}}}$ *is positive-definite,*

(2) f *has a local maximum at* $\hat{\mathbf{x}}$ *if* $d^k f|_{\hat{\mathbf{x}}}$ *is negative-definite,*

(3) *nothing can be said if* $d^k f|_{\hat{\mathbf{x}}}$ *is only semi-definite.*

If k *is odd, there is no extremum at* $\hat{\mathbf{x}}$.

When the Hessian matrix of a function of several variables is indefinite (i.e. neither positive- nor negative-definite) at a stationary point, the function has no extremum at that point, which is then called a *saddle point.* A characteristic feature of a saddle point of a function of two variables, say x and y, is that it is an extremum point with respect to each variables separately, but different in kind for each variable, e.g. a maximum with respect to x (with y held constant at $\hat{y}$) and a minimum with respect to y (with x kept constant at $\hat{x}$). A similar characterization can also be given of saddle points of functions of more than two variables, with x and y replaced by multidimensional vectors.

Necessary and sufficient conditions for a constrained extremum of a function of several variables

We first consider the case where an extremum of a function $f(\mathbf{x})$ is sought subject to equality constraints $g_j(\mathbf{x}) = 0$, $j = 1, 2, \ldots, m$. We assume that $m \leqslant n$; otherwise the problem will have no solution. There are a few methods by which one can seek the constrained extrema of a function. The best known is the method of Lagrange multipliers:

THEOREM 6. *A necessary condition for a function* $f(\mathbf{x})$ *to have a constrained extremum at a point* $\hat{\mathbf{x}}$, *the constraints being* $g_j(\mathbf{x}) = 0, j = 1, 2, \ldots, m$, *is that there are numbers* $\hat{\lambda}_1, \hat{\lambda}_2, \ldots, \hat{\lambda}_m$ *(Lagrange multipliers) such that the first partial derivatives of the Lagrange function*

$$L(\mathbf{x}, \boldsymbol{\lambda}) = f(\mathbf{x}) + \sum_{j=1}^{m} \lambda_j g_j(\mathbf{x}) \cdot \tag{5.15}$$

with respect to each of the variables $x_i, i = 1, 2, \ldots, n$ and $\lambda_j, j = 1, 2, \ldots, m$, *vanish at* $(\mathbf{x}, \boldsymbol{\lambda}) = (\hat{\mathbf{x}}, \hat{\boldsymbol{\lambda}})$, *where* $\hat{\boldsymbol{\lambda}} = (\hat{\lambda}_1, \hat{\lambda}_2, \ldots, \hat{\lambda}_m)$.

THEOREM 7. *A sufficient condition for a point* $\hat{\mathbf{x}}$ *satisfying the conditions of Theorem 6 to be a constrained minimum point for the function* $f(\mathbf{x})$ *is that the quadratic form*

$$Q = \sum_{i=1}^{n} \sum_{j=1}^{n} \frac{\partial^2 L}{\partial x_i \partial x_j} \mathrm{d}x_i \mathrm{d}x_j \tag{5.16}$$

be positive-definite at $\hat{\mathbf{x}}$ *for all* $\mathrm{d}\mathbf{x}$ *for which* $\hat{\mathbf{x}} + \mathrm{d}\mathbf{x}$ *satisfies the constraints.*

When a function $f(\mathbf{x})$ is to be minimized subject to inequality constraints $g_j(\mathbf{x}) \leqslant 0, j = 1, 2, \ldots, m$, the problem can first be transformed to an equality-constrained problem by introducing non-negative slack variables y_j^2 such that

$$g_j(\mathbf{x}) + y_j^2 = 0, \quad j = 1, 2, \ldots, m. \tag{5.17}$$

Then, the method of Lagrange multipliers can be used (Theorem 6), with the Lagrange function

$$L(\mathbf{x}, \mathbf{y}, \boldsymbol{\lambda}) = f(\mathbf{x}) + \sum_{j=1}^{m} \lambda_j [g_j(\mathbf{x}) + y_j^2]. \tag{5.18}$$

As a result, one obtains the following

THEOREM 8. *A necessary condition for a function* $f(\mathbf{x})$ *to have a constrained minimum at a point* $\hat{\mathbf{x}}$, *the constraints being* $g_j(\mathbf{x}) \leqslant 0, j = 1, 2, \ldots, m$, *is that there exist* $\lambda_1, \ldots, \lambda_j$ *such that, at* $\hat{\mathbf{x}}$,

$$\begin{aligned} &\frac{\partial f}{\partial x_i} + \sum_{j=1}^{m} \lambda_j \frac{\partial g_j}{\partial x_1} = 0, \quad i = 1, 2, \ldots, n, \\ &\lambda_j g_j = 0, \\ &g_j \leqslant 0, \quad j = 1, 2, \ldots, m. \\ &\lambda_j \geqslant 0, \end{aligned} \tag{5.19}$$

If the function is to be maximized or if the constraints are of the form $g_j \geqslant 0$, *all* λ_j must be non-positive. Conditions (5.19) are known as *Kuhn-Tucker conditions.*

All the extremum conditions given above are conditions of local extrema. There is a class of extremum problems, however, called *convex problems,* for which they become conditions of global extrema. The basic definitions and theorems relating to the property of convexity are the subject of the next section.

5.3 Convex regions and convex functions

A set P in a Euclidean space E^n is said to be *convex* if, for any $\mathbf{x}_1, \mathbf{x}_2 \in P$ and $0 < \theta < 1$, the vector

$$\mathbf{x} = \theta \mathbf{x}_1 + (1-\theta)\mathbf{x}_2 \tag{5.20}$$

also belongs to P. Geometrically, this means that if the ends of a line segment belong to P then the entire segment is contained in P (Fig. 5.3).

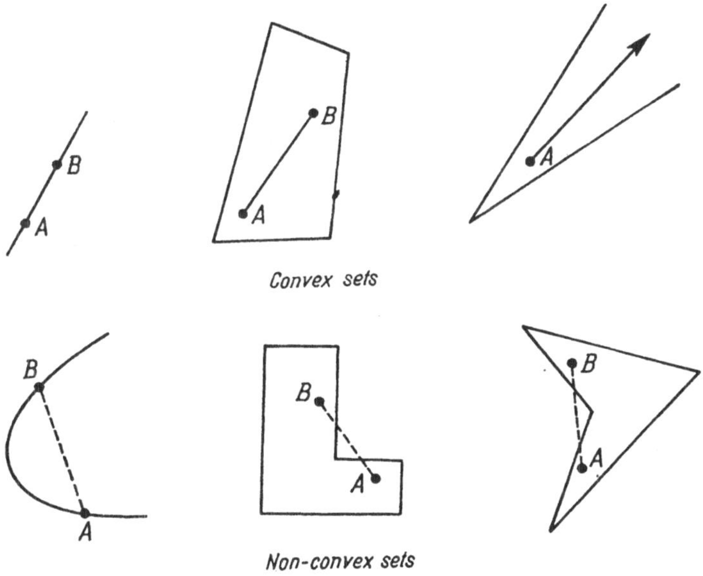

Fig. 5.3. Examples of convex and non-convex sets

A closed convex set will be called a *convex domain.* A convex domain may be unbounded and may lie in a linear manifold of a dimension lower than n. The intersection of a finite number of convex domains is also a convex domain (Fig. 5.4).

If a convex domain P is unbounded, then for every point $A \in P$ there is a vector $\mathbf{t} \in E^n$ such that for arbitrary $\lambda \geqslant 0$ the point $\mathbf{x}_A + \lambda \mathbf{t}$ also belongs to P.

A function $F(\mathbf{x})$ defined on a convex domain $P \subset E^n$ is said to be *convex* if for every two points $\mathbf{x}_1$ and $\mathbf{x}_2$ of P and arbitrary $0 < \theta < 1$

$$F(\theta \mathbf{x}_1 + (1-\theta)\mathbf{x}_2) \leqslant \theta F(\mathbf{x}_1) + (1-\theta)F(\mathbf{x}_2). \tag{5.21}$$

The function F is called *strictly convex* if the above condition holds as a sharp inequality.

The values that a convex function takes along a straight line segment between two arbitrary points cannot be larger than the corresponding values of the linear function coinciding with the given function at those points (Fig. 5.5).

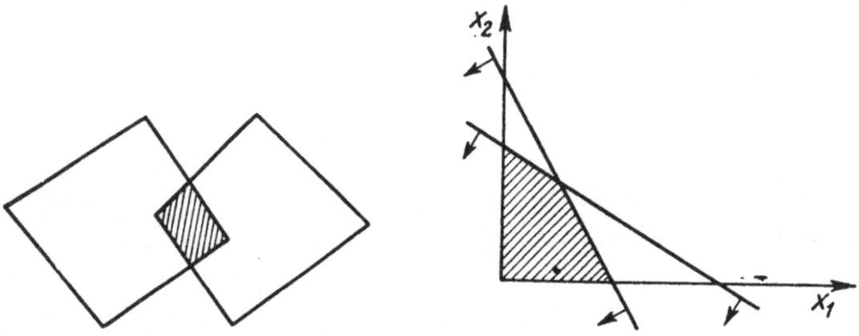

Fig. 5.4. Intersection of convex sets

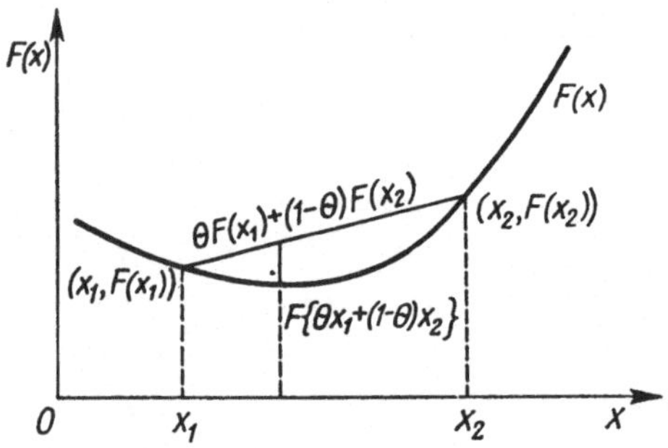

Fig. 5.5. A convex function

A *concave* function is defined in a similar manner, with the sign $\leqslant$ replaced by $\geqslant$ in inequality (5.21).

The linear function is both convex and concave.

Below we recall without proofs[1] some properties of convex functions.

(1) $F(\mathbf{x}_2) - F(\mathbf{x}_1) \geqslant \text{grad}\, F(\mathbf{x}_1)^T(\mathbf{x}_2 - \mathbf{x}_1)$ for arbitrary $\mathbf{x}_1$ and $\mathbf{x}_2 \in P$.

(2) A linear combination with positive coefficients of two or more convex functions is also convex; if at least one of the original functions is strictly convex, the combination will also be strictly convex.

[1] For proofs see e.g. GASS, S. I., *Linear Programming, Methods and Applications*, Third Edition, McGraw-Hill, New York 1969.

(3) Every local minimum of a convex function defined on a convex set is a global minimum; the set of all the minima is convex.
(4) A strictly convex function attains a minimum at exactly one point or is unbounded from below.
(5) The Hessian matrix $\left[\dfrac{\partial^2 F(\mathbf{x})}{\partial x_i \, \partial x_j}\right]$ is positive-definite for all $\mathbf{x}$.

Convex programming problem

A constrained optimization problem stated as

Minimize $F(\mathbf{x})$ *subject to*

$$g_j(\mathbf{x}) \leqslant 0, \quad j = 1, 2, \ldots, m, \tag{5.22}$$

is said to be *convex* if both the objective function F and the constraint functions g_j, $j = 1, \ldots, m$, are convex.

As mentioned in the preceding section, the Lagrange function for the problem is constructed as

$$L(\mathbf{x}, \mathbf{y}, \boldsymbol{\lambda}) = F(\mathbf{x}) + \sum_{j=1}^{m} \lambda_j (g_j(\mathbf{x}) + y_j^2), \tag{5.23}$$

where $\mathbf{y}^T = (y_1, \ldots, y_m)$ is the *slack variable vector*. If $\lambda_j \geqslant 0$, then the products $\lambda_j g_j(\mathbf{x})$ are convex, and since the extremum condition requires that

$$\frac{\partial L}{\partial y_j}(\mathbf{x}, \mathbf{y}, \boldsymbol{\lambda}) = 2\lambda_j y_j = 0, \quad j = 1, \ldots, m, \tag{5.24}$$

we may conclude that L is also a convex function. A necessary condition for the function F to have a constrained minimum at a point $\hat{\mathbf{x}}$ is that $\hat{\mathbf{x}}$ is a stationary point of L. However, since L is a convex function, its derivative may vanish at one point only, which therefore must be the global constrained minimum point for the function F. Thus the Kuhn–Tucker conditions are both necessary and sufficient for the solution of a convex programming problem.

6

Linear Programming

6.1 The linear programming problem

We recall from Section 5.1 that a mathematical program "Minimize $F(\mathbf{x})$ subject to $g_j(\mathbf{x}) = 0$ $(j = 1, ..., r)$ and $g_j(\mathbf{x}) \leqslant 0$ $(j = r+1, ..., m)$" is called a *linear programming problem* if the objective function F and the constraint functions g_j are all linear.

Since it was first formulated by Dantzig in the late 1960s, the general linear programming problem has been investigated by many authors and efficient methods have been developed for either solving it or showing that a solution does not exist.

Linear programming has proved extremely useful in many fields, including transportation and production planning, storage control and network flow control, economics and management. Many optimization problems in engineering design, particularly in the design of bar structures, can also be reduced to linear programs or at least approximated by linear programming solutions. In addition, even when a non-linear problem has to be solved, an iterative procedure can sometimes be used, converting the problem into a series of linear programs.

As examples of optimum-structural-design problems which reduce to linear programming we may mention the following:

(i) Find a minimum-volume truss with given support conditions, configuration and loading. The constraints are the equilibrium equations at the joints and the allowable stresses for each bar.

(ii) Determine the prestressing forces in a multi-span beam from the condition of minimum volume of the prestressing cables, given the feasible cable layouts. The constraints are the allowable stresses under multiple loading conditions.

A general linear programming problem can be formulated as follows:[1]
Minimize the function

$$F(\mathbf{x}) = c_1 x_1 + \ldots + c_n x_n \tag{6.1}$$

[1] Kalichman, I. L., *Linear Algebra and Programming*, Moskva 1967 (in Russian).

subject to the equality and inequality constraints

$$\begin{aligned}
&a_{11}x_1+a_{12}x_2+ \dots a_{1n}x_n = b_1, \\
&\dots\dots\dots\dots\dots \\
&a_{h1}x_1+a_{h2}x_2+ \dots +a_{hn}x_n = b_h, \\
&a_{h+1,1}x_1+a_{h+1,2}x_2+ \dots +a_{h+1,n}x_n \leqslant b_{h+1}, \\
&\dots\dots\dots\dots\dots \\
&a_{m1}x_1+a_{m2}x_2+ \dots +a_{mn}x_n \leqslant b_m,
\end{aligned} \tag{6.2}$$

and

$$x_1 \geqslant 0, \dots, x_t \geqslant 0 \quad (t \leqslant n). \tag{6.3}$$

If all constraints (6.2) are equalities, the problem is said to be in the *standard form.* In order to put an arbitrary linear program (6.1)–(6.3) into this form, we convert the inequalities to equalities by introducing new variables, called *slack variables.* Given an inequality

$$a_1x_1+a_2x_2+ \dots +a_nx_n \leqslant b, \tag{6.4}$$

we define a new variable $x_{n+1} \geqslant 0$ such that

$$a_1x_1+a_2x_2+ \dots +a_nx_n+x_{n+1} = b. \tag{6.5}$$

6.2 Constrained extrema of a linear function

We now recall the basic definitions and state without proofs the basic theorems relating to the feasible region for a system of linear equations and inequalities and to constrained extrema of linear functions.[1] We assume that the reader is familiar with the fundamentals of linear algebra.

We consider a linear programming problem in the standard form:

Minimize

$$F = \mathbf{c}^T\mathbf{x} \tag{6.6}$$

subject to the set of simultaneous linear equations

$$\mathbf{A}\mathbf{x} = \mathbf{b} \tag{6.7}$$

and non-negativity conditions

$$\mathbf{x} \geqslant \mathbf{0}, \tag{6.8}$$

where $\mathbf{c}^T = (c_1, \dots, c_n)$, $\mathbf{x}^T = (x_1, \dots, x_n)$, $\mathbf{A} = [a_{ij}]$ and $\mathbf{b} = (b_1, \dots, b_m)$.

Basic definitions

A *feasible solution* of the linear programming problem (6.6)–(6.8) is a vector $\mathbf{x}$ that satisfies conditions (6.7) and (6.8).

[1] Gass, S. I., *Linear Programming, Methods and Applications*, Third Edition, McGraw-Hill, New York 1969. See also Kalichman, I. L., *loc. cit.*, p. 202.

A *basic solution* of system (6.7) is the solution of the system obtained from (6.7) by setting $n-m$ variables equal to zero; the determinant of the reduced system is assumed to be non-zero. The m variables involved in the basic solution are called the *basic variables*.

A *basic feasible solution* is a basic solution satisfying condition (6.3), i.e. one in which all the basic variables are non-negative.

A *non-degenerate basic feasible solution* is a basic feasible solution in which all the basic variables are positive.

An *optimum feasible solution* is a feasible solution which minimizes function (6.6).

Properties of the feasible solutions of a linear programming problem

The set of all feasible solutions of a linear programming problem is convex. A feasible solution $\mathbf{x}$ is an extreme (corner) point of the feasible set if and only if it is a basic feasible solution.

If the region of feasible solutions to problem (6.6)–(6.8) is bounded, then every feasible solution can be represented as a convex linear combination of all the basic feasible solutions $\mathbf{x}_1, \ldots, \mathbf{x}_s$, i.e.

$$\mathbf{x} = \sum_{i=1}^{s} t_i \mathbf{x}_i, \tag{6.9}$$

where $\sum_{i=1}^{s} t_i = 1,\ t_i \geqslant 0.$

If program (6.6)–(6.8) has at least one feasible solution, it has at least one basic solution.

If the feasible region is bounded, then at least one basic feasible solution is an optimum solution.

If the feasible region is unbounded, an optimum solution exists and is given by at least one basic solution if and only if the objective function F is bounded from below in this region.

If $\mathbf{x}_1, \ldots, \mathbf{x}_l$ are all different optimal basic solutions to a linear programming problem, then any optimal solution $\hat{\mathbf{x}}$ can be represented as a convex combination of $\mathbf{x}_1, \ldots, \mathbf{x}_l$.

6.3 The graphical method of linear programming

The graphical method can only be applied to problems involving two variables. Let us consider a plane ($n = 2$) linear problem with inequality constraints:

Find $\mathbf{x} = (x_1, x_2)$ *which minimizes*

$$z = c_1 x_1 + c_2 x_2$$

subject to the inequalities

$$\begin{aligned} a_{11}x_1 + a_{12}x_2 &\leqslant b_1, \\ a_{21}x_1 + a_{22}x_2 &\leqslant b_2, \\ &\cdots\cdots\cdots \\ a_{r1}x_1 + a_{r2}x_2 &\leqslant b_r, \end{aligned} \tag{6.10}$$

and the non-negativity conditions

$$x_1 \geqslant 0, \quad x_2 \geqslant 0. \tag{6.11}$$

To solve this problem graphically, we proceed as follows:

— find in the x_1x_2-plane the region of feasible solutions, i.e. the set of all points (x_1, x_2) satisfying inequalities (6.10) and (6.11);

— draw the straight line $c_1x_1 + c_2x_2 = 0$, which is the line of intersection of the plane $z = c_1x_1 + c_2x_2$ with the x_1x_2-plane, and indicate the vector

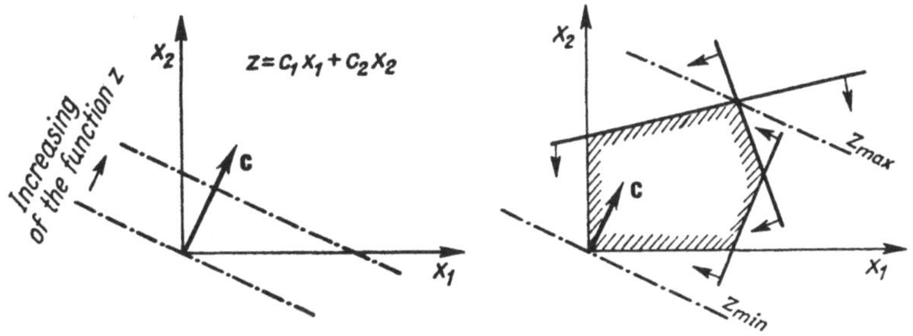

Fig. 6.1. Graphical linear programming

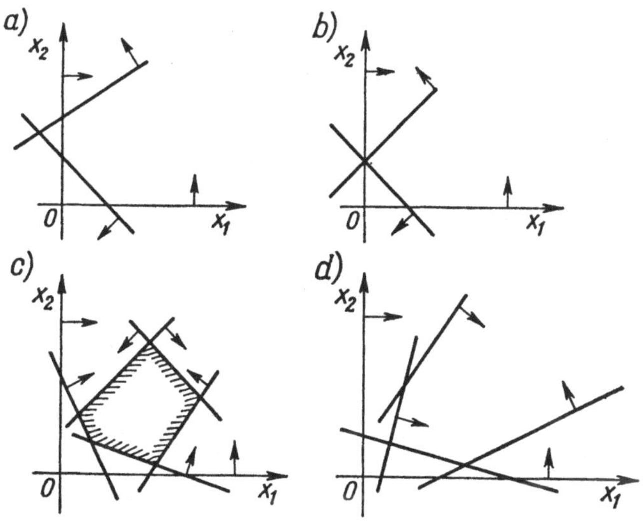

Fig. 6.2. Examples of feasible regions in linear programming

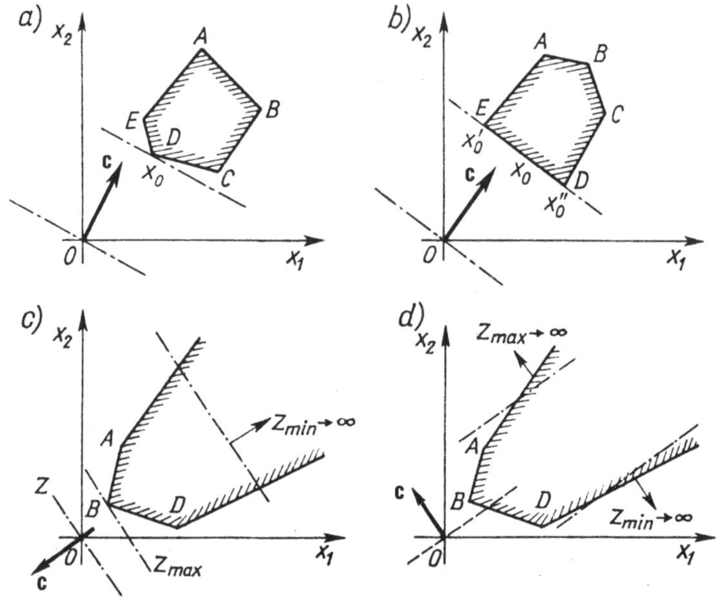

Fig. 6.3. Maximum and minimum values of the objective function for various feasible regions

$\mathbf{c} = (c_1, c_2)$, which points in the direction of steepest ascent of the function z;

— by moving the line $c_1x_1 + c_2x_2 = 0$ parallelly, determine the furthest extreme point of the feasible region in the direction opposite to the vector **c**. This point is the desired constrained minimum point for the function z (Fig. 6.1).

The feasible region described by inequalities (6.10) and (6.11) can be the empty set, a one-point set, a closed convex polygon or a convex unbounded polygonal region (Fig. 6.2).

When the feasible region is a closed convex polygon, the minimum of z in this region is attained at at least one vertex of the polygon (Fig. 6.3a). If the objective function attains a minimum at two vertices of the polygon, it takes the same value all along the edge joining those vertices (Fig. 6.3b).

When the feasible region is unbounded, then z may still take its minimum at a vertex or it may have no lower bound (Fig. 6.3c, d).

6.4 Determination of an initial basic feasible solution

The first systematic computational procedure for solving arbitrary linear programs was the simplex method developed by Dantzig[1]. It is an efficient scheme for obtaining an optimal solution in a finite number of iterations

[1] DANTZIG, G. B., *Linear Programming and Extensions*, Princeton, N. J., 1963.

once any basic feasible solution has been determined. Each iteration then gives a new basic feasible solution with a lower value of the objective function, and the process is continued until the minimum is reached. To discuss this procedure, we first need to know how to compute an initial basic feasible solution.

The sequential elimination method and simplex transformation

It is known from linear algebra that a system of r linear equations in n unknowns, where $n > r$, can always be reduced to a canonical form with the first r variables (after renumbering if necessary) as the basic variables:

$$\begin{aligned}
x_1 \qquad\qquad & +\bar{a}_{1,r+1}x_{r+1}+ \dots +\bar{a}_{1p}x_p+ \dots +\bar{a}_{1n}x_n = \bar{b}_1, \\
x_2 \qquad\quad & +\bar{a}_{2,r+1}x_{r+1}+ \dots +\bar{a}_{2p}x_p+ \dots +\bar{a}_{2n}x_n = \bar{b}_2, \\
& \dots\dots\dots\dots\dots\dots\dots\dots \\
x_i \quad & +\bar{a}_{i,r+1}x_{r+1}+ \dots +\bar{a}_{ip}x_p+ \dots +\bar{a}_{in}x_n = \bar{b}_i, \\
& \dots\dots\dots\dots\dots\dots\dots\dots \\
x_q \; & +\bar{a}_{q,r+1}x_{r+1}+ \dots +\bar{a}_{qp}x_p+ \dots +\bar{a}_{qn}x_n = \bar{b}_q, \\
& \dots\dots\dots\dots\dots\dots\dots\dots \\
x_r & +\bar{a}_{r,r+1}x_{r+1}+ \dots +\bar{a}_{rp}x_p+ \dots +\bar{a}_{rn}x_n = \bar{b}_r.
\end{aligned} \tag{6.12}$$

By performing a single pivot operation, known from the Gauss–Jordan reduction procedure, we can obtain another basic solution.

In linear programming we are concerned, however, with basic *feasible* solutions, i.e. basic solutions in which the basic variables are non-negative. The basic solution obtained from system (6.12) will clearly be feasible if all the free terms $\bar{b}_i \geqslant 0$.

If some of the free terms in (6.12) are negative, we transform the system in the following way. Among the equations with negative free terms we find one in which the free term has the largest absolute value. We multiply this equation, say it is the sth equation, by -1 and add to each of the remaining equations with negative free terms. As a result we obtain a system

$$\begin{aligned}
x_1+a_{1s}x_s+a_{1,r+1}x_{r+1}+ \dots +a_{1p}x_p+ \dots +a_{1n}x_n &= b_1, \\
x_2+a_{2s}x_s+a_{2,r+1}x_{r+1}+ \dots +a_{2p}x_p+ \dots +a_{2n}x_n &= b_2, \\
\dots\dots\dots\dots\dots\dots\dots\dots & \\
-x_s+a_{s,r+1}x_{r+1}+ \dots +a_{sp}x_p+ \dots +a_{sn}x_n &= b_s, \\
\dots\dots\dots\dots\dots\dots\dots\dots & \\
a_{rs}x_s+x_r+a_{r,r+1}x_{r+1}+ \dots +a_{rp}x_p+ \dots +a_{rn}x_n &= b_r,
\end{aligned} \tag{6.13}$$

in which all the free terms b_i are non-negative, $b_i \geqslant 0$, $i = 1, \dots, r$. However, the new system is no longer canonical and has to be reduced to one in such a manner as to maintain all the free terms non-negative. To this end, we use the *simplex transformation*, i.e. a pivot operation for which the pivot element is chosen as follows:

As the pivot column p we choose a column with a positive coefficient in the sth row, i.e. such that $a_{sp} > 0$. As the pivot row q we choose the one for which the ratio b_i/a_{ip} is the smallest.

If there were no positive coefficient a_{sp} in the sth equation, system (6.13) would have no feasible solutions.

If there is an $a_{sp} > 0$, then one of the following three cases must occur.

1. The pivot element a_{qp} belongs to row s $(q = s)$. On performing the pivot operation, we obtain the system in a canonical form with x_p as a new basic variable.
2. The pivot element a_{qp} is positive and does not belong to row s $(q \neq s)$. In this case, the pivot operation does not put equation s into a canonical form with respect to any of the variables; however, it will reduce the value of the free term b_s. If we repeat the whole procedure, i.e. pivot again following the same rules, b_s will further decrease. Since in each such simplex iteration all the free terms of the system remain non-negative, after a finite number of iterations we shall come to case 1 or to case 3 considered below, or the system will turn out inconsistent.
3. The pivot element a_{qp} does not belong to row s $(q \neq s)$ and, at the same time, $b_q = 0$, so that the pivot on a_{qp} does not affect the free term in equation s. In this case we try to find a different pivot column (starting from a different $a_{sp'} > 0$) for which the pivot row would have $b_{q'} > 0$. If this is impossible, we perform the original pivot and, after the set of the basic variables is changed, try again to find a new pivot element $a_{qp} > 0$ with $b_q > 0$. After a finite number of such attempts we come to case 2 or to case 1 or the system turns out inconsistent.

The artificial basis method

Determination of an initial basic feasible solution by the procedure described above may be very laborious. We can avoid that by using the artificial basis method, which is particularly effective when the number of variables is much larger than the number of constraint equations.

The problem we are considering is: *Minimize the function*

$$z = \sum_{i=1}^{n} c_i x_i$$

subject to the constraints

$$\begin{aligned} a_{11}x_1 + \ldots + a_{1n}x_n &= b_1, \\ a_{21}x_1 + \ldots + a_{2n}x_n &= b_2, \\ \cdots\cdots\cdots\cdots \\ a_{m1}x_1 + \ldots + a_{mn}x_n &= b_m, \\ x_i \geqslant 0, \quad i = 1, \ldots, n, \end{aligned} \tag{6.14}$$

where all b_i are assumed to be non-negative; this can always be assured by multiplying the equations by -1 where necessary.

For the artificial basis method, we augment the given constraint equations by adding a different new variable to each of them,

$$\begin{aligned} &a_{11}x_1 + \dots + a_{1n}x_n + x_{n+1} = b_1, \\ &a_{21}x_1 + \dots + a_{2n}x_n + x_{n+2} = b_2, \\ &\dots\dots\dots\dots\dots \\ &a_{m1}x_1 + \dots + a_{mn}x_n \dots + x_{m+n} = b_m, \end{aligned} \tag{6.15}$$

and form a new objective function

$$z^* = z + M \sum_{i=1}^{m} x_{n+i}, \tag{6.16}$$

where M is an unspecified large positive number. The new variables $x_{n+1}, \dots, x_{n+m}$ are called *artificial variables* and the problem of minimizing the function z^* subject to constraints (6.15) is called the *M-problem.*

An immediate basic feasible solution to system (6.15) is

$$\mathbf{x}^0(x_{n+1}^0, \dots, x_{n+m}^0) = (b_1, \dots, b_m).$$

Knowing this solution, we can find an optimum solution to the M-problem by the simplex method.

It can be shown that if no artificial variables appear in an optimal solution of the M-problem, then this solution is an optimal solution to the original problem.

If an optimal solution of the M-problem contains one or more non-zero artificial variables, the original constraint equations are inconsistent.

If in the M-problem the minimum of z^* is $-\infty$, then in the original problem either the constraint system is inconsistent or the function z is unbounded in the feasible region.

6.5 The simplex method

We now move on to describe the main iterative scheme of the simplex method. We assume that we are given a basic feasible solution, or, equivalently, that the program reads

$$\textit{Minimize} \quad z = \sum_{i=1}^{n} c_i x_i$$

subject to constraints

$$\begin{aligned} &x_1 \quad + a_{1,r+1}x_{r+1} + \dots + a_{1p}x_p + \dots + a_{1n}x_n = b_1, \\ &\dots\dots\dots\dots\dots\dots\dots \\ &\quad x_q \; + a_{q,r+1}x_{r+1} + \dots + a_{qp}x_p + \dots + a_{qn}x_n = b_q, \\ &\dots\dots\dots\dots\dots\dots\dots \\ &\qquad x_r + a_{r,r+1}x_{r+1} + \dots + a_{rp}x_p + \dots + a_{rn}x_n = b_r, \end{aligned} \tag{6.17}$$

and conditions

$$x_1 \geqslant 0, \ \ldots, \ x_p \geqslant 0, \ \ldots, \ x_n \geqslant 0,$$

where all the free terms are non-negative: $b_i \geqslant 0$, $i = 1, \ldots, r$. For system (6.17), the initial basic feasible solution is

$$\mathbf{x}^0 = (b_1, \ldots, b_r, 0, \ldots, 0), \tag{6.18}$$

with a corresponding value of the objective function

$$z^0 = z(\mathbf{x}^0) = \sum_{i=1}^{r} c_i b_i. \tag{6.19}$$

To find whether $\mathbf{x}^0$ is an optimum solution, let us compare the value $\mathbf{z}^0$ with the value $\mathbf{z}^1$ that the objective function takes at a different apex of the feasible polyhedron, or a different basic feasible solution. In the simplex method, a new basic feasible solution is obtained from the given one by performing a simplex transformation on the constraint system (6.17), as a result of which a new variable x_p is introduced into the basis in place of a variable x_q, where q is such that $b_q/a_{qp} = \min\limits_{a_{ip}>0} [b_i/a_{ip}]$. This is done by solving the qth equation for x_p,

$$x_p = \frac{1}{a_{qp}}(a_{q0} - x_q - a_{q,r+1}x_{r+1} - \ldots - a_{qn}x_n), \tag{6.20}$$

and eliminating x_p from all the other equations (6.17). The resulting system is

$$\begin{gathered}
x_1 + \left(a_{1,r+1} - \frac{a_{1p}}{a_{qp}} a_{q,r+1}\right) x_{r+1} + \ldots - \frac{a_{1p}}{a_{qp}} x_q + \ldots \\
\ldots + \left(a_{1n} - \frac{a_{1p}}{a_{qp}} a_{qn}\right) x_n = b_1 - \frac{a_{1p}}{a_{qp}} b_q, \\
\cdots\cdots\cdots\cdots\cdots\cdots\cdots\cdots\cdots\cdots \\
x_p + \frac{a_{q,r+1}}{a_{qp}} x_{r+1} + \ldots + \frac{1}{a_{qp}} x_q + \ldots + \frac{a_{qn}}{a_{qp}} x_n = \frac{b_q}{a_{qp}}, \\
\cdots\cdots\cdots\cdots\cdots\cdots\cdots\cdots\cdots\cdots \\
x_r + \left(a_{r,r+1} - \frac{a_{rp}}{a_{qp}} a_{q,r+1}\right) x_{r+1} + \ldots - \frac{a_{rp}}{a_{qp}} x_q + \ldots \\
\ldots + \left(a_{rn} - \frac{a_{rp}}{a_{qp}} a_{qn}\right) x_n = b_r - \frac{a_{rp}}{a_{qp}} b_q.
\end{gathered} \tag{6.21}$$

Hence we obtain a new basic feasible solution

$$\mathbf{x}^1 = (b_1^1, \ldots, b_r^1, \ 0, \ldots, \ 0) \tag{6.22}$$

for which the objective function is

$$z^1 = z(\mathbf{x}^1) = c_1\left(b_1 - \frac{a_{1p}}{a_{qp}} b_q\right) + \dots + c_p \frac{b_q}{a_{qp}} + \dots + c_r\left(b_r - \frac{a_{rp}}{a_{qp}} b_q\right)$$

$$= z^0 - \left(\sum_{i=1}^{r} c_i a_{ip} - c_p\right) \frac{b_q}{a_{qp}}, \tag{6.23}$$

where $b_q^1/a_{qp} > 0$. The expression $\sum_{i=1}^{r} c_i a_{ip} - c_p$ is called the *optimality index.*

The following statements can be inferred from equality (6.23):

(1) If the optimality index $\sum_{i=1}^{r} c_i a_{ij} - c_j \leqslant 0$ for all j, the solution is optimal.

(2) If the optimality index is positive for at least one column (j) and each such column contains at least one positive element, the value of the objective function can be decreased by another simplex iteration,

(3) If a column for which the optimality index is positive has no positive elements, the function z is unbounded from below in the feasible region.

The simplex computational algorithm. Simplex tableau

On the basis of what has been said so far, we can now briefly outline the step-by-step procedure for solving a linear programming problem by the simplex method.

Simplex computations can conveniently be arranged in a tabular form similar to that of a Gauss–Jordan table and called the *simplex tableau.*

Table 6.1 presents the simplex tableau for system (6.17). The right-hand part of the tableau contains the matrix $[a_{ij}]$ on which the simplex operations

Table 6.1

		$-c_0$	c_1	c_2	$\dots x_q \dots$	c_r	c_{r+1}	$\dots c_k$	$\dots c_p$	$\dots c_n$	
c_i	x_i										b_i/a_{ip}
		b_i	x_1	x_2	$\dots x_q \dots$	x_r	x_{r+1}	$\dots x_k$	$\dots x_p$	$\dots x_n$	
c_1	x_1	b_1	1				$a_{1,r+1}$	$\dots a_{1k}$	$\dots a_{1p}$	$\dots a_{1n}$	
c_2	x_2	b_2		1			$a_{2,r+1}$	$\dots a_{2k}$	$\dots a_{2p}$	$\dots a_{2n}$	
$\vdots$	$\vdots$	$\vdots$					$\vdots$	$\vdots$	$\vdots$	$\vdots$	
c_i	x_i	b_i					$a_{i,r+1}$	$\dots a_{ik}$	$\dots a_{ip}$	$\dots a_{in}$	
$\vdots$	$\vdots$	$\vdots$					$\vdots$	$\vdots$	$\vdots$	$\vdots$	
c_q	x_q	b_q			1		$a_{q,r+1}$	$\dots a_{qk}$	$\dots a_{qp}$	$\dots a_{qn}$	b_q/a_{qp}
$\vdots$	$\vdots$	$\vdots$					$\vdots$	$\vdots$	$\vdots$	$\vdots$	
c_r	x_r	b_r				1	$a_{r,r+1}$	$\dots a_{rk}$	$\dots a_{rp}$	$\dots a_{rn}$	
$\sum_{i=1}^{r} c_i a_{ij} - c_j$											

will be performed. In the last column we calculate the quotients b_i/a_{ip}, which we need for correct choice of the pivot row. In the last row of the tableau we compute the optimality indices.

The second column on the left-hand side lists the basic variables while the third gives the free terms of the constraint equations. This arrangement makes it easy to read off the basic solution from the tableau.

Having constructed an initial simplex tableau, we examine the optimality index. If case 1 or case 3 occurs (see statements (1), (2), (3) above), the process is ended: either the initial basic solution is optimal or the problem has no solution. If, which is more likely, case 2 occurs, we introduce a new variable into the basis. This is done in the following way:

1. Choose a column p for which the optimality index is positive as the pivot column.
2. Choose as the pivot row q the one for which
$$\frac{b'_q}{a_{qp}} = \min_{a_{ip}>0}\left\{\frac{b_i}{a_{ip}}\right\}.$$
3. Transform row q using the formula
$$a'_{qk} = \frac{a_{qk}}{a_{qp}}, \qquad k = 1, \ldots, n.$$
4. Transform the remaining rows $i \neq q$ using the formula
$$a'_{ik} = a_{ik} - a'_{qk}a_{ip}, \qquad i = 1, \ldots, q-1, q+1, \ldots, r.$$
5. Calculate the optimality indices.

On completion of this iteration we obtain a new simplex tableau for the problem, corresponding to a different canonical form of the constraint system, with a different basis.

We then return to the first step, i.e. examine the optimality indices, and repeat the whole procedure if necessary. After a finite number of such iterations we arrive at an optimal solution or show that there is no bounded solution to the problem.

A numerical example of solving a linear-programming problem by the simplex method is worked out in Sec. 6.10.

Alternative optima

A linear-programming problem may have several different optimal basic solutions, $\mathbf{x}_1, \ldots, \mathbf{x}_s$. In geometrical terms, this implies that a boundary face of the feasible region is parallel to the plane $z(\mathbf{x}) = \text{const}$. More precisely, every vector which can be represented as a convex combination of the optimal basic solutions, i.e. every vector of the form

$$\mathbf{x} = t_1\mathbf{x}_1 + t_2\mathbf{x}_2 + \ldots + t_s\mathbf{x}_s,$$

where

$$\sum_i t_i = 1 \quad \text{and} \quad t_i \geqslant 0, \quad i = 1, 2, \ldots, s,$$

is also an optimal solution.

When an optimal solution has been obtained by the simplex method, the information whether there are any alternative optima appears in the simplex tableau: an alternative optimal solution exists if the optimality index vanishes for some j not in the basis (cf. statement 1 above).

Degeneracy problem

It may happen that the free terms in some of the equations in the canonical form are zero. In this case, the values of the basic variables corresponding to the zero b_i are also zero.

A basic solution in which one or more basic variables are equal to zero (i.e. a solution with more than $n-r$ zero variables) is called a *degenerate basic solution*; if a linear-programming problem admits a degenerate solution, it is called a *degenerate problem.*

For a plane problem, degeneracy means that more than two constraint lines intersect at one point (Fig. 6.4).

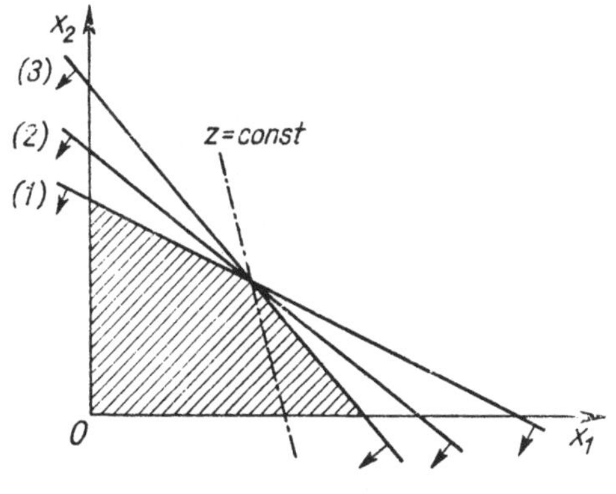

Fig. 6.4. A degenerate vertex

In solving a degenerate problem by the simplex method we may find ourselves in a situation where the same sequence of bases is repeated cyclically and so the iteration procedure does not lead anywhere. In the simplex tableau we then have a few identical minimum ratios b_i/a_{ip}, and hence a few alternative choices for the pivot row.

To avoid the possiblity of cycling, we adopt the following rule in the simplex algorithm. If there is a tie for the minimum ratio b_i/a_{ip}, we choose that one of the tied rows for which the ratio $a_{i,r+1}/a_{ip}$ is minimum. If a unique minimum is not obtained, we look for the smallest ratio $a_{i,r+2}/a_{ip}$. We continue the comparison until the tie is broken and the pivot row is uniquely determined.

6.6 Duality in linear programming

The dual problem

Associated with every linear-programming problem is an optimization problem called its *dual*. The optimal solution of either problem carries information about the optimal solution of the other. Simultaneous consideration of a linear-programming problem and its dual has proved very useful in the development of the theory and the numerical methods of linear programming.

Given any linear-programming problem (the primal problem):

$$\textit{Maximize } z = \mathbf{c}^T\mathbf{x} \textit{ subject to } \mathbf{A}\mathbf{x} \leqslant \mathbf{b} \textit{ and } \mathbf{x} \geqslant 0, \tag{6.24}$$

we define the dual problem as

$$\textit{Minimize } w = \mathbf{b}^T\mathbf{y} \textit{ subject to } \mathbf{A}^T\mathbf{y} \geqslant \mathbf{c} \textit{ and } \mathbf{y} \geqslant 0. \tag{6.25}$$

In the above, **x** and **c** are n-dimensional vectors, **y** and **b** are m-dimensional vectors and **A** is an $m \times n$-matrix.

Example of dual problems:

The primal problem	The dual problem
Maximize the function	*Minimize the function*
$z = x_1 + x_2$	$w = 2y_1 + 12y_2$
subject to	*subject to*
$-\frac{1}{2}x_1 + x_2 \leqslant 2,$	$-\frac{1}{2}y_1 + 2y_2 \geqslant 1,$
$2x_1 + x_2 \leqslant 12,$	$y_1 + y_2 \geqslant 1,$
$x_1 \geqslant 0,$	$y_1 \geqslant 0,$
$x_2 \geqslant 0.$	$y_2 \geqslant 0.$

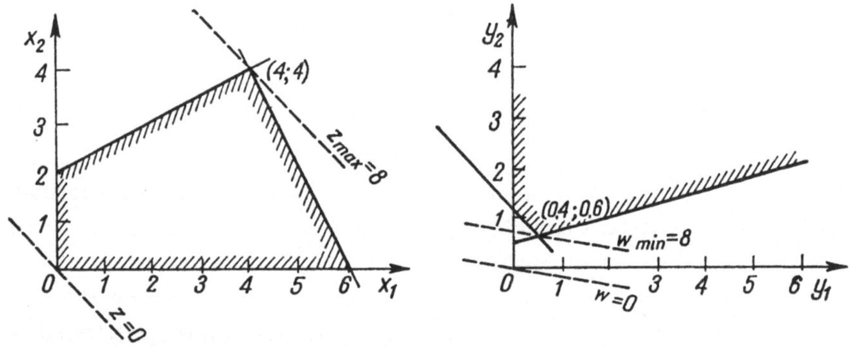

Fig. 6.5. Graphical solution of the original and the dual problems

Figure 6.5 shows graphical solutions to both problems. The solution of the primal problem is $\mathbf{x} = (4, 4)$ with $z_{max} = 8$, while that of the dual is $\mathbf{y} = (0.4, 0.6)$ with $w_{min} = 8$. Thus, $z_{max} = w_{min}$.

Fundamental theorems of duality

If one of the pair of dual problems (6.24) and (6.25) has an optimal solution, then the other also has an optimal solution and

$$z_{max} = w_{min}. \tag{6.26}$$

If one of the pair of dual problems has an unbounded solution (i.e. $z_{max} = \infty$ or $w_{min} = -\infty$), then the other problem has no feasible solutions (i.e. the constraint system is inconsistent).

If the ith constraint of one of the problems is satisfied by an optimal solution as a sharp inequality, then the ith variable of the corresponding optimal solution of the dual system is zero. Conversely, if the ith variable of an optimal solution of one of the problems is positive, then the ith constraint of the dual system is satisfied as an equality by any optimum solution of the dual problem.

Feasible solutions $\mathbf{x}$ and $\mathbf{y}$ of a pair of dual problems are optimal if and only if they satisfy the condition

$$x_k\left(c_k - \sum_{i=1}^{m} a_{ik} y_i\right) = 0. \tag{6.27}$$

The values of the variables y_i of an optimal solution of the dual problem provide estimates of the influence of a change in the free terms on the value z_{max} in the primal problem, i.e.

$$y_i = \frac{\partial z_{max}}{\partial b_i}, \quad i = 1, \ldots, m.$$

Based on the properties of dual problems and solutions, a programming algorithm called the *dual simplex method* has been devised by Lemke[1], in which the primal and the dual problems are computed simultaneously in a tableau of the same type as in the simplex method.

6.7 Other important methods of linear programming

Besides the simplex method, several other efficient methods for solving large linear-programming problems have been developed. They include

— the revised simplex method,[2]

[1] LEMKE, C. E., The dual method of solving the linear programming problem, *Naval Res. Logistics Quarterly*, **1**, 48–54, 1954.

[2] GASS, S. I., *loc. cit.* p. 232

— the duplex method,[1]
— the revised duplex method,[2]
— the symmetrical revised simplex method,[3]
— the multiphase simplex method,[4]
— the multiphase gradient projection method,[4]
— the Dantzig–Wolfe decomposition method,[5]

There is also a number of methods by which approximate solutions to linear-programming problems can be found effectively, for example,
— the multiplex method,[6,7]
— the method of logarithmic potential.[6]

6.8 Linear discrete programming

Discrete-programming problems are optimization problems in which the design variables are required to be discrete or, less generally, to be integers. The number of prestressing cables in a beam, for example, is obviously an integer variable, but also the member sections of a structure may happen to be available in discrete sizes, or, in the optimization of a steel beam, the only available steel sheets may be of a thickness being a multiple of, say, 2 mm.

According as all or only some of the design variables are discrete in a given problem, we call it a *discrete* or a *mixed discrete problem*. A special type of discrete programming is zero-one programming, in which the design variables can only be 0 or 1.

Sometimes a solution to an integer-programming problem can be obtained by rounding off the fractions occurring in an optimal solution of the corresponding continuous problem. In general, however, this operation may lead us outside the feasible region or to a point far from the optimal solution, and so other methods must be employed. In most of them, the solving procedure consists of two phases: Phase I to find a solution of the continuous problem and Phase II to modify that solution until an integer solution is

[1] Künzi, H. P. and Krelle, W., *Nichtlineare Programmierung*, Springer-Verlag, Heidelberg 1966.

[2] Künzi, H. P. and Tzshbach, T., *The duplex-algorithm*, Numer. Mathem. 7, 1965.

[3] Müller-Merbach, H., Die symmetrisch revidierte Simplex-Methode der linearen Planungsrechnung, *Elektron. Detenverarbeitung*, **7**, 1965.

[4] Künzi, H. P., Tzshbach, H. G. and Zehnder, C. A., *Numerische Methoden der Mathematischen Programmierung mit Algol- und Fortran-Programmen*, Stuttgart 1966.

[5] Dantzig, G. B. and Wolfe, Ph., Decomposition principle for linear programming, *Oper. Res.*, **8**, 1960.

[6] Frisch, R. A. K., *The multiplex method for linear and quadratic programming*, Mem. Univ. Social. Institute of Oslo, 1957.

[7] Bode, B., Multiplexmethode—ein Verfahren zur Lösung linearer und quadratischer Optimierungs-probleme, *Messen-Steuern-Regeln*, **12**, 1963.

reached. One such method, introduced by Gomory[1,2,3] is outlined below. A different approach is represented by branch-and-bound methods, which are combinatorial algorithms for systematic elimination of non-optimal solutions. The first method of this kind was given by Land and Doing[4] in 1960. A general procedure for branch-and-bound methods was developed by Harve[5] and Roy[6]. Zero-one programs can be solved by any branch-and-bound method or by a special method devised by Balas[7].

In Gomory's "cutting plane" method, we start by solving the problem in the usual manner (e.g. by the simplex method), treating the variables as continuous. If all the basic variables in the solution are integers, the task is completed. If any of them have fractional parts, we introduce an additional constraint, called *Gomory's cut*, which forces a move from the optimal non-integer solution towards an optimal integer solution.

To see how this is done, consider an optimal tableau for the continuous problem (Table 6.2), where for convenience the basic variables are denoted

Table 6.2

Basic variables	x_1	x_2	$\ldots x_l$	$\ldots x_m$	y_1	y_2	$\ldots y_j$	$\ldots y_n$	Object. function	s_l	Free terms
x_1	1	0	0	0	$\bar{a}_{11}$	$\bar{a}_{12}$ ...	$\bar{a}_{1j}$	$\bar{a}_{1n}$	0	0	$\bar{b}_1$
x_2	0	1	0	0	$\bar{a}_{21}$	$\bar{a}_{22}$ ...	$\bar{a}_{2j}$	$\bar{a}_{2n}$	0	0	$\bar{b}_2$
$\vdots$											
x_l	0	0	1	0	$\bar{a}_{l1}$	$\bar{a}_{l2}$ ...	$\bar{a}_{lj}$	$\bar{a}_{ln}$	0	0	$\bar{b}_l$
$\vdots$											
x_m	0	0	0	1	$\bar{a}_{m1}$	$\bar{a}_{m2}$ ...	$\bar{a}_{mj}$	$\bar{a}_{mn}$	0	0	$\bar{b}_n$
f	0	0	0	0	$\bar{c}_1$	$\bar{c}_2$ ...	$\bar{c}_j$	$\bar{c}_n$	1	0	$\bar{f}$
s_l	0	0	0	0	$-\alpha_{l1}$	$-\alpha_{l2}$ ...	$-\alpha_{lj}$	$-\alpha_{ln}$	0	1	$-\beta_l$

[1] Gomory, R. E., An algorithm for the mixed integer problem, *Rand Report, R. H.* 25797, July 1960.

[2] Gomory, R. E., *An all-integer programming algorithm. Chapter* 13, *Industrial Scheduling*, eds. Huth, J. F. and Thompson, G. L., Prentice Hall, Englewood Cliffs, New Jersey 1963.

[3] Rao, S. S., *Optimization*; *Theory and Application*, Wiley, New Delhi 1978.

[4] Land, A. H. and Doing. A. G., An automatic method for solving discrete programming problems, *Econometrica*, **28**, 1960.

[5] Herve, P., Les procédures arborescentes d'optimisation, *Revue d'Ing. et de Rech. Oper.*, **14**, 69–80, 1968.

[6] Roy, B., Procédure d'exploration par séparation et évaluation, *Revue d'Ing. et de Rech. Oper.*, **3**, V–1 (1966).

[7] Balas, E., An additive algorithm for solving linear programs with zero-one variables, *Operations Research*, **13**, 517–546, 1965.

by x_i ($i = 1, \ldots, m$) and the non-basic ones by y_j ($j = 1, \ldots, n$). The additional row and column (s_i) contain coefficients of the Gomory constraint, which we explain below.

Let x_i be the basic variable with the largest fractional part. If two or more basic variables are tied with respect to this property, we choose any one of them. From the ith equation, we have

$$x_i = \bar{b}_i - \sum_{j=1}^{n} \bar{a}_{ij} y_j. \tag{6.28}$$

The coefficients $\bar{b}_i$ and $\bar{a}_{ij}$ can be represented as

$$\bar{b}_i = \hat{b}_i + \beta_i, \qquad \bar{a}_{ij} = \hat{a}_{ij} + \alpha_{ij}, \tag{6.29}$$

where $\hat{b}_i$ and $\hat{a}_{ij}$ are the integral parts and β_i and α_{ij} the fractional parts of $\bar{b}_i$ and $\bar{a}_{ij}$, respectively. Equation (6.28) can now be written in the form

$$\beta_i - \sum_{j=1}^{n} \alpha_{ij} y_j = x_i - \hat{b}_i - \sum_{j=1}^{n} \hat{a}_{ij} y_j. \tag{6.30}$$

Since in the final solution all x_i and y_i must be integers, the right-hand side of equation (6.30) must be an integer number and so the left-hand side must also be one. Note that α_{ij} are non-negative fractions and y_j are non-negative integers, and therefore $\sum_{j=1}^{n} \alpha_{ij} y_j$ cannot be negative. As $0 < \beta_i < 1$, it follows that

$$\beta_i - \sum_{j=1}^{n} \alpha_{ij} y_j \leqslant \beta_i < 1, \tag{6.31}$$

and hence, if the left-hand side is an integer, it can only be a negative one or zero:

$$\beta_i - \sum_{j=1}^{n} \alpha_{ij} y_j \leqslant 0. \tag{6.32}$$

Adding a non-negative slack variable s_i, we obtain a new constraint equation, or a *Gomory cut*,

$$s_i - \sum_{j=1}^{n} \alpha_{ij} y_j = -\beta_i; \tag{6.33}$$

by definition, s_i is also an integer.

Should all y_j be equal to 0, equation (6.33) would give $s_i = -\beta_i$, i.e. a negative value, which is not feasible. To obtain a new, feasible solution we use the simplex method. If the variables in the new solution are not all integers, we construct a new Gomory cut and apply a simplex transformation again. The process is repeated until an integer solution is obtained or shown not to exist.

For more details regarding this method and also for a discussion of the mixed-discrete case, we refer the reader to Gomory's original papers or Rao's book (see footnote p. 217).

6.9 Examples of linear programming in optimum structural design

Optimization of a rectangular prestressed-concrete beam

A prestressed-concrete beam of rectangular cross-section with a constant depth h and a width $b(x)$ variable along the beam length l, subjected to a load producing a bending moment $M(x)$, is optimized for the minimum substitute volume of the concrete and steel used (Brandt 1964),

$$V^* = \int_0^l (A+kP)\mathrm{d}x \approx \frac{l}{n}\sum_{i=0}^{n} (A_i+kP_i).$$

where A is the cross-sectional area, P the prestressing force and k a coefficient which takes into account different steel and concrete costs. In the numerical example, k has been calculated as

$$k = \frac{9000 \cdot 78.5}{750 \cdot 1000000} \approx 0.0001 \ [\mathrm{m}^2/\mathrm{kN}];$$

it is assumed that the unit costs of the materials are 750 per $1\,\mathrm{m}^3$ for concrete and 9000 per 1kN for prestressing steel, the maximum allowable stress in the prestressing steel is 10^6 kN/m^2, and the weight density of steel is 78.5 kN/m^3.

The approximating sum extends over n uniformly spaced cross-sections of the beam. The subscript i indicates the value of a given quantity at the ith cross-section. In order to minimize the substitute volume V^*, we minimize the sum A_i+kP_i for each i.

The stresses at the lower and the upper edges of the beam in the ith section are constrained by the conditions

$$\text{(a)} \quad \frac{P_i}{b_i h} + \frac{6(M_i-P_i e)}{b_i h^2} \leqslant \bar{\sigma},$$

$$\text{(b)} \quad \frac{P_i}{b_i h} - \frac{6(M_i-P_i e)}{b_i h^2} \geqslant \underline{\sigma},$$

$$\text{(c)} \quad \frac{P_i}{b_i h} + \frac{6(M_i-P_i e)}{b_i h^2} \geqslant \underline{\sigma},$$

$$\text{(d)} \quad \frac{P_i}{b_i h} - \frac{6(M_i-P_i e)}{b_i h^2} \leqslant \bar{\sigma}.$$

Furthermore, we have constructional constraints (Fig. 6.6)

(e) $b_i \geqslant \underline{b}$,

(f) $h \leqslant \bar{h}$,

(g) $P_i \geqslant 0$,

(h) $e \leqslant \frac{1}{2}h-d$,

where $\underline{b}$ is the smallest allowed width of the beam, $\bar{h}$ the largest allowed depth and d the minimum distance of the prestressing cables to the beam edge.

Conditions (c) and (d) are active when $P_i e > M_i$.

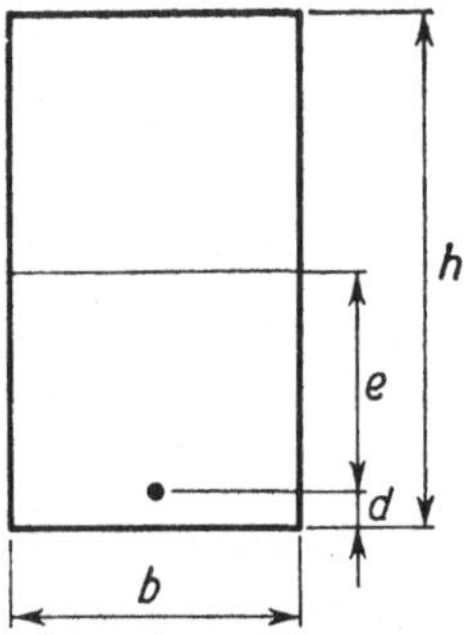

Fig. 6.6. The beam section under consideration

Considering the design variables to be the beam width b_i and the prestressing force P_i, and treating all the other quantities as preassigned, we obtain the following linear-programming problem for the ith section:

Minimize the function

$$A_i^* = hb_i+kP_i$$

subject to

(a) $(h-6e)P_i-\bar{\sigma}h^2b_i \leqslant -6M_i$,

(b) $(h+6e)P_i-\underline{\sigma}h^2b_i \geqslant 6M_i$,

(c) $(h-6e)P_i-\underline{\sigma}h^2b_i \geqslant -6M_i$,

(d) $(h+6e)P_i-\bar{\sigma}h^2b_i \leqslant 6M_i$,

(e) $b_i \geqslant \underline{b}$,

(f) $P_i \geqslant 0$,

(g) $e = \frac{1}{2}h-d$.

The problem has been solved graphically with the following numerical data: $M = 375$ kNm, $h = 0.70$ m, $d = 0.05$ m, $\bar{\sigma} = 10$ MPa, $\underline{\sigma} = 0.5$ MPa. The function F and constraints (a)–(f) then become

$$A^* = 0.7b+0.0001P;$$

(a) $1.1P+4900b \geqslant 2250$,

(b) $2.5P-225b \geqslant 2250$,

(c) $1.1P+245b \leqslant 2250$,

(d) $2.5P-4900b \leqslant 2250$,

(e) $b \geqslant 0.10$,

(f) $P \geqslant 0$.

The solution is shown in Fig. 6.7. The function A^* attains a minimum at the point K, at which $b = 0.178$ m, $P = 1250$ kN and $A^*_{min} = 0.250\ m^2$. In the cross-section so designed, the stress is constant throughout the section and equal to 10 MPa.

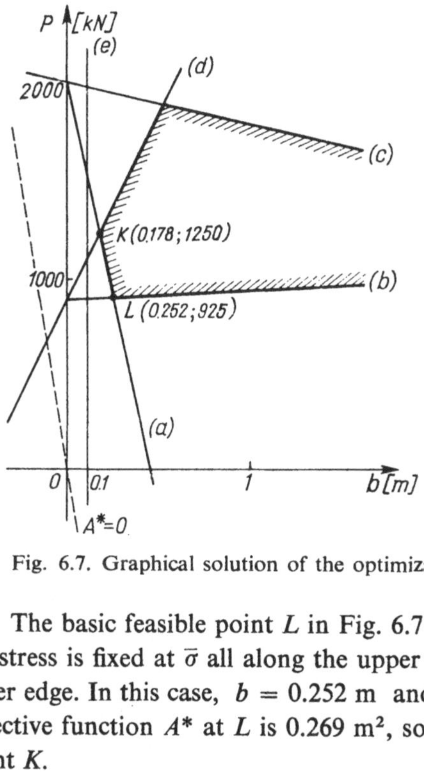

Fig. 6.7. Graphical solution of the optimization problem considered

The basic feasible point L in Fig. 6.7 corresponds to the design in which the stress is fixed at $\overline{\sigma}$ all along the upper edge of the section and at $\underline{\sigma}$ on the lower edge. In this case, $b = 0.252$ m and $P = 925$ kN. The value of the objective function A^* at L is 0.269 m^2, somewhat more than at the optimum point K.

Optimization of a plane truss

A plane truss with n members and a given configuration of nodes is to be designed for the minimum volume of material under a given external loading (Niemierko 1972). The volume of the truss is

$$V = \sum_{i=1}^{n} A_i l_i,$$

where A_i and l_i denote the cross-sectional area and the length of the ith bar.

The constraints for the problem are given by the equilibrium equations for the forces at each joint of the truss

$$\sum_{i=1}^{n} c_{ij} N_i = P_j, \quad j = 1, \ldots, r,$$

where P_j denote the external force components, N_i is the longitudinal force in the ith bar, c_{ij} are geometrical configuration constants, $r = 2m$ is the number of equations and m the number of nodes. Furthermore, the internal stresses must not exceed the allowable values in each bar:

$$-\bar{\sigma} A_i \leqslant N_i \leqslant \bar{\sigma} A_i.$$

Since a truss designed for minimum volume under a single loading state must be fully stressed, i.e. each bar must sustain a limiting allowable stress, we assume that $|N_i| = \bar{\sigma} A_i$, $i = 1, \ldots, n$, and so the problem can be restated as follows:

Minimize the function

$$V = \frac{1}{\bar{\sigma}} \sum_{i=1}^{n} |N_i| l_i,$$

subject to the constraints

$$\sum_{i=1}^{n} c_{ij} N_i = P_j, \quad j = 1, \ldots, r.$$

In order to solve this problem by the simplex method, we first have to modify it so that all the variables may be non-negative. To this end, we represent each force N_i as

$$N_i = N'_i - N'_{i+n}$$

where

$$N'_i = \max[N_i, 0] \quad \text{and} \quad N'_{i+n} = \max[-N_i, 0].$$

In this way we have doubled the number of variables but all of them are non-negative.

The new matrix $\mathbf{C}$ of the coefficients c_{ij}, the new vector $\mathbf{L}$ of the bar lengths l_i and the new vector $\mathbf{N}'$ of the bar forces N'_i are defined by

$$\mathbf{C} = \begin{bmatrix} c_{11}, \ldots, c_{1n}, -c_{11}, \ldots, & -c_{1n} \\ \cdot \quad \cdot \quad \cdot \quad \cdot \quad \cdot \quad \cdot \quad \cdot \quad \cdot \quad \cdot \quad \cdot & \cdot \\ c_{r1}, \ldots, c_{rn}, -c_{r1}, \ldots, & -c_{rn} \end{bmatrix},$$

$$\mathbf{L} = [l_1, \ldots, l_n, l_1, \ldots, l_n],$$

$$\mathbf{N}' = [N'_1, \ldots, N'_n, N'_{n+1}, \ldots, N'_{n+n}],$$

and the optimization problem takes the form:

Minimize

$$V = \mathbf{N'L}$$

subject to

$$\mathbf{CN'} = \mathbf{P}.$$

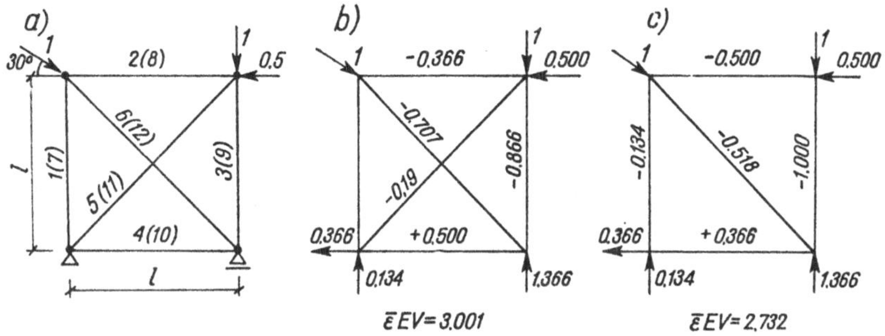

Fig. 6.8. Optimization of a truss

As a numerical example, let us consider a four-node truss loaded as shown in Fig. 6.8. The equilibrium equations for the nodes are

$$(1)\quad N_1+\frac{\sqrt{2}}{2}N_5 = -0.134, \qquad (5)\quad N_3+\frac{\sqrt{2}}{2}N_5 = -1.000,$$

$$(2)\quad N_4+\frac{\sqrt{2}}{2}N_5 = 0.366, \qquad (6)\quad N_2+\frac{\sqrt{2}}{2}N_5 = -0.500,$$

$$(3)\quad N_1+\frac{\sqrt{2}}{2}N_6 = -0.500, \qquad (7)\quad N_3+\frac{\sqrt{2}}{2}N_6 = -1.366,$$

$$(4)\quad N_2+\frac{\sqrt{2}}{2}N_6 = -0.866, \qquad (8)\quad N_4+\frac{\sqrt{2}}{2}N_6 = 0.$$

Since the truss is externally isostatic and the equilibrium equations already include the reactions calculated from the conditions of equilibrium for the whole truss (3 equations), 3 of the above equations can be left out as linearly dependent on the others.

Omitting equations (5), (6) and (8), and expressing the member forces N_i as differences of non-negative variables as shown before, we obtain the following constraint system

$$N_1'+\frac{\sqrt{2}}{2}N_5'-N_7' +\frac{\sqrt{2}}{2}N_{11}' = -0.134,$$

$$N_4'+\frac{\sqrt{2}}{2}N_5'-N_{10}'-\frac{\sqrt{2}}{2}N_{11}' = 0.366,$$

$$N_1'+\frac{\sqrt{2}}{2}N_6'-N_7' -\frac{\sqrt{2}}{2}N_{12}' = -0.500,$$

$$N_2' + \frac{\sqrt{2}}{2} N_6' - N_8' \quad - \frac{\sqrt{2}}{2} N_{12}' = \quad 0.866,$$

$$N_3' + \frac{\sqrt{2}}{2} N_6' - N_9' \quad + \frac{\sqrt{2}}{2} N_{12}' = \quad 1.366,$$

subject to which a minimum is to be found of the objective function

$$V = \frac{1}{\bar{\sigma}} (N_1' + N_2' + N_3' + N_4' + \sqrt{2} N_5' + \sqrt{2} N_6' + N_7' + N_8' + N_9' +$$
$$+ N_{10}' + \sqrt{2} N_{11}' + \sqrt{2} N_{12}').$$

To find an initial basic feasible solution, we have transformed the system in the manner described in Sec. 6.4. The first solution is displayed in the upper Table 6.3. It corresponds to the design shown in Fig. 6.8b, with the volume $V = 3.001\ LP/\bar{\sigma}$.

Table 6.3

C_i	N_i	0	1	1	1	1	1.41	1.41	1	1	1	1	1.41	1.41	a_{i0}
		a_{i0}	N_1	N_2	N_3	N_4	N_5	N_6	N_7	N_8	N_9	N_{10}	N_{11}	N_{12}	a_{ip}
1	N_4	0.500	−1			1			1						0.500
1	N_8	0.366	1	−1					−1	1					
1	N_9	0.866	1		−1				−1		1				
1.41	N_{11}	0.190	−1.41				−1		1.41				1		0.134
1.41	N_{12}	0.707	−1.41					−1	1.41					1	0.500
		3.001	−4	−2	−2		−2.83	−2.83	2			−2			
							1st iteration								
1	N_4	0.366	0			1	0.71					−1	−0.71		
1	N_7	0.134	−1				−0.71		1				0.71		
1	N_8	0.500	0	−1			−0.71			1			0.71		
1	N_9	1.000	0		−1		−0.71				1		0.71		
1.41	N_{12}	0.518	0				−0.71	−1					−1	1	
		2.732	−2	−2	−2		−1.41	−2.83				−2	−1.41		

Since the optimality index is positive in the N_7'-column, we perform a simplex iteration (introducing N_7 into the basis and removing N_{11} from it) and obtain an improved solution (the lower Table 6.3). The corresponding truss has the volume $V = 2.732\ LP/\bar{\sigma}$ (Fig. 6.8c). This time, the optimality index is negative throughout and therefore the new solution is an optimum solution, i.e. the design it represents is a minimum-volume design.

Niemierko's paper quoted above presents optimal solutions for several different trusses, obtained on an ODRA-1204 computer by the simplex method.

7

Non-linear Programming

7.1 Problem statement

If the objective function and the constraints in a mathematical programming problem are not all linear, we say it is a *non-linear programming problem.* The general non-linear programming problem can thus be stated as follows:

Minimize the function

$$F(\mathbf{x}), \quad \mathbf{x}^T = (x_1, \ldots, x_n), \tag{7.1}$$

subject to the constraints

$$g_j(\mathbf{x}) \leqslant 0, \quad j = 1, \ldots, m, \quad \mathbf{x} \geqslant \mathbf{0}, \tag{7.2}$$

where at least one of the functions $F(\mathbf{x})$ *and* $g_j(\mathbf{x})$ *is non-linear.*

Figure 7.1 shows a geometric interpretation of non-linear programming in the case where there are only two variables. The curves $g_j(\mathbf{x}) = 0$ delimit the feasible region (shaded) in the first quadrant. A minimum solution may be a point where the constraints have no influence, i.e. an interior point of the feasible region (Fig. 7.1a), but in most practical problems it lies where a contour of the objective function, i.e. a curve $F(\mathbf{x}) = \text{const}$, is tangent to the boundary (Fig. 7.1b).

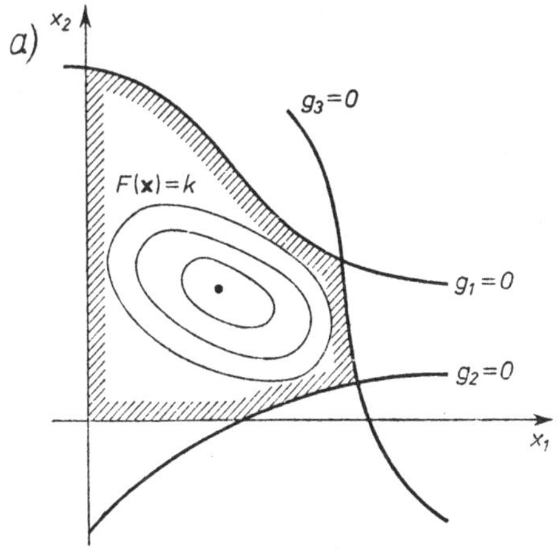

Fig. 7.1a. Non-linear programming problem

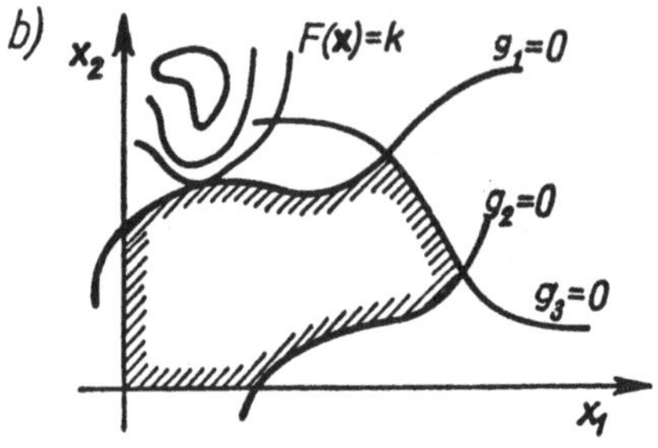

Fig. 7.1b. Non-linear programming problem

A general method for solving non-linear programs has not yet been found. There are numerous methods by which approximate minima can be found to a good accuracy. Determining an exact absolute minimum, however, is generally a difficult and very laborious task, if at all possible.

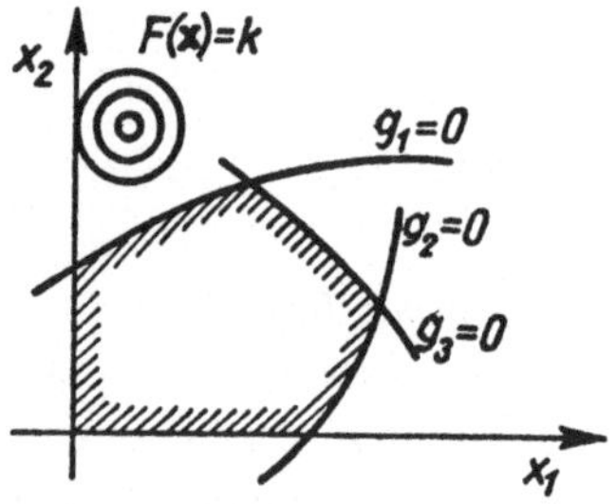

Fig. 7.2. A convex program

Efficient methods exist for solving non-linear programming problems in which the objective function and all the constraints are convex (see Chapter 5). Problems of this kind are called *convex programs*. A plane convex program is illustrated in Fig. 7.2. The feasible region bounded by the curves $g_j(\mathbf{x}) = 0$ is then convex.

7.2 Kuhn–Tucker conditions

The Kuhn–Tucker theorem, fundamental in non-linear programming, generalizes the classical method of Lagrange multipliers to inequality-constrained problems. In a general formulation (Sec. 5.3) it specifies the necessary conditions for a vector $\hat{\mathbf{x}}$ to be a solution of problem (7.1)–(7.2). The conditions involve a vector $\boldsymbol{\lambda}$ of m new variables $\lambda_1, \ldots, \lambda_m$ (Lagrange multipliers) and a new function L (Lagrange function) of $m+n$ variables $(\mathbf{x}, \boldsymbol{\lambda})$:

$$L(\mathbf{x}, \boldsymbol{\lambda}) = F(\mathbf{x}) + \sum_{j=1}^{m} \lambda_j g_j(\mathbf{x}). \tag{7.3}$$

As noted in Sec. 5.3, in the case of convex programming, the Kuhn–Tucker conditions are both necessary and sufficient for an absolute constrained minimum of the objective function. Here we present another formulation of the Kuhn–Tucker theorem for the convex case:

A vector $\hat{\mathbf{x}}$ is *a solution of problem* (7.1)–(7.2) *if and only if there is a vector* $\hat{\boldsymbol{\lambda}}$ *such that*

$$\hat{\mathbf{x}} \geqslant \mathbf{0}, \quad \hat{\boldsymbol{\lambda}} \geqslant \mathbf{0}, \tag{7.4}$$

$$L(\hat{\mathbf{x}}, \boldsymbol{\lambda}) \leqslant L(\hat{\mathbf{x}}, \hat{\boldsymbol{\lambda}}) \leqslant L(\mathbf{x}, \hat{\boldsymbol{\lambda}}). \tag{7.5}$$

Inequality (7.5) means that the function $L(\mathbf{x}, \boldsymbol{\lambda})$ attains at the point $(\hat{\mathbf{x}}, \hat{\boldsymbol{\lambda}})$ a global minimum with respect to $\mathbf{x}$ ($\mathbf{x} \geqslant \mathbf{0}$) and a global maximum with respect to $\boldsymbol{\lambda}$ ($\boldsymbol{\lambda} \geqslant \mathbf{0}$), and hence that $(\hat{\mathbf{x}}, \hat{\boldsymbol{\lambda}})$ is a saddle point for L. Thus, the problem of minimizing the objective function $F(\mathbf{x})$ is reduced to that of determining the saddle point of the Lagrange function L, or to a minimax problem.

When the functions $F(\mathbf{x})$ and $g_j(\mathbf{x})$ are differentiable, inequalities (7.4)–(7.5) are equivalent to the following set of conditions:

$$\left(\frac{\partial L}{\partial x_i}\right)_{\hat{\mathbf{x}}, \hat{\boldsymbol{\lambda}}} \geqslant 0, \tag{7.6}$$

$$\hat{x}_i \left(\frac{\partial L}{\partial x_i}\right)_{\hat{\mathbf{x}}, \hat{\boldsymbol{\lambda}}} = 0, \tag{7.7}$$

$$\hat{x}_i \geqslant 0, \tag{7.8}$$

$$\left(\frac{\partial L}{\partial \lambda_j}\right)_{\hat{\mathbf{x}}, \hat{\boldsymbol{\lambda}}} \leqslant 0, \tag{7.9}$$

$$\lambda_j \left(\frac{\partial L}{\partial \lambda_j}\right)_{\hat{\mathbf{x}}, \hat{\boldsymbol{\lambda}}} = 0, \tag{7.10}$$

$$\hat{\lambda}_j \geqslant 0. \tag{7.11}$$

They can easily be derived by using the fact that for a fixed non-negative vector $\hat{\boldsymbol{\lambda}}$, the function $L(\mathbf{x}, \hat{\boldsymbol{\lambda}})$ is convex in X. A proof of the Kuhn–Tucker theorem can be found in most books on mathematical programming.[1,2,3]

Example. Minimize the function

$$F(x_1, x_2) = (x_1 - 2)^2 + (x_2 - 2)^2$$

---

[1] KÜNZI, H. P. and KRELLE, W., *Nichtlineare Programmierung*, Springer-Verlag, Heidelberg 1966.

[2] GASS, S. I., *Loc. cit.* p. 203.

[3] KÜNZI, H. P., TZSCHACH, H. G. and ZEHNDER, C. A., *Numerische Methoden der Mathematischen Optimierung*, B. G. Teubner, Stuttgart 1966.

subject to the constraints

$$
\begin{aligned}
x_1+x_2-2 &\leqslant 0,\\
-x_1+x_2-1 &\leqslant 0,\\
x_1 &\geqslant 0,\\
x_2 &\geqslant 0.
\end{aligned}
$$

The Lagrange function for the problem is

$$
\begin{aligned}
&L(x_1, x_2, \lambda_1, \lambda_2)\\
&\quad = (x_1-2)^2+(x_2-2)^2+\lambda_1(x_1+x_2-2)+\lambda_2(-x_1+x_2-1).
\end{aligned}
$$

Since the objective function and constraints are all differentiable and convex, we use the Kuhn–Tucker conditions (7.6)–(7.11):

$$
\begin{array}{llll}
\text{(a)} & \dfrac{\partial L}{\partial x_1} \geqslant 0, & & 2(x_1-2)+\lambda_1-\lambda_2 \geqslant 0,\\[2ex]
\text{(b)} & \dfrac{\partial L}{\partial x_2} \geqslant 0, & & 2(x_2-2)+\lambda_1+\lambda_2 \geqslant 0,\\[2ex]
\text{(c)} & x_1\dfrac{\partial L}{\partial x_1} = 0, & & x_1[2(x_1-2)+\lambda_1-\lambda_2] = 0,\\[2ex]
\text{(d)} & x_2\dfrac{\partial L}{\partial x_2} = 0, & & x_2[2(x_2-2)+\lambda_1+\lambda_2] = 0.\\[2ex]
\text{(e)} & x_1 \geqslant 0, & &\\
\text{(f)} & x_2 \geqslant 0, & &\\[1ex]
\text{(g)} & \dfrac{\partial L}{\partial \lambda_1} \leqslant 0, & & x_1+x_2-2 \leqslant 0,\\[2ex]
\text{(h)} & \dfrac{\partial L}{\partial \lambda_2} \leqslant 0, & & -x_1+x_2-1 \leqslant 0,\\[2ex]
\text{(i)} & \lambda_1\dfrac{\partial L}{\partial \lambda_1} = 0, & & \lambda_1(x_1+x_2-2) = 0,\\[2ex]
\text{(j)} & \lambda_2\dfrac{\partial L}{\partial \lambda_2} = 0, & & \lambda_2(-x_1+x_2-1) = 0,\\[2ex]
\text{(k)} & \lambda_1 \geqslant 0, & &\\
\text{(l)} & \lambda_2 \geqslant 0. & &
\end{array}
$$

All solutions of the set of equations (c), (d), (i) and (j) are collected in Table 7.1, where each of them is checked for the remaining conditions. It can be seen that the only solution which satisfies all the Kuhn–Tucker conditions, and therefore the optimal solution to the problem, is $x_1 = 1$, $x_2 = 1$ (Fig. 7.3). At that point, the objective function is $F(1, 1) = 2$.

Table 7.1

	Solution								
	1	2	3	4	5	6	7	8	9
x_1	0	2	0	2	$\frac{3}{2}$	1	0	−1	$\frac{1}{2}$
x_2	0	0	2	2	$\frac{5}{2}$	1	1	0	$\frac{3}{2}$
λ_1	0	0	0	0	0	2	0	0	2
λ_2	0	0	0	0	−1	0	2	−6	−1
$x_1 \geqslant 0,\ x_2 \geqslant 0,$ $\lambda_1 \geqslant 0,\ \lambda_2 \geqslant 0$	yes	yes	yes	yes	no	yes	yes	no	no
$\frac{\partial L}{\partial x_1} \geqslant 0$	no	yes	no	yes		yes	no		
$\frac{\partial L}{\partial x_2} \geqslant 0$		no		yes		yes			
$\frac{\partial L}{\partial \lambda_1} \leqslant 0$				no		yes			
$\frac{\partial L}{\partial \lambda_2} \leqslant 0$						yes			

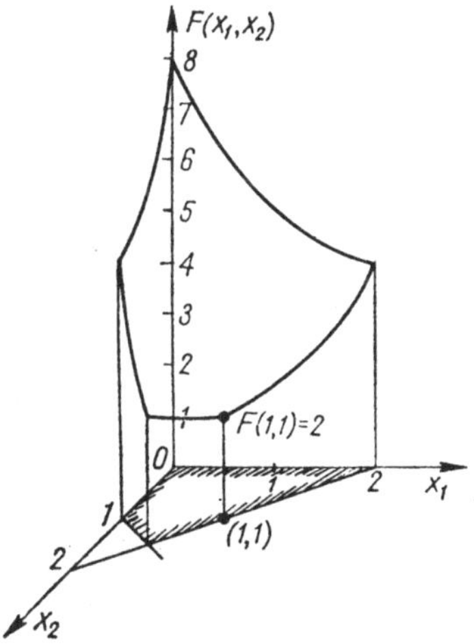

Fig. 7.3. Constrained minimum of the function F

The Kuhn–Tucker conditions admit various modifications according to the form of the original problem:

(a) When it is not required that $x \geqslant 0$, conditions (7.6)–(7.8) can be replaced by the single condition

$$\left(\frac{\partial L}{\partial x_i}\right)_{\hat{x},\hat{\lambda}} = 0. \tag{7.12}$$

(b) If $g_j(\mathbf{x})$ is linear, a constraint of the form $g_j(\mathbf{x}) = 0$ is feasible. It is evident from the form of the Lagrange function that in this case conditions (7.9)–(7.11) reduce to the equality

$$\left(\frac{\partial L}{\partial \lambda_j}\right)_{\hat{\mathbf{x}},\hat{\boldsymbol{\lambda}}} = 0, \tag{7.13}$$

where the sign of $\boldsymbol{\lambda}$ is no longer restricted.

(c) If the original problem is of the form

Minimize $F(\mathbf{x})$ *subject to* $g_j(\mathbf{x}) = 0$,

where g_j are linear, then there only remain conditions (7.12) and (7.13), i.e. we obtain the classical theorem of differential calculus on constrained extrema of a function of several variables.

7.3 Quadratic programming

Quadratic programming is convex programming in which the constraints are linear and the objective function is the sum of a linear and a quadratic form. The general quadratic problem can be stated as

Minimize

$$F(\mathbf{x}) = \mathbf{c}^T\mathbf{x}+\mathbf{x}^T\mathbf{D}\mathbf{x}, \tag{7.14}$$

subject to

$$\mathbf{A}\mathbf{x} \leqslant \mathbf{b}, \tag{7.15}$$

$$\mathbf{x} \geqslant \mathbf{0}, \tag{7.16}$$

where $\mathbf{x}$ and $\mathbf{c}$ are n-vectors, $\mathbf{b}$ is an m-vector, $\mathbf{A}$ is an $m \times n$ matrix and $\mathbf{D}$ is an $n \times n$ symmetric and positive definite or semidefinite matrix.

Two alternative formulations are also used:

$$\text{Min}\, F(\mathbf{x}) \textit{ subject to } A\mathbf{x} = \mathbf{b} \textit{ and } \mathbf{x} \geqslant \mathbf{0}, \tag{7.17}$$

$$\text{Min}\, F(\mathbf{x}) \textit{ subject to } A\mathbf{x} \leqslant \mathbf{b}. \tag{7.18}$$

The first is obtained by introducing appropriate slack variables into constraints (7.15), and the second by incorporating the non-negativity condition (7.16) into the constraint matrix.

The assumption that the matrix $\mathbf{D}$ is positive semidefinite ensures the convexity of the objective function. Geometrically, the solution to a quadratic programming problem with a positive definite $\mathbf{D}$ can be interpreted as shown in Fig. 7.4. The difference when compared with linear programming is evident. In a linear problem, the objective function attains a minimum at an apex of the constraint polyhedron, while in a quadratic problem a minimum may occur at an apex (Fig. 7.4a), but also at a non-apex boundary point (Fig. 7.4b) or even in the interior of the constraint polyhedron (Fig. 7.4c).

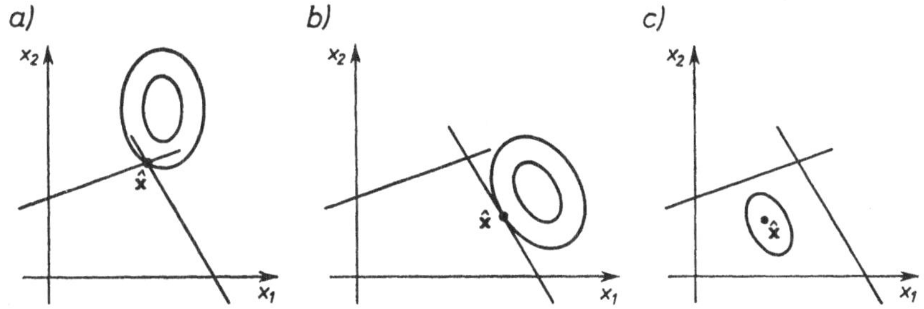

Fig. 7.4. Possible locations of the solution to a quadratic program

If a linear programming problem has a solution, then there is an optimal point at which exactly n of $m+n$ constraint inequalities are satisfied as equalities, while in quadratic programming usually less then n constraints (and never more) hold as equalities at an optimal point.

For each of the three problem statements ((7.14)–(7.16), (7.17) and (7.18)) the Lagrange function takes the form

$$L(\mathbf{x}, \boldsymbol{\lambda}) = \mathbf{c}^T\mathbf{x}+\mathbf{x}^T\mathbf{D}\mathbf{x}+\boldsymbol{\lambda}^T(\mathbf{A}\mathbf{x}-\mathbf{b}). \tag{7.19}$$

For brevity, let us write

$$\frac{\partial L}{\partial \mathbf{x}} = \mathbf{v} \quad \text{and} \quad -\frac{\partial L}{\partial \boldsymbol{\lambda}} = \mathbf{u}; \tag{7.20}$$

we have

$$\begin{aligned} \mathbf{v} &= \mathbf{c}+2\mathbf{D}\mathbf{x}+\mathbf{A}^T\boldsymbol{\lambda}, \\ \mathbf{u} &= -\mathbf{A}\mathbf{x}+\mathbf{b}. \end{aligned} \tag{7.21}$$

In this notation, the Kuhn–Tucker conditions for problem (7.17) can be written as

$$\begin{aligned} &\mathbf{A}\mathbf{x} = \mathbf{b}, \\ &2\mathbf{D}\mathbf{x}-\mathbf{v}+\mathbf{A}^T\boldsymbol{\lambda} = -\mathbf{c}, \\ &\mathbf{x} \geqslant \mathbf{0}, \qquad \mathbf{v} \geqslant \mathbf{0}, \\ &\mathbf{x}^T\mathbf{v} = 0. \end{aligned} \tag{7.22}$$

7.4 Duality in quadratic programming

In non-linear programming the concept of duality is much less important than it is in linear programming, mainly because the symmetry between a primal problem and its dual is here much less pronounced. We shall therefore restrict ourselves to introducing the dual problem and quoting the central result of the duality theory for quadratic programming.

As the primal problem we take a quadratic program in form (7.18):

$$\min F(\mathbf{x}) = \mathbf{c}^T\mathbf{x}+\mathbf{x}^T\mathbf{D}\mathbf{x}$$

subject to

$$\mathbf{Ax} \leqslant \mathbf{b}.$$

The dual problem can then be formulated as follows:

$$\max L(\mathbf{x}, \boldsymbol{\lambda}) = \mathbf{c}^T\mathbf{x}+\mathbf{x}^T\mathbf{Dx}+\boldsymbol{\lambda}^T(\mathbf{Ax}-\mathbf{b})$$

subject to the constraints

$$\frac{\partial L}{\partial \mathbf{x}} = \mathbf{0} \quad \text{and} \quad \boldsymbol{\lambda} \geqslant \mathbf{0}.$$

Using the equality

$$\frac{\partial L}{\partial \mathbf{x}} = \mathbf{c}+2\mathbf{Dx}+\mathbf{A}^T\boldsymbol{\lambda} = \mathbf{0},$$

we can restate the dual problem as

$$\max L(\mathbf{x}, \boldsymbol{\lambda}) = -\mathbf{x}^T\mathbf{Dx}-\mathbf{b}^T\boldsymbol{\lambda}$$

subject to

$$2\mathbf{Dx}+\mathbf{A}^T\boldsymbol{\lambda} = -\mathbf{c} \quad \text{and} \quad \boldsymbol{\lambda} \geqslant \mathbf{0}.$$

When $\mathbf{D} = \mathbf{0}$, we obtain a pair of dual linear programs (cf. Sec. 6.7).

Dorn[1] has given the following duality theorem for quadratic programming:

If $\hat{\mathbf{x}}$ is a solution to the primal problem, then there is a vector $\hat{\boldsymbol{\lambda}}$, such that $(\hat{\mathbf{x}}, \hat{\boldsymbol{\lambda}})$ is a solution to the dual problem. Conversely, if $(\hat{\mathbf{x}}, \hat{\boldsymbol{\lambda}})$ maximizes the function $L(\mathbf{x}, \boldsymbol{\lambda})$ of the dual problem, then $\hat{\mathbf{x}}$ is a solution to the primal problem. In either case $F(\hat{\mathbf{x}}) = L(\hat{\mathbf{x}}, \hat{\boldsymbol{\lambda}})$, i.e. the minimum of the primal objective function is equal to the maximum of the dual objective function.

As mentioned earlier, there is no general algorithm for non-linear programming. The class of problems best researched are quadratic programs, for which a number of effective iterative methods have been devised. Among the more important methods we may mention those of Beale[2], Wolfe[3], Frank and Wolfe[4], Hildreth[5] and Houthakker[6]. Below we discuss the methods of Beale and Wolfe.

---

[1] Dorn, W. S., Duality in quadratic programming, *Quart. Appl. Math.*, **18**, 1960.

[2] Beale, E. M. L., On quadratic programming, *Naval Res. Logist. Quart.*, **6**, 227–243, 1959.

[3] Wolfe, Ph., The simplex method for quadratic programming, *Econometrica*, **27**, 382–398, 1959.

[4] Frank, M., and Wolfe, Ph., An algorithm for quadratic programming, *Naval Res. Logist. Quart.*, **3**, 1–2, 15–110, 1956.

[5] Hildreth, C., A quadratic programming procedure, *Naval Res. Logist. Quart.*, **4**, **1**, 79–85, 1957.

[6] Houthakker, H. S., The capacity method of quadratic programming, *Econometrica*, **28**, **1**, 62–87, 1960.

7.5 Beale's method

Beale's method is a generalization of the linear simplex method to quadratic programming. It is adapted to quadratic programs of form (7.17), i.e. to the problem of minimizing a convex second-degree polynomial $F(\mathbf{x})$ subject to constraints

$$\begin{aligned} \mathbf{Ax} &= \mathbf{b}, \\ \mathbf{x} &\geqslant \mathbf{0}. \end{aligned} \tag{7.23}$$

where $\mathbf{x}$ is an n-vector, $\mathbf{b}$ an m-vector and $\mathbf{A}$ an $m \times n$ matrix ($m < n$).

The procedure starts from an arbitrary basic feasible solution (the first "trial point") of system (7.23). Let the basic variables be $x_1, x_2, \ldots, x_m$ and put $x_{m+h} = z_h$, $h = 1, 2, \ldots, n-m$, so that

$$x_g = d_{g0}^1 + \sum_{h=1}^{n-m} d_{gh}^1 z_h, \quad g = 1, \ldots, m, \tag{7.24}$$

where $d_{g0}^1 \geqslant 0$ (the upper index, 1, refers to the 1st trial point).

Using (7.24), we can represent the function $F(\mathbf{x})$ in terms of the variables $z_1, \ldots, z_{n-m}$ as

$$\begin{aligned} F(x_1, \ldots, x_n) &= F^1(z_1, \ldots, z_{n-m}) \\ &= c_{00}^1 + 2\sum_{i=1}^{n-m} c_{i0}^1 z_i + \sum_{h=1}^{n-m}\sum_{i=1}^{n-m} c_{hi}^1 z_i z_h \\ &= c_{00}^1 + \sum_{i=1}^{n-m} c_{0i}^1 z_i + \sum_{h=1}^{n-m}\Big(c_{h0}^1 + \sum_{i=1}^{n-m} c_{hi}^1 z_i\Big) z_h, \end{aligned} \tag{7.25}$$

where the coefficients c_{ih}^1 are symmetric, i.e. $c_{ih}^1 = c_{ih}^1$, $i, h = 0, 1, \ldots, n-m$, and

$$\frac{1}{2}\frac{\partial F^1}{\partial z_h} = c_{h0}^1, \quad h = 1, \ldots, n-m. \tag{7.26}$$

Of course, the value of the function $F(\mathbf{x})$ at the first trial point is c_{00}^1. It is apparent from representation (7.25) that the first trial point is a Kuhn–Tucker point, or the optimal solution, if and only if

$$\frac{\partial F^1}{\partial z_h} \geqslant 0 \tag{7.27}$$

for all $h = 1, 2, \ldots, n-m$ at that point. Indeed, an increase of any of the variables z_h would then increase the value of F. On the other hand, if for some h

$$\frac{\partial F^1}{\partial z_h} < 0 \quad (\text{i.e. } c_{h0}^1 < 0), \tag{7.28}$$

then an improved solution, i.e. one that yields smaller F, can be obtained by making z_h positive (Fig. 7.5: $\partial F^1/\partial z_h < 0$ at $z_h = 0$).

Suppose that inequality (7.28) occurs for $h = 1$ and let z_1 rise. An increase in z_1 will of course affect the basic variables $x_1, \ldots, x_m$. The question is how far z_1 should be increased. Two cases can occur:

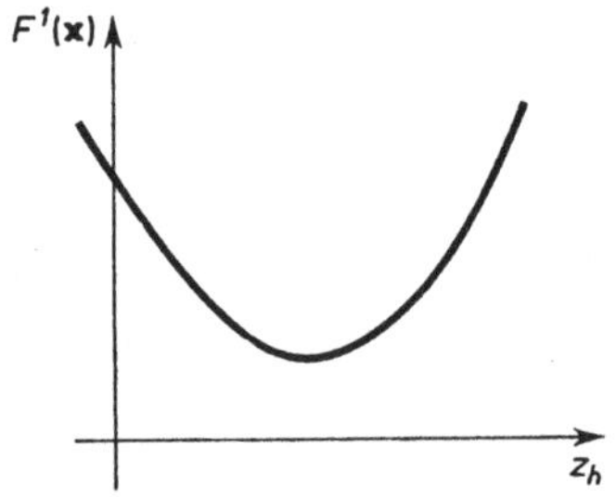

Fig. 7.5. $\partial F^1/\partial z_h < 0$ at $z_h = 0$

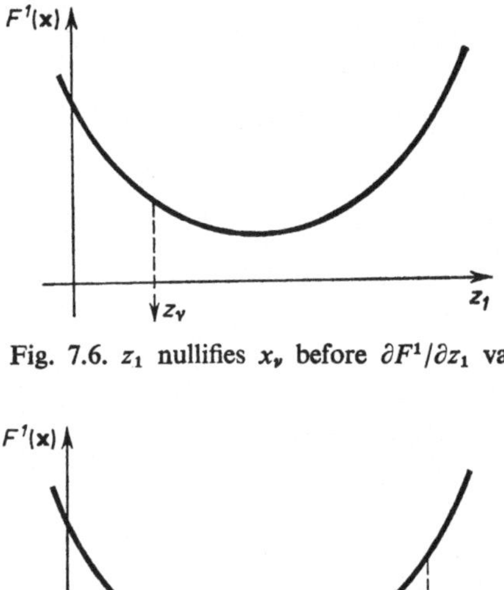

Fig. 7.6. z_1 nullifies x_v before $\partial F^1/\partial z_1$ vanishes

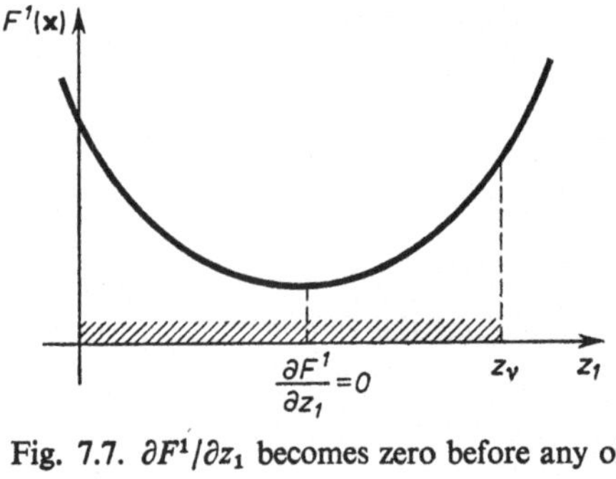

Fig. 7.7. $\partial F^1/\partial z_1$ becomes zero before any of the basic variables does

(1) z_1 reaches a point at which one of the variables $x_1, \ldots, x_n$, say x_v, vanishes before $\partial F^1/\partial z_1$ does (Fig. 7.6). We then introduce z_1 into the basis in place of x_v as in linear programming. The new basic solution is the second trial point.

(2) $\partial F^1/\partial z_1$ becomes zero before any of the basic variables x_g does (Fig. 7.7). In this case, z_1 can only be increased to the value at which $\partial F^1/\partial z_1 = 0$

(a further rise would increase F) and a new, unconstrained variable is introduced:

$$u_1 = \frac{1}{2}\frac{\partial F^1}{\partial z_1}. \tag{7.29}$$

As the second trial point we then take the point at which the first unconstrained variable u_1 and the non-basic variables $z_2, \ldots, z_{n-m}$ all vanish; z_1 becomes a new basic variable. Both the constraint system and the objective function are rearranged, with u_1 as a non-basic variable:

From

$$u_1 = c_{10}^1 + \sum_{h=1}^{n-m} c_{1h}^1 z_h = \frac{1}{2}\frac{\partial F^1}{\partial z_1} \tag{7.30}$$

we find

$$z_1 = -\frac{c_{10}^1}{c_{11}^1} + \frac{1}{c_{11}^1}u_1 - \sum_{h=2}^{n-m} \frac{c_{1h}^1}{c_{11}^1} z_h = d_{10}^2 + d_{11}^2 u_1 + \sum_{h=2}^{n-m} d_{1h}^2 z_h \tag{7.31}$$

and eliminate z_1 from equation (7.24). On renumbering the variables ($z_1 \to x_1, x_1 \to x_2, \ldots, x_m \to x_{m+1}$) and including (7.31) as the first equation in the system, we obtain a new system of $m+1$ constraint equations

$$x_g = d_{g0}^2 + d_{g1}^2 u_1 + \sum_{h=2}^{n-m} d_{gh}^2 z_h, \qquad g = 1, 2, \ldots, m, m+1. \tag{7.32}$$

Similarly, using (7.31), we eliminate z_1 from F^1:

$$F^1(z_1, \ldots, z_{n-m}) = F^2(u_1, z_2, \ldots, z_{n-m}). \tag{7.33}$$

We next consider the second trial point in the same way as the first one, except that the Kuhn–Tucker condition with respect to the variable u_1 takes the form

$$\frac{\partial F^2}{\partial u_1} = 0. \tag{7.34}$$

If this derivative is not zero at the trial point, we can decrease F by changing u_1 in the positive or in the negative direction according as $\partial F^2/\partial u_1$ is negative or positive. Once the unconstrained variable becomes a basic variable and is eliminated from the constraints and the objective function, we can ignore it: the constraints only check whether the basic variables are non-negative, and no such control is needed for the unconstrained variable.

Beale[1] has shown that the above procedure yields the optimum solution after a finite number of steps provided the following auxilliary rule is obeyed:

[1] Beale, E. M. L., On minimizing a convex function subject to linear inequalities, *J. Roy. Stat. Soc.*, **17**, 2, 173–184, 1955.

as the new basic variable in each step, select, if possible, an unconstrained variable, and only when the derivatives of F with respect to the unconstrained variables are all zero, i.e. when F cannot be decreased by changing an unconstrained variable, choose an original (constrained) variable.

7.6 Wolfe's method

The procedure developed by Wolfe[1] is largely based on the simplex method, and hence is suitable for computerized use. It can be applied to quadratic programs of form (7.17). As we have seen in Sec. 7.3, the Kuhn–Tucker conditions for such programs can be written

$$\begin{aligned} &\mathbf{Ax} = \mathbf{b}, \\ &2\mathbf{Dx}-\mathbf{v}+\mathbf{A}^T\boldsymbol{\lambda} = -\mathbf{c}, \quad \mathbf{x} \geqslant \mathbf{0}, \quad \mathbf{v} \geqslant \mathbf{0}, \end{aligned} \tag{7.35}$$

$$\mathbf{x}^T\mathbf{v} = \mathbf{0}. \tag{7.36}$$

The free terms b_j can be assumed to be non-negative. Let us recall that $\mathbf{x}$ and $\mathbf{v}$ are n-vectors and $\boldsymbol{\lambda}$ is an m-vector. System (7.35) consists of $m+n$ linear equations with $2n$ non-negative variables and m variables with an unrestricted sign. The condition $\mathbf{x}^T\mathbf{v} = \mathbf{0}$ requires that, for each i, x_i and v_i should not both be positive. Thus not more than n out of $2n$ variables $\mathbf{x}$ and $\mathbf{v}$ can be positive, i.e. at least n of the $2n+m$ variables of system (7.35) must vanish. As in linear programming, any solution to system (7.35) with at least n zero variables is called a *basic feasible solution.* To solve the problem it suffices to find a basic feasible solution satisfying condition (7.36). A set of artificial variables is first introduced into system (7.35) so that a basic feasible solution of the augmented system is available, obeying condition (7.36). The simplex method is then used to eliminate the new variables, but care must be taken that condition (7.36) is satisfied at each iteration.

There are two versions of Wolfe's algorithm, called the *short* and the *long forms.* The first is applicable to problems in which $\mathbf{c} = \mathbf{0}$ or $\mathbf{D}$ is positive definite. The long algorithm can be used for solving arbitrary quadratic programs.

The short form

To solve system (7.35) we introduce $m+2n$ non-negative artificial variables

$$\begin{aligned} \mathbf{w} &= w_1, \ldots, w_m, \\ \mathbf{z}^1 &= z_1^1, \ldots, z_n^1, \\ \mathbf{z}^2 &= z_1^2, \ldots, z_n^2. \end{aligned}$$

---

[1] Wolfe, Ph., *loc. cit.*, p. 232.

and form a new system containing $m+n$ equations in $4n+2m$ unknowns

$$\begin{aligned} &\mathbf{Ax}+\mathbf{w} = \mathbf{b}, \\ &2\mathbf{Dx}-\mathbf{v}+\mathbf{A}^T\boldsymbol{\lambda}+\mathbf{z}^1-\mathbf{z}^2 = -\mathbf{c}, \\ &\mathbf{x} \geqslant \mathbf{0}, \quad \mathbf{v} \geqslant \mathbf{0}, \quad \mathbf{w} \geqslant \mathbf{0}, \quad \mathbf{z}^1 \geqslant \mathbf{0}, \quad \mathbf{z}^2 \geqslant \mathbf{0}. \end{aligned} \tag{7.37}$$

A basic feasible solution of the augmented system, satisfying condition (7.36), is readily obtained by setting $(4n+2m)-(m+n) = 3n+m$ variables equal to zero, and namely

$$\mathbf{x} = \mathbf{0}, \quad \mathbf{v} = \mathbf{0}, \quad \boldsymbol{\lambda} = \mathbf{0},$$

and for each i one of the variables z_i^1 and z_i^2, according as c_i is positive or negative. Thus the initial basic variables are $w_j = b_j$ ($\geqslant 0$ by assumption), $j = 1, \ldots, m$, and for each i either

$$z_i^1 = -c_i \quad \text{if} \quad c_i \leqslant 0,$$

or

$$z_i^2 = +c_i \quad \text{if} \quad c_i > 0.$$

The elimination of the artificial variables from the basis is carried out at two stages. At stage 1, starting from the initial solution and using the simplex method, we minimize the linear function

$$\sum_{j=1}^{m} w_j$$

subject to constraints (7.37) and the additional conditions

$$\boldsymbol{\lambda} = \mathbf{0} \quad \text{and} \quad \mathbf{v} = \mathbf{0},$$

so that λ_j and v_j remain non-basic throughout stage 1. If the constraints are consistent, then, as proved by Wolfe, $\min \sum_{j=1}^{m} w_j = 0$.

When stage 1 is completed, we have a basic solution of the system

$$\begin{aligned} &\mathbf{Ax} = \mathbf{b}, \\ &2\mathbf{Dx}-\mathbf{v}+\mathbf{A}^T\boldsymbol{\lambda}+\mathbf{Bz} = -\mathbf{c}, \\ &\mathbf{x} \geqslant \mathbf{0}, \quad \mathbf{v} \geqslant \mathbf{0}, \quad \mathbf{z} \geqslant \mathbf{0}, \end{aligned} \tag{7.38}$$

where $\mathbf{B}$ is a diagonal matrix with $+1$ or -1 on the diagonal according as there remains z_i^1 or z_i^2. We next go to stage 2, in which the simplex method is used again to minimize the linear form

$$\sum_{i=1}^{n} z_i$$

subject to conditions (7.38).

It can be shown (see e.g. Künzi and Krelle[1]) that if $\mathbf{c} = \mathbf{0}$ or $\mathbf{D}$ is positive definite then the sum $\sum z_i$ can be reduced to zero with condition (7.36) maintained. The minimum point so obtained is the desired solution of the original problem.

The long form

The long algorithm consists of three stages, of which the first two are in fact short-form stages applied to the same problem but with the vector $\mathbf{c}$ replaced by a zero vector, i.e. with equation $(7.37)_2$ reading

$$2\mathbf{D}\mathbf{x}-\mathbf{v}+\mathbf{A}^T\boldsymbol{\lambda}+\mathbf{z}^1-\mathbf{z}^2 = \mathbf{0}.$$

According to the previous discussion, the short algorithm will produce a basic solution of the system

$$\begin{aligned} &\mathbf{A}\mathbf{x} = \mathbf{b},\\ &2\mathbf{D}\mathbf{x}-\mathbf{v}+\mathbf{A}^T\boldsymbol{\lambda} = \mathbf{0},\\ &\mathbf{x} \geqslant \mathbf{0}, \quad \mathbf{v} \geqslant \mathbf{0}, \end{aligned} \tag{7.39}$$

satisfying condition (7.36). We then introduce one more variable, μ, and consider the system

$$\begin{aligned} &\mathbf{A}\mathbf{x} = \mathbf{b},\\ &2\mathbf{D}\mathbf{x}-\mathbf{v}+\mathbf{A}^T\boldsymbol{\lambda}+\mu\mathbf{c} = \mathbf{0},\\ &\mathbf{x} \geqslant \mathbf{0}, \quad \mathbf{v} \geqslant \mathbf{0}, \quad \mu \geqslant 0. \end{aligned} \tag{7.40}$$

The basic solution of system (7.39) will with $\mu = 0$ be still valid, but what we need is a basic solution of (7.40) with $\mu = 1$ and satisfying condition (7.36). To find such a solution, we move on to the third stage of the procedure, in which we choose $(-\mu)$ as a new objective function and use the simplex method to minimize it subject to conditions (7.40) and (7.36), starting from the basic solution obtained in stage 2.

Two cases may occur:

(a) There is a bounded optimum solution to the linear problem defined, i.e. $-\mu$ cannot be decreased to $-\infty$. It can be shown (see e.g. Wolfe[2] or Künzi and Krelle[3]) that in this case it is not possible to execute even one simplex transformation without violating restriction (7.36).

(b) $-\mu$ can be decreased to $-\infty$. Performing simplex iterations, we obtain a finite sequence of basic solutions $\mathbf{x}^i$, $\mathbf{v}^i$, $\boldsymbol{\lambda}^i$, μ^i, $i = 1, \ldots, s$, with

$$0 = \mu^1 < \mu^2 < \ldots < \mu^s.$$

---

[1] Künzi, H. P. and Krelle, W., *loc. cit.*, p. 227.
[2] Wolfe, Ph., *loc. cit.* p. 232.
[3] Künzi, H. P. and Krelle, W., *loc. cit.* p. 227.

To obtain a solution with $\mu = 1$, we construct a suitable linear combination of two of the solutions found:

(1) If $\mu^s \geqslant 1$, we choose $1 \leqslant j \leqslant s-1$, such that $\mu^j < 1 \leqslant \mu^{j+1}$ and put

$$(\hat{\mathbf{x}}, \hat{\mathbf{v}}, \hat{\boldsymbol{\lambda}}, \hat{\mu}) = \frac{\mu^{j+1}-1}{\mu^{j+1}-\mu^j}(\mathbf{x}^j, \mathbf{v}^j, \boldsymbol{\lambda}^j, \mu^j) + \\ + \frac{1-\mu^j}{\mu^{j+1}-\mu^j}(\mathbf{x}^{j+1}, \mathbf{v}^{j+1}, \boldsymbol{\lambda}^{j+1}, \mu^{j+1}). \tag{7.41}$$

Since $\hat{\mu} = 1$, equation (4.41) represents the solution of the problem.

(2) If $\mu^s < 1$, the optimum solution with $\hat{\mu} = 1$ is given by

$$(\hat{\mathbf{x}}, \hat{\mathbf{v}}, \hat{\boldsymbol{\lambda}}, \hat{\mu}) = (\mathbf{x}^g, \mathbf{v}^g, \boldsymbol{\lambda}^g, \mu^g) + \frac{1-\mu^g}{\mu^{g+1}}(\mathbf{x}^{g+1}, \mathbf{v}^{g+1}, \boldsymbol{\lambda}^{g+1}, \mu^{g+1}). \tag{7.42}$$

7.7 Geometric programming: unconstrained problem

Geometric programming, a comparatively new method of mathematical programming developed by Duffin, Peterson and Zener,[1,2,3] is useful in dealing with optimization problems in which both the objective function and constraints are posynomials, i.e. sums of power functions with positive coefficients. The requirement that the coefficients be positive ensures the convexity of the problem, and hence the identity of a global minimum with a local minimum, if one exists. A feature which distinguishes geometric programming from other programming techniques is that it first yields the optimal value of the objective function and only then is the optimal design vector found if needed. Another advantage of the method is that it often reduces a complicated optimization problem to solving a set of simultaneous linear algebraic equations.

In the absence of constraints, a geometric programming problem can be stated as follows:[4]

Find a vector $\mathbf{x}^T = (x_1, x_2, \ldots, x_n)$ *which minimizes the function*

$$F(\mathbf{x}) = \sum_{j=1}^{N} U_j(\mathbf{x}) = \sum_{j=1}^{N}\left(c_j \prod_{i=1}^{n} x_i^{a_{ij}}\right) = \sum_{j=1}^{N}(c_j x_1^{a_{1j}} \ldots x_n^{a_{nj}}), \tag{7.43}$$

where $c_j > 0$, $x_i > 0$ *and* a_{ij} *are real constants.*

---

[1] Duffin, R. J., Peterson, E. L., and Zener, S., *Geometric Programming*, Wiley, New York 1967.

[2] Zener, C,. A mathematical aid in optimizing engineering designs, *Proc. National Acad. Sci.*, **47**, p. 537 (1961).

[3] Zener, C,. *Engineering Design by Geometric Programming*, Wiley-Interscience, New York 1971.

[4] Rao, S. S., *Optimization, Theory and Application*, Wiley Eastern Ltd, New Delhi 1978.

Among different procedures that can be used to solve this problem we shall outline two: one based on the differential calculus and one involving arithmetic-geometric inequalities.

Solution by way of differential calculus

The necessary conditions for a minimum of the function F are

$$\frac{\partial F}{\partial x_k} = \sum_{j=1}^{N} \frac{\partial U_j}{\partial x_k}$$

$$= \sum_{j=1}^{N} (c_j x_1^{a_{1j}} \dots x_{k-1}^{a_{k-1,j}} a_{kj} x_k^{a_{kj}-1} x_{k+1}^{a_{k+1,j}} \dots x_n^{a_{nj}}) = 0, \quad k = 1, 2, \dots, n. \tag{7.44}$$

Multiplying the kth equation by x_k, we get

$$x_k \frac{\partial F}{\partial x_k} = \sum_{j=1}^{N} a_{kj}(c_j x_1^{a_{1j}} x_2^{a_{2j}} \dots x_{k-1}^{a_{k-1,j}} x_k^{a_{kj}} x_{k+1}^{a_{k+1,j}} \dots x_n^{a_{nj}})$$

$$= \sum_{j=1}^{N} a_{kj} U_j(\mathbf{x}) = 0, \quad k = 1, 2, \dots, n. \tag{7.45}$$

It can be shown that the necessary conditions (7.44) or (7.45) are also sufficient for a minimum of F. To this end, it is enough to prove that the Hessian matrix of F is positive definite (cf. Theorem 4′, Sec. 5.2). We leave this task to the reader.

Suppose that $\hat{\mathbf{x}}$ is the optimal vector, i.e. the one that yields the minimum of F, $\hat{F} = F(\hat{\mathbf{x}})$. By (7.45) we have

$$\sum_{j=1}^{N} a_{kj} U_j(\hat{\mathbf{x}}) = 0, \quad k = 1, 2, \dots, n. \tag{7.46}$$

Dividing this equation by $\hat{F}$ gives

$$\sum_{j=1}^{N} \hat{\Delta}_j a_{kj} = 0, \quad k = 1, 2, \dots, n, \tag{7.47}$$

where the quantities $\hat{\Delta}_j$ are defined as

$$\hat{\Delta}_j = \frac{U_j(\hat{\mathbf{x}})}{\hat{F}} = \frac{\hat{U}_j}{\hat{F}}. \tag{7.48}$$

So that

$$\sum_{j=1}^{N} \hat{\Delta}_j = \frac{1}{\hat{F}} \sum_{j=1}^{N} \hat{U}_j = 1. \tag{7.49}$$

Equations (7.47) are termed the *orthogonality conditions* and equation (7.49) is called the *normality condition.*

The minimum value $\hat{F}$ can be determined by the following procedure. By the normality condition, $\hat{F}$ can be written as

$$\hat{F} = (\hat{F})^1 = \hat{F}^{\sum_{j=1}^{N}\hat{\Delta}_j} = (\hat{F})^{\hat{\Delta}_1}(\hat{F})^{\hat{\Delta}_2}\dots(\hat{F})^{\hat{\Delta}_N}. \tag{7.50}$$

Since, from (7.48),

$$\hat{F} = \frac{\hat{U}_1}{\hat{\Delta}_1} = \frac{\hat{U}_2}{\hat{\Delta}_2} = \dots = \frac{\hat{U}_N}{\hat{\Delta}_N}, \tag{7.51}$$

expression (7.50) can be written as

$$\hat{F} = \left(\frac{\hat{U}_1}{\hat{\Delta}_1}\right)^{\hat{\Delta}_1}\left(\frac{\hat{U}_2}{\hat{\Delta}_2}\right)^{\hat{\Delta}_2}\dots\left(\frac{\hat{U}_N}{\hat{\Delta}_N}\right)^{\hat{\Delta}_N}. \tag{7.52}$$

Now if we substitute

$$\hat{U}_j = c_j\prod_{i=1}^{n}(\hat{x}_i)^{a_{ij}}, \quad j = 1, 2, \dots, N,$$

equation (7.52) will become

$$\begin{aligned}\hat{F} &= \left\{\left(\frac{c_1}{\hat{\Delta}_1}\right)^{\hat{\Delta}_1}\left[\prod_{i=1}^{n}(\hat{x}_i)^{a_{i1}}\right]^{\hat{\Delta}_1}\right\}\cdot\dots\cdot\left\{\left(\frac{c_N}{\hat{\Delta}_N}\right)^{\hat{\Delta}_N}\left[\prod_{i=1}^{n}(\hat{x}_i)^{a_{iN}}\right]^{\hat{\Delta}_N}\right\}\\ &= \left\{\prod_{j=1}^{N}\left(\frac{c_j}{\hat{\Delta}_j}\right)^{\hat{\Delta}_j}\right\}\left\{\prod_{j=1}^{N}\left[\prod_{i=1}^{n}(\hat{x}_i)^{a_{ij}}\right]^{\hat{\Delta}_j}\right\}\\ &= \left\{\prod_{j=1}^{N}\left(\frac{c_j}{\hat{\Delta}_j}\right)^{\hat{\Delta}_j}\right\}\left\{\prod_{i=1}^{n}(\hat{x}_i)^{\sum_{j=1}^{N}a_{ij}\hat{\Delta}_j}\right\}\\ &= \prod_{j=1}^{N}\left(\frac{c_j}{\hat{\Delta}_j}\right)^{\hat{\Delta}_j},\end{aligned} \tag{7.53}$$

because $\sum_{j=1}^{N} a_{ij}\hat{\Delta}_j = 0$ for each i.

Thus, once $\hat{\Delta}_j$ are known, the minimum value $\hat{F}$ can be calculated from eqn. (7.53). To determine $\hat{\Delta}_j$, we have $n+1$ equations in N unknowns, viz. (7.47) and (7.49). The number $N-n-1$ is called the *degree of difficulty* of the program (in constrained geometric programming, N denotes the total number of terms in all the posynomials and n the number of design variables). If $N-$ $-n-1 = 0$, the problem is said to be of *zero degree* of difficulty. Indeed, the orthogonality and normality conditions then provide exactly as many equations as there are unknowns, and so $\hat{\Delta}_j$ can be determined uniquely. When

$N > n+1$, the number of unknowns exceeds the number of equations; in this case, the program should be solved by the Cauchy inequality method discussed below. Finally, when the degree of difficulty is negative, geometric programming is inapplicable.

Having found $\hat{\Delta}_j$ and $\hat{F}$, we can determine the optimal design vector $\hat{\mathbf{x}}$ by solving the system of non-linear equations

$$\hat{U}_j = \hat{\Delta}_j \hat{F} = c_j \prod_{i=1}^{n} (\hat{x}_i)^{a_{ij}}, \quad j = 1, 2, \ldots, N. \tag{7.54}$$

To this end, we divide equation (7.54) by c_j and take logarithms of both sides. The result is

$$\ln\left(\frac{\hat{\Delta}_j \hat{F}}{c_j}\right) = a_{1j} \ln \hat{x}_1 + a_{2j} \ln \hat{x}_2 + \ldots + a_{nj} \ln \hat{x}_n. \tag{7.55}$$

Putting $w_i = \ln \hat{x}_i$, $i = 1, 2, \ldots, n$ we obtain N linear equations in n unknowns w_i:

$$\begin{aligned} a_{11} w_1 + a_{21} w_2 + \ldots + a_{n1} w_n &= \ln\left(\frac{\hat{\Delta}_1 \hat{F}}{c_1}\right), \\ a_{12} w_1 + a_{22} w_2 + \ldots + a_{n2} w_n &= \ln\left(\frac{\hat{\Delta}_2 \hat{F}}{c_2}\right), \\ \ldots\ldots\ldots\ldots\ldots\ldots\ldots &\ldots \\ a_{1N} w_1 + a_{2N} w_2 + \ldots + a_{nN} w_n &= \ln\left(\frac{\hat{\Delta}_N \hat{F}}{c_N}\right). \end{aligned} \tag{7.56}$$

If there are more equations than unknowns, any subsystem of n linearly independent equations can be chosen to determine w_i, and hence

$$\hat{x}_i = e^{w_i}, \quad i = 1, 2, \ldots, n. \tag{7.57}$$

Solution based on Cauchy's inequality

Another important method for solving geometric-programming problems is a procedure involving Cauchy's inequality between the arithmetic and the geometric mean of n non-negative numbers,

$$\frac{x_1 + x_2 + \ldots + x_n}{n} \geqslant (x_1 x_2 \ldots x_n)^{1/n}. \tag{7.58}$$

The equality between the two means holds only when $x_1 = x_2 = \ldots = x_n$. An equivalent form of inequality (7.58), more useful in geometric programming, is

$$\Delta_1 u_1 + \Delta_2 u_2 + \ldots + \Delta_N u_N \geqslant u_1^{\Delta_1} \cdot u_2^{\Delta_2} \ldots u_N^{\Delta_N} \tag{7.59}$$

with

$$\Delta_1 + \Delta_2 + \ldots + \Delta_N = 1. \tag{7.60}$$

If we apply this inequality to the objective function (7.43), setting $U_i(\mathbf{x}) = u_i\Delta_i$, $i = 1, 2, \ldots, N$, with $\Delta_1+\Delta_2+ \ldots +\Delta_N = 1$ (the normality condition), we obtain

$$U_1+U_2+ \ldots +U_N \geqslant \left(\frac{U_1}{\Delta_1}\right)^{\Delta_1}\left(\frac{U_2}{\Delta_2}\right)^{\Delta_2} \ldots \left(\frac{U_N}{\Delta_N}\right)^{\Delta_N}. \tag{7.61}$$

We call the left-hand side of inequality (7.61), i.e. the original function $F(\mathbf{x})$, the *primal function*, and the right-hand side, the *predual function*.

Using the relations

$$U_j = c_j \prod_{i=1}^{n} x_i^{a_{ij}}, \quad j = 1, 2, \ldots, N, \tag{7.62}$$

we can represent the predual function as

$$\begin{aligned}\left(\frac{U_1}{\Delta_1}\right)^{\Delta_1} \ldots \left(\frac{U_N}{\Delta_N}\right)^{\Delta_N} &= \left(\frac{c_1 \prod_{i=1}^{n} x_i^{a_{i1}}}{\Delta_1}\right)^{\Delta_1} \ldots \left(\frac{c_N \prod_{i=1}^{n} x_i^{a_{iN}}}{\Delta_N}\right)^{\Delta_N} \\ &= \left(\frac{c_1}{\Delta_1}\right)^{\Delta_1} \ldots \left(\frac{c_N}{\Delta_N}\right)^{\Delta_N} \left\{\left(\prod_{i=1}^{n} x_i^{a_{i1}}\right)^{\Delta_1} \ldots \left(\prod_{i=1}^{n} x_i^{a_{iN}}\right)^{\Delta_N}\right\} \\ &= \left(\frac{c_1}{\Delta_1}\right)^{\Delta_1} \ldots \left(\frac{c_N}{\Delta_N}\right)^{\Delta_N} \left\{x_1^{\sum_{j=1}^{N} a_{1j}\Delta_j} \ldots x_n^{\sum_{j=1}^{N} a_{nj}\Delta_j}\right\} \\ &= \left(\frac{c_1}{\Delta_1}\right)^{\Delta_1} \ldots \left(\frac{c_N}{\Delta_N}\right)^{\Delta_N},\end{aligned} \tag{7.63}$$

provided that the weights Δ_j are selected so as to satisfy the orthogonality condition

$$\sum_{j=1}^{N} a_{ij}\Delta_j = 0, \quad i = 1, 2, \ldots, n. \tag{7.64}$$

Thus, inequality (7.61) becomes

$$U_1+U_2+ \ldots +U_N \geqslant \left(\frac{c_1}{\Delta_1}\right)^{\Delta_1}\left(\frac{c_2}{\Delta_2}\right)^{\Delta_2} \ldots \left(\frac{c_N}{\Delta_N}\right)^{\Delta_N}. \tag{7.65}$$

The expression on the right-hand side is termed the *dual function*. If we denote it by $\nu(\Delta_1, \ldots, \Delta_N)$, inequality (7.65) can be written as

$$F(\mathbf{x}) \geqslant \nu(\mathbf{\Delta}).$$

It follows that the maximum of the dual function is less than or equal to the minimum of the primal function. Now the central result of the theory, which we state here without proof, is that these two values are equal.

In this way the unconstrained minimization of the primal function can be accomplished by maximizing the dual function

$$\nu(\mathbf{\Delta}) = \prod_{j=1}^{N} \left(\frac{c_j}{\Delta_j} \right)^{\Delta_j} \tag{7.66}$$

or its logarithm, $\ln \nu(\mathbf{\Delta})$, whichever is convenient, subject to the constraints given by the normality condition

$$\sum_{j=1}^{N} \Delta_j = 1 \tag{7.67}$$

and the orthogonality conditions

$$\sum_{j=1}^{N} a_{ij} \Delta_j = 0, \quad i = 1, 2, \ldots, n. \tag{7.68}$$

If the degree of difficulty of the problem is zero, there will be a unique solution for Δ_j. In cases where $N > n+1$, the constraints can sometimes be used to eliminate $n+1$ variables Δ_j from the dual function, which can subsequently be maximized with respect to the remaining $N-n-1$ independent Δ_j's.

7.8 Geometric programming: constrained problem

A constrained minimization problem with the objective function and the constraints all in the form of posynomials can be stated as follows:

Find a vector $\mathbf{x}^T = (x_1, x_2, \ldots, x_n)$ which minimizes the objective function

$$F(\mathbf{x}) = \sum_{j=1}^{N_0} c_{0j} \prod_{i=1}^{n} x_i^{a_{0ij}} \tag{7.69}$$

and satisfies the constraints

$$g_k(\mathbf{x}) = \sum_{j=1}^{N_k} c_{kj} \prod_{i=1}^{n} x_i^{a_{kij}} \lesseqgtr 1, \quad k = 1, 2, \ldots, m, \tag{7.70}$$

where the coefficients c_{0j} $(j = 1, 2, \ldots, N_0)$ and c_{kj} $(k = 1, 2, \ldots, m; j = 1, 2, \ldots \ldots, N_k)$ are positive numbers, the exponents a_{kij} $(k = 0, 1, \ldots, m;\ i = 1, 2, \ldots \ldots, n;\ j = 1, 2, \ldots, N_k)$ are any real numbers, m is the number of constraints, N_0 the number of terms in the objective function and N_k $(k > 0)$ the number of terms in the kth constraint. The design variables $x_1, \ldots, x_n$ are assumed to be positive.

For a unified notation, it will be convenient to denote the objective function as

$$\mathbf{x}_0 = g_0(\mathbf{x}) = F(\mathbf{x}) \tag{7.71}$$

and rewrite the constraints as

$$f_k = \sigma_k(1-g_k(\mathbf{x})) \geqslant 0, \quad k = 1, 2, \ldots, m, \tag{7.72}$$

where σ_k equals 1 or -1 according as $g_k(\mathbf{x}) \leqslant 1$ or $g_k(\mathbf{x}) \geqslant 1$.

The problem formulated above will be referred to as the *primal problem.* To solve it, we construct an equivalent problem called its *dual,* which will have linear constraints and therefore will often be easier to solve. For this purpose, just as in the unconstrained case, we introduce new variables w_i by the relations

$$x_i = e^{w_i}, \quad i = 0, 1, 2, \ldots, n; \tag{7.73}$$

w_i can take any real values. Next, we define the variables Δ_{kj} as

$$\begin{aligned} \Delta_{0j} &= \frac{c_{0j}\prod_{i=1}^{n} x_i^{a_{0ij}}}{x_0} > 0, \quad j = 1, 2, \ldots, N_0, \\ \Delta_{kj} &= c_{kj}\prod_{i=1}^{n} x_i^{a_{kij}} > 0, \quad k = 1, 2, \ldots, m, \quad j = 1, 2, \ldots, N_k, \end{aligned} \tag{7.74}$$

so that, by the definition of x_0,

$$\sum_{j=1}^{N_0} \Delta_{0j} = 1 \tag{7.75}$$

and, if the kth constraint is active, also

$$\sum_{j=1}^{N_k} \Delta_{kj} = 1. \tag{7.76}$$

After applying logarithms to equation (7.74) we can restate the primal problem as

Minimize $w_0 = \ln x_0$, *subject to the equality constraints*

$$\begin{aligned} &\sum_{j=1}^{N_0} \Delta_{0j} = 1, \\ &\ln \frac{\Delta_{0j}}{c_{0j}} = -w_0 + \sum_{i=1}^{n} a_{0ij} w_i, \quad j = 1, 2, \ldots, N_0, \\ &\ln \frac{\Delta_{kj}}{c_{kj}} = \sum_{i=1}^{n} a_{kij} w_i, \quad j = 1, 2, \ldots, N_k, \quad k = 1, 2, \ldots, m, \end{aligned} \tag{7.77}$$

and the inequality constraints

$$f_k = \sigma_k\Big(1 - \sum_{j=1}^{N_k} \Delta_{kj}\Big) \geqslant 0, \quad k = 1, 2, \ldots, m; \tag{7.78}$$

the variables Δ_{kj} *are positive, whereas* $w_0, w_1, \ldots, w_n$ *can take any real values.*

The minimum of the function w_0 can be sought by the Lagrange multiplier method. Thus, the Lagrange function is formed and the Kuhn–Tucker conditions are applied, leading to a system of equations for a saddle point at which the primal function F attains a minimum and another function v—to be called the *dual function*—attains a maximum. If all the constraints are of the form $g_k(\mathbf{x}) \leqslant 1$,[1] which ensures strict convexity of the objective function g_0 with respect to the transformed variables $w_1, w_2, \ldots, w_n$, the primal-dual relation is

$$F(\mathbf{x}) \geqslant \min F = \max v \geqslant v(\boldsymbol{\lambda}).$$

Without going into the details of the derivation, let us just state the dual problem arrived at.

Find the vector of dual variables

$$\boldsymbol{\lambda}^T = (\lambda_{01} \ldots, \lambda_{0N_0}, \lambda_{11}, \ldots, \lambda_{1N_1}, \ldots, \lambda_{m1}, \ldots, \lambda_{mN_m})$$

which maximizes the dual function

$$v(\boldsymbol{\lambda}) = \prod_{k=0}^{m} \prod_{j=1}^{N_k} \left(\frac{c_{kj}}{\lambda_{kj}} \sum_{l=1}^{N_k} \lambda_{kl} \right)^{\lambda_{kj}} \tag{7.79}$$

subject to the normality constraint

$$\sum_{j=1}^{N_0} \lambda_{0j} = 1, \tag{7.80}$$

the orthogonality constraints

$$\sum_{k=0}^{m} \sum_{j=1}^{N_k} a_{kij} \lambda_{kj} = 0, \quad i = 1, 2, \ldots, n, \tag{7.81}$$

and the non-negativity conditions

$$\lambda_{kj} \geqslant 0, \quad j = 1, 2, \ldots, N_k, \quad k = 0, 1, \ldots, m. \tag{7.82}$$

Comparing the primal and the dual programs, we can note the following characteristics:

1. The coefficients c_{kj} occurring in the dual function $v(\boldsymbol{\lambda})$ are the coefficients of the posynomials $g_k(\mathbf{x})$, $k = 0, 1, \ldots, m$.
2. The vector $\boldsymbol{\lambda}$ has as many components as there are terms in all the posynomials $g_0, g_1, \ldots, g_m$. Associated with each term in g_k, there is a corresponding Δ_{kj}.
3. Every factor $\left(\sum_{l=1}^{N_k} \lambda_{kl}\right)^{\lambda_{kj}}$ of $v(\boldsymbol{\lambda})$ originates from an inequality constraint

[1] The following discussion is restricted to this case; for a more general treatment see Duffin *et al.*, Zener or Rao (cf. footnotes p. 239).

$g_k(\mathbf{x}) \leqslant 1$; none comes from the primal function $g_0(\mathbf{x})$ since by the normality condition $\sum_{j=1}^{N_0} \lambda_{0j} = 1$.

4. The coefficient matrix $[a_{kij}]$ of the orthogonality conditions is the exponent matrix for the primal program.

Concluding remarks

The theory of geometric programming presented above is applicable to programs with positive terms only. If we tried to construct a dual function to a primal in which some of the terms are negative, the result would involve fractional powers of negative numbers, and hence imaginary numbers. Passy and Wilde[1] developed a method by which an inequality-constrained polynomial problem with coefficients of an arbitrary sign can be transformed into a problem involving posynomials only, and hence a geometric programming problem susceptible to the primal-dual techniques described above. The transformation introduces some new variables but then an equal number of terms is also added so as to keep the degree of difficulty of the problem unchanged.

Avriel and Williams[2] extended the method of geometric programming to include an even larger class of problems, allowing the objective function and constraints to be rational functions of polynomial terms. The method, termed the *complementary geometric programming*, consists in the successive approximation of the rational functions involved by polynomials and solving the corresponding sequence of ordinary geometric programs. The successive solutions converge to a local minimum of the original problem. However, it should be noted that, unlike ordinary geometric programs, complementary geometric programs are not in general convex and therefore a local minimum need not be global.

7.9 Non-linear integer programming

A general *mixed-integer programming problem* can be formulated as follows (Rao, *loc. cit.*, p. 239):

Find a vector $\mathbf{x}^T = (\mathbf{x}_d, \mathbf{x}_c)$ *which minimizes the objective function* $F(\mathbf{x})$ *subject to the constraints*

$$g_j(\mathbf{x}) \geqslant 0, \quad j = 1, 2, \dots, m,$$

$$\mathbf{x}_c \in S_c \quad \textit{and} \quad \mathbf{x}_d \in S_d,$$

[1] Passy, U., and Wilde, D. J., Generalized polynomial optimization, *SIAM, J. Appl. Math.*, **15**, No 5, pp. 1344–1356 (1967).

[2] Avriel, M., and Williams, A. C., Complementary geometric programming, *SIAM J. Appl. Math.*, **10**, No 1, pp. 125–141 (1970).

where $\mathbf{x}_c$ *and* $\mathbf{x}_d$ *are the vectors of continuous and discrete design variables and* S_c *and* S_d *denote their respective feasible ranges.*

The problem can be solved by an extended interior penalty function method (see Sec. 7.11). To this end, the following transformed problem is defined:

Minimize the function

$$\Phi(\mathbf{x}, r, s) = F(\mathbf{x}) + r\sum_{j=1}^{m} G[g_j(\mathbf{x})] + sQ(\mathbf{x}_d), \tag{7.83}$$

where r and s are the penalty parameters of the continuous and the discrete variables, respectively, $G[g_j(\mathbf{x})]$ *can be taken as* $1/g_j(\mathbf{x})$ *and the function Q has the property*

$$Q(\mathbf{x}_d) = \begin{cases} 0 & \textit{if} \quad \mathbf{x}_d \in S_d, \\ \mu > 0 & \textit{if} \quad \mathbf{x}_d \notin S_d. \end{cases} \tag{7.84}$$

For example, as Q one can take a normalized symmetric beta function integrand

$$Q_\beta(\mathbf{x}_d) = \sum \left\{4\frac{x_i - y_i}{z_i - y_i}\left(1 - \frac{x_i - y_i}{z_i - y_i}\right)\right\}^\beta, \tag{7.85}$$

where $y_i \leqslant x_i$ and $z_i \geqslant x_i$ are the two neighbouring integers for x_i, and $\beta \geqslant 1$ is a constant.

The transformed problem is then solved for a sequence of values of the parameters r_k, s_k and β_k such that, as $k \to \infty$,

$$\begin{aligned} &\min \Phi(\mathbf{x}, r_k, s_k) \to \min F(x), \\ &g_j(\mathbf{x}) \geqslant 0, \quad j = 1, 2, \ldots, m \end{aligned} \tag{7.86}$$

and

$$Q_{\beta_k}(\mathbf{x}_d) \to 0.$$

The shape of the function Φ strongly depends on the numerical values of r_k, s_k and β_k, and so the choice of these parameters is crucial for the convergence of the process. No general and precise rule exists, however, by which a given set of values can be considered the best for a given problem, and besides certain indications based on experience, there remains in fact the trial and error method. Still, in most of the practical problems, 5 to 10 steps (k) are enough to reach a satisfactory solution.

A general proof of convergence of the penalty function method—also for integer programming—was given by Fiacco[1]. Some applications of this method to optimum-structural-design are discused by Gisvold and Moe[2].

[1] Fiacco, A. V., Penalty methods for mathematical programming in E^n with general constraint sets, *JOTA*, **6**, pp. 252–268 (1970).

[2] Gisvold, K. M., and Moe, I., A method for nonlinear mixed-integer programming and its application to design problems, *J. Eng. Industry, Trans. ASME*, **94**, No. 2, pp. 353–364 (1972).

A different approach to non-linear integer programming was earlier suggested by Reiter and Rice[1], who applied a modified gradient method, very similar to the methods used in the non-linear continuous programming.

7.10 Unconstrained optimization: direct search methods

In this section, we consider the methods of minimizing a function $F(\mathbf{x})$ of several unconstrained variables without the use of derivatives. They are all iterative methods in that each of them starts from an initial point $\mathbf{x}_1$ and proceeds towards the optimum point $\mathbf{x}$ in a sequential manner. What may differ from method to method is the way the sequence is generated and the testing of the successive points for optimality.

A Random search methods

The random search methods use the facility of random number generators which most digital computers are fitted up with. Three well-known methods of this kind are presented below.

Random jumping

To find the minimum of the objective function in an n-cube

$$l_i \leqslant x_i \leqslant u_i, \quad i = 1, 2, \ldots, n,$$

where l_i and u_i are the lower and the upper bounds on the variable x_i, we "jump" at random within the cube. Each jump is carried out by generating a set of n random numbers $r_1, r_2, \ldots, r_n$ between 0 and 1 and evaluating the function F at the corresponding point of the cube,

$$\mathbf{x} = \begin{Bmatrix} x_1 \\ \vdots \\ x_n \end{Bmatrix} = \begin{Bmatrix} l_1 + r_1(u_1 - l_1) \\ \vdots \\ l_n + r_n(u_n - l_n) \end{Bmatrix}.$$

After making a large number of jumps, we choose the least of the calculated values of the objective function as the desired minimum.

Random walk method

In this method, a sequence of approximations to the minimum is formed with every next approximation constructed from the preceding one as

$$\mathbf{x}_{i+1} = \mathbf{x}_i + \lambda \mathbf{u}_i, \tag{7.87}$$

where λ is a prescribed step length and $\mathbf{u}_i$ is a randomly generated unit vector, such that $f(\mathbf{x}_{i+1}) < f(\mathbf{x}_i)$. If an improved point cannot be obtained despite

[1] Reiter, S., and Rice, D. B., Discrete optimizing solution procedures for linear and non-linear integer programming problems, *Management Science*, **12**, pp. 829–850 (1966).

a large number of trials, the step length λ is reduced and random generation of a descent direction attempted again. The process is stopped when λ is reduced below a specified small number ε.

Random walk method with direction exploitation

This is an improved version of the random walk method, the improvement being that every successful direction $\mathbf{u}_i$ is exploited until it is no longer useful. This is done by increasing the step size along the direction $\mathbf{u}_i$ as long as it yields decreasing values of the objective function. Thus the new point, $\mathbf{x}_{i+1}$, is found as

$$\mathbf{x}_{i+1} = \mathbf{x}_i + \hat{\lambda}_i \mathbf{u}_i \tag{7.88}$$

where $\hat{\lambda}_i$ is the optimal step size in the direction $\mathbf{u}_i$, i.e. such that

$$F_{i+1} = F(\mathbf{x}_i + \hat{\lambda}_i \mathbf{u}_i) = \min_{\lambda_i} F(\mathbf{x}_i + \lambda_i \mathbf{u}_i). \tag{7.89}$$

Random search methods are generally not very efficient but have some advantages when compared with other methods. Indeed, they can be useful

- when the objective function is discontinuous or non-differentiable at some points,
- when the objective function has several local minima,
- when other methods fail owing to very rapid or very slow variation of the functions involved.

B Univariate method

In each iteration of the univariate method only one variable is allowed to vary, with the remaining $n-1$ variables fixed at the values of the preceding iteration. The one-dimensional minimization problem so obtained is solved, and hence a new approximation point is found. The search is then continued in a new coordinate direction by changing one of the $n-1$ variables that were fixed previously. After all the variables have been changed in turn, the first cycle is completed. The process can be repeated until no futher improvement is possible in the objective function in any of the n directions of a cycle.

The method can be applied to any continuously differentiable function but may converge very slowly or even fail if the function has a steep valley.

C Pattern search methods

The search for the minimum need not proceed exclusively along the coordinate directions, as in the univariate method. There are several methods known as *pattern search methods*, in which after m univariate steps ($m = n$ if there are n variables in the problem) a new direction $\mathbf{s}_i$ is chosen as

$$\mathbf{s}_i = \mathbf{x}_i - \mathbf{x}_{i-m},$$

where $\mathbf{x}_i$ is the point reached after the univariate steps and $\mathbf{x}_{i-m}$ is the starting point. The direction $\mathbf{s}_i$ is called a *pattern direction*, and hence the term: pattern search methods. Three methods of this kind are outlined below.

The method of Hooke and Jeeves[1]

Each iteration of the Hooke–Jeeves method involves two kinds of moves, exploratory moves and pattern moves (Fig.7. 8). The first explore the

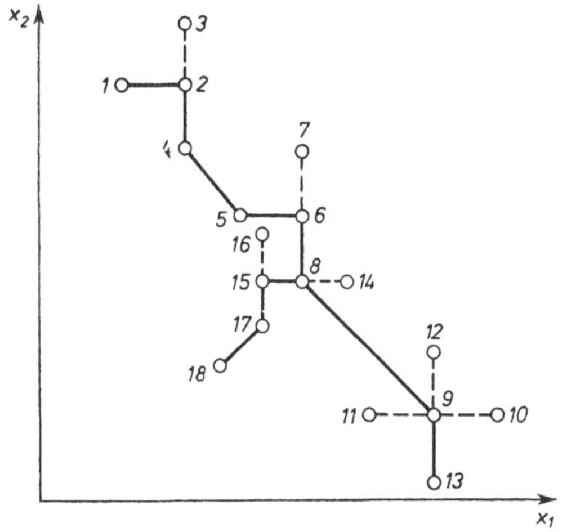

Fig. 7.8. The Hooke–Jeeves pattern search

local behaviour of the objective function along the coordinate directions within a prescribed step size from the current temporary base point. This leads to a new base point, whereupon a pattern direction is established and a move is made along it to an adjacent region. There, exploratory search is performed again, to be followed by a new pattern move, but only if at least one of the exploratory moves has been successful, i.e. has yielded a smaller value of the objective function. Otherwise, we return to the region explored before and repeat the procedure with a reduced step size. The process is terminated when the step size falls below a prescribed small number ε.

Powell's method[2]

Powell's method starts in exactly the same way, as that of Hooke and Jeeves, i.e. by exploratory moves along the coordinate directions followed by

---

[1] Hooke, R. and Jeeves, T. A., Direct search solution of numerical and statistical problems, *JACM*, **8**, 2, pp. 212–229 (1961).

[2] Powell, H. I. D., An efficient method for finding the minimum of a function of several variables without calculating derivatives, *Computer Journal*, **7**, 4, pp. 303–307 (1964).

a move in the corresponding pattern direction $\mathbf{s}_1$. For the next cycle, however, one of the coordinate directions, say x_1, is replaced by the pattern direction $\mathbf{s}_1$, and a new pattern direction $\mathbf{s}_2$ is generated after exploratory search in the directions $\mathbf{s}_1, x_2, \ldots, x_n$ (Fig. 7.9). In the next cycle, the x_2-direction is discarded in favour of $\mathbf{s}_2$, and so on. When all the coordinate directions are replaced by pattern directions, we restart the procedure with a univariate search.

Rosenbrock's method[1]

This method, sometimes called the *method of rotating coordinates*, can be considered a further development of the method of Hooke and Jeeves. The procedure, illustrated in Fig. 7.10, divides into stages and cycles. In a single cycle a trial move is made in each of the current search directions, the initial search directions being those of the coordinate axes. If a trial move leads to a smaller value of the objective function, it is considered a success and the step size in the same direction during the next cycle is taken as α times the present step size, with $\alpha > 1$. If, on the other hand, the objective function is increased, the move is deemed a failure and the step size for next cycle is taken as $-\beta$ times the present step size, with $0 < \beta < 1$. A single stage consists of as many cycles as is necessary for at least one success and one failure to occur in every search direction. The next stage begins by the formation of a new orthogonal set of search directions, the starting and the end points of the completed stage indicating the first direction of the new set. The remaining directions of the new set are obtained by a procedure known as the *Gram–Schmidt orthogonalization.* The process is continued until the distance between the starting and the final points of any particular stage becomes smaller than a given number ε.

Rosenbrock's method performs well in the case of strongly curved and steep-valleyed surfaces, because the orthogonal search-direction system is rotated according to the locally estimated direction of the valley. The method is particularly useful in locating early approximations to the minimum. Davies, Swann and Campey[2] modified Rosenbrock's procedure by incorporating a line search within it, i.e. replacing discrete steps by a complete one-dimensional minimization in each of the search directions. The modified method appears to be superior both to Hooke and Jeeves' method and to Rosenbrock's method.

---

[1] Rosenbrock, H. H., An automatic method for finding the greatest or least value of function, *Computer Journal*, **3**, 3, pp. 175–184 (1960).

[2] Box, M. I., Davies, D., and Swann, W. H., *Nonlinear Optimization Techniques*, ICI Ltd. Monograph No. 5, Oliver and Boyd, Edinburgh 1969.

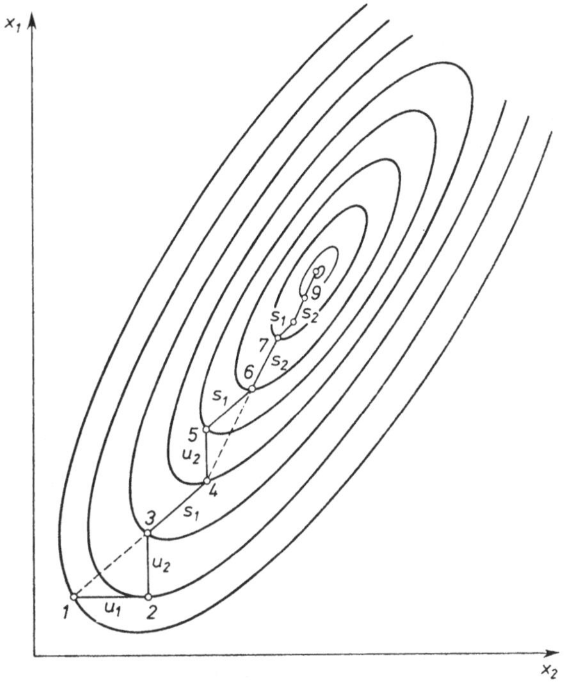

Fig. 7.9. Powell's method

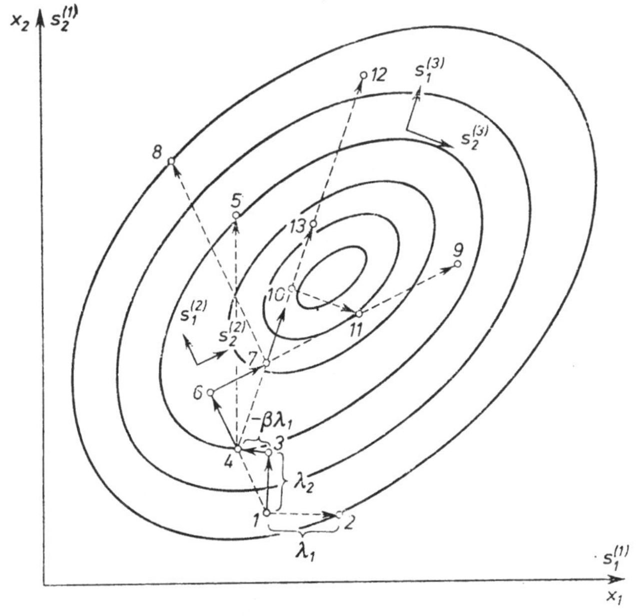

Fig. 7.10. Rosenbrock's method

D Simplex search

The method of sequential simplex search was originally proposed by Spendley, Hext and Himswirth[1] and later modified by Nelder and Mead[2].

Essentially, the method consists in evaluating the objective function at the vertices of a simplex (e.g. a triangle in a two-dimensional space, a tetra-

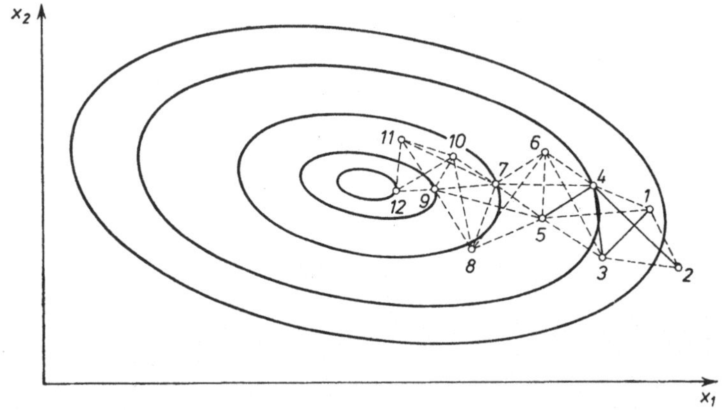

Fig. 7.11. Simplex search

hedron in a three-dimensional space; in general: a figure of $n+1$ points in an n-dimensional space). The "worst" vertex is rejected and replaced by a new point along the line joining that vertex and the centroid of the remaining points. The simplex is thus moved gradually towards the optimum point during the iterative process (Fig. 7.11).

7.11 Unconstrained optimization: descent methods

A Steepest descent method

The method of steepest descent makes use of the gradient ∇F of the objective function. The gradient vector has the property that it represents the direction of the highest rate of increase of the function at a given point, or the direction of steepest ascent at that point. Hence the opposite direction, i.e. that represented by $-\nabla F$, is the direction of steepest descent, and by moving in this direction we can expect to minimize the objective function.

In the method in question, we start from an initial point $\mathbf{x}_1$ and iteratively move towards the optimum point according to the rule

$$\mathbf{x}_{i+1} = \mathbf{x}_i + \hat{\lambda}_i \mathbf{s}_i = \mathbf{x}_i - \hat{\lambda}_i \nabla F(\mathbf{x}_i), \tag{7.90}$$

where $\hat{\lambda}_i$ is the optimal step size in the direction $\mathbf{s}_i = -\nabla F(\mathbf{x}_i)$, i.e. it is the

[1] Spendley, W., Hext, G. R., and Himsworth, F. R., Sequential application of simplex design in optimization and evolutionary operation, *Technometries*, 4, p. 441 (1962).

[2] Nelder, I. A., and Mead, R., A simplex method for function minimization, *Computer Journal*, 7, p. 308 (1965).

optimal solution to the line-search problem of minimizing $F(\mathbf{x}_i+\lambda_i\mathbf{s}_i)$ subject to $\lambda_i \geqslant 0$. To terminate this process, any of the following optimality tests can be used:

$$\left|\frac{F(\mathbf{x}_{i+1})-F(\mathbf{x}_i)}{F(\mathbf{x}_i)}\right| \leqslant \varepsilon_1, \tag{7.91}$$

$$\left|\frac{\partial F}{\partial x_i}\right| \leqslant \varepsilon_2, \tag{7.92}$$

$$|\mathbf{x}_{i+1}-\mathbf{x}_i| \leqslant \varepsilon_3. \tag{7.93}$$

The method usually performs quite well at the early stages of the optimization process. However, owing to the fact that the steepest descent direction is a local property, the process may become very slow as the minimum point is approached. In order to speed up the convergence, Forsythe and Motzkin[1] suggested to use the search direction

$$\mathbf{s}_i = \mathbf{x}_i-\mathbf{x}_{i-2}, \quad i \geqslant 2, \tag{7.94}$$

from time to time instead of $-\nabla F(\mathbf{x}_i)$. Shah *et al.*[2] proposed to take these two directions alternately.

B Conjugate gradient method

The conjugate gradient method, developed by Fletcher and Reeves[3], converges much more rapidly than that of steepest descent. This is achieved by deflecting the direction of steepest descent by adding to it a positive multiple of the direction used in the preceding step. The algorithm of the method is as follows:

1. Choose an arbitrary initial point $\mathbf{x}_1$.
2. Set the first search direction as $\mathbf{s}_1 = -\nabla F(\mathbf{x}_1) = -\nabla F_1$.
3. Find the point $\mathbf{x}_2$ according to the relation $\mathbf{x}_2 = \mathbf{x}_1+\hat{\lambda}_1\mathbf{s}_1$, where $\hat{\lambda}_1$ is the optimal step size in the direction $\mathbf{s}_1$. Let $i = 2$ and go to the next step.
4. Find $\nabla F_i = \nabla F(\mathbf{x}_i)$ and let
$$\mathbf{s}_i = -\nabla F_i+(|\nabla F_i|^2/|\nabla F_{i-1}|^2)\mathbf{s}_{i-1}. \tag{7.95}$$
5. Compute the optimal step size $\hat{\lambda}_i$ in the direction $\mathbf{s}_i$ and find the new point
$$\mathbf{x}_{i+1} = \mathbf{x}_i+\hat{\lambda}_i\mathbf{s}_i.$$
6. Test the point $\mathbf{x}_{i+1}$ for optimality. If the result is positive, stop the process. Otherwise replace i by $i+2$ and repeat steps 4, 5 and 6.

[1] Forsythe, G. E., and Motzkin, T. S. Asymptotic properties of the optimum gradient method (abstract), *American Math. Soc. Bull.*, **57** (1951).

[2] Shah, B. V., Buehler, R. I., and Kempthorne, O., Some algorithms for minimizing a function of several variables, *SIAM Journal*, **12**, p. 74 (1964).

[3] Fletcher, R., and Reeves, C. M., Function minimization by conjugate gradients, *Computer Journal*, **7**, 2, pp. 149–154 (1964).

Much more efficient than the steepest descent method, the Flecher–Reeves algorithm is rather inferior to the quasi-Newton and the variable metric methods, which we present next.

C Quasi-Newton methods

The class of quasi-Newton methods derive from Newton's method for solving a set of non-linear equations. All the local minima of a differentiable function $F(\mathbf{x})$ satisfy the equations

$$\mathbf{g}(\mathbf{x}) = \nabla F(\mathbf{x}) = 0. \tag{7.96}$$

In Newton's method, the equations are first linearized about some point $\mathbf{x}_i$, which can be considered the ith approximation to the solution $\hat{\mathbf{x}}$. If we put $\hat{\mathbf{x}} = \mathbf{x}_i + \mathbf{s}$, the Taylor series expansion of $\mathbf{g}$ gives

$$\mathbf{g}(\hat{\mathbf{x}}) = \mathbf{g}(\mathbf{x}_i+\mathbf{s}) = \mathbf{g}(\mathbf{x}_i)+\mathbf{J}(\mathbf{x}_i)\mathbf{s}+ \ldots = \mathbf{g}_i+\mathbf{J}_i\mathbf{s}+ \ldots \tag{7.97}$$

where $\mathbf{g}_i = \mathbf{g}(\mathbf{x}_i)$ and $\mathbf{J}_i = \mathbf{J}(\mathbf{x}_i)$ is the matrix of the second partial derivatives of F at $\mathbf{x}_i$. Neglecting higher order terms and setting $g(\hat{\mathbf{x}}) = \mathbf{0}$, we get

$$\mathbf{g}_i+\mathbf{J}_i\mathbf{s} = \mathbf{0}, \quad \text{or} \quad \mathbf{s} = -\mathbf{J}_i^{-1}\mathbf{g}_i \tag{7.98}$$

if $\mathbf{J}_i$ is non-singular. Since, in general, the above equality is only approximate, the point

$$\mathbf{x}_{i+1} = \mathbf{x}_i+\mathbf{s} = \mathbf{x}_i-\mathbf{J}_i^{-1}\mathbf{g}_i \tag{7.99}$$

will only approximate to the solution $\hat{\mathbf{x}}$. The sequence $\mathbf{x}_i$ of Newton's iterations so constructed can be shown to converge to $\hat{\mathbf{x}}$ if the initial point $\mathbf{x}_1$ is chosen sufficiently close to the solution.

When only first derivatives are available or when evaluating $\mathbf{J}_i$ and its inverse is too complicated, Newton's method cannot be implemented directly. In quasi-Newton methods, a positive definite symmetric matrix $\mathbf{H}_i$ is used in each step as an approximation to $\mathbf{J}_i^{-1}$, and each new point is found by linear search in the current search direction.

Let $\mathbf{x}_i$ and $\mathbf{x}_{i+1}$ be two successive points obtained by Newtons iterations. By equation (7.97), we can write

$$\mathbf{g}_{i+1}-\mathbf{g}_i = \mathbf{J}_i\mathbf{s} = \mathbf{J}_i(\mathbf{x}_{i+1}-\mathbf{x}_1). \tag{7.99}$$

If we define $\mathbf{G}_i = \mathbf{g}_{i+1}-\mathbf{g}_i$ and $\mathbf{s}_i = \mathbf{x}_{i+1}-\mathbf{x}_i$, equation (7.99) becomes

$$\mathbf{G}_i = \mathbf{J}_i\mathbf{s}_i \quad \text{or} \quad \mathbf{s}_i = \mathbf{J}_i^{-1}\mathbf{G}_i, \tag{7.100}$$

provided that $\mathbf{J}_i$ is non-singular. Thus, the approximation matrix $\mathbf{H}_i$ to $\mathbf{J}_i^{-1}$ should be constructed in each step so as to satisfy

$$\mathbf{s}_i = \mathbf{H}_i\mathbf{G}_i. \tag{7.101}$$

Now if the point found in a kth step, i.e. $\mathbf{x}_{k+1}$, were a stationary point, we would have $\mathbf{g}_{k+1} = \mathbf{0}$, and then

$$\mathbf{s}_k = \mathbf{H}_k\mathbf{G}_k = \mathbf{H}_k(\mathbf{g}_{k+1}-\mathbf{g}_k) = -\mathbf{H}_k\mathbf{g}_k. \tag{7.102}$$

Obviously, the point $\mathbf{x}_{k+1}$ may still be far from a stationary point, but we may use equation (7.102) as a direction of search and find a new point $\mathbf{x}_{k+1}$ as

$$\mathbf{x}_{k+1} = \mathbf{x}_k+\mathbf{s}_k, \tag{7.103}$$

where

$$\mathbf{s}_k = -\hat{\lambda}_k\mathbf{H}_k\mathbf{g}_k, \tag{7.104}$$

where $\hat{\lambda}_k$ is the optimal step size along the direction $-\mathbf{H}_k\mathbf{g}_k$.

Equations (7.101), (7.103) and (7.104) constitute a general iterative scheme of quasi-Newton methods. Particular methods of this class differ from one another in the ways the matrix $\mathbf{H}_k$ is constructed and $\hat{\lambda}_k$ determined.

D Variable metric method

The variable metric method, originally proposed by Davidon[1] and later developed by Fletcher and Powell[2], is one of the most efficient methods of unconstrained optimization. The iterative procedure of this method can be summarized as follows:

1. Choose an initial point $\mathbf{x}_1$ and a positive-definite symmetric $n\times n$ matrix $\mathbf{H}_1$ (usually $\mathbf{H}_1 = \mathbf{I}$, the identity matrix). Let $i = 1$ and go to step 2.
2. Compute the gradient $\nabla F_i = \nabla F(\mathbf{x}_i)$ and set

$$\mathbf{s}_i = -\mathbf{H}_i\nabla F_i. \tag{7.105}$$

3. Find the optimal step size $\hat{\lambda}_i$ in the direction $\mathbf{s}_i$ and set

$$\mathbf{x}_{i+1} = \mathbf{x}_i+\hat{\lambda}_i\mathbf{s}_i. \tag{7.106}$$

4. Test the point $\mathbf{x}_{i+1}$ for optimality. If $\mathbf{x}_{i+1}$ is optimal, stop. Otherwise go to step 5.
5. Construct the matrix $\mathbf{H}_{i+1}$ as

$$\mathbf{H}_{i+1} = \mathbf{H}_i+\hat{\lambda}_i\frac{\mathbf{s}_i\mathbf{s}_i^T}{\mathbf{s}_i^T\mathbf{Q}_i} - \frac{(\mathbf{H}_i\mathbf{Q}_i)(\mathbf{H}_i\mathbf{Q}_i)^T}{\mathbf{Q}_i^T\mathbf{H}_i\mathbf{Q}_i} \tag{7.107}$$

---

[1] DAVIDON, W. C., Variable metric method of minimization, *Argonne National Laboratory Report*, No. **ANL-5990**, 1959.

[2] FLETCHER, R., and POWELL, M. I. D., A rapidly convergent descent method for minimization, *Computer Journal*, **6**, **2**, pp. 163–168 (1963).

where

$$\mathbf{Q}_i = \nabla F(\mathbf{x}_{i+1}) - \nabla F(\mathbf{x}_i) = \nabla F_{i+1} - \nabla F_i. \tag{7.108}$$

6. Replace i by $i+1$ and go to step 2.

The variable metric method shows a high degree of stability in progressing towards the minimum, which can be attributed to the fact that after each iteration the new matrix $\mathbf{H}_{i+1}$ is obtained by correcting $\mathbf{H}_i$ so as to take into account the information that has become available on that iteration. It can be shown that if $\mathbf{H}_1$ is positive definite, then all subsequent $\mathbf{H}_i$ are also positive definite. Furthermore, $\mathbf{H}_i$ converges to the inverse Hessian matrix $\mathbf{J}^{-1}$ at the optimal point $\hat{\mathbf{x}}$.

7.12 Constrained optimization: direct methods

A Methods of feasible directions

Methods of this group solve a non-linear constrained minimization problem by approaching the optimum point in a succession of steps along usable feasible directions, i.e. directions which make it possible to reduce the value of the objective function and to stay within the feasible space. Each iteration consists of finding a usable feasible direction at a specified point and determining the proper size of the step to be made along that direction. The manner in which usable feasible directions are generated and step sizes determined is different in different methods.

Zoutendijk's method

In this method, if the point $\mathbf{x}_i$ from which we are to move lies in the interior of the feasible space, the usable feasible direction $\mathbf{s}$ is taken as the negative of the gradient direction, while if $\mathbf{x}_i$ is a boundary point, $\mathbf{s}$ is chosen so as to satisfy the equations

$$\begin{aligned} \frac{\mathrm{d}}{\mathrm{d}\lambda} F(\mathbf{x}_i + \lambda\mathbf{s})|_{\lambda=0} &= \mathbf{s}^T \nabla F(\mathbf{x}_i) \leqslant 0, \\ \frac{\mathrm{d}}{\mathrm{d}\lambda} g_j(\mathbf{x}_i + \lambda\mathbf{s})|_{\lambda=0} &= \mathbf{s}^T \nabla g_j(\mathbf{x}_i) \leqslant 0, \end{aligned} \tag{7.109}$$

where, as usual, F denotes the objective function and g_j $(j = 1, 2, \ldots, m)$ represent the inequality constraints. For a complete algorithm of the method, including a discussion of different ways of finding a suitable direction $\mathbf{s}$ and a proper step size λ along $\mathbf{s}$, and also of the convergence problem and process termination criteria, see e.g. books by Zoutendijk[1] and Rao[2]. Let us only

[1] ZOUTENDIJK, G., *Methods of Feasible Directions*, Elsevier, Amsterdam 1960.
[2] RAO, S. S., *loc. cit.*, p. 239.

note that in cases where the constraints are non-linear the locally best feasible direction in each iteration is found by solving an auxiliary linear programming problem, and so the procedure may involve much computational work.

Gradient projection method

Rosen[1] proposed to construct a usable feasible direction in a different way, and namely by projecting the negative gradient onto the tangent planes to the constraints that are currently active. Rosen's method proves to be highly efficient for solving problems with linear constraints. It can also be used when some of the constraints are non-linear, but then the projected gradient may not lead to feasible points and a correction move may be required to re-enter the feasible space. One other possibility is to transform the objective function as in the interior penalty function method (Sec. 7.13) so as to take care of the non-linear constraints, and solve a sequence of constrained optimization problems with a non-linear objective and linear constraints using the gradient projection method.

The method of Klingman and Himmelblaum[2]

This method gives a prescription how to "bounce" from the constraints so as to remain within the feasible space and reduce the objective function. Namely, on reaching a constraint surface g_i, the new usable feasible direction $\mathbf{s}$ is defined as a combination of the gradients of the objective function and the binding constraint:

$$\mathbf{s} = \frac{\nabla g_i(\mathbf{x})}{|\nabla g_i(\mathbf{x})|} - \frac{\nabla F(\mathbf{x})}{|\nabla F(\mathbf{x})|}. \tag{7.110}$$

B Cutting plane method

The cutting plane method introduced by Cheney and Goldstein[3] and by Kelly[4] can conveniently be applied to solve convex programming problems with linear objective functions. The basic idea of the method is to linearize the non-linear constraints via the Taylor series expansions about an initial point and thus approximate the problem with a linear programming problem. The linear program is solved, e.g. by the simplex method, and the new point is

---

[1] ROSEN, J. B., The gradient projection method for non-linear programming. Part I: Linear constraints, *J. Soc. Ind. Appl. Math.*, **8** (1960); Part II: Non-linear constraints, *J. Soc. Ind. Appl. Maths*, **9** (1961).

[2] KLINGMAN, W. R., and HIMMELBLAUM, D. M., *Methods for Unconstrained Optimization problems*, American Elsevier Publ. Co., 1968.

[3] CHENEY, E. W., and GOLDSTEIN, A. A., Newton's method of convex programming and Tchebycheff approximation, *Numersiche Mathematik*, **1**, pp. 253–268 (1959).

[4] Kelly, J. E., The cutting plane method for solving convex programs, *Journal of SIAM*, **3**, 4, pp. 703–712 (1960).

used to construct a further linearization, which is added to the linear program. The augmented linear program is again solved, and so on. The cutting plane algorithm can be given in the following form:

1. Choose an initial point $\mathbf{x}_1$ (not necessarily feasible), let $i = 1$ and go to step 2.
2. Linearize the constraint functions g_j about $\mathbf{x}_1$:

$$g_j(\mathbf{x}) \cong g_j(\mathbf{x}_i) + [\nabla g_j(\mathbf{x}_i)]^T(\mathbf{x}-\mathbf{x}_i), \quad j = 1, 2, \ldots, m. \tag{7.111}$$

3. Formulate the approximating linear program:
   Minimize

$$F(\mathbf{x}) = \mathbf{c}^T\mathbf{x} = \sum_{i=1}^{n} c_i x_i$$

   subject to the constraints

$$g_j(\mathbf{x}_i) + [\nabla g_j(\mathbf{x}_i)]^T(\mathbf{x}-\mathbf{x}_i) \leqslant 0, \quad j = 1, 2, \ldots, m. \tag{7.112}$$

4. Solve the approximating LP problem to obtain the solution $\mathbf{x}_{i+1}$.
5. If $g_j(\mathbf{x}_{i+1}) \leqslant \varepsilon$ for $j = 1, 2, \ldots, m$, where ε is a prescribed tolerance (a small positive number), all the original constraints can be considered satisfied. Take $\mathbf{x}_{\text{opt}} = \mathbf{x}_{i+1}$ and stop. If $g_j(\mathbf{x}_{i+1}) > \varepsilon$ for some j, let g_k be the most violated constraint, i.e.

$$g_k(\mathbf{x}_{i+1}) = \max_j g_j(\mathbf{x}_{i+1}), \tag{7.113}$$

   Relinearize the constraint $g_k(\mathbf{x}) \leqslant 0$ about the point $\mathbf{x}_{i+1}$ as

$$g_k(\mathbf{x}) \cong g_k(\mathbf{x}_{i+1}) + [\nabla g_k(\mathbf{x}_{i+1})]^T(\mathbf{x}-\mathbf{x}_{i+1}) \leqslant 0 \tag{7.114}$$

   and add this as the $(m+1)$st constraint to the previous LP problem.
6. Set the new iteration number as $i = i+1$, the number of constraints as $m = m+1$ and return to step 4.

C The complex method

The complex method developed by Box[1] is a constrained version of the simplex method of unconstrained optimization (Sec. 7.10.D). To minimize an objective function $F(\mathbf{x})$ subject to the constraints

$$\begin{aligned} &a_i \leqslant x_i \leqslant b_i, \quad i = 1, 2, \ldots, n, \\ &g_j(\mathbf{x}) \leqslant 0, \quad j = 1, 2, \ldots, m, \end{aligned} \tag{7.115}$$

a complex with $k \geqslant n+1$ feasible vertices is formed and iteratively moved towards the optimum point by successive modifications. In each step, the vertex $\mathbf{x}_l$ which yields the largest value of F is selected and a new point $\mathbf{x}_r$

[1] Box, M. J., A new method of constrained optimization and a comparison with other methods, *Computer Journal*, **8** (1965).

giving a smaller F is found by reflecting $\mathbf{x}_l$ in the complex of the remaining vertices:

$$\mathbf{x}_r = \mathbf{x}_0 + \alpha(\mathbf{x}_0 - \mathbf{x}_l), \tag{7.116}$$

where $\alpha > 0$ and $\mathbf{x}_0$ is the centroid of all the vertices except $\mathbf{x}_l$. If the reflected point violates any of the constraints, it is moved half way towards the centroid until it becomes feasible. In this way, the complex is rolled over and over toward the minimum, remaining within the feasible space. The process is stopped when the complex shrinks to a specified small size or the standard deviation from the centroid of the function values at the vertices becomes sufficiently small.

The complex method is thus a gradient-free method and a very simple one from computational point of view. However, it becomes inefficient for a large number of variables and may fail if the feasible space is non-convex.

7.13 Constrained optimization: indirect methods

A Transformation of independent variables

The idea of this method is to convert a constrained minimization problem into an unconstrained one by transforming the variables in such a way that the constraints may be satisfied automatically.[1] This is possible when the constraints have certain simple forms, for example when the variables x_i are just bounded from below and from above by constants,

$$a_i \leqslant x_i \leqslant b_i.$$

To ensure the fulfilment of these constraints, it is enough to set

$$x_i = a_i + (b_i - a_i)\sin^2 x_i', \tag{7.117}$$

where the new variables x_i' can take any real values. After appropriate substitution, we seek the unconstrained minimum of the objective function with respect to the new variables x_i'.

A similar idea is utilized in the non-classical calculus of variations (Chapter 11).

B Penalty function methods

The problem of minimizing an objective function $F(\mathbf{x})$ subject to inequality constraints $g_j(\mathbf{x}) \leqslant 0$, $j = 1, 2, \ldots, m$, can be converted into one of unconstrained minimization by adding a penalty term to the objective function for any violation of the constraints.

[1] Box, H. J., A comparison of several current optimization methods and the use of transformations in constrained problems, *Comp. J.*, **9** (1966).

In a general form, the new function can be written

$$\Phi(\mathbf{x}, r) = F(\mathbf{x})+r\sum_{j=1}^{m} G_i\big(g_j(\mathbf{x})\big), \tag{7.118}$$

where G_i is an operator on the constraint g_j and r is a positive constant called the *penalty parameter*. The unconstrained minimum of the function Φ can be sought by any of the methods of unconstrained optimization discussed earlier. The unconstrained problem is solved for a sequence r_k ($k = 1, 2, \dots$) of the values of the penalty parameter, such that the resulting minima of the functions $\Phi(\mathbf{x}, r_k)$ converge to the solution of the original problem.

The penalty function formulations for inequality-constrained problems can be divided into two categories: interior and exterior (Fig. 7.12). In the

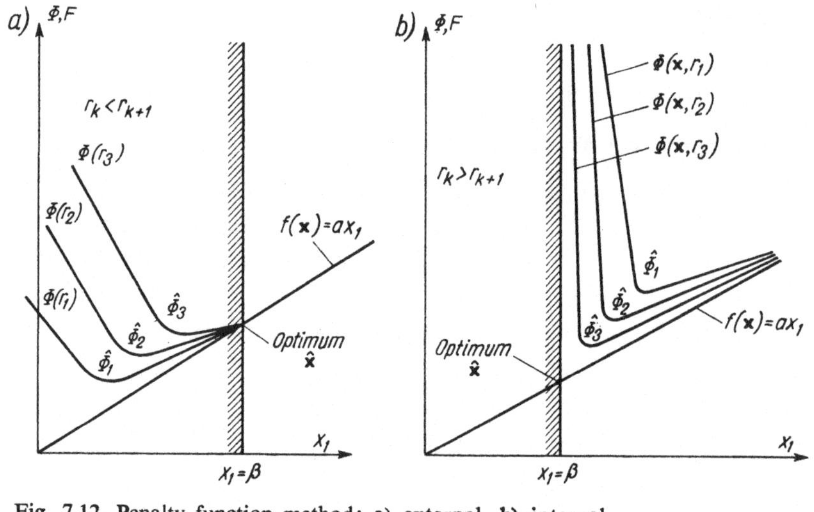

Fig. 7.12. Penalty function method: a) external, b) internal

interior formulations the unconstrained minima of $\Phi(\mathbf{x}, r_k)$ all lie in the feasible region and converge to the solution as $r_k \to 0$. Typical forms of G in the interior methods are

$$G(g) = -\frac{1}{g} \tag{7.119}$$

and

$$G(g) = \log(-g). \tag{7.120}$$

In the external penalty function methods, the unconstrained minima converge to the solution from outside the feasible region as $r_k \to \infty$. The most common forms of the function G in the external methods are

$$G(g) = \max[g, 0] \tag{7.121}$$

and

$$G(g) = (\max[g, 0])^2. \tag{7.122}$$

The internal penalty function method

The *internal method*, first proposed by Carroll[1], is also called the *barrier method*, the reason being that the penalty term in the function Φ is chosen so as to increase unboundedly as the constraints are approached and thus sets a barrier against leaving the feasible region. With G_j defined by equation (7.119), the auxiliary function Φ becomes

$$\Phi(\mathbf{x}, r) = F(\mathbf{x}) - r\sum_{j=1}^{m} \frac{1}{g_j(\mathbf{x})}. \tag{7.123}$$

The algorithm of the method can be summarized as follows:

1. Choose an initial feasible point $\mathbf{x}_1$ satisfying all the constraints as sharp inequalities, i.e. $g_j(\mathbf{x}_1) < 0$, $j = 1, 2, \ldots, m$, and an initial value $r_1 > 0$. Let $k = 1$.
2. Minimize $\Phi(\mathbf{x}, r_k)$ by any of the methods of unconstrained optimization to obtain $\mathbf{x}_k$.
3. Test $\mathbf{x}_k$ for convergence to the optimal solution of the original problem. If the convergence criterion is satisfied, stop the process. Otherwise go to the next step.
4. Reduce the penalty parameter by setting $r_{k+1} = cr_k$, where $c < 1$.
5. Replace k by $k+1$, take $\mathbf{x}_k$ as the new starting point and go to step 2.

The external penalty function method

The usual form of the function Φ in this method is

$$\Phi(\mathbf{x}, r) = F(\mathbf{x}) + r\sum_{j=1}^{m} \langle g_j(\mathbf{x})\rangle^q, \tag{7.124}$$

where the brackets $\langle \cdot \rangle$ denote $\max[\cdot, 0]$ and q is a non-negative constant. It can be seen that if any of the constraints is violated, a penalty proportional to the qth power of the amount of violation is imposed on the objective function. The minimum of Φ as a function of $\mathbf{x}$ is usually in the infeasible region but converges to the optimal solution of the original problem as $r \to \infty$. The exponent q is important for the behaviour of the auxiliary function. For $q = 0$, the function Φ is discontinuous on the boundary of the feasible region and it would be difficult to search for a minimum. For $0 < q < 1$, the function Φ is continuous but may fail to penalize strongly enough to ensure

[1] Carroll, C. W., The created response surface technique for optimizing nonlinear restrained systems, *Operations Research*, **9** 1961.

the desired behaviour. In addition, the function possesses discontinuous first derivatives along the boundary and hence would also be difficult to minimize. For $q = 1$, as has been shown by Zangwill[1], there exists an $\bar{r}$ large enough that the minimum of Φ is exactly the constrained minimum of the original problem for all $r \geqslant \bar{r}$. The most convenient choice, however, is a $q > 1$, for which the function Φ will have continuous first derivatives; $q = 2$ is a popular value.

The algorithm of the method consists of the following steps:

1. Choose an arbitrary initial point $\mathbf{x}_1$ and a suitable value of r_1. Let $k = 1$.
2. Find a vector $\mathbf{x}_k$ that minimizes the function

$$\Phi(\mathbf{x}, r_k) = F(\mathbf{x})+r_k \sum_{j=1}^{m} \langle g_j(\mathbf{x})\rangle^q.$$

3. Investigate the constraints to determine whether the point $\mathbf{x}_k$ is feasible. If it is feasible, it is the desired optimum. If it is not, go to step 4.
4. Choose a new penalty parameter $r_{k+1} > r_k$, set $k = k+1$ and go to step 2.

C Penalty function methods for mixed equality and inequality constrained problems

The algorithms given above cannot be applied directly to problems involving equality constraints. Below we present briefly some of the penalty functions proposed by various authors to deal with this case.

To solve a non-linear programming problem with equality constraints $h_j(\mathbf{x}) = 0, j = 1, 2, \ldots, p$, Courant[2] used a quadratic penalty function of the form

$$\Phi(\mathbf{x}, r) = F(\mathbf{x})+r \sum_{j=1}^{p} h_j^2(\mathbf{x}). \tag{7.125}$$

If the original problem has an optimal solution, the unconstrained minima of the function Φ obtained for a sequence $r_k \to \infty$ can be shown to converge to that solution, i.e.

$$\lim_{r_k\to\infty} \left(\min_{\mathbf{x}} \Phi(\mathbf{x}, r_k)\right) = \min_{h_j(\mathbf{x})=0} F(\mathbf{x}). \tag{7.126}$$

In following this idea, a mixed problem with inequality and equality constraints

$$\begin{aligned} g_j(\mathbf{x}) &\leqslant 0, \quad j = 1, 2, \ldots, m, \\ h_j(\mathbf{x}) &= 0, \quad j = 1, 2, \ldots, p, \end{aligned} \tag{7.127}$$

[1] ZANGWILL, W. I., Nonlinear programming via penalty function, *Management Science*, **13**, 5, pp. 344–358 (1967).

[2] COURANT, R., Variational methods for the solution of problems of equilibrium and vibrations, *Bull. Am. Soc.*, **48** (1943).

can be handled by an external penalty function of the form

$$\Phi(\mathbf{x}, r) = F(\mathbf{x})+r\sum_{j=1}^{m}\langle g_j(\mathbf{x})\rangle^2+r\sum_{j=1}^{p}h_j^2(\mathbf{x}) \tag{7.128}$$

to be minimized for an increasing sequence of values of r.

An internal penalty function including a barrier term that takes care of the inequality constraint and a penalty term that handles the equality constraints has been proposed[1] as

$$\Phi(\mathbf{x}, r) = F(\mathbf{x})-r\sum_{j=1}^{m}g_j^{-1}(\mathbf{x})+r^{-1/2}\sum_{j=1}^{p}h_j^2(\mathbf{x}). \tag{7.129}$$

If Φ is minimized for a decreasing sequence $r_k \to 0$, the resulting unconstrained minima $\mathbf{x}_k$ will converge to the solution $\hat{\mathbf{x}}$ of the constrained problem (7.127).

Although the method has been used successfully in a number of problems, it may present extreme difficulties due to "ill-conditioning" of the auxiliary function as r approaches zero. To avoid that, several penalty parameter-free methods have been developed.

Schmit and Fox (1965) proposed to reach the solution of the mixed equality-inequality constrained problem (7.127) by sequential unconstrained minimization of the auxiliary function

$$\Phi(\mathbf{x}) = \langle F(\mathbf{x})-F_k\rangle^2+\sum_{j=1}^{m}\langle g_j(\mathbf{x})\rangle^2+\sum_{j=1}^{p}h_j^2(\mathbf{x}) \tag{7.130}$$

where F_k is a constant selected as a goal for the objective function in the kth iteration. Every next value F_k is chosen below the minimum of Φ calculated in the preceding step.

A different approach, originally proposed by Powell[2] for solving equality-constrained problems and later extended to mixed equality-inequality constrained problems by Wierzbicki[3] and Szymanowski *et al.*[4] is to shift the penalty function during the iterative process according to the extent of constraint violations. The auxiliary function in this method has the form

$$\Phi(\mathbf{x}, \boldsymbol{\sigma}, \boldsymbol{\theta}) = F(\mathbf{x})+\sum_{j=1}^{m}\sigma_j\langle g_j(\mathbf{x})+\theta_j\rangle^2, \tag{7.131}$$

---

[1] Fiacco, A. V., and McCormick, G. P., *Nonlinear programming: Sequential Unconstrained Minimization Techniques*, J. Wiley, 1968.

[2] Powell. M. J. D, A method for non-linear constraints in minimization problems, *AERE Harwell Report*, **TP 310**, 1967.

[3] Wierzbicki, A., A penalty function method with shifts, *National Automatics Conference*, Gdańsk 1971 (in Polish).

[4] Szymanowski, J., *et al.*, *Program Library for Static Optimization*, Warsaw Technical University, 1970 (in Polish).

where $\boldsymbol{\sigma} = (\sigma_1, \sigma_2, \ldots, \sigma_m)$ is the penalty parameter vector ($\sigma_j > 0$) and $\boldsymbol{\theta} = (\theta_1, \ldots, \theta_m)$ is the penalty shift vector, with $\theta_j \leqslant 0$.

Rosenbrock[1] considered non-linear programs with constraints

$$a_i(\mathbf{x}) \leqslant g_i(\mathbf{x}) \leqslant b_i(\mathbf{x}). \tag{7.132}$$

In his method, the feasible space is divided into three subregions: an inner zone and two boundary zones of width $\alpha = 10^{-4}(b_i(\mathbf{x})-a_i(\mathbf{x}))$. A point $\mathbf{x}$ lies in the inner zone if

$$a_i(\mathbf{x})+\alpha \leqslant g_i(\mathbf{x}) \leqslant b_i(\mathbf{x})-\alpha, \tag{7.133}$$

and in a boundary zone if

$$\begin{aligned} a_i(\mathbf{x}) &\leqslant g_i(\mathbf{x}) < a_i(\mathbf{x})+\alpha, \\ b_i(\mathbf{x})-\alpha &< g_i(\mathbf{x}) \leqslant b_i(\mathbf{x}). \end{aligned} \tag{7.134}$$

The problem is solved by sequential unconstrained minimization of an auxiliary function Φ which is the same as the original objective function F in the inner zone and defined by

$$\Phi(\mathbf{x}) = F(\mathbf{x}) - \big(F(\mathbf{x}) - F^*\big)(3\eta - 4\eta^2 + 2\eta^3) \tag{7.135}$$

in the boundary zones, where F^* is the last value of F before the boundary zone is entered and η is given by

$$\eta = \frac{a(\mathbf{x})+\alpha-F(\mathbf{x})}{\alpha} \tag{7.136}$$

or by

$$\eta = \frac{F(\mathbf{x})-b(\mathbf{x})+\alpha}{\alpha}. \tag{7.137}$$

If $\eta = 0$, we have $\Phi(\mathbf{x}) = F(\mathbf{x})$, while if $\eta = 1$, then $\Phi(\mathbf{x}) = F^*$, with $F^* \geqslant F(\mathbf{x})$.

7.14 Choosing a method

So far we have reviewed several methods of unconstrained and constrained optimization. Which method should be chosen for solving a particular optimum-design problem is an important question, which a designer may find difficult to answer. No single method will perform best in all the situations to which it is applicable and not all the applicable methods may be realizable under given conditions. Often several methods have to be tried before the most suitable one is finally adopted. In the present section, we give some directions which may be useful in selecting a method.

---

[1] Rosenbrock, H. H., and Storey, C., *Computational Techniques for Chemical Engineers*, Pergamon Press (1966).

Here are some of the factors to be considered in choosing a method:

- — type of the problem to be solved (linear, non-linear, quadratic, geometric, etc. programming);
- — availability of a computer program or the time required for preparing one;
- — facility of calculating derivatives of the objective function and constraints;
- — available record of the efficiency of the method;
- — required accuracy of the solution;
- — reliability of the method in finding the true minimum;
- — adaptability of the method to the given problem and facility in interpreting the results.

Table 7.2 presents a decision tree proposed by Fletcher[1] for the selection of a method appropriate for a given optimization problem. Of necessity, the scheme is not a detailed one and does not account for the various devices that can be used to simplify the formulation of the problem before deciding on a solving algorithm. It may be possible, for example, to use linear or even non-linear[2] constraints to eliminate some of the design variables. Another possibility, to be recalled from Sec. 7.13.A, is that constraints can sometimes be eliminated by making a suitable transformation of variables.[3] Also, in solving sums-of-squares problems, i.e. problems in which the objective function is a sum of squares of a number of other functions, only non-linear variables should be supplied to the optimization algorithm, because those involved linearly in the objective function can be determined by a least-squares routine every time this function is calculated. In this way, the number of variables, and hence the complexity of the procedure, is reduced. However, particular care is required in evaluating derivatives in the modified formulation.[4]

The choice of a method will often be resolved by the local availability of a computer program. A notable collection of FORTRAN programs for solving linear, quadratic, geometric, dynamic and general non-linear programming problems was prepared by Kuester and Mize.[5]

---

[1] FLETCHER, R., Methods for the solution of optimization problems, *Study No. 5, Computer-aided Engineering, Solid Mechanics Division*, University of Waterloo, Ontario, Canada 1971.

[2] DOMMEL, M. W., and TINNEY, W. P., Optimal power flow solutions, *IEEE Trans. PAS*, **87**, 1866 (1968).

[3] For more examples of such transformations see e.g. BOX, M. J., DAVIES, D., and SWANN, W. H., *Nonlinear Optimization Techniques*, ICI Monograph No. 5, Oliver and Boyd, London 1969.

[4] FLETCHER, R.: see 23*, References to Table 7.2.

[5] KUESTER, J. L., and MIZE, J. H., *Optimization Techniques with Fortran*, McGraw-Hill, New York 1973.

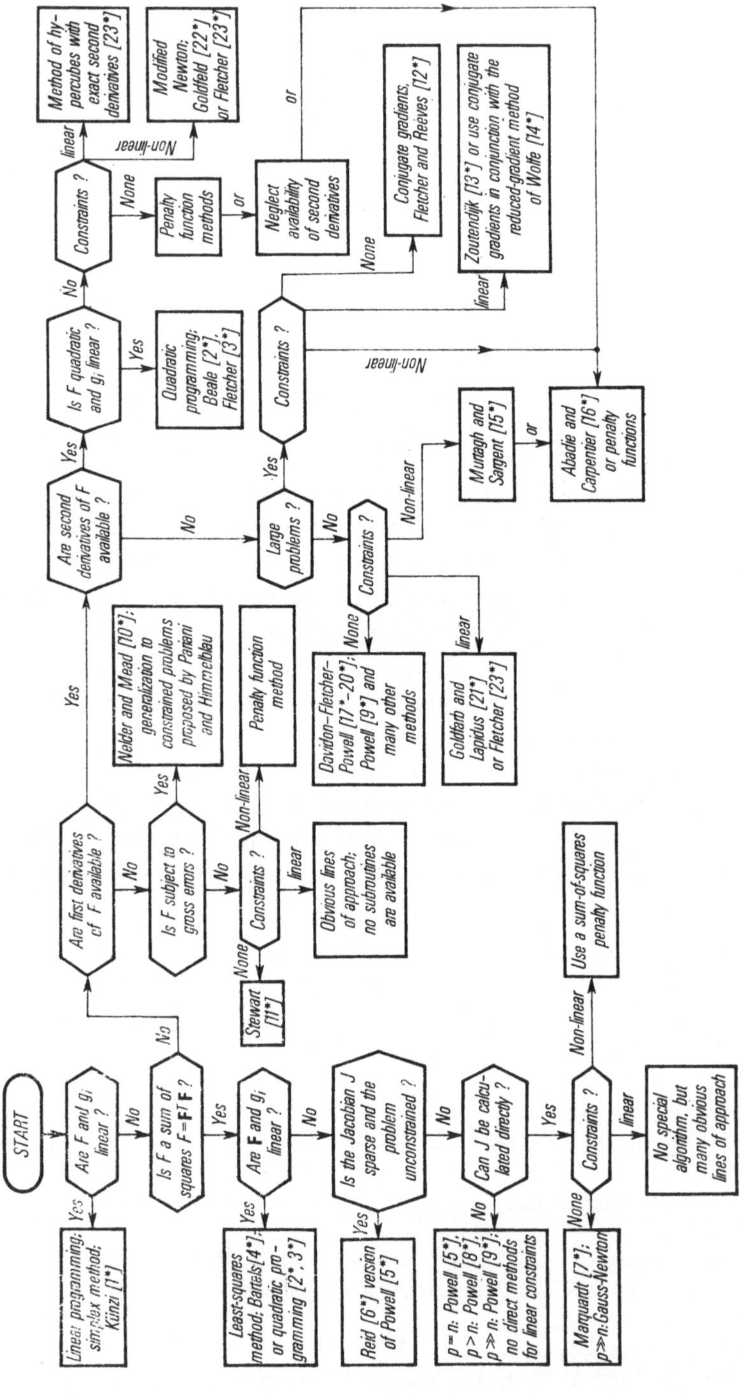
START
Are F and g_i linear ?
Yes
Linear programming; simplex method; Künzi [1*]
No
Is F a sum of squares $F=\mathbf{F}^T\mathbf{F}$?
Yes
Are $\mathbf{F}$ and g_i linear ?
Yes
Least-squares method; Bartels [4*]; or quadratic programming [2*, 3*]
No
Is the Jacobian J sparse and the problem unconstrained ?
Yes
Reid [6*] version of Powell [5*]
No
Can J be calculated directly ?
No
$p = n$: Powell [5*]; $p > n$: Powell [8*]; $p \gg n$: Powell [9*]; no direct methods for linear constraints
Yes
Constraints ?
None
Marquardt [7*]; $p \gg n$: Gauss-Newton
linear
No special algorithm, but many obvious lines of approach
Non-linear
Use a sum-of-squares penalty function
No
Are first derivatives of F available ?
No
Is F subject to gross errors ?
Yes
Nelder and Mead [10*]; generalization to constrained problems proposed by Paviani and Himmelblau
No
Constraints ?
None
Stewart [11*]
linear
Obvious lines of approach; no subroutines are available
Non-linear
Penalty function method
Yes
Are second derivatives of F available ?
Yes
Is F quadratic and g_i linear ?
Yes
Quadratic programming; Beale [2*], Fletcher [3*]
No
Constraints ?
linear
Method of hypercubes with exact second derivatives [23*]
Non-linear
Modified Newton; Goldfeld [22*] or Fletcher [23*]
None
Penalty function methods
or
Neglect availability of second derivatives
or
No
Large problems ?
Yes
Constraints ?
None
Conjugate gradients, Fletcher and Reeves [12*]
linear
Zoutendijk [13*] or use conjugate gradients in conjunction with the reduced-gradient method of Wolfe [14*]
Non-linear
No
Constraints ?
None
Davidon–Fletcher–Powell [17*–20*]; Powell [9*] and many other methods
linear
Goldfarb and Lapidus [21*] or Fletcher [23*]
Non-linear
Murtagh and Sargent [15*]
or
Abadie and Carpentier [16*] or penalty functions

Table 7.2

References to Table 7.2

1*. H. P. Künzi, H. G. Tzschach, and C. A. Zehnder, *Numerical methods of mathematical optimization*, Academic Press, New York 1968.

2*. E. M. L. Beale, On quadratic programming, *Naval Res. Logistics Quarterly*, **6**, 227, 1959.

3*. R. Fletcher, A Fortran subroutine for general quadratic programming, *UKAEA Research Group Report, AERE Report* **R. 6370**, 1970 (Available from HMSO).

4*. R. H. Bartels, G. H. Golub and M. A. Saunders, *Numerical techniques in mathematical programming*, presented at the *7th International Mathematical Programming Symposium*, The Hague 1970.

5*. M. J. D. Powell, A hybrid method for non-linear equations, and A Fortran subroutine for solving systems of non-linear algebraic equations, In: *Numerical Methods for Non-Linear Algebraic Equations* (Ed. P. Rabinowitz), Gordon and Breach, London 1970.

6*. J. K. Reid, A Fortran subroutine for the solution of a sparse system of non-linear equations, *UKAEA Research Group Report AERE Library subroutine* **NS02A**, 1970.

7*. D. W. Marquardt, An algorithm for least squares estimation of non-linear parameters, *Jour. SIAM*, **11**, 431, 1963.

8*. M. J. D. Powel, *AERE library subroutine VA05A*, 1970, available from HMSO.

9*. M. J. D. Powell, A Fortran subroutine for unconstrained minimization, requiring first derivatives of the objective function, *UKAEA Research Group Report, AERE* **R. 6469**, 1970, available from HMSO.

10*. J. A. Nelder, and R. Mead, A simplex method for function minimization, *Computer Jour.* **7**, 308, 1965.

11*. G. W. Stewart III, A modification of Davidon's minimization method to accept difference approximations to derivatives, *Jour. ACM*, **14**, 72, 1967.

12*. R. Fletcher, and C. M. Reeves, Function minimization by conjugate gradients, *Computer Jour.*, **7**, 149, 1964 (includes an Algol procedure).

13*. G. Zoutendijk, *Methods of feasible directions*, Elsevier, Amsterdam 1960.

14*. P. Wolfe, Methods of non-linear programming, in: *Non-linear Programming* (Ed. J. Abadie), North-Holland, Amsterdam 1967.

15*. B. A. Murtagh, and R. W. H. Sargent, *Projection methods for non-linear programming*, presented at the *7th International Mathematical Programming Symposium*, The Hague 1970.

16*. D. Davies, Some practical methods of optimization, in: *Integer and Non-linear Programming* (Ed. J. Abadie), North-Holland, Amsterdam 1970.

17*. W. C. Davidon, Variable metric method for minimization, *Argonne Nat. Lab.*, **ANL-5990** *Rev.*, 1959.

18*. R. Fletcher and M. J. D. Powell, A rapidly convergent descent method for minimization, *Computer Journ.*, **6**, 163, 1963. (An Algol procedure given in References 19* and 20*.)

19*. M. Wells, Algorithm 251, function minimization, *Comm. ACM*, **8**, 169, 1965.

20*. R. Fletcher, Certification of Algorithm 251, *Comm. ACM*, **9**, 686, 1966.

21*. D. Goldfarb, and L. Lapidus, Conjugate gradient method for non-linear programming problems with linear constraints, *IEC Fundamentals*, **1**, 142, 1966 (AERE Library subroutine VEOIA).

22*. S. M. Goldfeld, R. E. Quandt, and H. F. Trotter, Maximization by quadratic hill climbing, *Econometrica*, **34**, 541, 1966.

23*. R. Fletcher, An efficient globally convergent algorithm for unconstrained and linearly constrained optimization problems, *AERE Report TP* **431**, presented at the *7th International Mathematical Programming Symposium*, The Hague 1970.

In comparative studies of different optimization techniques, it is essential to have an appropriate efficiency criterion. Depending on what features of the method tested are considered the most important, the criterion can be formulated in various ways. For iterative methods of unconstrained optimization, the criterion proposed by Box[1] uses the number of calculations of the objective function needed to achieve the required accuracy. This provides a fair measure of the relative speed with which a given procedure converges. For constrained methods, also the number of calculations of the constraint functions should be included. Comparing the efficiencies of different methods is pointless of course if different computers, different test-problems or different starting points are used for testing each method. The speed with which a given method will converge when applied to a specific problem will obviously depend on the form of the objective function and constraints in that problem but also on the position of the minimum—whether it is inside the feasible region, at its boundary or at a vertex—and on the required accuracy of the solution.

Optimization methods can broadly be divided into two groups: those which use derivatives and those which do not. Generally, it can be said that the gradient methods are more efficient. In particular, this is true of sequential methods, which, in their own right, are usually more efficient than non-sequential methods.

Comparisons of various optimizations methods have been conducted by many authors, e.g. by Swann[2], Fletcher and Reeves[3], Box[1,4], Fletcher[5], Colville[6] and Stocker[7]. Some general ıemarks on the various classes of methods are presented below, separately for unconstrained and constrained optimization.

Unconstrained methods

The available data show that if the first derivatives of the objective function

---

[1] Box, M. J., A comparison of several current optimization methods and the use of transformations in constrained problems, *Comp. J.*, **9** (1966).

[2] Swann, W. H., Report on the development of a new direct search method of optimization, *ICI Ltd., Central Instr. Res. Lab., Res. Note* **64/3**, London 1964.

[3] R. Fletcher and C. H. Reeves, Function minimization by conjugate gradients, *Computer J.*, **7**, 149, 1964.

[4] Box, M. J., A new method of constrained optimization and a comparison with other hods, *Computer J.*, **8**, pp. 42–52 (1965).

[5] Fletcher, R., Function minimization without evaluting derivatives: a review, *Computer J.* **8**, 33–41 (1965).

[6] Colville, A. R., A comparative study on nonlinear programming codes, *Report No.* **320-2949**, IBM Scientific Center, 1968.

[7] Stocker, D. C., *A comparative study on nonlinear programming codes*, M. S. Thesis, University of Texas, Austin, Texas 1969.

are easy to evaluate either in closed form or by finite differences and if the number of design variables does not exceed 50, the best results are obtained by the quasi-Newton methods. However, as the number of variables increases the evaluation of the gradients of the objective function and the approximation of the Hessian matrix become exceedingly time-consuming. Also, the convergence of these methods is strongly influenced by the choice of the starting point. The least sensitive to the choice of the starting point is Rosenbrock's method, which, although much slower than the quasi-Newton methods, gives satisfactory results in almost every situation.

When the number of design variables exceeds 50, the variable metric methods are the best, particularly the Davidon–Fletcher–Powell method DFP. Their superiority over other gradient methods was confirmed by Leon[1]. Generally, it can be said that if the evaluation of the objective function and its gradient is very time-consuming, the problem is so far best solved by the variable metric methods, regardless of the number of variables. When the function is less complicated (e.g. a polynomial) and the calculation of its values and gradients is not so laborious, the Fletcher–Reeves method can work equally well or even prove more efficient than DFP. In addition, it does not require so large an operating store; when the number of variables is greater than 100 and the storage becomes a problem, the conjugate gradient method will be the most suitable.

Constrained methods

The penalty function methods can be employed successfully when the objective function and constraints are in explicit forms and the number of variables is small or moderate. The interior penalty function method appears to be more effective than the exterior method, because the latter may lead to an infeasible solution. The exterior method is useful when finding a feasible starting point is difficult.

In problems in which the objective function and constraints are given implicitly, so that their derivatives are not available in closed form but can be approximated by finite-difference expressions, Zoutendijk's method of feasible directions will be more efficient than the penalty function methods. However, if approximations are used to evaluate the objective function and the constraints themselves, the penalty methods should perform better. If the evaluation of the objective function and constraints is very complicated and if only an approximate minimum is required, the interior penalty function method, which produces an improving sequence of feasible designs, is the preferred choice.

---

[1] LEON, A., A comparison among eight known optimizing procedures, in: *Recent Advances in Optimization Techniques*, ed. by T. P. VOGL and A. LAVI, Wiley, New York 1966.

In cases where all the constraints are linear, the gradient projection method of Rosen can be expected to work most efficiently.

Clearly, the actual efficiency of a penalty function method in solving a particular problem depends on the efficiency of the minimization technique used to solve the penalized unconstrained problem. Ramamoorty and Rao[1] compared the performance of several unconstrained techniques used in conjunction with the exterior penalty function method in Zangwill's formulation in a real optimum design problem, namely in designing a three-phase reluctance motor. The problem involved 10 design variables and 8 constraints. The unconstrained techniques used were: random search, the simplex method, Hooke and Jeeves' method, Powell's method and the Davidon–Fletcher–Powell method. The results of calculations, which were carried out on an IBM 7044/1401 computer, are shown in Table 7.3. For each method, the initial penalty factor was 0.1 and was increased by a factor of 10 at every successive stage. The step lengths used for the 10 variables were different in the first three methods.

Table 7.3 Comparative performance of different unconstrained optimization methods

	Random search	Simplex	Hooke and Jeeve's	Powell's	DFP
1. Initial value of objective function	2.44076	2.44076	2.44076	2.44076	2.44076
2. Optimized value of objective function	1.08806	1.06247	1.06948	1.05727	1.30145
3. Computer time (including compilation)	7 min 38 s	9 min 18 s	4 min 49 s	1 min 40 s	30 min
4. Initial value of penalty factor	0.10	0.10	0.10	0.10	0.10
5. Final value of penalty factor	10^4	10^4	10^{15}	10^5	10^2
6. Maximum constraint violation	nil	nil	2.037×10^{-5}	nil	4.157×10^{-3}

In the random search method, the initial step length was 10% of the respective variables and was decreased gradually to a value of 0.01%. The computer time taken, including 2.52 s compilation time, was 7 min 38 s.

In the simplex method, the length of the simplex edge was initially taken to be 0.01 and then reduced to 0.0025 and 0.000625. The calculated optimum value of the objective function compared well with the lowest value obtained by Powell's method but the computing time, inclusive of 2 min 58 s compilation time, was 9 min 18 s.

The initial step lengths used in Hooke and Jeeves' method were 1% of the magnitudes of the respective variables and were diminished successively,

[1] Ramamoorty, M., and Rao, P. J., Comparative study of optimization methods for the design of polyphase reluctance motors, *Engineering Optimization*, **3**, 51–60, 1977.

with a reduction factor of 0.5. The computer time taken, including 2 min 42 s compilation time, was 4 min 48 s.

Powell's method took the least time, namely 1 min 40 s including compilation. It also gave the smallest value of the objective function.

The DFP method performed worst. Convergence was not reached even after 500 iterations in each minimization cycle and a total computer time of 30 minutes. The failure could be due to the fact that the objective function and some of the constraints were discontinuous at some points. Furthermore, the gradient was calculated by a central difference scheme and cubic interpolation was used for one-dimensional minimization. As noted by Lasdon and Fox[1], cubic interpolation used in conjunction with the exterior penalty function method may fail to work in certain optimization problems. Cumulative errors in numerical differentiation and cubic interpolation could also impede convergence. Finally, as observed by Ramarathnam *et al.*[2], the descent methods which make use of the first and second derivatives have inherently poor convergence properties for the types of functions encountered in the problem considered.

7.15 Applications of non-linear programming methods to structural optimization

The literature of structural optimization offers a great variety of examples of application of non-linear programming to optimum design problems. Reinschmidt, Cornell and Brotchie (1966) and Bigelow and Gaylord (1967) used non-linear programming methods in optimum plastic design of multistorey frames without the simplifying assumptions that the objective function (weight) is linear in the limit plastic moments. Brown and Ang (1966) used Rosen's gradient projection method to optimize steel frames in the elastic and plastic ranges. They found that an elastically designed optimal frame is about 10% more expensive than its plastically designed counterpart.

Some general formulations of optimum structural design involving non-linear programming were considered by Schmit (1960) and Vinogradov (1965_2). Vinogradov pointed out that in many problems the sufficient conditions for an absolute extremum to be attainable by the methods of non-linear programming are not satisfied. Also, the objective function is often non-convex. For convexity, the Hessian matrix $[\partial^2 F/\partial x_i \partial x_j]$ should be positive defi-

---

[1] Lasdon, L. S., and Fox, R. L., Efficient algorithm for one-dimensional minimization problems, *Man. Sci*, **22**, 42–51, 1975.

[2] Ramarathnam, R., Desai, B. G., and Rao, V. S., A comparative study of minimization techniques for optimization of induction motor design, *IEEE Transactions on Power Apparatus and Systems*, **PAS-92**, 1448–1454, 1973.

nite. A simple example where it is not positive definite is provided by the objective function in the minimum weight design of a frame if the members of the frame are chosen from a set of typical I-bars. On the other hand, the problem of an optimum reinforcement of a frame with given sections is convex.

Further examples of non-linear programming in application to structural optimization were given by Schmit, Kicher and Morrow (1963), Gellatly, Gallagher and Luberacki (1964), Leśniak (1965) and Bogatyriev (1968). Peredy (1964) and Abramov (1966) used non-linear programming for the optimum elastic design of ferroconcrete frames. Achmadaliev (1967) determined the optimum height of a truss with given span lengths. Karihaloo, Pathare and Ramesh (1967) applied the methods of non-linear programming to the optimum elastic design problem for spatial bar structures. Moses (1967) used similar methods to optimize a frame and a truss. Schmit, Kicher and Morrow (1963) calculated the optimum reinforcement of a perforated plate and Kavlie, Kowalik, Lund and Moe (1966) optimized elastic ribbed plates.

As an illustration, we present below three simple examples of application of non-linear programming methods to optimum design of engineering structures. The examples concern optimization of a box beam, a three-bar truss and a portal frame.

Optimization of a box beam

Consider a box beam of a given length L subject to a pure bending moment M (Fig. 7.13). The problem is to determine the parameters of the beam section, such that the volume of the beam is a minimum (Akita and

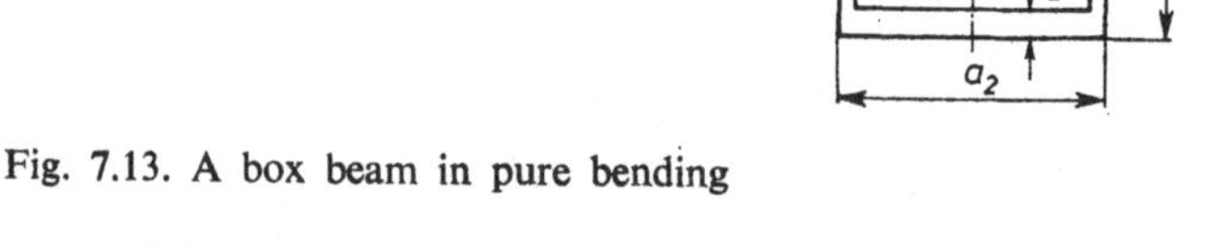

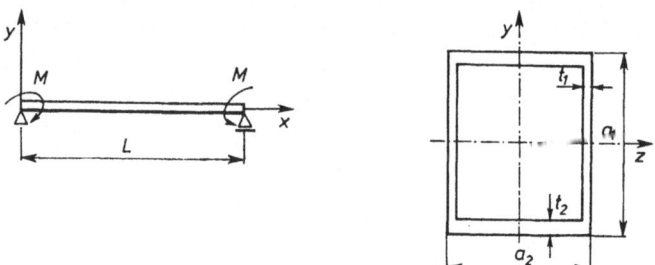

Fig. 7.13. A box beam in pure bending

Kitamura 1970). The design variables representing the sectional form are the thicknesses t_1 and t_2 of the web and the flange, the depth a_1 of the web and the width a_2 of the flange. The constraints for the problem follow from the global and local buckling restrictions in the elastic and plastic ranges, respectively.

Accordingly, the problem under consideration can be formulated as follows:

Find the section variables t_1, t_2, a_1 and a_2 that minimize the cross-sectional area, and hence the volume, of the beam,

$$V = AL = 2L(a_1 t_1 + a_2 t_2), \tag{7.138}$$

subject to the constraints

$$n\sigma \leqslant \sigma_c^*, \quad n\sigma \leqslant \sigma_{p1}^*, \quad n\sigma \leqslant \sigma_{p2}^*. \tag{7.139}$$

The moments of inertia with respect to the z- and y-axis of the beam, i.e. J_z and J_y, the polar moment of inertia K and the section modulus Z are given by

$$J_z = \tfrac{1}{2}a_1^2 a_2 t_2 + \tfrac{1}{6}a_1^3 t_1; \quad J_y = \tfrac{1}{6}a_2^3 t_2 + \tfrac{1}{2}a_2^2 a_1 t_1;$$
$$K = \frac{2a_1^2 a_2^2}{a_1/t_1 + a_2/t_2}; \quad Z = \frac{2I_z}{a_1}. \tag{7.140}$$

The buckling analysis of the beam in the elastic and plastic ranges gives the following relations:

$$\sigma = \frac{M}{Z}, \quad \sigma_c^m = \frac{M_c^*}{Z} = \frac{\pi}{ZL}\sqrt{EJ_y GK},$$

$$\sigma_c^* = \begin{cases} \sigma_c^m, & \text{if} \quad \sigma_c^m \leqslant \dfrac{\sigma_y}{2}, \\[2ex] \sigma_y - \dfrac{1}{4}\dfrac{\sigma_y^2}{\sigma_c^m}, & \text{if} \quad \sigma_c^m > \dfrac{\sigma_y}{2}, \end{cases}$$

$$\sigma_{p1}^* = \begin{cases} \dfrac{2E\pi^2}{1-\nu^2}\left(\dfrac{t_1}{a_1}\right)^2, & \text{if} \quad \dfrac{2E\pi^2}{1-\nu^2}\left(\dfrac{t_1}{a_1}\right)^2 \leqslant \dfrac{\sigma_y}{2}, \\[2ex] \sigma_y - \dfrac{1}{4}\dfrac{\sigma_y^2}{\dfrac{2E\pi^2}{1-\nu^2}\left(\dfrac{t_1}{a_1}\right)^2}, & \text{if} \quad \dfrac{2E\pi^2}{1-\nu^2}\left(\dfrac{t_1}{a_1}\right)^2 > \dfrac{\sigma_y}{2}, \end{cases} \tag{7.141}$$

$$\sigma_{p2}^* = \begin{cases} \dfrac{E\pi^2}{3(1-\nu^2)}\left(\dfrac{t_2}{a_2}\right)^2, & \text{if} \quad \dfrac{E\pi^2}{3(1-\nu^2)}\left(\dfrac{t_2}{a_2}\right)^2 \leqslant \dfrac{\sigma_y}{2}, \\[2ex] \sigma_y - \dfrac{1}{4}\dfrac{\sigma_y^2}{\dfrac{E\pi^2}{3(1-\nu^2)}\left(\dfrac{t_2}{a_2}\right)^2}, & \text{if} \quad \dfrac{E\pi^2}{3(1-\nu^2)}\left(\dfrac{t_2}{a_2}\right)^2 > \dfrac{\sigma_y}{2}, \end{cases}$$

The following notation are used:

σ —working stress,
σ_y—yielding point,
σ_c—total buckling stress in the beam treated as a column,
σ_c^m—lateral buckling stress due to the bending moment,
σ_{p1}^*—buckling stress of the web treated as a plate,
σ_{p2}^*—buckling stress of the flanges treated as plates,
M_c^*—lateral buckling bending moment,
n—safety factor to buckling.

In deriving equation (7.141), the following *Johnson–Ostenfeld formulae* for plastic buckling were used:

$$\begin{aligned} \sigma_b &= \sigma_E = \frac{\pi^2 E}{\lambda^2} \quad \text{if} \quad \sigma_E \leqslant \frac{\sigma_y}{2}, \\ \sigma_b &= \sigma_y\left(1-\frac{\sigma_y}{4\sigma_E}\right) \quad \text{if} \quad \sigma_E > \frac{\sigma_y}{2}, \end{aligned} \tag{7.142}$$

where $\lambda = l_b/i$.

Let x_1, x_2, x_3 and x_4 be dimensionless design variables defined by

$$\begin{aligned} x_1 &= \alpha\left(\frac{a_1}{L}\right), \quad x_2 = \beta\left(\frac{t_1}{a_1}\right)^2, \\ x_3 &= \gamma\left(\frac{t_2}{a_2}\right), \quad x_4 = \delta\left(\frac{a_2}{a_1}\right)^2. \end{aligned} \tag{7.143}$$

The parameters α, β, γ and δ in these relations express the material and load properties and are given by

$$\begin{aligned} \alpha &= \sqrt{\frac{3}{2(1+\nu)}}\,\pi\left(\frac{E}{\sigma_y}\right), \quad \beta = \frac{2\pi^2}{1-\nu^2}\frac{E}{\sigma_y}, \\ \gamma &= \frac{\pi^2}{3(1-\nu^2)}\frac{E}{\sigma_y}, \quad \delta = 3n\left(\frac{M}{L^3\sigma_y}\right)\alpha^3. \end{aligned} \tag{7.144}$$

For numerical facility, let us also put

$$\begin{aligned} x_1'' &= x_1 f_1, \quad x_2' = \left(\frac{x_2}{6}\right)^{1/2}, \quad x_3' = x_3^{1/2}, \\ f_1 &= \frac{x_4}{x_2'+3x_3'x_4}\sqrt{\frac{x_2'x_3'(3x_2'+x_3'x_4)}{x_2'+x_3'x_4}}, \\ f_2 &= \gamma^{-1/2}x_1^3(x_2'+3x_3'x_4). \end{aligned} \tag{7.145}$$

After all the substitutions, the optimum-design problem for the beam takes the form

Minimize the objective function (*volume*)

$$V = \frac{2L^3}{\alpha^2\gamma^{1/2}}\Phi, \quad \Phi = x_1^2(x_2'+x_3'x_4) \tag{7.146}$$

subject to the constraints

$$g_1 = \begin{cases} x_1'' - \dfrac{\delta}{f_2} \geqslant 0 & \text{if} \quad x_1'' \leqslant \dfrac{1}{2}, \\ 1-\dfrac{1}{4}\dfrac{1}{x_1''} - \dfrac{\delta}{f_2} \geqslant 0 & \text{if} \quad x_1'' > \dfrac{1}{2}, \end{cases} \tag{7.147$_1$}$$

$$g_2 = \begin{cases} x_2 - \dfrac{\delta}{f_2} \geqslant 0 & \text{if} \quad x_2 \leqslant \dfrac{1}{2}, \\ 1 - \dfrac{1}{4}\dfrac{1}{x_2} - \dfrac{\delta}{f_2} \geqslant 0 & \text{if} \quad x_2 > \dfrac{1}{2}, \end{cases} \tag{7.147$_2$}$$

$$g_3 = \begin{cases} x_3 - \dfrac{\delta}{f_2} \geqslant 0 & \text{if} \quad x_3 \leqslant \dfrac{1}{2}, \\ 1 - \dfrac{1}{4}\dfrac{1}{x_3} - \dfrac{\delta}{f_2} \geqslant 0 & \text{if} \quad x_3 > \dfrac{1}{2}. \end{cases} \tag{7.147$_3$}$$

To solve this problem, Akita and Kitamura (1970) used an internal penalty function (IPF) method coupled with the variable metric method (or DFP method—after Davidon, Fletcher and Powell).

The material of the beam was assumed to be steel with $E = 200$ GPa, $\nu = 0.3$, $\sigma_y = 200$ MPa and $n = 1.5$, and the parameters α, β, γ and δ were calculated as

$$\begin{aligned} \alpha &= 3.37\times 10^3, \quad \beta = 2.18\times 10^4, \\ \gamma &= 3.53\times 10^3, \quad \delta = 8.65\times 10^9\left(\frac{M}{L^3}\right). \end{aligned} \tag{7.148}$$

The optimum solutions were calculated at $M/L^3 = 10^{-11}$ to 10^{-8} kNmm/ /mm^3. Table 7.4 gives the results of optimization obtained for ten different initial points (x_1', x_2, x_3, x_4). For convenience, the variables x_1 and Φ are scaled down to

$$x_1' = \frac{x_1}{10} \quad \text{and} \quad \Phi' = \frac{\Phi}{100}. \tag{7.149}$$

Table 7.4 Results of optimization of a box beam by IPF+DFP method

No.	Initial point				Goal				
	x_1'	x_2	x_3	x_4	x_1'	x_2	x_3	x_4	Φ'
1	10.0	10.0	10.0	10.0	3.558	5.135	5.140	1.648	59.03
2	8.0	8.0	8.0	8.0	3.428	4.539	4.372	2.008	60.33
3	6.0	6.0	6.0	6.0	5.553	2.610	2.468	0.5647	47.94
4	4.0	4.0	4.0	4,0	5.450	2.650	2.074	0.6273	47.83
5	2.0	2.0	2.0	2.0	infeasible initial point				
6	10.0	1.0	1.0	1.0	7.041	2.192	1.292	0.3113	45.89
7	8.0	1.0	1.0	1.0	6.222	2.926	1.783	0.3830	46.50
8	6.0	1.0	1.0	1.0	6.392	1.538	1.338	0.5311	46.24
9	4.0	1.0	1.0	1.0	infeasible initial point				
10	2.0	1.0	1.0	1.0					

Figure 7.14 shows the optimum solutions x_1', x_2, x_3, x_4 and Φ' obtained for various values of M/L^3. The values x_1', x_2, $x_3 = 0.5$ are the limiting values between the elastic and the plastic buckling. At the optimal point, the local bucklings of the flanges and webs are plastic bucklings, while the total

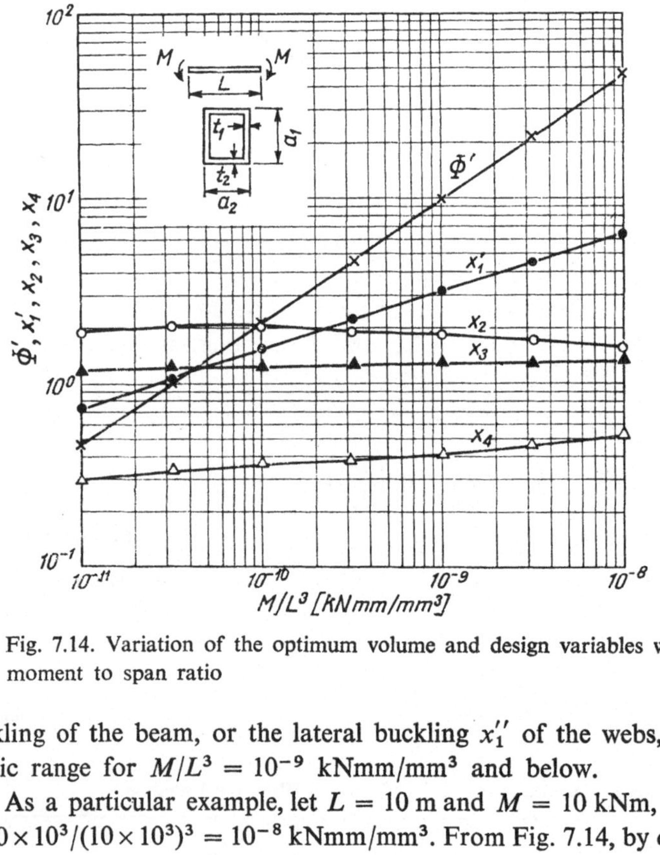

Fig. 7.14. Variation of the optimum volume and design variables with the bending moment to span ratio

buckling of the beam, or the lateral buckling x_1'' of the webs, is within the elastic range for $M/L^3 = 10^{-9}$ kNmm/mm^3 and below.

As a particular example, let $L = 10$ m and $M = 10$ kNm, so that $M/L^3 = 10 \times 10^3/(10 \times 10^3)^3 = 10^{-8}$ kNmm/mm^3. From Fig. 7.14, by ordinate interpolation, we obtain $x_1' = 6.392$, $x_2 = 1.538$, $x_3 = 1.338$, $x_4 = 0.5311$ and $\Phi' = 46.24$. Equations (7.143)–(7.149) then give $a_1/L = 1.90 \times 10^{-2}$, $t_1/a_1 = 8.39 \times 10^{-3}$, $t_2/a_2 = 1.92 \times 10^{-2}$, $a_2/a_1 = 0.730$, and hence $a_1 = 190$ mm, $a_2 = 139$ mm, $t_1 = 1.59$ mm, $t_2 = 2.67$ mm, $V = 1.35 \times 10^{-5} L^3 = 1.35 \times \times 10^7$ mm^3. Assuming the density of steel to be 7.8×10^{-8} kN/mm^3, we find that the unit weight of the beam, W/L, is 0.105 kN/m.

Optimization of a three-bar truss

Consider a three-bar planar truss shown in Fig. 7.15 (Schmit 1960; see Sec. 4.3). The configuration and the material of the truss are preassigned, with

$$N = 254 \text{ mm} \quad \beta_1 = 135°, \quad \beta_2 = 90°, \quad \beta_3 = 45°,$$
$$\varrho = 2.77 \times 10^{-8} \text{ kN/mm}^3 \quad E = 70.3 \text{ GPa}.$$

In view of the symmetry, we assume $A_1 = A_3$, so that the only independent design variables are the cross-sectional areas A_1 and A_2. The truss is designed for two different loading conditions. The first loading state is that of a load

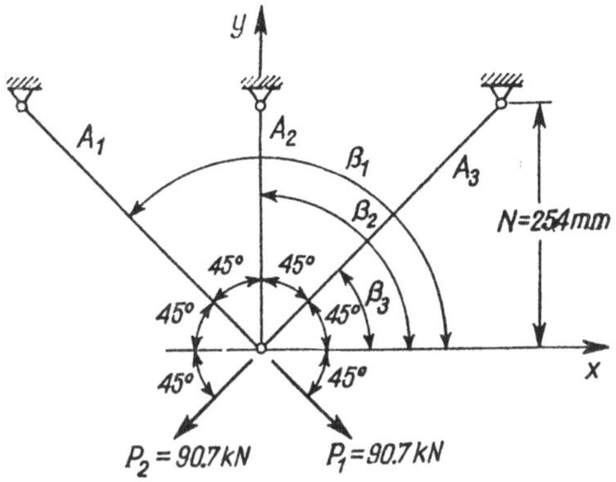

Fig. 7.15. Schematic of a three-bar truss

$P_1 = 90.7$ kN acting at an angle of 45° to the x-axis, and the second is due to a load $P_2 = 90.7$ kN acting at 135° to the x-axis. The objective is to minimize the weight of the truss

$$W(\mathbf{x}) = \varrho N(2\sqrt{2}A_1 + A_2), \tag{7.150}$$

where $\mathbf{x}^T = (A_1, A_2)$, subject to the requirement that the compressive and the tensile stresses do not exceed certain allowable limit values in either loading state. To keep the bar cross-sections non-negative, we also require that $A_1 \geqslant 0$ and $A_2 \geqslant 0$.

Let σ_{ij} denote the stress in the ith member under the jth load condition. By symmetry, we have $\sigma_{11} = \sigma_{32}$, $\sigma_{21} = \sigma_{22}$ and $\sigma_{31} = \sigma_{12}$, and so it is enough to consider the stresses σ_{11}, σ_{21} and σ_{31}. With 140 MPa taken as the limit for tensile stresses and 105 MPa taken as the limit for compressive stresses, all the stress constraints can be written

$$\begin{aligned} g_1(\mathbf{x}) &= \sigma_{11} - 0.14 \leqslant 0, \\ g_2(\mathbf{x}) &= \sigma_{21} - 0.14 \leqslant 0, \\ g_3(\mathbf{x}) &= \sigma_{31} - 0.14 \leqslant 0. \\ g_4(\mathbf{x}) &= -\sigma_{11} - 0.105 \leqslant 0, \\ g_5(\mathbf{x}) &= -\sigma_{21} - 0.105 \leqslant 0, \\ g_6(\mathbf{x}) &= -\sigma_{31} - 0.105 \leqslant 0. \end{aligned} \tag{7.151}$$

In addition, we have the non-negativity conditions

$$\begin{aligned} g_7(\mathbf{x}) &= -A_1 \leqslant 0, \\ g_8(\mathbf{x}) &= -A_2 \leqslant 0. \end{aligned} \tag{7.152}$$

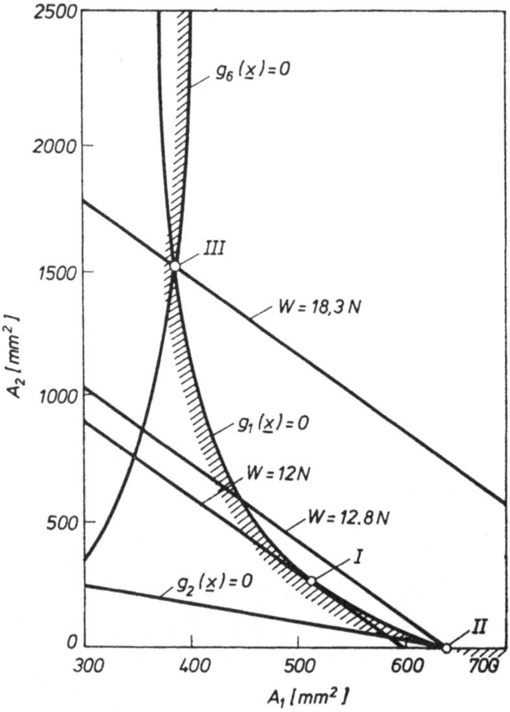

Fig. 7.16. The design space and objective contours

From elementary statical analysis of the structure, we obtain

$$\begin{aligned}\sigma_{11} &= 0.14\left(\frac{1}{A_1} - \frac{A_2}{2A_1A_2+\sqrt{2}A_1^2}\right),\\ \sigma_{21} &= \frac{0.14\sqrt{2}A_1}{2A_1A_2+\sqrt{2}A_1^2},\\ \sigma_{31} &= -\frac{0.14A_2}{2A_1A_2+\sqrt{2}A_1^2}.\end{aligned} \tag{7.153}$$

The solution to the constrained non-linear problem (7.150)–(7.153) was computed by Liu (1972) by a few different techniques, each using a linear or quadratic external penalty function and either the Hooke–Jeeves or Powell's search methods (Sec. 7.10.C) to solve the penalized unconstrained problem. Computationally, the Hooke–Jeeves algorithm turned out to be a better choice.

As there are only two design variables in the problem, it allows a lucid geometrical interpretation. The design space with the constraint curves and few objective contours is shown in Fig. 7.16. It can be seen that the feasible region is bounded by the constraint curves $g_1 = 0$ (the limit on tension in bar 1 under load P_1), $g_6 = 0$ (the limit on compression in bar 3 under load P_1) and $g_8 = 0$ ($A_2 = 0$). Note that the condition $g_2(\mathbf{x}) \leqslant 0$, setting the limit on tension in bar 2 in the first loading state, is automatically satisfied if the constraint $g_1(\mathbf{x}) \leqslant 0$ is satisfied. The objective contours displayed are in fact the contours of $W/\varrho N$ ($\varrho N = 2.77\times10^{-8}\times254 = 7\times10^{-6}$).

The minimum-weight design is represented by point *I*, where $A_1 = A_3 = 508$ mm², $A_2 = 265$ mm² and $W = 0.012$ kN. Thus the optimum does not lie at a vertex of the feasible space, i.e. not at a point where constraint curves intersect. It is a statically indeterminate structure, in which bar 2 is not fully stressed in either loading state.

A statically determinate and fully stressed design is given by point *II*, at which $A_1 = A_3 = 645$ mm², $A_2 = 0$ (the middle bar vanishes) and $W = 0.0128$ kN. This design is somewhat heavier than design *I*. The well-known conclusion can hence be drawn that a fully stressed structure, i.e. one in which every member sustains a limiting allowable stress under at least one load condition, need not be a minimum-volume structure. This problem was also investigated by Gerard (1965), Gellatly (1966) and Dayaratnem and Partaik (1969).

Optimization of a frame

One more example computed by Liu (1972) with the use of the externa penalty function method concerned optimization of a frame as shown in Fig 7.17. The optimization aimed at determining the cross-section characteristics

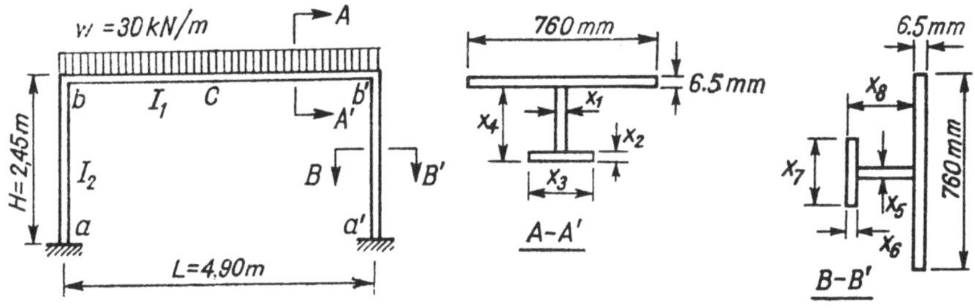

Fig. 7.17. The frame and the member sections

for the bars of the frame so that the weight of the frame is minimum. The cross-sections are assumed to be *T*-shaped, with the thickness and width of the upper flange fixed and identical in all the bars. The design variables are the thicknesses and the depths of the webs and the thicknesses and widths of the lower flanges.

The objective function for the problem can be written

$$W = \gamma V = \gamma(760\times6.5+x_1x_4+x_2x_3)L+2\gamma(760\times6.5+x_5x_8+x_6x_7)H. \tag{7.154}$$

The constraints on the shear stresses are

$$\begin{aligned} g_1 &= \sigma_y-\tau_1 n_1 \geqslant 0,\\ g_2 &= \sigma_y-\tau_2 n_1 \geqslant 0, \end{aligned} \tag{7.155}$$

where

$$\tau_1 = \frac{V_b}{A_{w1}}, \qquad \tau_2 = \frac{H_b}{A_{w2}};$$

σ_y denotes the yield stress and n_1 the safety factor for shear stresses.

The constraints on the normal stresses are

$$\begin{aligned} g_3 &= \sigma_{b1} - \sigma_1 n_1 \geqslant 0, \\ g_4 &= \sigma_{b1} - \sigma_2 n_2 \geqslant 0, \\ g_5 &= \sigma_{b1} - \sigma_3 n_2 \geqslant 0, \\ g_6 &= \sigma_{b1} - \sigma_4 n_2 \geqslant 0, \\ g_7 &= \sigma_{b2} - \sigma_5 n_2 \geqslant 0, \\ g_8 &= \sigma_{b2} - \sigma_6 n_2 \geqslant 0. \end{aligned} \tag{7.156}$$

The normal stresses involved are given by

$$\begin{aligned} \sigma_1 &= \left|\frac{M_{bc}}{Z_{11}} + \sigma_7\right|, & \sigma_5 &= \left|\frac{M_{ba}}{Z_{12}} + \sigma_8\right|, \\ \sigma_2 &= \left|\frac{M_{bc}}{Z_{21}} - \sigma_7\right|, & \sigma_6 &= \left|\frac{M_{ba}}{Z_{22}} - \sigma_8\right|, \\ \sigma_3 &= \left|\frac{M_{cb}}{Z_{11}} - \sigma_7\right|, & \sigma_7 &= \frac{H_b}{A_1}, \\ \sigma_4 &= \left|\frac{M_{cb}}{Z_{21}} + \sigma_7\right|, & \sigma_8 &= \frac{V_b}{A_2}. \end{aligned} \tag{7.157}$$

The bending moments and shear forces occurring in these equations are found from the statical analysis of the structure to be (Fig. 7.18)

$$\begin{aligned} k &= \frac{I_1}{I_2}\frac{H}{L}, & M_{cb} &= \frac{wL^2}{8} + M_{bc}, \\ N_1 &= k+2, & V_a &= V_{a'} = \frac{wL}{2} = V_b = V_{b'}, \\ M_{ab} &= M_{a'b'} = \frac{wL^2}{12N_1}, & H_a &= H_{a'} = \frac{3M_{ab}}{H} = H_b = H_{b'}. \\ M_{bc} &= M_{b'c'} = -\frac{wL^2}{6N_1} = -2M_{ab}, \end{aligned} \tag{7.158}$$

In terms of the design variables $x_1, \dots, x_8$, the relevant section properties can be expressed as (Fig. 7.19)

$$\begin{aligned} A_{w1} &= x_1 x_4, & A_{f2} &= x_6 x_7, \\ A_{f1} &= x_2 x_3, & A_1 &= 760 \times 6.5 + A_{w1} + A_{f1}, \\ A_{w2} &= x_5 x_8, & A_2 &= 760 \times 6.5 + A_{w2} + A_{f2}, \end{aligned} \tag{7.159}$$

$$
\begin{aligned}
I_{xk} &= x_{1+j}^2 760\times 6.5 + x_{1+j}x_{4+j}^3/3 - A_k y_k^2,\\
y_k &= (x_{4+j}760\times 6.5 + x_{4+j}^2 x_{1+j}/2)/A_k,\\
I_{yk} &= (x_{4+j}x_{1+j}^3 + 760^3\times 6.5 + x_{2+j}x_{3+j}^3)/12,\\
I_{tk} &= (x_{4+j}x_{1+j}^3 + 760\times 6.5^3 + x_{3+j}x_{2+j}^3)/3, \qquad (7.159)_2\\
I_{pk} &= I_{xk} + I_{yk},\\
C_{wk} &= \frac{x_{4+j}^2}{12}\,\frac{6.5\times 760^3 x_{2+j}x_{3+j}^3}{6.5\times 760^3 + x_{2+j}x_{3+j}^3},\\
Z_{2k} &= \frac{I_{xk}}{x_{4+j} - y_k},\\
Z_{k1} &= \frac{I_{xk}}{y_k}, \qquad k = 1, 2, \quad j = 0, 4.
\end{aligned}
$$

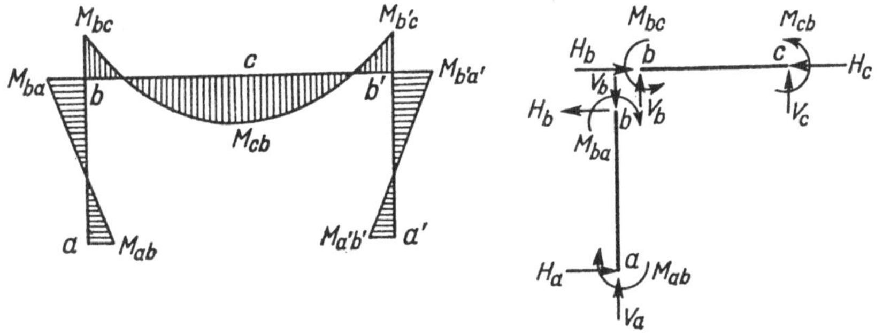

Fig. 7.18. Bending moments and internal forces in the frame

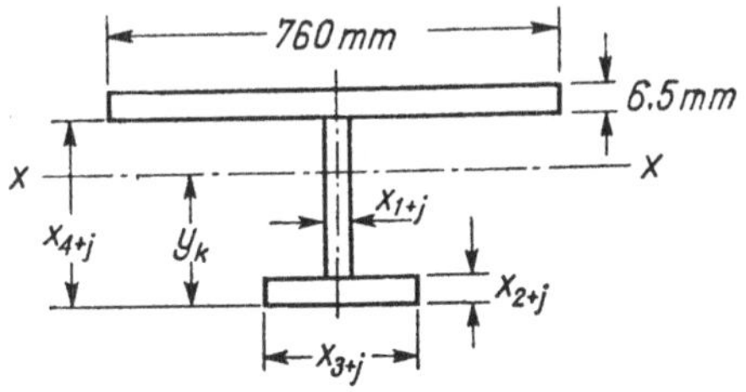

Fig. 7.19. Generalized cross-section: *1*—upper beam, $k = 1$, $j = 0$; 2—columns, $k = 2$, $j = 4$

The buckling criteria are

$$\sigma_{b1} = \min(\sigma_{bx1}, \sigma_{by1}, \sigma_{bt1}, \sigma_{bw1}, \sigma_{bf1}),$$

$$\sigma_{b2} = \min(\sigma_{bx2}, \sigma_{by2}, \sigma_{bt2}, \sigma_{bw2}, \sigma_{bf2}).$$

In the buckling analysis of the frame, the Johnson–Ostenfeld empirical formulae for elastic and plastic buckling stresses are

— global x-wise buckling:

$$\sigma_{bxk} = \frac{\pi^2 E}{\lambda_{xk}^2} \qquad \text{if} \quad \sigma_{bxk} \leqslant \frac{\sigma_y}{2},$$

$$\sigma_{bxk} = \sigma_y\left(1 - \frac{\sigma_y}{4\pi^2 E/\lambda_{xk}^2}\right) \qquad \text{if} \quad \frac{\pi^2 E}{\lambda_{xk}^2} > \frac{\sigma_y}{2},$$

where

$$\lambda_{xk} = \frac{L_{bk}}{i_{xx}}, \quad i_{xk} = \sqrt{\frac{I_{xk}}{A_k}},$$

— global y-wise buckling:

$$\sigma_{byk} = \frac{\pi^2 E}{\lambda_{yk}^2} \qquad \text{if} \quad \frac{\pi^2 E}{\lambda_{yk}^2} \leqslant \frac{\sigma_y}{2},$$

$$\sigma_{byk} = \sigma_y\left(1 - \frac{\sigma_y}{4\pi^2 E/\lambda_{yk}^2}\right) \qquad \text{if} \quad \frac{\pi^2 E}{\lambda_{yk}^2} > \frac{\sigma_y}{2},$$

where

$$\lambda_{yk} = \frac{L_{bk}}{i_{yk}}, \quad i_{yk} = \sqrt{\frac{I_{yk}}{A_k}},$$

— torsion buckling:

$$\sigma_{btk} = \frac{\pi^2 E}{\lambda_{tk}^2} \qquad \text{if} \quad \frac{\pi^2 E}{\lambda_{tk}^2} \leqslant \frac{\sigma_y}{2},$$

$$\sigma_{btk} = \sigma_y\left(1 - \frac{\sigma_y}{4\pi^2 E/\lambda_{tk}^2}\right) \qquad \text{if} \quad \frac{\pi^2 E}{\lambda_{tk}^2} \geqslant \frac{\sigma_y}{2},$$

where

$$\lambda_{tk} = \frac{L_{bk}}{i_{\beta k}}, \quad i_{\beta k} = \sqrt{\frac{C_{wk}}{I_{pk}} + \frac{G L_{bk}^2 I_{tk}}{\pi^2 E I_{pk}}},$$

— local buckling of webs:

$$\sigma_{bwk} = \frac{4\pi^2 E}{12(1-\nu^2)}\left(\frac{t_k}{b_k}\right)^2 \qquad \text{if} \quad \frac{4\pi^2 E}{12(1-\nu^2)}\left(\frac{t_k}{b_k}\right)^2 \leqslant \frac{\sigma_y}{2},$$

$$\sigma_{bwk} = \sigma_y\left[1 - \frac{\sigma_y}{4\,\dfrac{4\pi^2 E}{12(1-\nu^2)}\left(\dfrac{t_k}{b_k}\right)^2}\right] \qquad \text{if} \quad \frac{4\pi^2 E}{12(1-\nu^2)}\left(\frac{t_k}{b_k}\right)^2 > \frac{\sigma_y}{2},$$

where

$$\begin{aligned} t_k &= x_{1+j} \\ b_k &= x_{4+j} \end{aligned} \quad \text{for} \quad \begin{cases} k = 1, & j = 0, \\ k = 2, & j = 4, \end{cases}$$

Table 7.5 Results of optimization of a frame

No.	Quantity	Initial point	Optimal point EPF+ HJ	Optimal point EPF+ PDS	No.	Quantity	Initial point	Optimal point EPF+ HJ	Optimal point EPF+ PDS	No.	Quantity	Initial point	Optimal point EPF+ HJ	Optimal point EPF+ PDS
1	x_1 (mm)	12.7	2.4	3.0	2	x_1(mm)	20.3	4.3	2.9	3	x_1 (mm)	10.2	3.9	3.6
	x_2 (mm)	12.7	17.7	25.0		x_2 (mm)	7.6	7.4	14.1		x_2 (mm)	15.2	16.6	5.5
	x_3 (mm)	127.0	149.2	76.5		x_3 (mm)	152.4	294.0	169.4		x_3 (mm)	152.4	176.5	255.0
	x_4 (mm)	177.8	174.0	230.3		x_4 (mm)	174.8	174.8	207.7		x_4 (mm)	203.2	204.1	258.1
	x_5 (mm)	5.8	3.1	4.5		x_5 (mm)	3.8	3.8	3.8		x_5 (mm)	10.2	9.5	4.6
	x_6 (mm)	12.7	14.1	7.2		x_6 (mm)	12.7	9.4	7.1		x_6 (mm)	15.2	9.5	6.2
	x_7 (mm)	177.8	144.4	177.4		x_7 (mm)	152.4	188.2	214.7		x_7 (mm)	177.8	173.7	192.5
	x_8 (mm)	177.8	157.7	204.6		x_8 (mm)	203.2	164.0	182.7		x_8 (mm)	203.2	137.5	218.9
	V (m^3)		0.074453	0.070645		V (m^3)		0.073171	0.072631		V (m^3)		0.079882	0.069384
	M_{bc} (kNm)		41.77	40.50		M_{bc} (kNm)		42.63	39.70		M_{bc} (kNm)		32.43	40.95
	M_{cb} (kNm)		43.17	44.44		M_{cb} (kNm)		42.30	45.23		M_{cb} (kNm)		52.51	43.99
	V_b (kN)		69.67	69.67		V_b (kN)		69.67	69.67		V_b (kN)		69.67	69.67
	H (kN)		25.70	24.92		H (kN)		26.23	24.43		H (kN)		19.95	25.19
Number of function evaluations			1564	4938				2295	9355				3136	7933

— local buckling of flanges:

$$\sigma_{bfk} = \frac{0.425\pi^2 E}{12(1-\nu^2)}\left(\frac{t_k}{b_k}\right)^2 \qquad \text{if} \quad \frac{0.425\pi^2 E}{12(1-\nu^2)}\left(\frac{t_k}{b_k}\right)^2 \leqslant \frac{\sigma_y}{2},$$

$$\sigma_{bfk} = \sigma_y\left[1 - \frac{\sigma_y}{4\dfrac{0.425\pi^2 E}{12(1-\nu^2)}\left(\dfrac{t_k}{b_k}\right)^2}\right] \qquad \text{if} \quad \frac{0.425\pi^2 E}{12(1-\nu^2)}\left(\frac{t_k}{b_k}\right)^2 > \frac{\sigma_y}{2},$$

where

$$t_k = x_{2+j}, \qquad \begin{cases} k = 1, & j = 0, \\ k = 2, & j = 4, \end{cases}$$
$$b_k = x_{3+j},$$

$k = 1$ for top beam,

$k = 2$ for side columns.

In the above formulae

L_b = buckling length, $L_{b1} = L$, $L_{b2} = 0.7\text{H}$,

G = modulus of elasticity = 80.8 GPa,

ν = Poisson's ratio = 0.3

For computations, the material of the frame was assumed to be ordinary structural steel. The optimum solution was obtained by the *external penalty function method* (EPF) with two different unconstrained minimization techniques used for comparison: *Hooke and Jeeves's method* (HJ) and *Powell's Direct Search* (PDS). The results obtained for three different initial points are presented in Table 7.5.

The comparison of the two methods shows that EPF+HJ is superior to EPF+PDS in computational efficiency.

8

Dynamic Programming

8.1 Introduction

The theory of dynamic programming was developed by an American mathematician, Richard Bellman[1], in the early 1950s. Based on a recursion principle known as the *Bellman principle of optimality*, it has proved very successful in dealing with optimization problems which can be represented as *multistage decision processes*, i.e. processes in which a sequence of decisions has to be made, each decision affecting the state of the system and the final result.

In structural design this kind of problems is often encountered, for example when a structure to be optimized consists of a number of components in a serial interaction scheme, as in the case of a bridge whose spans are carried by supports which in turn rest on foundations. For a truss or a frame, the components may be individual bars or groups of bars.

The dynamic programming technique exploits this sequential feature of a structure to solve the optimization problem in a sequence of steps. In the first step only the last component of the n-component serial structure is considered, in the second the last two are optimized, and so on. The solution of every successive step uses the result of the preceding step. The optimal solution to the original problem is obtained in the nth step, in which all the components are included.

In this way an n-variable design problem can be reduced to a sequence of n single-variable subproblems, which are generally much easier to solve than the original problem.

The optimization technique to be used in different steps depends on the form of the objective function and constraints. It may be a method of differential calculus, linear or non-linear programming, or any other technique.

An important advantage of dynamic programming is that it can handle discrete variables and allows objective functions to take complex forms, not necessarily convex, differentiable or continuous. In such cases, an integer-

[1] Bellman, R. E., *Dynamic Programming*, Princeton University Press, Princeton, New Jersey 1957.

programming or random-search method is used to solve the successive subproblems.

For a comprehensive exposition of the theory of dynamic programming and examples of its application the reader may be referred to many books, among them to those of Bellman and Dreyfus[1], Hadley[2], Nemhauser[3] and Jacobs[4]. Considerable chapters on this subject are also included in the recent books of Aguilar[5] and Rao[6].

8.2 Multi-stage decision process

A typical multi-stage decision process considered in dynamic programming is a series of single-stage processes representable as shown in Fig. 8.1. A single-stage decision process is characterized by *input data*—or *input state vari-*

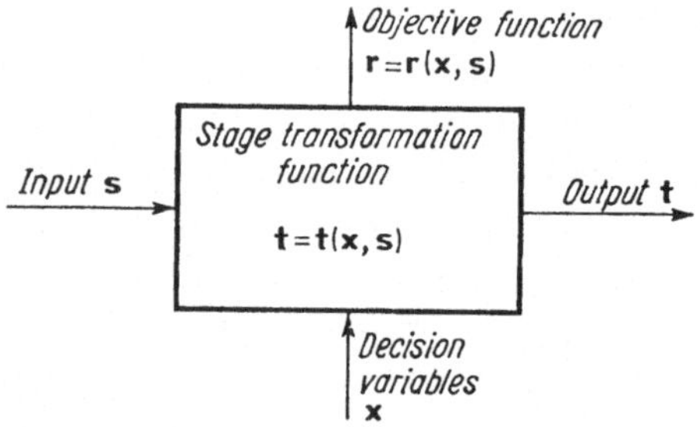

Fig. 8.1. Variables in a single stage decision process

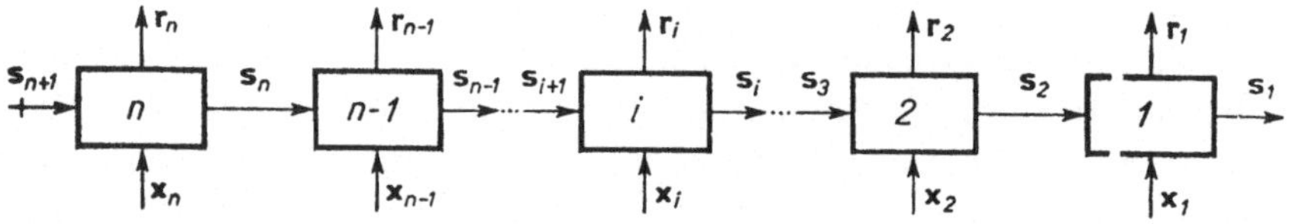

Fig. 8.2. Multi-stage decision process

ables—**s**, *decision* (*design*) *variables* **x**, *output state variables* **t** and a *return* (*objective*) *function* $r = r(\mathbf{x}, \mathbf{s})$ which measures the effectiveness of the decision. The output is related to the input by a *stage transformation function* (design equation) $\mathbf{t} = \mathbf{t}(\mathbf{x}, \mathbf{s})$.

[1] Bellman, R. E., and Dreyfus, S. E., *Applied Dynamic Programming*, Princeton University Press, Princeton, New Jersey 1962.

[2] Hadley, G., *Nonlinear and Dynamic Programming*, Addison-Wesley, Reading Mass. 1964.

[3] Nemhauser, G. L., *Introduction to Dynamic Programming*, Wiley, New York 1966.

[4] Jacobs, O. L. R., *An Introduction to Dynamic Programming*, Chapman and Hall, London 1967.

[5] Aguilar, R. J., *System Analysis and Design in Engineering, Architecture Construction and Planning*, Prentice-Hall, Englewood Cliffs, New Jersey 1973.

[6] Rao, S. S., *Optimization, Theory and Applications*, Wiley Eastern Limited, New Delhi 1978.

A scheme of a sequential multi-stage decision process is shown in Fig. 8.2. The input and the output state vectors in the ith stage are denoted by $\mathbf{s}_{i+1}$ and $\mathbf{s}_i$, respectively (the arrangement being sequential, the output of stage $i+1$ is the input of stage i) and therefore the stage transformation and the return function for the ith stage can be written

$$\mathbf{s}_i = \mathbf{t}_i(\mathbf{x}_i, \mathbf{s}_{i+1}), \tag{8.1}$$

$$r_i = r_i(\mathbf{x}_i, \mathbf{s}_{i+1}), \tag{8.2}$$

where $\mathbf{x}_i$ is the decision vector at stage i.

In an optimization problem associated with this process, an overall objective function may be expressed as a function of the individual returns, viz. $F(r_1, r_2, \ldots, r_n)$, and the problem is to find the decision vectors $\mathbf{x}_1, \ldots, \mathbf{x}_n$ which minimize F and satisfy equations (8.1) and (8.2). The form of the function F with regard to $r_1, \ldots, r_n$ is crucial for the feasibility of the dynamic programming method, whose essential idea is to decompose the original multi-stage problem into a series of single-variable problems. This will indeed be possible if the objective function F is separable with respect to r_i, for example when F takes the form of a sum

$$F = \sum_{i=1}^{n} r_i(\mathbf{x}_i, \mathbf{s}_{i+1}) \tag{8.3}$$

or a product

$$F = \prod_{i=1}^{n} r_i(\mathbf{x}_i, \mathbf{s}_{i+1}). \tag{8.4}$$

With regard to the preassigned data, a multi-stage decision problem may be

- an initial value problem—if the initial state vector $\mathbf{s}_{n+1}$ is given;
- a final value problem—if the final state vector $\mathbf{s}_1$ is given;
- a boundary value problem—if both the initial and the final state vectors are prescribed.

8.3 Bellman's principle of optimality

Different components in a serial structure are not independent of each other and cannot be optimized individually in an arbitrary order. A classical approach to the problem is to consider all the components jointly and to optimize the objective function with respect to all the variables involved. As mentioned in the introduction, dynamic programming solves the problem in a sequence of steps, beginning from an optimization of the last component of the series and then gradually expanding to involve—one by one—all the components (Fig. 8.3). At each step, and this is the pivotal point of the method, only those designs are considered which contain optimal substructures found in the preceding steps. In terms of a decision process, this principle of optimality was originally stated by Bellman as follows:

"An optimal policy (i.e. an optimal set of decisions) has the property that whatever the initial state and initial decision are, the remaining decisions

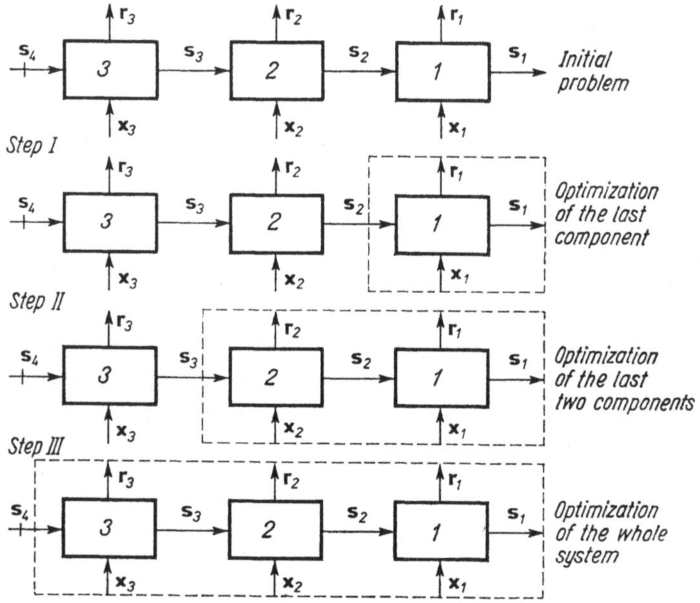

Fig. 8.3. The dynamic programming method

must constitute an optimal policy with regard to the state resulting from the first decision."

To show how this principle governs the process of optimization, let us consider an initial-value multi-stage problem with the objective function of form (8.3). We shall assume for simplicity that the decision (design) variables x_i and state variables s_i are all scalars, although the theory is equally applicable when they are vectors.

We thus want to find $x_1, \ldots, x_n$ which minimize

$$F = r_n(x_n, s_{n+1}) + r_{n-1}(x_{n-1}, s_n) + \ldots + r_1(x_1, s_2), \tag{8.5}$$

where the state and the decision variables are related by

$$s_i = t_i(x_i, s_{i+1}), \quad i = 1, 2, \ldots, n, \tag{8.6}$$

and the initial input s_{n+1} is given.

We start by minimizing the last component with respect to x_1, i.e. by finding

$$f_1^*(s_2) = \min_{x_1} [r_1(x_1, s_2)], \tag{8.7}$$

which is a function of the yet unknown input s_2.

Next, we optimize the last two components jointly, i.e. seek a minimum of the sum r_2+r_1:

$$f_2^*(s_3) = \min_{x_1, x_2} [r_2(x_2, s_3)+r_1(x_1, s_2)]. \tag{8.8}$$

By the principle of optimality, x_1 must be chosen so as to minimize r_1 for a given s_2. Hence, according to equation (8.7), we can write

$$f_2^*(s_3) = \min_{x_2} [r_2(x_2, s_3)+f_1^*(s_2)]. \tag{8.9}$$

Using the equation of state (8.6), we can rewrite (8.9) as

$$f_2^*(s_3) = \min_{x^2} [r_2(x_2, s_3)+f_1^*(t_2(x_2, s_3))] \tag{8.10}$$

from which it is apparent that for a specified input s_3, the second-step optimization is also a single-variable problem.

We continue in a similar manner, involving one by one the successive components and solving a single-variable optimization problem in each step. In the *i*th step we seek the minimum

$$f_i^*(s_{i+1}) = \min_{x_i, x_{i-1}, \ldots, x_1} [r_i(x_i, s_{i+1})+r_{i-1}(x_{i-1}, s_i)+ \ldots +r_1(x_1, s_2)], \tag{8.11}$$

which by the Bellman principle can be written

$$f_i^*(s_{i+1}) = \min_{x_i} [r_i(x_i, s_{i+1})+f_{i-1}^*(s_i)], \tag{8.12}$$

where f_{i-1}^* denotes the optimum found in step $i-1$ and s_i is the input to stage $i-1$. In the last step, the recursion formula (8.12) gives the optimal value $f_n^* = \min F$ and the optimal decision x_n in terms of the initial input s_{n+1}, which is known (in an initial value problem). By retracing the steps down to the last component, we then find the complete set of optimum decisions, x_n^*, $x_{n-1}^*, \ldots, x_1^*$ (see next section).

Let us stress again that, owing to the Bellman principle of optimality, the original problem (8.5)–(8.6), which would normally require a simultaneous variation of n design variables, can be solved in a sequence of n steps, each involving only one decision variable.

8.4 The variant method of dynamic programming

The chief computational problem in implementing the dynamic programming procedure is the construction of the succesive functions $f_i^*(s_{i+1})$. In particularly simple problems f_i^* can be represented analytically, but even a moderate complexity of the return functions r_i soon leads to extremely complex expressions for f_i^*. Besides, the functions r_i may be given in a tabular form or by graphs, in which case an exact analytical treatment is impossible.

A simple alternative is what we call the *variant method.* As we have seen, dynamic programming begins by optimizing the last component, i.e. finding

$$f_1^*(s_2) = \min_{x_1} [r_1(x_1, s_2)]. \tag{8.13}$$

Both f_1^* and the value x_1^* which realizes this minimum depend on the input state variable s_2, which is unknown at this stage. In the variant methods, instead of expressing f_1^* and x_1^* as functions of s_2, we limit s_2 to a set of possible values, or variants, and tabulate or plot $f_1^*(s_2)$ and $x_1^*(s_2)$ for each of them. In the second step, we similarly treat s_3 and tabulate or plot $f_2^*(s_3)$ and $x_2^*(s_3)$, using the results of the first step and the optimality principle (8.9). We proceed in this manner until stage n is reached. Since the initial input s_{n+1} is assumed to be given, we obtain a definite optimal value $f_n^*(s_{n+1}) = \min F$ and a definite $x_n^*(s_{n+1})$.

It remains to return "downstream" and "collect" the optimum decision variables x_i^*. This can be done as follows. Optimization in the nth step has determined x_n^* and f_n^* for the given s_{n+1}. Using the design equation

$$s_n = t_n(s_{n+1}, x_n^*) \tag{8.14}$$

we can compute the input s_n^* to stage $n-1$. The tabulated values of $x_{n-1}(s_n)$ then show the correct variant x_{n-1}^*. We next use the design equation

$$s_{n-1} = t_{n-1}(s_n, x_{n-1}^*) \tag{8.15}$$

to calculate the input s_{n-1}^* to stage $n-2$, and again find the decision variant x_{n-2}^* corresponding to s_{n-1}^* from the tabulated values of $x_{n-2}(s_{n-1})$. This procedure is continued until s_2^* and the corresponding x_1^* are determined. The vector $(x_1^*, x_2^*, \ldots, x_n^*)$ so found constitutes the optimal solution of the problem.

8.5 Applications of dynamic programming to structural optimization

The first examples of application of dynamic programming to structural optimization are due to Kalaba (1962), Goble and DeSantis (1966) and Goff (1966). Further applications were given by Palmer (1968, 1969, 1973), Palmer and Sheppard (1970, 1972), Rejtman (1968), Swiszczowski (1971, 1974), Distefano (1972), Distefano and Ratha (1975) and also by Twisdale and Khachaturian (1973, 1975), Kirch (1975) and Ho (1975). Pochtman and Baranenko (1975) were the first to write a monographic account of applications of dynamic programming to structural mechanics and specifically to optimum design of bar structures, membranes and plates. Raj and Durrant (1976) demonstrated the usefulness of dynamic programming in the very practical problem of optimizing transmission towers that support overhead electric lines. Similar problems had earlier been investigated by Palmer and Sheppard

(1972) and were later considered by Howell and Doyle (1978). Blachut (1977) used dynamic programming to optimize a flexible bar in compression, and Vitiello (1978) determined in this way the optimal number of uni-section bar families is a large structural system. New possibilities of application of dynamic programming were shown by Distefano and Samartin (1975), who used this approach to formulate and solve finite element equations. The works of Distefano (1975) and Distefano and Bellows (1977) on optimization of two-point boundary value systems have opened yet another area of application of the dynamic programming method.

We give below a simple example of a structural application of dynamic programming, in which a seven-bar symmetric truss is optimized for minimum volume. At each step of the procedure an analytic representation of the functions involved is possible and differential calculus gives the required minima.

Example. It is required to determine the cross-sectional areas of the bars of the truss shown in Fig. 8.4, such that the volume of the truss is minimum and the displacement at point A is 5 mm. The load P is given as 250 kN, the Young modulus of the material is $E = 200\ \mathrm{kN/mm^2}$ and the member lengths are $l_1 = l_3 = l_5 = l_7 = 1.0$ m and $l_2 = l_4 = l_6 = 1.2$ m.

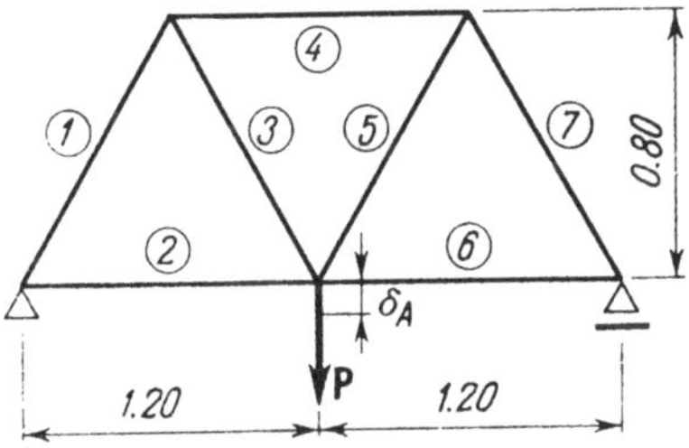

Fig. 8.4. The truss scheme

The forces induced in the bars of the truss by the load P are readily calculated as

$$N_1 = N_7 = -\frac{P}{1.6}, \qquad N_3 = N_5 = \frac{P}{1.6},$$

$$N_2 = N_6 = \frac{3}{8}P, \qquad N_4 = -\frac{3}{4}P.$$

Hence, if x_i ($i = 1, \ldots, 7$) denotes the cross-sectional area of bar i, the deflection at point A is given by

$$\delta_A = \sum_{i=1}^{7} \frac{\partial}{\partial P}\left(\frac{N_i^2 l_i}{E x_i}\right) = 2\frac{488}{x_1} + 2\frac{211}{x_2} + 2\frac{488}{x_3} + \frac{844}{x_4}.$$

The optimization problem can now be formulated as follows: *Find* x_1, x_2, x_3

and x_4 *which minimize the volume* $V(x_1, x_2, x_3, x_4) = 1000\ F(x_1, x_2, x_3, x_4) = 1000(2x_1+2.4x_2+2x_3+1.2x_4)$ *subject to the constraints*

$$\frac{976}{x_1}+\frac{422}{x_2}+\frac{976}{x_3}+\frac{844}{x_4} = 5$$

and

$$x_i \geqslant 0, \quad i = 1, 2, 3, 4.$$

Owing to the form of the objective function F and the constraint equation, in which the successive terms represent the "contributions" from the pairs of symmetrical members 1 and 7,2 and 6, 3 and 5, and from member 4, the problem can be regarded as a multi-stage decision process as shown in Fig. 8.5, where $s_5 = 5$ mm is the total displacement at A available for "dis-

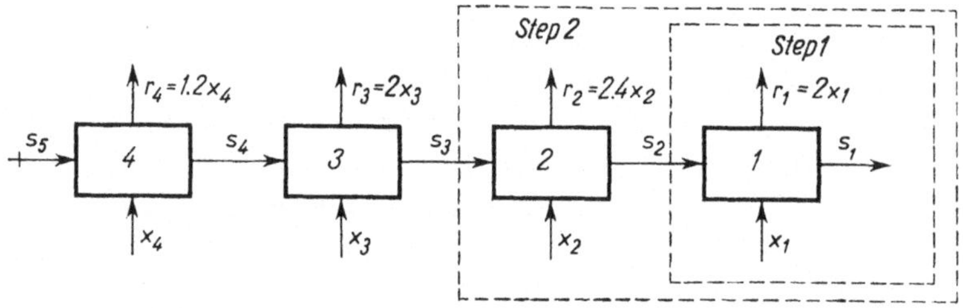

Fig. 8.5. Successive stages of the truss problem

tribution" among all the components, s_4 is the displacement left after allocation at stage 4, etc.

The solution proceeds as follows. In the first step, we consider component 1 (bars 1 and 7) and find the minimum of the return function r_1:

$$f_1^*(s_2) = \min_{x_1 \geqslant 0} [r_1(x_1, s_2)] = \min_{x_1 \geqslant 0} [2x_1(s_2)],$$

where

$$s_2 = \frac{976}{x_1}$$

is the unknown displacement at A to be produced by the displacements of bars 1 and 7. We obtain

$$x_1^* = \frac{976}{s_2} \quad \text{and} \quad f_1^* = \frac{1952}{s_2}.$$

In the second step, we consider components 1 and 2 and seek the minimum of the sum r_1+r_2. According to the principle of optimality, it can be found as

$$f_2^*(s_3) = \min_{x_2 \geqslant 0} [r_2(x_2, s_3)+f_1^*(s_2)] = \min_{x_2 \geqslant 0}\left[2.4x_2(s_3)+\frac{1952}{s_2}\right],$$

where s_3, the unknown displacement at A to come from bars 1, 7 (component 1) and 2 and 6 (component 2), is related to s_2 and x_2 by

$$s_3 = s_2 + \frac{422}{x_2}.$$

Hence we have

$$s_2 = s_3 - \frac{422}{x_2}$$

and so $f_2^*(s_3)$ can be written

$$f_2^*(s_3) = \min_{x_2 \geq 0}\left[2.4x_2 + \frac{1952}{s_3 - 422/x_2}\right].$$

For any fixed s_3, differential calculus gives the minimum point

$$x_2^* = \frac{1008}{s_3}$$

so that

$$f_2^* = \frac{5777}{s_3}.$$

A similar procedure at steps 3 and 4 yields

$$x_3^* = \frac{2655}{s_4}, \quad f_3^* = \frac{14445}{s_4}$$

and

$$x_4^* = \frac{4031}{s_5}, \quad f_4^* = \frac{23107}{s_5}.$$

Since the initial state variable s_5, or the displacement due to all the members, is given as 5 mm, we find

$$x_4^*(5) = 806.2 \text{ mm}^2 \quad \text{and} \quad \min F = f_4^*(5) = 4622 \text{ mm}^3.$$

The optimum values of the remaining design variables are now calculated in turns:

$$s_4 = 5 - \frac{844}{806.2} = 3.95 \text{ mm},$$

$$x_3^* = 672.2 \text{ mm}^2, \quad s_3 = 3.95 - \frac{976}{672.2} = 2.50 \text{ mm},$$

$$x_2^* = 403.2 \text{ mm}^2, \quad s_2 = 2.50 - \frac{422}{403.2} = 1.45 \text{ mm},$$

$$x_1^* = 673.1 \text{ mm}^2.$$

The minimum volume of the truss is $V = 1000 \times \min F = 4622000 \text{ mm}^3$.

8.6 Final to initial value problem conversion

The theory presented in Sections 8.3 and 8.4 dealt with an initial value problem, i.e. one in which the input s_{n+1} to stage n is known. In a final value problem, on the other hand, we are given the final output, s_1. To solve such a problem, we convert it into an equivalent initial value problem by interchanging the troles of inpu and output state variables. This can be done if the design equation of the original problem,

$$s_i = t_i(x_i, s_{i+1}), \quad i = 1, 2, \ldots, n \tag{8.16}$$

has an inverse relation

$$s_{i+1} = \bar{t}_i(x_i, s_i), \quad i = 1, 2, \ldots, n, \tag{8.17}$$

which expresses the input to stage i as a function of its output and the decision variable. Replacing equation (8.16) by (8.17) is called *state inversion.* After inversion, the return function of stage i

$$r_i = r_i(x_i, s_{n+1}) \tag{8.18}$$

takes the form

$$r_i = r_i(x_i, \bar{t}_i(x_i, s_i)) = \bar{r}_i(x_i, s_i). \tag{8.19}$$

The optimization problem can now be formulated as follows:

Find $x_1, x_2, \ldots, x_n$ which minimize

$$F(x_1, \ldots, x_n) = \sum_{i=1}^{n} \bar{r}_i(x_i, s_i), \tag{8.20}$$

where the state and decision variables are related by (8.17).

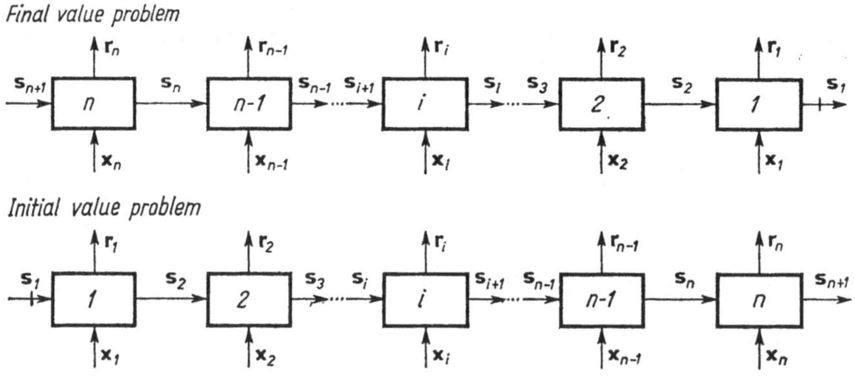

Fig. 8.6. Reversing the flow of information

The use of (8.17) amounts to reversing the flow of information through the state variables (Fig. 8.6). The optimization process can therefore start at stage n and proceed through stages $n-1$, $n-2$, etc., reaching in the last step stage 1, where the known value s_1 will permit the determination of the required optimum. On changing the stage numbers $1, 2, \dots, n$ to $n, n-1, \dots, 1$, respectively, we obtain the standard initial value procedure.

8.7 Continuous dynamic programming

Dynamic programming can also be used with success to extremize functionals. The classical method of approach to such problems is the calculus of variations, but it is often impossible to obtain an analytical solution and then dynamic programming may provide an efficient numerical approximation procedure.

As an illustration of the method, we shall consider a simple example. Suppose that it is required to find a function $y(x)$ which minimizes the integral

$$F = \int_a^b r\left(\frac{\mathrm{d}y}{\mathrm{d}x}, y, x\right)\mathrm{d}x \tag{8.21}$$

subject to the end conditions $y(a) = \alpha$ and $y(b) = \beta$. By applying dynamic programming we can determine the values of $y(x)$ at a finite number of points in the interval (a, b). To this end, let us divide this interval into n segments of length Δx each, using the points

$$x_1 = a, x_2 = a+\Delta x, \dots, x_i = a+(i-1)\Delta x, \dots, x_{n+1} = a+n\Delta x = b.$$

If Δx is small, the derivative $\mathrm{d}y/\mathrm{d}x$ at x_i can be approximated by the difference quotient

$$\frac{\mathrm{d}y}{\mathrm{d}x}(x_i) \approx \frac{y_{i+1}-y_i}{\Delta x}, \tag{8.22}$$

where $y_i = y(x_i)$, $i = 1, 2, \dots, n+1$. Hence the integral (8.21) can be approximated by the sum

$$F \cong \sum_{i=1}^{n} r\left(\frac{y_{i+1}-y_i}{\Delta x}, y_i, x_i\right)\Delta x \tag{8.23}$$

and the optimization problem can be restated as follows:

Find $y_2, \dots, y_n$ which minimize sum (8.23), *where $y_1 = \alpha$ and $y_{n+1} = \beta$.*

The problem can be solved as a final value problem. Let

$$f_i^*(y_i) = \min_{y_{i+1}, \dots, y_n} \left[\sum_{k=1}^{n} r\left(\frac{y_{k+1}-y_k}{\Delta x}, y_k, x_k\right)\Delta x\right]. \tag{8.24}$$

By Bellman's principle of optimality f_i^* can be written as

$$f_i^*(y_i) = \min_{y_{i+1}} \left[r\left(\frac{y_{i+1}-y_i}{\Delta x}, y_i, x_i\right)\Delta x + f_{i+1}^*(y_{i+1})\right]. \tag{8.25}$$

This relation is valid for $i = 1, 2, \ldots, n-1$. For $i = n$, we have

$$f_n^*(y_n) = r\left(\frac{\beta - y_n}{\Delta x}, y_n, x_n\right)\Delta x. \tag{8.26}$$

The desired minimum is given by $f_1^*(y_1) = f_1^*(\alpha)$.

If, for simplicity, y_i are treated as discrete variables, the solution can be obtained by the variant method (Section 8.4). For each i a set of "possible" values of y_i is established and the corresponding values of f_i^* are tabulated. In the last step, $f_1^*(\alpha)$ and y_2^* are found and then the optimal values $y_3^*, \ldots, y_n^*$ can easily be tracked down.

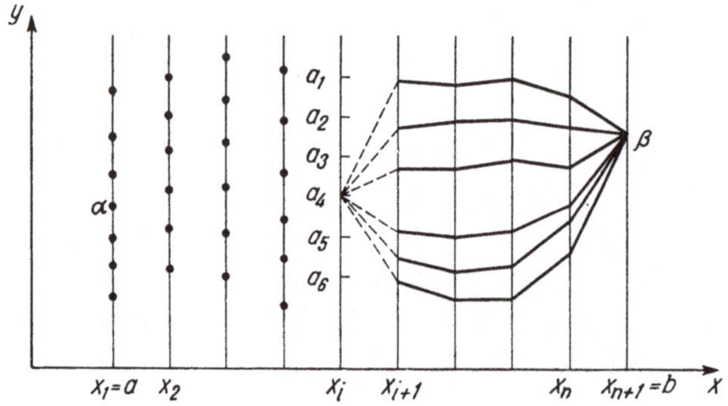

Fig. 8.7. A continuous dynamic programming process

Thus the solution of our continuous decision problem by dynamic programming involves the determination of a whole family of extremal trajectories as we move from b to a. In the last step, we find the particular extremal that passes through both the given points, (a, α) and (b, β). The process is illustrated in Fig. 8.7.

Let us note that a similar procedure can be used to solve problems involving several unknown functions constrained by differential equations.

9

Stochastic Programming

9.1 Linear stochastic programming

A linear stochastic programming problem can be formulated as follows:

Minimize the objective function

$$F(x) = \sum_{j=1}^{n} c_j x_j \tag{9.1}$$

subject to the constraints

$$\begin{aligned} &\sum_{j=1}^{n} a_{ij} x_j \leqslant b_j, \quad i = 1, 2, \ldots, m, \\ &\qquad x_j \geqslant 0, \quad j = 1, 2, \ldots, n \end{aligned} \tag{9.2}$$

where c_j, a_{ij} and b_j are random variables with known probability distributions. For simplicity, the design variables x_j are assumed to be deterministic. We shall also assume that all the random variables involved are statistically independent and normally distributed with known mean values and standard deviations.

There are several methods for solving problem (9.1)–(9.2). Below, basing ourselves on a discussion by Rao[1], we outline a *chance-constrained programming technique* developed by Charnes and Cooper[2].

Chance-constrained programming permits constraints (9.2) to be violated with some small prescribed probabilities $1-p_i$. The objective function (9.1) is thus minimized subject to the modified conditions

$$\begin{aligned} &P\Big[\sum_{j=1}^{n} a_{ij} x_j \leqslant b_i\Big] \geqslant p_i, \quad i = 1, 2, \ldots, m, \\ &\qquad\qquad x_j \geqslant 0, \quad j = 1, 2, \ldots, n. \end{aligned} \tag{9.3}$$

Before considering the general case where c_j, a_{ij} and b_i are all random variables, we first look at the special cases in which only a_{ij} or only b_i or only c_j are random variables.

---

[1] Rao, S. S., *loc. cit.*, p. 288.

[2] Charnes, A., and Cooper, W. W., Chance constrained programming, *Management Science*, **6**, 73–79, 1959.

1 Only a_{ij} are random variables

Let $\bar{a}_{ij}$ and $\sigma_{a_{ij}}$ be the mean value and the standard deviation of the normally distributed random variable a_{ij}. Recall that the variables a_{ij} are assumed to be independent. Let us assume

$$d_i = \sum_{j=1}^{n} a_{ij} x_j, \quad i = 1, 2, \ldots, m. \tag{9.4}$$

Since a_{ij} are normally distributed and x_i are deterministic quantities, d_i are also normally distributed with the mean values

$$\bar{d}_i = \sum_{j=1}^{n} \bar{a}_{ij} x_j, \quad i = 1, 2, \ldots, m, \tag{9.5}$$

and the variances

$$\sigma_{d_i}^2 = \sum_{j=1}^{n} \sigma_{a_{ij}}^2 x_j^2 . \tag{9.6}$$

Constraints (9.3) can now be expressed as

$$P[d_i \leqslant b_i] \geqslant p_i \tag{9.7}$$

or

$$P\left[\frac{d_i - \bar{d}_i}{\sigma_{d_i}} \leqslant \frac{b_i - \bar{d}_i}{\sigma_{d_i}}\right] \geqslant p_i, \quad i = 1, 2, \ldots, m \tag{9.8}$$

where $(d_i - \bar{d}_i)/\sigma_{d_i}$ is a standard normal variable with a zero mean value and a unity variance. Hence the probability that $d_i \leqslant b_i$ is equal to

$$P[d_i \leqslant b_i] = \Phi\left(\frac{b_i - \bar{d}_i}{\sigma_{d_i}}\right), \tag{9.9}$$

where Φ is the cumulative distribution function of the standard normal distribution. If e_i is the value for which

$$\Phi(e_i) = p_i, \tag{9.10}$$

then constraints (9.8) can be written

$$\Phi\left(\frac{b_i - \bar{d}_i}{\sigma_{d_i}}\right) \geqslant \Phi(e_i), \quad i = 1, 2, \ldots, m. \tag{9.11}$$

These inequalities will only be satisfied if

$$\frac{b_i - \bar{d}_i}{\sigma_{d_i}} \geqslant e_i, \tag{9.12}$$

i.e.

$$\bar{d}_i + e_i \sigma_{d_i} - b_i \leqslant 0, \quad i = 1, 2, \ldots, m. \tag{9.13}$$

Substituting equations (9.5) and (9.6) gives the inequalities

$$\sum_{j=1}^{n} \bar{a}_{ij} x_j + e_i \Big(\sum_{j=1}^{n} \sigma_{a_{ij}}^2 x_j^2 \Big)^{1/2} - b_i \leqslant 0, \qquad i = 1, 2, \dots, m, \tag{9.14}$$

which represent non-linear deterministic constraints equivalent to the original linear stochastic constraints. In this way, the original stochastic linear problem has been reduced to that of minimizing the linear objective function (9.1) subject to non-linear deterministic constraints (9.14).

2 Only b_i are random variables

Let $\bar{b}_i$ and $\sigma_{b_i}^2$ denote the mean value and the variance of the normally distributed random variable b_i. Constraints (9.3) can be rewritten as

$$P\Big[\sum_{j=1}^{n} a_{ij} x_j \leqslant b_i\Big] = P\left[\frac{\sum_{j=1}^{n} a_{ij} x_j - \bar{b}_i}{\sigma_{b_i}} \leqslant \frac{b_i - \bar{b}_i}{\sigma_{b_i}}\right] \geqslant p_i,$$
$$i = 1, 2, \dots, m, \tag{9.15}$$

where $(b_i - \bar{b}_i)/\sigma_{b_i}$ is a standard normal variable with a zero mean and a unity variance. Instead of inequality (9.15) we can consider the inequality

$$P\left[\frac{b_i - \bar{b}_i}{\sigma_{b_i}} \leqslant \frac{\sum_{j=1}^{n} a_{ij} x_j - \bar{b}_i}{\sigma_{b_i}}\right] \leqslant 1 - p_i. \tag{9.16}$$

If E_i is the value of the standard variable such that

$$\Phi(E_i) = 1 - p_i, \tag{9.17}$$

constraints (9.16) can be expressed as

$$\Phi\left[\frac{\sum_{j=1}^{n} a_{ij} x_j - \bar{b}_i}{\sigma_{b_i}}\right] \leqslant \Phi(E_i), \qquad i = 1, 2, \dots, m. \tag{9.18}$$

Hence we obtain the condition

$$\frac{\sum_{j=1}^{n} a_{ij} x_j - \bar{b}_i}{\sigma_{b_i}} \leqslant E_i, \qquad i = 1, 2, \dots, m \tag{9.19}$$

or

$$\sum_{j=1}^{n} a_{ij} x_j - \bar{b}_i - E_i \sigma_{b_i} \leqslant 0, \qquad i = 1, 2, \dots, m. \tag{9.20}$$

Thus the original stochastic program (9.1)–(9.2) is equivalent—in the case considered—to a deterministic linear programming problem of minimizing function (9.1) subject to constraints (9.20) and $x_j \geqslant 0$.

3 Only c_j are random variables

The assumption that c_j are normally distributed implies that the objective function $F(x)$ is also a normally distributed random variable. The mean and the variance of F are given by

$$\bar{F} = \sum_{j=1}^{n} \bar{c}_j x_j \tag{9.21}$$

and

$$\sigma_F^2 = \sum_{j=1}^{n} \sigma_{c_j}^2 x_j^2, \tag{9.22}$$

where $\bar{c}_j$ denotes the mean value of c_j.

A new deterministic objective function can be defined as

$$\mathscr{F}(x) = k_1 \bar{F} + k_2 \sigma_F, \tag{9.23}$$

where k_1 and k_2 are non-negative weights indicating the relative importance for minimization of the mean and the standard deviation. Setting $k_2 = 0$ would mean that the expected value of F is to be minimized with no regard to the standard deviation, while the choice $k_1 = 0$ would imply that we are only interested in minimizing the dispersion of F about an arbitrary mean value. The case $k_1 = k_2 = 1$ attaches equal importance to both characteristics of F.

Thus the original stochastic program has been reduced to the non-linear deterministic problem of minimizing the function

$$\mathscr{F}(x) = k_1 \sum_{j=1}^{n} \bar{c}_j x_j + k_2 \Big(\sum_{j=1}^{n} \sigma_{c_j}^2 x_j^2\Big)^{1/2} \tag{9.24}$$

subject to the constraints

$$\begin{aligned} &\sum_{j=1}^{n} a_{ij} x_j - b_i \leqslant 0, \quad i = 1, 2, \ldots, m, \\ &\qquad x_j \geqslant 0, \quad j = 1, 2, \ldots, n. \end{aligned} \tag{9.25}$$

4 General case: c_j, a_{ij} and b_i are all random variables

Since the random variables c_j only occur in the objective function F, a new deterministic objective function can be formed as in (9.24).

The original constraints

$$P\Big[\sum_{j=1}^{n} a_{ij} x_j \leqslant b_i\Big] \geqslant p_i, \quad i = 1, 2, \ldots, m,$$

can be written

$$P[h_i \leqslant 0] \geqslant p_i, \quad i = 1, 2, \ldots, m, \tag{9.26}$$

where h_i are new random variables defined by

$$h_i = \sum_{j=1}^{n} a_{ij}x_j - b_i = \sum_{k=1}^{n+1} q_{ik}y_k, \tag{9.27}$$

where

$$q_{ik} = \begin{cases} a_{ik}, & k = 1, 2, \dots, n, \\ b_i, & k = n+1; \end{cases} \quad \text{and} \quad y_k = \begin{cases} x_k, & k = 1, \dots, n, \\ -1, & k = n+1. \end{cases} \tag{9.28}$$

Being a linear combination of the normally distributed random variables q_{ik}, h_i will also be a normal variable. The mean value of h_i is

$$\bar{h}_i = \sum_{k=1}^{n+1} \bar{q}_{ik}y_k = \sum_{i=1}^{n} \bar{a}_{ij}x_j - \bar{b}_i \tag{9.29}$$

and its variance is

$$\sigma_{h_i}^2 = \sum_{k=1}^{n+1} y_k^2 \sigma_{q_{ik}}^2 = \sum_{k=1}^{n} x_k^2 \sigma_{a_{ik}}^2 + \sigma_{b_i}^2. \tag{9.30}$$

Constraints (9.26) can be rewritten as

$$P\left[\frac{h_i - \bar{h}_i}{\sigma_{h_i}} \leqslant \frac{-\bar{h}_i}{\sigma_{h_i}}\right] \geqslant p_i, \quad i = 1, 2, \dots, m, \tag{9.31}$$

where $(h_i - \bar{h}_i)/\sigma_{h_i}$ is a standard normal variable with a zero mean value and a unity variance. If e_i is a number such that $\Phi(e_i) = p_i$, this inequality can be represented in the form

$$\Phi\left(\frac{-\bar{h}_i}{\sigma_{h_i}}\right) \geqslant \Phi(e_i), \quad i = 1, 2, \dots, m, \tag{9.32}$$

which is equivalent to the non-linear deterministic condition

$$\frac{-\bar{h}_i}{\sigma_{h_i}} \geqslant e_i, \quad i = 1, 2, \dots, m, \tag{9.33}$$

i.e.

$$\bar{h}_i + e_i\sigma_{h_i} \leqslant 0, \quad i = 1, 2, \dots, m. \tag{9.34}$$

Thus the original linear stochastic problem is reduced to an equivalent non-linear deterministic problem of minimizing the function

$$\mathscr{F}(x) = k_1 \sum_{j=1}^{n} \bar{c}_j x_j + k_2 \left(\sum_{j=1}^{n} \sigma_{c_j}^2 x_j^2\right)^{1/2} \tag{9.35}$$

subject to the constraints

$$\begin{aligned} \bar{h}_i + e_i\sigma_{h_i} &\leqslant 0, \quad i = 1, 2, \dots, m, \\ x_j &\geqslant 0, \quad j = 1, 2, \dots, n. \end{aligned} \tag{9.36}$$

9.2 Non-linear stochastic programming

A general non-linear stochastic programming problem can be stated as follows.

Find a vector $\mathbf{x} = (x_1, x_2, \ldots, x_n)$ *which minimizes the objective function* $F(\mathbf{y})$ *subject to the constraints*

$$P[g_j(\mathbf{y}) \geqslant 0] \geqslant p_j, \quad j = 1, 2, \ldots, m, \tag{9.37}$$

where $\mathbf{y}$ is the vector of N random variables $y_1, y_2, \ldots, y_N$, which include the design variables $x_1, \ldots, x_n$ and all other parameters involved in the problem and considered to be random variables. The case where $\mathbf{x}$ is deterministic is a special case of the present formulation.

As before, we shall assume that all the random variables are independent and normally distributed.

The stochastic problem stated above can be converted into an equivalent deterministic problem by a chance-constrained programming technique.

We begin by resolving the objective function into a Taylor series about the expected values of y_i:

$$F(\mathbf{y}) = F(\bar{\mathbf{y}}) + \sum_{i=1}^{N} \left(\frac{\partial F}{\partial y_i}\bigg|_{\bar{\mathbf{y}}} \right)(y_i - \bar{y}_i) + \text{higher order derivative terms.} \tag{9.38}$$

If the standard deviations of y_i, σ_{y_i} are small, $F(\mathbf{y})$ can be approximated by the first two terms of equation (9.38), i.e.

$$F(\mathbf{y}) \cong F(\bar{\mathbf{y}}) - \sum_{i=1}^{N} \left(\frac{\partial F}{\partial y_i}\bigg|_{\bar{\mathbf{y}}} \right)\bar{y}_i + \sum_{i=1}^{N} \left(\frac{\partial F}{\partial y_i}\bigg|_{\bar{\mathbf{y}}} \right) y_i = \Psi(\mathbf{y}). \tag{9.39}$$

Being a linear function of normally distributed variables y_i, $\Psi(\mathbf{y})$ also follows normal distribution. The mean value and the variance of Ψ are given by

$$\bar{\Psi} = \Psi(\bar{\mathbf{y}})$$

and

$$\sigma_\Psi^2 = \sum_{i=1}^{N} \left(\frac{\partial F}{\partial y_i}\bigg|_{\bar{\mathbf{y}}} \right)^2 \sigma_{y_i}^2, \tag{9.40}$$

because all y_i are independent.

As in linear stochastic programming, a new deterministic objective function can be defined as

$$\mathscr{F}(\mathbf{y}) = k_1 \Psi + k_2 \sigma_\Psi, \tag{9.41}$$

where k_1 and k_2 are non-negative weights indicating the relative importance of $\bar{\Psi}$ and σ_Ψ for minimization.

An alternative possibility is to choose the mean value Ψ as the objective function and to introduce the additional constraint $\sigma_\Psi \leqslant k_3 \bar{\Psi}$, where k_3 is a constant.

Constraint inequalities (9.37) can be written

$$\int_0^\infty f_{g_j}(g_j)\,\mathrm{d}g_j \geqslant p_j, \quad j = 1, 2, \ldots, m, \tag{9.42}$$

where $f_{g_j}(g_j)$ is the probability density function of the random variable g_j, whose range is assumed to be $-\infty$ to $+\infty$. To within the first order terms of its Taylor series expansion about the mean $\bar{\mathbf{y}}$, the constraint function $g_j(\bar{\mathbf{y}})$ can be represented as

$$g_j(\mathbf{y}) \cong g_j(\bar{\mathbf{y}}) + \sum_{i=1}^{N} \left(\frac{\partial g_j}{\partial y_i}\bigg|_{\bar{\mathbf{y}}} \right)(y_i - \bar{y}_i). \tag{9.43}$$

Hence, the mean value $\bar{g}_j$ and the standard deviation σ_{g_j} are calculated as

$$\bar{g}_j = g_j(\bar{\mathbf{y}}),$$

$$\sigma_{g_j} = \left[\sum_{i=1}^{n} \left(\frac{\partial g_j}{\partial y_i}\bigg|_{\bar{\mathbf{y}}} \right)^2 \sigma_{y_i}^2 \right]^{1/2}. \tag{9.44}$$

Introducing the standard normal variable

$$\theta = \frac{g_j - \bar{g}_j}{\sigma_{g_j}} \tag{9.45}$$

and using a similar standardization procedure as in the case of linear stochastic programming, it has been obtained

$$\Phi\left(-\frac{\bar{g}_j}{\sigma_{g_j}} \right) \geqslant \Phi(e_j), \quad j = 1, 2, \ldots, m, \tag{9.46}$$

where e_j is the value of the standard normal variable corresponding to probability p_j, i.e. such that

$$\Phi(e_j) = p_j. \tag{9.47}$$

Inequalities (9.46) will be satisfied only if

$$-\bar{g}_j/\sigma_{g_j} \geqslant e_j \tag{9.48}$$

or, by substituting equation (9.44), this condition can be written

$$\bar{g}_j + e_j \left[\sum_{i=1}^{N} \left(\frac{\partial g_j}{\partial y_i}\bigg|_{\bar{\mathbf{y}}} \right)^2 \sigma_{y_i}^2 \right]^{1/2} \leqslant 0, \quad j = 1, 2, \ldots, m \tag{9.49}$$

Thus the original stochastic optimization problem has been transformed into the deterministic problem of minimizing the objective function (9.41) subject to constraints (9.49).

As an illustration of the above stochastic programming method let us consider the following

Example. Design the cross-section of the steel *I*-beam shown in Fig. 9.1, so that the beam may carry the load P and have a minimum volume. All the dimensions of the beam as well as the load and the allowable stresses are assumed to be random variables with normal distributions.

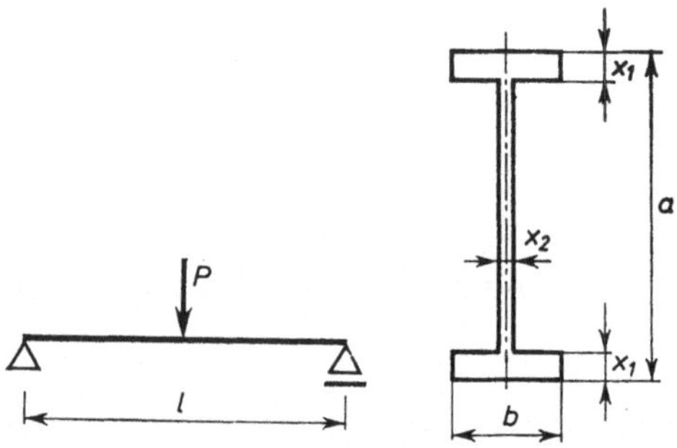

Fig. 9.1. Loading and cross-section of the beam under consideration

The objective function, i.e. the cross-sectional area, is given by

$$A = 2x_1 b + (a - 2x_1)x_2,$$

where a and b are the depth and the width of the beam, and x_1 and x_2 denote the flange and web thicknesses. The design variables are the mean thicknesses $\bar{x}_1$ and $\bar{x}_2$. The standard deviations of x_1 and x_2 are assumed to be related to the mean values by

$$\sigma_{x_1} = \alpha_1 \bar{x}_1 \quad \text{and} \quad \sigma_{x_2} = \alpha_2 \bar{x}_2,$$

where α_1 and α_2 are constants.

A minimum of the function A is sought subject to the following constraints:

— stress constraint at the cross-section edges

$$g_1 = \frac{Ma}{2J} - f \leqslant 0,$$

i.e.

$$\frac{Pla}{8J} - f \leqslant 0,$$

where

$$J = \frac{1}{12}a^3 b - \frac{1}{12}(b - x_2)(a - 2x_1)^3$$

and f is the allowable stress;

— combined stress constraint under the flange

$$g_2 = \left[\frac{M}{J}\left(\frac{a}{2}-x_1\right)\right]^2+3\left(\frac{TS}{Jx_2}\right)^2-f^2 \leqslant 0,$$

i.e.

$$\left[\frac{Pl}{4J}\left(\frac{a}{2}-x_1\right)\right]^2+3\left[\frac{Px_1 b(a-x_1)}{4Jx_2}\right]^2-f^2 \leqslant 0;$$

— side constraints, i.e. limitations on the minimum and maximum thicknesses of the web and the flanges

$$g_3 = -x_2+x_{2\,\min} \leqslant 0,$$
$$g_4 = x_2-x_{2\,\max} \leqslant 0,$$
$$g_5 = -x_1+x_{1\,\min} \leqslant 0,$$
$$g_6 = x_1-x_{1\,\max} \leqslant 0,$$

Each constraint is required to hold with a probability not less than a prescribed probability p_i.

The random variable vector $\mathbf{y}$ has the following components, each assumed to be normally distributed with the mean and standard deviation as indicated:

loading force	$y_1 = (\bar{P}, \sigma_P)$,
allowable stress	$y_2 = (\bar{f}, \sigma_f)$,
beam length	$y_3 = (\bar{l}, \sigma_l)$,
beam depth	$y_4 = (\bar{a}, \sigma_a)$,
beam width	$y_5 = (\bar{b}, \sigma_b)$,
flange thickness	$y_6 = (\bar{x}_1, \sigma_{x_1})$,
web thickness	$y_7 = (\bar{x}_2, \sigma_{x_2})$.

To construct a deterministic objective function (cf. (9.41)), we find the derivatives at $\mathbf{y} = \bar{\mathbf{y}}$ of the function

$$\bar{\Psi} = A(\bar{\mathbf{y}}) = 2\bar{x}_1\bar{b}+(\bar{a}-2\bar{x}_1)\bar{x}_2$$

with respect to the variables y_i:

$$\frac{\partial A}{\partial y_1} = \frac{\partial A}{\partial y_2} = \frac{\partial A}{\partial y_3} = 0,$$

$$\frac{\partial A}{\partial y_4} = \bar{x}_2, \qquad \frac{\partial A}{\partial y_5} = 2\bar{x}_1,$$

$$\frac{\partial A}{\partial y_6} = 2(\bar{b}-\bar{x}_2), \qquad \frac{\partial A}{\partial y_7} = \bar{a}-2\bar{x}_1.$$

Thus the new objective function takes the form

$$\mathscr{A} = k_1\bar{\Psi}+k_2\sigma_\Psi = k_1[2\bar{x}_1\bar{b}+(\bar{a}-2\bar{x}_1)\bar{x}_2]+$$
$$+k_2[\bar{x}_2^2\sigma_a^2+4(\bar{b}-\bar{x}_2)^2(\alpha_1\bar{x}_1)^2+4\bar{x}_1^2\sigma_b^2+(\bar{a}-2\bar{x}_1)^2(\alpha_2\bar{x}_2)^2]^{1/2}.$$

To obtain the deterministic constraints for the function $\mathscr{A}$ we compute the partial derivatives of the constraint functions, viz. $\left.\frac{\partial g_i}{\partial y_i}\right|_{\bar{y}}$ and use formula (9.49), i.e.

$$\bar{g}_j+e_j\left[\sum_{i=1}^{7}\left(\left.\frac{\partial g_j}{\partial y_i}\right|_{\bar{y}}\right)^2\sigma_{y_i}^2\right]^{1/2} \leqslant 0, \quad j = 1,2,\ldots,6.$$

The resulting constraints are:
For $j = 1$:

$$\frac{\bar{P}\bar{l}\bar{a}}{8\bar{J}}-\bar{f}+e_1\frac{\bar{P}\bar{l}}{8\bar{J}}\left\{\left(\frac{\bar{a}}{\bar{P}}\sigma_P\right)^2+\left(\frac{8\bar{J}}{\bar{P}\bar{l}}\sigma_f\right)^2+\left(\frac{\bar{a}}{\bar{l}}\sigma_l\right)^2+\right.$$
$$+\left[\frac{1}{2}(\bar{b}-\bar{x}_2)(\bar{a}-2\bar{x}_1)^2\left(\frac{1}{3}\bar{a}-\bar{x}_1\right)-\frac{1}{6}\bar{a}^3\bar{b}\right]^2\left(\frac{\sigma_a}{\bar{J}}\right)^2+$$
$$+\left[\frac{\bar{a}}{2\bar{J}}(\bar{b}-\bar{x}_2)(\bar{a}-2\bar{x}_1)\alpha_1\bar{x}_1\right]^2+\left(\frac{\bar{a}}{12\bar{J}}\right)^2+[\bar{a}^3-(\bar{a}-2\bar{x}_1)^3]^2\sigma_b^2+$$
$$\left.+\left[\frac{\bar{a}}{12\bar{J}}(\bar{a}-2\bar{x}_1)^3\alpha_2\bar{x}_2\right]^2\right\}^{1/2}\leqslant 0,$$

For $j = 2$:

$$\left[\frac{\bar{P}\bar{l}}{4\bar{J}}\left(\frac{\bar{a}}{2}-\bar{x}_1\right)\right]^2+3\left[\frac{\bar{P}\bar{x}_1\bar{b}(\bar{a}-\bar{x}_1)}{4\bar{J}\bar{x}_2}\right]^2-\bar{f}^2+$$
$$+e_2\left\{\left(\frac{\bar{P}}{8\bar{J}^2}\right)^2\left[\bar{l}^2\left(\frac{\bar{a}}{2}-\bar{x}_1\right)^2+3\frac{\bar{x}_1^2\bar{b}^2(\bar{a}-\bar{x}_1)^2}{\bar{x}_2^2}\right]^2\sigma_P^2+4\bar{f}^2\sigma_f^2+\right.$$
$$+\left[\frac{\bar{P}^2\bar{l}}{8\bar{J}^2}\left(\frac{\bar{a}}{2}-\bar{x}_1\right)^2\right]^2\sigma_l^2+\left(\frac{\bar{P}^2}{8\bar{J}^3}\right)^2\left[\frac{1}{12}\bar{l}^2\left(\frac{\bar{a}}{2}-\bar{x}_1\right)(3\bar{a}^2x_1\bar{b}-\bar{a}^3\bar{b}+\right.$$
$$+(\bar{b}-\bar{x}_2)(\bar{a}-2\bar{x}_1)^3)+\frac{3\bar{x}_1^2\bar{b}^2(\bar{a}-\bar{x}_1)}{\bar{x}_2^2}\left(\frac{1}{4}\bar{a}^2\bar{x}_1\bar{b}-\frac{1}{6}\bar{a}^3\bar{b}+\right.$$
$$\left.\left.+\frac{1}{12}(\bar{b}-\bar{x}_2)(\bar{a}-2\bar{x}_1)^2(2\bar{a}-\bar{x}_1)\right)\right]^2\sigma_a^2+$$
$$+\left(\frac{\bar{P}^2}{8\bar{J}^3}\right)^2\left[-\frac{1}{12}\bar{l}^2\left(\frac{\bar{a}}{2}-\bar{x}_1\right)(\bar{a}^3\bar{b}+2(\bar{b}-\bar{x}_2)(\bar{a}-2\bar{x}_1)^3)+\right.$$

$$+\frac{3\bar{x}_1\bar{b}^2(\bar{a}-\bar{x}_1)(\bar{a}-2\bar{x}_1)}{\bar{x}_2^2}\left(\frac{\bar{a}^3\bar{b}}{12}-\frac{(\bar{b}-\bar{x}_2)(\bar{a}-2\bar{x}_1)^3}{12}-\right.$$

$$\left.\left.-\frac{1}{2}\bar{x}_1(\bar{a}-\bar{x}_1)(\bar{a}-2\bar{x}_1)(\bar{b}-\bar{x}_2)\right)\right](\alpha_1\bar{x}_1)^2+\left(\frac{\bar{P}}{8\bar{J}^3}\right)^2\times$$

$$\times\left[-\frac{1}{12}\bar{l}^2\left(\frac{\bar{a}}{2}-\bar{x}_1\right)(\bar{a}^3-(\bar{a}-2\bar{x}_1)^3)+\frac{\bar{x}_1^2(\bar{a}-\bar{x}_1)^2\bar{b}}{4\bar{x}_2^3}(\bar{a}-2\bar{x}_1)^3\right]\sigma_b^2+$$

$$+\left(\frac{\bar{P}}{8\bar{J}^3}\right)^2\left[-\frac{2}{3}\bar{l}^2\left(\frac{\bar{a}}{2}-\bar{x}_1\right)^5+\right.$$

$$\left.\left.+\frac{\bar{x}_1^2\bar{b}^3(\bar{a}-\bar{x}_1)^2}{4\bar{x}_2^3}(\bar{a}^3-(\bar{a}-2\bar{x}_1)^3)\right)\right](\alpha_2\bar{x}_2)^2\right\}\leqslant 0.$$

For $j = 3$:

$$g_3 = -\bar{x}_2+x_{2\,\min}+e_3\alpha_2\bar{x}_2 \leqslant 0.$$

For $j = 4$:

$$q_4 = \bar{x}_2-x_{2\,\max}+e_4\alpha_2\bar{x}_2 \leqslant 0.$$

For $j = 5$:

$$g_5 = -\bar{x}_1+x_{1\,\min}+e_5\alpha_1\bar{x}_1 \leqslant 0.$$

For $j = 6$:

$$g_6 = \bar{x}_1-x_{1\,\max}+e_6\alpha_1\bar{x}_1 \leqslant 0.$$

The deterministic problem so formulated has been solved with the following numerical data:

$\bar{P} = 4000$ kN, $\sigma_P = 40$ kN, $\bar{f} = 200$ MPa, $\sigma_f = 20$ MPa,
$\bar{l} = 5$ m, $\sigma_l = 0.05$ m, $\bar{a} = 0.60$ *m*, $\sigma_a = 0.006$ m, $\bar{b} = 0.20$ m,
$\sigma_b = 0.002$ m, $\alpha_1 = \alpha_2 = 0.05$, and $p_j = 0.95$ $(j = 1, \ldots, 6)$

with the corresponding value (from normal distribution tables) $e_j = 1.645$.

The presence of only two design variables has allowed a graphical solution. The feasible region and objective contours are shown in Figs 9.2–9.4 for three cases:

$k_1 = 1$, $k_2 = 0$ (Fig. 9.2),
$k_1 = 0$, $k_2 = 1$ (Fig. 9.3),
$k_1 = 1$, $k_2 = 1$ (Fig. 9.4).

In each case, the minimum of the objective function is realized at the intersection of the g_1 and g_3 constraints, where the mean web thickness $\bar{x}_2$ is at an allowable minimum of 10.90 mm and the flange thickness is chosen for full stress, at $\bar{x}_1 = 23.84$ mm. The values of the objective function $\mathscr{A}$ in the three cases considered are 0.01556 m², 0.00054 m² and 0.01610 m².

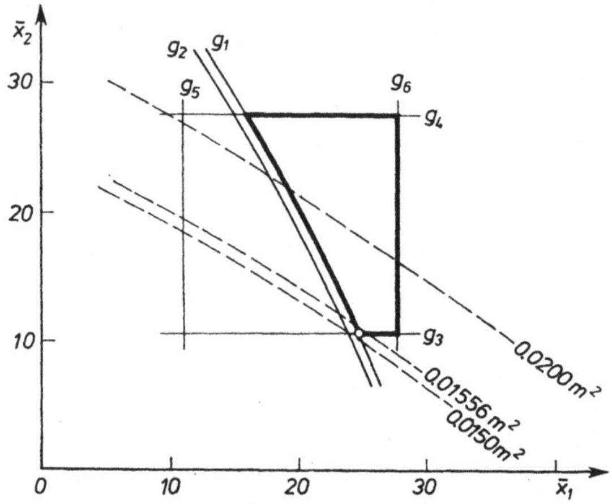

Fig. 9.2. The feasible region and objective contours for $k_1 = 1$ and k_2 =

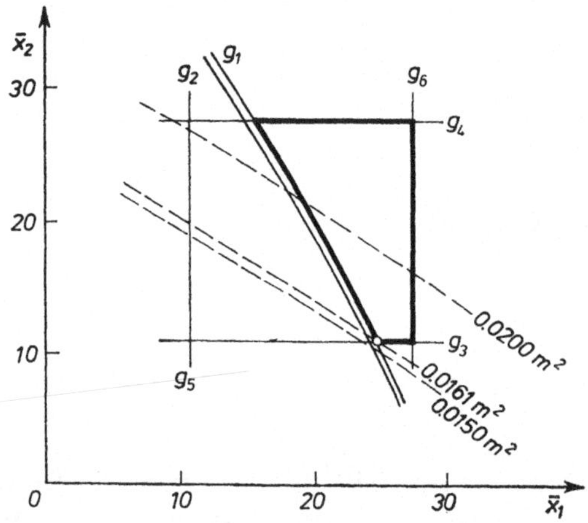

Fig. 9.3. The feasible region and objective contours for $k_1 = 0$ and $k_2 = 1$

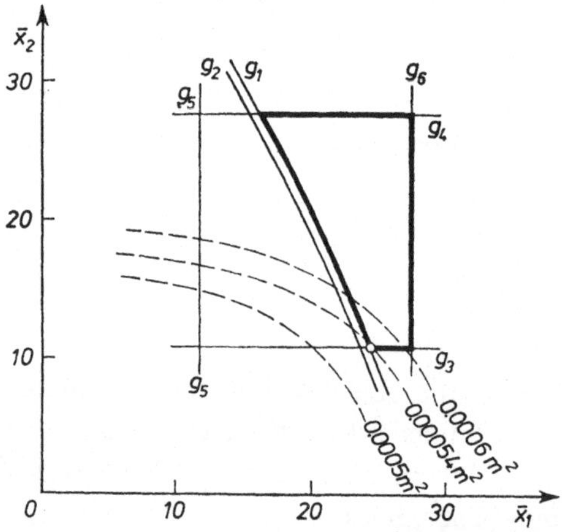

Fig. 9.4. The feasible region and objective contours for $k_1 = 1$ and $k_2 = 1$

If all the quantities involved in the problem are treated as deterministic, the optimum solution is found at the intersection of the same two constraints, with $x_1 = 17.7$ mm, $x_2 = 10$ mm and the minimum cross-sectional area equal to 0.0127 m^2.

9.3 Reliability-based structural optimization: a stochastic programming approach

As discussed in Section 4.6, the basic reliability-based approach to an optimum-structural-design problem is to replace the numerous behaviour constraints by a single constraint on the overall probability of failure of the structure. In the present section, we consider, following Davidson, Felton and Hart[1], a formulation which includes constraints on individual failure modes as well.

The minimum-weight optimization problem for a structure with random parameters may be stated as follows:

Minimize the objective function

$$F(\mathbf{x}) \tag{9.50}$$

subject to the constraints

$$P_0(\mathbf{x}) \leqslant p_0, \tag{9.51}$$

$$P_i(\mathbf{x}) = P[g_i(\mathbf{x}) > G_i(\mathbf{x})] \leqslant p_i, \quad i = 1, \ldots, M, \tag{9.52}$$

where $\mathbf{x}^T = (x_1, x_2, \ldots, x_n, \ldots, x_N)$ is a vector of N random variables the first n of which are the design variables; $P_0(\mathbf{x})$ denotes the overall probability of failure and p_0 is the allowable limit for this quantity; $g_i(\mathbf{x})$ is the ith response quantity of the structure (e.g. displacement or stress) and $G_i(\mathbf{x})$ its allowable limit. Note that, in many cases, the allowable response G_i may be independent of $\mathbf{x}$. The expression $P[\ldots] \leqslant p_i$ in equation (9.52) means that the probability $P_i(x)$ of the ith response being greater than the allowable limit must not exceed a specified failure probability p_i. The probabilities p_0 and p_i will generally have to be very small for constraints (9.51) and (9.52) to be active and sensitive to the particular distributions of the individual variables.

Assuming all the random variables to be normally distributed, we may relate p_i to response quantities by

$$p_i = 1 - \Phi(e_i) \tag{9.53}$$

where $\Phi(e_i)$ is the cumulative normal distribution function for which

$$e_i = \frac{|\bar{G}_i(\mathbf{x}) - \bar{g}_i(\mathbf{x})|}{[\sigma^2_{G_i}(\mathbf{x}) + \sigma^2_{g_i}(\mathbf{x})]^{1/2}} \tag{9.54}$$

[1] Davidson, J. W., Felton, L. P., and Hart, G. C., Optimum design of structures with random parameters, *Computers and Structures*, **7**, 3, 481–486, 1977.

is the coupling equation between response and allowable response. For a given p_i, e_i may be obtained from Φ-tables and the original constraint (9.52) replaced by the equivalent condition

$$\bar{g}_i(\mathbf{x})+e_i[\sigma^2_{\bar{G}_i}(\mathbf{x})+\sigma^2_{\bar{g}_i}(\mathbf{x})]^{1/2}-G_i(\mathbf{x}) \leqslant 0 \tag{9.55}$$

where e_i is required to be positive.

An important factor in the reliability analysis of a structure is the relationship between the overall probability of failure and the probabilities of the individual failure modes. For multiply-loaded hyperstatic structures which are assumed to fail when any constraint is violated ("weakest-link" model), failure modes are usually neither completely statistically dependent nor completely statistically independent and exact correlations are difficult to determine. However, these two extreme cases provide a lower and an upper bound on the overall reliabity:

$$\max P_i(\mathbf{x}) \leqslant P_0(\mathbf{x}) \leqslant \sum_{i=1}^{M} P_i(\mathbf{x}). \tag{9.56}$$

If the design involves relatively few active failure modes, which is often the case, the upper bound can reasonably be used for actual evaluation of P_0:

$$P_0(\mathbf{x}) = \sum_{i=1}^{M} P_i(\mathbf{x}). \tag{9.57}$$

In this relation, $P_i(\mathbf{x})$ is obtained from a modified form of (9.53),

$$P_i(\mathbf{x}) = 1-\Phi(\bar{e}_i), \tag{9.58}$$

where $\bar{e}_i$ is the value of e_i satisfying relation (9.55) as an equality.

The above considerations allow the original optimization problem to be restated as:

Minimize

$$F(x) \tag{9.59}$$

subject to

$$\sum_{i=1}^{M} P_i(\mathbf{x})-p_0 \leqslant 0, \tag{9.60}$$

$$\bar{g}_i(x)+e_i[\sigma^2_{\bar{G}_i}(\mathbf{x})+\sigma^2_{\bar{g}_i}(\mathbf{x})]^{1/2}-\bar{G}_i(\mathbf{x}) \leqslant 0, \quad i = 1, \ldots, M. \tag{9.61}$$

In order to find appropriate approximate expressions for the standard deviations of the response quantities, we replace the responses by the first order terms of their Taylor series expansions about means:

$$g_i(\mathbf{x}) = \bar{g}_i(\mathbf{x})+\sum_{j=1}^{N} \left.\frac{\partial g_i}{\partial x_j}\right|_{\bar{\mathbf{x}}}(x_j-\bar{x}_j). \tag{9.62}$$

The variance of the ith response is defined by

$$\sigma^2_{g_i}(\mathbf{x}) = E[g_i(\mathbf{x}) - \bar{g}_i(\mathbf{x})]^2 \tag{9.63}$$

where $E[\ldots]$ denotes the expected value. Substituting $g_i(\mathbf{x})$ from (9.62), we get

$$\sigma^2_{g_i}(\mathbf{x}) = E\left[\sum_{j=1}^{N}\sum_{k=1}^{N} \frac{\partial g_i}{\partial x_j}\bigg|_{\bar{\mathbf{x}}} \frac{\partial g_i}{\partial x_k}\bigg|_{\bar{\mathbf{x}}} (x_j - \bar{x}_j)(x_k - \bar{x}_k)\right], \tag{9.64}$$

i.e.

$$\sigma^2_{g_i}(\mathbf{x}) = \sum_{i=1}^{N}\sum_{k=1}^{N} \frac{\partial g_i}{\partial x_j}\bigg|_{\bar{\mathbf{x}}} \frac{\partial g_i}{\partial x_k}\bigg|_{\bar{\mathbf{x}}} E[(x_j - \bar{x}_j)(x_k - \bar{x}_k)]. \tag{9.65}$$

The expected-value term in this equation is by definition the covariance of x_j and x_k:

$$E[(x_j - \bar{x}_j)(x_k - \bar{x}_k)] = \mathrm{cov}(x_j, x_k). \tag{9.66}$$

For numerical purposes, the covariance is conveniently represented in terms of the correlation coefficient ϱ_{jk}, $-1 \leqslant \varrho_{jk} \leqslant 1$, as

$$\mathrm{cov}(x_j, x_k) = \varrho_{jk}\sigma_j\sigma_k, \tag{9.67}$$

where σ_j and σ_k are standard deviations of the random variables x_j and x_k, respectively. A value $\varrho_{jk} = \pm 1$ implies complete statistical dependence for x_j and x_k while $\varrho_{jk} = 0$ implies statistical independence.

Furthermore, the standard deviations σ_j and σ_k can be related to the mean values $\bar{x}_j$ and $\bar{x}_k$ through the coefficients of variation, γ_j and γ_k, defined by

$$\gamma_l = \frac{\sigma_l}{\bar{x}_l}, \quad l = 1, 2, \ldots, N. \tag{9.68}$$

Substituting equations (9.66)–(9.68) into (9.65) gives

$$\sigma_{g_i}(\mathbf{x}) = \left[\sum_{i=1}^{N}\sum_{k=1}^{N} \frac{\partial g_i}{\partial x_j}\bigg|_{\bar{\mathbf{x}}} \frac{\partial g_i}{\partial x_k}\bigg|_{\bar{\mathbf{x}}} \varrho_{jk}\gamma_j\gamma_k\bar{x}_j\bar{x}_k\right]^{1/2}. \tag{9.69}$$

In this way, the standard deviations of the responses have been expressed in terms of the mean values of the random variables, the correlation coefficients and the coefficients of variation. The values of ϱ_{jk}, γ_j and γ_k must be specified before beginning the design process.

10

Classical Variational Methods. Examples of Application to Optimum Structural Design

10.1 Introductory remarks and fundamental concepts

The classical methods of the calculus of variations allow determination of the extrema of certain types of functionals of one or more function variables. The extrema can be sought without any side conditions but can also be required to satisfy some functional or isoperimetric constraints. Such variational problems are very often obtained in the optimization of engineering structures, particularly in optimization for minimum volume (or cost) or minimum potential.

Excellent expositions of the classical calculus of variations are available in many textbooks.[1] In the present chapter we recall some basic notions, formulations and relations which are useful in structural optimization, and give several examples of application of variational methods to the optimum design of columns, beams and plates.

Basic notions

In the simplest case, by a *functional* we mean a corespondence which assigns a real number to each function belonging to some class. In more complicated cases, a functional may depend on a number of function variables. The elastic strain energy of an elastic body and the volume of a structure are important examples of functionals dealt with in structural optimization.

More generally, a functional is a real-valued function defined on a subset of a vector space, usually a normed vector space. Let us recall that a vector space E is said to be a *normed vector space* if to each x in E there corresponds a non-negative number $||x||$, called the *norm* of x, such that

[1] See e.g. Fox, C., *An Introduction to the Calculus of Variations*, Oxford University Press, London 1950; Gelfand, I. M., and Fomin, S. V., *Calculus of Variations*, trans. by Silverman, R. A., Prentice-Hall, Englewood Cliffs, N. J. 1963; Elsgolc, L. E., *Calculus of Variations*, Pergamon Press, New York–Oxford–London–Paris 1961.

(1) $||x|| = 0$ if and only if $x = 0$,

(2) $||\alpha x|| = |\alpha| \, ||x||$ for every $\alpha \in R^1$,

(3) $||x+y|| \leqslant ||x||+||y||$.

The set C of all continuous real-valued functions $y(x)$ defined on an interval $[a, b]$ is a vector space with ordinary addition of functions and multiplication by numbers. It can be made into a normed vector space with norm defined by

$$||y|| = \max_{a \leqslant x \leqslant b} |y(x)|. \tag{10.1}$$

In this norm, the distance between two functions $y(x)$ and $y_0(x)$, measured as $||y-y_0||$, is less than ε if the curve $y = y(x)$ lies entirely within the $2\,\varepsilon$-strip about the graph of $y_0(x)$.

Another important example of a normed vector space is the set D_1 of all functions continuously differentiable on an interval $[a, b]$, with addition and multiplication by numbers defined as before and norm given by

$$||y||_1 = \max_{a \leqslant x \leqslant b} |y(x)| + \max_{a \leqslant x \leqslant b} |y'(x)|. \tag{10.2}$$

Thus, two functions are close in the space D_1 whenever the functions themselves and their first derivatives are uniformly close on $[a, b]$.

Let E be a normed vector space and V a functional defined in E. V is *continuous* at a point $y_0 \in E$ if for every $\varepsilon > 0$ there is a $\delta > 0$ such that $|V(y)-V(y_0)| < \varepsilon$ whenever $||y-y_0|| < \delta$.

A functional V is said to be *linear* if it is continuous and satisfies the condition $V(y_1+y_2) = V(y_1)+V(y_2)$ for arbitrary $y_1, y_2 \in E$.

For instance, the functional

$$V(y) = \int_a^b h(x)y(x)\mathrm{d}x, \tag{10.3}$$

where $h(x)$ is a given continuous function on $[a, b]$, is a linear functional in C and in D_1.

Functional (10.3) plays a central role in the following Lagrange lemma which is essential for the derivation of Euler's equations in the classical calculus of variations: If $h(x)$ is continuous on the interval $[a, b]$ and

$$\int_a^b h(x)y(x)\mathrm{d}x = 0 \tag{10.4}$$

for any continuously differentiable function $y(x)$ which vanishes at a and b, i.e. $y(a) = y(b) = 0$, then $h(x) \equiv 0$.

Let $V(y)$ be a functional defined on a normed vector space E and let

$$\Delta V(\eta) = V(y+\eta)-V(y) \tag{10.5}$$

be the increment of V corresponding to an increment η of y; if y is fixed ΔV is a functional of η. If ΔV can be represented in the form

$$\Delta V(\eta) = L(\eta)+\varepsilon\|\eta\|, \tag{10.6}$$

where $L(\eta)$ is a linear functional and $\varepsilon \to 0$ as $\|\eta\| \to 0$, the functional $V(y)$ is said to be *differentiable* at y and the principal linear part of the increment, i.e. $L(\eta)$, is called the *variation* (or *differential*) of V at y and is denoted by δV.

A functional $V(y, z)$, where y and z belong to a normed vector space E, is called *bilinear* if it is linear in either of its arguments. Setting $y = z$ in a bilinear functional, we obtain what is termed a quadratic functional. For example, the expression $\int_a^b A(x)y(x)z(x)\mathrm{d}x$, where A is a given function, defines a bilinear functional in C, and so the functional $\int_a^b A(x)y^2(x)\mathrm{d}x$ is quadratic in C.

By analogy to the first variation, we can now define the second variation of a functional:

If the increment ΔV of a functional $V(y)$ can be represented in the form

$$\Delta V = L(\eta)+Q(\eta)+\varepsilon\|\eta\|^2, \tag{10.7}$$

where $L(\eta)$ is a linear functional, $Q(\eta)$ a quadratic functional, and $\varepsilon \to 0$ as $\|\eta\| \to 0$, then V is said to be *twice differentiable* and the quadratic part of its increment, i.e. $Q(\eta)$, is called the *second variation* (*second differential*) of V and is denoted by $\delta^2 V$.

As an example, consider in D_1 the functional

$$V(y) = \int_a^b F(x, y, y')\mathrm{d}x, \quad y(a) = A, \quad y(b) = B, \tag{10.8}$$

where the function F is twice continuously differentiable with respect to all its arguments. By using Taylor's formula, we can represent the increment of V as

$$\Delta V = V(y+\eta)-V(y) = \int_a^b \left(\frac{\partial F}{\partial y}\eta+\frac{\partial F}{\partial y'}\eta'\right)\mathrm{d}x+$$

$$+\frac{1}{2}\int_a^b \left(\frac{\partial^2 F}{\partial y^2}\eta^2+2\frac{\partial^2 F}{\partial y\,\partial y'}\eta\eta'+\frac{\partial^2 F}{\partial y'^2}\eta'^2\right)\mathrm{d}x+\theta, \tag{10.9}$$

where θ is an infinitesimal of order higher than $||\eta||^2$. Hence the first variation of V, i.e. the linear part of ΔV, is given by

$$\delta V = \int_a^b \left(\frac{\partial F}{\partial y}\eta + \frac{\partial F}{\partial y'}\eta'\right)\mathrm{d}x = \int_a^b \left[\frac{\partial F}{\partial y} - \frac{\mathrm{d}}{\mathrm{d}x}\left(\frac{\partial F}{\partial y'}\right)\right]\eta(x)\mathrm{d}x \tag{10.10}$$

and the second variation of V, quadratic in η, is

$$\delta^2 V = \frac{1}{2}\int_a^b \left(\frac{\partial^2 F}{\partial y^2}\eta^2 + 2\frac{\partial^2 F}{\partial y\,\partial y'}\eta\eta' + \frac{\partial^2 F}{\partial y'^2}\eta'^2\right)\mathrm{d}x$$

$$= \frac{1}{2}\int_a^b \left\{\left[\frac{\partial^2 F}{\partial y^2} - \frac{\mathrm{d}}{\mathrm{d}x}\left(\frac{\partial^2 F}{\partial y\,\partial y'}\right)\right]\eta^2(x) + \frac{\partial^2 F}{\partial y'^2}\eta'^2(x)\right\}\mathrm{d}x. \tag{10.11}$$

A functional $V(y)$ defined on a normed vector space E is said to have a *local extremum* at $y = \hat{y}$ if there is a neighbourhood of $\hat{y}$ in which the difference $V(y) - V(\hat{y})$ does not change its sign.

In the classical calculus of variations, functionals are always considered in some sets of smooth, at least continuously differentiable functions. Such functions can be regarded as elements of different normed spaces, for example of the spaces C and D_1. Traditionally, a local extremum of a functional $V(y)$ in the space D_1 is called a *weak extremum*, and an extremum of the same functional in the uniform norm of the space C a *strong extremum*. More precisely, the functional $V(y)$ has a *weak local minimum (maximum)* at $\hat{y}$ if there exists a number $\varepsilon > 0$ such that the inequality $V(y) - V(\hat{y}) \geqslant 0$ ($\leqslant 0$) holds for all admissible y such that $||y-\hat{y}||_1 \leqslant \varepsilon$. V is said to have a *strong minimum (maximum)* at $\hat{y}$ if the same inequality holds for all admissible y such that $||y-\hat{y}|| \leqslant \varepsilon$. Every strong extremum is a weak extremum; the converse statement is not true in general.

The following theorem is fundamental for the theory of extremum problems:

A necessary condition for a differentiable functional $V(y)$ to have an extremum at $y = \hat{y}$ is that its variation at $\hat{y}$ vanishes, i.e.

$$\delta V(\eta) = 0 \tag{10.12}$$

for $y = \hat{y}$ and all admissible η.

Let us also note a necessary condition involving the second variation: *If $\hat{y}$ is a local minimum (maximum) vector for a functional V, then*

$$\delta^2 V(\eta) \geqslant 0 \quad (\leqslant 0) \tag{10.13}$$

for $y = \hat{y}$ and all admissible η.

The necessary conditions for an extremum in the various problems of the classical calculus of variations are summarized in the subsequent sections

10.2 Necessary conditions for an extremum of a functional

The fixed end point problem. Extremum of a single integral

Let F be a function of $mn+m+1$ variables, of class C^2 with respect to each of those variables, and consider a functional of the form

$$V(y_1, \ldots, y_m) = \int_a^b F(x, y_1, \ldots, y_m, y_1', \ldots, y_m', \ldots, y_1^{(n)}, \ldots, y_m^{(n)})\mathrm{d}x, \quad (10.14)$$

which depends on m functions of x of class C^n satisfying the boundary conditions

$$y_i^{(k)}(a) = A_{ik}, \quad y_i^{(k)}(b) = B_{ik}, \quad i = 1, 2, \ldots, m, \quad k = 0, 1, \ldots, n-1. \quad (10.15)$$

A necessary condition for the functional $V(y_1, \ldots, y_m)$ to have an extremum for $y_i = \hat{y}_i(x)$, $i = 1, \ldots, m$, is that the functions $\hat{y}_i$ satisfy the following Euler equations:

$$\frac{\partial F}{\partial y_i} - \frac{\mathrm{d}}{\mathrm{d}x}\left(\frac{\partial F}{\partial y_i'}\right) + \frac{\mathrm{d}^2}{\mathrm{d}x^2}\left(\frac{\partial F}{\partial y_i''}\right) + \ldots + (-1)^n \frac{\mathrm{d}^n}{\mathrm{d}x^n}\left(\frac{\partial F}{\partial y_i^{(n)}}\right) = 0. \quad (10.16)$$

Each of these equations is a differential equation of order $2n$ and so its general solution contains $2n$ arbitrary constants. To determine them, we use the boundary conditions (10.15).

When the functional (10.14) depends on only one function $y(x)$ and its first-order derivative, the Euler equation takes the form

$$\frac{\partial F}{\partial y} - \frac{\mathrm{d}}{\mathrm{d}x}\left(\frac{\partial F}{\partial y'}\right) = \frac{\partial F}{\partial y} - \frac{\partial^2 F}{\partial y'^2} y'' - \frac{\partial^2 F}{\partial y \partial y'} y' - \frac{\partial^2 F}{\partial y' \partial x} = 0. \quad (10.17)$$

In particular, we can distinguish the following special cases:

$$\begin{array}{ll} \text{Integrand} & \text{Euler's equation} \\ F(y, y'), & F - \dfrac{\partial F}{\partial y'} y' = C_1, \\ F(x, y'), & \dfrac{\partial F}{\partial y'} = C_1, \\ F(x, y), & \dfrac{\partial F}{\partial y} = 0, \\ f(x, y)\sqrt{1+y'^2}, & \dfrac{\partial f}{\partial y} - \dfrac{\partial f}{\partial x} y' - f\dfrac{y''}{1+y'^2} = 0. \end{array} \quad (10.18)$$

The fixed boundary problem. Extremum of a double integral

Consider a functional of the form

$$V(u) = \iint_B F(x, y, u, u_x, u_y)\mathrm{d}x\mathrm{d}y, \quad (10.19)$$

where F is of class C^2 with respect to all its arguments, B is some closed region in the plane x, y and u_x and u_y denote the partial derivatives of $u = u(x, y)$. The function u is assumed to be of class C^2 in B and to take given values on the boundary $\mathscr{C}$ of B.

A necessary condition for the functional $V(u)$ to attain an extremum for $u = \hat{u}(x, y)$ is that $\hat{u}$ satisfies the following Euler–Ostrogradski equation:

$$F_u - \frac{\partial}{\partial x} F_{u_x} - \frac{\partial}{\partial y} F_{u_y} = 0. \tag{10.20}$$

After performing the differentiation, we obtain

$$F_{u_x u_x} u_{xx} + 2F_{u_x u_y} u_{xy} + F_{u_y u_y} u_{yy} + F_{u_x u} u_x + F_{u_y u} u_y + F_{x u_x} + F_{y u_y} - F_u = 0. \tag{10.21}$$

In the more general case where the integrand function F depends on the partial derivatives of $u(x, y)$ to order n, the Euler–Ostrogradski equation becomes

$$F_u - \frac{\partial}{\partial x} F_{u_x} - \frac{\partial}{\partial y} F_{u_y} + \frac{\partial^2}{\partial x^2} F_{u_{xx}} + 2 \frac{\partial^2}{\partial x \partial y} F_{u_{xy}} + $$
$$+ \frac{\partial^2}{\partial y^2} F_{u_{yy}} - \dots + (-1)^n \frac{\partial^n}{\partial y^n} F_{u_{yy\dots y}} = 0. \tag{10.22}$$

The variable end point problem

In the fixed end point problem an extremal is sought among curves which connect two given fixed points. We now suppose that one or both end points

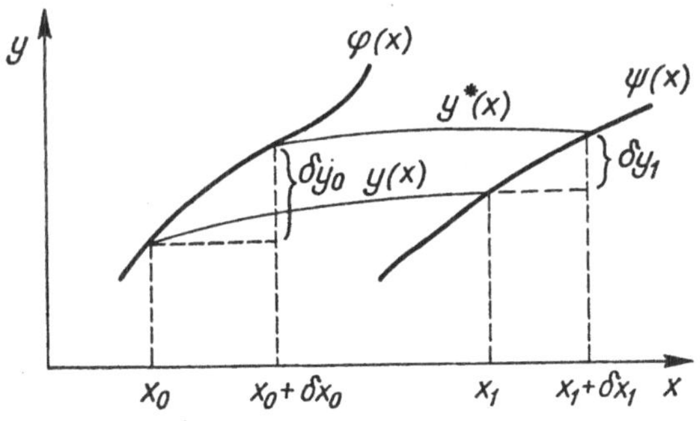

Fig. 10.1. Geometrical interpretation of a variable end point problem

of the curves considered are allowed to vary along two given curves (Fig. 10.1). Thus, we seek an extremum of the functional

$$V(y) = \int_{x_0}^{x_1} F(x, y, y')\mathrm{d}x \tag{10.23}$$

in the class of all smooth curves whose end points lie on two given curves $y = \varphi(x)$ and $y = \psi(x)$.

The variation of the functional V can be represented as

$$\delta V(\eta) = \int_{x_0}^{x_1} \left[\frac{\partial F}{\partial y} - \frac{\mathrm{d}}{\mathrm{d}x}\left(\frac{\partial F}{\partial y'}\right)\right]\eta(x)\mathrm{d}x + \left.\frac{\partial F}{\partial y'}\right|_{x=x_1} \delta y_1 +$$
$$+\left.\left(F - \frac{\partial F}{\partial y'}y'\right)\right|_{x=x_1} \delta x_1 - \left.\frac{\partial F}{\partial y'}\right|_{x=x_0} \delta y_0 - \left.\left(F - \frac{\partial F}{\partial y'}y'\right)\right|_{x=x_0} \delta x_0. \tag{10.24}$$

For an extremum, δV must vanish. It follows that the extremal must satisfy the Euler equation (10.17) and the boundary conditions

$$\begin{aligned} F + \frac{\partial F}{\partial y'}(\psi' - y')|_{x=x_1} &= 0, \\ F + \frac{\partial F}{\partial y'}(\varphi' - y')|_{x=x_0} &= 0, \end{aligned} \tag{10.25}$$

which are called the *transversality conditions.* Thus, to solve a variable end point problem, we must first solve the corresponding Euler equation and then use the transversality conditions to determine the values of the arbitrary constants appearing in its general solution.

Extremum problems with broken extremals

In some variational problems the class of admissible curves may be extended to include "broken", or piecewise smooth, curves. For example, consider the fixed-end point problem of finding an extremum of the functional

$$V(y) = \int_a^b F(x, y, y')\mathrm{d}x \tag{10.26}$$

in the class of functions $y(x)$ which are smooth in the interval $[a, b]$ except possibly at some interior point c, where they can have a cusp. It is clear that on each of the intervals $[a, c]$ and $[c, b]$ the required extremal must satisfy the Euler equation

$$\frac{\partial F}{\partial y} - \frac{\mathrm{d}}{\mathrm{d}x}\left(\frac{\partial F}{\partial y'}\right) = 0. \tag{10.27}$$

To use this fact, we write the functional $V(y)$ in the form

$$V(y) = V_1(y) + V_2(y) = \int_a^c F(x, y, y')\mathrm{d}x + \int_c^b F(x, y, y')\mathrm{d}x \tag{10.28}$$

and calculate the variations of the two terms separately, assuming that the two pieces of the curve $y = y(x)$ have fixed end points at $x = a$ and $x = b$ but join freely at $x = c$. At an extremum the sum of the two variations must be zero, which gives

$$\left(\frac{\partial F}{\partial y'}\bigg|_{x=c-0} - \frac{\partial F}{\partial y'}\bigg|_{x=c+0}\right)\delta y_1 + \\ + \left[\left(F - \frac{\partial F}{\partial y'}y'\right)\bigg|_{x=c-0} - \left(F - \frac{\partial F}{\partial y'}y'\right)\bigg|_{x=c+0}\right]\delta x_1 = 0. \qquad (10.29)$$

Hence we obtain the conditions

$$\begin{aligned} &\frac{\partial F}{\partial y'}\bigg|_{x=c-0} - \frac{\partial F}{\partial y'}\bigg|_{x=c+0} = 0, \\ &\left(F - \frac{\partial F}{\partial y'}y'\right)\bigg|_{x=c-0} - \left(F - \frac{\partial F}{\partial y'}y'\right)\bigg|_{x=c+0} = 0, \end{aligned} \qquad (10.30)$$

which are known as the *Weierstrass–Erdmann conditions.*

In each of the intervals $[a, c]$ and $[c, b]$, the extremal $y = \hat{y}(x)$ must satisfy the Euler equation, which is a second-order differential equation. The four arbitrary constants which appear in the corresponding general solutions can be found from the boundary conditions $y(a) = A$, $y(b) = B$ and the Weierstrass–Erdmann conditions (10.30).

10.3 Variational problems with side conditions

Functional conditions (constraints of the 1st kind).

The problem is to find an extremum of a functional of the form

$$V(y_1, \ldots, y_m) = \int_a^b F(x, y_1, \ldots, y_m, y_1', \ldots, y_m', \ldots, y_1^{(n)}, \ldots, y_m^{(n)})\mathrm{d}x, \qquad (10.31)$$

which depends on m functions of class C^n satisfying the usual boundary conditions at the points $x = a$ and $x = b$ (cf. (10.15)), subject to the additional requirement that the functions y_i must satisfy r side conditions

$$G_j(x, y_1, \ldots, y_m, y_1', \ldots, y_m') = 0, \quad j = 1, 2, \ldots, r \leqslant m. \qquad (10.32)$$

To solve this problem, we seek extrema of the auxiliary functional

$$\int_a^b \left[F + \sum_{j=1}^{r} \lambda_j(x) G_j\right]\mathrm{d}x = \int_a^b H(x, y_1, \ldots, y_m, \ldots, y_1^{(n)}, \ldots, y_m^{(n)})\mathrm{d}x. \qquad (10.33)$$

The corresponding Euler equations take the form

$$\frac{\partial H}{\partial y_i} - \frac{\mathrm{d}}{\mathrm{d}x}\left(\frac{\partial H}{\partial y_i'}\right) + \frac{\mathrm{d}^2}{\mathrm{d}x^2}\left(\frac{\partial H}{\partial y_i''}\right) + \ldots + (-1)^n \frac{\mathrm{d}^n}{\mathrm{d}x^n}\left(\frac{\partial H}{\partial y_i^n}\right) = 0. \qquad (10.34)$$

We thus obtain m Euler equations and r conditions (10.32) from which we can determine the m required functions $y_i(x)$ and r Lagrange multipliers $\lambda_j(x)$.

If the constraints (10.32) do not involve derivatives y_i' and can be solved for r functions y_i, then we can substitute these solutions into the functional and seek its unconstrained extrema with respect to the remaining $m-r$ variables $y_i(x)$.

Isoperimetric conditions (constraints of the 2nd kind)

In isoperimetric problems, extrema of a functional (10.31) are sought subject to constraints of the form

$$\int_a^b G_j(x, y_1, \ldots, y_m, \ldots, y_1^{(n)}, \ldots, y_m^{(n)})\mathrm{d}x = k_j, \quad j = 1, \ldots, r \leqslant m. \tag{10.35}$$

To solve such a problem, we set up a new functional

$$\int_a^b \Big[F + \sum_{j=1}^{r} \lambda_j G_j\Big]\, \mathrm{d}x = \int_a^b H(x, y_1, \ldots, y_m, \ldots, y_1^{(n)}, \ldots, y_m^{(n)})\mathrm{d}x, \tag{10.36}$$

where λ_j are constants, and write out its Euler equations

$$\frac{\partial H}{\partial y_i} - \frac{\mathrm{d}}{\mathrm{d}x}\left(\frac{\partial H}{\partial y_i'}\right) + \frac{\mathrm{d}^2}{\mathrm{d}x^2}\left(\frac{\partial H}{\partial y_i''}\right) + \ldots + (-1)^n \frac{\mathrm{d}^n}{\mathrm{d}x^n}\left(\frac{\partial H}{\partial y_i^{(n)}}\right) = 0. \tag{10.37}$$

By solving this system of m differential equations we find the required functions $y_i(x)$ in terms of λ_j and then use the isoperimetric conditions (10.35) to determine the constants λ_j.

10.4 Sufficient conditions for a minimum of a functional

Sufficient conditions for a weak minimum

The functional

$$V(y) = \int_a^b F(x, y, y')\mathrm{d}x, \quad y(a) = \bar{a}, \quad y(b) = \bar{b}, \tag{10.38}$$

has a weak minimum for a curve $y = \hat{y}(x)$ if $\hat{y}$ satisfies the following conditions:

1. The curve $y = \hat{y}(x)$ is an extremal, i.e. satisfies the Euler equation

$$\frac{\partial F}{\partial y} - \frac{\mathrm{d}}{\mathrm{d}x}\left(\frac{\partial F}{\partial y'}\right) = 0.$$

2. Along this curve we have

$$\frac{\partial^2 F}{\partial y'^2} > 0.$$

3. The interval $[a, b]$ contains no points conjugate to the point a (the Jacobi condition).

A point $\tilde{x}$ is said to be *conjugate* to the point a if the equation

$$-\frac{\mathrm{d}}{\mathrm{d}x}\left(\frac{\partial^2 F}{\partial y'^2}\eta'(x)\right)+\left[\frac{\partial^2 F}{\partial y^2}-\frac{\mathrm{d}}{\mathrm{d}x}\left(\frac{\partial^2 F}{\partial y\,\partial y'}\right)\right]\eta(x)=0 \tag{10.39}$$

has a solution which is not identically zero but vanishes at $x = a$ and $x = \tilde{x}$. Note that condition 2 above is necessary for a minimum of functional (10.38).

Sufficient conditions for a strong minimum

The functional

$$V(y) = \int_a^b F(x, y, y')\mathrm{d}x, \quad y(a) = \bar{a}, \quad y(b) = \bar{b}, \tag{10.40}$$

has a strong minimum for $y = \hat{y}(x)$ if this curve satisfies the conditions for a weak minimum and if the following inequality (the Weierstrass–Erdmann condition) holds along the curve in question for every finite $z \neq y'$:

$$F(x, y, z) - F(x, y, y') + (y' - z)\frac{\partial F}{\partial y'} > 0. \tag{10.41}$$

For proofs of the above sufficient conditions for weak and strong minima and also for the derivation of similar conditions for functionals of more complicated forms see e.g. Gelfand and Fomin, *loc. cit.* p. 314.

10.5 Direct methods in the calculus of variations

In many variational problems the solution of the corresponding Euler or Euler–Ostrogradski equation is very difficult or even impossible to obtain. This fact has led to the development of variational methods of a different kind, known as *direct methods*, which do not reduce variational problems to complicated differential equations or systems of such equations but consider them as limit problems for some extremum problems for functions of several variables, which can be solved by the usual methods.

The best known direct method is a *method of Ritz*. To find by this method a minimum of functional (10.40) we do not consider it along arbitrary admissible curves but only along all possible linear combinations

$$y_n = c_1\varphi_1(x) + \ldots + c_n\varphi_n(x) \tag{10.42}$$

of the first n functions of a certain sequence $\varphi_1, \ldots, \varphi_n, \ldots$ such that each function φ_i and combination (10.42) are admissible (i.e. are of class C^1 and satisfy the boundary conditions). We then seek a minimum of the integral

$$\int_a^b [c_1\varphi_1(x)+c_2\varphi_2(x)+ \dots +c_n\varphi_n(x)]\mathrm{d}x \tag{10.43}$$

with respect to n unknown coefficients c_i. This is a minimum problem for a function of several variables and can be solved by using the ordinary conditions $\partial V_n/\partial c_i = 0$. In this way we can obtain a solution $\hat{y}_n$ and a minimum value μ_n for each n. Clearly, the sequence $\{\mu_n\}$ cannot be increasing, i.e. we have $\mu_1 \geqslant \mu_2 \geqslant \dots$, because for each n the linear combinations of the first $n+1$ functions φ_i include all combinations of the first n functions. If the sequence $\hat{y}_1, \hat{y}_2, \dots$ is convergent to an admissible function, then $\hat{y} = \lim_{n\to\infty} \hat{y}_n$ is an exact solution of the problem in question and $\mu = \lim_{n\to\infty} \mu_n$ is the minimum value of the functional V. To ensure the existence of this last limit, the sequence φ_i should be complete in the admissible space, i.e. should be such that every admissible function can be approximated arbitrarily close (in uniform norm) by combinations of form (10.42). In practice, however, the limiting process need not be carried out in full and often a relatively small number of functions φ_i (3 or 4) suffice to obtain a satisfactory approximate solution.

10.6 Examples of application of the calculus of variations to structural optimization

Below we give several examples of the application of the classical variational methods to the optimum design of columns, beams and plates. We restrict ourselves to formulating the problems and writing out the corresponding Euler or Euler–Ostrogradski equations. Complete solutions are available in the references indicated in the text.

Optimization of a column for minimum potential

A column with length $2h$ and rectangular cross-section with dimensions $a(x)$ and b, fixed at both ends and loaded by a longitudinal force N_0 and a bending moment M_0 is optimized for minimum elastic strain energy at constant volume (Fig. 10.2). The only quantity to be determined in the design process is the cross-section depth $a(x)$, variable along the column (Owczarek 1969).

The minimum-energy condition for the column takes the form

$$U = \int_{-h}^{h} \frac{M^2}{2EJ}\mathrm{d}x + \int_{-h}^{h} \frac{N^2}{2EA}\mathrm{d}x = \min,$$

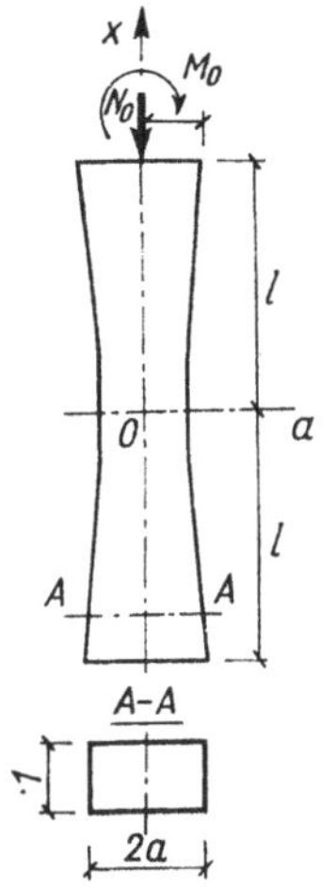

Fig. 10.2. The column under consideration

with

$$V = \int_{-h}^{h} A\,\mathrm{d}x = C,$$

and so the problem is an isoperimetric variational problem. The auxiliary functional (cf. (10.36)) can be written

$$U^* = \frac{1}{2E}\int_{-h}^{h}\left(\frac{M^2}{J}+\frac{N^2}{A}+2\lambda EA\right)\mathrm{d}x.$$

Substituting appropriate expressions for A and J with $b = 1$, we get

$$U^* = \frac{1}{2E}\int_{-h}^{h}\left(\frac{3M^2}{2a^3}+\frac{N^2}{2a}+4\lambda Ea\right)\mathrm{d}x.$$

The Euler equation for this functional is

$$8\lambda Ea^4 - N^2a^2 - 9M^2 = 0.$$

It is an algebraic equation from which we can find the unknown a. The Lagrange multiplier λ can be determined directly from the condition

$$b\int_{-h}^{h} a\,\mathrm{d}x = C,$$

if the material volume C is given *a priori*, or in some other way, for example from the condition that the maximum normal stress in the column, σ_{max}, be equal to an allowable value:

$$\sigma_{max} = \frac{6M_0}{ba^2}+\frac{N_0}{ba} = \bar{\sigma}.$$

Optimization of a beam for minimum potential

A simply supported rectangular beam of length l is loaded by a concentrated force P at the mid-span. The depth a of the beam is assumed to be given and constant. It is required to determine the variable width of the beam, $b(x)$, so that the elastic strain energy of the beam is minimum at constant volume.

With shearing forces taken into account, the minimum-energy condition for the beam can be written

$$U = \int_0^l \frac{M^2}{2EJ}\,dx + \int_0^l \frac{mT^2}{2GA}\,dx$$

with

$$V = \int_0^l A\,dx = C;$$

m is a coefficient depending on the cross-sectional shape. The auxiliary functional for this isoperimetric problem takes the form

$$U^* = \int_0^l \left[\frac{3P^2x^2}{2Ea^3b(x)} + \frac{m(1+\nu)P^2}{4Eab(x)} + \lambda ab(x)\right]dx$$

and the corresponding Euler equation is

$$4E\lambda a^4b(x)^2 - 6P^2x^2 - m(1+\nu)P^2a^2 = 0.$$

Hence we find $b(x)$. The multiplier λ can be determined as in the previous example.

Optimization of the cross-section of a thin-walled bar in bending

The cross-section is assumed to have two axes of symmetry (Fig. 10.3). The wall thickness h and the cross-section width b are constant. The function $y(x)$ defining the shape of the cross-section is unknown.

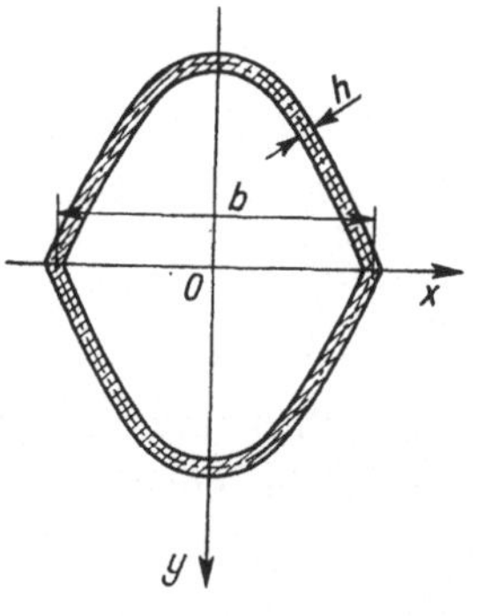

Fig. 10.3. Bar cross-section with constant wall thickness

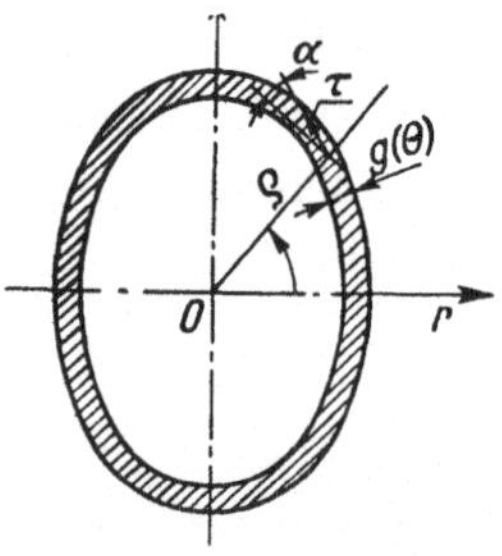

Fig. 10.4. A thin-walled bar section

The bar is optimized for minimum cross-sectional area,

$$A = 4h \int_0^{b/2} \sqrt{1+y'^2}\, dx = \min,$$

subject to the isoperimetric condition

$$W = \frac{4h}{\bar{y}} \int_0^{b/2} y^2 \sqrt{1+y'^2}\, dx = \text{const},$$

which expresses the requirement that the maximum stress in the section should be constant. The auxiliary functional for this problem is

$$A^* = 4h \int_0^{b/2} \left[\sqrt{1+y'^2} + \frac{\lambda y^2}{\bar{y}} \sqrt{1+y'^2}\right] dx.$$

The corresponding Euler equation is a second-order differential equation:

$$2\lambda y(1+y'^2) - (\bar{y} + \lambda y^2) y'' = 0.$$

A single integration gives the relation

$$\frac{dy}{dx} = \sqrt{\left(C + \frac{C\lambda}{\bar{y}} y^2\right)^2 - 1},$$

from which the inverse solution can be found, i.e. x as a function of y, in terms of elliptic functions. A complete solution, including determination of all the constants involved, is given in a paper by Życzkowski (1968_2).

Optimization of a thin-walled bar in torsion and bending

It is required to design the cross-sectional shape of a thin-walled bar subject to torsion and bending. The cross-section of the bar can be described by means of two functions of a single variable: $\varrho = \varrho(\theta)$, representing the middle line of the bar wall, and $g = g(\theta)$, defining the wall thickness (Fig. 10.4). It is assumed that the Huber–Mises–Hencky combined stress will be uniform over the cross-section, i.e.

$$\sigma^2 + 3\tau^2 = \bar{\sigma}^2.$$

It is further assumed that the normal and the shearing stresses can be calculated from the formulae

$$\sigma = C_1 \varrho \sin\theta, \qquad \tau = \frac{C_2}{g},$$

where C_1 and C_2 are constants depending on the bending moment M_b and the torsional moment M_t. Substituting these expressions in the Huber–Mises–Hencky condition gives

$$g = \frac{C_2\sqrt{3}}{\sqrt{\sigma^2 - C_1^2\varrho^2\sin^2\theta}}.$$

The cross-sectional area A and the moments M_t and M_b can be expressed as

$$A = C_2\sqrt{3}\int_0^{2\pi}\frac{\sqrt{\varrho^2+\varrho'^2}}{\sqrt{\bar{\sigma}^2 - C_1^2\varrho^2\sin^2\theta}}\,d\theta,$$

$$M_t = C_2\int_0^{2\pi}\varrho^2 d\theta,$$

$$M_b = C_1C_2\sqrt{3}\int_0^{2\pi}\frac{\varrho^2\sqrt{\varrho^2+\varrho'^2\sin^2\theta}}{\sqrt{\bar{\sigma}^2 - C_1^2\varrho^2\sin^2\theta}}\,d\theta.$$

For an optimum design, the cross-sectional area A is minimized subject to the side requirements that the moments M_t and M_b be constant.

The auxiliary functional takes the form

$$A^* = C_2\sqrt{3}\int_0^{2\pi}\frac{\sqrt{\varrho^2+\varrho'^2}}{\sqrt{\bar{\sigma}^2 - C_1^2\varrho^2\sin^2\theta}}\,d\theta +$$
$$+\lambda_2C_1C_2\sqrt{3}\int_0^{2\pi}\frac{\varrho^2\sqrt{\varrho^2+\varrho'^2\sin^2\theta}}{\sqrt{\bar{\sigma}^2 - C_1^2\varrho^2\sin^2\theta}}\,d\theta + \lambda_1C_2\int_0^{2\pi}\varrho^2 d\theta.$$

The necessary condition for a minimum of the functional A^* leads to the following second-order differential equation:

$$\frac{2}{\sqrt{3}}\lambda_1\varrho(\bar{\sigma}^2 - C_1^2\varrho^2\sin^2\theta)^{3/2}(\varrho^2+\varrho'^2)^{3/2} + [(3\varrho\sin\theta - 2\varrho'\cos\theta - \varrho''\sin\theta)\times$$
$$\times C_1\lambda_2\varrho^2\sin^2\theta + \varrho - \varrho''](\bar{\sigma}^2 - C_1^2\varrho^2\sin^2\theta)(\varrho^2+\varrho'^2) +$$
$$+[C_1^2\varrho\sin^2\theta(\varrho^2+\varrho'^2)^2 - C_1^2\varrho\varrho'\sin\theta(\varrho'\sin\theta+\varrho\cos\theta)(\varrho^2+\varrho'^2) +$$
$$+\varrho'^2(\varrho+\varrho'')(\bar{\sigma}^2 - C_1\varrho^2\sin^2\theta)](1 + C_1\lambda_2\varrho^2\sin^2\theta) = 0.$$

An exact solution of this equation does not seem possible. An approximate solution obtained by a small-parameter method was given by Życzkowski (1968_2).

Optimization of a prestressed plate for minimum potential

We consider the problem of designing a thin prestressed plate with a variable thickness $h(x, y)$ and an arbitrary boundary $\Gamma(s)$, and supported in an arbitrary manner (Dzieniszewski 1965, 1968). The plate is loaded by its own weight γh, an arbitrary vertical load $p(x, y)$ and boundary forces and moments, $P^*(s)$ and $M^*(s)$. It is prestressed by two orthogonal families of cables with the prestressing force densities $S_x(x, y)$ and $S_y(x, y)$, distributed within the plate on two surfaces $e_x(x, y)$ and $e_y(x, y)$ (Fig. 10.5). It is assumed that the material of the plate is homogeneous and isotropic and that the thickness $h(x, y)$ is sufficiently small relative to other dimensions of the plate.

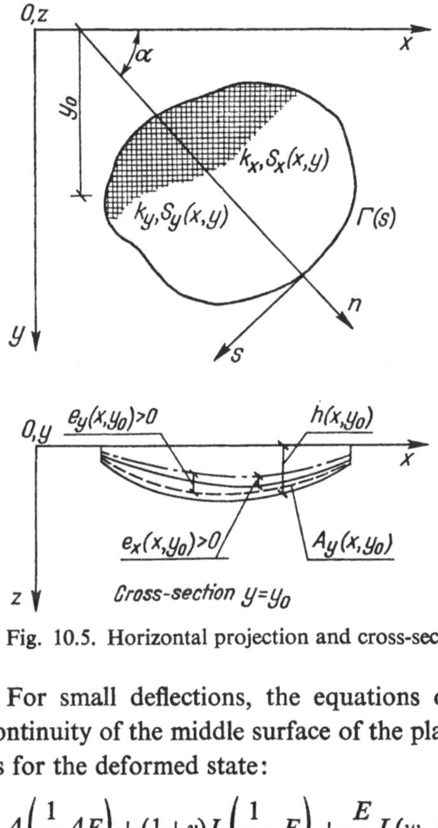

Fig. 10.5. Horizontal projection and cross-section of the plate

For small deflections, the equations of equilibrium and the condition of continuity of the middle surface of the plate lead to the following two equations for the deformed state:

$$\Delta\left(\frac{1}{h}\Delta F\right)+(1+\nu)L\left(\frac{1}{h}, F\right)+\frac{E}{2}L(w, h)$$

$$=\Delta\left[\frac{1}{h}(S_x+S_y)\right]-(1+\nu)\left[\left(\frac{S_x}{h}\right)_{,xx}+\left(\frac{S_y}{h}\right)_{,yy}\right], \tag{10.44}$$

$$\Delta(D\Delta w)+(1-\nu)L(D, w)+\frac{1}{2}L(h, F)=p+\gamma h+(S_x e_x)_{,xx}+(S_y e_y)_{,yy};$$

here D denotes the expression

$$D=\frac{Eh^3}{12(1-\nu^2)},$$

F is Airy's function defined by the relations

$$N_x+S_x=F_{,yy}, \quad N_y+S_y=F_{,xx}, \quad N_{xy}=-F_{,xy},$$

where N_x and N_y are the normal forces and N_{xy} the shearing force in the plane of the plate, Δ denotes the Laplace operator and

$$L(\varphi, \psi) = 2\varphi_{,xy}\psi_{,xy} - \varphi_{,xx}\psi_{,yy} - \varphi_{,yy}\psi_{,xx}.$$

As the design criterion we choose the condition of minimum elastic strain energy in the plate and the prestressing cables at constant volume of concrete and prestressing steel:

$$U_c = \min \quad \text{and} \quad U_s = \min$$

for

$$V_c = \text{const} \quad \text{and} \quad V_s = \text{const}.$$

On the basis of this criterion it is possible to determine
— the plate thickness $h(x, y)$,
— the prestressing forces $S_x(x, y)$ and $S_y(x, y)$,
— the eccentricities of the prestressing cables $e_x(x, y)$ and $e_y(x, y)$,
— the cross-sectional areas of the cables, $A_x(x, y)$ and $A_y(x, y)$.

The elastic energy of the plate can be calculated from the formula

$$U_c = \frac{1}{2E_c}\iint_B \Big\{\frac{1}{h}[(\Delta F - S_x - S_y)^2 + (1+\nu)L(F, F) + 2(1+\nu)(S_x F_{,xx} + S_y F_{,yy} - S_x S_y)] + D[(\Delta w)^2 + (1-\nu)L(w, w)]\Big\} \mathrm{d}x\mathrm{d}y \tag{10.45}$$

and that of the cables from

$$U_{si} = \frac{1}{2E_s}\iint_B \frac{S_i^2}{A_{si}} \mathrm{d}x\, \mathrm{d}y, \quad i = x, y. \tag{10.46}$$

On the other hand, if the deflection is small compared with h, i.e. $w \ll h$ the potential energy of the prestressed plate due to bending is given by

$$W_c = -\iint_B \Big[S_x e_x w_{,xx} + S_y e_y w_{,yy} + w\Big(\gamma h + p - \frac{1}{2}L(F, h)\Big)\Big] \mathrm{d}x\mathrm{d}y + \oint (M^* w_{,n} - P^* w)\mathrm{d}s.$$

The volumes of concrete and of prestressing cables are

$$V_c = \iint_B h(x, y)\mathrm{d}x\mathrm{d}y$$

and

$$V_{si} = \iint_B A_i(x, y)\mathrm{d}x\mathrm{d}y,$$

respectively.

Our design problem can now be formulated as one of minimizing functional (10.45) with respect to the functions $h(x, y)$, $S_x(x, y)$, $S_y(x, y)$, $e_x(x, y)$ and $e_y(x, y)$ subject to the constraint $V_c = \text{const}$, and minimizing functionals (10.46) with respect to the functions $A_x(x, y)$ and $A_y(x, y)$ subject to the constraints $V_{si} = \text{const}$. Furthermore, the required functions must satisfy conditions (10.44), which together with appropriate boundary conditions are the Euler–Ostrogradski equations for the functional $U_c + W_c$ and constitute a necessary condition for its minimum. From the general theorems about minima of the potential energy of an elastic body in equilibrium it can be inferred that the requirement of a minimum of the functional U_c or $U_c + \lambda V_c$ with respect to the functions h, S_x, S_y, e_x, and e_y subject to constraints (10.44) is equivalent to the requirement of an unconstrained minimum of the functional $U_c + W_c + \lambda V_c$ with respect to the functions h, S_x, S_y, e_x, e_y, w and F.

Eventually, we obtain the following auxiliary functionals for the problem:

$$
\begin{aligned}
U^* &= U_c + W_c + \lambda_c V_c + \lambda_{sx} V_{sx} + \lambda_{sy} V_{sy} \\
&= \frac{1}{2} \iint_B \left\{ \frac{1}{E_c h} [(\Delta F - S_x - S_y)^2 + (1+\nu) L(F, F) + 2(1+\nu)(S_x F_{,xx} + \right. \\
&\quad \left. + S_y F_{,yy} - S_x S_y)] + D[(\Delta w)^2 + (1-\nu) L(w, w)] \right\} \mathrm{d}x\mathrm{d}y - \\
&\quad - \iint_B \left[S_x e_x w_{,xx} + S_y e_y w_{,yy} + w\left(\gamma h + p - \frac{1}{2} L(F, h)\right)\right] \mathrm{d}x\mathrm{d}y + \\
&\quad + \oint (M^* w_{,n} - P^* w) \mathrm{d}s + \lambda_c \iint_B h \,\mathrm{d}x\mathrm{d}y + \lambda_{sx} \iint_B A_x \mathrm{d}x\mathrm{d}y + \\
&\quad + \lambda_{sy} \iint_B A_y \mathrm{d}x\mathrm{d}y = \min,
\end{aligned}
$$

$$
U^*_{sx} = U_{sx} + \lambda'_{sx} V_{sx} = \frac{1}{2E_s} \iint_B \frac{S_y^2}{A_x} \mathrm{d}x\mathrm{d}y + \lambda'_{sx} \iint_B A_x \mathrm{d}x\mathrm{d}y = \min,
$$

$$
U^*_{sy} = U_{sy} + \lambda'_{sy} V_{sy} = \frac{1}{2E_s} \iint_B \frac{S_y^2}{A_y} \mathrm{d}x\mathrm{d}y + \lambda'_{sy} \iint_B A_y \mathrm{d}x\mathrm{d}y = \min.
$$

The conditions for minima of these functionals take the form

$$
\begin{aligned}
&-\frac{1}{2E_c h^2} [(\Delta F - S_x - S_y)^2 + (1+\nu) L(F, F) + 2(1+\nu)(S_x F_{,xx} + \\
&\quad + S_y F_{,yy} - S_x S_y)] + \frac{E_c h^2}{8(1-\nu^2)} [(\Delta w)^2 + (1-\nu) L(w, w)] - \gamma w + \\
&\quad + \tfrac{1}{2} L(F, w) + \lambda_c = 0,
\end{aligned}
$$

$$-\frac{1}{E_c h}(\Delta F - S_x - S_y) - \frac{1+\nu}{E_c h}(F_{,xx} - S_y) - e_x w_{,xx} + \lambda_{sx}\frac{1}{\bar{\sigma}_s} = 0,$$

$$-\frac{1}{E_c h}(\Delta F - S_x - S_y) - \frac{1+\nu}{E_c h}(F_{,yy} - S_x) - e_y w_{,yy} + \lambda_{sy}\frac{1}{\bar{\sigma}_s} = 0,$$

$$S_x w_{,xx} = 0, \quad S_y w_{,yy} = 0,$$

$$\Delta\left(\frac{1}{h}\Delta F\right) + (1+\nu)L\left(\frac{1}{h}, F\right) - \frac{E_c}{2}L(w, h)$$

$$= \Delta\left[\frac{1}{h}(S_x + S_y)\right] - (1+\nu)\left[\left(\frac{S_x}{h}\right)_{,xx} + \left(\frac{S_y}{h}\right)_{,yy}\right],$$

$$\Delta(D\Delta w) + (1-\nu)L(D, w) + \tfrac{1}{2}L(h, F) = p + \gamma h + (S_x e_x)_{,xx} + (S_y e_y)_{,yy},$$

$$\frac{S_x^2}{A_x^2} = 2E_s \lambda'_{sx}, \quad \frac{S_y^2}{A_y} = 2E_s \lambda'_{sy}.$$

Supplemented by suitable boundary conditions, the above system of equations can be solved for the unknown functions. The boundary conditions depend on the way in which the plate is supported. Solution of the above system for various boundary conditions is given by Dzieniszewski (1968).

11

Non-classical Variational Methods of Optimization

11.1 Assumptions and problem formulations

The principal distinguishing feature of non-classical variational problems is the presence of constraints in the form of inequalities, either directly imposed on the unknown functions or involving functionals of these functions. Constraints encountered in practical engineering optimization problems are often of just such a form and therefore the knowledge of non-classical variational methods may be of considerable help to the structural engineer, who otherwise may tend to simplify or neglect various side conditions and reach an idealized solution of little practical value.

The following discussion of non-classical variational methods aims at popularizing these methods and stimulating their direct implementation in structural optimization. The material presented is largely based on a book by Trukhaev and Khomenyuk[1] and should only be regarded as an introduction to the subject.

We shall consider functionals of the form

$$V_0(\boldsymbol{\varphi}) = \int_P f_0(x, \boldsymbol{\varphi}(x))\mathrm{d}x, \tag{11.1}$$

defined in the linear normed space $E = L_2^n(P)$ of all vector-valued functions $\boldsymbol{\varphi}(x) = (\varphi_1(x), \ldots, \varphi_n(x))$ of several variables $x = (x_1, \ldots, x_k)$, square-integrable on a bounded interval $P \subset R^k$, i.e. such that $||\boldsymbol{\varphi}||^2 = \int_P \sum_{i=1}^{n} \varphi_i^2(x)\mathrm{d}x < \infty$. The function $f_0(x, \boldsymbol{\varphi})$ is assumed to be continuous in x and continuously differentiable with respect to $\boldsymbol{\varphi}$.

A general optimization problem for functional (11.1) can be formulated as follows:

Let Q be a subset of E. Find an element $\boldsymbol{\varphi}^0 \in Q$ (an optimal element) such that

[1] Trukhaev, R. I., and Khomenyuk, V. V., *Theory of Nonclassical Variational Problems* (in Russian), Izd. Lenin. Univ., 1971.

$$V_0(\boldsymbol{\varphi}^0) = \min_{\boldsymbol{\varphi} \in Q} V_0(\boldsymbol{\varphi}). \tag{11.2}$$

In classical variational problems the set Q is open or coincides with the whole space E (unconstrained problem). The classical calculus of variations also deals with problems in which the set Q is defined by functional and/or integral equality constraints

$$\begin{aligned} &\Phi_k(x, \boldsymbol{\varphi}(x)) = c_k', \qquad k = 1, \ldots, r, \quad r \leqslant n, \\ &\int_P f_i(x, \boldsymbol{\varphi}(x))\mathrm{d}x = c_i'', \quad i = 1, \ldots, m, \quad m \leqslant n, \end{aligned} \tag{11.3}$$

where c_k' $(k = 1, \ldots, r)$ and c_i'' $(i = 1, \ldots, m)$ are given constants (*Euler–Lagrange constrained problem*).

When the constraint set Q is not open, we call problem (11.2) *non-classical*. Our interest here will be confined to those non-classical variational problems in which Q is closed and convex.

Recall that a set Q is said to be *convex* if, for any $\boldsymbol{\varphi}_1, \boldsymbol{\varphi}_2 \in Q$ and $0 \leqslant \lambda \leqslant 1$, the element

$$\boldsymbol{\varphi}^\lambda = 2\boldsymbol{\varphi}_1 + (1-\lambda)\boldsymbol{\varphi}_2 \tag{11.4}$$

also belongs to Q.

In particular, we shall consider sets $Q \subset L_2^n(P)$ in the form $Q = Q_1 \cap Q_2$, where Q_1 and Q_2 are defined by inequality constraints of the first and the second kind, respectively:

$$\begin{aligned} Q_1 &= \{\boldsymbol{\varphi}(x)\colon 0 \leqslant \varphi_j(x) \leqslant b_j,\ j = 1, \ldots, n\}, \\ Q_2 &= \{\boldsymbol{\varphi}(x)\colon V_i(\boldsymbol{\varphi}) \leqslant c_i,\ i = 1, \ldots, m\}; \end{aligned} \tag{11.5}$$

$b_1, \ldots, b_n, c_1, \ldots, c_m$ are given real numbers and $V_1, \ldots, V_m$ are given functionals of integral form

$$V_i(\boldsymbol{\varphi}) = \int_P f_i((x, \boldsymbol{\varphi}(x))\mathrm{d}x, \quad i - 1, \ldots, m, \tag{11.6}$$

where all f_i are continuous in $x \in P$ and continuously differentiable in $\boldsymbol{\varphi}$, $\mathbf{0} \leqslant \boldsymbol{\varphi} \leqslant \mathbf{b}$. It can be shown that the set Q_2 is convex and, since the convexity of Q_1 is obvious, so is the set $Q = Q_1 \cap Q_2$.

In practical optimization problems, an admissible solution which is not necessarily optimal but which is known to give a value of the objective functional sufficiently close to the minimum is often accepted as satisfactory. It may also happen that an exact solution does not exist and then an approximate solution is the only possibility. Bearing this in mind, we shall also consider the following "ε-modification" of problem (11.2):

For a given $\varepsilon > 0$, find an element $\boldsymbol{\varphi}_\varepsilon^0 \in Q$ (an ε-optimal element) such that

$$V_0(\boldsymbol{\varphi}_\varepsilon^0) \leqslant \inf_{\boldsymbol{\varphi}_\varepsilon \in Q} V_0(\boldsymbol{\varphi}) + \varepsilon. \tag{11.7}$$

In this formulation, ε specifies the degree of accuracy required in approaching the possible minimum value of the objective functional. Note that an ε-optimal solution needs not be close in norm to the exact solution, if one exists.

11.2 Optimality conditions: necessary and sufficient

We first consider a classical problem for functional (11.1), assuming Q to be the whole space E. Let $\boldsymbol{\varphi}^0$ be a solution to this problem. For an arbitrary function $\boldsymbol{\varphi} \in E$ and a real number λ, define

$$\boldsymbol{\varphi}^\lambda = \boldsymbol{\varphi}^0 + \lambda(\boldsymbol{\varphi} - \boldsymbol{\varphi}^0). \tag{11.8}$$

Clearly, $\boldsymbol{\varphi}^\lambda \in E$ for any λ and, by optimality of $\boldsymbol{\varphi}^0$,

$$V_0(\boldsymbol{\varphi}^\lambda) \geqslant V_0(\boldsymbol{\varphi}^0). \tag{11.9}$$

Denote the gradient of the functional V_0 by $\mathbf{g}_0$:

$$\mathbf{g}_0(\boldsymbol{\varphi}) \equiv \operatorname{grad} V_0(\boldsymbol{\varphi}) = \left(\frac{\partial f_0(x, \boldsymbol{\varphi})}{\partial \varphi_1}, \ldots, \frac{\partial f_0(x, \boldsymbol{\varphi})}{\partial \varphi_m} \right). \tag{11.10}$$

Using Taylor's formula, we can write $V_0(\boldsymbol{\varphi}^\lambda)$ as

$$V_0(\boldsymbol{\varphi}^\lambda(x)) = V_0(\boldsymbol{\varphi}^0(x)) + \lambda \int_P [\mathbf{g}_0(\boldsymbol{\varphi}^0(x)), \boldsymbol{\varphi}(x) - \boldsymbol{\varphi}^0(x)]\mathrm{d}x + \mathbf{r}(\lambda), \tag{11.11}$$

where $\lim\limits_{\lambda \to 0} \dfrac{\mathbf{r}(\lambda)}{\lambda} = 0$; the brackets $[\cdot, \cdot]$ denote the ordinary scalar product

$$[\mathbf{g}_0(\boldsymbol{\varphi}^0(x)), \boldsymbol{\varphi}(x) - \boldsymbol{\varphi}^0(x)]$$
$$= \mathbf{g}_0^1(\boldsymbol{\varphi})[\varphi_1(x) - \varphi_1^0(x)] + \ldots + \mathbf{g}_0^n(\boldsymbol{\varphi})[\varphi_n(x) - \varphi_n^0(x)]. \tag{11.12}$$

By (11.9), we have from equation (11.11)

$$\lambda \left\{ \int_P [\mathbf{g}_0(\boldsymbol{\varphi}^0(x)), \boldsymbol{\varphi}(x) - \boldsymbol{\varphi}^0(x)]\mathrm{d}x + \frac{\mathbf{r}(\lambda)}{\lambda} \right\} \geqslant 0. \tag{11.13}$$

Hence we obtain for $\lambda \to 0+$

$$\int_P [\mathbf{g}_0(\boldsymbol{\varphi}^0(x)), \boldsymbol{\varphi}(x) - \boldsymbol{\varphi}^0(x)]\mathrm{d}x \geqslant 0 \tag{11.14}$$

and for $\lambda \to 0-$

$$\int_P [\mathbf{g}_0(\boldsymbol{\varphi}^0(x)), \boldsymbol{\varphi}(x) - \boldsymbol{\varphi}^0(x)]\mathrm{d}x \leqslant 0. \tag{11.15}$$

Thus, for arbitrary $\varphi \in E$, we have the equality

$$\int_P [g_0(\varphi^0(x)), \varphi(x) - \varphi^0(x)] dx = 0. \tag{11.16}$$

In particular, setting $\varphi(x) = \varphi^0(x) + g_0(\varphi^0(x))$, we get

$$\int_P [g_0(\varphi^0(x)), g_0(\varphi^0(x))] dx = 0. \tag{11.17}$$

For this equality to hold, the integrand, being non-negative, must vanish almost everywhere in P, and so

$$g_0(\varphi^0(x)) = 0 \quad \text{a.e. in } P \tag{11.18}$$

Condition (11.18) or, equivalently, (11.16), is necessary for a function φ^0 to be an optimal solution. If the objective functional V_0 is convex, the same condition proves also sufficient.

Recall that a functional V_0 is said to be *convex* if, for arbitrary $\varphi_1, \varphi_2 \in E$ and $0 \leqslant \lambda \leqslant 1$,

$$V_0(\varphi_1 + \lambda(\varphi_2 - \varphi_1)) \geqslant V_0(\varphi_1) + \lambda[V_0(\varphi_2) - V_0(\varphi_1)]. \tag{11.19}$$

It can be shown that if V_0 is convex, then for arbitrary φ and φ_0 in E

$$V_0(\varphi(x)) - V_0(\varphi^0(x)) \geqslant \int_P [g_0(\varphi^0(x)), \varphi(x) - \varphi^0(x)] dx. \tag{11.20}$$

If φ_0 satisfies condition (11.18) or (11.16), the right-hand side of the above inequality will vanish for arbitrary φ in E, and so $V_0(\varphi^0) \leqslant V_0(\varphi)$, which means that φ^0 is an optimal solution, as required.

Let us note that conditions (11.16) and (11.18) remain valid when Q is an arbitrary open set in E.

We now move on to the case where Q is an arbitrary convex set not necessarily open. To derive optimality conditions for this non-classical variational problem, we shall follow an approach similar to the one used before.

Suppose that $\varphi^0 \in Q$ is an optimal solution. By the convexity of Q, the function φ^λ belongs to Q for any $0 \leqslant \lambda \leqslant 1$ and, since φ^0 is optimal, we have $V_0(\varphi^\lambda) \geqslant V(\varphi^0)$. From Taylor's expansion (11.11) of $V_0(\varphi^\lambda)$, with positive λ, we can then infer that

$$\int_P [g_0(\varphi^0), \varphi - \varphi^0] dx + \frac{r(\lambda)}{\lambda} \geqslant 0. \tag{11.21}$$

In the limit, as λ tends to 0, we obtain

$$\int_P [g_0(\varphi^0), \varphi - \varphi^0] dx \geqslant 0. \tag{11.22}$$

Since this inequality holds for arbitrary $\boldsymbol{\varphi} \in Q$, we also have

$$\min_{\boldsymbol{\varphi} \in Q} \int_P [\mathbf{g}_0(\boldsymbol{\varphi}^0), \boldsymbol{\varphi}(x) - \boldsymbol{\varphi}^0(x)] \mathrm{d}x \geqslant 0. \tag{11.23}$$

On the other hand,

$$\min_{\boldsymbol{\varphi} \in Q} \int_P [\mathbf{g}_0(\boldsymbol{\varphi}^0), \boldsymbol{\varphi}(x) - \boldsymbol{\varphi}^0(x)] \mathrm{d}x$$

$$\leqslant \int_P [\mathbf{g}_0(\boldsymbol{\varphi}^0(x)), \boldsymbol{\varphi}(x) - \boldsymbol{\varphi}^0(x)]_{\boldsymbol{\varphi}(x) = \boldsymbol{\varphi}^0(x)} \mathrm{d}x = 0. \tag{11.24}$$

Together, inequalities (11.23) and (11.24) give

$$\min_{\boldsymbol{\varphi}(x) \in Q} \int_P [\mathbf{g}_0(\boldsymbol{\varphi}^0(x)), \boldsymbol{\varphi}(x) - \boldsymbol{\varphi}^0(x)] \mathrm{d}x = 0. \tag{11.25}$$

The above equality is a necessary condition for a function $\boldsymbol{\varphi}^0$ to yield a minimum of the functional V_0 in a convex set Q. Comparison with the optimality condition (11.16) obtained for the classical problem shows that the present condition holds whenever (11.16) does and so includes (11.16) as a particular case.

If the functional V_0 is additionally assumed to be convex in Q, the necessary condition (11.25) becomes also sufficient. Indeed, if $\boldsymbol{\varphi}_0$ satisfies equation (11.25) and V_0 is convex, then, using inequality (11.20), we can write

$$V_0(\boldsymbol{\varphi}(x)) - V_0(\boldsymbol{\varphi}^0(x)) \geqslant \int_P [\mathbf{g}_0(\boldsymbol{\varphi}^0(x)), \boldsymbol{\varphi}(x) - \boldsymbol{\varphi}^0(x)] \mathrm{d}x$$

$$\geqslant \min_{\boldsymbol{\varphi}(x) \in Q} \int_P [\mathbf{g}_0(\boldsymbol{\varphi}^0(x)), \boldsymbol{\varphi}(x) - \boldsymbol{\varphi}^0(x)] \mathrm{d}x = 0, \tag{11.26}$$

i.e. $V_0(\boldsymbol{\varphi}^0) \leqslant V_0(\boldsymbol{\varphi})$ for any $\boldsymbol{\varphi} \in Q$, as required.

Conditions (11.16) and (11.25) provide the basis for the derivation of operator equations for optimal elements in particular classes of problems.

11.3 Trukhaev–Khomenyuk operator equations

As we have seen, if the unconstrained problem for a functional of form (11.1) has a solution, it is given by the Euler equation

$$\mathbf{g}_0(\boldsymbol{\varphi}^0(x)) = 0, \tag{11.27}$$

i.e.

$$\left[\frac{\partial f_0(x, \boldsymbol{\varphi})}{\partial \varphi_j}\right]_{\varphi_j = \varphi_j^0} = 0, \quad j = 1, \ldots, n. \tag{11.28}$$

For solving the classical equality-constrained problem (11.2)–(11.3) the calculus of variations offers the well-known Lagrange multipliers method.

According to this method, a solution to the original constrained problem can be obtained by unconstrained minimization of the auxiliary functional

$$\int_P H(x, \boldsymbol{\varphi}(x), \psi_1(x), \ldots, \psi_r(x), \psi_{r+1}, \ldots, \psi_{r+m})\mathrm{d}x, \tag{11.29}$$

where H is the Lagrange function defined by

$$H = f_0(x, \boldsymbol{\varphi}(x)) + \sum_{k=1}^{r} \psi_k(x)\big(\Phi_k(x, \boldsymbol{\varphi}(x)) - c_k'\big) + \\ + \sum_{i=1}^{m} \psi_{r+i}\big(f_i(x, \boldsymbol{\varphi}(x)) - c_i''\big). \tag{11.30}$$

The functions $\psi_1(x), \ldots, \psi_r(x)$ and constants $\psi_{r+1}, \ldots, \psi_{r+m}$ are called the *Lagrange multipliers.* A necessary condition for an element $\boldsymbol{\varphi}^0$ to be optimal for problem (11.2)–(11.3) is that there exist Lagrange multipliers $\boldsymbol{\psi}^0 = (\psi_1^0, \ldots \ldots, \psi_{r+m}^0)$, not all zero, which together with $\boldsymbol{\varphi}^0$ satisfy the Euler–Lagrange equations:

$$\left.\frac{\partial H}{\partial \varphi_j}\right|_{\substack{\boldsymbol{\varphi}=\boldsymbol{\varphi}^0\\ \boldsymbol{\psi}=\boldsymbol{\psi}^0}} = 0, \qquad \left.\frac{\partial H}{\partial \psi_s}\right|_{\substack{\boldsymbol{\varphi}=\boldsymbol{\varphi}^0\\ \boldsymbol{\psi}=\boldsymbol{\psi}^0}} = 0. \tag{11.31}$$

$j = 1, \ldots, n \qquad s = 1, \ldots, r+m.$

We will now derive operator equations for the solution $\boldsymbol{\varphi}^0(x)$ of a non-classical problem with inequality constraints of the first kind:

$$V(\boldsymbol{\varphi}^0(x)) = \min_{\boldsymbol{\varphi} \in Q_1} V(\boldsymbol{\varphi}(x)), \\ Q_1 = \{\boldsymbol{\varphi}(x)\colon 0 \leqslant \varphi_j(x) \leqslant b_j,\ j = 1, \ldots, n\}. \tag{11.32}$$

In order that the operator which will appear in the right-hand side of the said equations may be defined uniquely, we shall assume that the functional V satisfies the condition of strong regularity.

We shall say that the functional V satisfies the *condition of regularity* if

$$\sum_{j=1}^{n} [g^j(\boldsymbol{\varphi}^0(x))]^2 \neq 0 \tag{11.33}$$

almost everywhere in P. This means that the optimal vector-function $\boldsymbol{\varphi}^0$ is not an Euler extremal for V. An obvious example of a functional satisfying the condition of regularity is provided by the linear functional $\int_P [\mathbf{p}(x), \boldsymbol{\varphi}(x)]\,\mathrm{d}x$, where $\mathbf{p}$ is a given non-zero vector-function.

The functional V will be said to satisfy the *condition of strong regularity* if

$$g^j(\boldsymbol{\varphi}^0(x)) \neq 0, \quad j = 1, \ldots, n, \tag{11.34}$$

almost everywhere in P. Strong regularity implies that none of the Euler equations for unconstrained minimization of V may hold for the optimal vector-function $\boldsymbol{\varphi}^0$ on a set of non-zero measure in P. In the one-dimensional case ($n = 1$), the conditions of regularity and strong regularity coincide.

According to the optimality condition for non-classical problems derived in the previous section, if $\boldsymbol{\varphi}^0(x)$ is a solution to problem (11.32), then

$$\min_{\boldsymbol{\varphi}(x)\in Q_1} \int_P [g(\boldsymbol{\varphi}^0(x)), \boldsymbol{\varphi}(x)-\boldsymbol{\varphi}^0(x)]\mathrm{d}x = 0. \tag{11.35}$$

Let us consider the linear functional

$$\int_P [g(\boldsymbol{\varphi}^0(x)), \boldsymbol{\varphi}(x)]\mathrm{d}x. \tag{11.36}$$

The minimum of this functional in the set Q_1 (regarded as a subset of $L_2^n(P)$) is attained at any function $\boldsymbol{\varphi}^*$ such that

$$g^j(\boldsymbol{\varphi}^0(x))\varphi_j^*(x) = \min_{0\leqslant z_j\leqslant b_j} g^j(\boldsymbol{\varphi}^0(x))z_j, \quad j = 1,\dots,n, \tag{11.37}$$

for almost all $x \in P$. It follows that

$$\varphi_j^*(x) = \begin{cases} 0, & \text{when } g^j(\boldsymbol{\varphi}^0(x)) > 0, \\ b_j, & \text{when } g^j(\boldsymbol{\varphi}^0(x)) < 0. \end{cases} \tag{11.38}$$

If $g^j(\boldsymbol{\varphi}^0(x)) = 0$, which may only hold on a set of measure zero in P in view of the condition of strong regularity (11.34), the value $\varphi_j^*(x)$ can be chosen arbitrarily, e.g. 0. Then, $\boldsymbol{\varphi}^*(x)$ can be represented as

$$\varphi_j^*(x) = \frac{b_j}{2}[1-\operatorname{sign} g^j(\varphi^0(x))], \quad j = 1,\dots,n. \tag{11.39}$$

On the other hand, from (11.35) we obtain

$$\int_P [g(\boldsymbol{\varphi}^0(x)), \boldsymbol{\varphi}^0(x)]\mathrm{d}x = \min_{\boldsymbol{\varphi}\in Q_1} \int_P [g(\boldsymbol{\varphi}^0(x)), \boldsymbol{\varphi}(x)]\mathrm{d}x, \tag{11.40}$$

which means that functional (11.36) attains its minimum in Q_1 for $\boldsymbol{\varphi} = \boldsymbol{\varphi}^0$. Consequently, we have $\boldsymbol{\varphi}^0(x) \equiv \boldsymbol{\varphi}^*(x)$ for almost all x, i.e. $\boldsymbol{\varphi}^0$ may be found from the Trukhaev–Khomenyuk operator equation

$$\varphi_j^0(x) = \frac{b_j}{2}[1-\operatorname{sign} g^j(\boldsymbol{\varphi}^0(x))], \quad j = 1,\dots,n. \tag{11.41}$$

When only regularity condition (11.33) is satisfied, the optimal element $\boldsymbol{\varphi}^0$ is not defined uniquely. In this case, the operator equation for $\boldsymbol{\varphi}^0$ can be written in the form

$$g_j(\boldsymbol{\varphi}^0(x))\left\{\boldsymbol{\varphi}_j^0(x) - \frac{b_j}{2}[1-\operatorname{sign} g^j(\boldsymbol{\varphi}^0(x))]\right\} = 0, \quad j = 1, 2, \dots, n, \tag{11.42}$$

which may be regarded as a generalization of the classical Euler equations.

In the special case where there is no upper limit on the admissible functions $\boldsymbol{\varphi}$, i.e. if $Q_1 = \{\boldsymbol{\varphi}: \boldsymbol{\varphi}(x) \geqslant 0\}$, an optimal element $\boldsymbol{\varphi}^0$ may be sought from the equation

$$g_j(\boldsymbol{\varphi}^0(x))\varphi_j^0(x) = 0, \quad j = 1, \dots, n. \tag{11.43}$$

11.4 Elimination of inequality constraints

In this section we shall consider a few methods for reducing variational problems with inequality constraints of the first and second kind

$$\begin{aligned} Q_1 &= \{\boldsymbol{\varphi}(x):\ 0 \leqslant \varphi_j(x) \leqslant b_j\ (j = 1, \dots, n)\}, \\ Q_2 &= \{\boldsymbol{\varphi}(x):\ V_i(\boldsymbol{\varphi}) \leqslant c_i\ (i = 1, \dots, m)\}, \end{aligned} \tag{11.44}$$

to unconstrained or equality-constrained problems.

The transformation method

The general idea of the transformation method is the replacement of the functions φ_j by new unconstrained function variables $d_j(x)$, $-\infty < d_j(x) < \infty$, by means of a transformation

$$\varphi_j(x) = F_j(d_j), \quad j = 1, \dots, n, \tag{11.45}$$

where the functions F_j are chosen so as to ensure the satisfaction of the original constraints for arbitrary values of d_j. The choice of the transformation functions F_j is not unique; using the known properties of elementary or special functions, it is possible to construct many suitable combinations.

Constraints of the form

$$|\varphi_j| \leqslant B_j, \quad j = 1, \dots, n \tag{11.46}$$

will be satisfied automatically if we set, for example,

$$F_j = B_j \sin d_j, \quad j = 1, \dots, n. \tag{11.47}$$

To eliminate inequality constraints of the first kind,

$$0 \leqslant \varphi_j(x) \leqslant b_j, \quad j = 1, \dots, n, \tag{11.48}$$

we can choose F_j to be, among many other possibilities, of any of the forms

$$\begin{aligned} F_j &= \frac{b_j}{2}(1+\sin d_j), \quad & F_j &= b_j \mathrm{e}^{-(d_j)^2}, \\ F_j &= \frac{b_j}{2}(1+\cos d_j), \quad & F_j &= \frac{b_j}{2}(1+\Phi(d_j)); \end{aligned} \tag{11.49}$$

in the last expression, Φ denotes the Gauss probability integral, $\Phi(d_j) = 2/\sqrt{2\pi} \int_0^{d_j} \mathrm{e}^{-t^2/2}\,\mathrm{d}t$.

Choosing an appropriate transformation for constraints of the second kind is much more difficult. In a particular case where V are linear functionals of the form

$$\int_R p_i(x)\varphi_i(x)\mathrm{d}x \leqslant c_i, \quad i = 1, \ldots, m \leqslant n, \tag{11.50}$$

where $p_i(x)$ are given functions, we may apply the substitution

$$\varphi_i(x) = \frac{c_i}{\mu(P)p_i(x)} G_i\big(d_i(x)\big), \quad i = 1, \ldots, m,$$

$$\varphi_i(x) = d_i(x), \quad i = m+1, \ldots, n \quad (\text{if } m < n),$$

where $\mu(P) = \int_P \mathrm{d}x$ denotes the measure of the set $P \subset R^k$ and G_i can be taken any of the functions (11.49) divided by b_i, i.e. $G_i = F_i/b_i$. With this substitution, constraints (11.50) will be satisfied automatically.

In problems with inequality constraints of both kinds, a transformation eliminating constraints of one of the kinds will often complicate constraints of the other kind and may lead to a problem of increased difficulty. Also, the transformed objective function in the resulting unconstrained problem will generally be much more complex than the original one owing to the non-linearity of the transformation. These disadvantages considerably limit the usefulness of the transformation method.

Reduction of inequality constraints to equalities

Inequality constraints of the first or the second kind can be replaced by equalities with an increased number of unknown functions. This approach was first used by Valentine[1] who solved a variational problem with side conditions of the form

$$\varphi_{j\,\min}(x) \leqslant \varphi_j(x) \leqslant \varphi_{j\,\max}(x), \quad j = 1, \ldots, n. \tag{11.51}$$

Instead of these inequalities, Valentine considered the equivalent set of equalities

$$[\varphi_j(x) - \varphi_{j\,\min}(x)][\varphi_{j\,\max}(x) - \varphi_j(x)] = \Psi_j^2(x), \quad j = 1, \ldots, n, \tag{11.52}$$

where $\Psi_j(x)$ are additional unknown functions, $-\infty < \Psi_j(x) < \infty$, to be determined in the optimization process.

In the same manner, constraints of the first kind, $0 \leqslant \varphi_j(x) \leqslant b_j$, can be replaced by the equalities

$$\varphi_j(x)\big(b_j - \varphi_j(x)\big) = \Psi_j^2(x), \quad j = 1, \ldots, n. \tag{11.53}$$

Equivalently, we may introduce two additional sets of unknown functions, $\Psi_j(x)$ and $\eta_j(x)$, such that

[1] VALENTINE, F. A., *The problem of Lagrange with differential inequalities as added side conditions*, Diss. Depart., Chicago, Illinois, 1937.

$$\begin{aligned} &\varphi_j(x)+\Psi_j^2(x)-b_j = 0, \\ &\varphi_j(x)-\eta_j^2(x) = 0, \end{aligned} \qquad j = 1, \ldots, n. \tag{11.54}$$

Another possibility is to introduce new functions $\xi_j(x)$ and $\zeta_j(x)$ such that

$$\begin{aligned} &\varphi_j(x)+\Psi_j(x)-b_j = 0, \\ &\varphi_j(x)-\eta_j(x) = 0, \end{aligned} \qquad j = 1, \ldots, n, \tag{11.55}$$

where

$$\begin{aligned} \Psi_j(x) &= \sigma\big(\xi_j(x)\big)\xi_j^p(x), \\ \eta_j(x) &= \sigma\big(\zeta_j(x)\big)\zeta_j^p(x), \end{aligned} \qquad \sigma(a) = \begin{cases} 0, & a < 0, \\ 1, & a \geqslant 0, \end{cases} \tag{11.56}$$

p—positive integer.

A similar approach can be applied to constraints of the second kind. Inequalities $V_i \leqslant c_i$ $(i = 1, \ldots, m)$ can be replaced by the equalities

$$V_i(\boldsymbol{\varphi})\big(c_i-V_i(\boldsymbol{\varphi})\big) = \Psi_j^2, \quad i = 1, \ldots, m, \tag{11.57}$$

where Ψ_i, unknown real numbers $(-\infty < \Psi_i < \infty)$, are introduced as additional parameters of optimization.

Alternatively, we may use the equalities

$$v_i(\boldsymbol{\varphi})+\Psi_i-c_i = 0, \quad i = 1, \ldots, m, \tag{11.58}$$

where $\Psi_i = \sigma(\Xi_i)\Xi_i^p$, with a vector $\boldsymbol{\Xi}$ as an additional unknown of optimization.

In all the above instances of replacing inequality constraints by equalities, the number of unknowns, and hence the dimensionality of the problem, has been increased relative to the original. There are also many ways, however, in which the same end may be reached without introducing new unknowns.

For example, constraints of the first kind, $0 \leqslant \varphi_j(x) \leqslant b_j$, $j = 1, \ldots, n$, may be represented as

$$\varphi_j(x)[1-\operatorname{sign}\varphi_j(x)] = 0, \quad \big(\varphi_j(x)-b_j\big)[1+\operatorname{sign}\big(\varphi_j(x)-b_j\big)] = 0 \tag{11.59}$$

or as

$$\varphi_j(x)\big(\varphi_j(x)-b_j\big)+|\varphi_j(x)\big(\varphi_j(x)-b_j\big)| = 0. \tag{11.60}$$

Inequalities of the second kind, $V_i \leqslant c_i$, $i = 1, \ldots, m$, may be replaced by the equalities

$$\big(V_i(\boldsymbol{\varphi})-c_i\big)[1-\operatorname{sign}\big(V_i(\boldsymbol{\varphi})-c_i\big)] = 0 \tag{11.61}$$

or by

$$V_i(\boldsymbol{\varphi})-c_i-|V_i(\boldsymbol{\varphi})-c_i| = 0. \tag{11.62}$$

Reduction of the number of constraints

The number of inequality constraints (of the first or the second kind) in a given optimization problem can be reduced by the following simple method. Let

$$F(\alpha, \beta) = \alpha+\beta-\sqrt{\alpha^2+\beta^2} \tag{11.63}$$

or, alternatively,

$$F(\alpha+\beta) = \alpha+\beta-|\beta-\alpha|. \tag{11.64}$$

It is evident that the condition $F(\alpha, \beta) \geqslant 0$ is equivalent to the inequalities $\alpha \geqslant 0$, $\beta \geqslant 0$. Using this property of the function $F(\alpha, \beta)$, we can replace n inequalities $0 \leqslant \varphi_j \leqslant b_j$, or $\varphi_j \geqslant 0$ and $b_j-\varphi_j \geqslant 0$, by a single inequality. To this end, in the first step, we write these inequalities in the form

$$F(j) = F[\varphi_j(x), b_j-\varphi_j(x)] \geqslant 0, \quad j = 1, \ldots, n. \tag{11.65}$$

In the second step, if n is even, we replace inequalities (11.65) by the conditions

$$F(j,j+1) = F[F(j), F(j+1)] \geqslant 0, \quad j = 1, 3, 5, \ldots, n-1, \tag{11.66}$$

or, if n is odd, by the inequalities

$$F(j,j+1) = F[F(j), F(j+1)] \geqslant 0, \quad j = 1, 3, 5, \ldots, n-2, \tag{11.67}$$
$$F(n) \geqslant 0.$$

In this way, the number of constraints is reduced to $n/2$ if n is even and to $(n-1)/2$ if it is odd. By repeating the same procedure sufficiently many times, we eventually obtain a single inequality equivalent to the original system of n inequalities. Specifically, the system will be reduced to a single inequality after $n-E(n/2)+1$ steps, where $E(n/2)$ denotes the integral part of $n/2$.

Constraints of the second kind, $V_i(\boldsymbol{\varphi}) \leqslant c_i$ $(i = 1, \ldots, m)$, can be replaced by a single inequality in a similar manner. Instead of (11.65) we then have for an even m

$$\hat{F}(i) = F\big(c_i-V_i(\boldsymbol{\varphi}), c_{i+1}-V_{i+1}(\boldsymbol{\varphi})\big) \geqslant 0, \quad i = 1, 3, 5, \ldots, m-1, \tag{11.68}$$

and for an odd m

$$\hat{F}(i) = F\big(c_i-V_i(\boldsymbol{\varphi}), c_{i+1}-V_{i+1}(\boldsymbol{\varphi})\big) \geqslant 0,$$
$$i = 1, 3, 5, \ldots, m-2, \quad V_m(\boldsymbol{\varphi}) \leqslant c_m. \tag{11.69}$$

After $m-E(m/2)$ analogous steps, the original system is reduced to a single inequality.

Let us note that a similar method can be applied to problems with equality constraints or with mixed equality-inequality constraints.

11.5 Minimax formulation for non-classical variational problems

In this section we show how to reduce the variational problem (11.2) with inequality constraints of the first and the second kinds (11.5) to a minimax problem in which only constraints of the first kind are present.

Let $\boldsymbol{\psi}$ denote an m-dimensional vector and define

$$h(\boldsymbol{\varphi}(x), \boldsymbol{\Psi}) = V_0(\boldsymbol{\varphi}(x)) + [\boldsymbol{\Psi}, \mathbf{V}(\boldsymbol{\varphi}(x)) - \mathbf{c}], \tag{11.70}$$

where

$$[\boldsymbol{\Psi}, \mathbf{V}(\boldsymbol{\varphi}) - \mathbf{c}] = \sum_{i=1}^{m} \Psi_i (V_i(\boldsymbol{\varphi}) - c_i). \tag{11.71}$$

We shall call a pair $\{\boldsymbol{\varphi}^0(x), \boldsymbol{\Psi}^0\}$ a *saddle point* for the functional h in the region $\{\boldsymbol{\varphi} \in Q_1, \boldsymbol{\Psi} \geqslant \mathbf{0}\}$ if $\boldsymbol{\varphi}^0 \in Q_1, \boldsymbol{\Psi}^0 \geqslant \mathbf{0}$ and

$$h(\boldsymbol{\varphi}^0, \boldsymbol{\Psi}) \leqslant h(\boldsymbol{\varphi}^0, \boldsymbol{\Psi}^0) \leqslant h(\boldsymbol{\varphi}, \boldsymbol{\Psi}^0) \tag{11.72}$$

for all $\boldsymbol{\varphi} \in Q_1$ and $\boldsymbol{\Psi} \geqslant \mathbf{0}$. The same condition can be written

$$h(\boldsymbol{\varphi}^0(x), \boldsymbol{\Psi}^0) = \min_{\varphi(x) \in Q_1} \max_{\Psi \geqslant 0} h(\boldsymbol{\varphi}(x), \boldsymbol{\Psi}) = \max_{\Psi \geqslant 0} \min_{\varphi(x) \in Q_1} h(\boldsymbol{\varphi}(x), \boldsymbol{\Psi}). \tag{11.73}$$

We shall show that if $\{\boldsymbol{\varphi}^0(x), \boldsymbol{\Psi}^0\}$ is a solution of problem (11.73), then

$$[\boldsymbol{\Psi}^0, \mathbf{V}(\boldsymbol{\varphi}^0(x)) - \mathbf{c}] = 0 \tag{11.74}$$

and the function $\boldsymbol{\varphi}^0$ is a solution of the original problem.

With h defined by equation (11.70), condition (11.72) reads

$$\begin{aligned}\mathbf{V}_0(\boldsymbol{\varphi}^0) - [\boldsymbol{\Psi}, \mathbf{V}(\boldsymbol{\varphi}^0) - \mathbf{c}] &\leqslant \mathbf{V}_0(\boldsymbol{\varphi}^0) + [\boldsymbol{\Psi}^0, \mathbf{V}(\boldsymbol{\varphi}^0) - \mathbf{c}] \\ &\leqslant \mathbf{V}_0(\boldsymbol{\varphi}) + [\boldsymbol{\Psi}^0, \mathbf{V}(\boldsymbol{\varphi}) - \mathbf{c}]\end{aligned} \tag{11.75}$$

for all $\boldsymbol{\varphi} \in Q_1$ and $\boldsymbol{\Psi} \geqslant \mathbf{0}$. The fact that the left-hand side inequality holds for arbitrary $\boldsymbol{\Psi} \geqslant \mathbf{0}$ implies that all the components of the vector $\mathbf{V}(\boldsymbol{\varphi}^0) - \mathbf{c}$ must be non-positive and, consequently, that the scalar product $[\boldsymbol{\Psi}^0, \mathbf{V}(\boldsymbol{\varphi}^0) - \mathbf{c}]$ must be zero:

$$\mathbf{V}(\boldsymbol{\varphi}^0) - \mathbf{c} \leqslant \mathbf{0}, \quad [\boldsymbol{\Psi}^0, \mathbf{V}(\boldsymbol{\varphi}^0) - \mathbf{c}] = 0. \tag{11.76}$$

Hence, the right-hand-side inequality (11.75) becomes

$$V_0(\boldsymbol{\varphi}^0) \leqslant V_0(\boldsymbol{\varphi}) + [\boldsymbol{\Psi}^0, \mathbf{V}(\boldsymbol{\varphi}) - \mathbf{c}] \tag{11.77}$$

for all $\boldsymbol{\varphi} \in Q_1$. Since, in particular, for any $\boldsymbol{\varphi} \in Q_2$ we have $[\boldsymbol{\psi}^0, \mathbf{V}(\boldsymbol{\varphi}) - \mathbf{c}] \leqslant 0$, it follows that for all $\boldsymbol{\varphi} \in Q = Q_1 \cap Q_2$

$$V_0(\boldsymbol{\varphi}^0) \leqslant V_0(\boldsymbol{\varphi}) + [\boldsymbol{\Psi}^0, \mathbf{V}(\boldsymbol{\varphi}) - \mathbf{c}] \leqslant V_0(\boldsymbol{\varphi}), \tag{11.77}$$

which proves the optimality of $\boldsymbol{\varphi}^0$.

11.6 Approximate methods for non-classical variational problems

The penalty function method

An approximate solution to a variational problem $V_0(\boldsymbol{\varphi}^0) = \min_{\boldsymbol{\varphi} \in Q} V_0(\boldsymbol{\varphi})$, where, as before, the set $Q = Q_1 \cap Q_2$ is given by inequality constraints of the first and the second kinds (11.5), can be obtained by solving the following problem, free of constraints of the second kind:

Find a vector-function $\boldsymbol{\varphi}_\psi^0 \in Q_1$ such that

$$V_\psi(\boldsymbol{\varphi}_\psi^0) = \min_{\boldsymbol{\varphi} \in Q_1} V_\psi(\boldsymbol{\varphi}), \tag{11.78}$$

where

$$V_\psi(\boldsymbol{\varphi}) = V_0(\boldsymbol{\varphi}) + \frac{1}{2}\sum_{i=1}^{m} \Psi_i [V_i(\boldsymbol{\varphi}) - c_i]^2 [1 + \operatorname{sign}(V_i(\boldsymbol{\varphi}) - c_i)]. \tag{11.79}$$

The second term in (11.79), called the *penalty function,*

$$\frac{1}{2}\sum_{i=1}^{m} \Psi_i [V_i(\boldsymbol{\varphi}) - c_i]^2 [1 + \operatorname{sign}(V_i(\boldsymbol{\varphi}) - c_i)], \tag{11.80}$$

is non-negative for arbitrary $\boldsymbol{\varphi} \in Q_1$ and $\Psi_i \geqslant 0$ and vanishes only when $V_i(\boldsymbol{\varphi}) \leqslant c_i$, $i = 1, \ldots, m$. When any of the constraints is violated, function (11.80) imposes a "penalty" on the objective functional. For sufficiently large positive coefficients Ψ_i, which means imposing a large penalty for any violation of the constraints, the solution of problem (11.78) can be shown to satisfy the constraints of the second kind to any required accuracy and yield a value of the original objective functional V_0 not greater than its minimum on Q. More precisely, the following theorem holds:

Assume that for every $\boldsymbol{\varphi} \in Q_1$

$$V_0(\boldsymbol{\varphi}) \geqslant \bar{c} > -\infty, \tag{11.81}$$

and that there exists a vector-function $\overline{\boldsymbol{\varphi}} \in Q_1$ such that $V_i(\overline{\boldsymbol{\varphi}}) \leqslant c_i, i = 1, \ldots, m$, i.e. the admissible space Q is non-empty.

Then, the solution $\boldsymbol{\varphi}_\psi^0$ of problem (11.78) *with*

$$\Psi_1 = \Psi_2 = \ldots = \Psi_m = \Psi \geqslant \Psi_\varepsilon = \frac{2}{\varepsilon^2}\left(V_0(\overline{\boldsymbol{\varphi}}) - \bar{c}\right) \geqslant 0 \tag{11.82}$$

satisfies the optimality condition

$$V_0(\boldsymbol{\varphi}_\psi^0) \leqslant \min_{\boldsymbol{\varphi} \in Q} V_0(\boldsymbol{\varphi}) \tag{11.83}$$

and the inequalities

$$V_i(\boldsymbol{\varphi}_\psi^0) \leqslant c_i + \varepsilon, \quad i = 1, \ldots, m, \tag{11.84}$$

where $\varepsilon > 0$ is an arbitrarily small given number.

In the particular case where $V_0(\overline{\boldsymbol{\varphi}}) = \bar{c}$, the function $\overline{\boldsymbol{\varphi}} = \boldsymbol{\varphi}^0 = \boldsymbol{\varphi}^0_\psi$ is an exact solution to the original problem.

Inequality (11.83) follows directly from the definition of the functional V_φ (11.79). Indeed, if $\Psi_i \geqslant 0$, $i = 1, \ldots, m$, then

$$V_0(\boldsymbol{\varphi}^0_\psi) \leqslant V_\psi(\boldsymbol{\varphi}^0_\psi) = \min_{\varphi \in Q_1} V_\psi(\boldsymbol{\varphi}) \leqslant \min_{\varphi \in Q_1 \cap Q_2} V_\psi(\boldsymbol{\varphi}) = \min_{\varphi \in Q} V_0(\boldsymbol{\varphi}). \tag{11.85}$$

To prove inequalities (11.84), let us first note that since $V_\psi(\boldsymbol{\varphi}^0_\psi) \leqslant \min\limits_{\varphi \in Q} V_0(\boldsymbol{\varphi})$ and $\overline{\boldsymbol{\varphi}} \in Q$, we have

$$V_\psi(\boldsymbol{\varphi}^0_\psi) \leqslant V_0(\overline{\boldsymbol{\varphi}}). \tag{11.86}$$

Hence, using expression (11.79) with $\Psi_i = \Psi$, $i = 1, \ldots, m$, we obtain

$$\frac{\Psi}{2} \sum_{i=1}^{m} (V_i(\boldsymbol{\varphi}^0_\psi) - c_i)^2 [1 + \operatorname{sign}(V_i(\boldsymbol{\varphi}^0_\psi) - c_i)] \leqslant V_0(\overline{\boldsymbol{\varphi}}) - V_0(\boldsymbol{\varphi}^0_\psi). \tag{11.87}$$

By assumption (11.81), the right-hand side is less than or equal to $V_0(\overline{\boldsymbol{\varphi}}) - \bar{c}$ and therefore

$$\sum_{i=1}^{m} (V_i(\boldsymbol{\varphi}^0_\psi) - c_i)^2 [1 + \operatorname{sign}(V_i(\boldsymbol{\varphi}^0_\psi) - c_i)]$$
$$\leqslant \frac{V_0(\overline{\boldsymbol{\varphi}}) - \bar{c}}{\Psi} 2 \leqslant 2 \frac{V_0(\overline{\boldsymbol{\varphi}}) - \bar{c}}{\Psi_\varepsilon} = \varepsilon^2. \tag{11.88}$$

Since each summand in the left-hand side of this inequality is non-negative, we may conclude that

$$(V_i(\boldsymbol{\varphi}^0_\psi) - c_i)^2 [1 + \operatorname{sign}(V_i(\boldsymbol{\varphi}^0_\psi) - c_i)] \leqslant \varepsilon^2 \tag{11.89}$$

and so

$$V_i(\boldsymbol{\varphi}^0_\psi) \leqslant c_i + \varepsilon, \tag{11.90}$$

which completes the proof.

In problem (11.78), the penalized functional V_ψ is minimized subject to constraints of the first kind only, which permits us to use the results of Section 11.3 to derive an operator equation for the optimal function $\boldsymbol{\varphi}^0_\psi$.

Let $\boldsymbol{\Psi} = \boldsymbol{\Psi}_\varepsilon$ as in (11.82), and let $\boldsymbol{\varphi}^0_{\psi_\varepsilon}$ be a solution of the corresponding problem (11.78), i.e.

$$V_{\psi_\varepsilon}(\boldsymbol{\varphi}^0_{\psi_\varepsilon}(x)) = \min_{\varphi \in Q_1} V_{\psi_\varepsilon}(\boldsymbol{\varphi}(x)), \tag{11.91}$$

where

$$V_{\psi_\varepsilon}(\boldsymbol{\varphi}) = V_0(\boldsymbol{\varphi}) + \frac{\Psi_\varepsilon}{2} \sum_{i=1}^{m} [V_i(\boldsymbol{\varphi}) - c_i]^2 [1 + \operatorname{sign}(V_i(\boldsymbol{\varphi}) - c_i)]. \tag{11.92}$$

Assuming that the functions $f_0(x, \boldsymbol{\varphi}), f_1(x, \boldsymbol{\varphi}), \ldots, f_m(x, \boldsymbol{\varphi})$ are continuously differentiable with respect to $\varphi_1, \ldots, \varphi_n$, we define the gradients $\mathbf{g}_i(\boldsymbol{\varphi})$ of the functionals $V_i(\boldsymbol{\varphi})$, $i = 0, 1, \ldots, m$, in the usual manner

$$\mathbf{g}_i(\boldsymbol{\varphi}) = \operatorname{grad} V_i(\boldsymbol{\varphi}) = \left(\frac{\partial f_i(x, \boldsymbol{\varphi}(x))}{\partial \varphi_1}, \ldots, \frac{\partial f_i(x, \boldsymbol{\varphi}(x))}{\partial \varphi_n}\right), \quad i = 0, 1, \ldots, m. \tag{11.93}$$

Then, the functional $V_{\psi\varepsilon}$ has a continuous gradient, $\mathbf{g}_{\psi\varepsilon}(\boldsymbol{\varphi}) = (g^1_{\psi\varepsilon}(\boldsymbol{\varphi}), \ldots$ $\ldots, g^n_{\psi\varepsilon}(\boldsymbol{\varphi}))$, which can be represented as

$$\mathbf{g}_{\psi\varepsilon}(\boldsymbol{\varphi}) = \operatorname{grad} V_{\psi\varepsilon}(\boldsymbol{\varphi})$$

$$= \mathbf{g}_0(\boldsymbol{\varphi}) + \Psi_\varepsilon \sum_{j=1}^{n} [V_i(\boldsymbol{\varphi}) - c_i][1 + \operatorname{sign}(V_i(\boldsymbol{\varphi}) - c_i)] g_i(\boldsymbol{\varphi}). \tag{11.94}$$

According to equation (11.25), the necessary (and sufficient if V_i are convex, $i = 0, 1, \ldots, m$) condition for $\boldsymbol{\varphi}^0_{\psi\varepsilon}$ to be optimal for $V_{\psi\varepsilon}$ is that

$$\min_{\boldsymbol{\varphi} \in Q_1} \int_P [\mathbf{g}_{\psi\varepsilon}(\boldsymbol{\varphi}^0_{\psi\varepsilon}(x)), \boldsymbol{\varphi}(x) - \boldsymbol{\varphi}^0_{\psi\varepsilon}(x)] \mathrm{d}x = 0. \tag{11.95}$$

As shown in Sec. 11.3, if the functional $V_{\psi\varepsilon}$ satisfies the condition of regularity, this equation leads to the Trukhaev–Khomenyuk operator equation

$$g^j_{\psi\varepsilon}(\boldsymbol{\varphi}^0_{\psi\varepsilon}(x))\left\{\varphi^0_{j\psi\varepsilon} - \frac{b_j}{2}[1 - \operatorname{sign} g^j_\psi(\boldsymbol{\varphi}^0(x))]\right\} = 0, \quad j = 1, \ldots, n. \tag{11.96}$$

The ε-optimization method

Let ε be an arbitrary (small) positive number. A vector-function $\boldsymbol{\varphi}^0_\varepsilon(x)$ is called an *ε-optimal solution* to the problem

$$V(\boldsymbol{\varphi}^0) = \min_{\boldsymbol{\varphi} \in Q_1} V(\boldsymbol{\varphi}),$$
$$Q_1 = \{\boldsymbol{\varphi}\colon 0 \leqslant \varphi_i(x) \leqslant b_i,\ i = 1, \ldots, n\}, \tag{11.97}$$

if $\boldsymbol{\varphi}^0_\varepsilon \in Q_1$ and

$$V(\boldsymbol{\varphi}^0_\varepsilon) \leqslant \min_{\boldsymbol{\varphi} \in Q_1} V(\boldsymbol{\varphi}) + \varepsilon. \tag{11.98}$$

It is evident that $\boldsymbol{\varphi}^0_\varepsilon$ is not unique.

Let $\Omega \subset Q_1 \times Q_1$ be the set of all pairs $(\boldsymbol{\varphi}, \boldsymbol{\eta})$, $\boldsymbol{\varphi} \in Q_1, \boldsymbol{\eta} \in Q_1$, such that

$$\left(\varphi_j(x) - \frac{b_j}{2}\right)^2 + \left(\eta_j(x) - \frac{b_j}{2}\right)^2 \leqslant \frac{b_j^2}{4}, \quad j = 1, \ldots, n. \tag{11.99}$$

We will show that if a pair $(\boldsymbol{\varphi}_\varepsilon^0, \boldsymbol{\eta}_\varepsilon^0)$ minimizes in Ω the following auxiliary functional

$$V_\varepsilon(\boldsymbol{\varphi}; \boldsymbol{\eta}) = V(\boldsymbol{\varphi}) + \frac{2\varepsilon}{n\mu(P)} \sum_{j=1}^{n} \frac{1}{b_j} \int_P \left(\eta_j(x) - \frac{b_j}{2}\right) \mathrm{d}x, \tag{11.100}$$

i.e. if

$$V_\varepsilon(\boldsymbol{\varphi}_\varepsilon^0, \boldsymbol{\eta}_\varepsilon^0) = \min_{(\boldsymbol{\varphi}, \boldsymbol{\eta}) \in \Omega} V_\varepsilon(\boldsymbol{\varphi}, \boldsymbol{\eta}), \tag{11.101}$$

then $\boldsymbol{\varphi}_\varepsilon^0$ is an ε-optimal solution to problem (11.97).

Indeed, since for each $\boldsymbol{\varphi} \in Q_1$ the pair $(\boldsymbol{\varphi}, \mathbf{b}/2)$ belongs to Ω and $V_\varepsilon(\boldsymbol{\varphi}, \mathbf{b}/2) = V(\boldsymbol{\varphi})$, we can write

$$\begin{aligned} V_\varepsilon(\boldsymbol{\varphi}_\varepsilon^0, \boldsymbol{\eta}_\varepsilon^0) &= \min_{(\boldsymbol{\varphi}, \boldsymbol{\eta}) \in \Omega} V_\varepsilon(\boldsymbol{\varphi}, \boldsymbol{\eta}) \\ &\leqslant \min_{\boldsymbol{\varphi} \in Q_1} V_\varepsilon(\boldsymbol{\varphi}, \mathbf{b}/2) = \min_{\boldsymbol{\varphi} \in Q_1} V(\boldsymbol{\varphi}). \end{aligned} \tag{11.102}$$

Using this inequality, we obtain from the definition of V_ε

$$\begin{aligned} V(\boldsymbol{\varphi}_\varepsilon^0) &= V_\varepsilon(\boldsymbol{\varphi}_\varepsilon^0, \boldsymbol{\eta}_\varepsilon^0) - \frac{2\varepsilon}{n\mu(P)} \sum_{j=1}^{n} \frac{1}{b_j} \int_P \left(\eta_{\varepsilon j}^0 - \frac{b_j}{2}\right) \mathrm{d}x \\ &\leqslant V_\varepsilon(\boldsymbol{\varphi}_\varepsilon^0, \boldsymbol{\eta}_\varepsilon^0) + \varepsilon \leqslant \min_{\boldsymbol{\varphi} \in Q_1} V(\boldsymbol{\varphi}) + \varepsilon \end{aligned} \tag{11.103}$$

as required.

To derive an operator equation for $\boldsymbol{\varphi}_\varepsilon^0$, we use the optimality condition (11.25). Applied to problem (11.101), it takes the form

$$\begin{aligned} \min_{(\boldsymbol{\varphi}, \boldsymbol{\eta}) \in \Omega} \Bigg\{ \sum_{j=1}^{n} \int_P \Bigg[& g^j(\boldsymbol{\varphi}_\varepsilon^0(x))(\varphi_j(x) - \varphi_{\varepsilon j}^0(x)) + \\ & + \frac{2\varepsilon}{n\mu(P) b_j} (\eta_j(x) - \eta_{\varepsilon j}^0(x)) \Bigg] \mathrm{d}x \Bigg\} = 0. \end{aligned} \tag{11.104}$$

It follows that, for almost all $x \in P$,

$$\begin{aligned} \min_{(\boldsymbol{\varphi}, \boldsymbol{\eta}) \in \Omega} \Bigg\{ & g^j(\boldsymbol{\varphi}_\varepsilon^0(x))(\varphi_j(x) - \varphi_{\varepsilon j}^0(x)) + \\ & + \frac{2\varepsilon}{n\mu(P) b_j} (\eta_j(x) - \eta_{\varepsilon j}^0(x)) \Bigg\} = 0. \end{aligned} \tag{11.105}$$

Since the minimized expression is linear and Ω is convex, the minimum will be realized on the boundary of Ω, i.e. in the set of pairs $(\boldsymbol{\varphi}, \boldsymbol{\eta})$ satisfying the equalities

$$\left(\varphi_j(x) - \frac{b_j}{2}\right)^2 + \left(\eta_j(x) - \frac{b_j}{2}\right)^2 = \frac{b_j^2}{4}, \quad j = 1, \ldots, n. \tag{11.106}$$

Hence, by applying the Lagrange multipiers method, one will obtain the following equations for the ε-optimal solution $\boldsymbol{\varphi}_{\varepsilon}^{0}$:

$$\varphi_{j\varepsilon}^{0}(x) = \frac{b_j}{2}\left\{1 - \frac{g^j(\boldsymbol{\varphi}_{\varepsilon}^{0}(x))}{\sqrt{[g^j(\boldsymbol{\varphi}_{\varepsilon}^{0}(x))]^2 + \dfrac{4\varepsilon^2}{n^2[\mu(R)]^2 b_j^2}}}\right\}, \quad j = 1, \dots, n. \tag{11.107}$$

It can be seen that if the gradient $\mathbf{g}$ of the functional V is sufficiently smooth, then the operator defined by the right-hand side of equation (11.107) is also smooth.

12

Mathematical Theory of Extremum Problems

12.1 Introduction

Extremum problems have been among the main interests of mathematics from the very earliest stages of its development and can be traced back to the works of Euclid and Archimedes. In the 17th and 18th centuries the works of Galileo, Newton, Johann and Jacob Bernoulli, de l'Hospital, Leibniz and, above all, Euler, laid the foundations of what is now called the calculus of variations. The classical results of this theory are mostly concerned with the extrema of smooth functions or functionals defined over an entire space or restricted to some smooth manifold. The extremum conditions in this case are provided by Euler equations with or without Lagrange multipliers, according as the problem is or is not constrained.

Technological progress has created the need for solving extremum problems of a new type: the control of objects whose control parameters are varied within some closed sets. Investigation of such problems led Pontryagin, Boltyanski, Gamkrelidze and Mishchenko to formulating a new necessary condition for an extremum, which is known as the *Pontryagin maximum principle.* The nature of this condition and the form of the optimal solution differed so much from the classical results of the calculus of variations that some writers began to speak of a "new" calculus of variations.

At the same time the demands of economics have stimulated the search for new methods of determining extrema of smooth functions of a large number of variables on closed domains with piecewise-smooth boundaries. As a result, a whole new field of special methods has emerged, known as mathematical programming.

The above classes of extremum problems do not exhaust all possibilities of classification in extremum theory. Divisions are also made with respect to the differentiability of functionals, the form of constraints, functions of state or the number of variables.

As an illustration, let us list a few different types of problems.

1. Let $F_0(x)$ be a smooth function in R^m and $Q \subset R^m$ a set defined by

a system of equations $F_i(x) = 0$, $x \in R^m$, $i = 1, \dots, n$.[1] Find the minimum of F_0 on Q. The necessary conditions for the function F_0 to take its minimum on Q at a point $x_0 \in Q$ are formulated in classical analysis with the use of Lagrange multipliers.

2. Let Φ be a smooth function of three variables. Find the minimum of the functional

$$F = \int_{t_0}^{t_1} \Phi\left[x(t), \frac{dx(t)}{dt}, t\right] dt,$$

in the class of differentiable functions $x(t)$ satisfying the condition $x(t_0) = c$, $x(t_1) = d$. This is the fundamental problem of the calculus of variations. Necessary conditions for an extremum are given by the Euler equations.

3. Minimize the functional

$$\int_{t_0}^{t_1} \Phi\left[x_1(t), x_2(t), x_3(t), \frac{dx_1(t)}{dt}, \frac{dx_2(t)}{dt}, \frac{dx_3(t)}{dt}, t\right] dt,$$

in the set of all curves $x_1(t), x_2(t), x_3(t)$ lying on the surface $G(x_1, x_2, x_3, t) = 0$. This problem is known as the *Lagrange problem*. Necessary conditions for its solution are given by the Euler equation with additional terms to allow for the constraint $G = 0$.

4. Find the minimum of the functional

$$\int_{t_0}^{t_1} \Phi[x(t), u(t), t] \, dt,$$

where $x \in R^n$, $u(t) \in M \subset R^r$ for $t_0 \leqslant t \leqslant t_1$, and

$$\frac{dx(t)}{dt} = \varphi(x(t), u(t), t),$$

$$x(t_0) = c, \quad x(t_1) = d.$$

This problem is known as the *optimal control problem*. The necessary conditions to be satisfied by its solution are provided by the Pontryagin maximum principle.

5. Let $F_0(x)$ be a smooth function in R^m and $Q \subset R^m$ a set defined by a system of equations $F_i(x) = 0$, $i = 1, \dots, k$ and inequalities $F_i(x) \leqslant 0$, $i = k+1, \dots, n$. Find the minimum of F_0 on Q. This is the general problem

[1] Notation used in this chapter corresponds with that of topology and functional analysis: scalars will be denoted by Greek letters, the elements of a topological space E by small Latin letters and sets in E by Latin capitals; bold-faced type of vector notation will not be used.

of non-linear programming. A special case is the problem of linear programming: find $\min(c, x)$ subject to $Ax \leqslant b$, where $x, c \in R^m$, $b \in R^n$ and A is an $n \times m$ matrix. In linear programming the necessary conditions are also sufficient.

Of problems 1–5 above, problem 1 is solved in classical analysis, problems 2 and 3 are dealt with in the classical calculus of variations, problem 4 in what is sometimes called the non-classical calculus of variations, and problem 5 in the field of non-linear or linear programming.

Thus, owing to the great diversity of extremum problems, the theory developed into several separate branches dealing with different types of problems and for a long time no attempts were made to produce a unified treatment. Finally, in 1962, Dubovitskii and Milyutin found a necessary condition for an extremum in the form of an equation set down in the language of functional analysis. From that equation it was possible to derive—as special cases—almost all previously known extremum conditions and thus to unify all the earlier theories into one modern calculus of variations.

The universality of the Dubovitskii–Milyutin method makes it a powerful theoretical tool in the realm of optimization. In particular, it may provide a basis for a future synthetic approach to structural optimization. It is this possibility of a future application of the Dubovitskii–Milyutin theory to optimum design of engineering structures which has led us to include an outline of that theory in the present book. The following exposition is based on a book by Girsanov[1].

12.2 Cones and dual cones

The extremum conditions derived by Dubovitskii and Milyutin make use of a certain special type of sets in a linear topological space and its dual, namely of cones and dual cones. Below we give the most important definitions and theorems relating to this kind of sets.

A set K in a linear space E is called a *cone with apex at* 0 if $x \in K$ implies that $\lambda x \in K$ for all $\lambda > 0$. If K is a cone with apex at 0, then the set $x_0 + K$ is called a *cone with apex at* x_0.

A cone which is a convex set is called a *convex cone*. A convex cone with apex at 0 may be defined as a set which, together with any two points x, y, contains all points of the form $\alpha x + \beta y$, $\alpha > 0$, $\beta > 0$.

The following sets are convex cones in the plane: a ray, a straight line, an angle not greater than π, a half plane and the entire plane; an angle greater than π is a cone but is not a convex cone.

[1] Girsanov, I. V., *Lectures on Mathematical Theory of Extremum Problems*, Lecture Notes in Economics and Mathematical Systems, 67, Springer-Verlag, Berlin–Heidelberg–New York 1972.

Let K be a cone with apex at 0 in a linear topological space E. The set K^* of all continuous linear functionals non-negative on K is called the *dual cone* of K. Thus, $K^* = \{f \in E': f(x) \geqslant 0 \text{ for all } x \in K\}$, where E' denotes the dual space of E, i.e. the set of all bounded linear functionals on E. It follows from the definition that K^* is a convex cone with apex at 0.

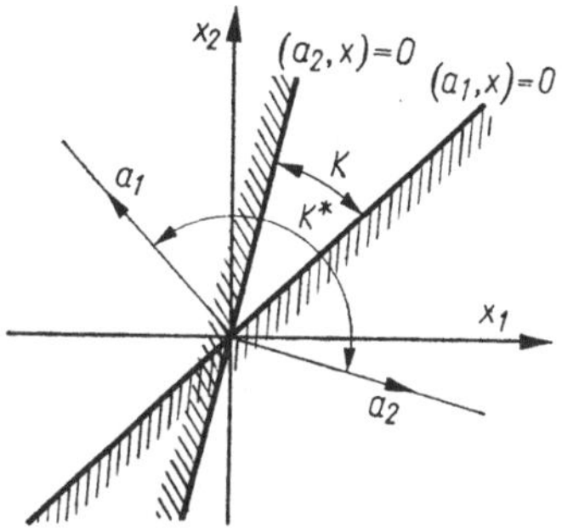

Fig. 12.1. A cone and its dual

The set of all x in E such that $f(x) \geqslant 0$ for all linear functionals f in the dual cone K^* will be denoted by K^{**}:

$$K^{**} = \{x \in E: f(x) \geqslant 0 \text{ for all } f \in K^*\}.$$

Here are two simple examples of cones and their duals:

Example 1.

$$E = R^2, \quad K = \{x: (a_1, x) \geqslant 0, (a_2, x) \geqslant 0\}, \quad a_1 \in R^2, a_2 \in R^2,$$

$$K^* = \{f: f = \lambda_1 a_1 + \lambda_2 a_2, \lambda_1 \geqslant 0, \lambda_2 \geqslant 0\};$$

(a, x) denotes the scalar product of the vectors a and x. Vector f is identified with a linear functional on R^2. The sets K and K^* are shown in Fig, 12.1.

Example 2.

$$E = R^2, \quad K = \{x = (x_1, x_2), x_1 \geqslant 0, x_2 \geqslant 0\}, \quad K^* = K.$$

Let A and B be two sets in a linear topological space E. A non-zero continuous linear functional f is said to *separate* the sets A and B if there exists a number α such that $f(x) \geqslant \alpha$ for $x \in A$ and $f(x) \leqslant \alpha$ for $x \in B$ (if both inequalities in this definition are sharp, f is said to *strongly separate* A and B). The closed hyperplane $H = \{x: f(x) = \alpha\}$ is then called a *separating hyperplane* for A and B, and the sets A and B themselves are said to be *separable.*

The fundamental theorem on separability was given by Hahn and Banach. In a geometric formulation the Hahn–Banach theorem reads:

Any two disjoint convex subsets of a linear topological space, one of which contains an interior point, are separable.

A non-zero continuous linear functional f is said to be a *supporting functional* for a set $A \subset E$ at a point $x_0 \in A$ if $f(x) \geqslant f(x_0)$ for all $x \in A$. Under this condition, the closed hyperplane $H = \{x: f(x) = f(x_0)\}$ is called a *supporting hyperplane* for A at x_0.

LEMMA 1. *If K is a cone with apex at x_0 and $f(x)$ a linear functional such that $f(x) \geqslant \alpha$ for $x \in K$, then $f(x) \geqslant f(x_0)$ for $x \in K$, i.e. the functional f is a supporting functional for the cone K at the apex.*

COROLLARY. *If two cones K_1 and K_2 with apex at 0 are separable by a hyperplane, then it must pass through 0. In particular, if a cone with apex at 0 and a subspace L are separable by a hyperplane, then L lies entirely in that hyperplane.*

THEOREM 1 (on extensions of linear functionals) (Krein). *If K is a convex cone with apex at 0 and non-empty interior, L a subspace such that $(\operatorname{int} K) \cap \cap L \neq \varnothing$, and $\bar{f}(x)$ a linear functional on L such that $\bar{f}(x) \geqslant 0$ on $K \cap L$, then there is a continuous linear functional $f(x)$ on E such that $f(x) = \bar{f}(x)$ for $x \in L$ and $f(x) \geqslant 0$ for all $x \in K$.*

LEMMA 2. *The dual of a union of cones is equal to the intersection of the corresponding dual cones:*

$$\Big(\bigcup_{\alpha \in A} K_\alpha\Big)^* = \bigcap_{\alpha \in A} K_\alpha^*.$$

LEMMA 3. *If $K_1 \subset K_2$, then $K_2^* \subset K_1^*$.*

LEMMA 4. *If $K_1, \ldots, K_n$ are convex open cones and their common part is non-empty, i.e. $\bigcap_{i=1}^{n} K_i \neq \varnothing$, then*

$$\Big(\bigcap_{i=1}^{n} K_i\Big)^* = \sum_{i=1}^{n} K_i^*.$$

LEMMA 5. (Dubovitskii–Milyutin). *Let $K_1, \ldots, K_n, K_{n+1}$ be convex cones with apex at 0 and let $K_1, \ldots, K_n$ be open. The common part of all the cones is empty, $\bigcap_{i=1}^{n+1} K_i = \varnothing$, if and only if there exist linear functionals $f_i \in K_i^*$, $i = 1, \ldots$ $\ldots, n+1$, not all zero, such that*

$$f_1 + f_2 + \ldots + f_n + f_{n+1} = 0.$$

The above lemma provides the basis for the proof of the fundamental necessary conditions for an extremum of a functional.

12.3 Necessary extremum conditions

Let $F(x)$ be a functional defined in a linear locally convex topological space E and consider the problem of finding a minimum of F subject to constraints of two types:

1. $x \in Q_i$, $i = 1, \ldots, n$, where Q_i are sets with non-empty interior;
2. $x \in Q_{n+1}$, where Q_{n+1} has no interior points.

In the usual formulation, the sets Q_i, $i = 1, \ldots, n$, are given by inequality constraints and the set Q_{n+1} by a system of equality constraints, so that Q_{n+1} is a manifold of lower dimension than the space.

The problem amounts to determining a local minimum of the functional F on the common part Q of the sets Q_i, $Q = \bigcap_{i=1}^{n+1} Q_i$, i.e. finding a point $x_0 \in Q$ such that

$$F(x_0) = \min_{Q \cap U} F(x),$$

where U is some neighbourhood of x_0.

To formulate the conditions which must hold at x_0, we first introduce the concepts of

— directions of decrease of the funcional F,
— feasible directions for Q_i $(i = 1, \ldots, n)$ at x_0, and
— tangent directions to Q_{n+1} at x_0.

A vector h will be called a *direction of decrease* of the functional $F(x)$ at the point x_0 if there is a neighbourhood U of h and a number $\alpha = \alpha(F, x_0, h) < 0$ such that

$$F(x_0 + \varepsilon \bar{h}) \leqslant F(x_0) + \varepsilon \alpha$$

for every $0 < \varepsilon < \varepsilon_0$ and every $\bar{h} \in U$.

The following lemma is readily obtained from the above definition:

LEMMA 6. *The set of all directions of decrease of F at x_0 is an open cone with apex at* 0.

The functional $F(x)$ will be said to *decrease regularly* at x_0 if its cone of decrease at x_0 is convex.

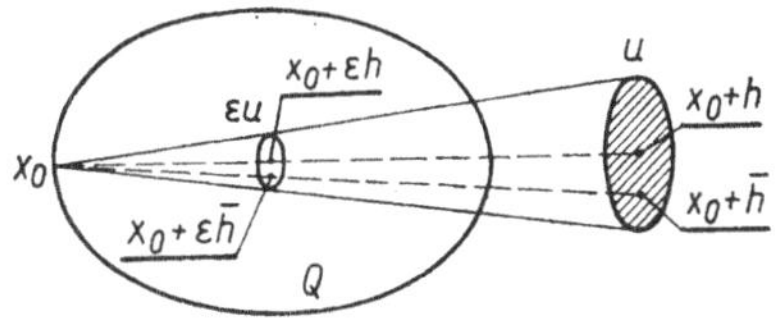

Fig. 12.2. A feasible direction for Q at x_0

Given an inequality constraint represented by a set Q, we shall say that a vector h is a *feasible direction* for Q at x_0 if there is a neighbourhood U of h such that for every $0 < \varepsilon < \varepsilon_0$ and all $\bar{h} \in U$ the vector $x_0 + \varepsilon \bar{h}$ is in Q (Fig. 12.2). As before, we have

LEMMA 7. *The set of all feasible directions for Q at x_0 is an open cone with apex at* 0.

We shall say that the inequality constraint represented by Q is *regular at* x_0 if the cone of feasible directions for Q at x_0 is convex.

For equality constraints we introduce the concept of a tangent direction:

A vector h will be called a *tangent direction to* Q at a point x_0 if for every $0 < \varepsilon < \varepsilon_0$ there exists a point $x(\varepsilon) \in Q$ such that if we set $x(\varepsilon) = x_0 + \varepsilon h +$

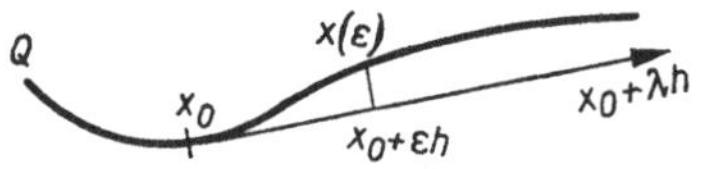

Fig. 12.3. A tangent direction to Q at x_0

$+r(\varepsilon)$ (Fig. 12.3), then the vector $r(\varepsilon) \in E$ has the property that, for any neighbourhood U of zero, $\frac{1}{\varepsilon} r(\varepsilon) \in U$ for all small $\varepsilon > 0$; in a Banach space this is equivalent to the requirement $||r(\varepsilon)|| = O(\varepsilon)$.

LEMMA 8. *The set of all tangent directions to Q at x_0 is a cone with apex at* 0.

Note that, unlike the cone of feasible directions, the cone of tangent directions is in general neither open nor closed.

An equality constraint Q will be said to be *regular at a point* x_0 if the cone of tangent directions to Q at x_0 is convex.

We are now ready to formulate the fundamental theorem:

THEOREM 2 (Dubovitskii–Milyutin). *If the functional $F(x)$ attains a local minimum on the set $Q = \bigcap_{i=1}^{n+1} Q_i$ at a point $x_0 \in Q$ and, in addition,*

(a) *$F(x)$ is regularly decreasing at x_0, with the cone of decrease K_0,*
(b) *the inequality constraints Q_i, $i = 1, \ldots, n$, are regular at x_0, with the corresponding cones of feasible directions K_i, and*
(c) *the equality constraint Q_{n+1} is also regular at x_0, with the cone of tangent* directions K_{n+1},

then there exist continuous linear functionals $f_i \in K_i^*$, $i = 0, 1, \ldots, n+1$, not all identically zero, which satisfy the Euler–Lagrange equation

$$f_0 + f_1 + \ldots + f_n + f_{n+1} = 0.$$

To prove this theorem, it is first shown that a necessary condition for a minimum to occur at x_0 is that $\bigcap_{i=0}^{n+1} K_i = \emptyset$, i.e. that no direction of decrease of the functional $F(x)$ be feasible for all constraints. This fact and Lemma 5 then imply Theorem 2.

The above extremum condition is a generalization of the rule of Lagrange multipliers in classical analysis and the Euler equation in the classical calculus of variations, which is why it is sometimes called the *Euler–Lagrange equation.*

It follows from the Dubovitskii–Milyutin theorem that in order to write out the necessary conditions for an extremum in some specific problem, one must first know how to

(1) determine the directions of decrease,
(2) calculate the feasible directions,
(3) find the tangent directions,
(4) construct the dual cones.

12.4 Directions of decrease

The directions of decrease of a functional can be calculated from its directional derivative (Gateau derivative) or Fréchet derivative.

A functional $F(x)$ in a linear space E is said to have a *derivative* $F'(x_0, h)$ at a point x_0 in the direction h if

$$\lim_{\varepsilon\to+0} \frac{F(x_0+\varepsilon h)-F(x_0)}{\varepsilon} = F'(x_0, h).$$

If $F(x)$ is a function of one variable, i.e. $E = R^1$, then the existence of $F'(x_0, h)$ is equivalent to the existence of the right (for $h > 0$) or left (for $h < 0$) derivative of $F(x)$ at the point x_0.

Let K denote the cone of directions of decrease for the functional $F(x)$ at the point x_0. The following theorems establish the relationship between the directional derivatives of F at x_0 and K.

THEOREM 3. *If $h \in K$ and $F'(x_0, h)$ exists, then $F'(x_0, h) < 0$.*

THEOREM 4. *Let E be a Banach space. If $F(x)$ satisfies the Lipschitz condition in the neighbourhood of x_0, i.e. there exist $\varepsilon_0 > 0$ and $\beta > 0$ such that $|F(x_1)-F(x_2)| \leqslant \beta\|x_1-x_2\|$ for all $\|x_1-x_0\| \leqslant \varepsilon_0$ and $\|x_2-x_0\| \leqslant \varepsilon_0$, and $F'(x_0, h) < 0$, then $h \in K$.*

THEOREM 5. *Let E be a Banach space and let $F(x)$ satisfy the Lipschitz condition in the neighbourhood of a point $x_0 \in E$. If $F(x)$ is differentiable at x_0 in any direction and the derivative $F'(x_0, h)$ is convex as a functional of h, then $F(x)$ is regularly decreasing at x_0 and $K = \{h\colon F'(x_0, h) < 0\}$.*

If F is convex, some of the assumptions of Theorem 5 can be relaxed:

THEOREM 6. *Let $F(x)$ be a convex continuous functional in a linear topological space E. Then*

(a) *$F(x)$ is differentiable at any point in any direction,*

(b) $F(x_0+h) \geqslant F(x_0)+F'(x_0, h)$,
(c) *$F(x)$ is regularly decreasing at any point, and*
(d) $K = \{h: F'(x_0, h) < 0\}$.

Thus, if the assumptions of Theorem 5 or Theorem 6 are satisfied, the cone of directions of decrease can be calculated in terms of the directional derivative as

$$K = \{h:\ F'(x_0, h) < 0\}.$$

An important class of functionals for which the cones of directions of decrease are easily determined are functionals differentiable in the sense of Fréchet.

A functional $F(x)$ defined on a Banach space E is said to be *Fréchet-differentiable* (or, simply, *differentiable*) at a point x_0 if there exists a linear functional $f \in E'$ such that for all $h \in E$

$$F(x_0+h) = F(x_0)+(f, h)+o(||h||).$$

The linear functional f is denoted by $F'(x_0)$ and called the *derivative, Fréchet-derivative* or *gradient* of the functional $F(x)$ at x_0. When $E = R^n$, i.e. F is a function of several variables, Fréchet-differentiability is equivalent to differentiability in the classical sense and the Fréchet derivative becomes the ordinary gradient vector $\left(\frac{\partial F}{\partial x_1}, \ldots, \frac{\partial F}{\partial x_n}\right)$,

LEMMA 9. *If $F(x)$ is differentiable at x_0, then it is differentiable in any direction and $F'(x_0, h) = (F'(x_0), h)$.*

THEOREM 7. *If $F(x)$ is differentiable at x_0, then $F(x)$ is regularly decreasing at x_0 and*

$$K = \{h:\ (F'(x_0), h) < 0\}.$$

Theorems 5, 6 and 7 describe certain classes of regularly decreasing functionals and show how to determine the directions of decrease by using directional derivatives and Fréchet derivatives.

Example 3. Let F be a linear functional defined by $F(x) = (f, x)$, where $f \in E'$. Then F is differentiable, $F'(x) \equiv f$, and by Theorem 7 its cone of decrease at any point x_0 is given by

$$K = \{h:\ (f, h) < 0\}.$$

Example 4. Let $E = C^{(n)}(0, T) = C$ (the Banach space of n-tuples of functions continuous on $[0, T]$, with the norm $||x|| = \max_{[0,T]} |x(t)|$) and consider the integral functional

$$F(x) = \int_0^T \Phi(x(t), t)dt,$$

where the function $\Phi(x, t)$ is continuous in x and t and differentiable with respect to x, with the partial derivative $\Phi_x(x, t)$ continuous in x and t. It is not difficult to show that $F(x)$ is differentiable and

$$[F'(x_0), h] = \int_0^T [\Phi_x(x_0(t), t), h(t)]\mathrm{d}t.$$

Therefore, by Theorem 7, the cone of decrease of F at a point x_0 is given by

$$K = \left\{h \in C: \int_0^T [\Phi_x(x_0(t), t), h(t)]\mathrm{d}t < 0\right\}.$$

Example 5. Consider the functional

$$F(x, u) = \int_0^T \Phi[x(t), u(t), t]\mathrm{d}t,$$

where $x \in C^{(n)}(0, T)$ and $u \in L_\infty^{(r)}(0, T)$ (the Banach space of all r-tuples of essentially bounded functions on $[0, T]$ with the norm $||x|| = \operatorname*{ess\,sup}_{[0,T]}|x(t)|$), Φ is continuous in all its arguments and differentiable with respect to x and u, with Φ_x and Φ_u continuous in x, u and t. It can be shown that

$$F'(x_0, u_0)(\bar{x}, \bar{u})$$
$$= \int_0^T [(\Phi_x(x_0(t), u_0(t), t), \bar{x}(t)) + (\Phi_u(x_0(t), u_0(t), t), \bar{u}(t))]\mathrm{d}t,$$

and so

$$K = \{\bar{x} \in C, \bar{u} \in L_\infty^r: F'(x_0, u_0)(\bar{x}, \bar{u}) < 0\}.$$

Example 6. Let F be given by

$$F(x) = \int_0^T G(x(t), t))\mathrm{d}t, \qquad x \in C^n(0, T),$$

where, contrary to Example 4, the function $G(x, t)$ is not assumed to be differentiable with respect to x or continuous in t. Instead, we assume that $G(x, t)$ satisfies the Lipschitz condition with respect to x in any bounded region (i.e. $|G(x_1, t) - G(x_2, t)| \leqslant L|x_1 - x_2|$ for $|x_1| \leqslant \varrho$, $|x_2| \leqslant \varrho$, $L = L(\varrho)$) and is measurable in t for any x and differentiable with respect to x in any direction for almost all $0 \leqslant t \leqslant T$, i.e. there exists the limit

$$G_x(x, h, t) = \lim_{\varepsilon \to 0+} \frac{G(x + \varepsilon h, t) - G(x, t)}{\varepsilon}.$$

Then $F(x)$ is differentiable at any point in any direction, and

$$F'(x_0, h) = \int_0^T G_x(x_0(t), h(t), t)\mathrm{d}t.$$

In particular, if $G(x, t)$ is measurable in t and differentiable with respect to x and $G_x(x, t)$ is bounded for bounded x, then

$$F'(x_0, h) = \int_0^T \big(G\big(x_0(t), t\big), h(t)\big)\mathrm{d}t,$$

and so

$$K = \Big\{h\colon \int_0^T \big(G_x\big(x_0(t), t\big), h(t)\big)\mathrm{d}t < 0\Big\}.$$

12.5 Feasible directions

Before we show how to find the cone K_f of feasible directions at a point x_0 for a set Q in a topological space E, let us first note that if x_0 is an interior point of Q then every direction is feasible, i.e. $K_f = E$, and so the only interesting situation is that in which x_0 is a boundary point of Q. Also, the set Q should have a non-empty interior or else the cone K_f will be empty.

Let Q represent an inequality constraint involving a functional F defined on E,

$$Q = \{x \in E\colon F(x) \leqslant F(x_0)\},$$

and let K_d denote the cone of directions of decrease for the functional $F(x)$ at the point x_0. We then have

LEMMA 10. $K_d \subset K_f$.

The following theorem shows that under certain conditions the two cones coincide.

THEOREM 8. *If $F(x)$ is differentiable at x_0 in any direction, $F'(x_0, h)$ is convex in h and there exists $\bar{h}$ such that $F'(x_0, \bar{h}) < 0$, then*

$$K_d \subset \{h\colon F'(x_0, h) < 0\}.$$

COROLLARY. *Assume that any one of the following conditions holds:*

(a) *E is a Banach space, $F(x)$ satisfies the Lipschitz condition in the neighbourhood of x_0, $F(x)$ is differentiable at x_0 in any direction, $F'(x_0, h)$ is convex in h, and there exists $\bar{h}$ such that $F'(x_0, \bar{h}) < 0$;*

(b) *$F(x)$ is a convex continuous functional and there exists $\bar{x}$ such that $F(\bar{x}) < F(x_0)$;*

(c) *E is a Banach space, $F(x)$ is differentiable at x_0 and $F'(x_0) \neq 0$.*

Then

$$K_f = K_d = \{h\colon F'(x_0, h) < 0\}.$$

Thus, in many important cases the problem of determining the feasible directions for a set is reduced to that of finding the directions of decrease of a functional.

When the set Q is not defined by a functional but is convex, its cone of feasible directions can be described very simply:

THEOREM 9. *Let Q be a convex set. Then the cone of feasible directions for Q at a point x_0 is given by*

$$K_f = \{\lambda(Q^0 - x_0),\ \lambda > 0\},$$

i.e.

$$K_f = \{h : h = \lambda(x - x_0),\ x \in Q^0,\ \lambda > 0\}.$$

The proof follows directly from the definition of a cone of feasible directions.

12.6 Tangent directions

Our next question is that of finding tangent directions to a set Q defined by a system of equality constraints. More generally, we can assume that Q is of the form $Q = \{x : P(x) = 0\}$ where P is an operator. To formulate the fundamental theorem on tangent directions, we need the concept of a differentiable operator and its derivative in the sense of Fréchet.

Let E_1 and E_2 be Banach spaces and $P(x)$ an operator with domain in E_1 and range in E_2. The operator P is said to be *differentiable* at a point $x_0 \in E_1$ if there exists a continuous linear operator A mapping E_1 into E_2, such that for all $h \in E_1$

$$P(x_0 + h) = P(x_0) + Ah + r(x_0, h),$$

where $\|r(x_0, h)\| = o(\|h\|)$. The operator A is called the *Fréchet derivative* of P and is denoted by $A = P'(x_0)$. When $E_2 = R^1$, i.e. $P(x)$ is a functional, this definition coincides with the definition of the derivative of a functional introduced before.

The following theorem shows how to find the tangent directions to an equality-constraint set defined by a differentiable operator.

THEOREM 10 (Lyusternik). *Let $P(x)$ be an operator mapping E_1 into E_2, differentiable in the neighbourhood of a point x_0, $P(x_0) = 0$. If $P'(x)$ is continuous in the neighbourhood of x_0 and $P'(x_0)$ maps E_1 onto E_2 (i.e. the linear equation $P'(x_0)h = b$ has a solution h for any $b \in E_2$), then the cone K of tangent directions to the set $Q = \{x : P(x) = 0\}$ at the point x_0 is the subspace*

$$K = \{h : P'(x_0)h = 0\}.$$

The proof of this theorem may be found in the well-known book of Lyusternik and Sobolev[1].

---

[1] LYUSTERNIK, L. A., and SOBOLEV, V. I., *Elements of Functional Analysis*, I. Wiley, New York 1975.

Below we present examples of tangent subspaces for various sets, determined on the basis of Lyusternik's theorem.

Example 7. Let $Q = \{x: G_i(x) = 0, i = 1, \ldots, n\}$, where

(a) $x \in R^m$,

(b) $G_i(x)$ are functions continuously differentiable at the point x_0, $G_i(x_0) = 0$, $i = 1, \ldots, n$,

(c) the vectors $G_i'(x_0)$, $i = 1, \ldots, n$, are linearly independent. The cone of tangent directions to the set Q at the point x_0 is given by

$$K = \{h \in R^m: (G_i'(x_0), h) = 0,\ i = 1, \ldots, n\}.$$

Example 8. Let $E_1 = C^{(n)}(0, T) \times L^r_\infty(0, T)$, and consider the set

$$Q = \left\{x, u \in E_1, x(t) = c + \int_0^t \varphi(x(\tau), u(\tau), \tau)\mathrm{d}\tau\right\}.$$

In other words, x and u are required to satisfy the differential equation

$$\frac{\mathrm{d}x(t)}{\mathrm{d}t} = \varphi(x(t), u(t), t), \qquad x(0) = c.$$

To determine the tangent subspace $K = \{\bar{x}(t), \bar{u}(t)\}$ for this constraint, we introduce the operator

$$P(x, u) = x(t) - c - \int_0^t \varphi(x(\tau), u(\tau), \tau)\mathrm{d}\tau$$

which maps E_1 into $C^{(n)}(0, T)$, so that the set Q can be written

$$Q = \{x, u: P(x, u) = 0\}.$$

It can be shown that, under appropriate conditions on φ, the operator $P(x, u)$ is differentiable, and

$$P'(x, u)(\bar{x}, \bar{u}) = \bar{x}(t) - \int_0^t [\varphi_x(x, u, \tau)\bar{x}(\tau) + \varphi_u(x, u, \tau)\bar{u}(\tau)]\mathrm{d}\tau;$$

here $\varphi_x(x, u, \tau)$ and $\varphi_u(x, u, \tau)$ denote the matrices of the partial derivatives

$$\frac{\partial \varphi_i(x, u, \tau)}{\partial x_j}, \qquad i = 1, \ldots, n, \quad j = 1, \ldots, r,$$

$$\frac{\partial \varphi_i(x, u, \tau)}{\partial u_j}, \qquad i = 1, \ldots, n, \quad j = 1, \ldots, r.$$

From the theory of integral equations, it can be found that the equation

$$\bar{x}(t) - \int_0^t [\varphi_x(x, u, \tau)\bar{x}(\tau) + \varphi_u(x, u, \tau)\bar{u}(\tau)]\mathrm{d}\tau = a(t)$$

has a solution $\bar{x}, \bar{u}$ for arbitrary $a(t) \in C^{(n)}(0, T)$, so that the operator $P'(x, u)$ maps E_1 onto $E_2 = C^{(n)}(0, T)$.

Thus, since all the assumptions of Lyusternik's theorem hold for the operator $P(x, u)$, the cone of tangent directions to Q at a point x, u is given by

$$K = \{\bar{x}, \bar{u}:\ P'(x, u)(\bar{x}, \bar{u}) = 0\},$$

i.e. consists of all pairs $\bar{x}(t), \bar{u}(t)$ which satisfy the integral equation

$$\bar{x}(t) = \int_0^t [\varphi_x(x, u, \tau)\bar{x}(\tau) + \varphi_u(x, u, \tau)\bar{u}(\tau)]\mathrm{d}\tau$$

or the equivalent differential equation

$$\frac{\mathrm{d}\bar{x}(t)}{\mathrm{d}t} = \varphi_x(x, u, t)\bar{x}(t) + \varphi_u(x, u, t)\bar{u}(t), \qquad \bar{x}(0) = 0.$$

Example 9. Consider the constraint set of the preceding example with an additional condition $x(T) = d$, $d \in R^n$:

$$Q = \left\{x, u \in E_1:\ x(t) = c + \int_0^t \varphi(x,(\tau), u(\tau), t)\mathrm{d}\tau,\ 0 \leqslant t \leqslant T,\ x(T) = d\right\}.$$

In order to describe a cone of tangent directions to this set, we need the following definition:

The differential equation

$$\frac{\mathrm{d}\bar{x}(t)}{\mathrm{d}t} = A(t)\bar{x}(t) + B(t)\bar{u}(t), \qquad \bar{x}(0) = 0,$$

where $A(t)$ and $B(t)$ are $n \times n$ and $n \times r$ matrices, respectively, is said to be *non-degenerate* if the set $D \subset R^n$ of all vectors $\bar{x}(T)$, where $\bar{x}(t)$ satisfies the above equation for some $\bar{u}(t) \in L_\infty^{(r)}(0, T)$, covers the entire space R^n, i.e. $D = R^n$.

Lyusternik's theorem leads to the following result:

If the system

$$\frac{\mathrm{d}\bar{x}}{\mathrm{d}t} = \varphi_x(x, u, t)\bar{x}(t) + \varphi_u(x, u, t)\bar{u}(t), \qquad \bar{x}(0) = 0,$$

is non-degenerate, then the cone of tangent directions to the set Q is given by

$$K = \left\{\bar{x}, \bar{u}:\ \bar{x}(t) = \int_0^t [\varphi_x(x, u, \tau)\bar{x}(\tau) + \varphi_u(x, u, \tau)\bar{u}(\tau)]\mathrm{d}\tau,\ \bar{x}(T) = 0\right\}$$

or, in a differential form, by

$$K = \left\{\bar{x}, \bar{u}\colon \frac{\mathrm{d}\bar{x}(t)}{\mathrm{d}t} = \varphi_x(x,u,t)\bar{x}(t) + \varphi_u(x,u,t)\bar{u}(t),\ \bar{x}(0) = \bar{x}(T) = 0\right\}.$$

12.7 Calculation of dual cones

So far we have seen how to determine the cones of directions of decrease, feasible directions and tangent directions. Before we can use the Dubovitskii–Milyutin theorem to formulate the necessary conditions for an extremum in various situations, we need to know how to construct dual cones. We begin with the simplest cases, where the given cone is a subspace.

THEOREM 11. *If K is a subspace, then the dual cone K^* is the annihilator of K, i.e.*

$$K^* = \{f \in E'\colon f(x) = 0 \text{ for all } x \in K\}.$$

THEOREM 12. *Let*

$$K_1 = \{x\colon f(x) = 0\}, \quad K_2 = \{x\colon f(x) \geqslant 0\}, \quad K_3 = \{x\colon f(x) > 0\},$$

where $f \in E'$. Then

$$K_1^* = \{\lambda f,\ -\infty < \lambda < +\infty\}, \quad K_2^* = \{\lambda f, 0 \leqslant \lambda < \infty\},$$
$$K_3^* = E' \quad \text{for} \quad f = 0 \quad \text{and} \quad K_3^* = K_2^* \quad \text{for} \quad f \neq 0.$$

THEOREM 13. *Let $E = E_1 \times E_2$, where E_1 and E_2 are linear topological spaces, and let A be a linear operator on E_1 into E_2. If*

$$K = \{x \in E,\ x = (x_1, x_2)\colon Ax_1 = x_2\},$$

then

$$K^* = \{f \in E',\ f = (f_1, f_2)\colon f_1 = -A^*f_2\},$$

where A^* denotes the adjoint operator.

The following important theorem of Minkowski and Farkas specifies the form of the dual cone for cones of a very general type.

THEOREM 14 (Minkowski–Farkas). *Let K_2 be a convex cone in E_2 with apex at 0, and let $K_1 = \{x_1 \in E_1\colon Ax_1 \in K_2\}$. If either of the following conditions holds:*

(a) *there exists $\bar{x}_1 \in E_1$ such that $A\bar{x}_1 \in \operatorname{int} K_2$,*

(b) *E_1 and E_2 are finite-dimensional and K_2 is the positive octant of E_2, i.e.*

$$K_1 = \{x \in R^m\colon (a^i, x) \geqslant 0,\ a^i \in R^m,\ i = 1, \dots, n\},$$

or, in matrix notation,

$$K_1 = \{x \in R^m\colon Ax \geqslant 0\}, \text{ where } A \text{ is an } n \times m \text{ matrix},$$

then

$$K_1^* = A^*K_2,$$

i.e., in case (b), $K_1^* = \left\{\sum a^i y_i\colon y_i \geqslant 0,\ i = 1, \dots, n\right\} = \{A^*y\colon y \geqslant 0\}$.

Example 10. *Let* K be given as

$$K = \{x \in R^m\colon (a^i, x) \geqslant 0,\ i = 1, \dots, k,\ (a^i, x) = 0,\ i = k+1, \dots, n\},$$

where $a^i \in R^m$, $i = 1, \dots, n$. Find the dual cone K^*.

The solution follows directly from Theorem 14 (case (b)):

$$K^* = \left\{f \in R^m, f = \sum_{i=1}^{n} \lambda_i a^i,\ \lambda_i \geqslant 0,\ i = 1, \dots, k\right\}.$$

Let Q be a set in a linear topological space E, and let K_f and K_t be the cones of feasible directions and of tangent directions for Q at a point $x_0 \in Q$, respectively. It can be shown that in some cases the duals of these cones coincide with the set of all supporting functionals for Q at x_0 (see Sec. 12.2),

$$Q^* = \{f \in E'\colon f(x) \geqslant f(x_0) \text{ for all } x \in Q\}.$$

THEOREM 15. *If Q is closed and convex, then $K_t^* = Q^*$. If, moreover, the interior of Q is non-empty, then $K_f^* = Q^*$.*

Thus, in many cases, determining dual cones is equivalent to finding supporting functionals.

Example 11. Let Q be a polyhedron in R^m,

$$Q = \{x\colon (a^i, x) \geqslant b_i,\ i = 1, \dots, n\},$$

$$a^i \in R^m,\ b_i \in R^1,\ i = 1, \dots, n.$$

Then, by Theorem 15,

$$K_s^* = Q^* = \left\{\sum_{i=1}^{n} \lambda_i a^i\colon \lambda_i \geqslant 0,\ \lambda_i[(a^i, x_0) - b_i] = 0\right\}.$$

Sometimes, determining all supporting functionals for a given set may be an extremely difficult task. In the next example we consider a subset Q of the space $L_\infty^{(r)}$ and give a necessary condition for an integral functional to be a support to Q at a point x_0.

Example 12. Let $Q = \{x \in L_\infty^{(r)}(0, T)\colon x(t) \in M$ for almost all $0 \leqslant t \leqslant T\}$, where $M \subset R^r$, and let $x_0 \in Q$. If the linear functional defined by

$$f(x) = \int_0^T (a(t), x(t))dt, \qquad a(t) \in L_1^r(0, T),$$

is a support to Q at the point x_0, then

$$(a(t), x - x_0(t)) \geqslant 0$$

for all $x \in M$ and almost all $0 \leqslant t \leqslant T$ (i.e., for almost all $0 \leqslant t \leqslant T$ the vector $a(t) \in R^r$ is a support to M at the point $x_0(t)$).

12.8 The Lagrange multiplier rule and the Kuhn–Tucker theorem

In this section, basing ourselves on the Dubovitskii–Milyutin theorem, we derive the necessary conditions for an extremum in a few important extremum problems.

PROBLEM 1. Equality-constrained minimum:

$$\min F_0(x), \quad F_i(x) = 0, \quad i = 1, \ldots, n; \tag{12.1}$$

here $x \in E$, where E is a Banach space, and $F_0, \ldots, F_n$ are functionals on E.

THEOREM 16. *If x_0 is a solution to Problem 1 and the functionals $F_i(x)$, $i = 0, \ldots, n$, are continuously differentiable in a neighbourhood of x_0, then there exist numbers $\lambda_0, \ldots, \lambda_n$, not all zero, such that*

$$\lambda_0 F_0'(x_0) + \lambda_1 F_0'(x_0) + \ldots + \lambda_n F_n'(x_0) = 0. \tag{12.2}$$

Proof: If $F_i'(x_0)$, $i = 1, \ldots, n$, are linearly dependent, i.e.

$$\sum_{i=1}^{n} \gamma_i F_i'(x_0) = 0, \quad \sum_{i=1}^{n} \gamma_i^2 \neq 0,$$

then condition (12.2) will hold with $\lambda_i = \gamma_i$, $i = 1, \ldots, n$, and $\lambda_0 = 0$. Now assume that $F_i'(x_0)$, $i = 1, \ldots, n$, are linearly independent. In this case, as shown in Example 7, the tangent subspace K_1 to the set $Q = \{x \colon F_i'(x) = 0, i = 1, \ldots, n\}$ is given by

$$K_1 = \{h \colon (F_i'(x), h) = 0, \ i = 1, \ldots, n\}.$$

The dual of this cone was determined in Example 10 as

$$K_1^* = \Big\{\sum_{i=1}^{n} \lambda_i F_i'(x_0)\Big\}.$$

By Theorem 7, the cone of decrease of the functional F_0 at x_0 can be written

$$K_0 = \{h \colon (F_0'(x_0), h) < 0\},$$

and by Theorem 12 its dual K_0^* is

$$K_0^* = \{-\lambda_0 F_0'(x_0), \ \lambda_0 \geqslant 0\}.$$

Application of the Dubovitskii–Milyutin theorem then gives condition (12.2). Note that if $F_i'(x_0)$, $i = 1, \ldots, n$, are linearly independent, then $\lambda_0 \neq 0$ and equation (12.2) can be rewritten as

$$F_0'(x_0) + \lambda_1 F_1'(x_0) + \ldots + \lambda_n F_n'(x_0) = 0. \tag{12.3}$$

In particular, if there are no constraints, we obtain the usual condition for an unconstrained minimum,

$$F_0'(x_0) = 0.$$

The numbers $\lambda_1, \ldots, \lambda_n$ are usually called the *Lagrange multipliers.* In the finite-dimensional case, condition (12.3) is well known from classical analysis as the Lagrange multiplier rule.

PROBLEM 2. Inequality-constrained minimum (non-linear programming):

$$\begin{aligned} &\min F_0(x), \\ &F_i(x) \leqslant 0, \quad i = 1, \ldots, n, \end{aligned} \tag{12.4}$$

where, as before, $F_0, \ldots, F_n$ are functionals on a Banach space E.

THEOREM 17. *If x_0 is a solution to Problem 2 and the functionals F_i, $i = 0, \ldots, n$, are differentiable at x_0, then there exist numbers λ_i, $i = 0, \ldots, n$, not all zero, such that*

$$\lambda_i \geqslant 0 \ (i = 0, \ldots, n); \qquad \lambda_i F_i(x_0) = 0 \ (i = 1, \ldots, n), \tag{12.5a}$$

$$\lambda_0 F_0'(x_0) + \lambda_1 F_1'(x_0) + \ldots + \lambda_n F_n'(x_0) = 0. \tag{12.5b}$$

Proof: We first determine the cone of decrease for the functional F_0 at x_0, the cones of feasible directions for the constraint sets and the dual cones, and then apply the Dubovitskii–Milyutin theorem.

According to Theorem 7, the cone K_0 of directions of decrease for $F_0(x)$ at x_0 is

$$K_0 = \{h\colon \ (F_0'(x_0), h) < 0\}.$$

It follows from Theorem 12 that if $F_0'(x_0) \neq 0$ then the dual cone K_0^* is given by

$$K_0^* = \{-\lambda_0 F_0'(x_0), \ \lambda_0 \geqslant 0\}.$$

In considering the feasible directions for a constraint set $Q_i = \{x\colon F_i(x) \leqslant 0\}$, $i = 1, \ldots, n$, let us distinguish two cases:

(1) *h_0 is an interior point of Q_i*, i.e. $F_i(x_0) < 0$. Then every direction is feasible, and so $K_i = E$. In this case $K_i^* = \{0\}$.

(2) *x_0 belongs to the boundary of Q_i*, i.e. $F_i(x_0) = 0$. We can assume that $F'(x_0) \neq 0$; otherwise conditions (12.5a, b) will hold with $\lambda_i = 1$ and $\lambda_j = 0$ for $j \neq i$. From Corollary (c) to Theorem 8 we then conclude that the cone of feasible directions is

$$K_i = \{h\colon \ (F_i'(x_0), h) < 0\}$$

and from Theorem 12 we find that the dual cone will be

$$K_i^* = \{-\lambda_i F_i'(x_0), \quad \lambda_i \geqslant 0\}.$$

Hence, using the Dubovitskii–Milyutin theorem, we obtain equation (12.5b). Condition (12.5a) is also satisfied, because whenever $F_i(x_0)$ is not zero (case (2) above), we can set $\lambda_i = 0$.

It may be asked under what assumptions the multiplier λ_0 in conditions (12.5) is non-zero. An answer to this question is given by the following theorem, which we state without proof.

THEOREM 18. *If besides the assumptions of Theorem* 17 *one of the following conditions holds*:

(a) *The functionals* $F_i(x)$, $i = 1, \ldots, n$, *are convex and there is* $\bar{x}$ *such that* $F_i(\bar{x}) < 0, i = 1, \ldots, n$;

(b) *the constraints are linear*:

$$F_i(x) = (a^i, x) - b_i, \quad a^i \in E', \quad b_i \in R^1, \quad i = 1, \ldots, n,$$

then $\lambda_0 \neq 0$ *and the necessary condition for a minimum at* x_0 *takes the form*

$$\begin{aligned} &\lambda_i \geqslant 0, \quad i = 1, \ldots, n, \quad \lambda_i F_i(x_0) = 0, \quad i = 1, \ldots, n, \\ &F_0'(x_0) + \lambda_1 F_1'(x_0) + \ldots + \lambda_n F_n'(x_0) = 0. \end{aligned} \tag{12.6}$$

After the names of mathematicians who were among the first to investigate the non-linear programming problem in the finite-dimensional case, the above conditions are often referred to as the Kuhn–Tucker theorem.

PROBLEM 3. *Mixed constraints.* Consider a generalization of the two preceding formulations,

$$\begin{aligned} &\min F_0(x), \\ &F_i(x) \leqslant 0, \quad i = 1, \ldots, k, \\ &F_i(x) = 0, \quad i = k+1, \ldots, n, \\ &x \in Q, \end{aligned} \tag{12.7}$$

where besides equality and inequality constraints given by smooth functionals there is an additional constraint which is not necessarily expressed by a functional.

Necessary conditions for a point x_0 to be a solution to this problem are provided by the following

THEOREM 19. *Let* x_0 *be a solution of Problem* 3. *If the functionals* F_i, $i = 0, \ldots, n$, *are differentiable in the neighbourhood of* x_0 *and* Q *is a convex set with non-empty interior, then there exist numbers* λ_i, $i = 0, \ldots, n$, *not all zero, such that*

$$\lambda_i \geqslant 0, \quad i = 0, \ldots, k, \quad \lambda_i F_i(x_0) = 0, \quad i = 1, \ldots, k,$$

and

$$f = \sum_{i=0}^{n} \lambda_i F_i'(x_0)$$

is a supporting functional for Q at x_0 (i.e., $(f, x_0) \leqslant (f, x)$ for all $x \in Q$). If $k = n$ or Q is a polyhedron, then the requirement that Q must contain interior points may be dropped.

The proof uses Theorem 15 and the result of Example 11.

12.9. Optimal control problems. Local maximum principle. Pontryagin maximum principle

PROBLEM 4. Determine functions $x(t)$ (phase trajectory) and $u(t)$ (control), $x(t) \in R^n$, $u(t) \in R^r$, which minimize the integral functional

$$\int_0^T \Phi(x(t), u(t), t)\mathrm{d}t \tag{12.8}$$

subject to the following constraints:

$$u(t) \in M \quad \text{for almost all } 0 \leqslant t \leqslant T; \tag{12.9}$$

$x(t)$ and $u(t)$ satisfy the equation of state

$$\frac{\mathrm{d}x(t)}{\mathrm{d}t} = \varphi(x(t), u(t), t) \tag{12.10}$$

with the boundary conditions

$$x(0) = c, \tag{12.11}$$

$$x(T) = d; \tag{12.12}$$

$\varphi(x, u, t)$ is a vector-valued function, $\Phi(x, u, t)$ a scalar function and M a subset of R^r.

Necessary conditions for a solution of Problem 4 are known as the local maximum principle:

THEOREM 20 (Local maximum principle). *Let $\varphi(x, u, t)$ and $\Phi(x, u, t)$ be continuous in x and u, measurable in t and continuously differentiable with respect to x and u, and let the derivatives*

$$\varphi_x(x, u, t), \quad \varphi_u(x, u, t), \quad \Phi_x(x, u, t), \quad \Phi_u(x, u, t)$$

be bounded for bounded x and u. Let M be a closed convex set in R^r such that $\operatorname{int} M \neq \emptyset$.

If $x^0(t)$ and $u^0(t)$ constitute a solution of Problem 4, then there exist a number $\lambda_0 \geqslant 0$ and a function $\psi(t)$, not both (identically) zero, satisfying the equation

$$\frac{\mathrm{d}\psi(t)}{\mathrm{d}t} = -\varphi_x^*(x^0(t), u^0(t), t)\psi(t) + \lambda_0 \Phi_x(x^0(t), u^0(t), t) \tag{12.13}$$

and such that the inequality

$$\big(-\varphi_u^*(x^0(t), u^0(t), t)\psi(t) + \lambda_0 \Phi_u(x^0(t), u^0(t), t), u - u^0(t)\big) \geqslant 0 \quad (12.14)$$

holds for almost all $0 \leqslant t \leqslant T$ *and all* $u \in M$; *the symbols* φ_x^* *and* φ_u^* *denote the transposes of the matrices* φ_x *and* φ_u, *respectively.*

We shall prove this theorem by constructing appropriate cones and dual cones for functional (12.8) and constraints (12.9)–(12.12), and applying the Dubovitskii–Milyutin theorem. As the initial space E we choose the set of all pairs

$$x \in C^{(n)}(0, T), \quad u \in L_\infty^{(r)}(0, T), \quad (\text{i.e. } E = C \times L_\infty).$$

(a) *Analysis of the functional*

Functionals of type (12.8) were considered in Example 5. We saw that a pair $\bar{x}, \bar{u}$ belongs to the cone of decrease K_0 if

$$\int_0^T [(\Phi_x(x^0, u^0, t), \bar{x}) + (\Phi_u(x^0, u^0, t), \bar{u})]\,\mathrm{d}t < 0.$$

By Theorem 12, if $K_0 \neq \varnothing$, then, for any $f_0 \in K_0^*$,

$$f_0(\bar{x}, \bar{u}) = -\lambda_0 \int_0^T [(\Phi_x(x^0, u^0, t), \bar{x}) + (\Phi_u(x^0, u^0, t), \bar{u})]\,\mathrm{d}t, \qquad \lambda_0 \geqslant 0. \quad (12.15)$$

(b) *Analysis of constraint* (12.9)

Let Q_1 denote the set of all $(x, u) \in E$ satisfying condition (12.9). The set Q_1' of functions $u(t)$ satisfying this condition is a closed convex set in the space L_∞ and, in virtue of the assumptions concerning M, $\operatorname{int} Q_1' \neq \varnothing$. Hence the set $Q_1 = C \times Q_1'$ is also closed and convex in E, and $\operatorname{int} Q_1 = C \times \operatorname{int} Q_1' \neq \varnothing$.

Let K_1 be the cone of feasible directions for Q_1 at the point x^0, u^0. Then, if $f_1 \in K_1^*$, it follows from Theorem 15 that $f_1 = (0, f_1')$, where $f_1' \in L_\infty^*$ is a supporting functional for Q_1' at the point u^0.

(c) *Analysis of constraints* (12.10)–(12.12)

Let Q_2 denote the set of all pairs $x, u \in E$ satisfying these constraints. We shall regard Q_2 as an equality constraint. A set of this type was considered in Example 9. According to the result stated there, the tangent subspace K_2 consists of all pairs $\bar{x}$, $\bar{u}$ such that

$$\frac{\mathrm{d}\bar{x}}{\mathrm{d}t} = \varphi_x(x^0, u^0, t)\bar{x} + \varphi_u(x^0, u^0, t)\bar{u}, \qquad \bar{x}(0) = 0, \quad (12.16)$$

$$\bar{x}(T) = 0. \quad (12.17)$$

Let $L_1 \subset E$ be the set of all pairs $\bar{x}, \bar{u}$ satisfying equation (12.16) and $L_2 \subset E$ the set of pairs satisfying (12.17), so that $K_2 = L_1 \cap L_2$. It can be shown that then $K_2^* = L_1^* + L_2^*$. Since L_1 is a subspace, it follows from Theorem 11 that if $f_2 \in L_1^*$ then $f_2(\bar{x}, \bar{u}) = 0$ for all $\bar{x}, \bar{u}$ satisfying (12.16). Also, it is clear that if $f_3 \in L_2^*$, then $f_3(\bar{x}, \bar{u}) = (\bar{x}(T), a)$, where $a \in R^n$, because L_2 is defined by n linear functionals: $\bar{x}_i(T) = 0$.

(d) *Euler equation*

By applying the Dubovitskii–Milyutin theorem to our problem, we find that there exist functionals $f_0, f_1, f_2, f_3 \in E^*$, not all zero, such that, for all $(\bar{x}, \bar{u}) \in E$,

$$f_0(\bar{x}, \bar{u}) + f_1(\bar{x}, \bar{u}) + f_2(\bar{x}, \bar{u}) + f_3(\bar{x}, \bar{u}) = 0, \tag{12.18}$$

where $f_0(\bar{x}, \bar{u})$ is given by (12.15), $f_1(\bar{x}, \bar{u}) = (0, f_1')(\bar{x}, \bar{u}) = f_1'(\bar{u})$ is a supporting functional to Q_1' at the point u^0, $f_2(\bar{x}, \bar{u})$ is only known to vanish for $\bar{x}, \bar{u}$ satisfying equation (12.16), and $f_3(\bar{x}, \bar{u}) = (a, \bar{x}(T))$.

To obtain a more specific form of equation (12.18), let $\bar{u}$ be arbitrary and $\bar{x} = \bar{x}(\bar{u})$ a solution of equation (12.16). Then $f_2(\bar{x}, \bar{u}) = 0$ and condition (12.18) becomes

$$f'(\bar{u}) = \lambda_0 \int_0^T [(\Phi_x(x^0, u^0, t), \bar{x}) + (\Phi_u(x^0, u^0, t), \bar{u})]\mathrm{d}t - (a, \bar{x}(T)). \tag{12.19}$$

We now transform the above expression so as to eliminate $\bar{x}$. Let $\psi(t)$ be a solution of equation (12.13) with the boundary condition $\psi(T) = a$. Then

$$\lambda_0 \int_0^T (\Phi_x(x^0, u^0, t), \bar{x})\,\mathrm{d}t = \int_0^T \left(\frac{d\psi}{dt} + \varphi_x^*(x^0, u^0, t)\psi(t), \bar{x}\right)\mathrm{d}t.$$

Integrating by parts and using the fact that $\bar{x}, \bar{u}$ satisfy equation (12.16), we get

$$\begin{aligned}
&\lambda_0 \int_0^T (\Phi_x(x^0, u^0, t), \bar{x})\mathrm{d}t \\
&= \psi\bar{x}|_0^T - \int_0^T \left(\psi, \frac{\mathrm{d}\bar{x}}{\mathrm{d}t}\right)\mathrm{d}t + \int_0^T (\psi, \varphi_x(x^0, u^0, t)\bar{x})\mathrm{d}t \\
&= (a, \bar{x}(T)) - \int_0^T (\psi, \varphi_u(x^0, u^0, t)\bar{u})\,\mathrm{d}t \\
&= (a, \bar{x}(T)) - \int_0^T (\varphi_u^*(x^0, u^0, t)\psi, \bar{u})\mathrm{d}t.
\end{aligned} \tag{12.20}$$

Using this result in (12.19), we obtain the condition

$$f_1'(\bar{u}) = \int_0^T (-\varphi_u^*(x^0, u^0, t)\psi + \lambda_0 \Phi_u(x^0, u^0, t), \bar{u})dt, \tag{12.21}$$

where $\bar{u}$ is arbitrary and $f_1'(\bar{u})$ is a support to Q_1' at u^0. Finally, according to the result of Example 12, which characterized integral supporting functionals of Q_1', we have

$$(-\varphi_u^*(x^0, u^0, t)\psi(t) + \lambda_0 \Phi_u(x^0, u^0, t), u - u^0(t)) \geqslant 0 \tag{12.22}$$

for almost all $0 \leqslant t \leqslant T$ and all $u \in M$, which completes the proof.

It is easy to see that the case $\lambda_0 = 0$ and $\psi(t) \equiv 0$ is impossible, because then we would have $f_0 = 0$, $f_1 = 0$, $f_2 = 0$ and $f_3 = 0$, which would contradict the assumption that there exist f_i, $i = 0, 1, 2, 3$, not all zero.

The local maximum principle is often formulated in a different, equivalent form, explaining the name of the principle. Let us introduce the function

$$H(x, u, \psi, t) = (\varphi(x, u, t), \psi(t)) - \lambda_0 \Phi(x, u, t). \tag{12.23}$$

A necessary condition for $H(x^0, u, \psi, t)$ to have a maximum on M at a point $u = u^0(t)$ is that $-H_u(x^0, u^0, \psi, t)$ be a supporting functional to M at u^0. Since

$$H_u(x^0, u^0, \psi, t) = (\varphi_u^*(x^0, u^0, t), \psi(t)) - \lambda_0 \Phi_u(x^0, u^0, t), \tag{12.24}$$

that condition is identical with (12.22). Hence the local maximum principle can be restated as follows:

If x^0, u^0 is a solution of Problem 4 *and the assumptions of Theorem* 20 *are satisfied, then $H(x^0, u, \psi, t)$, as a function of u on M, satisfies the necessary conditions for a maximum at the point $u^0(t)$ for almost all* $0 \leqslant t \leqslant T$.

The special case of Problem 4 in which $n = r = 1$ and the constraints are of the form

$$\frac{dx(t)}{dt} = u(t) = \varphi, \quad x(0) = C, \quad x(T) = d$$

is the classical problem of the calculus of variations. Since in this case $\varphi_u(x, u, t) \equiv 1$, inequality (12.14) becomes

$$(-\psi(t) + \Phi_u(x^0, u^0, t))(u - u^0(t)) \geqslant 0 \tag{12.25}$$

for all $u \in R^1$ and almost all $0 \leqslant t \leqslant T$. Satisfying this inequality for all u is possible only if $-\psi(t) + \Phi_u(x^0, u^0, t) = 0$ for all $0 \leqslant t \leqslant T$. Differentiating this equality with respect to t gives

$$-\frac{d\psi}{dt} + \frac{d}{dt}\Phi_u(x^0, u^0, t) = 0.$$

Since it follows from (12.13) that

$$\frac{d\psi}{dt} = \Phi_x(x^0, u^0, t),$$

we finally get

$$-\Phi_x(x^0, u^0, t) + \frac{\mathrm{d}}{\mathrm{d}t}\Phi_u(x^0, u^0, t) = 0. \tag{12.26}$$

Without differentiation with respect to t, we obtain the extremum condition in the following form:

$$\Phi_u(x^0, u^0, t) - \int_0^t \Phi_x(x^0, u^0, \tau)\mathrm{d}\tau = \psi(0). \tag{12.26a}$$

Equation (12.26) is the well-known Euler equation from the classical calculus of variation. Thus, inequality (12.14) is a generalization of the classical Euler equation.

PROBLEM 5. Find a minimum of the functional

$$F(x, u) = \int_{t_0}^{t_1} \Phi(x(t), u(t))\mathrm{d}t$$

over all continuous phase coordinates x, $x(t) \in R^n$, all bounded measurable controls u, $u(t) \in R^r$, and time $t_1 \geqslant t_0$, satisfying the constraints

$$u(t) \in M \subset R^r \quad \text{for almost all } t_0 \leqslant t \leqslant t_1, \tag{12.27}$$

$$\frac{\mathrm{d}x(t)}{\mathrm{d}t} = \varphi(x(t), u(t)), \qquad x(t_0) = C, \tag{12.28}$$

$$x(t_1) = d, \tag{12.29}$$

where the functions $\Phi(x, u)$ and $\varphi(x, u)$ are defined on R^{n+r}, with values in R^1 and R^n, respectively, continuous in u and continuously differentiable in x, and $\Phi_x(x, u)$ and $\varphi_x(x, u)$ are bounded for bounded x, u.

The present problem differs from Problem 4 in several respects:

(1) In Problem 4 the set M was convex and had non-empty interior; here M is an arbitrary set in R^r.

(2) $\Phi(x, u)$ and $\varphi(x, u)$ are not required to be differentiable in u, which is natural as M may consist of finitely many points.

(3) The functions Φ and φ do not depend explicitly on t.

(4) The time t_1 is not fixed.

The necessary conditions for a solution of Problem 5 are known as the *Pontryagin maximum principle.* We state it without proof.

THOOREM 21. (Pontryagin maximum principle). *Let $x^0(t), u^0(t), t_1$ be a solution of Problem 5. Then there exist $\psi(t)$ and $\lambda_0 \geqslant 0$, not both zero, such that*

$$\frac{d\psi}{dt} = -\varphi_x(x^0, u^0)\psi + \lambda_0 \Phi_x(x^0, u^0), \tag{12.30}$$

$$H(x^0(t), u^0(t), \psi(t), t) = 0 \text{ for almost all } t_0 \leqslant t \leqslant t_1, \tag{12.31}$$

$$H(x^0(t), u, \psi(t), t) \leqslant 0 \text{ for all } u \in M \text{ and almost all } t_0 \leqslant t \leqslant t_1, \tag{12.32}$$

where

$$H(x, u, \psi, t) = (\varphi(x, u), \psi) - \lambda_0 \Phi(x, u).$$

In other words, *the function $H(x^0(t), u, \psi(t), t)$ has a maximum on M for $u = u^0(t)$, for almost all $t_0 \leqslant t \leqslant t_1$.*

Let us apply Theorem 21 to the classical problem of the calculus of variations. In this case, we can set $\lambda_0 = 1$, and so

$$H(x^0, u, \psi, t) = u\psi - \Phi(x^0, u, t), \tag{12.33}$$

where $\psi(t)$ satisfies the differential equation

$$\frac{d\psi}{dt} = \Phi_x(x^0, u^0, t). \tag{12.34}$$

Hence

$$\psi(t) = \psi(0) + \int_0^t \Phi_x(x^0, u^0, \tau) d\tau, \tag{12.35}$$

and using the Euler equation (12.26a), we get

$$\psi(t) = \Phi_u(x^0, u^0, t).$$

The maximum principle now gives

$$u^0\psi - \Phi(x^0, u^0, t) \geqslant u\psi - \Phi(x^0, u, t). \tag{12.36}$$

Alternatively, introducing the Weierstrass function

$$E(x^0, u^0, u, t) = \Phi(x^0, u, t) - \Phi(x^0, u^0, t) + (u^0 - u)\Phi_u(x^0, u^0, t),$$

we obtain the condition

$$E(x^0, u^0, u, t) \geqslant 0$$

for all u and t. This is the *Weierstrass condition* for a strong extremum, well known from the calculus of variations. The maximum principle is thus a generalization of the Weierstrass condition to the problem of optimal control.

12.10 Sufficient extremum conditions

In this section we consider the question of whether the necessary conditions for an extremum given by Dubovitskii and Milyutin (Theorem 2) are also sufficient. In general, the answer is negative, which can be shown by elementary examples. However, there is a large class of extremum problems, namely convex problems, for which, under certain additional conditions, the necessary extremum conditions are also sufficient. Sufficient conditions have also been derived for various types of non-convex problems, mostly with the use of the second variation, but as the theory is far from being complete on this point, we shall restrict ourselves to the convex case.

We recall that a convex problem is an extremum problem in which the objective functional and the constraints are all convex. An important property of convex problems, which we note first, is that' in such problems the local and the global minima coincide (x_0 is a global minimum point for $F(x)$ on Q if $F(x_0) \leqslant F(x)$ for all $x \in Q$; it is a local minimum point if the same inequality holds for all $x \in Q \cap U$, where U is a neighbourhood of x_0):

THEOREM 22. *If $F(x)$ is a convex functional and Q a convex set in a linear topological space E, then every local minimum point for $F(x)$ on Q is also a global minimum point for $F(x)$ on Q.*

Proof: Let x_0 be a local minimum point, so that $F(x_0) \leqslant F(x)$ for $x \in Q \cap U$, where U is a neighbourhood of x_0. Choose an arbitrary point x_1 in Q and let λ be a number between 0 and 1 such that the point $x_\lambda = x_1 + \lambda(x_0 - x_1) = \lambda x_0 + (1-\lambda)x_1$ belongs to U. Since the set Q is convex, we have also $x_\lambda \in Q$ and, by assumption, $F(x_\lambda) \geqslant F(x_0)$. The convexity of F implies that

$$F(x_\lambda) \leqslant \lambda F(x_0) + (1-\lambda) F(x_1),$$

and so

$$F(x_1) \geqslant \frac{1}{1-\lambda}[F(x_\lambda) - \lambda F(x_0)] \geqslant \frac{1}{1-\lambda}[F(x_0) - \lambda F(x_0)] = F(x_0),$$

which completes the proof.

We now state without proofs three theorems on sufficient conditions for a minimum in convex problems.

THEOREM 23. *Assume that*

(a) *$F(x)$ is a convex continuous functional,*

(b) *$Q_1, \ldots, Q_{n+1}$ are convex sets,*

(c) *there exists a point $\bar{x}$ such that $\bar{x} \in \operatorname{int} Q_i$, $i = 1, \ldots, n$, and $\bar{x} \in Q_{n+1}$,*

(d) $x_0 \in Q = \bigcap_{i=1}^{n+1} Q_i$,

(e) *K_0 is the cone of decrease of $F(x)$ at x_0,*
(f) *$K_1, \ldots, K_n$ are the cones of feasible directions for $Q_1, \ldots, Q_n$,*
(g) *K_{n+1} is the tangent cone for Q_{n+1} at x_0.*

Then the necessary and sufficient condition for x_0 to be a minimum point of $F(x)$ on Q is that there exist $f_i \in K_i^*$, $i = 0, 1, \ldots, n+1$, not all zero, such that

$$f_0+f_1+ \ldots +f_{n+1} = 0.$$

The next theorem deals with a case where the sets Q_i, $i = 1, \ldots, n$, are defined by smooth functionals. The problem to be solved is:

$$\min F_0(x),$$

$$F_i(x) \leqslant 0, \quad i = 1, \ldots, k,$$

$$F_i(x) = 0, \quad i = k+1, \ldots, n, x \in Q,$$

where $F_i(x)$, $i = 0, \ldots, k$, are convex differentiable functionals, $F_i(x)$, $i = k+1, \ldots, n$, are of the form

$$F_i(x) = f_i(x)+\alpha_i, \quad f_i \in E', \quad \alpha_i \in R^1, \quad i = k+1, \ldots, n,$$

and Q is a convex set.

THEOREM 24. *Under the above conditions, assume that there exists a point* $x \in \operatorname{int} Q$ such that

$$F_i(\bar{x}) < 0, \quad i = 1, \ldots, k, \qquad F_i(\bar{x}) = 0, \quad i = k+1, \ldots, n.$$

Then the necessary and sufficient condition for F to have a minimum at x_0 is that there exist numbers $\lambda_1, \ldots, \lambda_n$ such that $\lambda_i \geqslant 0$, $\lambda_i F_i(x_0) = 0$, $i = 1, \ldots, k$, and the functional

$$f = F_0'(x_0) + \sum_{i=1}^{n} \lambda_i F_i'(x_0)$$

is a supporting functional for Q at x_0.

The last theorem deals with the optimal control problem for a linear system, viz.

$$\min \int_0^T F(x(t), u(t), t)\,\mathrm{d}t,$$

$$\frac{\mathrm{d}x(t)}{\mathrm{d}t} = A(t)x(t)+B(t)u(t),$$

$$x(0) = c, \quad x(T) = d,$$

$u(t) \in M$ for almost all $0 \leqslant t \leqslant T$.

THEOREM 25. *In the above problem, assume that*

(a) *$F(x, u, t)$ is convex and continuously differentiable with respect to x and u, and continuous in t,*

(b) *$A(t)$ and $B(t)$ are matrices continuous in t,*

(c) *M is a closed convex set in R^r,* $\operatorname{int} M \neq \varnothing$,

(d) *the state equation is non-degenerate (see Example* 9),

(e) *there exists $\bar{u}(t)$ such that $\bar{u}(t) \in \operatorname{int} M$ for almost all $0 \leqslant t \leqslant T$ and $\bar{x}(T) = d$.*

Then the local maximum principle (Theorem 20) *is both a necessary and sufficient condition for an extremum.*

13

Iterative and Experimental Methods of Shape Optimization of Structures

13.1 Iterative and experimental approaches to shape optimization

Optimal or near optimal shapes of structures can often be determined by successive approximations involving iterative analytical methods or model observations and measurements. This approach makes it possible to obtain valuable information of how to improve a given design, often saving time and resources required by the exact methods of optimization described so far.

In iterative analytical methods an initial solution to the problem is adopted on the basis of some previous experience, by guess or intuition, and then used in a simple computational scheme leading to an improved solution. The successive solutions should be more and more accurate, the decreasing differences between them indicating the accuracy of the solution adopted as final. Of course, the better the initial solution and the faster the convergence of the scheme chosen, the more effective the whole procedure.

Experimental methods of structural optimization usually also incorporate an iterative procedure by which the model can be improved. Sometimes, however, the observations or measurements of the first model alone provide sufficient information for the design. As an example we can mention hanging structures, which from the very earliest times of engineering were modelled after the shapes assumed by freely hanging cables or chains subjected to various loads.

An important device of experimental optimization is the use of photo-elastic effects, which permit observation of characteristic lines of force and strain distribution in models. Applications of photo-elasticity to structural optimization are described in some detail in Section 13.6.

Observation of models made of elastic and easily deformable materials can provide very useful suggestions how to improve the shape of the modelled structures, for example joints in trusses or girders and, above all, shell roofs of factories and stores or public buildings. Thin elastic membranes used in such investigations are deformed by concentrated forces, by rigid elements representing the adjacent structures, or by liquid or air pressure. Examples of such studies are presented in Section 13.8.

13.2 Types of iterations

The term "iteration" is customarily used to mean the repetition of a pair of processes of which the first represents a certain supposition and the second verifies its correctness with a view to improving it in the next step. Either process may consist of many different operations to be performed in some specified order. These operations may lie in the sphere of theoretical reasoning, involving elements of design and analytical or numerical mathematics, or may be experimental: constructional, technological or physical.

By considering various types of the component processes in iterations one can distinguish different types of iterations. The diagrams shown in Fig. 13.1 illustrate several such possibilities. The lines joining the small circles representing the component processes may symbolize a passage from hypothesis to verification in a given pair or that from verification in one pair to

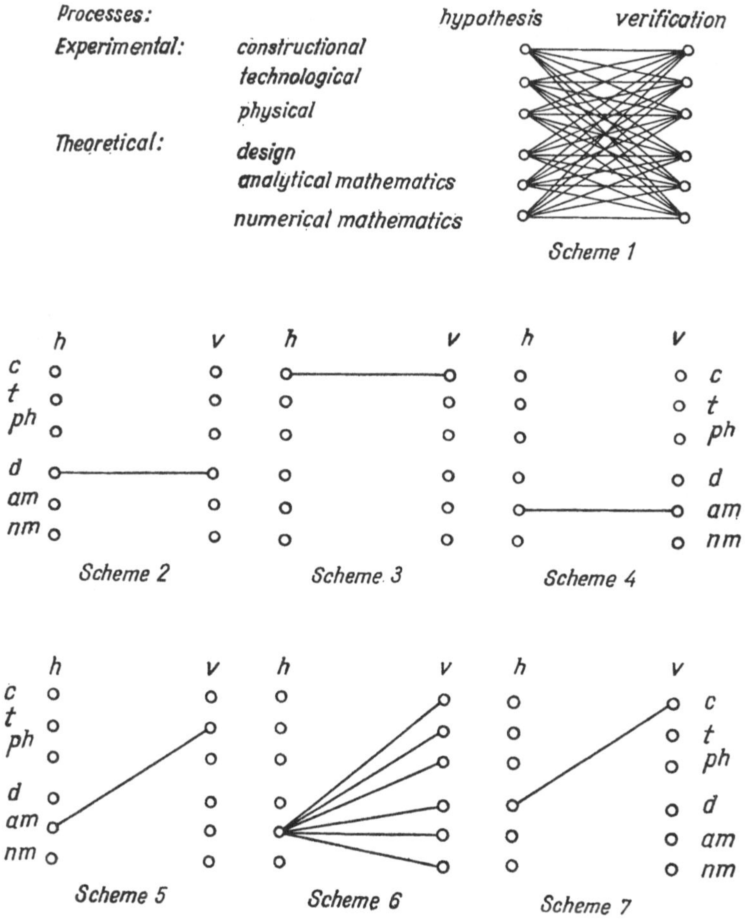

Fig. 13.1. Types of iterations

the hypothesis in the next pair. Scheme 1 shows how many types of iterations may be considered, depending on the kinds of operations that are taken into account. The remaining schemes represent the simplest or the most typical combinations.

13.3 Iteration vs. imitation and the trial and error method

A common approach which one tends to follow when trying to reach a certain objective is to apply procedures which are known from experience to have led to similar objectives under similar conditions. Conditions and objectives are never the same, however, and therefore imitating old solutions may be unsatisfactory. For this reason we abandon imitation, seeking solutions which would be better adapted to the changed conditions and better suit the new objectives. In all instances of this approach we observe two kinds of operations, of which the first presupposes a procedure as effective in reaching a given goal, and the second verifies the result of that procedure and evaluates the agreement between the prediction and the effect. This line of approach is known as the trial and error method.

Iterations resemble the scheme of the trial and error method but also differ from it in the following important respects.

Firstly, in an iterative approach, we do not perform any operations unless we are certain they will bring us sufficiently close to some definite solution, i.e., unless we know they converge. Secondly, the iterative operations are chosen so as to lead us to the desired solution. And thirdly, we always seek as rapidly convergent iterative procedures as possible; ideally, they should give a sufficiently accurate result in one step.

13.4 Designing a column for uniform strain

Objective, notation

It is required to determine the shape of a column so that the strain under the dead weight and a load uniformly distributed over the top cross-section may be constant throughout the column (Fig. 13.2).

Let

A_0, A_x, A_a [m²]—top, running and bottom cross-sectional areas of the column, respectively,

x, a [m] —abscissa along the column height and the height itself,

p [MPa] —vertical load at the top,

γ [kN/m³] —column weight per unit volume,

E [GPa] —deformation coefficient of column material,

ε [m/m] —equalized strain.

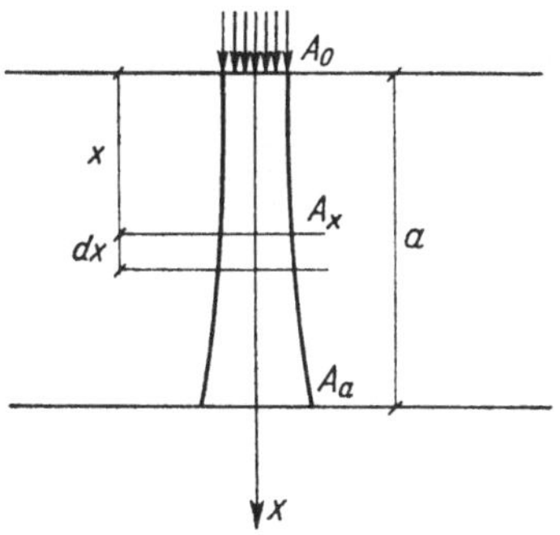

Fig. 13.2. A column designed for uniform strain

Solution by integration

The simplest way to solve the above problem is direct integration. The weight of the column layer between the sections x and $x + \mathrm{d}x$ can be calculated by using the area A_x as

$$A_x \gamma \,\mathrm{d}x,$$

or in terms of the area increment $\mathrm{d}A_x$ as

$$E\bar{\varepsilon}\,\mathrm{d}A_x.$$

Hence we obtain the equation

$$A_x \gamma \,\mathrm{d}x = E\bar{\varepsilon}\,\mathrm{d}A_x$$

or

$$\frac{\gamma}{E\bar{\varepsilon}}\,\mathrm{d}x = \frac{\mathrm{d}A_x}{A_x}.$$

Integration gives

$$\frac{\gamma}{E\bar{\varepsilon}}\,x = \log\frac{A_x}{A_0},$$

and so

$$A_x = A_0 \exp\left(\frac{\gamma}{E\bar{\varepsilon}}\,x\right) \quad \text{or} \quad A_x = A_0 \exp\left(\frac{\gamma}{p}\,x\right).$$

Solution by a physical iteration

The same problem can be solved by iterations. As explained before, an iterative procedure begins with the adoption of a simplified assumption which gives an initial approximate solution. Then a method is shown by which to improve the original assumption so as to obtain solutions approaching the exact solution. The simplifying assumption may concern some physical relations present in the problem or may be of a mathematical nature. Each of these two principal types may appear in an unlimited variety of specific forms.

A common iterative approach to designing a structure involves what is known as "removing weight outside the structure". The weight of a structure depends on its shape and therefore is unknown in the design problem. In many cases, this fact may considerably increase the difficulty of a direct solution. "Removing weight outside the structure" means replacing it by a system of external forces, approximately equal to the actual weight and independent of the unknown shape of the structure. In adopting this approach we are guided by experience: we choose an approximate initial value for the weight of the structure on the basis of the weight of structures designed in the past.

In the present problem, we shall begin by assuming that the weight of the column is zero. Every next iteration will be performed on the assumption that the weight of the column corresponds to the shape determined in the preceding iteration. In this way we obtain the following sequence.

Iteration 1. The load is $A_0 p$; the dead weight is zero; the strain is given everywhere in the column by

$$\bar{\varepsilon} = \frac{A_0 p}{A_0 E} = \frac{A_0 p}{A_x E};$$

the cross-sectional area is thus $A_x^{(1)} = A_0$. The uniformity of strain in the top section implies that $p = \bar{\varepsilon} E$.

Iteration 2. Dead weight in section x: $A_0 \gamma x$; load $A_0 p$; compressive force at x: $A_0(p+\gamma x)$; strain at x:

$$\bar{\varepsilon} = \frac{A_0(p+\gamma x)}{A_x^{(2)} E};$$

cross-sectional area at x: $A_x^{(2)} = A_0\left(1+\dfrac{\gamma}{p}x\right)$.

Iteration 3. Dead weight at x:

$$\int_0^x A_x^{(2)} \gamma \, dx = A_0 \gamma \int_0^x \left(1+\frac{\gamma}{p}x\right) dx = A_0 \gamma \left(x + \frac{1}{2}\frac{\gamma}{p}x^2\right),$$

compressive force at x: $A_0\left(p+\gamma x+\dfrac{1}{2}\dfrac{\gamma^2}{p}x^2\right)$,

cross-sectional area at x: $A_x^{(3)} = A_0\left(1+\dfrac{\gamma}{p}x+\dfrac{1}{2}\dfrac{\gamma^2}{p^2}x^2\right)$.

Iteration n. The cross-sectional area at x:

$$A_x^{(n)} = A_0\left[1+\left(\frac{\gamma}{p}x\right)+\frac{1}{2!}\left(\frac{\gamma}{p}x\right)^2+\ldots+\frac{1}{(n-1)!}\left(\frac{\gamma}{p}x\right)^{n-1}\right].$$

The above formula contains the first n terms of the series expansion of the exponential function occurring in the exact solution obtained previously, so our sequence of iterations converges to that solution.

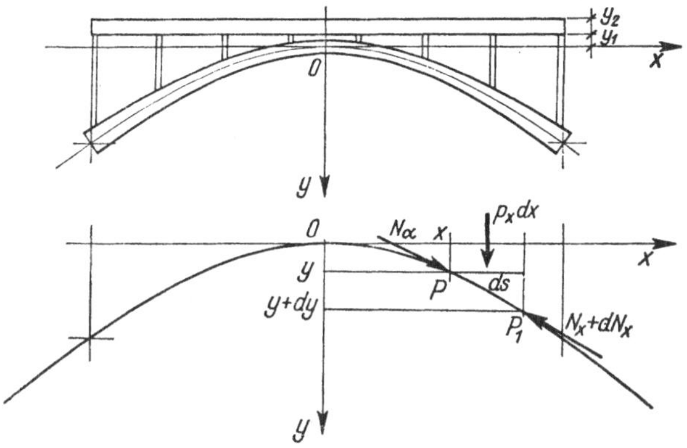

Fig. 13.3. An arch bridge designed for uniform strain

The starting assumption in the above sequence is that, in the first approximation, the weight of the column may be neglected. In each subsequent approximation it is assumed that the weight can be calculated on the basis of the preceding result. These assumptions can be included in the physical and rational categories. On the other hand, verifications in this example consist of mathematical operations. The scheme of this type of iteration is not shown in Fig. 13.1.

13.5 Designing an arch bridge for uniform strain

Objective, notation, assumptions

Consider the problem of determining the shape of an arch supporting a deck as shown in Fig. 13.3.

Let us denote:

O_x —horizontal coordinate axis tangent to the arch axis at the crown,

O_y —horizontal coordinate axis through the crown, perpendicular to Q_x,

$\overline{OP} = s$ —arc length between the crown and the point P,

$\overline{PP_1} = \mathrm{d}s$ —differential arc length,

x, y —coordinates of the point P,

$x+\mathrm{d}x, y+\mathrm{d}y$ —coordinates of the point P_1,

$p_x\mathrm{d}x = (p_1+p_2+p_3)\mathrm{d}x$—weight of the deck, supporting columns or walls and arch over $\mathrm{d}x$,

$p_1 = c_1 y_0$ —weight of the deck per unit of its length,

$p_2 = c_2(y_1+y)$ —weight of columns or walls per unit of deck length,

$p_3 = 2ab\gamma\dfrac{ds}{dx}$ —weight of arch per unit of deck length,

a_0, a_x —half depths of the arch cross-sections at O and P,

b —arch width, constant along the arch,

A_0, A_x —cross-sectional areas of the arch at O and P,

$\bar{\sigma}$ —allowable stress,

N_0, N_x —normal forces in the sections O and P of the arch.

Assumptions:

The arch is erected without initial bending moments and has a uniform-strength shape, i.e. $N_x = A_x\gamma = 2a_x b\gamma$.

The deck weight per unit length is constant along the O_x axis.

The column weight per unit deck length is proportional to the sum of the constant y_1 and the column ordinates y.

The arch weight is proportional to its volume and unit weight.

Solution by a design iteration

The projections onto the O_x and O_y axes of the forces acting on the segment PP_1 are described by the equations

$$\mathrm{d}\left(N_x\frac{\mathrm{d}x}{\mathrm{d}s}\right) = 0, \qquad \mathrm{d}\left(N_x\frac{\mathrm{d}y}{\mathrm{d}s}\right) = p_x\mathrm{d}x.$$

It is assumed that the bending moment is zero when the arch is closed and therefore the moment equation for the segment PP_1 is satisfied identically.

From the first equation we find

$$N_x\frac{\mathrm{d}x}{\mathrm{d}s} = N_0, \qquad N_x = N_0\frac{\mathrm{d}s}{\mathrm{d}x},$$

and so the second equation can be written as

$$N_0\frac{\mathrm{d}^2y}{\mathrm{d}x^2} = p_x$$

or

$$N_0\frac{\mathrm{d}^2y}{\mathrm{d}x^2} = c_1y_0+c_2(y_1+y)+2ab\gamma\frac{\mathrm{d}s}{\mathrm{d}x}. \tag{13.1}$$

We can avoid the task of solving this complex equation by applying an iterative procedure.

We first express the unknown function p_x as a power series with unknown coefficients:

$$p_xx = \sum_0^n ka_{2k}x^{2k}.$$

The equation for the arch axis then takes the form

$$N_0 \frac{d^2y}{dx^2} = \sum_0^n k a_{2k} x^{2k}.$$

Double integration gives

$$N_0 \frac{dy}{dx} = C_1 + \sum_0^n k \frac{a_{2k}}{2k+1} x^{2k+1},$$

$$N_0 y = C_2 + C_1 x + \sum_0^n k \frac{a_{2k}}{(2k+1)(2k+2)} x^{2k+2}.$$

Since the arch axis is tangent to the O_x axis at the coordinate origin, we obtain $C_1 = C_2 = 0$, and

$$N_0 y = \sum_0^n k \frac{a_{2k}}{(2k+1)(2k+2)} x^{2k+2}$$

$$= \frac{a_0}{1 \cdot 2} x^2 + \frac{a_2}{3 \cdot 4} x^4 + \ldots + \frac{a_{2k}}{(2n+1)(2n+2)} x^{2n+2}. \tag{13.2}$$

The unknown coefficients a_{2k} can be determined by sketching an initial shape for the arch and calculating the approximate values of the ordinate y and the function $p(x)$ from equation (13.1) at k different cross-sections of the arch; then equation (13.2) is used to find the shape of the arch axis. Having obtained new values of the ordinates, we introduce them in equation (13.1) and repeat the calculation. Usually the second iteration gives a sufficiently accurate result.

The above method was used to calculate the shape of the axis of an arch bridge with a span of 96.25 m. The weight of the bridge, i.e. the weight of the arch, the deck and the columns supporting the deck per unit length, varied according to the equation

$$p_x = 32.2 + 1.317 \left(\frac{x}{10}\right)^2 + 0.007\,777 \left(\frac{x}{10}\right)^4 \quad (x \text{ in meters}). \tag{13.3}$$

Hence, by formula (13.2), the shape of the arch axis was found by double integration to be

$$y = 0.47619 \left(\frac{x}{10}\right)^2 \left[1 + 0.006\,818\,2 \left(\frac{x}{10}\right)^2 + 0.000\,016\,101 \left(\frac{x}{10}\right)^4\right]. \tag{13.4}$$

The weight of a bridge so shaped is sufficiently close to the weight given by equation (13.3).

In the above example, the first, assumptive part of the iteration is a design operation, because it consists of drawing an approximate shape of the arch axis, arithmetic calculation of the bridge weight per unit length at a few cross-sections, representing that weight in the form of a power expansion, and determining the improved shape of the arch axis from equation (13.4). The second component of the iteration, verification, is also a design operation as it again involves drawing the arch axis, this time in accordance with equation (13.4), and a repeated calculation of the bridge weight, following equation (13.3). Thus, the present case corresponds to Scheme 2 of Fig. 13.1.

13.6 Determination of an optimum profile of the junction between a beam and a column by means of photo-elastic modelling

Problem formulation

Combinations of horizontal beams or plates with vertical columns appear in many structures. As an example, consider a viaduct whose deck rests on columns (Fig. 13.4). The deck can be supported by a number of single

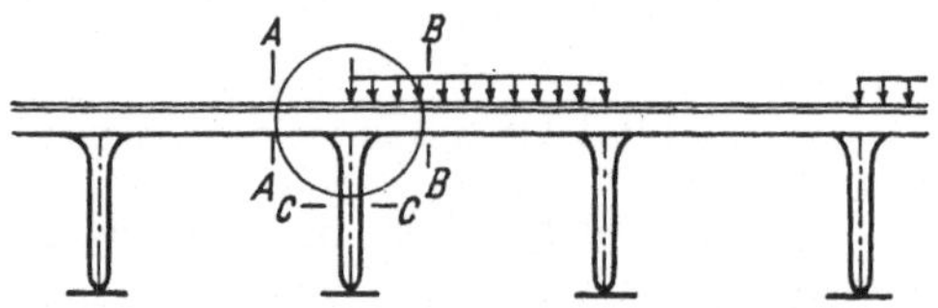

Fig. 13.4. A viaduct supported by columns

columns spaced along the viaduct, by pairs of columns or by rows of several columns each. The corners formed at the junctions between the deck and the columns give rise to concentrated stresses. To eliminate these concentrations, it is necessary to fill the corners, forming heads which join the column sides smoothly with the deck surface.

The above problem has been known in building engineering for a long time. Experience with structures existing in the past and at present provided the basis for establishing standards of junction profiles between beams and columns. It is recommended, for example, that the length of a straight-splay junction should be 1/6 to 1/5 of the beam span, while its height should equal 1/3 of the splay length. It is also indicated how to reinforce the splay with rods and stirrups. Similar recommendations accompany the standards for dimensioning and strength testing of flat slab floors. The recommendations specify the width of the column strip in relation to the column spacing. The sizing of truncated pyramid heads with square or rectangular bases is also given. Flat-surface heads do not eliminate stress concentrations but only reduce them by increasing the corner angles. Therefore the specifications also include design suggestions concerning curved head profiles.

We shall consider here the problem of reducing the strain in a beam-column corner to a sufficiently small value by inscribing into it a surface which will deform uniformly under a given load. After Doroszkiewicz *et al.* (1972), the problem will be stated as a plane stress problem under standard assumptions of the theory of elasticity. The optimized element of the column head and the interactions of the head with the remaining part of the structure are shown in Fig. 13.5. Thus given the boundaries $PGG'F'$ and EE' of

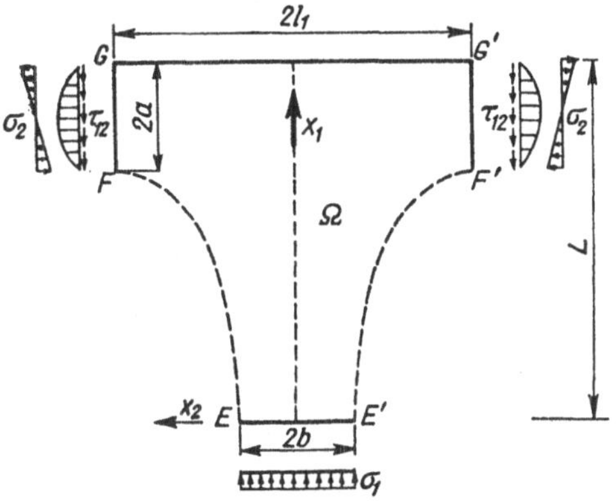

Fig. 13.5. The boundary stress conditions at the column head

the region Q and the stress along these boundaries we have to determine the load-free boundaries EF and $E'F'$ so that the non-zero principal stress, called in photoelasticity the *edge pressure*, be constant along them: $\sigma_b = \text{const}$.

With Airy's stress function Φ, the mathematical formulation of the problem is as follows:

Determine such boundaries EF and E'F' of the region Q that

(a) *The function $\Phi(x_1, x_2)$ satisfies inside Q the biharmonic equation*

$$\nabla^4 F = 0, \tag{13.5}$$

where ∇^4 denotes the biharmonic operator;

(b) *The function Φ and its first partial derivatives satisfy given boundary conditions on the boundaries EE', FG and G'F',*

$$F = G_1(s), \quad F_{,1} = G_2(s), \quad F_{,2} = G_3(s). \tag{13.6}$$

(c) *On the boundaries EF and E'F', Φ is a linear function of x_1 and x_2 and its parial derivatives are constant:*

$$\begin{aligned} F &= C_1 x_1 + C_2 x_2 + C_3, \\ F_{,1} &= C_1, \; F_{,2} = C_2. \end{aligned} \tag{13.7}$$

(d) *On the boundary EF, Φ satisfies the equation*

$$\Delta F = C, \tag{13.8}$$

where Δ denotes the Laplace operator.

Condition (13.5) replaces the equations of equilibrium and of continuity of deformations in the region Q. Conditions (13.6), and (13.7) follow from integration of stresses along the boundaries. Condition (13.8) is the design condition (criterion of optimization).

Thus, from mathematical viewpoint, we have to solve a boundary biharmonic problem with non-homogeneous boundary conditions and a part of the boundary not given *a prori*. The unknown boundary must then be chosen so as to satisfy the given criterion of optimization.

Iterative procedure

The solution proposed by Doroszkiewicz *et al.* (1972) used the following iterative procedure.

The first approximation of the free boundary was determined from one of the following strength formulae:

$$\sigma_b = \left[\frac{M(x_1)}{W(x_2)} + \frac{N(x_1)}{A(x_2)}\right]\left(\frac{l_1 - x_2}{l_1 - b}\right)\beta_1 + \left[\frac{M(\eta)}{W(x_1)} + \frac{N(\eta)}{A(x_1)}\right] \cdot \frac{x_1}{l-a}\beta_3, \tag{13.9}$$

$$\sigma_b = \left[\frac{M(x_1)}{W(x_2)} + \frac{N(x_1)}{A(x_2)}\right]\left(\frac{l_1 - x_2}{l_1 - b}\right)^2\beta_1 + \left[\frac{M(\eta)}{W(x_1)} + \frac{N(\eta)}{A(x_1)}\right]\left(\frac{x_1}{1-a}\right)^2\beta_3. \tag{13.10}$$

Notation:

x_1, x_2 —coordinates of an arbitrary point at the boundary EF (Fig. 13.5),

$M(x_1), N(x_1)$—bending moment and longitudinal force in the cross-section of the column; M and N are independent quantities,

$M(\eta), N(\eta)$ —bending moment and longitudinal force in the cross-section of the beam,

$W(x_2), A(x_2)$—modulus and area of the column cross-section,

$W(x_1), A(x_1)$—modulus and area of the beam cross-section,

σ_b —edge pressure,

β_1, β_3 —concentration indices at the points E and F of the model, equal to 1.0 and 1.3, respectively.

The second operation of the solving procedure was the checking up of the deformations along the boundary of the first approximation. The checking was performed on a suitable model by the photoelastic method. The photoelastic image and the corresponding stress diagram provided the basis for the third stage, a correction of the boundary calculated in the first step. This was done with the use of equations (13.11) given below.

The fourth step consisted of preparing a new photoelastic model, loading it, recording the photoelastic image and obtaining the stress diagrams along the boundaries. The uniformity of that stress system indicated that the corrected boundary answered satisfactorily the original requirement.

The first approximation of the boundary

Formulae (13.9) and (13.10) are based on simplified formulae for extreme stresses in plane elements subjected to bending and compression. It is assumed that the stress at the boundary of the junction is a sum of two components, of which the first depends on the stress in the beam and the second on that in the column. In formula (13.9) the coefficients of these components involve the coordinates x_1 and x_2 linearly, while in formula (13.10) they are assumed to depend on their second powers. For a constant σ_b, either equation defines the shape of the boundary.

The first check-up of stress along the boundary

The model examined and its loading are shown schematically in Fig. 13.6. The loading consists of three couples of forces: two couples of vertical forces P and one couple of oblique forces Q.

The actual model was made of epoxy resin VP 1527, the common material in photoelastic studies. Its dimensions, adapted to the photoelastic apparatus, were: height 0.22 cm, width 0.28 cm. The photoelastic image recorded is shown in Fig. 13.7 and the corresponding diagram of stresses along the tangents to the boundary in Fig. 13.8.

The second approximation of the boundary

The correction of the boundary was carried out according to the formulae

$$X_1 = x_1 + \frac{(\bar{\sigma}_b - \sigma_b^r)\,\sigma_{b,1}}{\bar{\sigma}_{b,1}^2 + \sigma_{b,2}^2}, \qquad X_2 = x_2 + \frac{(\bar{\sigma}_b - \sigma_b^r)\,\sigma_{b,2}}{\sigma_{b,1}^2 + \sigma_{b,2}^2}, \tag{13.11}$$

where

X_1, X_2 —coordinates at the corrected boundary,
x_1, x_2 —coordinates at the initial boundary,
σ_b^r —the postulated (constant) stress along the boundary,
$\bar{\sigma}_b$ —boundary pressure determined in the first check-up,
σ_{b1}, σ_{b2}—derivatives of the stress σ_b with respect to x_1 and x_2, calculated from formula (13.10).

The second check-up of stress along the boundary

As before, the verification was performed on a photoelastic model. That model differed from the first model only in the shape of the boundary. The fringe patterns obtained in this case are shown in Figs 13.9 and 13.10. The

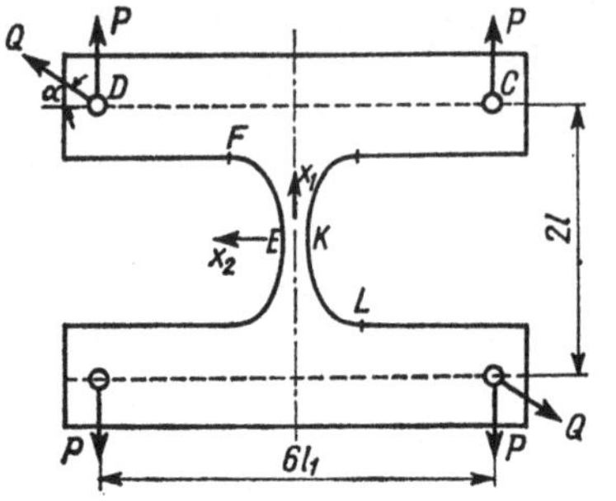

Fig. 13.6. Schematic ilustration of the model and its loading

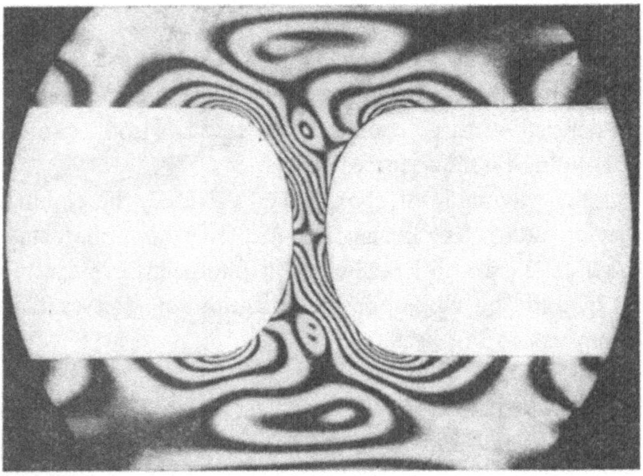

Fig. 13.7. Fringe pattern in the model for P = 750 N and Q = 300 N

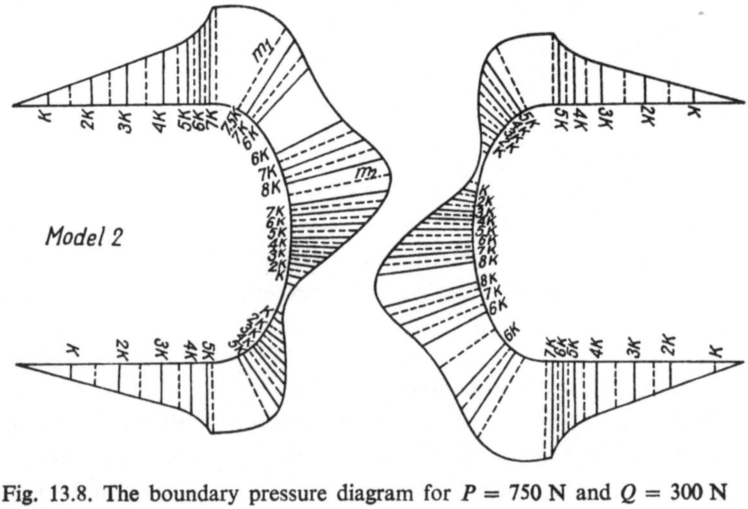

Fig. 13.8. The boundary pressure diagram for P = 750 N and Q = 300 N

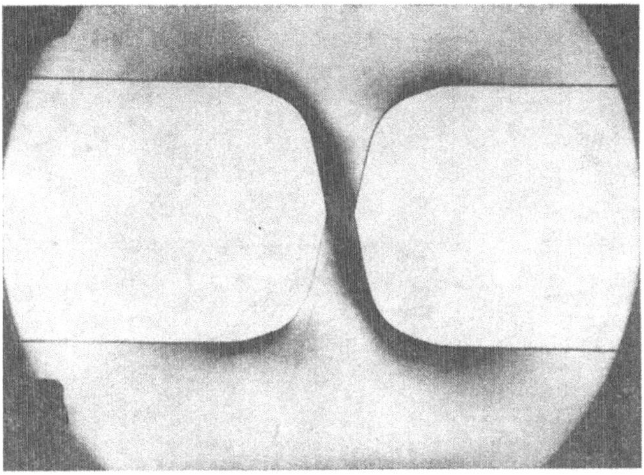

Fig. 13.9. The fringe pattern in the 2nd model loaded to produce a 0.5 K fringe at the boundary

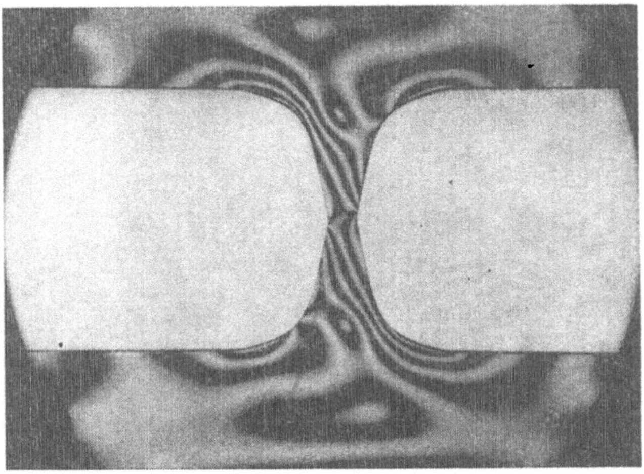

Fig. 13.10. The fringe pattern under a load corresponding to a boundary fringe of 3.5 K

extreme fringes are seen to be parallel to the boundary, which indicates an equalization of stresses along the tangents to the boundary. The fringe patterns when there is no bending in the column are shown in Figs 13.11 and 13.12.

13.7 Determination of the optimum shape of plane structures

The method of shape optimization of plane structures, for example truss nodes or wall beams, proposed by Burakiewicz *et al.* (1975) consists in drawing a network of orthogonal lines on a model made of a rubber membrane

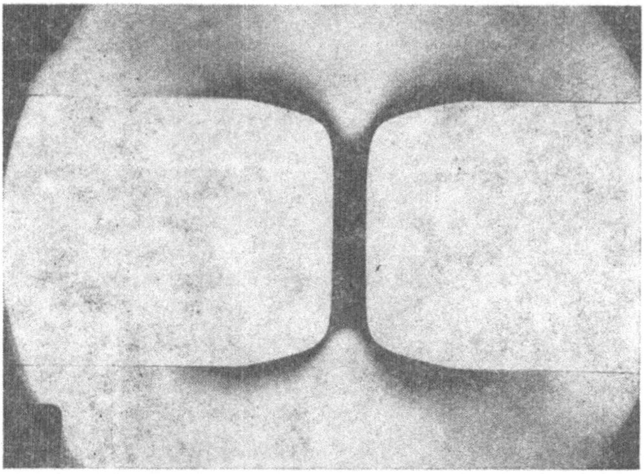

Fig. 13.11. The fringe pattern under an axial load corresponding to a 0.5 boundary fringe

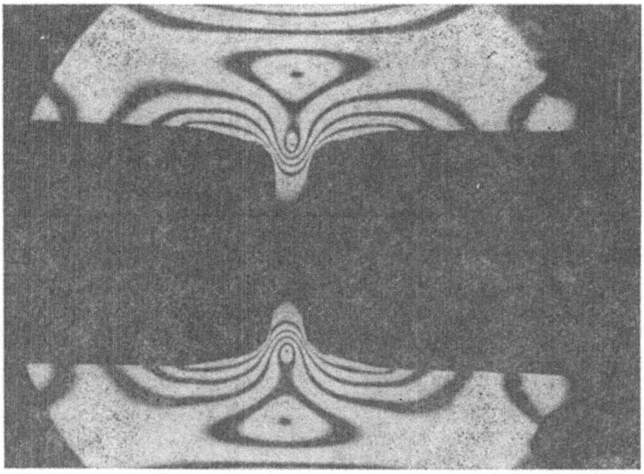

Fig. 13.12. The fringe pattern under an axial load corresponding to a 3.5 K boundary fringe

and observing its deformation under a given load. Analysis of the deformations provides indications as to the possible modifications of the model—by changing the shape of the boundary or the support elements—towards a uniform-strength shape. Depending on the conditions, visual observation of the networks may be sufficient, but it may also be necessary to measure their deformations under various loading states. The iterative scheme of the method can be summarized as follows:

(1) Construct a rubber-membrane model of the real structure, taking account of its support and loading conditions.

(2) Plot orthogonal networks in the regions required to deform uniformly
(3) Load the model and analyse deformations in the networks.
(4) Improve the shape of the model.

Operations (2), (3) and (4) are repeated successively as many times as necessary to obtain a shape sufficiently close to a uniform-strength shape.

The above method was applied to the optimization of a truss node between four identical bars in tension (Fig. 13.13). Sufficiently far from the node

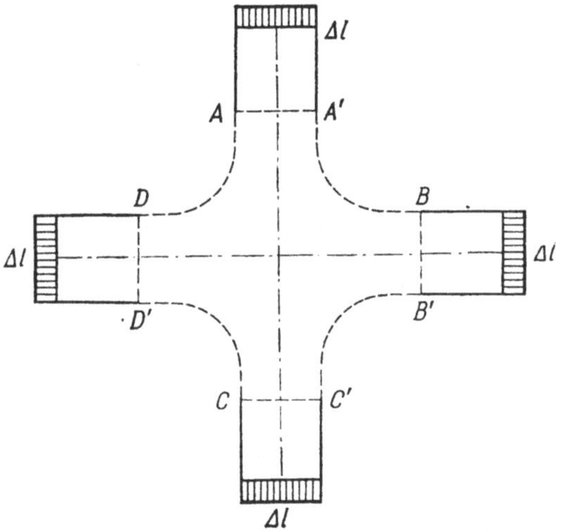

Fig. 13.13. Schematic illustration of Model 1 of a four-bar node in tension

centre, deformations are distributed uniformly and the condition of optimality is satisfied. The problem consists in determining such boundary curves $A'B$, $B'C'$, CD' and DA that the deformations along these boundaries are equalized. To this end, the bars were cut off along the lines AA', BB', CC' and DD', with an identical displacement Δl applied to each of them, and the procedure (1)–(4) described above was carried out. In the first stage, the model adopted was in the form of an octagon, $AA'BB'C'CD'D$, with orthogonal nets drawn along the free boundaries (Fig. 13.14). Model 1 subjected to tension in a special frame is shown in Fig. 13.15. Model 2, in the second stage, was designed after the shape of model 1 in the deformed state, with the dimensions reduced in the scale of deformation (Fig. 13.16). The successive models 3, 4 and 5 are shown in Figs. 13.17–13.19, and model 5 in the deformed state in Fig. 13.20. This last model was accepted as sufficiently close to an optimum solution, because its deformation was very nearly uniform. The strain curves for the boundaries of all the models are presented in Fig. 13.21.

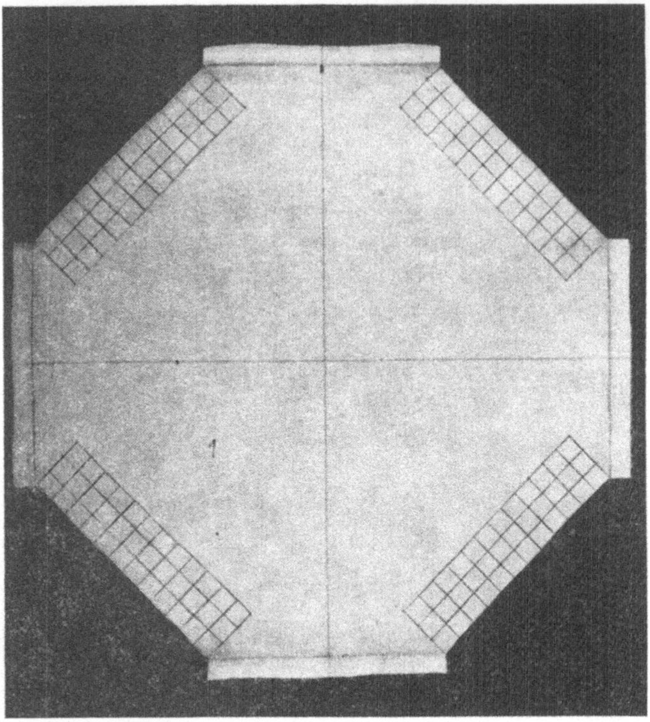

Fig. 13.14. Model 1: an octagon

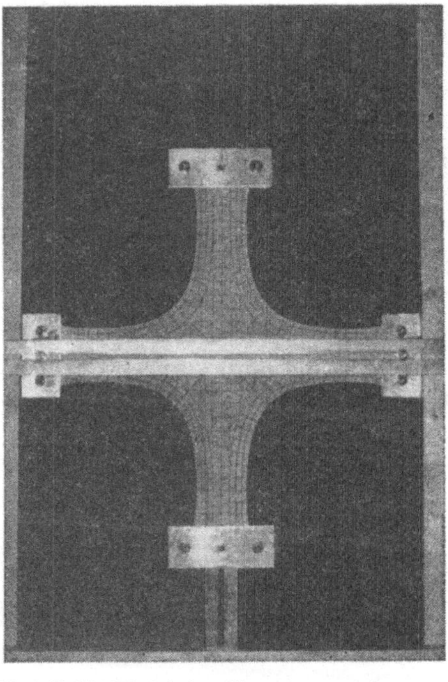

Fig. 13.15. Model 1 subjected to tension

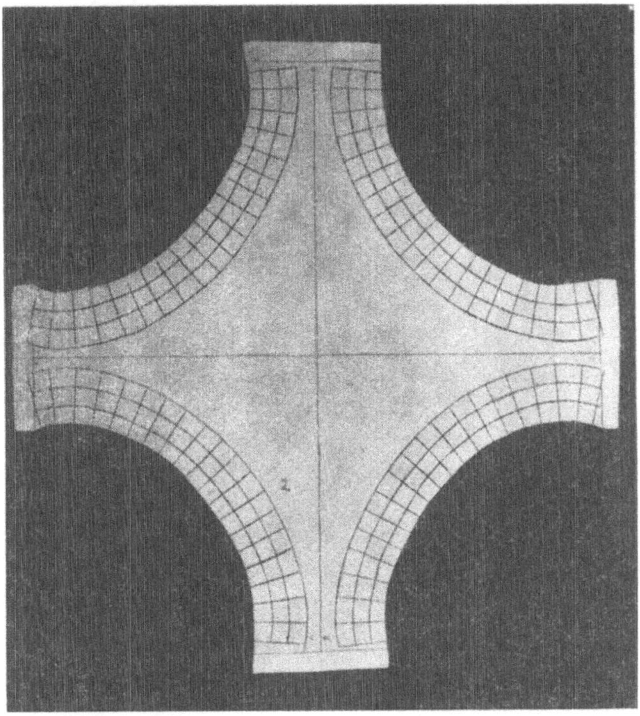
Fig. 13.16. Model 2

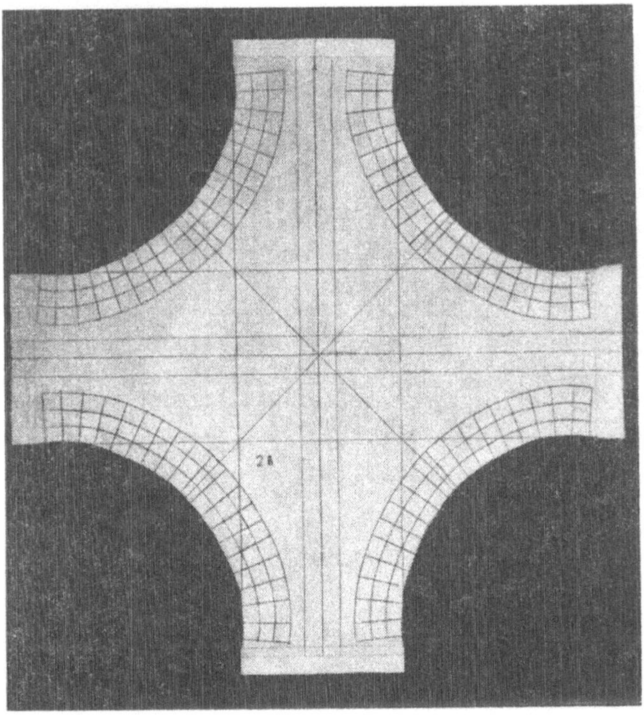
Fig. 13.17. Model 3

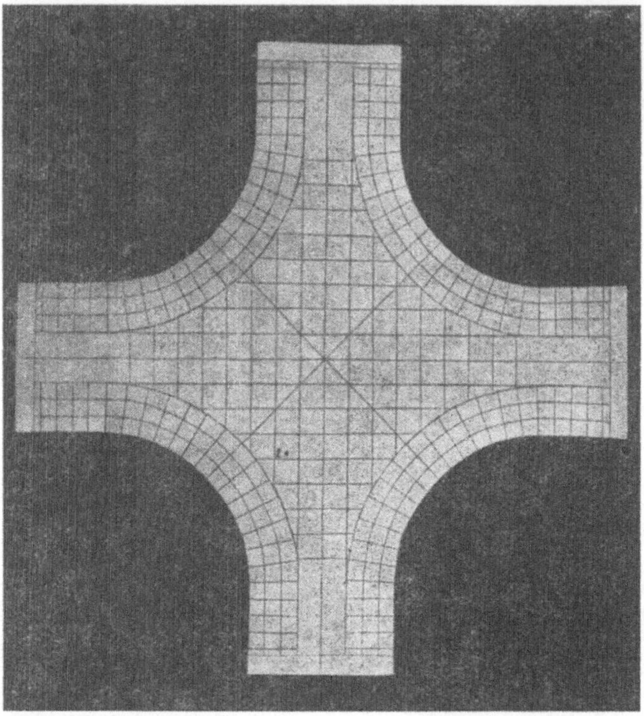

Fig. 13.18. Model 4

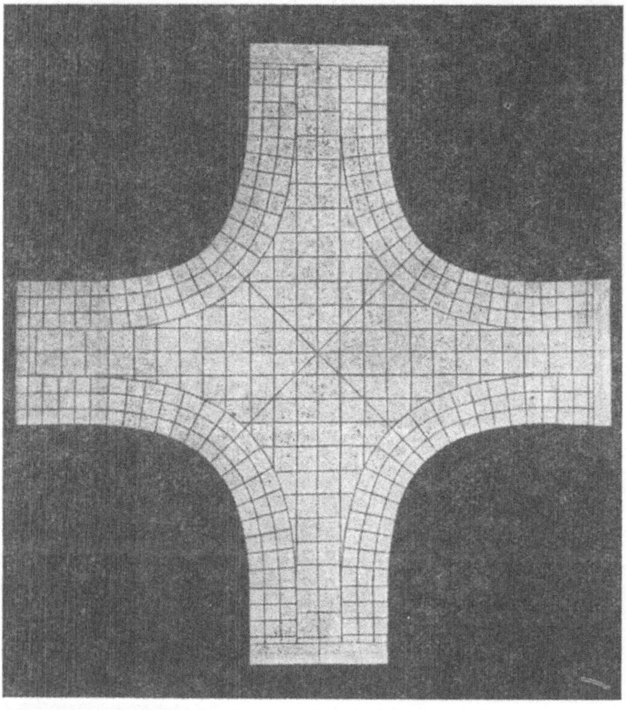

Fig. 13.19. Model 5

Fig. 13.20. Model 5 subjected to tension

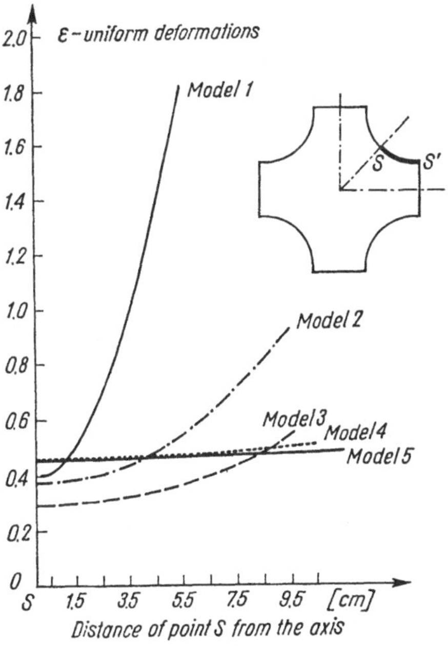

Fig. 13.21. The strain curves for the successive models (average strain from all the boundaries)

The shapes of the free boundaries in the sequence of improving models described above allow certain general conclusions:

— load-free optimal boundaries are given by curves with curvature of constant sign;

— curvatures and functions describing optimal boundaries, and the derivatives of these functions, are continuous.

It may be supposed that the above conlusions also hold for optimal surfaces and not only for plane curves, and, furthermore, that they may be useful in the design of loaded boundaries. In this case, the problem is reduced to the choice of load distribution for an optimum distribution of deformations.

Let us note that the elastic-membrane method of optimization has not been elaborated theoretically and is largely based on intuitive premises. Its convergence has been confirmed experimentally.

13.8 Shape optimization of shells

Optimum shapes of shells, understood here as shapes ensuring a uniform distribution of deformation, can also be obtained by means of elastic-membrane models with plotted line networks. In observing the network deformations after the model is loaded, one should be guided by the following features indicative of optimum shapes:

— invariance of angles between the network lines,

— invariance of side ratios in each mesh,

— invariance of side ratios between neighbouring meshes.

The above method was used by Burakiewicz *et al.* (1975) to optimize a shell with a rectangular plan and a centrally placed rectangular or square support, and a shell with a circular plan and a circular support. Figures 13.22–13.24 show deformed rubber-membrane models for the three cases An inspection of the deformed network systems indicates that

— shells with rectangular or square support elements exhibit none of the optimal shape characteristics mentioned above;

— distribution of deformation in the shell with a circular support is much more uniform than in the other two cases;

— non-uniformity of deformations in shells with rectangular and square supports is due to stress concentrations at the corners of the supports.

It follows that, of the three shapes considered, the shape of the model in Fig. 13.24 should be preferred in design.

Another experimental method of designing optimum shapes of shells is to spread an elastic membrane over a contour representing the plan of the shell to be designed, and fix it at the points corresponding to the actual supports. The membrane is then subjected to a loading in the form of air pressure or a large number of concentrated forces, which simulates the action

Fig. 13.22. Deformed rectangular rubber membrane

Fig. 13.23. Deformed square shell

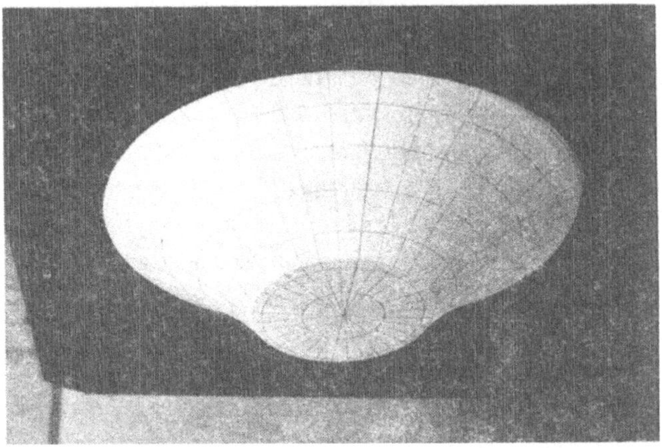

Fig. 13.24. Deformed circular shell

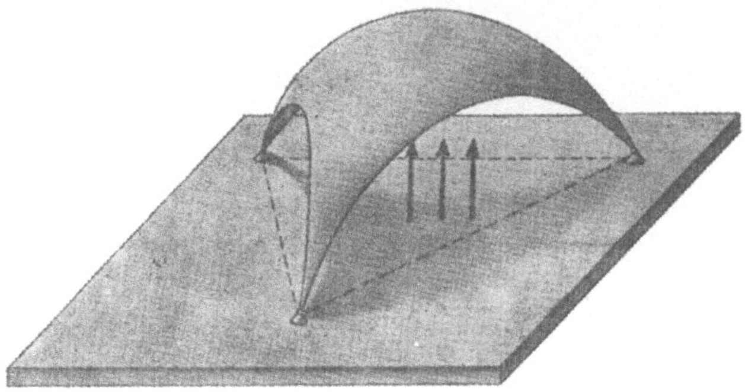

Fig. 13.25. Triangular membrane subjected to upward pressure

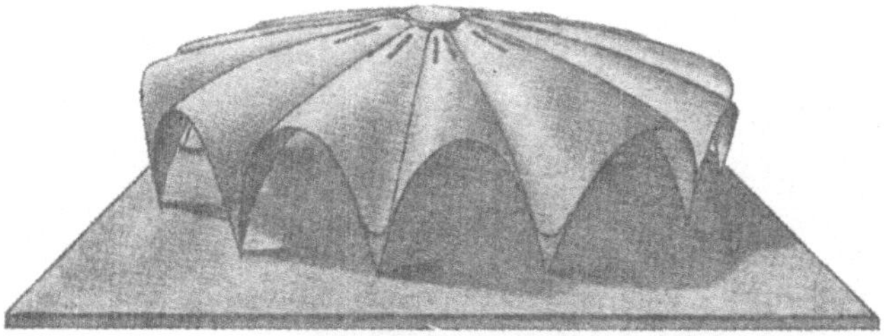

Fig. 13.26. A model of the market hall in Royan designed by Sobrero

of the dead weight of the structure and of a uniformly distributed loading, e.g. snow, with the internal forces having opposite directions to those which would occur in reality. After the deformation, the membrane corresponds to an optimum shell, because the shape it tends to assume is free from stress concentrations, bending moments and compressive forces, and also shows maximum stability.

Figure 13.25 shows a membrane stretched over a triangle and subjected to upward wind pressure. An obvious example of a real shell structure inspired by an elastic membrane model loaded by upward wind pressure is the market hall in Royan, designed by Sobrero in 1955 (Fig. 13.26).

In some instances the above method has made it possible to design novel and unconventional shell roofs. For example, the rubber-membrane model shown in Fig. 13.27 was used in the design of a basilica in Syracuse (Sobrero).

Copying the shapes taken by deformed thin elastic membranes amounts to adopting the criterion of minimum elastic strain energy as the criterion of optimization. This formulation was used by Koy (1963, 1965), who investigated the optimum shapes of shells supported by a single column. He showed that a reduction in the elastic potential corresponds to a decrease in the tangential forces and that a minimum potential occurs where the shell is in a membrane state of stress, i.e. where only normal stresses, n_1 and n_2, are present, whose values—in the absence of tangential stress—are equal, regardless of direction. It is impossible to bring the entire shell to such a state and to eliminate stresses due to bending at all its points, but any intervention in the design aiming in this direction undoubtedly improves the shape of the

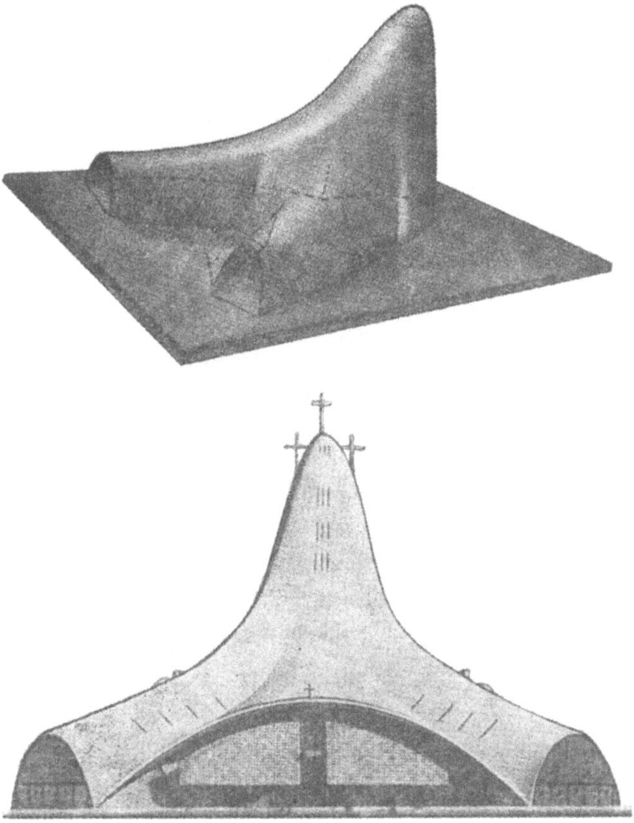

Fig. 13.27. a) A rubber-membrane model of a basilica in Syracuse; b) Design of the basilica

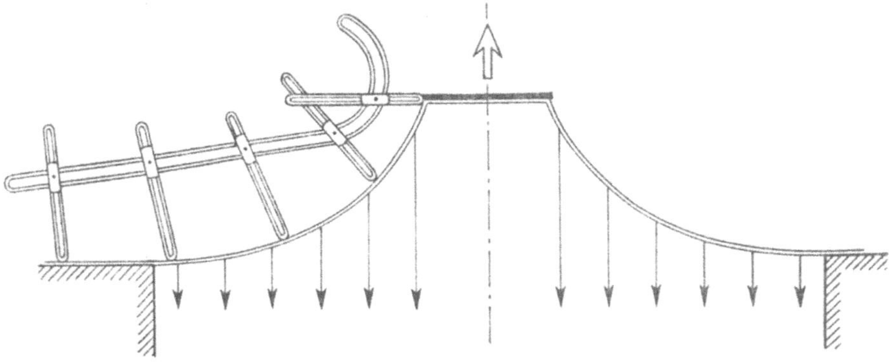

Fig. 13.28. Recording the curvature of a membrane by means of a templet

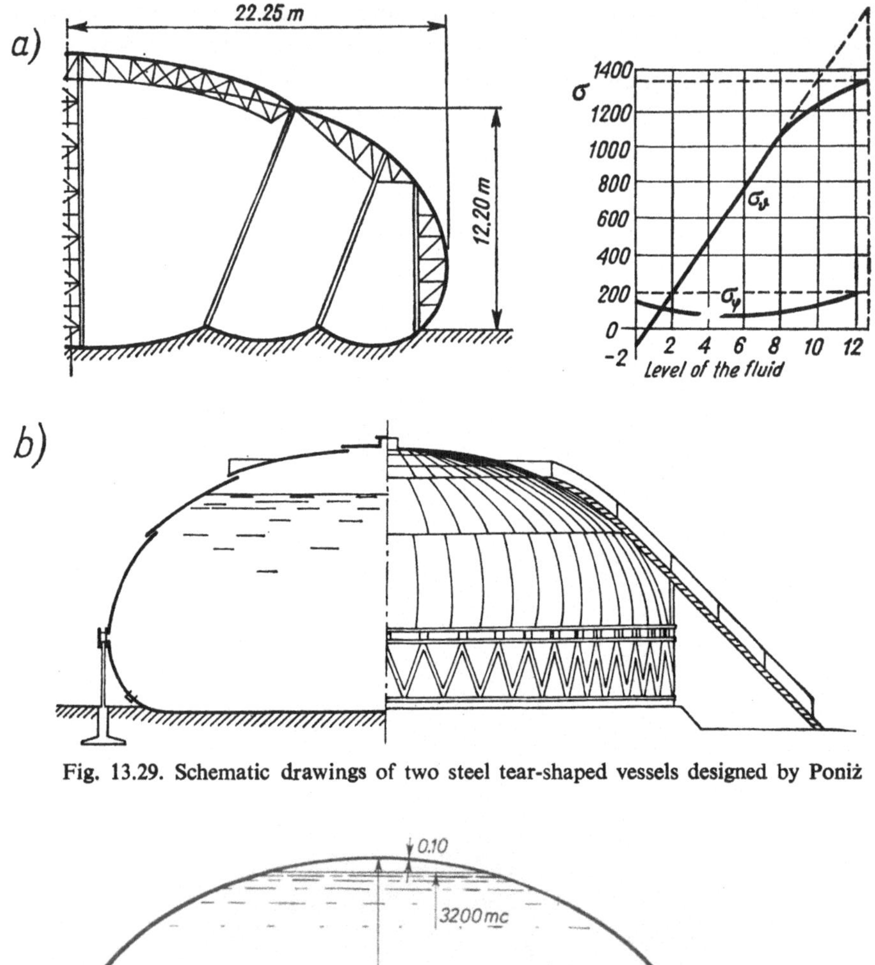

Fig. 13.29. Schematic drawings of two steel tear-shaped vessels designed by Poniż

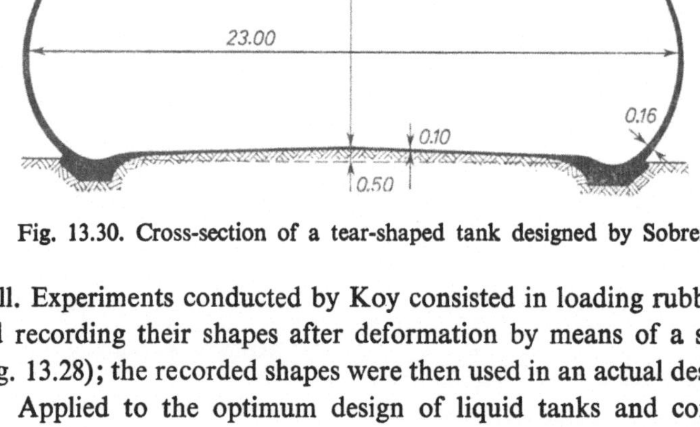

Fig. 13.30. Cross-section of a tear-shaped tank designed by Sobrero

shell. Experiments conducted by Koy consisted in loading rubber membranes and recording their shapes after deformation by means of a special templet (Fig. 13.28); the recorded shapes were then used in an actual design.

Applied to the optimum design of liquid tanks and containers, these ideas have led to shapes similar to those assumed by a free liquid drop or by

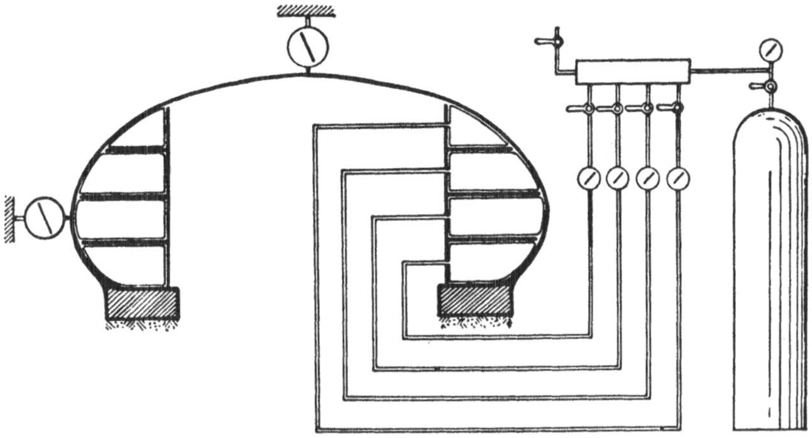

Fig. 13.31. Test loading of a model of the tank

liquid enclosed in a thin elastic membrane on a flat support. Bendings and stress concentrations observed in the containers designed in this way are considerably reduced, and the strain distribution is close to uniform. This has allowed application of exceptionally thin steel and reinforced concrete shells, with a very economical and uniformly distributed reinforcement in the latter case.

Figure 13.29 shows schematic drawings of two steel tear-shaped containers designed by Poniż (1961). A cross-section of a reinforced-concrete mazout tank designed by Sobrero (1955) and built in Sicily in 1957 is presented in Fig. 13.30. A model of this tank was tested in the Institute of Mechanics in Triest. Figure 13.31 schematically shows the way the model was loaded to simulate the liquid pressure from within.

PART III

BIBLIOGRAPHICAL SURVEY AND BIBLIOGRAPHY

14

Survey of the Literature of Structural Optimization

14.1 Scope and aim of the survey

Problems of optimum structural design have been formulated and solved by numerous authors for several hundred years. They used many different criteria and conditions of design and optimality and most varied methods for solving the relations they led to. Studies and solutions published up to the early 20th century, now only of historical interest, are not included in the present review. They are listed in the Bibliography, however, and many of them were noted in several earlier surveys of the field, e.g. by Wasiutyński (1939), Wasiutyński and Brandt (1963) or Sheu and Prager (1968_3). The aim of the present chapter is to review the principal trends and results in the research of structural optimization for different general types of structures. Short descriptions of the problems studied and solutions found include references to the basic sources. Publications in the field of mathematical methods of optimization are omitted, both in this chapter and in the Bibliography; whenever necessary, they are cited in Chapters 5–13.

The literature of structural optimization shows a continuous evolution of its methods, needs and research trends.

The early works concerned the simplest structural elements and employed mostly elementary mathematics. Solutions obtained in that period extended the knowledge of structures but had no significant influence on engineering practice.

The subsequent period brought a large number of publications in which optimization problems were formulated for generally applied types of structures and structural elements. Solutions were obtained by means of the classical methods of analysis and calculus of variations. However, the actual practical applications of the optimum solutions were still very limited. The reason was the omission in the formulated problems of various properties or parameters of design which, although not directly related to the strength of the structure, were relevant to the conditions of its execution und use. Exact solutions of such simplified problems could only provide certain general suggestions as to the optimum shape and permit estimation of changes in weight or stiffness if the design variables could not be chosen in accordance with the solution.

In the present development of structural optimization a new situation is observed. Depending on the purpose they are to serve or the material they are to be made of, many structures must be optimized, whether for minimum weight, for maximum stiffness, or for maximum load carrying capacity. An incorrect shape of a nuclear reactor containment renders the structure unsafe under an emergency pressure growth. A non-optimum shape of an element to be produced in large numbers leads to serious material losses. Most of these problems cannot be solved by classical mathematical methods. The effectiveness of structural optimization, its ability to solve more and more complex design problems, giving better and better approximations of real structures and loadings, depends on employing the latest achievements of modern mathematics, above all of numerical methods. Mathematical programming methods have become widely applied tools in optimization and almost all recent work on specific solutions is done with the use of computers.

14.2 Survey works and general studies on structural optimization

One of the first reviews of the methods of optimum strength design was published by Wasiutyński in 1939. The review, starting from the works of Galileo in the first half of the 17th century, is based on source materials and includes numerous quotations of the original texts and proofs. Besides presenting historical references, it also gives a comprehensive account of later contributions which influenced the development of strength design and, in general, of structural optimization.

Of monographic publications which appeared in the post-war period we should note a textbook by Shanley (1952), which provides an extensive exposition of the principles and methods of minimum weight design. In spite of the limitation of the subject to aircraft structures, as suggested in its title, the book includes a great deal of material of general significance. Similarly, the methods of optimum design presented in Johnson's monograph (1961), which deals primarily with mechanical elements, can be applied to structures of all kinds.

In the Polish literature, a survey of modern trends in the optimum design of building structures was presented by Wasiutyński and Brandt (1962), in a sequel to two earlier papers by Brandt (1961_1, 1961_2). The survey appeared in English in the Applied Mechanics Reviews in 1963 and 1966, each time in a new and revised edition.

A review of more recent developments in optimization, covering the period 1963–1968, was published by Sheu and Prager (1968). The review reflects a considerable advance in numerical methods and their increasing role in optimum design as effective supplements or alternatives to the classical analytical methods. A similar range of results was reviewed by Reitman and Shapiro (1966). The survey articles of Radtsig (1968) and Volkova (1971) are restricted to the investigations conducted by a very active group of Soviet researchers led by Radtsig in Kazan.

A unified view of the field of optimum structural design was presented by Niordson and Pedersen (1973). Besides general problem formulations and description of solution methods, their report gives a review of the state of research separately for different structural elements. However, only very little reference is made to Polish authors and none to the papers published in the Soviet Union.

A survey of optimality conditions for minimum cost of various multi-member and multiple-load systems was carried out by Rozvany (1973). More recently, an extensive survey of optimum design methods for civil engineering structures was published by Krishnamoorthy and Mosi (1979).

General problems of optimum structural design are treated extensively in the numerous papers of Wasiutyński. His theorems on optimization for minimum potential are formulated in parts II and III of his work of 1939. Further fundamental information is given in his papers of $1956_{1,2}$ and 1960_2. Convergence and equivalence of different criteria of optimization is analysed in Wasiutyński's papers of 1960_1 and 1966_3. Optimality conditions for structures subjected to n different loading states are discussed in his papers of 1966_2 and 1967_1.

A classification of strength-optimization problems with respect to the kinds of structures and design variables was proposed by Krzyś and Życzkowski (1963_1).

Drucker and Shield (1957_1) discussed general problems of minimum-weight design of structures of elastic-perfectly-plastic materials and derived bounds on the minimum design based on a limit analysis. The same authors formulated the principle of uniform dissipation of energy in minimum structures (1957_2). Drucker (1958) developed minimum-weight design methods for non-homogeneous plastic bodies, considering irregular and regular distributions of inserts in a homogeneous medium.

A probabilistic approach to optimum design was first formulated by Hilton (1958) and developed by Hilton and Feigen (1960), who used the concept of the reliability of a structure. Statistical methods were also employed by Hsuan–Loh Su (1959). Probabilistic or reliability-based design was subsequently studied by many authors, who contributed new formulations and solutions; we should note three papers on this subject in Study No. 1, SM

Division, University of Waterloo, Canada, by Khachaturian (1969), Moses (1969) and Lind (1969). More recent solutions in probabilistic design are summarized in a short article by Frangopol and Rondal (1978). These and other papers are presented in more detail in Section 4.6 of the present book.

Sufficient conditions for minimum-volume design of structures made of isotropic and perfectly plastic materials were discussed by Shield (1960) and by Mróz (1960, 1961). They pointed out that in various cases both minimum-volume and maximum-volume structures can be obtained. A similar conclusion with regard to hyperstatic frames was reached by Megarefs and Hodge (1964).

Fundamental research into the ways of formulating optimum design problems has been of considerable importance for the development of structural optimization. It is worth while to note here a paper by Radtsig (1962) on applications of mirror functions to the optimization of hyperstatic structures. The theory of mirror functions had been developed by Radtsig and his collaborators and presented in numerous publications.

A method of parametric optimization, i.e. optimization in which the selection of functions is replaced by the selection of a number of parameters for the functions assumed *a priori*, was proposed by Krzyś and Życzkowski ($1963_{2,3}$). The method involves linearization of non-linear inequality constraints, considerably simplifying the investigation of the extrema of the objective function.

The aim of designing a structure which would be optimum with regard to some criterion adopted *a priori* is often realized by modifying an initial design. To avoid repetition of calculations for the whole structure, methods of modification have been proposed which require only some additional calculations to be carried out, corresponding to the modification introduced, whereby different variants of the structures can be examined and the optimal one chosen. A review of the methods of structural modifications, with most relations in a matrix form adapted to computer treatment, was given by Gierliński (1974).

In general, optimum design problems have been considered for structures under single or multiple loading conditions. There are several papers, however, dealing with optimum design of structures subjected to moving loads. Gross and Prager (1962) presented a method of minimum weight design for hyperstatic beams under a moving concentrated load, using an example of a beam fixed at one end and simply supported at the other. Several more examples of optimum-plastic-design of hyperstatic beams under such loads were given by Prager and Save (1963). Radtsig (1960_2) took up the problem of designing a hyperstatic truss loaded by a single force moving from node to node. Vinogradov (1954_1) considered beams and frames under various combinations of several independent loads applied at different points. The structures were optimized for minimum volume of material subject to stress constraints. Optimum-elastic-design of beams under moving loads and multiple load conditions was investigated by Marks (1966_2, 1971). He used the criterion of equalizing maximum unit potentials at as many points of the structure as possible. Another paper by Marks (1968_4) considers optimization of an isostatic beam subjected to a moving load and shows that for each cross-section of the beam there are two specific load positions which produce an extreme combined stress, due to a maximum bending moment or to a maximum shearing force.

A separate group of papers deal with optimum design problems in which constraints are also imposed on certain dynamic parameters, e.g. frequency of free vibrations or vibration amplitude; let us note here a paper by Niordson (1965). Icerman (1969) and Mróz (1970_2) optimized beams with forced transverse vibration. Similar problems were also solved by Brach (1968_1) and Plaut (1970, 1971_2). Of Polish authors who studied optimum structural design under dynamic constraints, let us also note Olszowski (1970_1). A survey of the work done within this subject was presented by Pierson (1972).

14.3 Optimization of beams

Homogeneous beams

Optimum design of beams has been studied by many authors; beams have also been used frequently as example structures on which to illustrate or test more general methods of optimization.

Wasiutyński applied his theorems (see Sec. 2.1) to minimum volume problems for beams with rectangular cross-sections and to the optimization of steel I-beams (1939, part III). In a later paper (1960_3) he considered the design of steel I-beams and box beams with regard to various criteria of optimization, e.g. design for minimum total potential at constant volume or for uniform strength in the sense of the Huber hypothesis, and under various conditions. According to his general theorems, the condition of minimum potential at constant volume of material can be replaced by that of uniform unit potential, which fact may simplify calculations considerably.

The question of how to account for the dead weight in designing a beam was also examined by Wasiutyński (1960_3), and by Grycz (1962_3).

Kosmowski ($1960_{3,4}$) determined shape parameters for cantilever beams of uniform strength in the sense of the Huber–Mises hypothesis and investigated the dependence of the beam volume on various parameters. In continuous hinged beams the distribution of supports and hinges influences the volume of the material; a search for a distribution which yields a minimum volume is the subject of yet another paper by the same author (1960_2).

Leśniak (1970_2) discussed optimization of rolled parallel-flange beams

IPE, wide-flange welded beams and welded plate girders—all for a minimum cross-sectional area.

Optimum design of beams made of homogeneous materials with equal tensile and compressive strength (e.g. steel) has beem investigated by several authors in the Soviet Union. The studies initiated by Rabinovich (1933) were later continued by Radtsig in Kazan, Khuberyan in Tbilisi and Vinogradov in Kharkov, and their numerous collaborators. Most of the work concerned hyperstatic beams, the internal forces being determined from strength of materials formulae and little or no attention being paid to stability, the influence of tangential forces and other secondary factors in the design. Dead weight was usually treated as an external loading, while both steady and changing loads—concentrated or continuous—were assumed to be distributed arbitrarily, so that extreme values of forces could be obtained in arbitrary cross-sections. The design variables were cross-sectional parameters, with the general shape of cross-section assumed *a priori*, e.g. flanged, box or rectangular.

The early papers of Radtsig (e.g. 1941, 1946) concerned a wider range of structures, not only beams. Later (1958_3) he considered applications of the minimum-volume approach to the design of hyperstatic beams and, in a paper of 1963 (Radtsig 1963_1), formulated a general optimum design problem for such beams, gave a historical overview of the problem and pointed out the significance of the location of the points where the bending moment curves change their sign. He also introduced and developed the theory of mirror, or modular, functions (e.g. Radtsig 1958_2, 1961_1, 1966_5, 1969_4), which became an important mathematical tool for solving optimization problems where the investigated functions may be discontinuous and change sign while the required design variables must be positive. Mirror functions were used by researchers of Radtsig's group in numerous investigations (e.g. Volkova 1965_{1-4}, Gajnullina 1956). Khuberyan's main interests were frames and shells, and only his paper of 1938 takes up an optimum design problem for hyperstatic beams.

Among the numerous writings of Vinogradov we note two papers published in 1954 and particularly the monograph of 1955, which mainly deals with hyperstatic trusses but in a large proportion of its general considerations is also relevant to the design of beams.

Studies of the optimum design of beams were pursued by several coworkers of Radtsig, including Borisov ($1959_{1,2}$), who derived relations between the zero points of the bending moment curves for continuous beams and frames. Hyperstatic beams were also investigated by Izraelit (1956) and Fesik ($1961_{1,2}$). We should also mention Sofronov (1969_4), who introduced deflection constraints for minimum volume beams, and Sungatullin (1969_1), who used inequality constraints for stresses and rotations of support cross-sections, reducing optimization of isostatic beams to non-linear programming. Gajnullina (1966) formulated an isoperimetric variational problem for hyperstatic beams under various loading systems; some of her earlier papers (e.g. 1958 and 1963_2) were also devoted to optimization of beams.

Hyperstatic beams were also the subject of papers by Peredy (1959, $1960_{1,2}$) and, to a lesser extent, of those by Readey (e.g. 1954).

Optimization of isostatic beams was studied by several Soviet authors: Kiselev (1956), Korshunov (1959), Yuryev (1959) and Tsentsevitskaya (1961).

An important work in this field, frequently quoted in the literature, was published by Barnett (1961). It concerned steel beams with I- or box-section. The design variables were flange width and web thickness regarded as variable along the beam length. Barnett's solution was superior to earlier solutions in that it imposed deflection constraints on the minimum-weight beam, including constraints on deflections due to tangential forces. An interesting conclusion obtained by Barnett pointed to the expediency of using different materials for flanges and for webs. A similar view had been presented in 1951 by Vakhurkin, who solved a minimum cost problem for a steel I-beam made of two different materials, defining the cost functions as an appropriate combination of the two corresponding volumes.

Optimum proportions for minimum-weight symmetrical steel girders without web stiffeners were calculated by Holt and Heithecker (1969). Practical techniques for optimizing box girders and wide-flanged beams were described by Goble and Moses (1975). Chern (1971) discussed optimization of beams for a given deflection; as an example he used a cantilever beam loaded by a concentrated force. Hornbuckle, Neville and Boykin (1975) applied mathematical programming and the finite element method to optimize continuous beams. Mathematical programming, combined with an iterative technique, was also applied by Bronowicki and Felton (1975) to the optimum design of continuous thin-walled beams. Plate girders under multiple loading conditions were considered by Brown and Ang (1966) and by Martin (1970).

Optimization of thin-walled beams with stability constraints was studied by Saelman (1962), Życzkowski ($1968_{1,2}$) and Krzyś; (1964, 1966); these authors sought minimum-volume beams at the plastic limit. Demokritov (1968) solved similar problems with regard to deflection and local stability constraints.

Minimum-weight design of beams subject to moving loads was investigated by Save and Prager (1963).

A minimum-weight beam problem with inequality constraints on stress and deflection was solved by Haug, Jr., and Kirmser (1967). Their considerations are restricted to isostatic beams made of isotropic and elastic materials with equal compressive and tensile strengths and subjected to arbitrary loads. The solution provides necessary conditions for an optimum, obtained by classical variational methods; no sufficient conditions are given.

Optimization of beams in the elastic range with inequality constraints on stress was also considered by Masur (1970).

Optimization of prestressed and reinforced concrete beams

Optimization problems for concrete beams differ from those reviewed above in that they must take into account the presence of two different materials: concrete and steel, and—in the case of prestressed beams—in that the distribution and the magnitude of the prestressing force are additional very important parameters of design.

The influence of the dead weight of a reinforced beam on the design for uniform extreme stress was considered by Brandt and Ignaczak (1958) and by Ignaczak (1959). Optimization of beams with respect to the same condition under various assumptions on shape parameters was discussed by Brandt in several papers (1958, 1960_1, 1964), the last of which uses the concept of auxiliary volume, i.e. the sum of the volume of concrete and the volume of steel multiplied by coefficients depending on the relative strength properties and unit costs of both materials. The same paper (Brandt 1964) presents further solutions concerning prestressed beams designed for minimum elastic strain energy—total or only due to useful loads—and also examines the influence of the tangential forces on the optimum shape of I-beams. Brandt also discussed the optimality conditions implied by the equalization of stresses at the lower and upper edges of the beam (1960_2) and formulated a theorem on the effect of the dead weight of the beam on its shape parameters (1962). According to that theorem, among the cross-sectional parameters determined from the condition of minimum elastic strain energy and equalized stress at constant volume of material, only the function describing the location of the prestressing cable depends on the dead weight of the beam. The theorem was later presented in a modified form (Brandt 1963, 1967).

Jaworski and Marks (1968) used a method of successive approximations to determine cross-sectional parameters for a prestressed beam of uniform strength in the sense of Caquot's hypothesis. A subsequent paper of Marks (1968_3) concerned prestressed beams designed with respect to the same condition but under moving loads. Touma and Wilson (1973) sought optimum designs for prestressed concrete spans carrying high speed ground transportation, using the direct search method.

A few papers have been devoted to the optimum design of beams prestressed with thin wires continuously distributed within the beam. The problem was originally formulated by Wasiutyński *et al.* (1962) and later solved in several variants by Marks (1966, 1967).

A separate trend in the design of concrete beams, reinforced or prestressed, is represented by all those works in which the principal objective is to determine the basic parameters of a structure so as to satisfy certain requirements of strength and economy. Investigations of this kind do not involve optimization problems in the strict sense, i.e. do not seek extrema of functions or functionals derived from some criterion of optimization. Instead, they provide systematic design procedures or formulae which make it possible to avoid the laborious way of successive trials in reaching a satisfactory design.

Mames (1957) took up the problem of designing a post-tensioned continuous I-beam with a cross-section non-symmetrical with respect to the horizontal axis. The shape and prestressing of the section were defined by eight parameters. The problem was solved under two loading states and supplemented by a discussion concerning the limit value of the dead weight of the beam beyond which the fulfilment of all the conditions of the design was impossible.

Bilge (1964) presented a method for designing pre-tensioned and post-tensioned prestressed concrete beams and discussed the differences between the corresponding designs. As in Mames' work, the design variables were determined from the strength requirements for prestressed concrete and from structural requirements. The structural requirements differ in the two cases; in particular, in pre-tensioned beams with wires of constant cross-section it is impossible to equilibriate the dead weight, while it is possible to do so in post-tensioned beams. The magnitudes of the coefficients characterizing the structural conditions for the beams were chosen by Bilge on the basis of their frequencies in the bridge structures designed by the Société Technique pour l'Utilisation de la Précontrainte (STUP) and built in France.

Gawęcki (1965) studied the design of simply supported post-tensioned prestressed concrete beams with respect to the following three criteria:

— minimum beam height,
— minimum cross-sectional area,
— minimum prestressing force.

Croci (1963) considered the problem of determining the eight cross-sectional parameters for a prestressed I-beam from the structural and strength requirements. He presented two methods for finding an optimum solution characterized by a minimum cross-sectional area and the least possible prestressing force in the cross-section. Another paper by Croci (1964) proposes a method for an optimum design of a cable system in a prestressed beam, emphasizing the possibility of non-uniform prestressing and also of concreting by stages.

Optimization of prestressed concrete beams for minimum cost was investigated by Ramaswamy and Raman (1970). The problem was formulated so as to allow for partial prestressing and to comply with the FIP-CEB recommendations concerning stress conditions in structures designed in accordance with class I, II and III requirements. The beam cost per unit length was defined as the combination

$$C = C_c B_p + C_a A_a + C_{ap} A_{ap},$$

where C_c, C_a and C_{ap} denoted the costs per unit volume of the concrete, reinforcing steel and prestressing steel, respectively, and B_p, A_a and A_{ap} represented the areas of the component materials in the cross-section of the beam.

An extensive analysis of the problem of choice of the most economical cross-section for a prestressed beam was carried out by Kaufman and Hop (1959). They considered rectangular sections, symmetrical and unsymmetrical I-sections, and T-sections. The cost of a section was estimated by means of a special performance coefficient combined with the section depth, area and prestressing force. By minimizing the cost, it was possible to determine the beam height and the compressive stress in the concrete, which was required not to exceed an allowable value.

Design of reinforced concrete structures for minimum joint cost of the reinforcement and concrete was studied by Kaliszky (1968). Earlier, minimum cost design of beams and frames in reinforced concrete was taken up by Massonnet and Save (1963). In either case a rigid-plastic model was adopted, i.e. one of a simultaneous yield of the reinforcement in tension and the concrete in compression under the limit load.

Continuous prestressed beams were optimized by Kirsch (1972) with the use of linear programming. Goble and Moses (1975) optimized a reinforced concrete beam using a penalty function method. Geometric programming was successfully employed for the same purpose by Templeman and Winterbottom (1973). Much earlier, Hill (1966) was one of the first to apply dynamic programming to similar problems. Non-linear programming methods were also used by Marks to optimize prestressed beams (1974) and, prestressed concrete beams with postcompressed reinforcement (1976, 1977, 1978).

14.4 Optimization of plates

Publications on the optimum design of plates are not as numerous as those dealing with beams. In most papers, optimization problems for plates are formulated under the assumptions of plasticity theory; plates are designed for a minimum volume of material in a limit equilibrium state under a given external load. A variety of shapes and both static and dynamic loads have been considered. The limit state has been defined via the Coulomb–Tresca or Huber–Mises–Hencky yield conditions. In a few cases the limit state was taken as corresponding to the appearance of plastic deformations on the outer surfaces of the plate, i.e. the plate was considered in the elastic range.

A separate group of papers can be distinguished which deal with prestressed concrete plates, using minimum elastic strain energy at a constant volume of material as the criterion of optimization.

Among the older solutions of the first group we should note those of Hopkins and Prager (1955), who calculated optimum circular plates of variable thickness and discussed limit cases. Prager (1955) also considered non-circular plates. Solid and sandwich circular plates were optimized by Freiberger and Tekinalp (1956). A sandwich plate consists of two supporting face sheets of variable thickness, with a constant-thickness filler layer inside; the principal design variable is the face thickness.

Drucker and Shield (1957_1) showed that a plate has a minimum volume when it is designed to be just at the point of collapse under the given loads and has a continuous collapse mode such that the dissipation of energy per unit volume is constant over the faces of the plate. Mróz (1961) extended the theorem of Drucker and Shield, showing that every plate designed to collapse in a mode such that the dissipation rate is constant over the entire surface has an extremum volume.

Minimum weight design of plates was further studied by Onat, Schuman and Shield (1957), who indicated the possibility of the existence of local minima in the volume of a plate, and also by Prager and Shield (1959), and by Eason (1960), who considered various support and loading conditions. Optimum design methods for plates under multiple loadings were presented by Shield (1963).

Various optimum plate design problems were also presented as illustrations to more general considerations of minimum-weight plastic optimization of structures. We may note here the works of Mróz (1960), Marçal and Prager (1964), Lackman and Ault (1966), Marçal (1967) and Megarefs (1967, 1968). Optimization of plates with constraints on the fundamental frequency of free vibrations was investigated by Olhoff (1970) and by Armand (1971).

Minimum weight design of annular plates, simply supported or fixed at the inner or outer edge, or at both edges, was studied by Mróz (1958_2). Further solutions to similar problems were given by Shamiev (1962, $1963_{1,2}$).

All the works quoted so far concerned isotropic plates. The optimum design problem for technically orthotropic plates was analysed by Mróz (1958_1). The material of the plates considered by Mróz was assumed to be brittle in tension and perfectly plastic in compression, as in concrete plates reinforced orthogonally.

Reinforced concrete plates, as made of steel which is elastic-perfectly plastic in tension and of concrete which has the same properties in compression, were considered by Morley (1966) and Kaliszky (1968). Optimization of reinforced concrete plates was also studied by Mróz (1965) and by Mróz and Shamyev (1970). The problem of minimizing the weight of reinforcing steel in reinforced concrete slabs was solved by Dutta and Ratnalikar (1973) for simply supported slabs and for slabs with clamped edges.

Ribbing may be regarded as another kind of orthotropy. An analysis of optimum ribbing for a square slab with respect to buckling strength was presented by Kloppel and Moller (1965). A similar problem was considered by Rockey and Cook (1965).

Ramesh and Karve (1976) and Keshava Rao (1976) studied optimization of orthotropic plates and grid systems under various support and load conditions.

Optimization of multilayer plates was the subject of several papers by Mikeladze (1957, 1960, 1961). An extensive discussion of the minimum volume design of perfectly plastic discs was presented by Hu and Shield (1961).

Uniform strength design of three-layer plates was investigated by Dzieniszewski ($1969_{2,3}$). He considered simply supported plates with a neutral middle layer and solved the problem for several different shapes of the plate.

Optimization of circular plates at limit states corresponding to the appearance of plastic stresses at the external surfaces was studied by Brzoska (1954) and by Tadjbakhsh (1961). The plates were assumed to obey Tresca's yield condition. In a later paper on minimal design of sandwich axisymmetric plates, Reiss and Megarefs (1971) used the Mises criterion.

Extensive studies of the optimum design of prestressed plates were carried out by Rozvany and Hampson (1963) and by Rozvany (1964), who presented a method for finding plates with minimum thickness or with minimum weight of the prestressing cables.

Dzieniszewski (1964_2, 1965) took up the problem of optimizing prestressed plates for a minimum elastic strain potential at a constant volume of the material. He considered two-way prestressed solid plates and obtained solutions for the variable plate thickness, cable shape and prestressing forces in both directions. Another study of Dzieniszewski (1969_1) deals with prestressed concrete plates of uniform strength in two loading states. A solution is obtained for plates with arbitrary contours. An energy approach to the optimum design of prestressed plates was also followed by Taylor (1971).

Finally, let us note an extensive comparison between the minimum-weight and the minimum-cost designs of prestressed plates presented by Naaman (1976).

14.5 Optimization of trusses

Optimum design of trusses has been treated in a great number of publications. In the last decade interest in this field has somewhat declined, mainly because trusses—in the classical sense of the word—are being more and more rarely used in present day engineering. However, owing to the simplicity with which optimization problems can be formulated for trusses (only axial member forces need be considered!), they continue to serve as test examples for the developing methods of optimization.

In the Soviet Union, the first to formulate a general optimization problem for trusses was Rabinovich (1933). He considered sets of fully stressed trusses, including both isostatic and hyperstatic designs; optimization consisted in choosing a minimum weight design.

Radtsig (1941) optimized a singly statically indeterminate truss, considering it jointly with the corresponding isostatic truss. A subsequent paper by the same author (1946), showing how to transform a hyperstatic truss into an isostatic one by reducing member forces to zero, was of considerable importance for further research. The transformation aims at finding a minimum volume truss. In his later investigations of hyperstatic trusses (1956_1) Radtsig employed the theory of mirror functions, which enabled him to show new possibilities for obtaining exact solutions, and hence for verifying approximate solutions used in practical design. Yet another paper by Radtsig (1960_1) surveys the earlier work on optimization of isostatic and hyperstatic trusses and offers a method for determining an optimum configuration of bars in a single-span truss. Among the numerous publications of Radtsig let us also note a monograph (1969_1) covering the whole of optimum design problems for hyperstatic trusses with applications of the theory of mirror functions.

Kozyrev (1953) proposed an analytical and a graphical method for determining isostatic trusses of minimum weight.

Korshunov (1961) presented a method of successive approximations for designing indeterminate trusses which takes into account the dead weight of the structure. In a later paper (1966), Korshunov observed that optimal height of a truss depends on structural requirements and on redundant bars with finite cross-sections.

Volkova (1966) considered minimum-volume elastic design of hyperstatic trusses with members having different strength properties, and observed that such structures cannot be fully stressed.

A new algorithm for selecting a minimum weight design from the set of hyperstatic trusses satisfying given strength and structural requirements, using methods of linear programming, was proposed by Vinogradov (1967).

In Poland, studies of the optimum design of trusses were initiated by Wasiutyński's papers on transformations of trusses through the introduction of new nodes (1950_1) or new bars (1950_2). Optimization of a truss by means of shifting its unloaded node joining three members was later considered by Brandt, Wasiutyński and Kosmowski (1957). Their solutions concerned isostatic trusses under static loads; the criterion of optimization used was that of minimum elastic strain potential, equivalent in this case to the criterion of minimum weight.

The minimum-potential design of four-node trusses was worked out in detail by Kosmowski (1960_1). Four-node trusses were also optimized by Grycz (1960) by means of bar exchanges.

The work of Biernawski and Grochowski (1960) on optimization of four-node trusses, using methods of analytical geometry, confirmed in a more general form the results of Brandt, Wasiutyński and Kosmowski and created the possibility of extending them to trusses with a larger number of nodes. An applications of such methods to a spatial grating with four loaded nodes was presented by Biernawski (1962).

Optimization of member layout for an isostatic truss with a given configuration of nodes was also analysed by Fleron (1964), who proved the existence of an optimal solution. To obtain such solutions, Dorn *et al.* (1964) and Pedersen (1970) used methods of linear programming.

Toakley (1968_2) optimized isostatic trusses subject to displacement constraints.

The simultaneous optimization of member sizes and structural configuration, following the pioneer work of Michell (1904), was undertaken by several authors, e.g. Hemp (1966), Richards and Chan (1966) and Hegemier and Prager (1969).

The optimum design of trusses with stability conditions for bars, in the elastic range and beyond it, was analysed by Wojdanowska-Zając and Życzkowski (1969).

Optimization of hyperstatic trusses leads to non-linear problems, which are often solved by numerical methods. Solutions of this kind were published e.g. by Brown and Ang (1966) and Pope (1968). Pope also reviewed various procedures for the reduction of non-linear problems to a sequence of linear programs (Pope 1969). Rao (1975) used methods of optimal control. Geometric programming was applied by Morris (1972) to the minimum weight design of a three-bas truss. Similar methods were used by Templeman (1973) to optimize a complex statically indeterminate truss.

A unified mathematical programming approach to the optimization of elastic trusses was presented by Niemierko (1978). His work deals both with linear and with non-linear problems, including discrete cases and allowing for multiple conditions. Elastic trusses were also optimized by Templeman (1976), who proposed a new dual formulation in which constraint activity levels are used as variables.

14.6 Optimization of columns, arches and frames

Optimization problems for columns have been formulated in various ways. Most authors have sought an optimum distribution of material along the column axis for a given type of cross-section, but problems of determining an optimum-shaped cross-section have also been considered. The criteria used include those of minimum volume of material at a given load carrying capacity, minimum elastic strain energy at constant volume, uniform strength according to various strength hypotheses, and maximum buckling load. Columns can be solid or tubular, variously shaped, subjected to static or dynamic loads—axial or eccentric.

The least-weight problem for a column in compression was first formulated by Lagrange (1770) in a variational form. The problem was solved by Clausen (1851), who showed that the most efficient tapered column has a volume $\sqrt{3}/2$ times the volume of a cylindrical column of the same strength. Nikolai (1907) studied Lagrange's problem for circular columns with an additional constraint on the compressive stress. Chentsov (1936) derived a simple form of the Euler–Lagrange differential equation to be satisfied by a minimum-volume column.

In papers by Życzkowski (1956) and by Krzyś and Życzkowski ($1963_{1,2}$) the problem is restricted to tapered columns, with the cross-sectional shape assumed *a priori* and the column taper used as a design variable. The corresponding variational problem is thus reduced to determining a function of a single variable.

Wasiutyński (1951) proposed a method for designing columns of uniform strength under axial or eccentric compression.

The problem of determining the optimum-shaped cross-section among all possible convex cross-sections was studied by Keller (1960_1), who showed that the equilateral triangular section increases the critical buckling load by 20.9 per cent over that of a circular cylinder. Similar conclusions were obtained by Riggs (1964), who investigated circular, square and triangular sections for elastic and inelastic buckling.

Tadjbakhsh and Keller (1962) derived a necessary condition for a maximum buckling strength solution and obtained such solutions for columns under various support conditions. Column designs for which the buckling load is a maximum were also obtained by Taylor and Liu (1968).

Stability problems are usually taken into account in the design of thin-walled compressed elements with box sections, for example in the papers by Krzyś (1964, 1966) and by Krzyś and Życzkowski (1964) dealing with the minimum-weight design of beams in bending. Gajewski and Życzkowski (1969_1) gave an optimal solution, including numerical tables, for a bar compressed by a polar force.

Optimization of thin-walled tubular columns was studied by Kirste (1958), who pointed to the advantages of the annular cross-section over sections in the form of regular polygons. Krzyś (1967_2, 1968) also considered annular columns in axial compression, which he optimized for minimum weight with stability constraints for the walls, both in the elastic and in the inelastic range. Design of thin-walled columns was also investigated by Janiczek (1964, 1968) and by Spunt (1974), and that of tubular columns supported by tension ties by Mauch and Felton (1967).

Kowalski (1967) analysed the stability of bars with discretely varying cross-sections, loaded by a compressive force whose point of application and direction changed with deformation of the bar ("follow-up" force) and showed how to determine the cross-section jump points from the condition of minimum weight.

Eccentrically loaded columns were optimized for a minimum elastic potential at a constant volume of the material by Owczarek (1967, $1969_{1,2}$). The same author investigated, both analytically and experimentally—by photoelastic methods, optimum shapes of joints between columns and beams ($1968_{1,2}$, $1970_{1,2}$).

Most of the work on the design of arches and vaults dates back to the last century and to the early years of the present century, stimulated—as it was mentioned previously—by the rapid development of civil engineering and the demand for stone and concrete bridges. Arch designs were obtained by determining the axis shape, the rise to span ratio and the variation of the cross-section along the axis. The principal aim was usually to find an axis shape such that the bending moments could be reduced. The cross-sections were chosen so as to minimize the amount of material used, and hence to reduce the dead weight.

A comprehensive exposition of arch design is available in a book by Wasiutyński (1959).

Since the basic problems in arch design have long been solved and, more importantly, arches have all but disappeared from modern building structures, new publications in this field are rare. We may note a paper by Grycz (1968_3) on maximum-stiffness design of moment-free arches and two by Dzhaparidze ($1968_{1,2}$) on the optimum design of hyperstatic reinforced concrete arches under limit loads. A necessary condition for an arch to be optimal was given by Budiansky *et al.* (1969).

Optimization of frames has been investigated by many authors, mostly in the USA. Optimum shapes have been sought for frames in the limit state, i.e. at the point of appearance of plastic hinges, when the geometry of the structure becomes variable. In some formulations, the configuration of a frame was optimized with respect to the load it was to carry; more often, the configuration was assumed to be known and only the member cross-sections were sought, variable or constant along the lengths of the members.

Plastic design of plane frames for minimum weight was studied by Heyman (1951, 1953). A general method for optimizing frames and conditions which should be satisfied by the distribution of the bending moments at collapse were given by Foulkes (1953, 1954).

Schmidt (1962) recalled the classical solution of Michell (1904) for plane elastic frames of minimum volume, reviewed the most important work in this field and calculated minimum weight layouts of such frames under alternative load systems. Michell's theorem was also used by Ghista (1966) to obtain minimum-volume frameworks for six different load cases.

Khuberyan (1963) considered a minimum-weight problem for hyperstatic frames with inequality constraints on stresses, deflections and slenderness ratios; the frames were analysed in two loading states.

Plastic design of steel frames for minimum weight with structural constraints (side sway limitation) was studied by Bigelow and Gaylord (1967); a non-linear relationship was assumed between the plastic moment and the unit weight.

Megarefs and Hodge (1964) analysed a singular case in which the optimal design did not correspond to a stationary point of the objective function. The structure considered was a single-span frame with a given configuration and loading; the member cross-sections, defined by a single parameter, were determined from the condition of minimum cost for a given strength.

Further investigations of the optimum plastic design of frames were made by Ridha and Wright (1967), who arrived at minimum-cost frames by including the connection cost in the objective function, by Megarefs and Sidhu (1968), and by Yokoo and Nakamura (1970), who considered Vierendeel frames.

Kavlie and Moe (1969) optimized grids and frames in the elastic range using non-linear programming methods.

Gajnullina (1963) showed that the effect of longitudinal forces on indeterminate quantities in minimum-weight frames may only reach a few percent and can be neglected, particularly in initial calculations.

Kokorin (1969) compared uniform-strength and minimum-weight designs of hyperstatic frames subjected to changing loads. In general, it was found that a minimum-weight solution was not fully stressed. Optimization of singly hyperstatic frames for minimum weight was further developed by Safin (1969_1). Previously, minimum-volume hyperstatic frames were calculated by Khasanshin (1966) with the use of Radtsig's mirror functions and certain approximate methods.

Kirsch, Reiss and Shamir (1972) proposed a technique for dividing a framed structure into substructures in problems with a large number of variables and constraints. A similar technique combined with the plastic moment distrubution method was also used by Kuzmanovic and Willems (1972) to optimize multistorey steel frames. An algorithm for stability-constrained optimum design of frames was presented by Khot, Venkayya and Berke (1976). A solution to the optimum design problem for plane frames obtained by non-linear programming methods has been given by Jendo (1979). The solution is illustrated with an example design of a supporting frame structure of a large ship.

Optimization of concrete frames has been studied by few authors; let us note two papers: by Grierson and Cohn (1970) and by Rozvany and Cohn (1970). Reinforced concrete frames have attracted much more attention. Grierson and Cohn obtained optimal elastic designs of such frames by using linear programming techniques (1968) and presented a general formulation for the optimal design of reinforced concrete frames in which plastic moments, member stiffness and frame geometry are all treated as design variables (1970). Munro, Krishnamoorty and Yu (1972) and Krishnamoorty and Munro (1973) sought minimum reinforcements satisfying various design requirements, applying the simplex method to solve the resulting linear programs. Rozvany and Cohn (1970) optimized reinforced concrete frames under several alternative loading conditions, analysing the effect of axial forces in the members and the non-linearity of the moment-steel area relation.

14.7 Optimization of shells, hanging structures and lattice structures

Shell structures have long been an object of interest for optimum design researchers. This interest probably follows from the fact that the shape of a shell can often be chosen within wide limits independently of the constructional and functional requirements and, at the same time, is crucial for the stiffness of the structure and for material consumption.

One of the first papers in this field, concerning optimum shapes of cylindrical shells, was published in 1946 by Khuberyan. Several other papers by Khuberyan (1945, 1947, 1955_3) dealt with the optimum design of supporting shell elements in loose material containers. The same author investigated the problem of designing thick-walled shells for underground tubing (Khuberyan 1951, 1955_2, 1956). Let us also note Khuberyan's paper on optimization of arch dams (1969), presenting fundamental equations and describing a numerical procedure for computer use.

An example of optimization of a cylindrical shell for a minimum volume of material was given by Onat and Prager (1955). Optimum design and uniqueness problems for layer shells of perfectly plastic materials were investigated by Hu and Shield (1961).

Saelman (1961) studied the minimum-weight elastic design of spherical and cylindrical high-pressure shell elements of aircraft, rocket and vessel structures, with the radii of curvature as the principal design variables.

Pipkin and Rivlin (1963) optimized for minimum weight axisymmetric shells reinforced with inextensible fibres; in the minimal solution the fibres are equally stressed.

Reinforced shells were also investigated by Crawford and Schwartz (1965), who showed that ribbed shells exhibit a much higher stiffness than equal-volume shells without ribs. A special type of reinforcement of cylindrical shells was considered by Życzkowski (1967_2). The reinforcement, increasing the stability of the shell, consists of a number of point connections with a rigid surface; the distribution of these connections is a design variable to be determined by optimization for maximum stability.

An extensive survey of the optimum design problems for shells, including bibliography, was also published by Życzkowski (1974).

Optimization of hanging structures has been studied by comparatively few authors. Lessaer (1963) determined the optimum shapes of carrying cables in axisymmetric structures. Optimum design methods for axisymmetric roofs, using the minimum weight and the minimum potential criteria, were developed by Jendo ($1969_{2,3}$). In one of his subsequent papers, Jendo (1970_2) solved an optimization problem for prestressed two-surface cable structures under two alternative loading conditions. Further results in the optimum design of hanging structures, using methods of the calculus of variations, were obtained by Jendo and Stachowicz (1974, 1975) and again by Jendo (1979). Let us also note a paper by Mukhadze (1968) considering prestressed spatial hanging structures under two kinds of loading: prestressing forces and useful loads. The criterion of optimization adopted by the author was that of minimum difference between the elastic potentials induced by the two loads.

Lattice structures are relatively new and rare in engineering practice but have also been considered by optimum-design theorists. We may note here a paper by Lessaer (1967) dealing with axisymmetric lattices and several papers by Dzieniszewski ($1970_{1,2}$, $1971_{1,2}$) on the optimization of lattice rod structures for minimum deformability under a given load at a constant volume of the material. Dzieniszewski's results were further developed by Gierliński (1973_2, $1974_{1,2}$), Related problems of optimization of stress fields in plane elastic media by means of an initial prestressing were studied by Holnicki-Szulc (1971_{1-4}).

Bibliography of Structural Optimization

Presented below is what we believe to be a fairly complete list of publications in the field of structural optimization. More accurately, the list has been meant to cover all publications dealing with the fundamental problems of optimization and with the optimum design of building and civil engeneering structures. It has also been thought useful to include all those works on optimum design of mechanisms and mechanical elements which are of more general significance. The bibliography does not contain any mathematical or mechanical references, even those dealing directly with problems and methods of optimization; in the present book, they are given in page footnotes wherever required. This restriction has seemed necessary to maintain a uniform character of the selection. The first fifty or so references are included in recognition of their historical value.

Except in a few cases, every title in the Bibliography is accompanied by complete bibliographical data, which makes it possible to find any required publication. Whenever possible, information about reviews in Polska Bibliografia Mechaniki (PBAM) (Polish Mechanics Reviews), Applied Mechanics Reviews (AMR) and Referativnyj Zhurnal—Mekhanika (RŻM) (Russian Mechanics Reviews) is also included. This additional information is provided in the belief that a review can often be interesting to the prospective reader of the original paper, and may even help him to understand the original; sometimes, reading a review may save the effort of finding the original.

The Bibliography is set out in chronological order. Publications which appeared in a single year are arranged alphabetically according to the authors' names.

1638–1700

Galileo Galilei Linceo, *Discorsi e dimonstrazioni matematiche, interno, a due nuove scienze attenenti alla mecanica et i movimenti locali*, Leida 1638.

Blondel, F., *Epistola ad P. Wurtius in qua famosa Galilei propositio discutitur circa naturam lineae qua trabes secari debeant ut suit aequalis ubique resistentiae et in qua lineam illam non quidem parabolicam ut ipse Galileus arbitratus est, sed ellipticam esse demonstratur*, Data Parisiis, 1661.

Marchetti, A., *De resistentia solidorum*, 1669.

LEIBNIZ, G. W., Demonstrationes novae de resistentia solidorum, *Acta Eruditorum*, Lipsiae, Julii, 319, 1684.

JACQUES BERNOULLI, *A letter to Leibniz on designing a beam of uniform resistance*, 1687.

JACQUES BERNOULLI, *Equation of a catenary*, 1690.

1701–1800

LA HIRE, P., Remarques sur la forme de quelques arcs dont on se sert dans L'architecture, *Mém. Acad. Roy. Sci.*, 1702.

PARENT, A., Des résistances des poutres ... et des poutres de plus grande résistance, indépendamment de tout système physique, *Mém. Acad. Roy. Sci.*, 1708.

PARENT, A., Des points de la rupture des figures. D'en déduire celles qui sont partout d'une résistance égale, *Mém. Acad. Roy. Sci.*, 1710.

MARCHETTI, A., *Lettera nella si ribattono l'ingiuste acouse dati dal P. D. G. G.*, **18** and **27**, 1711.

LAGRANGE, J. L., Sur la figure des collonnes, *Miscellanea Taurinensia*, **5**, 1770–1773.

LAGRANGE, J. L., Sur la force des ressorts pliés, *Mém. Acad.*, Berlin 1771.

GAUTHEY, E. M., *Mémoire sur l'application de la mécanique à la construction des voûtes et des dômes*, 1772.

COULOMB, C. A., *Mémoires sur la stabilité des voûtes*, Académie des Sciences, 1776.

PERRONET, J. R., Mémoire sur la réduction de l'épaisseur des piles et sur la courbure qu'il convient donner aux voûtes pour que l'eau puisse passer plus librement sous les ponts, *Histoire de L'Académie des Sciences*, **264**, 1777.

GIRARD, P. S., *Traité analitique de la résistance des solides d'égale résistance*, 1798.

1801–1900

YOUNG, T., *A course of lectures on natural philosophy and the mechanical arts*, London 1807.

YOUNG, T., *Strongest form of columns and walls* (unpublished).

BERARD, J. B., *Théorie de l'équilibre des voûtes*, 1810.

ROBINSON, J. T., *Mechanical Philosophy*, 1822.

NAVIER, L. M. H., *Sur les ponts suspendus*, 1823.

SEGUIN, M., *Pont sur le Rhône entre Tain et Touron*, 1824.

NAVIER, L. M. H., *Résumé des leçons ...*, 1833.

POLONCEAU, A. R., *Mémoire sur le nouveau système des ponts en fonte suivi dans la construction du pont du Carrousel*, 1839.

ROBINSON, J. T. R., *Description of Col. Long's bridges, together with a series of directions to bridge builders*, Philadelphia 1841.

ARDANT, P. J., *Nouvelles recherches sur le profil de revêtement le plus économique*, 1848.

CLAUSEN, T., Über die Form architectonischer Säulen, *Bulletin physico-math. de l'Académie de St. Petersbourg*, vol. **IX**, 1851.

SEGUIN, J., *Ann. Ponts et Chaus.*, 1851.

SCHWEDLER, P. W., *Z. Bauwesen*, 1851–1859, *ZVDI*, 1859–1880.

MORIN, A. J., *Résistance des Matériaux*, 1853.

PAULI, F. A., *Grosshesselcher Brücke über die Isar*, 1853.

VILLARCEAU, Y., *Sur l'établissement des arches de pont envisagé au point de vue de la plus grande stabilité*, 1853.

BRESSE, J. A. C., *Recherches analytiques sur la fléxion et la résistance des pièces courbes*, 1854.

BARTON, I., On the economic distribution of material in the sides or vertical portion of wrought-iron beams, *Institution of Civil Engineers, Minutes of Proceedings*, 443, 1854–1855.

ALBARET, The form of trusses of minimum volume, *Nouvelles Annales de la Construction*, 1859.

ALBARET, The form of section irons, *Hannoversche Bauzeitung*, 523, 1860.

FINK, A., Allgemaine Betrachtungen über die Biegungsfestigkeit und Biegunswiderstand zur Erzielung eines einheitlichen Standpunktes für die Bauteilung verschiedener Brücken-Systeme, *Z. Ost. Ing. Ver.*, 1860.

DE ST. VENANT, B., *Notes et Appendices, au résumé des leçons sur application de la mécanique, de Navier*, 1864.

CULMANN, *Die graphische Statik*, 1886.

BARLOW, P., *A treatise on the strength of materials*, 1867.

LÉVY, M., *Sur la recherche des tensions dans les systèmes de barres et sur les systèmes qui, à volume égal de matière, présentent la plus grande résistance possible, présenté à l'Académie des Sciences*, 1873.

FÖPPL, A., *Theorie des Fachwerke*, 1880.

1901–1940

MICHELL, A. G. M., The limit of economy of material in frame structures, *Phil. Mag.*, **8**, **47**, 589–597, 1904.

BELZETSKIJ, S. I., Linear elastic arch of uniform strength (in Russian), *SPB*, 1904.

NIKOLAJ, J. L., Lagrange's problem of the most efficient forms of columns (in Russian), *Izdat. Peterb. Politekhn. In-ta*, vol. **8**, 1907.

MILLER, S., A method for arch design (in Polish), *Sprawozdanie z posiedzeń Koła Inżynierów przy Politechnice Warszawskiej*, 1919.

KAUFMAN, S., *On graphical design of special trusses* (in Polish), Ph.D. Thesis, Lwów 1920.

KEFELI, A. I., On theoretical weights of structures (in Russian), *So. Len. Inzh. P. S.*, **96**, 1927.

SHTAERMAN, I. YA., Beam of uniform strength in bending (in Russian), *Vesti KPI*, 1928.

BLINOV, V. S., On rational design of vaults (in Russian), *Sb. Leningrad. Inst. Inzh. Putej Soobshch.*, 1929.

RUDNEV, V. I., On a rational form of a continuous elastic arch and modern construction methods (in Russian), *Trudy MIIT*, 1930.

RABINOVICH, I. M., On the theory of statically indeterminate trusses. Laws of distribution of forces; the method of allowable stresses; initial forces in statically indeterminate trusses (in Russian), *TsNIITS*, 1933.

SWIDA, W., Über die Form eines Pfeilers gleicher Festigkeit bei gleichzeitiger Druck- und Biegebeanspruchung, *Zeitschr. Angew. Math. Mech.*, **378**, 1934.

CHENTSOV, H. R., Pillars of minimum weight (in Russian), *Trudy TsLGI*, **265**, 1936.

ZANABONI, O., *Atti Academiae dei Lincei*, **25**, 117–121, 535–601, 1937.

WASIUTYŃSKI, Z., Heights of lattice girders (in Polish), *Przegl. Techn.*, **21/22**, 1937.

KHUBERYAN, K. M., On the design of statically indeterminate trusses (in Russian), *Tbilis. Nauchno-Issl. Ins. Sooruzh. TNIS*, **32**, Tbilisi 1938.

LUDWIG, K., Über Träger mit Eigengewicht und gleichem Widerstand gegen Biegung, *Zeitscher. Angew. Math. Mech.*, 1938.

SIKORSKIJ, J. S., On equilibrium of a catenary of uniform strength on a surface of revolution (in Russian), *Prikladnaya Matematika i Mekhanika*, **1**, 1, 1938.

SIKORSKIJ, J. S., On equilibrium form of uniform-strength filament (in Russian), *Prikladnaya Matematika i Mekhanika*, **1**, 1, 1938.

SIKORSKIJ, J. S., On equilibrium form of a heavy filament of uniform strength under vertical forces (in Russian), *Prikladnaya Matematika i Mekhanika*, **2**, **1**, 1938.

FEDERHOFER, K., Über Schalen gleicher Festigkeit, *Bauingenieur*, **20**, 366, 1939.

WASIUTYŃSKI, Z., *Strength Design. Part I: Methods of strength design. Part II: Design for minimum potential. Part III: On the design of I-beams* (in Polish), Akademia Nauk Technicznych, Warszawa 1939.

ISHLINSKII, A. YU., Beam with cross-section of uniform strength (in Russian), *Uchenye zapiski Moskovskogo gosudarstvennogo universiteta*, vyp. **39**, Mekhanika Izd-vo MGU, 1940.

KISELEV, V. A., *Funicular curve theory and its applications to the design of arches and suspended systems in engineering structures* (in Russian), Dokt. diss. Moskov. Gidr. Inst., 1940.

1941–1950

SIKORSKIJ, D. S., On heavy girders of uniform strength in bending (in Russian), *Vest. Inzh. i Tekh.*, **1**, 29–33, 1941.

KISELEV, V. A., On the design of suspended cylindrical bunkers and hydraulic channels (in Russian), *Inzh. Sb.*, **1**, AN SSSR, 1941.

RADTSIG, YU. A., The principle of optimality and the basic problems of structural mechanics (in Russian), *Trudy Kazanskogo Aviatsionnogo Instituta (KAI)*, **13**, 1941.

DRYMAEL, J., The design of trusses and its influence on weight and stiffness. *J. Roy. Aeronaut. Soc.*, **46**, 297–308, 1942.

COX, H. L., and SMITH, H. E., Structures of minimum weight, *Aeronaut. Res. Co. Rep. and Mem.*, **1923**, 1943.

RADTSIG, YU. A., The principles of the design of plane trusses (in Russian), *Tezisy k nauchnoj konferentsii Kazanskogo aviatsionnogo instituta*, Kazan', Tatknigoizdat, 1945.

KHUBERYAN, K. M., Flexible shells under pressure of liquid or loose materials (in Russian), *Soobshch. AN GSSR*, **6**, No. 1, 11–20, 1945.

NORMAN, D. G., Economic aspects in the design of some reinforced concrete structural members, *J. Amer. Concr. Inst.*, **61**, 4, 419–440, 1946; RŻM, 1965, 1B428.

RABINOVICH, I. M., Design of continuous beams on stiff and elastic supports by the permissible moment and stress method (in Russian), *Issl. po Teorii Sooruzh.*, 1946.

RADTSIG, YU. A., Determination of a minimum volume for statically indeterminate trusses (in Russian), *Trudy KAI*, **17**, 55–59, 1946.

SAGATELOVA, E. S, *Moment-free reinforced concrete aqueducts* (in Russian), Kand. diss., Tbilisi 1946.

KHUBERYAN, K. M., Determination of the funicular curve for rational design of cylindrical shells (in Russian), *Soobshch. Tbilis. Nauchno-issl. Inst. Sooruzh. i Gidroenerg.*, **1**, 15–26, Tbilisi 1946.

PROTASOV, K. G., Design of statically indeterminate bridge trusses with allowance for plastic deformations (in Russian), *Transzheldorizdat*, **13**, 1947.

KHUBERYAN, K. M., Moment-free suspended cylindrical bunkers (in Russian), *Izv. Tbilis. Nauchno-issl. Inst. Sooruzh. i Gidroenerg.*, **1**, 167–182, 1947.

BECKER, H., The optimum proportions of a long unstiffened circular cylinder in pure bending, *J. Aeronaut. Sc.*, **15**, 10, 616–624, 1948.

LAASONEN, P., On the selection of the most advantageous cross-section of a buckling bar (in Finnish), *Teknilinen Aikakauslehti*, **38**, 2, 49–52, 1948.

PROTASOV, K. G., A method for selecting statically indeterminate trusses of uniform strength (in Russian), *Sbornik LIIZhT*, **137**, 1948.

RADTSIG, YU. A., Mirror functions and their application to structural mechanics (in Russian), *Trudy KAI*, **20**, 1948.

KHUBERYAN, K. M., On the design of moment-free suspended cylindrical bunkers (in Russian), *Izv. Tbilis. Nauchno-issl. Inst. Sooruzh. i Gidroenerg.*, **2**, 1948.

FARRAV, D. J., The design of compression structures for minimum weight, *J. Roy. Aeronaut. Soc.*, **53**, 467, 1041–1052, 1949.

KHUBERYAN, K. M., On flexible shells loaded with liquid or loose materials (in Russian), *Issl. po Teorii Sooruzh.*, **4**, 1949.

KHUBERYAN, K. M., The permissible stress method (in Russian), *Issl. po Teorii Sooruzh.*, **4**, 1949.

PARKES, E. W., The design of redundant structures for minimum weight, *Aircraft Engng.*, **21**, 243, 162–163, 1949.

RADTSIG, YU. A., Solution of a system of linear algebraic equations by the methods of multidimensional geometry (in Russian), *Trudy KAI*, **23**, 3–9, 1949.

RZHANITSYN, A. R., On theoretical weight of bar structures (in Russian), *Issl. po Teorii Sooruzh.*, **4**, 1949.

VAKHURKIN, V. M., The most efficient form of a flanged beam (in Russian), *Byull. Stroit. Tekh.*, 1949.

VINOGRADOV, A. I., *The design of roofs for given stress* (in Russian), Avtoref. Dokt. Diss. (Ph.D. Thesis) MIIT, 1949.

VOROB'EV, L. N., The most efficient form of a pillar (in Russian), *Trudy Novocherkas. Polit. Inst.*, **21/35**, 1949.

SMIRNOV, A. F., Rods and arches of minimum weight in buckling (in Russian), *Trudy MIIT*, 1950.

VALETTE, R., Utilisation maximum de la matière dans les constructions, *Ass. Int. Ponts et Charp., Mém.*, Zürich 1950.

WASIUTYŃSKI, Z., Transformation of trusses by introduction of new nodes (in Polish), *Księga jubileuszowa dla uczczenia zasług naukowych prof. M. T. Hubera*, 1–19, Gdańsk 1950.

WASIUTYŃSKI, Z., Transformation of trusses by exchange of members, (in Polish), *Inż. i Bud.* **7**, 11, 542–553, 1950.

1951

FILIN, A. N., Problems of efficient design of bridge arches (in Russian), *Trudy Khabarovsk. Inst. Inzh. Zh. D. Trans.*, 1951.

HEYMAN, J., Plastic design of beams and plane frames for minimum material consumption. *Quart. Appl. Math.*, **8**, 373, 1951.

KHUBERYAN, K. M., The design of trusses for static loads by the permissible stress method (in Russian), *Issl. po Teorii Sooruzh.*, **5**, 1951.

KHUBERYAN, K. M., A pipeline with the cross-section designed after funicular curves (in Russian), *Izv. Tbilis. Nauchno-Issl. Inst. Sooruzh. i Gidroen.*, **4**, 3–21, 1951.

RADTSIG, YU. A., The principal errors in the design of statically indeterminate trusses (in Russian), *Vestnik Inzhenerov i Tekhnikov*, **5**, 1951.

RADTSIG, YU. A., On the errors found in the book by Professor N. K. Snitko, "Statika sooruzhenij v zadachakh i primerakh", part 3, edited 1941, concerning the design of statically indeterminate trusses (in Russian), *Trudy KAI*, **26**, 1951.

RADTSIG, YU. A., The design of statically indeterminate trusses by the minimum volume method (in Russian), *Trudy Kazan. Aviats. Inst.*, **25**, 1951.

SZELĄGOWSKI, F., On the efficient forms of bridge vaults (in Polish), *Arch. Mech. Stos.*, **3**, 3/4, 271–291, 1951; *AMR*, 1954, **2**, 489.

TIETIAJEW, J., and WASIUTYŃSKI, Z., Utility of single-span frames in the construction of bridges (in Polish), *Inż. i Bud.*, **8**, 7–8, 1951.

VAKHURKIN, V. M., The least-cost form of a flanged beam (in Russian), *Vest. Inzh. i Tekh.*, 1951.

VINOGRADOV, A. I., Some invariants of a statically indeterminate system and their application to rational design (in Russian), *Trudy Khark. Inst. Zh. D. Trans.*, 1951.

WASIUTYŃSKI, Z., On the design of cross-sections in columns (in Polish), *Arch. Mech. Stos.*, **3**, 3/4, 345–370, 1951; *AMR*, 1954, **2**, 463.

1952

BIJLAARD, P. P., On the optimum distribution of material in sandwich plates loaded in their plane, *Proc. I US Nat. Congr. Appl. Mech., June 1951*, J. W. Edwards, Ann Arbor Mich., 373–380, 1952; *AMR*, 1953, **8**, 2483.

FEIGEN, M., Minimum weight of tapered round thin-walled columns, *J. Appl. Mech.*, **19**, 3, 1375–1380, 1952; *AMR*, 1953, **2**, 426.

PAUL, T., Das Eigengewicht idealer Tragwerke, *Der Stahlbau*, **4**, 1952.

SHANLEY, F. R., *Weight-strength analysis of aircraft structures*, McGraw-Hill Book Co., New York 1952.

SLYUSARCHUK, F. I., *On the design of statically indeterminate trusses of uniform strength* (in Russian), 1952.

SLYUSARCHUK, F. I., Systematic design of statically indeterminate trusses by the permissible stress method (in Russian), *Tr. Novosibir. Inst. Inzh. Zh. D. Trans.*, **8**, 34–71, 1952.

1953

BILLINGTON, D. P., Economical design of prestressed concrete beams, *J. Amer. Concr. Inst.*, **25**, 73–87, 1953; *RŻM*, 1954, 3858.

FOULKES, J., Minimum-weight design and the theory of plastic collapse, *Quart. Appl. Math.*, **4**, 10, 347–358, 1953; *AMR*, 1953, **9**, 2787.

FRANCIS, A. I., Direct design of elastic statically indeterminate triangulated frameworks for single systems of loads, *Austral. J. Appl. Sci.*, **4**, 2, 1953.

HEYMAN, J., Plastic design of plane frames for minimum weight, *Struct. Engng.*, **5**, 31, 125–129, 1953; *AMR*, 1954, **3**, 768.

HORNE, M. R., Determination of the shape of fixed-ended beams for maximum economy according to the plastic theory, *Final Rep. IABSE, 4th Congress*, Cambridge and London 1953.

HOWARD, H. B., Tube of least weight for given torsional stiffness, *J. Roy. Aeronaut. Soc.*, **57**, 505, 1953.

JOHNSON, A. I., Strength, safety and economical dimensions of structures, *Bull. Div. of Build. Stat. and Struct. Engng. of the Royal Inst. of Tech.*, Stockholm 1953.

KISELEV, V. A., Rational forms of arches and suspended systems (in Russian), *M., Gos. Izdat. po stroit. i arkhit.*, 1953.

KOLOM, A. L., Optimum design considerations for aircraft wing structures, *Aeronaut. Engng. Rev.*, **12**, 10, 31–41, 1953; *AMR*, 1954, **8**, 2466.

KOPYCIŃSKI, B., Design of the cross-section of a pre-tensioned prestressed beam from economical viewpoint (in Polish), *Inż. i Bud.* **11**, 1953.

KOZYREV, YA. F., On the theoretical weight of statically determinate trusses (in Russian), *Nauchn. Tr. Kazans. Inst. Inzh.-Str. Neft. Prom.*, **1**, 1953.

MACDONALD, E. P., The minimum weight design of wings for flutter conditions, *J. Aeronaut. Sci.*, **20**, 8, 573–574, 1953; *AMR*, 1954, **4**, 1086.

MOUGENOT, E., La poutre la plus économique, *Travaux*, **37**, 225, 339–347, 1953; *RŻM*, 1954, 2543.

RADTSIG, YU. A., Symmetrical properties of mirror functions (in Russian), *Trudy KAI*, **28**, 109–134, 1953.

WALING, J. L., Least-weight proportions of the bridge trusses, *Univ. III, Eng. Exp. Sta. Bull.*, 417, 1953.

1954

BRZOSKA, Z., Circular plates of uniform strength under axisymmetric loading (in Polish) *Arch. Bud. Masz.*, **1**, 3, 1954.

CATCHPOLE, E. I., The optimum design of compression surfaces having unflanged integral stiffness, *J. Roy. Aeronaut. Soc.*, **58**, 527, 765–768, 1954; *AMR*, 1957, **11**, 3617.

CORBETTA, G., Le travi di uniforme resistenza a flessione nelle structure aeronautiche, *Ingegnere*, **28**, 8, 867–874, 1954; *AMR*, 1955, **6**, 1663; *RŻM*, 1956, 4773.

FOULKES, J., The minimum-weight design of structural frames, *Proc. Roy. Soc. London, ser. A*, **223**, 482–494, 1954; *AMR*, 1955, **1**, 75.

GRIGOR'EV, A. S., *On plates of uniform bending strength* (in Russian), *Tezisy dokladov, izd. AN SSSR*, 1954.

GUYON, Y., *Béton précontraint*, Eyrolles, Paris 1954.

GUREVICH, YA. I., On the problem of optimum variation of cross-sections along the members in statically indeterminate bar systems (in Russian), *Trudy Khabarov. Inst. Inzh. Zhel. Trans.*, 1954.

HORNE, M. R., Shells with zero bending stresses, *J. Mech. and Phys. Solids*, **2, 2**, 117–126, 1954.

KAUFMAN, S., Determination of the shape of a continuous prestressed beam with rectilinear cable routing (in Polish), *Inż. i Bud.*, **4**, 1954.

KHUBERYAN, K. M., Force surfaces of given stress under hydrostatic loads (in Russian) *Iss. po Teorii Sooruzh.*, **4**, 347–355, 1954.

KLYAVIN, EH. N., *Selection of rational forms for stone arches* (in Russian), Gosudarstvennyj Universitet, Riga 1954.

PLAINEVAUX, J. E., Sur le profil optimum à donner aux tôles ondulées et aux palplanches, *Acad. Roy. Belgique, Bull. Cl. Sci.*, **40**, 9, 962–969, 1954; *AMR*, 1955, **9**, 2682.

RABINOVICH, I. M., *A course of the structural mechanics of bar systems* (in Russian), Part 2. 1954.

READEY, W., Optimum design of indeterminate frames, *J. Aeronaut. Sci.*, **21**, 9, 615–620, 1954; *AMR*, 1955, **4**, 976.

ROBERTSON, R. G., Prestressed concrete beams: the economical shape of section, *Proc. Inst. Civ. Engrs.*, Part III, 1, 1954.

SEROV, N. A., *Minimum-weight design methods for statically indeterminate trusses* (in Russian), Ph.D. Thesis, 1954.

SVED, G., The minimum weight of certain redundant structures, *Austral. J. Appl. Sci.*, **5**, 1–8, March 1954; *RŻM*, 1955, 4589.

VINOGRADOV, A. I., Some problems in the design of bar systems for given stress (in Russian), *Issl. po Teorii Sooruzh.*, **6**, 357–379, 1954.

VINOGRADOV, A. I., On the statical indeterminacy of minimum-weight bar systems (in Russian), *Issl. po Teorii Sooruzh.*, **6**, 381–387, 1954.

1955

ALEKSANDROV, A. V., Another proof of a certain theorem on the stress method (in Russian), *Trudy Novosibir. Inst. Inzh. Zh. D. Trans.*, 1955.

CADAMBE, V., and KRISHNAN, S., Minimum weight design of thin-walled cells in torsion, *J. Roy. Aeronaut. Soc.*, **59**, 530, 120–126, 1955; *AMR*, 1957, 1, 52.

FOULKES, J., Linear programming and structural design, *Proc. II Symp. in Linear Programming, Washington National Bureau of Standards*, 177–184, 1955; *AMR*, 1956, **5**, 1433.

FREIBERGER, W., Minimum weight design of cylindrical shells subjected to axial loading and arbitrary internal or external pressure, *Brown Univ., Div. Appl. Math., Tech. Rep.*, **20**, 1955.

GAJNULLINA, S. KH., *Design of transverse frames for minimum weight* (in Russian), Diss. Kand. KAI, 1955. (Thesis).

HOP, T., *Influence of the geometrical and mechanical properties of the cross-section on the economy of prestressed structures* (in Polish), Thesis, Gliwice 1955.

HOPKINS, H. G., and PRAGER, W., Limits of economy of material in plates, *J. Appl. Mech.*, **22**, 372–374, 1955.

KHUBERYAN, K. M., *The stress method and its application to statically indeterminate trusses* (in Russian), Dokt. Diss. (Ph.D. Thesis), 1955.

KHUBERYAN, K. M., Determination of efficient cross-sections for underground pressure tunnels (in Russian), *Izv. Tbilis. Nauchno-issl. Inst. Sooruzh. i Gidroenerg.*, **8**, 65–74, 1955.

KHUBERYAN, K. M., Design of a suspended cylindrical bunker with flexible walls and rigid bottom (in Russian), *Izv. Tbilis. Nauchno-issl. Inst. Sooruzh. i Gidroenerg.*, **8**, 75–80, 1955.

KISELEV, V.A., On the theoretical form of a beam of uniform bending strength with account for dead weight (in Russian), *Sb. Trudy Mosk. Avtomobil. Inst.*, 1955.

KLEIN, B., Direct use of extremal principles in solving certain optimization problems involving inequalities, *Operations Research*, **3**, 168–175, 1955.

KRISHNAN, S., and CADAMBE, V., A note on the minimum weight design of a thin-walled stiffened rectangular cell subjected to torsion, *J. Aeronaut. Soc. India*, **7**, 3, 43–48, 1955; *AMR*, 1956, **5**, 1436.

LOZHKIN, A. G., *On efficient shapes of steel profiles* (in Russian), 1955.

MAMES, J., Calculation of deflections in a post-tensioned prestressed concrete beam (in Polish), *Inż. i Bud.* **5**, 1955.

ONAT, E. T. and PRAGER, W., Limits of economy of material in shells, *Ingen.*, **10**, 11–3, 1955.

PRAGER, W., Minimum weight design of plates, *De Ingen.*, **67**, 48, 141–142, 1955; *AMR*, 1956, **4**, 1095.

SAHMEL, P., Optimale Formgebing bei torsionsbeanspruchten I- und *C*-Querschnitten, *Bauingenieur*, **30**, 11, 403–404, 1955.

SLYUSARCHUK, F. I., Investigation of volume of a statically indeterminate truss carrying a single load (in Russian), *Trudy Novosibir. Inst. Inzh. Zh. D. Trans.*, **11**, 1955.

SLYUSARCHUK, F. I., On the existence of uniform-strength solutions for statically indeterminate trusses (in Russian), *Trudy Novosibir. Inst. Inzh. Zh. D. Trans.*, **11**, 1955.

1956

AYERS, K. B., Struts of minimum weight. Theoretical properties of struts of maximum efficiency and practical approximation, *Aircraft Engng.*, **28**, 324, 43–45, 1956; *RŻM*, 1957, 4837.

CADAMBE, V., and KRISHNAN, S., Note on the minimum weight of thin-walled cells in combined bending and torsion, *J. Roy. Aeronaut. Soc.*, **60**, 65–66, 1956.

FREIBERGER, W., Minimum weight design of cylindrical shells, *J. Appl. Mech.*, **4**, 23, 576–580, 1956.

FREIBERGER, W., and TEKINALP, B., Minimum weight design of circular plates, *J. Mech. Phys. Solids*, **4**, 4, 294–299, 1956; *AMR*, 1957, **2**, 420.

GAJNULLINA, S. KH., *Design of transverse frames for minimum weight* (in Russian), Avtoref. kand. diss. (Thesis), KAI, 1956.

GERARD, G., *Minimum weight analysis of compressive structures*, New York Univ. Press, pp. 194, 1956; *AMR*, 1957, **3**, 695.

GROSS, W. A., and LI, J. P., Beams of uniform strength subjected to uniformly distributed loading, *ASME, Ann. Meet., New York*, Pap. 56-A-9, p. 4, 1956; *AMR*, 1957, **8**, 2461.

IZRAELIT, A. B., On guarantees for positive solutions in the design of statically indeterminate beams and frames by the method of given forces (in Russian), *Trudy Vsesoyuz. Zaoch. Lesotech. Inst.*, **2**, 136–138, 1956.

IZRAELIT, A. B., Design of statically indeterminate thin-walled structures by the method of given forces and stresses (in Russian), *Trudy Vsesoyuz. Zaoch. Lesotech. Inst.*, **2**, 167–183, 1956.

KAUFMAN, S., On efficient design of prestressed sections in bending (in Polish), *Inż. i Bud.*, **4**, 1956.

KHUBERYAN, K. M., *Efficient shapes of pipelines, reservoirs and pressure roofs* (in Russian), M. Gosstrojizdat, 1956.

KISELEV, V. A., On the theoretical form of a beam of uniform bending strength with account for dead weight (in Russian), *Trudy MADI*, **16**, 1956.

LIVESLEY, R. K., The automatic design of structural frames, *Quart. J. Mech. Appl. Math.*, **9**, 257–258, 1956.

MANSFIELD, E. H., Optimum design for reinforced circular holes, *Aeronaut. Res. Counc. London, Curr. Pap.*, **239**, 15, 1956; *AMR*, 1956, 11, 3560.

PAUL, T., Das Eigengewicht idealer Tragwerke, *Stahlbau*, **25**, 4, 97–100, 1956; *RŻM*, 1957, 4867.

PRAGER, W., Minimum weight design of a portal frame, *Proc. ASCE*, **82**, 4, *J. Engng. Mech. Div.*, 1073/1–1073/10, 1956; *RŻM*, 1957, **11**, 13158.

RADTSIG, YU. A., Design of statically indeterminate trusses under transient loading (in Russian), *Trudy KAI*, **32**, 3–43, 1956.

RADTSIG, YU. A., Application of the least-volume method to the design of statically indeterminate beams (in Russian), *Trudy KAI*, **32**, 44–53, 1956.

RADTSIG, YU. A., *Application of the least-volume method to the design of statically indeterminate trusses and beams* (in Russian, Thesis) Avtoref. Dokt. Diss. Moskov. Inst. Inzh. Zh. D. Transp., 1956.

SAELMAN, B., Round tubes in bending. How to select optimum diameter for minimum weight, *Machine Design.* **28**, 19, 141–144, 1956; *RŻM*, 1957, 7171.

SEROV, N. A., *Design methods for statically indeterminate trusses and beams* (in Russian, Thesis), Avtoref. Dokt. Diss. Moskov. Inst. Inzh. Zh. D. Transp., 1956.

VARGO, L., Nonlinear minimum weight design of planar structures, *J. Aeronaut. Sci.*, **23**, 10, 956–960, 1956; *AMR*, 1957, **8**, 2466; *RŻM*, 1957, 8246.

WASIUTYŃSKI, Z., Fundamentals of strength design (in Polish), *Arch. Inż. Ląd.*, **2**, 1, 1956.

WASIUTYŃSKI, Z., On the ways and methods of strength design (in Polish), *Podst. Probl. Współcz. Tech.*, 44–110, PWN, 1956.

WASIUTYŃSKI, Z., Determination of the proper rise to span ratio in complex arch structures (in Polish), *Inż. i Bud.*, **2**, 1956.

ŻYCZKOWSKI, M., On the question of choice and optimum shape of axially compressed bars (in Polish), *Rozpr. Inż.*, **4**, 4, 441–456, 1956; *AMR*, 1958, **2**, 422.

1957

Barta, J., On the minimum weight of certain redundant structures, *Acta Tech. Acad. Sci. Hungaricae*, **18**, 1/2, 67–76, 1957.

Brandt, A., and Wasiutyński, Z., Strength design of prestressed concrete plates (in Polish), *Inż. i Bud.*, **2**, 1957.

Brandt, A., Wasiutyński, Z., and Kosmowski, J., Minimum potential design of trusses by shifting unloaded nodes joining three members (in Polish), *Rozpr. Inż.*, **2**, 157–205, 1957; *AMR*, 1958, **6**, 2113; *RŻM*, 1959, 4422.

Clarkson, J., Design for minimum weight of simply supported flat grillages to withstand a single concentrated load, *Trans. North East Coast Inst. Engrs. Shipbuilders, Newcastle-on-Tyne*, **73**, 1, 145–155, 1957.

Drucker, D. C., On minimum weight design and strength of non-homogeneous plastic bodies, *Symp. IUTAM*, 139–146, Warszawa 1958.

Drucker, D. C., and Shield, R. T., Bounds on minimum weight design, *Quart. Appl. Math.*, **3**, 15, 269–281, 1957; *AMR*, 1958, **6**, 2125.

Drucker, D. C., and Shield, R. T., Design for minimum weight, *IX Congr. Appl. Mech. Univ. Bruxelles*, **5**, 212–222, 1957; *AMR*, 1959, **10**, 5024.

Freiberger, W., On the minimum weight design problem for cylindrical sandwich shells, *J. Aeronaut. Sci.*, **24**, 847–848; *AMR*, 1957, **8**, 3078.

Gross, W. A., and Li, J. P., Beams of uniform strength subjected to uniformly distributed loading, *J. Appl. Mech.*, **2**, 1, 1957.

Hodge, P. G., Discussion on a paper by W. Freiberger; Minimum weight design of cylindrical shells, *J. Appl. Mech.*, **24**, **3**, 486–497, 1957.

Kaufman, S., Design of prestressed flexure sections with the full use of eccentricity (in Polish), *Inż. i Bud.*, **14**, 2, 7–17, 1957.

Khuberyan, K. H., *The stress method in application to statically indeterminate trusses* (in Russian, Thesis), Avtoref. Dokt. Diss., Moskov. Inzh. Stroit. Inst., Moskva 1957.

Kopyciński, B., Economical design of the cross-section of a prestressed beam (in Polish), *Zesz. Nauk. P. Krak., Bud. Ląd.*, **1**, 1957.

Kuznetsov, Yö. I., *Bar of uniform strength under transverse-longitudinal bending* (in Russian) Inf. Mat. Inst. Stroit. Mekhan. AN SSSR, 1957.

Mames, J., Prestressed continuous beam—analysis and design (in Polish), *Arch. Inż. Ląd.*, **4**, 1957.

Mikeladze, M. Sh., On minimum weight of anisotropic shells (in Russian), *Soobshch. AN Trudy SSR*, **19**, 1, 11–19, 1957; *RŻM*, **10**, 11504; *AMR*, 1960, **4**, 1696.

Onat, E. T., Schumann, W., and Shield, R. T., Design of circular plates for minimum weight, *Z. Ang. Math. Phys.*, **8**, 6, 485–499, 1957; *AMR*, 1959, **10**, 4919; *RŻM*, 1961, 12B233.

Pelikan, J., Theory of highly economical reinforced concrete beams, *Acta Techn. Acad. Sci. Hung.*, **17**, 39–56, 1957; *AMR*, 1959, **3**, 1293.

Wasiutyński, Z., Ordering the internal force system as a guiding principle in the development of concrete arch bridges (in Polish), *Inż. i Bud.*, **14**, 1, 8–10, 1957.

Vinogradov, A. I., On the design of closed-contour systems for given stresses (in Russian), *Issl. po Teorii Sooruzh.*, **7**, 373–400, 1957.

1958

Brandt, A., Prestressed beam design (in Polish), *Inż. i Bud.*, **15**, 1, 17–21, 1958.

Brandt, A., and Ignaczak, J., Strength design of a cantilever beam (in Polish), *Rozpr. Inż.*, **1**, 6, 167–180, 1958.

COX, H. L., Structures of minimum weight: the basic theory of design applied to the beam under pure bending, *Aeronaut. Res. Council*, Rep. No, **19785**, 1958.

COX, H. L., The theory of design, *Aeronaut. Res. Conucil.*, Rep. No. **19791**, 1958.

COX, H. L., The application of the theory of stability in structural design, *J. Roy. Aeronaut. Soc.*, **62**, 571, 497–515, 1958.

HEMP, W. S., Notes on the problem of the optimum design of structures, *Coll. Aero. Cranfield, Note* **73**, **8**, 1958; *AMR*, 1958, **12**, 4952; *RŻM*, 1959, 4215.

GAJNULLINA, S. KH., Design of statically indeterminate beams and frames for minimum weight (in Russian), *Trudy KAI*, **33/34**, 439–459, 1958.

HEMP, W. S., Theory of structural design, *Coll. Aero. Cranfield, Rep.* **115**, 1958; *AMR*, 1959, **11**, 5580; *RŻM*, 1960, 9485.

HEYMAN, J., Minimum weight of frames under shake-down loading, *Proc. ASCE, EM4 J. Engng. Mech. Div.*, 1–25, 1958; *AMR*, 1959, **5**, 2410; *RŻM*, 1959, **9**, 10617.

HEYMAN, J., and PRAGER, W., Automatic minimum weight design of steel frames, *J. Franklin Inst.*, **266**, 5, 339–364, 1958; *AMR*, 1959, **5**, 2406; *RŻM*, 1961, 2B138.

HILTON, H. H., Minimum weight analysis for combined loads based on probability of failure, *Hughes Aircraft Comp. Rep. S. F.*, **1**, 3–16, 1958.

IZRAELIT, A. B., Design of beams and frames by the method of given forces and displacements (in Russian), *L. Trudy Vsesoyuz. Zaochn. Lesotekhn. Inst.*, **3**, 67–80, 1958.

KHUBERYAN, K. M., Design of statically indeterminate trusses by the stress method for given loading and temperature conditions (in Russian), *Izv. Tbilis. Nauchn. Issl. Inst-Sooruzh. i Gidroenerg.* **10**, 165–180, 1958.

KIRSTE, L., Druckstäbe geringsten Gewichts, *Öst. Ing-Arch.*, **12**, 1/2, 36–31, November 1958; *AMR*, 1960, **1**, 109; *RŻM*, 1959, 10621.

KOZYREV, YA. F., On the theoretical weight of statically determinate trusses (in Russian), *Trudy Kaz. Inst. Inzh. Stroit. Neft. Prom.*, **1**, 1958.

LAUSHEY, L. M., Direct design of optimum ideterminate trusses, *Proc. ASCE*, **84**, *ST* 8, *J. Struct. Div.*, 1958.

MAZURKIEWICZ, Z., Designing the arch axis against bending under a given load (in Polish), *Arch. Inż. Ląd.*, **4**, 1, 1958.

MRÓZ, Z., On the design of non-homogeneous technically orthotropic plates, *Symp. IUTAM*, 191–202, Warszawa 1958.

MRÓZ, Z., The limit load-carrying capacity and the strength design of annular plates (in Polish), *Rozpr. Inż.*, **6**, 4, 603–626, 1958; *AMR*, 1960, **2**, 626.

PEARSON, C., Structural design by high speed computing machines, *Proc. 1st Nat. Conf. on Electronic Computation*, ASCE, Kanzas 1958.

PRAGER, W., On a problem of optimal design, *Symp. IUTAM*, Warszawa 1958.

PRAGER, W., Minimum weight design of a portal frame, *Trans. ASCE*, vol. **123**, 1958.

RADTSIG, YU. A., The least volume method and its application to the design of statically indeterminate bar systems of structural mechanics (in Russian), *Trudy Kazanskogo Sel'skokhozyajstvennogo Instituta*, **1**, 37, 1958.

RADTSIG, YU. A., Foundations of the theory of mirror functions (in Russian), *Trudy KAI*, **33/34**, 139–199, 1958.

RADTSIG, YU. A., Application of the least volume method to the design of statically indeterminate trusses and beams (in Russian), *Diss. Mosk. Inst. Inzh. Zh. D. Transp.*, **37**, 101–114, 1958.

SAELMAN, B., A note on the optimum distribution of material in a beam for stiffness, *IASS*, **4**, 1958.

SAPOZHKOV, N. M., Investigation of optimum proportions of I- and T-sections (in Russian), *Sb. Rasch. Prochn. Mashinostr. MVTU*, **89**, 147–167, Moskva 1958; *RŻM*, 1960, **2**, 2556; *AMR*, 1962, **4**, 1982.

SCHMIDT, L. C., Fully-stressed design of elastic redundant trusses under alternative load system, *Austr. J. of Appl. Sci.*, **9**, 4, 1958.

SERGEEV, N. D., Certain inverse problems for beams on elastic foundation (in Russian), *Leningr. Inst. Inzh. Zh. D. Transp.*, 137–141, 1958.

STAROSOLSKI, W., Design of prestressed flexure sections with regard to all programmed loading states (in Polish), *Arch. Inż. Ląd.*, **4**, 3, 1958.

STRASSER, G., Optimization of multiweb beams under combined bending and torsional loading, *J. Aerospace Sci.*, **25**, 8, 529, 1958; *RŻM*, 1959, 9348.

VISY, Z., Iterative method for the calculation of moments on highly economical reinforced concrete beams, *Acta Techn. Acad. Sci. Hungaricae*, **22**, 1/2, 12–26, 1958.

VAKHURKIN, V. M., Prestressing and optimum shapes of flexural elements (in Russian), *Sb. Mater. po Stal'. Konstr.*, **3**, 87–111, Moskva 1958, 1960, I; *AMR* 1962, **3**, 1459.

VINOGRADOV, O., On minimum-weight trusses (in Ukrainian), *Prikladna Mekhanika AN USSR*, **4**, 3, 1958.

ZIEGLER, H., Kuppeln gleicher Festigkeit, *Ing. Arch.*, **26**, 378–382, 1958.

ZIELIŃSKI, Z., Problems relative to design and use of economical prestressed concrete beams, *III Congress FIP, Ser. III*, **17**, Berlin 1958.

1959

BALOVIEV, G. G., and TROFIMOV, G. S., On efficient forms of cross-sections of thin-walled flexural profiles (in Russian), *Vest. Mashin.*, **39**, 4, 3–10, 1959.

BORISOV, V. P., On the design of continuous beams of uniform bending strength (in Russian), *Trudy KAI*, **46**, 95–105, 1959.

BORISOV, V. P., On the design of frames of uniform bending strength (in Russian), *Trudy KAI*, **46**, 107–112, 1959; *RŻM* 1962, 9B333.

DAVIDON, W. C., Variable metric method for minimization, *Argonne National Lab. ANL-*5990 Rev., Univ. of Chicago 1959.

GAJNULLINA, S. KH., On the question of cross-sectional area limitations near the zero points of the bending moments (in Russian), *Trudy KAI*, **46**, 35-43, 1959.

GAJNULLINA, S. KH., An approximate design method for minimum volume frames (in Russian), *Trudy KAI*, **46**, 67–74, 1959.

GAJNULLINA, S. KH., On the question of stability check for frames of uniform bending strength (in Russian), *Trudy KAI*, **46**, 113–119, 1959.

GATEWOOD, B. E., and JONES, C. R., Optimum design of stiffened panels and sandwich panels at elevated temperature, *Proc. IV Midwest Conf. Solid Mech., Austin Texas*, 270–297, 1959; *AMR*, 1960, **3**, 1225.

GELLATLY, G. D., Optimum design of thin circular plates on an elastic foundation, *Proc. Inst. Mech. Engrs.*, **173**, 27, 687–698, 1959; *AMR* 1961, **8**, 4061.

GRANHOLM, O. A., Ekonomiska aluminium profiler, *Chalmers Tekn. Högskol. Handl.*, **220**, 1959; *RŻM*, 1960, 15220.

GRIGOR'EV, A. S., On plates of uniform bending strength (in Russian), *Inzh. Sb.*, **25**, 45–50, 1959; *RŻM*, 1961, 5B168.

HEYMAN, J., On the absolute minimum weight design of framed structures, *Quart. J. Mech. Appl. Math.*, **12**, 3, 314–324, 1959; *AMR*, 1960, **8**, 3997.

HEYMAN, J., Inverse design of beams and grillages, *Proc. Instn. Civ. Engrs.*, **13**, 339–352, 1959; *AMR*, 1960, **11**, 5668; *RŻM*, 1960, 16685.

HEYMAN, J., On the minimum weight design of a simple portal frame, *Inst. J. of Mech. Sci.*, **1**, 1959.

HU, T. C., and SHIELD, R. T., Uniqueness in the optimum design of structures, *Brown University Tech. Rep.* **DA-4795/2**, 1959.

IGNACZAK, J., On the spans of beams designed for uniform stress (in Polish), *Rozpr. Inż.*, **7**, 2, 181–189, 1959; *AMR*, 1963, **3**, 1533; *RŻM*, 1960, **9**, 12219.

ISSLER, W., Eine Kuppel gleicher Festigkeit, *Z. Angew. Math. Phys.*, **10**, 6, 1959.

KAUFMAN, S., and HOP, T., A study on the rational design of cross-section for prestressed beams (in Polish), *Arch. Inż. Ląd.*, **5**, 1, 81–127, 1959.

KORSHUNOV, A. I., A graphical-analytical method for the design of beams of uniform strength with allowance for dead weight (in Russian), *Trudy KAI*, 46, 75–86, 1959.

KRISHNAN, S., and SHETTY, K. V., On the optimum design of an I-section beam, *J. Aerospace Sci.*, **26**, 6, 393–399, 1959; *AMR*, 1960, **6**, 2741; *RŻM*, 1961, 6B315.

KRISHNAN, S., and SHETTY, K. V., Methods in optimum structural design for compression elements, *J. Aeronaut. Soc., India*, **11**, 2, 23–29, 1959; *AMR*, 1960, **7**, 3416.

LIVESLEY, R. K., Optimum design of structural frames for alternative systems of loads, *Civ. Engng. London*, **54**, 636, 732–740, 1959; *AMR* 1961, **6**, 3049.

MCWITNEY, R. R., Minimum-weight analysis of symmetrical-multiweb-beam structures subjected to thermal stress, *NASA Techn. Note*, No. **D-104**, Washington 1959; *RŻM*, 1961, 1B331.

MIKELADZE, M. SH., Weight and strength analysis of orthotropic rigid-plastic shells (in Russian), *Arch. Mech. Stos.*, **11**, 1, 1959.

PEREDY, J., Über eine neue Minimumaufgabe der technischen Festigkeitslehre, *Acta Techn. Acad. Sci. Hung.*, **24**, 3/4, 329–346, 1959; *AMR*, 1960, **7**, 3402.

PRAGER, W., Dimensionnement plastique et économie des matériaux, *Bull. CERES, Gén. Civ.*, **10**, 335–362, Liège 1959; *AMR*, 1962, **3**, 1449; *RŻM*, 1961, 9B139.

PRAGER, W., On a problem of optimal design. Nonhomogeneity in elasticity and plasticity, Pergamon Press, 127–132, 1959; *RŻM*, 1960, **10**, 13587.

PRAGER, W., and SHIELD, R. T., Minimum weight design of circular plates under arbitrary loading, *J. Appl. Math. Phys., ZAMP*, **10**, 4, 421–426, 1959; *RŻM*, 1960, **9**, 12131; *AMR*, 1961, **6**, 2920.

RADTSIG, YU. A., *Application of the least-volume method to the design of statically indeterminate trusses and beams* (in Russian, Thesis), Avtoref. Dokt. Diss., MIIZhG, 1959.

RADTSIG, YU. A., Characteristic features in the application of crosssection sizing constraints to the design of statically indeterminate trusses (in Russian), *Trudy KAI*, **46**, 15–33, 1959.

RADTSIG, YU. A., Design of a multiply statically indeterminate truss of least volume under moving concentrated forces (in Russian), *Trudy KAI*, **46**, 45–65, 1959.

SEGOV, N. A., Design of statically indeterminate trusses for minimum weight under multiple loading conditions and under moving loads (in Russian), *Sb. Nauch. Tr. Leningr. Inzh.-Str. Inst.* **30**, 26–40, 1959; *RŻM*, 1963, 3B359.

STAROSOLSKI, W., Prestressed continuous beam with a variable moment of inertia. Choice of prestressing force (in Polish), *Arch. Inż. Ląd.*, **5**, 3, 1959.

STAROSOLSKI, W., Design of prestressed continuous beams of variable moment of inertia allowing for the variation of the prestressing force along the beam (in Polish), *Arch. Inż. Ląd.*, **5**, 4, 1959.

STAROSOLSKI, W., Design of cable layout in a prestressed beam for all planned loading states (in Polish), *Inż. i Bud.*, **10**, 1959.

SU, HSUAN-LOH, Statistical approach to structural design, *Proc. Instn. Civ. Engrs.*, **13**, 353–362, 1959; *AMR*, 1960, **10**, 5186.

WASIUTYŃSKI, Z., *Arch Bridges* (in Polish), PWN, Warszawa 1959.

VASILIEV, V. G., On the most efficient variation of cross-section along uniform-strength arches (in Russian), *Stroitel'stvo*, **1**, 111–118, 1959.

VINOGRADOV, A. I., On the most effective elimination of redundant connections in trusses for full stress (in Russian), *Stroit. Mech. i Raschet Sooruzh.*, **5**, 7–13, 1959; *RŻM*, 1961, 7B365.

VINOGRADOV, A. I., On the design of bar systems of minimum weight (in Russian), *Issl. po Teorii Sooruzh.*, **8**, 499–521, 1959.

VINOGRADOV, A. I., Dead weight considerations in the fully-stressed design of bar systems (in Russian), *Issl. po Teorii Sooruzh.*, **8**, 523–533, 1959.

YURYEV, V. P., On the theoretical form of a beam of uniform bending strength (in Russian), *Trudy KAI*, **46**, 87–94, 1959.

1960

BALOVNEV, G. G., On the determination of efficient flexural profiles for frame structures (in Russian), *Stroit. Mekh. Rasch. Sooruzh.*, **1**, 38–42, 1960; *RŻM*, 1960, **11**, 15203; *AMR*, 1963, **8**, 4444.

BIERNAWSKI, A., and GROCHOWSKI, B., Design of trusses for minimum potential (in Polish), *Rozpr. Inż.*, **8**, 2, 137–165, 1960; *AMR*, 1961, **4**, 1936.

BRANDT, A., Design of prestressed beams by equalization of extreme stresses (in Polish), *Rozpr. Inż.*, **8**, 2, 231–252, 1960.

BRANDT, A., Détermination de la forme des poutres précontraintes par l'égalisation des contraintes, *Bull. Acad. Pol. Sci., Sér. Sci. Tech.*, **8**, **5**, 219–224, 1960.

CHAN, A. S. L., The design of Michell optimum structures, *Coll. Aeronaut. Cranfield, Rep.* **142**, 34, Dec. 1960; *AMR*, 1962, **3**, 1450.

EASON, G., The minimum weight design of circular sandwich plates, *ZAMP*, **11**, 5, 368–375, 1960; *AMR*, 1961, **9**, 4674.

EHRSHOV, N. F., Elastic-plastic buckling of bars (in Russian), *Tr. Gor'kovs. Inst. Inzh. Vodn. Transp.*, **27**, 30–61, 1960; *RŻM*, 1961, 11B210; *AMR*, 1963, **12**, 6901.

ENGLISH, J. M., Optimization of framing arrangements for large metal roof systems, *Publ. Int. Ass. Bridge Struct. Engng.*, **20**, 75–86, 1960; *AMR*, 1962, **6**, 3378; *RŻM*, 1963, 8B404.

GERARD, G., Minimum weight analysis of orthotropic plates under compressive loading, *J. Aerospace Sci.*, **27**, 1, 21–26, 64, 1960; *AMR*, 1960, **11**, 5673.

GRAHAM, J. D., Optimum design of reinforced concrete buildings, *Second Conf. on Electr. Comp., Struct. Div. ASCE*, **8–9**, Pittsburgh, Sept. 1960.

GRYCZ, J., On transformation of a four-node structure by member exchange (in Polish), *Rozpr. Inż.*, **8**, 1, 31–43, 1960; *AMR*, 1961, **3**, 1397.

GUSEV, K. P., Optimum design of a steel I-beam in buckling (in Russian), *Stroit. Mekh. i Raschet Sooruzh.*, **1**, 34–38, 1960.

HEYMAN, J., On the minimum weight design of a simple portal frame, *Int. J. Mech. Sci.*, **1**, 2, 121–134, 1960; *AMR*, 1961, **7**, 3617.

HILTON, H. H., and FEIGEN, M., Minimum weight analysis based on structural reliability, *J. Aerospace Sci.*, **27**, 9, 641–651, 1960; *RŻM*, 1961, 9B293; *AMR*, 1961, **5**, 2464.

HOFFMAN, G. A., Minimum-weight proportions of pressure vessel heads, *Rand. Corp.. Aero-Astronaut. Dept. Calif.*, **P-2137**, 29 Nov. 1960; *AMR*, 1961, **4**, 1952.

HU, T. C., *Optimum design for structures of perfectly plastic materials*, Ph.D. Thesis, Brown University, 1960.

KAPKOWSKI, J., Uniform-strength design of a disc for a concentrated force (in Polish), *Arch. Bud. Masz.*, **1**, 1960.

KELLER, J. B., The shape of the strongest column, *Arch. Rat. Mech. Anal.*, **5**, 4, 275–285, 1960.

KELLER, J. B., *Minimum weight and minimum stress design of beams*, Inst. Math. Sci., New York University, 1960.

KHUBERYAN, K. M., Forces in a statically indeterminate truss of minimum weight under multiple loading (in Russian), *Izv. AN SSSR, Mekh. i Mashinostr.*, **3**, 24–29, 1960; *RŻM*, 1961, 10B401.

KHUBERYAN, K. M., The regions of the existence of redundant unknowns in singly statically indeterminate trusses designed for a given load and irregular temperature effects (in Russian), *Izv. Akad. Nauk Armyan. SSR*, **13**. 1–15, Erevan' 1960.

KHUBERYAN, K. M., Design of statically indeterminate trusses by the method of stresses: general case (in Russian), *Issl. po Teorii Sooruzh.*, **9**, 285–296, 1960.

KIRSTE, L., Beitrag zum Problem des "Tragwerks, Mindestgewicht", *Z. Flugwiss.*, **8**, 12, 352–359, 1960; *RŻM*, 1963, 9B367; *AMR*, 1961, **8**, 4103.

KOSMOWSKI, J., Determination of member layout in a four-node truss field from the condition of minimum potential (in Polish), *Rozpr. Inż.*, **8**, 1, 3–30, 1960.

KOSMOWSKI, J., Determination of an optimum distribution of zero points of the bending moments in multi-span beams (in Polish), *Prace Katedry Budowy Mostów Politechniki Warszawskiej* (typescript), 1960.

KOSMOWSKI, J., Examples of uniform-strength design of a cantilever steel beam with allowance for dead weight (in Polish), *Prace Katedry Budowy Mostów Politechniki Warszawskiej* (typescript), 1960.

KOSMOWSKI, J., Investigations of the relationship between the volume of a steel cantilever and the load, allowable stresses and depth at the end of the cantilever (in Polish), *Prace Katedry Budowy Mostów Politechniki Warszawskiej* (typescript), 1960.

ŁUKASIEWICZ, S., Circular plates of uniform strength under axisymmetrical loading (in Polish), *Arch. Bud. Masz.*, **7**, 1, 75–85, 1960; *RŻM*, 1962, 4B218.

MIKELADZE, M. SH., On plastic shells of uniform strength (in Russian), *Soobshch. AN GSSR*, **25**, 4, 1960.

MRÓZ, Z., On a problem of minimum weight design, *Brown Univ. Div. Appl. Math.*, **TR 59**, 18, May 1960; *AMR*, 1961, **2**, 661.

PEREDY, J., Die Planung wirtschaftlicher Tragwerke, *Acta Techn. Acad. Sci. Hung.*, **29**, 3/4, 251–273, 1960; *RŻM*, 1961, 9B295.

PEREDY, J., Wirtschaftliche, statisch unbestimmte Konstruktionen, *Acta Techn. Acad. Sci. Hung.*, **31**, 3/4, 285–309, 1960.

RADTSIG, YU. A., On truss synthesis (in Russian), *Trudy KSKhI*, **3**, 42, 165–175, 1960.

RADTSIG, YU, A., Statically indeterminate trusses of minimum volume (in Russian), *Trudy KAI*, **51**, 1960; *RŻM*, 1962, 1B323.

ROBERTSON, R. G., The two-span haunched continuous beam in prestressed concrete, *Proc. Instn. Civ. Engrs.*, **15**, 255–272, 1960; *AMR*, 1960, **10**, 5177.

SCHMIT, L. A., Structural design by systematic synthesis, *Proc. Second National Conf. on Electronic Comp., Struct. Div. ASCE*, 105–132, Pittsburg, Sept. 1960.

SHIELD, R. T., On the optimum design of shells, *Trans. ASME 82E, J. Appl. Mech.*, **2**, 316–322, 1960; *AMR*, 1961, **4** 1933; 1961, **9**, 4634; *RŻM*, 1961, 7B186.

SHIELD, R. T., Plate design for minimum weight, *Quart. Appl. Math.*, **18**, 2, 131–144, 1960; *AMR*, 1961, **2**, 659; *RŻM*, 1961, 6B163.

SHIELD, R. T., *Optimum design methods for structures. Plasticity*, Pergamon Press, 1960.

SOBRERO, L., *Les Voûtes minces*, Ist. Mecc., Trieste 1960.

SOROKIN, P. I., On efficient cross-sectional dimensions in thin-walled frames (in Russian), *Sb. Nauch. Trudy Krivirozhsk. Gornorudn. Inst.*, **8**, 277–282, 1960; *RŻM*, 1961, **6**, 6B294; *AMR*, 1963, **6**, 3403.

STAROSOLSKI, W., and SULIMOWSKI, Z., Design of eccentrically prestressed sections (in Polish), *Zesz. Nauk. P. Śl., Bud.*, **3**, 1960.

STAROSOLSKI, W., Prestressed section: general case; Analysis and choice of the prestressing force (in Polish), *Arch. Inż. Ląd.*, **3**, 1960.

VINOGRADOV, A. I., On least-weight systems with elastic-deformable joints (in Russian), *Stroit. Mekh. i Rasch. Sooruzh.*, **5**, 1–6, 1960.

WASIUTYŃSKI, Z., On the congruency of the forming according to the minimum potential energy with that according to the equal strength, *Bull. Acad. Pol. Sci., Sér. Sci. Techn.*, **6**, 8, 259–268, 1960; *AMR*, 1962, **6**, 3189..

WASIUTYŃSKI, Z., *On the application of exact sciences to building design* (in Polish), *Nauka Polska*, **1**, 29, 1960.

WASIUTYŃSKI, Z., Design of steel beams with I- or box sections (in Polish), *Księga jubileuszowa W. Wierzbickiego*, PWN, Warszawa 1960.

YUSUFF, S., Design for minimum weight. Considerations based on the long wave instability of stiffened plates in compression, *Aircraft Engng.*, **32**, 380, 288–294, 1960; *RŻM*, 1962, 2B102; *AMR*, 1961, **11**, 6079.

1961

BARNETT, R. L., Minimum weight design of beams for deflection, *Proc. ASCE* 87, *EM 1*, *J. Engng. Mech. Div.*, 75–109, 1961.

BENDYUK, D. G., Tabular design of a hingeless arch of minimum weight (in Russian), *Trudy Khar'k. Inst. Inzh. Zh. D. Transp.*, **14**, 67–86, 1961.

BIERNAWSKI, A., Application of point calculus to truss design (in Polish), *VII Konf. PZITB i PAN w Krynicy 1961.*

BIERNAWSKI, A., A theorem on the potential of a spatial truss with four loaded nodes (in Polish), *VII Konf. PZITB i PAN w Krynicy 1961.*

BORISOV, V. P., Application of the method of foci to the design of thin-walled structures (in Russian), *Trudy KAI*, **62**, 1961.

BORISOV, V. P., Fully stressed design of a box girder with torsion constraints (in Russian), *Trudy KAI*, **62**, 19–29, 1961.

BRANDT, A., A historical survery of strength design (in Polish), *VII Konf. PZITB i PAN w Krynicy 1961.*

BRANDT, A., A comparison of the foundations and scopes of Minimum Weight Design and Minimum Potential Design (in Polish), *VII Konf. PZITB i PAN w Krynicy 1961.*

BRANDT, A., Design of prestressed beams (in Polish), *VII Konf. PZITB i PAN w Krynicy 1961.*

CZUBAKOWSKI, H., Determination of the most economical cross-section for a steel beam in bending (in Polish), *Inż. i Bud.*, **18**, 1, 1961; *RŻM*, 1962, 3B353.

DAVIDSON, J. R., and Dalby, J. F., Optimum design of insulated compression plates subjected to aerodynamic heating, *NASA TN D*-520, **53**, Jan. 1961; *AMR*, 1961, **4**, 1954.

DZIENISZEWSKI, W., The influence of the normal forces in links, arches and beams on the arrangement and magnitude of the internal forces in Langer and Maillart systems (in Polish), *VII Konf. PZITB i PAN w Krynicy 1961.*

DZIENISZEWSKI, W., Determination of an optimum variation of the moments of inertia of beams in Maillart systems (in Polish), *VII Konf. PZITB i PAN w Krynicy 1961.*

EHSTRIN, M. I., Plates of minimum weight in the plane state of stress (in Russian), *Trudy Centr. n.-issl. Inst. Stroit. Konstr. Akad. Stroit. i. Arkhit.*, **4**, 91–103, 1961; *RŻM*, 1962, 4B217.

FESIK, S. P., *Design of statically indeterminate frames of minimum weight* (in Russian), Kand. Diss., (Thesis) Khar'kov 1961.

FESIK, S. P., On the minimum weight of singly statically indeterminate frames (in Russian), *Stroit. Mekh. Rasch. Sooruzh.*, **1**, 32–37, 1961; *RŻM*, 1961, 10B316; *AMR*, 1964, 2, 852.

GAJNULLINA, S. KH., On the design of statically indeterminate systems of minimum weight with variable depth of cross-section (in Russian), *Trudy KAI*, **62**, 31–36, 1961.

GAJNULLINA, S. KH., Frame design with consideration of axial forces (in Russian), *Trudy KAI*, **62**, 1961.

GERARD, G., Minimum weight design of ring stiffened cylinders under external pressure, *J. Ship Res.*, **5**, 2, 44–49, 1961; *RŻM*, 1962, 8B396; *AMR*, 1962, **8**, 4551.

GRYCZ, J., Basic relations and solving procedure for the design of concrete beams prestressed arbitrarily densely by arbitrarily thin wires from the condition of uniform principal stresses (in Polish), *VII Konf. PZITB i PAN w Krynicy 1961.*

GRYCZ, J., Determination of member layout in extreme isostatic trusses with a given configuration of nodes under fixed loading (in Polish), *VII Konf. PZITB i PAN w Krynicy 1961.*

HOUGHTON, D. S., Optimum design of a band reinforced pressurised cylinder, *Coll. Aero. Cranfield, Note* **116**, 7, 1961; *AMR*, 1962, **2**, 730.

HU, T. C., and SHIELD, R. T., Uniqueness in the optimum design of structures, *Trans. ASME* 83*E*, *J. Appl. Mech.*, **28**, 284–287; *AMR*, 1963, **7**, 3934; *RŻM*, 1962, 7B121.

HU, T. C., and SHIELD, R. T., Minimum-volume design of discs, *ZAMP*, **12**, 414–433, 1961; *RŻM*, 1962, 7B112.

IZRAELIT, A. B., Rational design of continuous beams and fixed frames with given stresses (in Russian), *Trudy Vses. Zaoch. Lesotekh. Inst.*, **7**, 139–152, 1961; *RŻM*, 1962, 10B 365.

JOHNSON, R. C., *Optimum design of mechanical elements*, Wiley, New York 1961; *RŻM*, 1962, 10B519.

JOHNSON, W., An analogy between upper-bound solutions for plane-strain metal working and minimum-weight two-dimensional frames, *Int. J. Mech. Sci.*, **3**, 4, 239–246, 1961; *AMR*, 1963, **9**, 5225.

KAPKOWSKI, J., and ŁUKASIEWICZ, S., Effect of temperature on the uniform strength of rotating discs (in Polish), *Arch. Bud. Masz.*, **2**, 1961.

KIRSTE, L., Ein weiterer Beitrag zum Problem des "Tragwerks-Mindestgewicht", *Z. Flugwiss.*, **9**, 11, 343–347, 1961; *AMR*, 1962, **7**, 4011; *RŻM*, 1963, 9B368.

KHUBERYAN K. M., Rational forms of hydraulic engineering structures (in Russian), *Stroit. i Arkhit. Inf. Byull. Gosstroya Gruz. SSR*, **6–7**, 32–35, 1961.

KORSHUNOV, A. I., On frame design with consideration of dead weight (in Russian), *Trudy KAI*, **62**, 1961.

KOSMOWSKI, J., Design of a steel cantilever with constant depth and constant flange thickness (in Polish), *VII Konf. PZITB i PAN w Krynicy 1961.*

KOSMOWSKI, J., Determination of member layout in a four-node truss field from the condition of minimum potential and practical member layout correctness tests for plane trusses (in Polish), *VII Konf. PZITB i PAN w Krynicy 1961.*

KOSMOWSKI, J., Determination of an optimum distribution of the zero points of the bending moments in multi-span beams designed for equal extreme values of non-dilatational strain potential (in Polish), *VII Konf. PZITB i PAN w Krynicy 1961.*

KOSMOWSKI, J., and BIERNAWSKI, A., Minimal trusses with two and three fields (in Polish), *VII Konf. PZITB i PAN w Krynicy 1961.*

KRISHNAN, S., and SHETTY, K. V., A method of minimum weight design for thin-walled beams, *Struct. Eng.*, **39**, 5, 174–180, 1961; *RŻM*, 1962, 2B427; *AMR*, 1962, **2**, 832.

L'VIN, YA. B., Canonical formulae of the uniform stress method (in Russian), *Stroit. Mekhan. i Konstr. Sbornik Trudov*, **8**, 119–124, 1961.

MARKETOS, J. D., Optimum theoretical pressure vessel filament wound along geodesic lines, *AIAA J.*, **1**, 8, 1942–1944, 1961; *RŻM*, 1964, 4B45.

MIKELADZE, M. SH., On layer structures of uniform strength (in Russian), *Dokl. AN Gruz. SSR*, **26**, 4, 397–404, 1961.

MRÓZ, Z., On a problem of minimum weight design, *Quart. Appl. Math.*, **19**, 127–135, 1961.

OZELL, A. M., and CONYERS, A. L., Effect of geometry in the economical design of cylindrical shells, *J. Amer. Concr. Inst.*, **32**, 12, 1585–1592, 1961; *RŻM*, 1962, 2B77.

PONIŻ, D., Design of building structures (in Polish), *Biul. Tech. BSPTBP*, 1961.

PONIŻ, W., Design of a tear-shaped container (in Polish), 1961 (unpublished).

RADTSIG, YU. A., Determination of the least value of a mirror function of several independent variables (in Russian), *Trudy KAI*, **62**, 91–97, 1961; *RŻM*, 1963, 2B380.

RADTSIG, YU. A., On the design of prestressed frames (in Russian), *Trudy KAI*, **62**, 1961.

RUBINSHTEJN, G. M., Optimum supporting mode for a uniformly loaded thin circular plate on a continuous concentric support (in Russian), *Izv. Vyssh. Uchebn. Zaved.*, **4**, 6, 115–123, 1961; *RŻM*, 1963, 12B89.

SAELMAN, B., A note on the minimum-weight design of spherical and cylindrical pressure surfaces, *J. Aeronaut. Sci.*, **28**, 1, 72–73, 1961; *AMR*, 1961, **8**, 4236; *RŻM*, 1961, 12B89.

SAELMAN, B., and RUBIN, A. E., Designing minimum-section column, *Mach. Design*, **33**, 22, 167–169, 1961; *RŻM*, 1964, 1B473.

STAROSOLSKI, W., Design of prestressed continuous beams (in Polish), *VII Konf. PZITB i PAN w Krynicy 1961.*

SUHUBI, E., Design of plates for minimum weight, *Bull. Techn. Univ. Istambul*, **14**, 1, 11–30, 1961; *AMR*, 9962, **5**, 2582; *RŻM*, 1962, 6B161.

SULIMOWSKI, Z., Analysis of a simply supported beam prestressed by a variable force (in Polish), *VII Konf. PZITB i PAN w Krynicy 1961.*

TADJBAKHSH, I. G., Elastic optimum design of circular plates, Development in Mechanics, *Proc. VII Midwestern Mech. Conf., Mich. State Univ.*, vol. **1**, Sept. 1961; *RŻM*, 1963, 12B90.

TSENTSEVITSKAYA, E. S., On the design of beams of minimum weight (in Russian), *Trudy KAI*, **62**, 45–58, 1961; *RŻM*, 1962, 7B287.

VASIL'EV, A. A., Optimum parameters of prestressed steel beams (in Russian), *Stroit. Mekh. i Raschet Sooruzh.*, **4**, 47–51, 1961.

VINOGRADOV, A. I., and FESIK, S. P., On the most efficient distribution of forces in composite structures (in Russian), *Prikl. Mekhanika*, **2**, 7, 157–163, 1961.

VOLKOVA, N. S., On the design of statically indeterminate trusses with the least volume of the 2nd kind (in Russian), *Trudy KAI*, **62**, 1961.

VOLKOVA, N. S., Prestressed trusses with the least volume of the 2nd kind (in Russian), *Trudy Kazansk. S. Kh. Inst.*, **45**, 120–131, 1961; *RŻM*, 1963, 11B450.

VOLKOVA, N. S., RADTSIG, YU. A., and KHASANSHCHIN, N. KH., On the design of statically indeterminate trusses of minimum weight (in Russian), *Stroit. Mekh. Raschet Sooruzh.*, **4**, 10–14, 1961; *RŻM*, 1962, **5**, 5B373; *AMR*, 1964, 2, 851.

VOLODARSKIJ, B. YA., Strength analysis of prestressed steel trusses of minimum cost with the use of mirror functions (in Russian), *Proekt. i Stroit. Proizv. Zdanij. Akad. Stroit. i Arkhit. SSSR*, 152–165, 1961.

WASIUTYŃSKI, Z., The problem of designing a uniform-strength concrete beam prestressed arbitrarily densely by arbitrarily thin wires (in Polish), *VII Konf. PZITB i PAN w Krynicy 1961.*

WIDMANN, R., Zur Berechnung and wirtschaftlichen Formgebung von Bogengewichtsmauren, *Öst. Ingenieur Z.*, **4**, 8, 274–280, 1961.

ZHURAVLEV, V. N., *Reducing the weight of mechanical engineering structures* (in Russian), Mashgiz. M., Sverdlovsk 1961.

YUR'EV, V. K., On the design of continuous beams of uniform bending strength with allowance for dead weight (in Russian), *Trudy KAI*, **62**, 39–43, 1961.

1962

ALEKSANDROV, A. YA., *Optimum parameters of three-layer plates and shells* (in Russian), Kiev, AN USSR, 463–466, 1962; *RŻM*, 1963, **7**, 7B85; *AMR*, 1964, **9**, 5020.

ABGARYAN, K. A., On the theory of minimum-weight beams (in Russian), Raschety na prochn., *Mashgiz. M.*, **8**, 136–151, 1962; *RŻM*, 1963, 7B322.

BALOVNEV, G. G., Determination of optimum dimensions of thin-walled flexural profiles for compression elements (in Russian), *Stroit. Mekh. i Raschet Sooruzh.*, **6**, 40–44, 1962; *RŻM*, 1963, 2B440.

BAUSIC, V., and CLIMESCU, A., Solutio economica pentru dimensionarea barelor din tronsoane supuse la solicitari axiale, *Bull. Inst. Polit. Iasi*, **8**, 3/4, 289–294, 1962; *RŻM*, 1964, 6B382.

BIERNAWSKI, A., A theorem on the potential of a spatial grating with four loaded nodes, *Bull. Acad. Pol. Sci., Sér. Sci. Tech.*, **10**, 11, 61–70, 1962; *AMR*, 1962, **11**, 6479.

BRANDT, A., Some theorems on statically determinate prestressed beams designed for minimum potential, *Bull. Acad. Pol. Sci., Sér. Sci. Tech.*, 10, 2, 57–62, 1962; *AMR*, 1962, **12**, 7066.

BROTCHIE, J. F., Direct design of plate and shell structures, *Proc. ASCE* 88, *ST 6, J. Struct. Div.*, 127–148, 1962; *AMR*, 1963, **9**, 5220.

BROWN, E. H., The minimum weight design of closed shells of revolution, *Quart. J. Mech. Appl. Math.*, **15**, 1, 109–128, 1962; *RŻM*, 1963, 5B55; *AMR*, 1963, **2**, 745.

BURT, M. E., Structural weight estimations for novel configurations, *J. Roy. Aeronaut. Soc.*, **66**, 613, 15–30, 1962; *RŻM*, 1963, 5B368.

COHN, M. Z., Limit design of reinforced concrete structures for maximum yield safety, *India Concr. J.*, **36**, 6, 214–224, 1962; *RŻM*, 1963, 2B423.

COLE, B. M., and DAVIES, R., Optimum design methods for short duplex cylinders, *Engineer*, **214**, 5566, 528–531, 1962; *AMR*, 1963, **7**, 3931; *RŻM*, 1963, 7B477.

FESIK, S. P., On the design of frames of minimum weight under transient loads (in Russian), *Nauchn. Tr. Khar'k. Inst. Inzh. Zh. D. Transp.*, **58**, 33–46, 1962; *RŻM*, 1963, 11B386.

FESIK, S. P., On the inverse problem of the theory of structures (in Russian), *Trudy KhIIZhT*, **58**, Khar'kov, 1962.

FILIP, A. P., and GUREVICH, YA. I., Application of the calculus of variations to the rational design of structures (in Russian), *LIIZhT, L.*, **190**, 161–187, 1962; *AMR*, 1964, 7B435.

FLEJSHMAN, N. P., Circular and annular plates of minimum weight. (in Russian), *Raschety na prochn., Mashgiz. M.*, **8**, 127–135, 1962; *RŻM*, 1963, 5B94; *AMR*, 1964, **8**, 4504.

GALLAGHER, R. H., RATTINGER, I., and KRIVETSKY, A., Minimum weight shells in bending, *Aerospace Engng.*, **21**, 58–82, 1962; *RŻM*, 1962, 12B99.

GROSS, O., and PRAGER, W., Minimum weight design for moving loads, *Proc. IV US Nat. Congr. Appl. Mech. Berkeley* 1962. *Amer. Soc. Mech. Engineers*, **2**, 1047–1051, 1962; *AMR*, 1963, **9**, 5042; *RŻM*, 1964, 2B367.

GRYCZ, J., Design of beams with rectangular cross-section (in Polish), *Zesz. Nauk. P. Warsz.*, **68**, *Bud.*, 19, 53–60, 1962; *RŻM*, 1964, 2B371.

GRYCZ, J., Design of a simply supported prestressed beam with rectangular cross-section (in Polish), *Zesz. Nauk. P. Warsz.*, **68**, *Bud.*, 19, 61–68, 1962.

GRYCZ, J., Determining shapes of isostatic box beams with variable flange and web thicknesses (in Polish), *Zesz. Nauk. P. Warsz.*, **68**, *Bud.*, 19, 69–77, 1962; *RŻM*, 1964, 2B372.

GUREVICH, YA. I., Design of double- and triple-hinged arches of minimum volume with given axes by variational methods (in Russian), *Sb. Tr. Leningr. Inst. Inzh. Zh. D. Transp.*, **190**, 189–218, 1962; *RŻM*, 1964, 7B410.

GUSEJNOV, T. D., Equations of rational profiles of cylindrical shells under various loading conditions (in Russian), *Trudy Azerb. S. Kh. Inst.*, **16**, 111–118, 1962; *RŻM*, 1964, 4B350.

HARRIS, R. S., JR., and DAVIDSON, J. R., Methods for determining the optimum design of structures protected from aerodynamic heating and application to typical boost-glide or reentry flight paths, *NASA TN, D-990*, **37**, 1962; *AMR*, 1962, **11**, 6494.

HEDGEPETH, J. M., Design of stiffened cylinders in axial compression, *NASA Techn. Note, No.* **D-1510**, 77–83, 1962.

HOFFMAN, G. A., Minimum-weight proportions of pressure vessel heads, *Trans. ASME, E* **29**, 662–668, 1962; *RŻM*, 1963, 10B686.

HOFFMAN, G. A., Optimum proportions of pressure vessel heads, *J. Aerospace Sci.*, **29**, 12, 1471–1475, 1962; *AMR*, 1963, **10**, 5722; *RŻM*, 1964, 8B337.

KALABA, R., Design of minimal-weight structures for given reliability and cost, *J. Aerospace Sci.*, **29**, 3, 355–356, 1962; *RŻM*, 1963, 1B428.

KHACHATURIAN, M., ALI, I., and THORPE, L. T., Analytical studies of relations among various design criteria for prestressed concrete. Investigation of prestressed concrete for highway bridges, *III Engng. Exp. Sta. Bull.* No. **463**, 68, 1962; *AMR*, 1968, **12**, 7068.

KITOV, YU. P., On the minimum weight of complex frames (in Russian), *Nauchn. Tr. Khar'k. Inst. Inzh. Zh. D. Transp.*, **58**, 47–64, 1962; *RŻM*, 1963, 10B525.

KORACH, M., Economy of reinforced concrete slab structures, *Magyar Tud. Akad. Müsz. Tud. Oszt. Közl.*, **30**, 1/4, 97–119, 1962; *AMR*, 1963, **9**, 5219.

MAYERJAK, R. I., On the weight and design of a redundant truss, *Aeronaut. Res. Lab., Rep.* **62–332**, USAF, Apr. 1962.

MEGAREFS, G. J., and HODGE, P. G., JR., Singular cases in the optimum design of frames, *III Inst. Technol. Dept. Mech. Rep. No.* **1-19**, 28, Mai 1962; *AMR*, 1962, **11**, 6486; *RŻM*, 1964, 8B558.

PADUART, A., Recherche des charactéristiques optimales d'une section fléchie en béton précontraint, *Ind. Ital. Cem.*, **32**, **3**, 141–146, 1962; *RŻM*, 1963, 2B446.

PANCHOVSKI, D., On the optimum design of the front spar in a double-spar wing (in Bulgarian), *Godishnik Khim.-Tekhnol. Inst.*, **8**, 263–273, 1961; *RŻM*, 1964, 2B401.

PETCU, V., Momente plastica optime in grinzi continue de beton armat cu sectiunea variabila, *Studi Cerc. Mec. Apl. Acad. APR*, **13**, 2, 497–506, 1962; *RŻM*, 1963, 8B464.

PETCU, V., Les bases du calcul plastique des structures hyperstatiques en béton armé, *Rev. Méc. Appl.*, **7**, 2, 285–296, 1962; *AMR*, 1963, **7**, 3929.

PRAGER, W., Lineare Ungleichungen in der Baustatik, *Schweizerische Bauzeitung*, **80**, 315–320, 1962.

RADTSIG, YU. A., Application of modular forms to the modelling problems of structural mechanics (in Russian), *Trudy Mezhvuzovskoj Nauchno-tekhnicheskoj Konferentsii po Elektricheskomu Modelirovaniyu Zadach Stroitel'noj Mekhaniki, Teorii Uprugosti i Soprotivleniya Materialov*, Novocherkask 1962.

RAO, K. V., Economical design of beams, *J. Instn. Engrs. (India), Civil Engng. Div.*, **42**, 11, 608–615, 1962; *RŻM*, 1963, 11B413.

SAELMAN, B., Some notes on the optimum design of box beams for combined strength and stiffness, *J. Aerospace Sci.*, **29**, 4, 498–499, 1962; *AMR*, 1962, **9**, 5164; *RŻM*, 1962, 11B413.

SCHMIDT, L. C., Minimum weight layouts of elastic, statically determinate, triangulated frames under alternative load systems, *J. Mech. Phys. Solids*, **10**, 139–149, 1962; *AMR*, 1962, **11**, 6480; *RŻM*, 1963, 10B578.

SCHMIT, JR., L. A., and KICHER, T. P., Synthesis of material and configuration selection, *Proc. ASCE*, 88, 3, *Part 1, J. Struct. Div.* 79–102, 1962; *RŻM*, 1964, 6B448.

SHAMIEV, F. G., On the minimum-weight design of plates (in Russian), *Izv. AN Azerbejdzh. SSR*, **6**, 41–47, 1962; *RŻM*, 1964, 8B342.

SUHUBI, E., The bending of rectangular plates of varying thickness simply supported along two opposite sides, *Bul. Istanbul. Tekn. Univ.*, **15**, 63–70, 1962.

SULIMOWSKI, Z., Determination of an optimum height of a post-tensioned prestressed beam (in Polish), *Zesz. Nauk. P. Śl.*, **60**. *Bud.*, 8, Gliwice 1962.

SWITZKY, H., The minimum weight design of structures operating in an aerospace environment, *ASD-TDR*-62-763, *Rep. Av. Corp., Farmingdale, L. I., N. Y.*, **124**, 1962; *AMR*, 1963, **3**, 1555.

TADJBAKHSH, I. G., and KELLER, J. B., Strongest columns and isoperimetric inequalities for eigenvalue, *Trans. ASME 84E, J. Appl. Mech.*, **1**, 159–164, 1962; *AMR*, 1962, **10**, 5774.

TANAKA, H., Automatic analysis and design of plastic frames, *Rep. Inst. Industr. Sci. Tokyo Univ.*, **12**, 3, 108–168, 1962; *AMR*, 1963, **6**, 3410; *RŻM*, 1963, 4B194.

TURKSTRA, C. J., *A formulation of structural design decisions*, Dissertation, Waterloo Univ. 1962.

WASIUTYŃSKI, Z., Fundamental theorems of minimum potential design and applications to the design of beams (in Polish), *Súcasné problémy mostov*, *SAV* 393–433, Smolenice–Bratislava 1962.

WASIUTYŃSKI, Z., and BRANDT, A., The current state of knowledge in structural strength design (in Polish), *Rozpr. Inż.*, **10**, 2, 307–332, 1962.

WASIUTYŃSKI, Z., BRANDT, A., GRYCZ, J., and IGNACZAK, J., Recherches des formes des poutres précontraintes, *Proc. IV Congr. FIP, thème III*, 9, 76–81, Rome–Naples 1962.

VINOGRADOV, A. I., and FESIK, S. P., On statically indeterminate frames of minimum weight (in Russian), *Stroit. Mekhan. i Raschet Sooruzh.*, **3**, 11–14, 1962; *RŻM*, 1963, 2B318.

1963

BABUSKA, I., and KAUTSKY, J., Über die Optiemierung von Kerbformen, *ZAMM*, 43, 1/2, 47–54, 1963; *AMR*, 1964, **4**, 1979.

BARNETT, R. L., Minimum deflection design of a uniformly accelerating cantilever beam, *J. Appl. Mech.*, **30**, 466–467, 1963; *AMR*, 1964, **5**, 2574; *RŻM*, 1964, 8B582.

BAUS, R., Considérations sur le dimensionnement économique du béton armé en flexion, *Ann. Trav. Pub. Belg.*, **6**, 489–504, 1963; *RŻM*, 1965, 1B427.

BERT, C. W., Optimum elastic restraint for pressurised circular plates, *J. Roy. Aeronaut. Soc.*, **67**, 632, 525–526, 1963; *RŻM*, 1964, 9B99.

BEST, G., A method of structural weight minimization suitable for high-speed computers, *AIAA J.*, **1**, 478–479, 1963; *AMR*, 1963, **7**, 3953.

BORISOV, V. P., The influence of stiffness constraints on the minimum volume of fully stressed systems (in Russian), *Trudy KAI*, **77**, 6–70, 1963; *RŻM*, 1964, 5B355.

BORISOV, V. P., On the design of statically indeterminate beams and frames of minimum weight (in Russian), *Tezisy dokladov, Pervaya Nauchnaya Sessiya Povolozhskogo Soveta*, Kazan' 1963.

BORISOV, V. P., *Some problems in the design of optimum bar systems of beam and frame types* (in Russian), Avtoref. Kand. Diss. (Thesis) KhAI, 1963.

BRABNER, R., About the direct design of prestressed concrete beams with arbitrary cross-section, *Gest. Ingen. Z.*, **6**, 4, 109–118, 1963; *AMR*, 1964, **5**, 2705.

BRANDT, A., A theorem on the strength design of prestressed concrete beams (in Polish), *Rozpr. Inż.*, **11**, 4, 559–566, 1963; *AMR*, 1965, **4**, 2169; *RŻM*, 1965, 2B464.

BRODING, W. C., DIEDERICH, F. W., and PARKER, P. S., Structural optimization and design based on a reliability design criterion, *J. Spacecraft and Rockets*, **1**, 1, 56–61, 1964; *AMR*, 1964, **12**, 7173; *RŻM*, 1964, 12B324.

BURNS, A. B., and SHOGH, J., Combined loads minimum weight analysis of stiffened plates and shells, *J. Spacecrafts and Rockets*, **3**, 2, 235–240, 1963.

BURT, M., Structural design for minimum weight, *Engng. Mater. Design*, **6**, 10, 712–715, 1963; *RŻM*, 1964, 6B387.

CALKOVSKY, A., Minimum-weight plates under shearing (in Czech), *Sb. Vojen. Akad.*, **B 11**, 3, 55–61, Praha 1963; *RŻM*, 1964, 6B116.

CHAN, H. S. Y., Optimum Michell frameworks for three parallel forces, *Coll. Aeronautics, Cranfield*, Rep. **167**, 15, 1963; *AMR*, 1964, **10**, 5825; *RŻM*, 1966, 12B213.

COHEN, G. A., Optimum design of truss-core sandwich cylinders under axial compression, *AIAA J.*, **1**, 7, 1626–1630, July 1963; *AMR*, 1964, **1**, 141.

CRAWFORD, R. F., and BURNS, A. B., Minimum weight potentials for stiffened plates and shells, *AIAA J.*, **1**, 4, 879–886, 1963; *AMR*, 1964, **2**, 743; *RŻM*, 1964, 6B68.

CROCI, G., Due metodi per il dimensionamento rapido di una sezione presollecitata soggetta a flessione, *Ingegneria Civile*, **8**, 3–59, 1963.

CSONKA, P., Hyperboloid shaped cooling tower with a mantlewall of equal strength, *Acta Techn. Acad. Sci. Hung.*, **44**, 1/2, 215–221,

DZHAPARIDZE, G. S., On the rational forms of statically indeterminate reinforced concrete arches designed for a given load-carrying capacity (in Russian), *Trudy Inst. Str. Mekh. i Sejsmostojkosti AN Gruz. SSR*, **9**, 51–56, 1963; *RŻM*, 1964, 5B474.

EHDOR, V. S., Selection of a rational variation rule for the stiffness of continuous beams and frames under moving loads (in Russian), *Trudy Mosk. Inst. Inzh. Zh. D. Transp.*, **174**, 147–152, 1963; *RŻM*, 1964, 10B396.

EVANS, J. H., and KHOUSHY, D., Optimized design of midship section structure, *Trans. SNAME*, **71**, 144–191, 1963.

FOULKERS, J. S., Systematical determination of the geometric properties of a built girder, *Int. Shipbuilding Progress*, **10**, 105, 159–169, 1963; *RŻM*, 1964, 10B432.

GAJNULLINA, S. KH., Design of frames of minimum weight with consideration of longitudinal forces (in Russian), *Trudy KAI*, **77**, 71–82, 1963, *RŻM*, 1964, 4B359.

GAJNULLINA, S. KH., On the selection of cross-sections for uniform-bending-strength systems (in Russian), *Trudy KAI*, **77**, 83–88, 1963; *RŻM*, 1964, 6B390.

GERASIMENKO, T. E., On the determination of an optimum weight of a statically indeterminate reinforced concrete beam (in Russian), *Tezisy Dokladov, Pervaya Nauchnaya Sessiya Povolzhskogo Soveta*, Kazan' 1963.

HESS, T. E., BAILEY, J. C., and MOST, H. J., Reentry vehicle optimization and synthesis program, *Proc. ASCE*, **89**, 4, *Part I, J. Struct. Div.* 249–268, 1963; *RŻM*, 1964, 8B634; *AMR*, 1964, **4**, 2126.

JANICZEK, R., Design of bars and beams from the criterion of minimum material (in Polish), *Zesz. Nauk. P. Częst.*, **2**, 1963.

KHUBERYAN, K. M., The stresses in a statically indeterminate truss corresponding to the minimum weight under fixed forces (in Russian), Part 1, *Izv. AN Arm. SSR*, **16**, 1, 13–19, 1963; Part 2, *Izv. AN Arm. SSR*, **16**, 4, 15–22, 1963; *RŻM*, 1963, 11B449.

KOY, B., Analysis and experimental design of single-column roofs (in Polish), *Inż. i Bud.*, **20**, 4, 125–131, 1963.

KRATOCHVIL, J., Determination of the optimum profile of a buttress dam (in Czech), *Knižn. Odbor, Vysok. Uceni. Techn. Brno*, **2**, 213–221, 1963; *RŻM*, 1965, 7B453.

KRZYŚ, W., and ŻYCZKOWSKI, M. Classification of strength design problems (in Polish), *Czasop. Tech.*, **68**, 2, 1–3, 1963.

KRZYŚ, W., and ŻYCZKOWSKI, M., A certain method of parametrical structural optimum shape-design, *Bull. Acad. Pol. Sci., Sér. Sci. Tech.*, **10**, 335–345, 1963; *AMR*, 1964, **12**, 7174; *RŻM*, 1964, 8B577.

KRZYŚ, W., and ŻYCZKOWSKI, M., A certain method of parametric strength design (in Polish), *Rozpr. Inż.*, **11**, **4**, 643–666, 1963; *RŻM*, 1965, 2B394.

KUAZIS, A. P., Optimum reinforcement of flexural reinforced concrete elements of annular cross-section (in Russian), *Beton i Zhelezobeton*, **4**, 179–181, 1963; *AMR*, 1965, **4**, 2161.

LESSAER, S., Determination of the carrying cable curves in suspended axisymmetrical surface structures loaded by self-weight (in Polish), *Arch. Inż. Ląd.*, **9**, 4, 1963.

LEŚNIAK, Z. K., Computer-aided application of nonlinear programming to the strength design of structures (in Polish), *IX Konf. PZITB* i *PAN w Krynicy*, 139–243, 1963.

MASSONNET, Ch., and SAVE, M., Optimum design of beams and frames in reinforced concrete, *Progr. Appl. Mech., Prager Anniversary Volume*, Macmillan Ltd., New York–London, 297–294, 1963; *RŻM*, 1965, 5B545.

MANEVICH, A., Optimum design of a supported cylindrical shell loaded by external pressure (in Ukrainian), *Dopovidi AN USSR*, **7**, 875–878, 1963; *RŻM*, 1964, 7B85.

MCKAY, C., A method of designing bridge wing walls which are of minimum length and of minimum area, *Struct. Engr.*, **41**, 5, 163–165, 1963; *AMR*, 1964, **3**, 1500; *RŻM*, 1964, 1B459.

MEGAREFS, G. J., and HODGE, P. G., JR., Singular cases in the optimum design of frames, *Quart. Appl. Math.*, **21**, 2, 91–103, 1963.

MEGAREFS, G. J., and HODGE, P. G., Jr., Method for plastic design of frames, *Proc. ASCE*, **89**, 1, Part I, *J. Struct. Div.* 197–214, 1963; *RŻM*, 1963, 12B190.

MRÓZ, Z., Limit analysis of plastic structures subject to boundary variations, *Arch. Mech. Stos.*, **15**, 1, 63–76, 1963; *AMR*, 1964, **11**, 6291.

PIPKIN, A. C., and RIVLIN, R. S., Minimum weight design for pressure vessels reinforced with inextensible fibres, *Trans. ASME, 85E, J. Appl. Mech.*, **30**, 1, 103–108, 1963; *AMR*, 1963, **10**, 5726; *RŻM*, 1963, 12B539.

PRAGER, W., Optimization in structural design, *Math. Optimiz. Techn. Univ. Calif. Press*, 279–289, 1963; *RŻM*, 1965, 3B184.

RADTSIG, E. A., Calcul sistemeloz static nedeterminate de wolum minim, *Buletinul Institutului Politechnic Bucuresti*, tamul **25**, fascicola 2, 1963.

RADTSIG, YU. A., Development of practical design methods for statically indeterminate beams of minimum volume (in Russian), *Trudy KAI*, **77**, 89–105, 1963; *RŻM*, 1964, 10B417.

RADTSIG, YU. A., Review of the research work carried out in Kazan' in the field of minimum-weight structural design (in Russian), *Pervaya nauchnaya sessiya Povolzhskogo Soveta*, Kazan' 1963.

RADTSIG, YU. A., Computer-aided rational strength design of statically indeterminate trusses of minimum volume (in Russian), *Tezisy doklada na Leningradskoe soveshchanie po primeneniyu EhTsVM v stroitel'noj mekhanike*, Leningrad 1963.

ROZVANY, G. I. N., and HAMPSON, A. I. K., Optimum design of prestressed plates, *J. Amer. Concr. Inst.*, **60**, 8, 1065–1082, 1963; *AMR*, 1964, **4**, 2112.

SATO, T., Automatic minimum weight design of plastic frames used for moment distribution method, *Proc. 12 Japan Nat. Congr. Appl. Mech.*, 1962, 17–21, Tokyo 1963; *RŻM*, 1965, 6B173.

SAVE, M., Minimum-weight design of beams subjected to fixed and moving loads, *Brown Univ. Div. Appl. Math. TR*, **86**, 21, 1963; *AMR*, 1963, **7**, 3935.

SAVE, M., and PRAGER, W., Minimum-weight design of beams subject to fixed and moving loads, *J. Mech. Phys. Solids*, **11**, 255–267, 1963; *RŻM*, 1964, 5B364.

SCHMIT, L. A., JR., KICHER, T. P., and MORROW, W. M., Structural synthesis capability for integrally stiffened waffle plates, *AIAA J.*, **1**, 12, 2820–2836, 1936; *RŻM*, 1964, 10B102.

SCHMIT, L. A., JR., and MALLET, R. H., Structural synthesis and design parameter hierarchy, *Proc. ASCE*, **89**, *ST* 4, *J. Struct. Div.*, 269–299, 1963; *AMR*, 1964, **3**, 1488; *RŻM*, 1964, 5B425.

SCHMIT, L. A., Jr., and MORROW, W. M., Structural synthesis with buckling constraints, *Proc. ASCE* **89**, *ST* 2, *J. Struct. Div.*, 107–126, 1963; *AMR*, 1965, **4**, 2043; *RŻM*, 1964, 4B426.

SHAMIEV, F. G., Minimum-weight design of annular plates (in Russian), *Izv. AN Azerbajdzh. SSR*, **3**, 13–20, 1963; *RŻM*, 1964, 8B343.

SHAMIEV, F. G., Design of plates of minimum weight (in Russian), *Izv. AN Azerbajdzh. SSR*, **5**, 37–41, 1963; *RŻM*, 1965, 1B173; *AMR*, 1965, **6**, 3370.

SHIELD, R. T., Optimum design methods for multiple loading, *ZAMP*, **14**, 1, 38–45, 1963; *AMR*, 1964, **10**, 5812; *RŻM*, 1963, 12B177.

SULIMOWSKI, Z., Analysis of a simply supported beam prestressed by a varying force (in Polish), *Inż. i Bud.*, **5**, 1963.

SWITZKY, H., and CARY, J. W., Minimum weight design of cylindrical structures, *AIAA J.*, **1**, 10, 2330–2337, 1963; *AMR*, 1964, **3**, 1385; *RŻM*, 1964, 6B83.

TRAUM, E., Economical design of reinforced concrete slabs using ultimate strength theory, *J. Amer. Concr. Inst.*, **6**, Part 1, 60, 763–774, 1963; *RŻM*, 1964, 5B465.

VARVAK, P. M., and VARVAK, O. P., Moment-free uniform-strength shell of small rise with a rectangular plan (in Ukrainian), *Dopovidi AN USSR*, **8**, 1021–1025, 1963; *RŻM*, 1964, 6B70.

VOLKOVA, N. S., Certain general propositions concerning singly statically indeterminate trusses of minimum volume of the 2nd kind (in Russian), *Trudy KAI*, **77**, 109–120, 1963; *RŻM*, 1964, 5B393.

VOLKOVA, N. S., Multiply statically indeterminate trusses of minimum volume of the $(n+1)$st kind (in Russian), *Trudy KAI*, **77**, 121–129, 1963; *RŻM*, 1964, 5B392.

WASIUTYŃSKI, Z., and BRANDT, A., The present state of knowledge in the field of optimum design of structures, *Appl. Mech. Rev.*, **16**, 5, 341–350, 1963; *RŻM*, 1964, 4B436.

WERNER, G., and FRANKE, W., Tonnenschalen mit optimalen Trageingenschaften, *Bauplan.-Bautechn.*, **17**, 8, 376–379, 1963; *RŻM*, 1964, 7B529.

1964

BAUS, R., Considérations sur le dimensionnement économique du béton armé en flexion, *Ann. Trav. Pub. Belg.*, **1**, 43–60, 1964; *RŻM*, 1964, 10B563.

BECKER, H., Design of cylinder-cone intersections, *J. Spacecrafts and Rockets*, **1**, 1, 1964.

BELEN'KIJ, L. M., Design of crossed-beam systems of minimum weight under limit loads (in Russian), *Izv. Vyssh. Uchebn. Mashinostrojenie*, **2**, 31–55, 1964. *RŻM*, 1965, 2B189.

BIGELOW, R. H., and GAYLORD, E. H., Design of steel frames for minimum weight, *Proc. ASCE*, **93**, *ST* 6, *J. Struct. Div.*, 109–131, 1964; *AMR*, 1968, **8**, 5788.

BILGE, M., Sur l'étude de la forme des sections des poutres précontraintes à fils adhérents, *Ann. Inst. Tech. Bât. Trav. Pub.*, **194**, 255–274, 1964.

BRANDT, A., Examples of prestressed beam design (in Polish), *Rozpr. Inż.*, **12**, 1, 101–113, 1964; *AMR*, 1965, **9**, 5447; *RŻM*, 1964. 12B358.

BRODING, W. C., DIEDRICH, F. W., and PARKER, P. S., Structural optimization and design based on a reliability design criterion, *J. Spacecraft and Rockets*, **1**, 56–61, 1964; *AMR*, 1964, **12**, 7173; *RŻM*, 1964, 12B324.

BROTCHIE, J. F., Direct design of framed structures, *Proc. ASCE*, **90**, *ST* **6**, *J. Struct. Div.*, 243–257, 1964; *AMR*, 1965, **5**, 2840.

CERONI, E., Realizzazione economica delle travi in cemento armato, *Ingegnere*, **38**, 6, 571–575, 1964; *RŻM*, 1965, 3B411.

CHAN, H. S. Y., Optimum structural design and linear programming, *Coll. Aeronaut., Cranfield Rep. Aero.*, **175**, 7, 1964; *AMR*, 1965, **3**, 1567; *RŻM*, 1966, 10B493.

CHAN, H. S. Y., Tabulation of some layouts and virtual displacement fields in the theory of Michell optimum structures, *Coll. Aeronaut. Cranfield, Note Aero.*, **161**, 1964; *AMR*, 1964, **11**, 6474.

CROCI, G., Tracciamento dei cavi in una struttura presollecitata in cui il getto del calcestruzzo ed il tiro dei cavi possa essere effettuato in piu fasi successive, *Ingegneria Civile*, No. **10**, 3–28, 1964.

CHELNOKOV, R. V., Design of minimum-weight arches according to the funicular curve with stability and elastic fastening considerations (in Russian), *Vtoraya Nauchnaya Sessiya Povolzhskogo Soveta, Saratov* 1964.

DORN, W. S., GOMORY, R. E., and GREENBERG, J. J., Automatic design of optimal structures *J. de Mécanique*, **3**, 1, 25–52, 1964; *RŻM*, 1965, 4B496.

DZIENISZEWSKI, W., Equations of deformation of prestressed plates of variable thickness *Bull. Acad. Pol. Sci., Sér. Sci. Tech.*, **12**, 10, 23–39, 1964.

DZIENISZEWSKI, W., Strength design of prestressed plates of varying thickness (in Polish), *Konf. ZMOC PAN, Zakopane* 1964.

FESIK, S. P., On the most efficient force distribution in statically indeterminate beams and frames with members of constant cross-sections (in Russian), *2-j Vses. S'ezd po Teor. i Prikl. Mekh.*, 1964; *RŻM*, 1964, 9B320.

FIACCO, A. V., and MCCORMICK, G. P., The sequential unconstrained minimization technique for nonlinear programming. A primal dual method, *Management Sc.*, **10**, **2**, 360–366, 1964.

FIACCO, A. V., and MCCORMICK, G. P., Computational algorithm for the sequential unconstrained minimization technique for nonlinear programming, *Management Sc.*, **10**, 4, 601–617, 1964.

FLERON, P., The minimum weight of trusses, *Bygningsstatiske Meddelser*, **35**, 3, 81–96, 1964.

FLETCHER, R., and REEVES, C. M., Function minimization by conjugate gradients, *Computer J.*, **7**, 149–145, 1964.

FRENCH, M. J., The design of rotors to have the minimum weight consistent with a required first critical speed, *J. Mech. Engng. Sci.*, **1**, 99–100, 1964; *AMR*, 1964, **10**, 5688.

GAJNULLINA, S. KH., Design of optimum bar structures in bending (in Russian), *Vtoraya Nauchnaya Sessiya Povolzhskogo Soveta, Saratov*, 1964.

GELLATLY, R. A., GALLAGHER, R. H., and LUBERACKI, W. A., Development of a procedure for automated synthesis of minimum structures, *FDL-TDR*-64-141, 1964.

GOLIŃSKI, J., and LEŚNIAK, Z. K., Selection of the optimum dimensions of structures by the Monte Carlo method (in Polish), *Arch. Inż. Ląd.*, **10**, 3, 1964.

HACKMAN, L. E., and RICHARDSON, J. E., Design optimization of aircraft structures with thermal gradients, *J. Aircraft*, **1**, 27, 32, 1964; *AMR*, 1964, **10**, 5835.

ISSLER, W., Membranschalen gleicher Festigkeit, *Ingen. Archiv.*, **33**, 5, 330–345, 1964; *AMR*, 1965, **10**, 5997; *RŻM*, 1965, 7B178.

JANICZEK, R., Minimum-weight thin-walled tube in axial compression (in Polish), *Zesz. Nauk. P. Częst.*, **28**, *Nauki Podst.*, 6, 51–72, 1964.

JUNG, F. W., Direct design of compression members and economical optimum of steel structures, *Div. Paper Engng. Inst. Canada*, **2**, 12, 1964; *RŻM*, 1966, 11B558.

KALISZKY, S., Optimum limit design for reinforced concrete structures, *Rep. Univ. Southampton, Dept. Civ. Eng.*, 1964.

KARAIVANOV, P., PRODANOV, M., and CHAVUSHYAN, N., Determination of the optimum parameters of the main truss of a cantilever lattice girder for a load moving on the upper chord (in Russian), *Mashinostroenie*, **11**, 8–12, 1964; *RŻM*, 1965, 5B437.

KOLUPAEV, A. N., Design of statically indeterminate trusses with the use of electronic computers (in Russian), *Vtoraya Nauchnaya Sessiya Povolzhskogo Soveta, Saratov* 1964.

KORSHUN, L. I., Minimum volume conditions for flexible elements of composite systems in the design using the mixed problem of structural mechanics (in Russian), **2**-*j Vses. S'ezd po Teor. i Prikl. Mekh., Moskva 1964*; *RŻM*, 1964, 9B332.

KORSHUNOV, A. I., Optimum height and design scheme of a statically determinate truss (in Russian), *Vtoraya Nauchnaya Sessiya Povolzhskogo Soveta, Saratov* 1964.

KRZYŚ, W., Optimum design of the box-section of a beam bent in elastic-plastic range, *Bull. Acad. Pol. Sci., Sér. Sci. Tech.*, **12**, 5, 261–271, 1964; *AMR*, 1965, **4**, 2018; *RŻM*, 1965, 7B190.

KRZYŚ, W., and ŻYCZKOWSKI, M., Optimum design of the box section of a beam in elastic range, *Bull. Acad. Pol. Sci., Sér. Sci. Tech.*, **12**, 5, 309–312, 1964; *RŻM*, 1965, 9B399.

KUPRYANOV, V. V., Shell of revolution roof (suspension structure) of minimum weight under axisymmetric load (in Russian), *Raschet Prostr. Konstr.*, Moskva 1964.

KUZNETSOV, E. N., Geometry of suspension roofs with radial cable nets and moment-free boundary rings (in Russian), *Teoria Obolochek i Plastin, AN SSSR, Erevan* 1964, 613–620.

LEŚNIAK, Z. K., Optimum rolled I-sections (in Polish), *X Konf. PZITB i PAN w Krynicy* 1964.

LEŚNIAK, Z. K., Application of the nonlinear programming to the optimum design of structures with the use of digital computers, *Doc. P1, Int. Inst. of Welding*, 1964.

LESSAER, S., Economical design of axisymmetric suspended surface structures (in Polish), *Zesz. Nauk. Pol. Śląskiej, Budown.*, **14**, 1964.

LIPKA, I., On the determination of the optimum bearing distance of cantilever shafts with two supports, *Acta Tech. Acad. Sci. Hung.*, **45**, 65–81, Budapest 1964; AMR, 1965. 1, 128.

MACIULEVIČIUS, D. A., Some properties of the configuration of elastic statically determinate structures of minimum weight (in Russian), *Stroitel'naya Mekhanika i Konstruktsii, Doklady XI Nauchno-Tekhnicheskoj Konferentsii, Vil'nyus*, Izd. MINTIS, 1964.

MACIULEVIČIUS, D. A., Linear programming algorithms for the synthesis of statically determinate bar structures of minimum weight (in Russian), *Stroit. Mekh. i Konstr., Vil'nyus*, 33–49, 1964; *RŻM*, 1965, 5B433.

MAKSIMOV, L. YU., Rational design of cylindrical structures loaded by high internal pressure (in Russian), *Vest. Mashinostr.*, **44**, 5, 9–13, 1964; *AMR*, 1965, **4**, 2186.

MARÇAL, P. V., and PRAGER, W., A method of optimal plastic design, *J. Méc.*, **3**, 4, 509–530, 1964; *AMR*, 1965, **9**, 5288; *RŻM*, 1965, 9B172.

MARKS, W., Design of prestressed beams subjected to moving loads (in Polish), *Konf. ZMOC PAN, Zakopane* 1964.

MEGAREFS, G. J., and HODGE, P. G., Singular case in the optimum design of frames, *Quart. Appl. Math.*, **21**, 2, 91–103, 1964.

MORTENSEN, M., Bestimmung des optimalen Querschnitts vorgespannter stählerner Vollwandräger, *Stahlbau*, **33**, 8, 249–254, 1964; *RŻM*, 1965, 4B455.

MOSES, F., Optimum structural design using linear programming, *Proc. ASCE*, **90**, *ST* 6, *Part I, J. Struct. Div.*, 89–104, 1964; *AMR*, 1965, 5, 2839; *RŻM*, 1965, 7B440.

MRÓZ, Z., Optimum design of reinforced shells of revolution, *Proc. IASS, Symp. Non-Class. Shell Problems, Warszawa* 1963, North Holland–PWN, 732–748, 1964.

PEREDY, J., Optimierungsprogramme für statisch unbestimmte stabwerke, *Wissenschaftliche Zeitschrift der Hochschule für Architektur und Bauwesen*, **11**, 1, Weimar 1964.

PEREDY, J., and VISY, Z., The economic design of members subjected to shearing (ultimate-load method), Part 1: *Concr. Constr. Engng.* **59**, 2, 57–66, 1964; Part 2: *Concr. Constr. Engng.*, **59**, 3, 97–106, 1964; *AMR*, 1964, **11**, 6448 and 6449; *RŻM*, 1965, 2B434.

PETROV, A. M., Optimum parameters of prestressed aluminium beams with allowance for elastico-plastic deformations (in Russian), *Tr. Toms. Inzh.-Str. Inst.*, **11**, 90–100, 1964; *RŻM*, 1965, 7B193.

PETROV, A. M., Optimum parameters of prestressed steel beams with allowance for plastic deformations at the prestressing stage (in Russian), *Izv. Vyssh. Uchebn. Zaved., Stroit. i Arkhit.*, **3**, 9–16, 1964.

POWELL, M. J. D., An efficient method for finding the minimum of a function of several variables without calculating derivatives, *Computer J.*, **7**, 155–162, 1964.

RADTSIG, YU. A., Application of discontinuous forms in the strength analysis of minimum weight structures (in Russian), *Vtoraya Nauchnaya Sessiya Vuzov Powolzh'ya, Saratov*, Povolzhskoe Knizhn. Izd., 1964.

REJTMAN, M. I., Optimum design of anisotropic rigid-plastic shells (in Russian), *Raschet Prostr. Konstr.*, Moskva, Gosstorojizdat, 1964.

RIGGS, B., Optimization of columns under axial instability, Ph.D. Thesis, Univ. Southl Calif., 1964.

ROZVANY, G. I. N., Optimum synthesis of prestressed structures, *Proc. ASCE*, **90**, *ST* 6, *J. Struct. Div.*, Part 1, 189–211, 1964; *AMR* 1966, **2**, 900; *RŻM*, 1964, 3B437.

SAWYER, D. A., and GRINTER, L. E., Minimum weight plastic design of continuous frames, *Proc. ASCE*, **90**, *EM* **3**, Part 1, *J. Engng. Mech. Div.*, 39–42, 1964; *RŻM*, 1965, 4B447.

SCHMIDT, R., Sandwich shells of arbitrary shape, *Trans. ASME*, **31**, *E, J. Appl. Mech.*, **2**, 239–244, 1964; *AMR*, 1964, **10**, 5648.

SCHUERCH, H. U., and BURGGRAF, O. R., Analytical design for optimum filamentary pressure vessels, *AIAA J.*, **2**, 5, 809–820, 1964; *RŻM*, 1965, 2B494; *AMR*, 1964, 10, 5831.

SWITZKY, H., Minimum weight design with structural stability, *AIAA 5th Annual Struct. and Mat. Conf.*, 316–322, 1964.

TOCHACEK, M., Der Entwurf optimaler vorgespannter Fachwerke, *Wiss. Z. Tech. Univ. Dresden*, **13**, 3, 788–796, 1964.

TSYTOVICH, O. A., Optimum parameters of prestressed metal beams (in Russian), *Stal'n. Predvar. Napryazh. i Tros. Konstr.*, Moskva, Strojizdat, 21–28, 1964; *RŻM*, 1965, 7B442.

VARVAK, P. M., Optimum contours of sloping moment-free shells (in Russian), *Rasch. Prostr. Konstr.*, **9**, 187–200, Moskva 1964; *AMR*, 1965, **12**, 7461.

VARVAK, P. M., and VARVAK, O. P., Moment-free uniform-strength shells of small rise with an oblique plan (in Ukrainian), *Dopovidi AN USSR*, **1**, 47–49, 1964; *RŻM*, 1964, 9B69.

VINOGRADOV, A. I., On the optimum distribution of forces in bar systems and the properties of optimum systems (in Russian), *2-j Vses. S'ezd po Teor. i Prikl. Mekh., M.*, 1964; *RŻM*, 1964, 9B312.

VOLKOVA, N. S., Synthesis of statically indeterminate trusses of minimum volume of the $(n+1)$st kind (in Russian), *Vtoraya Nauchnaya Sessija Povolzhskogo Soveta, Saratov* 1964.

WANG, DU-CHING, Optimum design of the cross-section of an initially curved bar subjected to bending, *Konf. Nauk. ZMOC PAN, Zakopane* 1964.

WOLFENSBERGER, R., *Traglast und optimale Bemessung von Platten*, Dokt. Dissert., ETH Zürich 1964; *RŻM*, 1965, 1B434 D.

1965

BORISOV, V. P., On the design of continuous beams and frames of minimum weight by the four-parameter method (in Russian), *Materialy Nauchno-Tekhnicheskoj Konferentsii Sektsii Stroitel'noj Mekhaniki Tatarskogo Pravleniya NTO Strojindustrii, Kazan'* 1965,

BOX, M. J., A new method of constrained optimization and a comparison with other methods, *Comp. J.*, **8**, 42–52, 1965.

BRACH, R. M., *Optimum design of beams for transient loading*, Doct. Diss., Univ. Wisconsin 1965; *RŻM*, 1967, 11B236.

BROWN, D. M., and ANG, A. H. S., *A nonlinear programming approach to the minimum weight elastic design of shell structures*, Struct. Res., Ser. No. 298, Univ. of Illinois, Urbana Ill, 1965.

BURNETT, E. F. P., YU, C. W., Reinforced concrete linear structures at ultimate load, *Proc. Int. Symp. on Flexural Mech. of Reinf. Concr., Miami* 1964, *ASCE* New York, 29–52, 1965.

CHELNOKOV, R. V., Design of beams of uniform longitudinal-transverse bending strength (in Russian), *Materialy Vtoroj Konferentsii Molodykh Uchenykh Kazani, Tatknigoizdat, Kazan'* 1965.

CHELNOKOV, R. V., Reduced-weight curved bar in compression and bending (in Russian), *Mater. Nauch.-Tekhn. Konf. Sektsii Stroit. Mekh. Tatarskogo Pravleniya NTO Strojindustrii, Kazan'* 1965.

CHELNOKOV, R. V., and KUTIKOV, A. I., The influence of the elasticity curve of a compression-bent beam of uniform strength on the beam's volume (in Russian), *Mater. Nauch.-Tekhn. Konf. Sektsii Stroit. Mekh. Tatarskogo Pravleniya NTO Strojindustrii, Kazan'* 1965.

CLOUD, R. L., Minimum weight design of a radial nozzle in a spherical shell, *Trans. ASME*, 32 *E. J. Appl. Mech.*, **2**, 448–449, 1965; *AMR*, 1966, **11**, 6890,

COHN, M. Z., Optimum limit design for reinforced concrete continuous beams, *Proc. Instn. Civ. Engrs.*, **30**, 675–707, 1965; *AMR*, 1965, **11**, 6801; *RŻM*, 1965, 8B562.

CORNELL, C. A., REINSCHMIDT, K. F., and BROTCHIE, J. F., Structural optimization, *Res. Rep.* **R65-26**, Part **2**, Dept. of Civ. Eng., Mass. Inst. of Tech., Sept. 1965.

Cox, H. L., *The design of structures for least weight*, Pergamon Press, New York 1965; *RŻM*, 1967, 4B352 K.

Crawford, R. F., and Schwartz, D. B., General instability and design of grid stiffened spherical domes, *AIAA, J.*, **3**, 511–515, 1965; *AMR*, 1965, **10**, 6023.

Doroshenko, O. P., On the design of optimum systems (in Russian), *Sb. Tr. Khar'k. Inzh. Zh. D. Transp.*, **74**, 1965.

Dzieniszewski, W., Optimum design of plates of variables thickness for minimum potential energy, *Bull. Acad. Pol. Sc., Sér. Sc. Techn.* **13**, 6, 45–52, 1965; *AMR*, 1967, **1**, 77.

Fox, R. L., Constraint surface normals for structural synthesis techniques based on matrix analysis, *AIAA J.*, **3**, 1517–1518, 1965.

Garrooq, C. A., Optimum solidity, *ASME* No. **AV-21**, New York 1965; *RŻM*, 1966, 11B618.

Gajnullina, S. Kh., Design of closed curved frames of uniform bending strength (in Russian), *Materialy Nauch.-Tekhn. Konf. Sektsii Stroit. Mekh. Tatarskogo Pravleniya NTO Strojindustrii, Kazan'* 1965.

Gawęcki, A., Rational design of post-tensioned prestressed beams (in Polish), *Arch. Inż. Ląd.*, **11**, 503–567, 1965; *AMR*, 1967, **5**, 3350; *RŻM*, 1966, 8B511.

Gellatly, R. A., and Gallagher, R. H., Development of advanced structural optimization programs and their application to large order systems, *1st Conf. Matrix Math. Analysis, Wright-Patterson AF Base, AFFDL-TR*, **66-80**, 231–251, 1965.

Gerard, G., Optimum structural design concepts for aerospace vehicles: bibliography and assessment, *TR AFFDL-TR.*, **65-9** (*All. Res. Ass. Inc., ARA Div. TR No. 272-2*), **62**, 1965; *AMR*, 1965, **10**, 6147.

Girfanov, I. S., On the design of composite systems of optimum volume (in Russian), *Mater. Nauch.-Tekhn. Konf. Sektsii Stroit. Mekh. Tatarskogo Pravleniya NTO Strojindustrii, Kazan'* 1965.

Girfanov, I. S., On the selection of optimum cross-sections for bars in central compression (in Russian), *Mater. Nauch.-Tekhn. Konf. Sektsii Stroit. Mekh. Tatarskogo Pravleniya NTO Strojindustrii, Kazan'* 1965.

Girfanov, I. S., On the design of rigid trusses and rigid composite systems of optimum volume (in Russian), *Mater. Vtoroj Konf. Molodykh Nauchnykh Rabotnikov G. Kazanii*, Tatknigoizdat, Kazan' 1965.

Goliński, J., and Leśniak, Z. K., Optimales Entwerfen von Konstruktionen mit Hilfe der Monte-Carlo Methode, *Bautechnik*, **9**, 307–311, 1965; *RŻM*, 1967, 3B365.

Holmes, M., and Steel, K. A., Design of continuous concrete slabs supported on steel beams, *Concr. Constr. Engng.*, **60**, 8, 303–312, 1965; *AMR*, 1966, **5**, 2963.

Ibragimov, M. R., Design of a circular plate of minimum weight under two independent load systems (in Russian), *Izv. AN Azerb. SSR, Ser. Fiz.-Tekhn. i Matem.*, **3/33-40**, 1965; *RŻM*, 1966, 4B226; *AMR*, 1966, **12**, 7689.

Icerman, L. J., Optimal structural design for given dynamic deflection, *Int. J. Solids and Struct.*, **5**, 5, 473–490, 1965.

Kaliszky, S., Economic design by the ultimate-load method, *Concr. Constr. Engng.*, **60**, 10/12, 1965; *RŻM*, 1965, 6B629.

Kaliszky, S., *Optimum limit design for reinforced concrete structures*, Res. Rep. Univ. of Southhampton, 1965.

Kapkowski, J., Discs of uniform strength under axisymmetric load (in Polish), *Arch. Bud. Masz.*, **12**, 2, 249–260, 1965; *AMR*, 1966, **6**, 3508.

Kloppel, K., and Molier, K. H., Ein Beitrag zur Erhöhung der Boulwerte Längsausgesteifter Rechteckplatten durch eine entsprechende Verteilung der Steifen, *Stahlbau*, **34**, 10, 303–311, 1965; *AMR*, 1967, **3**, 1639.

KOLUPAEV, A. N., Programming of the design of statically indeterminate trusses of minimum volume (in Russian), *Mater. Nauch.-Techn. Konf. Sektsii Stroit. Mekh. Tatarskogo Pravleniya NTO Strojindustrii, Kazan'* 1965.

KOLUPAEV, A. N., and RADTSIG, YU. A., Programming of the design of statically indeterminate trusses of minimum volume (in Russian), *Tezisy Dokladov IV Vses. Konf. po Primen. EhVM v Stroit. Mekh., Kiev* 1965.

KOMAROV, W. A., On the rational distribution of material in structures (in Russian), *Izv. AN SSR, Mekhanika*, 5, 85–87, 1965.

KOROLEV, V. I., Some problems concerning the choice of structure for glass-reinforced plastics (in Russian), *Inzh. Zhurnal*, **5**, 2, 306–315, 1965.

KORSHUNOV, A. I., On the choice of design scheme for trusses (in Russian), *Mater. Nauch.-Techn. Konf. Sektsii Stroit. Mekh. Tatarskogo Pravleniya NTO Strojindustrii, Kazan'* 1965.

KOY, B., *Shape optimization of single-column shell structures* (in Polish), Ph.D. Thesis, Politechnika Warszawska 1965.

LEŚNIAK, Z. K., Application of nonlinear programming to the optimum design of structures by using digital computers, *Building Sc.*, 105–108, Oxford 1965.

LEŚNIAK, Z. K., Optimum design of welded structures by nonlinear programming methods with the use of computers (in Polish), *XIII Nauk. Tech. Konf. Spaw., Warszawa 1965.*

MACIULEVIČIUS, D. A., On the synthesis of configuration of statically determinate bar structures of minimum weight (in Russian), *Izv. AN SSSR, Tekhn. Kiberb.*, **1**, 1965.

MARTIN, G., Optimierung kastenförmiger-Querschnitte bei excentrisch angreifenden Kräften, *Konstruktion*, **17**, 12, 486–488, 1965; *RŻM*, 1966, 7B470.

MRÓZ, Z., On the optimum design of reinforced slabs, *Acta Mech.*, **3**, 1, 34–35, 1965.

MURPHY, R. D., SABAT, D. J., and TAYLOR, R. J., Least cost ship characteristics by computer techniques, *Marine Technol.*, **2**, 2, 174–202, 1965.

NIORDSON, F. I., On the optimal design of a vibrating beam, *Quart. Appl. Math.*, **23**, **1**, 47–53, 1965; *AMR*, 1965, 9, 5343.

OWEN, J. B. B., *The analysis and design of light structures*, E. Arnold, London 1965.

PORTER GOFF, R. F. D., Decision theory and the shape of structures, *J. Roy. Aeronaut. Soc.*, **70**, 663, 448–452, 1966; *AMR*, 1966, **10**, 6364.

RADTSIG, YU. A., The current state of the minimum-weight design of statically indeterminate systems (in Russian), *Mater. Nauch.-Tekhn. Konf. Sektsii Stroit. Mekh. Tatarskogo NTO Strojindustrii, Kazan'* 1965.

RADTSIG, YU. A., Characteristic features of the modern methods of minimum weight design for statically indeterminate systems (in Russian), *Rezyume Dokladov Nauchnoj Konferentsii "Otdeleniya Mekhaniki Sploshnykh Sred", Augustuv*, Izd. Pol'skoj Akad. Nauk, 1965.

RAZANI, R., Behavior of fully stressed design of structures and its relationship to minimum-weight design, *AIAA J.*, **3**, 12, 2262–2268, 1965; *AMR*, 1966, 9, 5652.

ROCKEY, K. C., and COOK, I. T., Optimum reinforcement by two longitudinal stiffeners of a plate subjected to pure bending, *Int. J. Solids Struct.*, **1**, 79–92, 1965; *AMR*, 1965, **12**, 7338.

SAFIN, R. K., On the design of statically indeterminate steel trusses for minimum weight at the limit state (in Russian), *Trudy Nauchnoj Konferentsii LISI*, 1965.

SAFIN, R. K., On the limit design of statically indeterminate trusses (in Russian), *Mater. Nauch.-Tekhn. Konf. Sektsii Stroit. Mekh. Tatarskogo Pravleniya NTO Strojindustrii, Kazan'* 1965.

SAFIN, R. K., Minimum-weight limit design of statically indeterminate steel trusses (in Russian), *Mater. Vtoroj Konf. Molod. Uchenykh Kazanii*, Tatknigoizdat, Kazan' 1965.

SAWYER, H. A., JR., Status and potentialities of nonlinear design of concrete frames, *Proc. Int. Symp. Flexural Mech. Reinf. Concr., Miami, Florida, Nov.* 1964, *ASCE* New York 1965.

SCHMIT, L. A., JR., and FOX, R. L., An intergated approach to structural synthesis and analysis, *AIAA J.*, **3**, 6, 1104–1112, 1965; *AMR*, 1966, **4**, 2279.

RUDNAI, G., and MICHELBERGER, P., Some principal aspects of lightweight design, *2nd Conf. on Dimensioning and Strength Calc., Budapest 1965*; *RŻM*, 1966, 6B234.

SEMENETS, G. L., Solution of least-volume and equilibrium equations by the method of successive approximations (in Russian), *Tr. Khar'k. Inst. Inzh. Zh. D. Transp.*, **74**, 1965.

SHIRKO, I. V., On the shape of a uniform-strength plate (in Russian), *AN SSSR, Inzh. Zhurn.*, **5**, 2, 293–298, 1965; *AMR*, 1967, **10**, 7703.

STERN, M., WANG, H. C., and WORLEY, W. J., A method for determining an optimum shape of a class of thin shells of revolution, *Dept. Theor. Appl. Mech., Univ. III, Urbana, Rep.* **281**, 50, 1965; *AMR*, 1966, **5**, 2832.

TANABASHI, R., and NAKAMURA, T., The minimum weight design of a class of tall multistorey frames subjected to large lateral forces, Part 1, *Trans. Arch. Inst. of Japan*, **118**, 10–16, 1965; *AMR*, 1969, **5**, 3394.

TUL'CHIJ, V. I., On the optimum stiffening of holes in plates (in Russian), *Prikl. Mekh.*, **1**, 3, 77–83, 1965; *AMR*, 1966, **3**, 1390.

VINOGRADOV, A. I., Investigation of the cost function in the optimum structural design (in Russian), *Issled. po Teorii Sooruzh.*, **14**, 143–154, 1965.

VINOGRADOV, A. I., On the optimum distribution of forces in bar systems and the properties of optimum systems (in Russian), *Prikl. Mekh.*, **1**, 1, 81–91, 1965; *RŻM*, 1965, 7B439.

VOLKOVA, N. S., Design of statically indeterminate trusses of minimum volume of the 2nd kind under transient loading (in Russian), *Mater. Vtoroj Konf. Molod. Uchenykh Kazanii*, Tatknigoizdat, Kazan' 1965.

VOLKOVA, N. S., *Synthesis of statically indeterminate trusses of minimum constant volume* (in Russian, Thesis), Avtoreferat Kandidatskoj Dissertatsii, UPI, 1965.

VOLKOVA, N. S., The practical part of the design of statically indeterminate trusses of minimum constant volume (in Russian), *Mater. Nauch.-Tekhn. Konf. Sektsii Stroit. Mekh. Tatarskogo Pravleniya NTO Strojindustrii, Kazan' 1965.*

VOLKOVA, N. S., The variety of indeterminate trusses of minimum volume of the second kind (in Russian), *Trudy Kazanskogo Sel'skokhozyajstvennogo Instituta*, **47**, 1965.

1966

ABRAMOV, N. I., Application of computers to the optimum design of reinforced concrete frame structures (in Russian), *Sbor. Tr. Len. Inst.*, **49**, 1966.

ANDERHEGGEN, E., and THURLIMANN, B., Optimum design using linear programming, *Publication IABSE*, **26**, 555–571, 1966.

BARNETT, R. L., Survey of optimum structural design, *Experimental Mech.*, **6**, 12, 19A-26A, Dec. 1966; *AMR*, 1967, **5**, 3361.

BIGELOW, R. H., and E. H. GAYLORD, Minimum weight of plastically designed steel frames, *Univ. Illinois Eng. Exp. Station, Bull.*, **485**, 1966; *AMR*, 1967, **4**, 2507.

BORISOV, V. R., A solution algorithm for the inverse problem of the theory of structures (in Russian), *Stroitel'naya Mekhanika i Raschet Sooruzhenij*, **3**, 1966.

BORISOV, V. R., On the limits of variation of redundant unknowns in statically indeterminate fully stressed trusses (in Russian), *Stroitel'nye konstruktsii* (*issledovaniya i metody rascheta*), Tatknigoizdat. Kazan' 1966.

BOX, M. J., A comparison of several current optimization methods and the use of transformations in constrained problems, *Computer J.*, **9**, 67–68, 1966.

BROWN, D. M., and ANG, A. H. S., Structural optimization by nonlinear programming, *Proc. ASCE*, **92**, *ST* 6, *J. Struct. Div.*, 319–340, 1966; *AMR*, 1967, **10**, 7578.

BURNS, A. B., Minimum weight of hydrostatically compressed ring-stiffened cones, *J. Spacecrafts and Rockets*, **3**, 3, 387–392, 1966.

BURNS, A. B., and ALMROTH, B. O., Structural optimization of axially compressed ring-stringer stiffened cylinders, *J. of Spacecraft*, **3**, 1, 19–25, 1966.

CELLA, A., and LOGCHER, R. D., Automated optimum design from discrete components, *Proc. ASCE* **92**, *ST*6, *J. Struct. Div.*, 1966.

CHELNOKOV, R. V., Design of arches of minimum weight according to the funicular curve with elastic fastening considerations (in Russian), *Trudy KAI*, **91**, 101–112, 1966.

CHELNOKOV, R. V., Reduced-weight curvilinear rod in longitudinal-transverse bending (in Russian), *Trudy KAI*, **91**, 113–125, 1966.

CHELNOKOV, R. V., and KUTIKOV, A. I., On the effect of tangent stresses in uniform-strength beams under longitudinal-transverse bending (in Russian), *Mater. Nauch.-Tekh. Konf. po Stroit. Mekh. i Stroit. Konstr.*, *KISI*, 1966.

COHN, M. Z., Limit design of continuous reinforced concrete crane girders, *Proc. ASCE*, **92**, *ST* 3, *J. Struct. Div.* 161–177, 1966; *AMR*, 1967, **2**, 853.

CORNELL, C. A., REINSCHMIDT, K. F., and BROTCHIE, J. F., A method for the optimum design of structures, *Proc. Int. Symp. on Use of Computers in Struct. Engng.*, Newcastle upon-Tyne, 1966.

DÖKMECI, M. C., A shell of constant strength, *Z. Angew. Math. Phys.*, **17**, 545–547, 1966.

FROYTON, G., Design aids-minimum depth and allowable bending stress, *Engng. J. Amer, Inst. Steel Constr.*, **3**, 2, 86–87, 1966; *RŻM*, 1969, 5B768.

GAJNULLINA, S. KH., Application of variational methods to the design of minimum-weight systems (in Russian), *Trudy KAI*, **91**, 69–77, 1966.

GAJEWSKI, A., Calculation of elastic stability of circular plates with variable thickness by an inverse method, *Bull. Acad. Pol. Sci. Sér. Sci. Techn.*, **14**, 5, 303–312. 1966.

GELLATLY R. A., Development of procedures for large scale automated minimum weight structural design, *AFFDL-TR*, **66-180**, *December* 1966.

GELLATLY, R. A., and GALLAGHER, R. H., A procedure for automated minimum weight structural design; Part I: Theoretical basis. Part II: Applications, *Aeronaut. Quart.*, **17**, 216–230 and 332–342, 1966; *AMR*, 1967, **8**, 6517; *RŻM*, 1967, 3B364, 6B540.

GERARD, G., Optimum structural design concepts for aerospace vehicles, *J. Spacecraft*, **3**, 1, 5–18, 1966.

GERARD, G., and LAKSHMIKANTHAM, C., Optimum thinwall pressure vessels of anisotropic materials, *J. Appl. Mech.*, **33**, 623–628, 1966; *AMR*, 1967, **3**, 1787.

GHISTA, D. N., Fully-stressed design for alternative loads, *Proc. ASCE*, **92**, *ST* 5, *J. Struct. Div.*, 237–260, 1966; *AMR*, 1967, 4, 2495; *RŻM*, 1967, 8B511.

GHISTA, D. N., Optimum frameworks under single load system, *Proc. ASCE*, **92**, *ST* 5, *J. Struct. Div.*, 261–286, 1966; *AMR*, 1967, 4, 2505; *RŻM*, 1967, 12B675.

GHISTA, D. N., Structural optimization with probability of failure constraint, *NASA TN D*-3777, **15**, 1966; *AMR*, 1967, **5**, 3344.

GHISTA, D. N., On the optimum stress level method for alternative load systems, *J. Sc. Eng. Res.*, *India*, **10**, 1, 1–18, 1966; *AMR*, 1967, **7**, 4814; *RŻM*, 1967, 7B462.

GHISTA, D. N., Weight optimization by structural synthesis, *J. Sc. Eng. Res. India*, **10**, Part 2, 199–220, 1966; *AMR*, 1068, 2, 1966; *RŻM*, 1968, 6B719.

GIRFANOV, I. S., The mixed problem of structural mechanics and the design of combined lattice structures of optimum volume (in Russian), *Trudy KAI*, **91**, 78–81, 1966.

GIRFANOV, I. S., Design of statically indeterminate trusses for given stress with allowance for plastic deformations (in Russian), *Mater. Nauch.-Tekhn. Konf. po Stroit. Mekh. i Stroit. Konstr., Kazan'* 1966.

GIRFANOV, I. S., Experimental-theoretical investigation of statically indeterminate welded trusses of optimum volume (in Russian), *Inzhenernye konstruktsii, Doklady k XXIV Nauchnoj Konf. LISI*, Leningrad 1966.

GIRFANOV, I. S., Theoretical investigation of rigid bar systems of optimum volume (in Russian), *Stroitel'nye Konstruktsii* (*issledovaniya i metody rascheta*), Tatknigoizdat, Kazan' 1966.

GIRFANOV, I. S., Design of multiply statically indeterminate trusses of optimum volume (in Russian), *Mater. Nauch.-Tekhn. Konf. po Stroit. Mekh. i Stroit. Konstr.*, Kazan' 1966.

GOBLE, G. G., and DE SANTIS, P. V., Optimum design of mixed steel composite girders, *Proc. ASCE*, **92**, *ST* 6, *J. Struct. Div.*, 25–43, 1966; *AMR*, 1968, **4**, 2640; *RŻM*, 1967 10B536.

GOFF, R. F. D., Decision theory and the shape of structures, *J. Roy. Aeronaut. Soc.*, **70**, 3, 1966.

GREEN, A. P., and LANCASTER, P. R., Design of a composite drawing die with a brittle insert, *Int. J. Mech. Sc.*, **8**, 281–294, 1966.

GUZMAN-BARRON, T., BROTCHIE, J. F., and CORNELL, C. A., A program for the optimum design of prestressed concrete highway bridges, *J. Prestr. Concr. Inst. Chicago* III, 166.

HAUG, E. J., JR., *Minimum weight design of beams with inequality constraints on stress and deflection*, Doct. Diss., Kanzas State Univ., 1966; *RŻM*, 1968, 6B724D.

HEMP, W. S., Studies in theory of Michell structures, *Proc. XI Int. Congr. Appl. Mech., München* 1964, Springer Verlag, 621–682, 1966; *AMR*, 1967, 7, 4826.

HEMP, W. S. and CHAN, H. S. Y., Optimum design of pin-jointed frame-works, *Aeronaut. Res. Council, Rep. and Mem. No.* 3632, London 1966.

HILL L. A., JR., Automated optimum cost building design, *Proc. ASCE*, **92**, *ST* 6, *J. Struct. Div.*, 247–263, 1966.

IZRAELIT, A. V., Regular design of continuous beams (in Russian), *Trudy Vses. Lesotekh. Inst.*, **2**, 1966.

KARPENKO, N. I., and REJTMAN, M. I., The lower limit of the load-carrying capacity and the optimum design of reinforced concrete plates (in Russian), *Trudy VI Vses. Konf. po Teorii Obol. i Plast.*, Nauka, Moskva 1966.

KAVLIE, D., KOWALIK, J., IUND, S., and MOE, J., Design optimization using general nonlinear programming method, *European Shipbuilding*, **4**, 1966; *RŻM*, 1968, 1B624,

KAVLIE, D., KOWALIK, J., and MOE, J., Structural optimization by means of nonlinear programming, *Meddelelse SKB II/M4*, Trondheim 1966.

KELLER, J. B., and NIORDSON, F. I., The tallest column, *J. Math. Mech.*, **16**, 433–446, 1966; *AMR*, 1967, 6, 3975.

KHACHATURIAN, N., and HAIDER, G. S., Probabilistic design of determinate structures, *Proc. Spec. Conf. ASCE*, 623–641, New York 1966.

KHACHATURIAN, N., and PREISS, K., Least area design of noncomposite prestressed concrete beams, *Civ. Engr. in South Africa*, **8**, 11, 321–329, 1966; *AMR*, 1968, **1**, 296; *RŻM.* 1967, 9B661.

KHASANSHIN, N. KH., Some approximate methods for the design of statically indeterminate systems of minimum weight (in Russian), *Trudy KAI*, **91**, 126–138, 1966.

KICHER, T. P., Optimum design—minimum weight versus fully stressed, *Proc. ASCE* **92**, *ST* 6, *J. Struct. Div.*, 265–279, 1966; *RŻM*, 1967, 7B512.

KORSHUNOV, A. I., Optimum height of statically determinate trusses (in Russian), *Trudy KAI*, **91**, 82–90, 1966.

KOWALIK, J., Nonlinear programming procedures and design optimization, *ACTA Polytech. Scand., Math. Comp. Mech., Ser.* **13**, Trondheim 1966.

KRZYŚ, W., Optimum design of the thin-walled box profile of a bar bent in the elastic-plastic range (in Polish), *Zesz. Nauk. P. Krak. Księga ku czci 600-lecia UJ*, 39–50, 1966.

LACKMAN, L. M., and AULT, R. M., Influence of plasticity correction factor in minimum weight analysis, *AIAA J.*, **4**, 714–715, 1966; *AMR*, 1966, 11, 6914.

LEE, H., *Optimum modeling of structure to predict dynamic elastic response*, Doct. Diss., Pa State Univ., 1966; *RŻM*, 1968, 5B290D.

LEON, A., *A classified bibliography on optimization, Recent Advances in Optimization Technique*, John Wiley and Sons, New York 1966.

LEŚNIAK, Z. K., Methods of optimum structural design using computers (in Polish), *III Konf. Nauk.-Techn. Konstrukcje Metalowe*, Vol. **2**, 213–215, Warszawa 1966.

LEŚNIAK, Z. K., Optimum design of welded structures by nonlinear programming methods with the use of computers (in Polish), *Prz. Spaw.*, **1**, 29–32, 1966.

LEŚNIAK, Z. K., and BAŁUCH, H., Optimum sizing of ordinary frogs (in Polish), *Arch. Inż. Ląd.*, **3**, 281–292, Warszawa 1966.

LEŚNIAK, Z. K., KWAŚNIEWSKI, M., and PARNIAK, H., Optimization of steel structures of belt conveyors (in Polish), *XII Konf. PZITB i PAN w Krynicy*, 201–207, 1966.

MANDEL, P., and LEOPOLD, R., Optimization methods applied to ship design, *Trans. SŃAME*, **74**, 477–521, 1966.

MARKS, W., Cable layout in prestressed beams (in Polish), *Konf. ZMOC PAN w Kołobrzegu* 1966.

MARKS, W., A certain case of beam design under moving load (in Polish), *Rozpr. Inż.*, **14**, 1, 49–68, 1966; *AMR*, 1967, **7**, 4823.

MAZURKIEWICZ, S., and ŻYCZKOWSKI, M., Optimum design of cross-section of thin-walled bar under combined torsion and bending, *Bull. Acad. Pol. Sc., Sér. Sc. Tech.*, **14**, 4, 273–281, 1966; *AMR*, 1967, **2**, 854; *RŻM*, 1967, 3B338.

MAZURKIEWICZ, S., and ŻYCZKOWSKI, M., Optimum design of the cross-section of a thin-walled bar under combined torsion and bending (in Polish), *Rozpr. Inż.*, **14**, 2, 199–213, 1966; *AMR*, 1967, **11**, 8596.

MEGAREFS, G. J., Method for minimal design of axisymmetric plates, *Proc. ASCE*, 92, *EM* 6, *J. Engng. Mech. Div.*; 79–99, 1966.

MIKELADZE, M. S., Some problems of the theory of spherical plastic shells, *Rev. Roum. Sc. Tech., Sér. Méc. Appl.*, **11**, 6, 1283–1294, 1966; *AMR*, 1968, **3**, 1675.

MORLEY, C. T., The minimum reinforcement of concrete slabs, *Int. J. Mech. Sc.*, **8**, 4, 305–319, 1966; *AMR*, 1966, 12, 7849.

MUKHADZE, L. G., Approximate design of two-surface prestressed cable systems (in Russian), *Voprosy Rascheta Slozhn. Staticheski Neopred. Sistem, Metsnereba, Tbilisi* 1966, 5–14.

MUKHADZE, L. G., TSKUASELI, N. P., and SULABERIDZE, O. G., Some optimum forms of hanging roofs (in Russian), *Voprosy Rascheta Slozh. Staticheski Neopred. Sistem, Metsnereba, Tbilisi* 1966, 15–21.

NESMEYANOV, A. S., Design of frames of minimum weight under dynamic loads (in Russian), *Mater. Nauch.-Tekhn. Konf. po Stroit. Mekhan. i Stroit. Konstr., Kazan'* 1966.

PEREDY, J., Optimum design of statically indeterminate structures, *Az opitoipari ez kozlekedesi muszaki egyetem tudomanycz kozlemenyeu*, **13**, 3, 99–123, Budapest 1966.

RABINOVICH, I. M., Bar systems of minimum weight (in Russian), *Mekhanika Tverdogo Tela*, Izd. Nauka, 265–275, Moskva 1966.

RADTSIG, YU. A., M. Levy's trusses (in Russian), *Mater. Nauch.-Tekhn. Konf. po Stroit. Mekh. i Stroit. Konstr., KISI*, 1966.

RADTSIG, YU. A., Invariant properties of statically indeterminate trusses of minimum volume (in Russian), *Mater. Nauch.-Tekhn. Konf. po Stroit. Mekh. i Stroit. Konstr., Kazan'* 1966.

RADTSIG, YU. A., Optimum design of statically indeterminate trusses in the elastic and plastic ranges (in Russian), *Tezisy Dokladov na IV Vsesoyuznoj Konferentsii po Prochnosti i Plastichnosti v Moskve*, M., Nauka, 1966.

RADTSIG, YU. A., Rational strength design of statically indeterminate trusses of minimum volume with the use of electronic computers (in Russian), *Trudy 1-go Vsesoyuznogo Soveshchaniya po Primeneniyu EhTsVM v Stroitel'noj Mekhanike, M.-D.*, Gosstrojizdat, 1966.

RADTSIG, YU. A., Mirror equations with one unknown, applicable to the problems of structural mechanics (in Russian), *Trudy KAI*, **91**, 91–100, 1966.

RADTSIG, YU. A., ARSLANOV, A. S., GIRFANOV, I. S., and SAFIN, R. K., Experimental investigation of the work of statically indeterminate steel trusses of optimum volume loaded to failure (in Russian), *Stroitel'nye Konstruktsii (issledovaniya i metody rascheta)*, Tatknigoizdat, Kazan' 1966.

RAZANI, R., and GOBLE, G. G., Optimum design of constant depth plate girders, *Proc. ASCE* **92**, *ST* 2, *J. Struct. Div.* 253–281, 1966; *RŻM*, 1966, 12B448.

REINSCHMIDT, K. F., CORNELL, C. A., and BROTCHIE, J. F., Iterative design and structural optimization, *Proc. ASCE*, **92**, *ST* 6, *J. Struct. Div.* 281–318, 1966; *RŻM*, 1967, 7B510.

REJTMAN, M. I., and SHAPIRO, G. S., Optimum design theory in structural mechanics and elasticity and plasticity theories (in Russian). *Mekhanika, Uprug. i Plast.*, 83–113, Moskva 1966.

RICHARDS, D. M., and CHAN, H. S. Y., Developments in the theory of Michell optimum structures, *AGARD Report* No. **543**, 1966.

ROZVANY, G. I. N., Minimum volume of uncurtailed orthogonal reinforcement in freely-supported slabs, *Concr. Constr. Engng.*, **61**, 8, 281–286, 1966; *AMR*, 1967, **6**, 4100.

ROZENKRANZ, B., Weight optimization of a lattice girder (in Czech), *Inż. Stavby*, **14**, 12, 527–533, 1966; *RŻM*, 1968, 3B643.

ROZVANY, G. I. N., Analysis versus synthesis in structural engineering, *Inst. Engrs., Australia, Civ. Eng. Trans.*, *CE* **8**, 2, 158–166, 1966; *AMR*, 1967, **8**, 6497.

ROZVANY, G. I. N., Rational approach to plate design, *J. Amer. Concr. Inst.*, **63**, 10, 1077–1094, 1966; *AMR*, 1967, **7**, 4816.

RUBINSTEIN, M. F., and KARAGOZIAN, J. J., Building design using linear programming, *Proc.* **92**, *ST* 6, *J. Struct. Div.* 223–245, 1966.

RYABCHENKO, V. M., Determination of the framework scheme in the synthesis of optimum thin-walled systems (in Russian), *Samoletostr. Tekhn. Vozd. Flota*, **6**, 50–55, 1966; *RŻM*, 1967, 6B531.

SAFIN, R. K., Design of statically indeterminate trusses allowing for plastic deformations (in Russian), *Stroitel'nye Konstruktsii (issledovaniya i metody rascheta)*, Tatknigoizdat, Kazan' 1966.

SAFIN, R. K., On the design of statically indeterminate trusses (in Russian), *Mater. Nauch.-Tekhn. Konf. po Stroit. Mekh. i Stroit. Konstr., Kazan'* 1966.

SIRAZUTDINOV, YU. K., On the design of structures of minimum volume allowing for dead weight (in Russian), *Mater. Nauch.-Tekhn. Konf. po Stroit. Mekh. i Stroit. Konstr.*, KISI, 1966.

SAVE, M. A., and SHIELD, R. T., Minimum-weight design of sandwich shells subjected to fixed and moving loads, *Proc. XI int. Congr. Appl. Mech. München, 1964*, Springer Verlag, 341–349, 1966; *AMR*, 1967, **5**, 3210.

SCORDELIS, A. C., BARON, F., and LIN, T. Y., Optimum design of two large span shells of post-tensioned precast concrete, *Proc. IASS Congress on Large Span Shells*, vol. **1**, Leningrad 1966.

SIRAZUTDINOV, YU. K., On optimum I-sections (in Russian), *Mater. Nauch.–Tekhn. Konf. po Stroit. Mekh. i Stroit. Konstr.*, KISI, 1966.

SOFRONOV, YU. D., On the design of a uniformly stressed and uniformly strong beam in longitudinal vibrations (in Russian), *IVUZ, Mashinostroenie*, **3**, 1966.

SOFRONOV, YU. D., Design of beams of least weight allowing for the scale factor (in Russian), *Mater. Nauch.-Tekhn. Konf. po Stroit. Mekh. i Stroit. Konstr., Kazan'* 1966.

SOFRONOV, YU. D., Design of beams of minimum weight under axisymmetric repeated loads (in Russian), *Mater. Nauchno-Tekhn. Konf. po Stroit. Mekh. i Stroit. Konstr.*, KISI, 1966.

SOFRONOV, YU. D., On the propagation of fatigue cracks in statically indeterminate bar structures (in Russian), *Trudy KAI*, **91**, 1966.

SOFRONOV, YU. D., Castigliano's theorem and certain special features of its applications (in Russian), *Trudy KAI*, 1966.

TEMPLEMAN, A. B., A note on light structures of the Maxwell/Michell type, *Inst. Civ. Engrs. Proc.*, **35**, 111–120, 1966; *AMR*, 1967, **10**, 7867.

TODOROV, T., and KOLEV, D., An optimum problem for an infinite beam on an elastic foundation (in Bulgarian), *Stroitel'stvo*, **13**, 8, 11–12, 1966.

TURNER, M. J., Design of minimum mass structures with specified natural frequencies, *AIAA J.*, **5**, 3, 406–412, 1966.

VOLKOVA, N. S., On the design of light non-uniformly stressed statically indeterminate trusses of constant volume (in Russian), *Trudy KAI*, **91**, 63–68, 1966.

WASIUTYŃSKI, Z., On the criterion of minimum deformability design of elastic structures. Effect of own weight of the material, *Bull. Acad. Sc., Sér. Sc., Tech.*, **14**, 9, 529–532, 1966; *AMR*, 1967, **8**, 6509; *PBAM*, 1967, **4**, 7126.

WASIUTYŃSKI, Z., On the criterion of optimum design of elastic structures subjected to ***n*** various systems of solicitations, *Bull. Acad. Pol. Sc., Sér. Sc. Tech.*, **14**, 9, 533–535, 1966; *AMR*, 1967, **8**, 6509; *PBAM*, 1967, **4**, 7127.

WASIUTYŃSKI, Z., On the equivalence of design principles: Minimum potential-constant volume and minimum volume-constant potential, *Bull. Acad. Pol. Sc., Sér. Sc. Tech.*, **14**, **9**, 537–539, 1966,; *AMR*, 1967, **10**, 7868; *PBAM*, 1967, **4**, 7128; *RŻM*, 1967, 6B575.

WASIUTYŃSKI, Z., and BRANDT, A., The present state of knowledge in the field of optimum design of structures, *Appl. Mech. Surveys*, 435–450, Spartan Books, 1966.

1967

AKHMADALIEV, M., Computer calculation of an optimum design of a statically determinate truss by a nonlinear programming method (in Russian), *Izv. AN UZSSR, Seriya Tekhn. Nauk*, **1**, 1967.

ARASLANOV, A. M., Minimum-weight beams under random loads (in Russian), *Materialy tret'ej konferentsii molodykh nauchnykh rabotnikov Kazani, Sektsiya mekhaniko-matematicheskaya i fiziko-tekhnicheskaya*, Tatknigoizdat, Kazan' 1967; *AMR*, 1969, **3**, 1757.

ARASLANOV, A. M., On the design of beams and frames of minimum weight under random loading with account taken of the accumulation of fatigue damage (in Russian), *Ma-*

terialy k yubilejnoj respublikanskoj konferentsii sektsii stroitel'noj mekhaniki Tatarskogo pravleniya NTO Strojindustrii (Tezisy dokladov), Kazan' 1967.

ARASLANOV, A. M., Rational structures under random loads (in Russian), *Materialy k yubilejnoj respublikanskoj konferentsii sektsii stroitel'noj mekhaniki Tatarskogo pravleniya NTO Strojindustrii* (Tezisy dokladov), Kazan' 1967.

ARSLANOV, A. S., Design of statically indeterminate trusses of minimum volume with stability constraints for compression members (in Russian), *Tezisy Vsesoyuznoj Konferentsii po Problemam Ustojchivosti v Stroitel'noj Mekhanike*, Kaunas 1967.

ARSLANOV, A. S., Application of the theory of discontinuous functions to the design of statically indeterminate trusses of minimum volume subject to member stability constraints (in Russian), *Materialy k jubilejnoj respublikanskoj konferentsii sektsii stroitel'noj mekhaniki Tatarskogo pravleniya NTO Strojindustrii*, Kazan' 1967.

ARSLANOV, A. S., Application of the theory of discontinuous out-functions to the design of statically indeterminate trusses of minimum volume subject to member stability constraints under multiple loading (in Russian), *Materialy k jubilejnoj respublikanskoj konferentsii sektsii stroitel'noj mekhaniki Tatarskogo pravleniya NTO Strojindustrii*, Kazan' 1967.

ARSLANOV, A. S., and GIRFANOV, I. S., On the selection of cross-sections for compression members of metal trusses (in Russian), *Trudy Kazanskogo Inzh.-Stroit. Instituta*, Vyp. **10**, 1967.

BASU, A. K., and CHAPMAN, J. C., Optimum design of plates with symmetrical trapezoidal corrugations subjected to lateral pressure, *Quart. Trans. Roy. Inst. Nav. Arch.*, **109**, 2, 209–221, April 1967; *AMR*, 1968, **3**, 1676.

BIGELOW, R. H., and GAYLORD, E. H., Design of steel frames for minimum weight, *Proc. ASCE*, **93**, *ST* 6, *J. Struct. Div.* 106–131, 1967.

BRAMSKI, C., Some problems of the design of steel tear-shaped vessels (in Polish), *XIII Konf. PZITB i PAN w Krynicy*, 349–357, 1967.

BRANDT, A., A theorem on the dependence of the prestressing cable path in a concrete beam on the dead weight of the beam (in Polish), *Konwersatorium PTMTS "Zagadnienia optymalizacji w mechanice"*, 7–14, Gliwice–Szczyrk 1967.

BROTCHIE, J. F., *Discussion Proc. ASCE, EM* **5**, *J. Engng. Mech. Div.*, 173–175, 1967.

CHAN, H. S. Y., Half-plane slip-line fields and Michell structures, *Quart. J. Mech. Appl. Math.*, **20**, **4**, 453–469, 1967; *AMR*, 1969, **5**, 3393.

COHN, M. Z., Limit-design solutions for concrete structures, *Proc. ASCE*, **93**, *ST* 1, *J. Struct. Div.*, 37–57, 1967.

CORNELL, C. A., HO, P. K., and EHRLICH, R. A., A method for the optimum design of simple span deck and stringer highway bridges, *Res. Rep. R67–22, Dept. Civ. Engng. MIT*, 1967.

DAYARATNEM, P., Minimum weight design of prestressed concrete beam, *Cem. and Concr.*, **8**, 1, 39–50, 1967; *RŻM*, 1968, 6B909.

DIXON, L. C. W., Pontryagin's maximum principle applied to the profile of a beam, *J. Roy. Aeronaut. Soc.*, **71**, 679, 513–515, July 1967.

FELTON, L. P., and DOBBS, M. W., Optimum design of tubes for bending and torsion, *Proc. ASCE*, **93**, *ST* 4, *J. Struct. Div.*, 185–200, 1967; *AMR*, 1968, **5**, 3245.

FEDOROV, I. A., On a minimum-weight truss (in Russian), *Stroit. Mekh. i Raschet Sooruzh.*, **6**, 11–14, 1967.

GIRFANOV, I. S., and KUTIKOV, A. I., A problem of piecewise-linear programming and its solution on an electronic computer (in Russian), *Trudy Kazanskogo Inzhenerno-Stroitel'nogo Instituta*, **10**, 1967.

GIRFANOV, I. S., On comparison standards for statically indeterminate trusses of optimum volume under some load combinations (in Russian), *Materialy k yubilejnoj respublikanskoj konferentsii sektsii stroitel'noj mekhaniki Tatarskogo Pravleniya NTO Strojindustrii, Kazan'* 1967.

GIRFANOV, I. S., On the design of statically indeterminate trusses with given member stresses and maximum stiffness (in Russian), *Materialy k yubilejnoj respublikanskoj konferentsii sektsii stroitel'noj mekhaniki Tatarskogo Pravleniya NTO Strojindustrii, Kazan'* 1967.

GIRFANOV, I. S., and SAFIN, R. K., Allowance for plastic deformations in the design of statically indeterminate trusses of optimum volume with given member stresses (in Russian), *Materialy k yubilejnoj respublikanskoj konferentsii sektsii stroitel'noj mekhaniki Tatarskogo Pravleniya NTO Strojindustrii, Kazan'* 1967.

HAUG, E. J., JR., and KIRMSER, P. G., Minimum-weight design of beams with inequality constraints on stress and deflection, *Trans. ASME, Ser. E*, **34**, 4, *J. Appl. Mech.*, 999–1004, 1967; *AMR*, 1968, **7**, 4981.

JAWORSKI, J., A successive approximation method for determining cross-sectional shape of prestressed beams of uniform strength according to Caquot's hypothesis (in Polish), *Konwersatorium PTMTS "Zagadnienia optymalizacji w mechanice"*, 25–36, Gliwice–Szczyrk 1967.

JENDO, S., On the design of cable lattices (in Polish), *Konwersatorium PTMTS "Zagadnienia optymalizacji w mechanice"*, 37–39, Gliwice–Szczyrk 1967.

KALISZKY, S., *Plastic design of reinforced concrete slabs* (in Hungarian), Müszaki Könyvkiadó, Budapest 1967.

KARIHALOO, B. L., PATHARE, P. R., and RAMESH, C. K., *The optimum design of space structures by linear programming using stiffness matrix method of analysis, Space Structures*, Oxford–Edinburgh 1967.

KOLUPAEV, A. N., Transformation of trusses by means of shifting their chords (in Russian), *Materialy k yubilejnoj respublikanskoj konferentsii sektsii stroitel'noj mekhaniki Tatarskogo pravleniya NTO Strojindustrii, Kazan'* 1967.

KOLUPAEV, A. N., Design of trusses for given stiffness with discretely varying cross-sectional areas of the members (in Russian), *Materialy k yubilejnoj respublikanskoj konferentsii sektsii stroitel'noj mekhaniki Tatarskogo pravleniya NTO Strojindustrii, Kazan'* 1967.

KOWALSKI, A., Stability of bars with discretely varying cross-section compressed by a follow-up force (in Polish), *Rozpr. Inż.*, **15**, 2, 197–209, 1967.

KRAVITZ, S., Optimum location for beam supports, *Design News*, **22**, 19, 100–103, 1967; *RŻM*, 1968, 6B726.

KRZYŚ, W., The optimum design of columns subject to stability constraints for the wall (in Polish), *Konwersatorium PTMTS "Zagadnienia optymalizacji w mechanice"*, 47–48, Gliwice–Szczyrk 1967.

KRZYŚ, W., Optimum design with regard to stability of compressed thin-walled columns with closed profiles (in Polish), *Zesz. Nauk. P. Krak., Mech.*, **4**, 24, 1967.

LESSAER, S., Economic design of axisymmetric lattice structures (in Polish), *Konwersatorium PTMTS "Zagadnienia optymalizacji w mechanice"*, 63–66, Gliwice–Szczyrk 1967.

LEŚNIAK, Z. K., Optimum wide-flanged welded I-beams (in Polish), *XIII Konf. PZITB i PAN w Krynicy*, 323–339, 1967.

LEŚNIAK, Z. K., Optimale Konstruktion von Bandförden ZIS-Mitteilungen, *Halle-Saale*, **1**, 154–159, 1967.

LEŚNIAK, Z. K., Optimum truss structures (in Czech), *VIII Celostatni konference Optimalizace ocelowych konstrukci*, 54–63, Brno 1967.

LEŚNIAK, Z. K., Optimization of a fabrication program for welded I-beams (in Polish), *Biul. Branżowy Mostostalu*, **2**, 5–8, Warszawa 1967.

LEŚNIAK, Z. K., Methods of optimizing dimensions of structures with the use of electronic computers (in Polish), *Prace COB i RTK*, **28**, 49–50, Warszawa 1967.

MACIULEVIČIUS, D. A., Duality principle in linear problems of the optimum synthesis of elastic hinged-rod systems (in Russian), *Lit. Mekh. Sb.*, **1**, 115–125, 1967; *AMR*, 1970, **7**, 5384.

MACIULEVIČIUS, D. A., On the synthesis of an elastic hinged-rod structure of minimum weight under multiple loading subject to a more general constraint system (in Russian), *Lit. Mekh. Sb.*, **1**, 126–131, 1967; *AMR*, 1970, **7**, 5385.

MANSFIELD, E. H., Optimum tapers for eccentrically loaded ties, *J. Roy. Aeronaut. Soc.*, **71**, 681, 647–650 (Tech. Not.), 1967; *AMR*, 1968, **4**, 2639.

MARÇAL, P. V., Optimal plastic design of circular plates, *Int. J. Sol. Struct.*, **3**, 3, 427–443, 1967; *AMR*, 1967, **10**, 7862.

MARKS, W., Design of a concrete beam of uniform strength prestressed with continuously distributed infinitely thin wires (in Polish), *Konwersatorium PTMTS "Zagadnienia optymalizacji w mechanice"*, 67–70, Gliwice–Szczyrk 1967.

MAUCH, H. R., and FELTON, L. P., Optimum design of columns supported by tension ties, *Proc. ASCE*, **93**, *ST* **3**, *J. Struct. Div.* 201–220, 1967; *AMR*, 1968, **5**, 3255; *RŻM*, 1968, 3B448.

MAYEDA, R., and PRAGER, W., Minimum-weight design of beams for multiple loading, *Int. J. Solids Struct.*, **3**, 6, 1001–1011, 1967; *AMR*, 1968, **5**, 3230.

MEGAREFS, G. J., Minimal design of sandwich axisymmetric plates, Part I, *Proc. ASCE*, **93**, *EM* 6, *J. Engng. Mech. Div.*, 245–269, 1967.

MELOSH, R. J., and LUIK, R., Approximate multiple configuration analysis and allocation for least weight structural design, *ASAF, AFFDL-TR*-67-59, 1967.

MOE, J., and LUND, S., Cost and weight minimization of structures with special emphasis on longitudinal strength members of tankers, Part 1, *Ingenieur*, **79**, 47, W243-249, Part 2, *Ingenieur* **79**, 49, W253–262; *AMR*, 1969, **9**, 6984.

MOSES, F., Some notes and ideas on mathematical programming methods for structural optimization, *Meddelese SKBII/M*8, Norges Tekniske Høgskole, Trondheim, January 1967.

MOSES, F., and KINSER, D. E., Optimum structural design with failure probability constraints, *AIAA J.*, **6**, 6, 1152–1158, 1967; *AMR*, 1968, **1**, 309.

MOSES, F., and KINSER, E. D., Analysis of structural reliability, *Proc. ASCE*, **93**, *ST 5 J. Struct. Div.*, 147–164, 1967.

MRÓZ, Z., On the optimum design of reinforced slabs, *Acta Mechanica*, **3**, 1, 34–35, 1967; *AMR*, 1967, **12**, 9372.

MUKHTARI, KH. M., On the theory of bending of optimum-weight plates of composite materials (in Russian), *Prikl. Mekh.*, **3**, 4, 1–7, 1967; *AMR*, 1968, **4**, 2493.

NESMEYANOV, A. S., Design of frames of uniform strength under repeated loading with allowance for dead weight (in Russian), *Materialy tret'ej konferentsii molodykh nauchnykh rabotnikov Kazani*, Tatknigoizdat, Kazan' 1967.

OWCZAREK, S., Design of eccentrically compressed columns (in Polish), *Konwersatorium PTMTS: Zagadnienia optymalizacji w mechanice*, Gliwice–Szczyrk 1967.

PRAGER, W., Optimum plastic design of a portal frame for alternative loads, *Trans. ASME, Ser. E*, **34**, *J. Appl. Mech.*, 772–774, 1967; *AMR*, 1968, **4**, 2653.

PRAGER, W., and SHIELD, R. T., A general theory of optimal plastic design, *Trans. ASME*, **34**, *J. Appl. Mech.*, 184–186, 1967; *RŻM*, 1967, 10B262.

PROTTE, W., and TROSS, W., Optimale Bemessung geschweisster Träger des Hochbaus, *Techn. Mitt. Krupp*, **25**, 3, 137–144, 1967; *RŻM*, 1968, 7B726.

RABINOVICH, I. M., On the design of trusses and beams of minimum weight subjected to dynamic loads and dead weight (in Russian), *Issledovaniya po Teorii Sooruzhenii*, Vyp. **15**, M. Strojizdat. 1967.

RADTSIG, YU. A., Prospects for the development of the theory of mirror functions in the light of their applications to the problems of structural mechanics (in Russian), *Materialy k yubilejnoj respublikanskoj konferentsii sektsii stroitel'noj mekhaniki Tatarskogo pravleniya NTO Strojindustrii, Kazan'* 1967.

RADTSIG, YU. A., and ARSLANOV, A. SH., Design of statically indeterminate trusses of minimum weight with member stability constraints (in Russian), *Tezisy dokladov Vsesoyuznoj konferentsii po problemam ustojchivosti v stroitel'noj mekhanike*, Kaunas 1967.

RAWAT, P., A note on the design synthesis of primary ship structure using iterative approach, *J. Ship. Res.*, **11**, 123–130, 1967; *AMR*, 1967, **11**, 8626.

REJTMAN, M. I., On the theory of optimal design of structures made of plastics with the time factor taken into account (in Russian), *Mekhanika Polimerov*, **2**, 357–360, Riga 1967.

REJTMAN, M. I., Optimum design of spatial reinforced concrete structures (in Russian), Vychisl. i Organiz. Tekh. v Stroit. i Proekt., Vyp. **II-3**, 40–44, Moskva 1967.

REJTMAN, M. I., On the analysis and optimum design of spatial reinforced concrete structures (in Russian), *Stroit. Mekh. i Rasch. Sooruzh.*, **3**, 1967.

REJTMAN, M. I., and YARIN, L. I., Selection of the cross-section of a reinforced concrete beam for minimum cost with deformation constraints (program A-2) (in Russian), *Sel'skoe Stroitel'stvo*, **9**, 13–22, 1967.

REJTMAN, M. I., and YARIN, L. I., Selection of the cross-section of a reinforced concrete I-beam for minimum cost with strength constraints (program A-3), (in Russian), *Sel'skoe Stroitel'stvo*, **9**, 23–31, 1967.

REJTMAN, M. I., Analysis of the equations of the theory of perfectly plastic shells, *Arch. Mech. Stos.*, **4**, 19, 595–601, 1967.

RIDHA, R. A., and WRIGHT, R. N., Minimum cost design of frames, *Proc. ASCE* **93**, *ST 4*, *J. Struct. Div.*, 165–183, 1967; *AMR*, 1968, **3**, 1828.

ROZVANY, G. I. N., A new calculus for optimum design, *Int. J. Mech. Sci.*, **9**, 885–886, 1967.

ROZVANY, G. I. N., The behaviour of optimized reinforced concrete slabs, *Inst. of Engrs., Australia CE9*, **2**, 283-294, 1967.

SAELMAN, B., A note on the optimum design of I-section beams, *J. Aircraft*, 4, 5, 1967; *RŻM*, 1968, 5B624.

SCHUERCH, H., Performance weight relations and shape parameters for Maxwell structures, *AIAA J.*, **5**, 2, 367–369, 1967; *AMR*, 1967, **10**, 7873.

SCIPIO, L. A., Structural design: some NASA contributions, *NASA SP 5039*, **174**, 1967; *AMR*, 1968, **8**, 5782.

SHANLEY, F. R., Optimum design of eccentrically loaded columns, *Proc. ASCE*, **93**, *ST* 4, *J. Struct. Div.* 201–326, 1967; *AMR*, 1968, **4**, 2496.

SHARMAN, P. W., Optimum stiffness-weight design of peripheral and ladder frames, *Proc. Inst. Mech. Engrs., Automob. Div., Part 2A*, **182**, 3, 61–70, 1967; *AMR*, 1969, **7**, 5236.

SIRAZUTDINOV, YU. K., On the design of structures of minimum volume by variational methods (in Russian), *Materialy k yubilejnoj respublikanskoj konferentsii sektsii stroitel'noj mekhaniki Tatarskogo pravleniya NTO Strojindustrii*, Kazan' 1967.

SIRAZUTDINOV, YU, K., On the design of beams of minimum volume in the plastic range (in Russian), *Materialy k yubilejnoj respublikanskoj konferentsii sektsii stroitel'noj mekhaniki Tatarskogo pravleniya NTO Strojindustrii*, Kazan' 1967.

SIRAZUTDINOV, YU. K., On the design of beams with allowance for dead weight (in Russian), *IVUZ, Stroitel'stvo i Arkhitektura,* **11**, 1967.

SOFRONOV, YU. D., and KRUCHININ, V. V., Studies of the propagation rate of fatigue cracks on the basis of the magnitude of deflection of test specimens (in Russian), *Prochnosti materialov pri tsiklicheskikh nagruzkakh,* M., Nauka, 1967.

SOFRONOV, YU. D., On the design of rods and beams of minimum weight subjected to repeated forces (in Russian), *Materialy k yubilejnoj respublikanskoj konferentsii sektsii stroitel'noj mekhaniki Tatarskogo pravleniya NTO Strojindustrii, Kazan'* 1967.

SOFRONOV, YU. D., On the design of a beam with a maximum damping factor (in Russian), *Materialy k yubilejnoj respublikanskoj konferentsii sektsii stroitel'noj mekhaniki Tatarskogo pravleniya NTO Strojindustrii, Kazan'* 1967.

SOFRONOV, YU. D., Design of beams of minimum weight with deformation constraints (in Russian), *Materialy k yubilejnoj respublikanskoj konferentsii sektsii stroitel'noj mekhaniki Tatarskogo pravleniya NTO Strojindustrii, Kazan'* 1967.

SOFRONOV, YU. D., Design of beams of minimum weight and with limited amplitude of vibrations (in Russian), *Materialy k yubilejnoj respublikanskoj konferentsii sektsii stroitel'noj mekhaniki Tatarskogo pravleniya NTO Strojindustrii, Kazan' 1967.*

SPERANSKIJ, B. A., *Prestressed trussed metal structures* (in Russian), Izd. Literatury po Stroit., Moskva 1967.

SPERANSKIJ, B. A., Design of steel prestressed lattice girders with external tie-rods by the prescribed-force method (in Russian), *Nauchno-Tekhn. Obshch. Stroit. Ind., Konf. Progress. Stroit. Konstr.* 54–58, Sverdlovsk 1967.

STEVENSON, J. D., Reliability analysis and optimum design of structural systems with applications to rigid frames, *Solid Mechanics, Struct. and Mech. Design Div., Rep. No.* **14**, Case Western University, November 1967.

SUNGATULLIN, R. Z., Application of the maximum principle to the optimum dynamic problems of structural mechanics (in Russian), *Materialy k yubilejnoj konferentsii sektsii stroitel'noj mekhaniki Tatarskogo pravleniya NTO Strojindustrii,* Kazan' 1967.

TADJBAKSH, I., An optimum design problem for the nonlinear elastic, *IBM Research Paper RC*-**1949**, 1967.

TAYLOR, J. E., Minimum-mass bar for axial vibration at specified natural frequency, *AIAA J.*, **5**, 1911–1913, 1967; *AMR*, 1968, **8**, 5628.

TAYLOR, J. E., The strongest column: An energy approach, *Trans. ASME, E,* **34**, 2, *J. Appl. Mech.* 486–487, 1967; *AMR*, 1968, **1**, 155.

TOAKLEY, A. R., The optimum design of triangulated frame-works, *Int. J. Mech. Sc.*, **9**, 1967.

TURKSTRA, C. J., Choice of failure probabilities, *Proc. ASCE,* **93**, *ST* **6**, *J. Struct. Div.*, Paper 5678, 1967.

TURNER, M. J., Design of minimum-mass structures with specified natural frequencies, *AIAA J.*, **5**, 406–412, 1967; *AMR*, 1967, **12**, 9278.

VASIL'EVA, EH. M., On the determination of an optimum force in the tie-bar of a prestressed truss (in Russian), *Stroitel'naya mekhanika i raschet sooruzhenij,* **2**, 1967.

VASIL'EVA, EH. M., Determination of an optimum force in the tie-bar of a prestressed truss (in Russian), *Materialy k yubilejnoj respublikanskoj konferentsii sektsii stroitel'noj mekhaniki Tatarskogo pravleniya NTO Strojindustrii,* Kazan' 1967.

VASIL'EVA, EH. M., On the design of prestressed trusses of minimum volume for moving loads (in Russian), *Tezisy dokladov nauchnoj konferentsii Chuvashskogo gosuniversiteta,* Cheboksary 1967.

VINOGRADOV, A. I., On a certain algorithm in the theory of optimum systems (in Russian), *Prikl. Mekh.*, **3**, 7, 8–13, 1967; *AMR* 1969, **2**, 1034.

VINOGRADOV, A. I., and DOROSHENKO, O. P., Investigations of optimum systems (in Russian), *Prikl. Mekh.*, **3**, 12, 1–9, 1967.

WANKE, J., Wirtschaftliche Bemessung von Vollwandträgern mit parallelen Gurten, *Stahlbau*, **36**, 11, 344–348, 1967; *RŻM*, 1968, 7B725.

WASIUTYŃSKI, Z., On a design criterion for structures subjected to *n* loading states (in Polish), *Konwersatorium PTMTS "Zagadnienia optymalizacji w mechanice"*, 105–107, Gliwice–Szczyrk 1967.

WASIUTYŃSKI, Z., On the criterion of minimum deformability for the design of elastic structures with allowance for dead weight (in Polish), *Konwersatorium PTMTS "Zagadnienia optymalizacji w mechanice"*, 109–113, Gliwice–Szczyrk 1967.

ŻYCZKOWSKI, M., Optimum point-reinforcement of cylindrical shells with respect to their stability (in Polish), *Konwersatorium PTMTS, "Zagadnienia optymalizacji w mechanice"*, 115–117, Gliwice–Szczyrk 1967.

ŻYCZKOWSKI, M., Optimum design of point-reinforcement of cylindrical shells with respect to their stability, *Arch. Mech. Stos.*, **19**, 5, 699–713, 1967; *PBAM*, 1968, **3**, 7592.

1968

ARASLANOV, A. M., On the design of beams and frames of minimum weight for random loading (in Russian), *Vtoraya nauchno-tekhnicheskaya konferentsiya Ural'skogo Politekhnicheskogo Instituta, Tezisy dokladov sektsii stroitel'nogo fakul'teta*, Sverdlovsk 1968.

ARASLANOV, A. M., Fully stressed structures under random loads represented by a stationary vector-function (in Russian), *Trudy KAI*, **101**, 1968.

ARIFKHODZHAEV, S. A., and TUJCHIEV, N., Determination of the optimum cross-sectional parameters for a plane frame (in Russian), *Izv. Akad. Nauk. Uzb. SSR*, **2**, 39–42, 1968; *AMR*, 1972, **2**, 1101.

ARSLANOV, A. S., Discrete mirror functions and their application to the design of statically indeterminate trusses of minimum volume with stability constraints for compression members (in Russian), *Trudy Kazanskogo Inzh.-Stroit. Instituta*, **7**, 1968.

AYER, F. N., and CORNELL, C. A., Grid moment maximization by mathematical programming, *Proc. ASCE*, **94**, *ST*2, *J. Struct. Div.*, 529–549, 1968.

BENJAMIN, J. R., Probabilistic structural analysis and design, *Proc. ASCE* **94**, *ST* 7, *J. Struct. Div.*, 1665–1680; *AMR*, 1969, **5**, 3392.

BOGATYREV, A. I., Design of minimum-weight forms by the methods of mathematical programming (in Russian), *Sbornik trudov instituta inzhenerov zheleznodorozhnogo transporta*, **284**, 132–146, 1968.

BORKAUSKAS, A. EH., and ČYRAS, A. A., Design of elastic-plastic plates of minimum weight using linear programming (in Russian), *Lit. Mekh. Sb.*, **1** (2), 136–150, 1968.

BRACH, R. M., Minimum dynamic response for a class of simply supported beams shapes, *Int. J. Mech. Sci.*, **10**, 429–439, 1968.

BRACH, R. M., Optimum design of beams for sudden loading, *Proc. ASCE* **94**, *EM* 6, *J. Engng, Mech. Div.*, 1395–1407, 1968; *AMR*, 1969, **9**, 6983.

CHAN, H. S. Y., Mathematical programming in optimal plastic design, *Int. J. Solids Structures*, **4**, 885–895, 1968.

COHN, M. Z., Limit design of reinforced concrete frames, *Proc. ASCE*, **94**, *ST* 10, *J. Struct. Div.*, 2467–2483, 1968.

COHN, M. Z., and GRIERSON, D. E., Optimal design of reinforced concrete beams and frames, *Final Publ. 8th IABSE Congress, Sept.* 1968.

CORNELL, C. A., MCGUIRE, R. K., and LATONA, R. W., Engineering user's manuals: simple span deck and girder optimization and span arrangement optimization, *Res. Rep.* **R68-28**, *Dept. Civ. Engng. MIT*, 1968.

CYPINAS, I. K., On the optimum synthesis of rod systems susceptible to stability loss (in Russian), *Lit. Mekh. Sb.*, **2** (3), 22–33, 1968.

ČYRAS, A. A., Duality in the problems of structural mechanics and elasticity and plasticity theories (in Russian), *Lit. Mekh. Sb.*, **2** (3), 34–54, 1968; *AMR*, 1972, **8**, 6289.

ČYRAS, A. A., Reliability theory and optimum design (in Russian), *Lit. Mekh. Sb.*, **2** (3), 115–122, 1968.

ČYRAS, A. A., On the rational design of elastic-plastic discrete systems for random loads (in Russian), *II Vsesoyuz. Konf. po Probl. Nadezhn. v Stroit. Mekh., Vil'nyus 1968.*

DEMOKRITOV, V. N., Form optimization for crane beams (in Russian), *Prikl. Mekh.*, **4**, **7**, 97–101, 1968; *AMR*, 1970, **4**, 2788.

DIXON, L. C. W., Further comments on Pontryagin's maximum principle applied to the profile of a beam, *Aeronautical*, **690**, 72, 518–519, June 1968.

DOMAŃSKI, S., JĘDRZEJEWSKI, W., and KOCA, J., Experimental design of a prestressed flexural element as a basis of an optimum AMC program (in Polish), *XIV Konf. PZITB i PAN w Krynicy*, 275–284, 1968.

DE DONATO, O., and SACCHI, G., *Soluzioni di minimo peso d'armatura metallica per piastre inflese in cemento armato*, Ist. Lomb. Scienza e Lettere, Milano 1968.

DOROSHENKO, O. P., On the selection of combined systems of minimum cost, *Prikl. Mekh.*, **4**, 5, 138–140, 1968.

DURELLI, A. J., PARKS, V. J., and URIBE, S., Optimization of a slot end configuration in a finite plate subjected to uniformly distributed load, *J. Appl. Mech.*, **35**, 403–406, 1968.

DYLĄG, Z., ORŁOŚ, Z., and PONIŻ, D., Optimization of a reinforcing steel profile (in Polish), *Konwersatorium PTMTS "Zagadnienia optymalizacji w mechanice"*, 35–36, Gliwice–Szczyrk 1968.

DZHAPARIDZE, G. S., Investigation of rational forms of reinforced concrete arches in terms of the limit equilibrium theory (in Russian), *Tbilis. Nauchn.-Issl. Inst. Sooruzh. i Gidroenerg.*, 1968.

DZHAPARIDZE, G. S., Investigation of rational forms of reinforced arches in terms of the limit equilibrium theory (in Russian), *Beton i Zhelezobeton, Sb. Trudov*, **11**, 107–118 Tbilisi 1968.

DZIENISZEWSKI, W., Plates of minimum potential energy of deformation (in Polish), *Konf. "Metody optymalizacji ustrojów odkształcalnych", Jabłonna*, cz. **I**, 69–131, Ossolineum 1968.

FELTON, L. P., and HOFMEISTER, L. D., Optimized components in truss synthesis, *J. AIAA*, **6**, 12, 235–237, 1968.

FOX, R. L., and STANTON, E. L., Developments in structural analysis by direct energy minimization, *AIAA J.*, **6**, 6, 1036–1042, 1968; *AMR*, **1969**, 6, 4258,

FRUITET, L., Optimisation d'un système de poutres croisées, *Constr. Mét.*, **5**, 4, 14–26, 1968; *AMR*, 1969, **12**, 9566; *RŻM*, 1969, 5B604.

GAJNULLINA, S. KH., Reliability analysis in the minimum-weight design of structures (in Russian), *II Vsesoyuz. Konf. po Probl. Nadezhn. v Stroit. Mekh., Vil'nyus* 1968.

GHISTA, D. N., and RESNIKOFF, M. M., Development of Michell minimum weight structures, *NASA Tech. Note* **D-4345**, 30, 1968; *AMR*, 1968, **4**, 2642.

GOLDSHTEIN, YU. B., and SOLOMESHCH, M. A., On the optimum design of beams under dynamic loads (in Russian), *Stroit. Mekh. i Raschet Sooruzh.*, **4**, 27–39, 1968.

GOLIŃSKI, J., A certain random search process for finding extrema in problems of applied mechanics (in Polish), *Konwersatorium PTMTS "Zagadnienia optymalizacji w mechanice"*, 37–41, Gliwice–Szczyrk 1968.

GÓRSKI, S., Optimum design of axially compressed two-branch steel bars (in Polish), *Inż. i Bud.*, **12**, 465–469, 1968; *PBAM*, 1969, **4**, 8264; *RŻM*, 1969, 5B638.

GRYCZ, J., Design for maximum stiffness of columns in axial compression (in Polish), *Konf. "Metody optymalizacji ustrojów odkształcalnych", Jabłonna*, cz. **I**, 132–139, Ossolineum, 1968.

GRYCZ, J., Design for maximum stiffness of arches in moment-free states (in Polish), *Konf. "Metody optymalizacji ustrojów odkształcalnych", Jabłonna*, cz. **I**, 140–157, Ossolineum, 1968.

GRYCZ, J., Maximum-stiffness design of moment-free arches (in Polish), *Arch. Inż. Ląd.*, **14**, 2, 287–298, 1968.

HAUGEN, E. B., *Probabilistic approaches to design*, London, Wiley, 1968.

HEGEMIER, G. A., and PRAGER, W., On Michell trusses, Univ. California, San Diego, *Report* 3 *to U.S. Army Research Office—Durham* 1968.

HUANG, N. C., Optimal design of elastic structures for maximum stiffness, *Int. J. Solids Struct.*, **4**, 7, 689–700, 1968; *AMR*, 1969, **10**, 7865.

HUANG, N. C., and SHEU, C. Y., Optimal design of an elastic column of thin-walled crosssection, *J. Appl. Mech.*, **35**, 285–288, 1968.

ICERMAN, L. J., *Optimal structural design for given dynamic deflection*, M.Sc. Thesis, Univ. California, San Diego 1968.

JANICZEK, R., Axially compressed bars of minimum mass (in Polish), *Konwersatorium PTMTS "Zagadnienia optymalizacji w mechanice"*, 43–48, Gliwice–Szczyrk 1968.

JAWORSKI, J., and MARKS, W., The method of successive approximations for the design of prestressed beams on the basis of Caquot's hypothesis (in Polish), *Rozpr. Inż.*. **16**, 3, 299–312, 1968; *AMR*, 1970, **4**, 2797.

JENDO, S., Minimum-weight design of axisymmetric surface cable structures (in Polish), *Konwersatorium PTMTS "Zagadnienia optymalizacji w mechanice"*, 49–80, Gliwice–Szczyrk 1968.

JENDO, S., An optimum design method for cable structures (in Polish), *Konf. "Metody optymalizacji ustrojów odkształcalnych", Jabłonna*, cz. I, 158–205, Ossolineum, 1968.

JENDO, S., Minimum-weight design of axisymmetric cable roofs (in Polish), *Konf. "Metody optymalizacji ustrojów odkształcalnych", Jabłonna*, cz. I, 206–231, Ossolineum, 1968.

JENDO, S., Design of prestressed cable roofs for uniform stress (in Polish), *Konf. "Metody optymalizacji ustrojów odkształcalnych", Jabłonna*, cz. **I**, 232–253, Ossolineum, 1968.

JOHNSON, L. G., *Practical design problems*, *Engineering Plasticity*, University Press, Cambridge, 363–383, 1968.

KALISZKY, S., On the optimum design for reinforced concrete structures, *Konf. "Metody optymalizacji ustrojów odkształcalnych", Jabłonna*, cz. I, 5–18, Ossolineum, 1968.

KASPRZYK, T., Some problems of optimization of car suspensions (in Polish), *Konwersatorium PTMTS "Zagadnienia optymalizacji w mechanice"*, 67–71, Gliwice–Szczyrk 1968.

KASPRZYK, S., and MYJAK, J., Optimum selection of elastic and damping characteristics for oscillation-free vibrations of a certain mechanical system (in Polish), *Konwersatorium PTMTS "Zagadnienia optymalizacji w mechanice"*, 61–66, Gliwice–Szczyrk 1968.

KICHER, T. P., Structural synthesis of integrally stiffened cylinders, *J. Spacecraft*, **5**, 1, 62–67, 1968.

KŁOSOWICZ, B., The non-homogeneous spherical pressure vessel of maximum rigidity; I Theoretical approach; II Practical applications; *Bull. Acad. Pol. Sc., Sér. Sc. Mech.*, **16**, 7, 1968.

KOLENDA, J., On parametric optimization of vibrohammers (in Polish), *Konwersatorium PTMTS "Zagadnienia optymalizacji w mechanice"*, 73–77, Gliwice–Szczyrk 1968.

KIM, T. Ts., Design of statically indeterminate trusses of minimum weight (in Russian), *Stroit. Mekh. i Raschet Sooruzh.*, 2, 20–22, 1968.

KOLUPAEV, A. N., Optimum search for a minimum-volume truss (in Russian), *Trudy KAI*, **101**, 1968.

KRZYŚ, W., Optimale Formen gedrückter dünnwandiger Stützen in elastisch-plastischen Bereich, *Wiss. Z. Tech., Univ. Dresden*, **17**, 2, 407–410, 1968.

LATONA, R. W., Optimization of span arrangement for highway bridges, *Res. Rep.* **R68-54**, *Dept. Civ. Engng. MIT*, 1968.

LEŚNIAK, Z. K., Optimum wide-flange welded I-beams (in Polish), *Inż. i Bud.*, **4**, 151–154, 1968.

LEŚNIAK, Z. K., Optimum cross-sections of welded plate girders (in Polish), *Prz. Spaw.*, **5**, 120–129, 1968.

LEŚNIAK, Z. K., Optimization of an arch shape regarded as a cross-section of vault girders (in Polish), *Mech. Teoret. Stos.*, **6**, **1**, 79–91, 1968; *PBAM*, 1969, **1**, 7841.

LIND, N. C., The relation of data to calculated failure probabilities, *8th Congress IABSE*, New York, Sept. 1968.

MACIULEVIČIUS, D. A., Some algorithms of the optimum synthesis of elastic bar structures of minimum weight (in Russian), *Primenenie EVM v Stroitel'noj Mekhanike*, Kiev 1968.

MACIULEVIČIUS, D. A., On degeneracy in linear problems of configuration synthesis for elastic hinged-rod systems (in Russian), *Lit. Mekh. Sb.*, **1** (2), 5–24, 1968; *AMR*, 1972, **8**, 6291.

MACIULEVIČIUS, D. A., Optimum synthesis of hinged-rod structures for a given material size range (partially an integer problem) (in Russian), *Lit. Mekh. Sb.*, **2** (3), 5–15, 1968.

MACIULEVIČIUS, D. A., On the problem of configuration synthesis for an elastic hinged-rod structure of maximum load-carrying capacity subject to weight and material size range limitations (in Russian), *Lit. Mekh. Sb.*, **2** (3), 16–21, 1968.

MARKS, W., On the optimization of structures for multiple loading states (in Polish), *Konwersatorium PTMTS "Zagadnienia optymalizacji w mechanice"*, 95–99, Gliwice–Szczyrk 1968.

MARKS, W., Optimum beam shapes (in Polish), *Konf. "Metody optymalizacji ustrojów odkształcalnych", Jabłonna*, cz. **I**, 254–311, Ossolineum, 1968.

MARKS, W., The design of prestressed concrete beams for uniform strength according to Caquot's hypothesis (in Polish), *Arch. Inż. Ląd.*, **14**, 3, 451–467, 1968; *PBAM*, 1969, **2**, 7971; *AMR*, 1969, **7**, 802.

MARKS, W., Determination of principal loading states in the design of girder bridge spans (in Polish), *Arch. Inż. Ląd.*, **14**, **4**, 1968; *PBAM*, 1969, **4**, 8256; *AMR*, 1970, **3**, 310.

MCINTOSH, S. C., and EASTEP, F. E., Design of minimum-mass structures with specified stiffness properties, *AIAA J.*, **6**, 5, 962–964, *Tech. Notes*, 1968; *AMR*, 1968, **9**, 6599.

MEGAREFS, G. J., Minimal design of sandwich axi-symmetric plates, Part II, *Proc. ASCE*, **94**, *EM* 1, *J. Engng. Mech. Div.* 177–198, 1968; *AMR*, 1968, **10**, 7220.

MEGAREFS, G. J., and SIDHU, H. S., Simplifications in minimal design of frames, *Proc. ASCE*, **94**, *ST* 12, *J. Struct. Div.* 2985–2998, 1968; *AMR*, 1969, **8**, 6166.

MIRZABEKYAN, B. YU., and REJTMAN, M. I., Determination of the load-carrying capacity and minimal reinforcement of reinforced concrete plates by linear programming methods (in Russian), *Issl. Konstr. Zdanij i Sooruzh. dlya Sel'sk. Stroit.*, **2-1**, 93–109, Moskva 1968.

MOE, J., and LUND, S., Cost and weight minimization of structures with special emphasis on longitudinal members of tankers, *Trans. RINA*, **110**, 1, 1968.

MORROW, W. M., and SCHMIT, L. A., Structural synthesis of a stiffened cylinder, *NASA* **CR-1217**, Dec. 1968.

MOSES, F., Optimum design for structural safety, *8th IABSE Congress*, New York 1968.

MOSES, F., and STEFENSON, J. D., Reliability based structural design, Case Western Reserve University, *DSMSMD Report* **16**, 1968.

MUKHADZE, L. G., Shape design of prestressed cable structures (in Polish), *Konwersatorium PTMTS "Zagadnienia optymalizacji w mechanice"*, 101–108, Gliwice–Szczyrk 1968.

MUKHADZE, L. G., Approximate solution of prestressed cable structures (in Polish), *Konf. "Metody optymalizacji ustrojów odkształcalnych"*, *Jabłonna*, cz. **I**, 19–35, Ossolineum, 1968.

MUKHADZE, L. G., Shape design of prestressed cable structures (in Polish), *Konf. "Metody optymalizacji ustrojów odkształcalnych"*, *Jabłonna*, cz. **I**, 36–46, Ossolineum, 1968.

MUKHADZE, L. G., On the determination of optimum shapes of spatial hanging systems (in Russian), *Arch. Inż. Ląd.*, **14**, 3, 367–371, 1968.

MURTHY, F. N., and SUBRAMANIAN, G., Minimum weight analysis based on structural reliability, *AIAA J.*, **6**, 10, 2037–2039, *Tech. Notes*, 1968.

NAGY, S., Determination of the optimum profile of I-beams (in Hungarian), *Melyepitestudomanyi Szemle*, **18**, 7, 327–332, 1968; *AMR*, 1970, **1**, 334.

NEMIROVSKIJ, YU. V., On weight estimates for optimum plastic structures (in Russian), *Mekhanika Twerd. Tela*, **4**, 159–162, 1968.

NICHOLIS, J. I. Linear programming and optimal structural design, *Trend in Engineering*, **20**, 3, 21–24, 1968; *AMR*, 1969, **4**, 2530.

OWCZAREK, S., Selection of an optimum shape for an eccentrically compressed column in monolithic junction with a beam on the basis of photo-elastic model studies (in Polish), *Konwersatorium PTMTS "Zagadnienia optymalizacji w mechanice"*, 109–118, Gliwice–Szczyrk 1968.

OWCZAREK, S., Determination of optimum shapes of columns monolithically joined with beams on the basis of photo-elastic model studies (in Polish), *Konf. "Metody optymalizacji ustrojów odkształcalnych"*, *Jabłonna*, cz. I, 312–342, Ossolineum, Warszawa 1968.

PALMER, A. C., Optimal structural design by dynamic programming, *Proc. ASCE* **94**, *ST* 8, *J. Struct. Div.*, 1887–1906, 1968; *AMR*, 1969, **5**, 3390.

POPE, G. G., The design of optimum structures of specified basic configuration, *Int. J. Mech. Sci.*, **10**, 251–263, 1968, *AMR*, 1969, **4**, 2528; *RŻM*, 1968, 10B463.

PRAGER, W., *Optimal plastic design of rings, Contributions to Mechanics*, Pergamon Press, 163–169, 1968.

PRAGER, W., Optimality criteria in structural design, *Proc. Nat. Acad. Sci.*, **61**, 3, 794–796, 1968; *RŻM*, 1969 7B704.

PRAGER, W., Optimal structural design for given stiffness in stationary creep, *J. Appl. Math. Phys. ZAMP*, **19**, 252–256, 1968.

PRAGER, W., Optimization of structural design, *Symp. Optim. National Meeting in Toronto, SIAM* 1968.

PRAGER, W., and SHIELD, R. T., Optimal design of multi-purpose structures, *Int. J. Solids Struct.*, **4**, 469–475, 1968; *AMR*, 1970, **4**, 2789; *RŻM*, 1968, 11B777.

PRAGER, W., and TAYLOR, J. E., Problems of optimal structural design, *Trans. ASME Ser.* **E 35**, 1, *J. Appl. Mech.*, 102–106, 1968; *AMR*, 1968, **9**, 6597.

RABENDA, M., and JEŻ, M., Strength optimization of a certain kind of heavily loaded screws (in Polish), *Tech. Lotn. Astronaut.*, **23**, 2, 9–12, 1968; *PBAM*, 1970, **1**, 8404.

RADTSIG, YU. A., Review of research work carried out in Kazan' in the field of optimum structural design (in Russian), *Konf. "Metody optymalizacji ustrojów odkształcalnych", Jabłonna*, cz. **I**, 64–68, Warszawa 1968.

RAUTU, S., and CHIROIU, V., L'automation du calcul des structures statiquement non déterminées par poids minime, *Konf. "Metody optymalizacji ustrojów odkształcalnych", Jabłonna*, cz. **I**, 47–63, Ossolineum, 1968.

REJTMAN, M. I., New methods of the design of spatial reinforced concrete structures (in Russian), *Issl. Konstr. Zdanij i Sooruzh. dlya Sel'sk. Stroit.*, vyp. **2-1**, 6–31, Moskva 1968.

REJTMAN, M. I., Optimum design of structures subjected to random loads (in Russian), *II Vsesoyuz. Konf. po Probl. Nadezhn. v Stroit. Mekh*, 50–55, Vil'nyus 1968.

REJTMAN, M. I., Optimum design and unification of structures by means of dynamic programming (in Russian), *Stroit. i Arkhitekt.*, **5**, 138–142, Novosibirsk 1968.

REJTMAN, M. I., *Optimum structural design by the methods of mathematical programming* (in Russian, Thesis), Avtoreferat Dissertatsii, M., TsNISK, 1968.

ROMSTAD, K. M., and WANG, C. K., Optimum design of framed structures, *Proc. ASCE*, **94**, *ST* **12**, *J. Struct. Div.*, 2817–2845, 1968; *RŻM*, 1969, 8B622.

ROZVANY, G. I. N., Optimal design of axisymmetric slabs, *Int. Engrs., Australia, Civ. Engng. Trans.*, **CE 10**, 1, 111–118, 1968; *AMR*, 1970, 4, 2785.

SALINAS, D., *On variational formulations for optimal structural design*, Doct. Thesis, Univ. California, Los Angeles 1968; *RŻM*, 1970, 4B801.

SAVE, M. A., Some aspects of minimum-weight design, *Engineering Plasticity*, University Press, Cambridge, 611–626, 1968; *AMR*, 1969, **6**, 4259.

SCHMIDLER, G. M., Natural geometry of orbiting structures, *AIAA J.*, **6**, 566–567, 1968.

SCHMIT, L. A., Automated design, *Int. Sc. and Techn.*, **54**, 63–78, 115–117, 1968.

SHEU, C. Y., Elastic minimum-weight design for specified fundamental frequency, *Int. J. Solids Struct.*, **4**, 10, 10, 953–958, 1968; *AMR*, 1968, **12**, 9567.

SHEU, C. Y., and PRAGER, W., Minimum-weight design with piecewise constant specific stiffness, *J. Optimization Theory Appl.*, **2**, 179–186, 1968.

SHEU, C. Y., and PRAGER, W., Optimal plastic design of circular and annular sandwich plates with piecewise constant cross section, *J. Mech. Phys. Solids*, **11-16**, 1968.

SHEU, C. Y., and PRAGER, W., Recent developments in optimal structural design, *Appl. Mech. Rev.*, **21**, 10, 985–992, 1968.

SIRAZUTDINOV, YU. K., On the design of structures of minimum volume (in Russian), *Trudy KAI*, **101**, 1968.

SIRAZUTDINOV, YU. K., On application of the calculus of variations to the design of minimum-volume structures (in Russian), *Trudy KAI*, **95**, 100–110, 1968.

SOFRONOV, YU. D., On minimum-weight beams under random loads (in Russian), *Stroit. Mekh. i Raschet Sooruzh.*, **6**, 1968.

SOFRONOV, YU. D., Design of a uniformly stressed and uniformly strong bar subjected to repeated longitudinal force (in Russian), *IVUZ, Mashinostroenie*, **2**, 1968.

SOFRONOV, YU. D., Design of minimum-weight beams for stationary random repeated loads (in Russian), *Trudy KAI*, **101**, 1968.

SOFRONOV, YU. D., Design of uniformly stressed and uniformly strong bars, beams and frames for repeated loads with allowance for hysteresis losses in the material (in Russian), *Rasseyanie energii pri kolebaniyakh mekhanicheskich sistem*, Kiev, Naukova Dumka, 76–82, 1968.

SOFRONOV, YU. D., and NESMEYANOV, A. S., Investigation of the scale factor by equal-size specimen testing (in Russian), *Trudy KAI*, **95**, 1968.

SOFRONOV, YU. D., and NESMEYANOV, A. S., Design of uniformly stressed frames for asymmetric repeated loads using electronic computers (in Russian), *Voprosy Prochnosti Mashinostroeniya Konstruktsii, Trudy Chelyabinskogo Politekhnicheskogo Instituta*, **45**, 1968.

SVED, C., and GINOS, Z., Structural optimization under multiple loading, *Int. J. Mech. Sc.*, **10**, 10, 803–805, 1968; *AMR*, 1969, **4**, 2529; *RŻM*, 1969, 2N764.

SWITZKY, H., Designing for minimum flexibility or weight, *J. Spacecrafts and Rockets*, **5**, 12, 1473–1476, 1968.

SZCZEPIŃSKI, W., *Design of machine elements by the limit load method* (in Polish), PWN, Warszawa 1968.

TAYLOR, J. E., Optimum design of a vibrating bar with specified minimum cross-section, *AIAA J.*, **6**, 1379–1381, 1968.

TAYLOR, J. E., and LIU, C. Y., Optimum design of columns, *AIAA J.*, **6**, 1497–1502, 1968; *AMR*, 1969, **5**, 3391.

TOAKLEY, A. R., Optimal design using available sections, *Proc. ASCE* **94**, *ST 5*, *J. Struct. Div.* 1219–1241, 1968.

TOAKLEY, A. R., The optimum design of triangulated frameworks, *Int. J. Mech. Sc.*, **10**, 2, 115–127, 1968; *AMR*, 1968, **9**, 6598; *RŻM*, 1968, 10B792.

TOAKLEY, A. R., Some computational aspects of optimum rigidplastic design, *Int. J. Mech. Sc.*, **10**, 531–537, 1968.

VASCO COSTA, F., Optimization of structures, *8th Congress IABSE*, New York 1968.

VASIL'EVA, EH. M., Interpolation technique in certain problems of truss design (in Russian), *IVUZ, Stroit. i Arkhitekt.*, **4**, 1968.

VASIL'EVA, EH. M., Design of prestressed trusses of minimum volume (in Russian), *Annotatsii dokladov III Vsesoyuznogo s'ezda po teoreticheskoj i prikladnoj mekhanike*, **M.**, Izd. AN SSSR, 1968.

VASIL'EVA, EH. M., On the design of prestressed braced trusses (in Russian), *Trudy KAI*, **101**, 1968.

VASIL'EVA, EH. M., On the design of prestressed trusses for specified stresses (in Russian), *Trudy Chuvashskogo Gosudarstvennogo Universiteta im. N. I. Ul'yanova*, **2**, 1968.

WARDASZKO, E., Distribution of nodes of a lattice structure in the orthogonal projection (in Polish), *Inż. i Bud.*, **25**, 12, 1968.

WASIUTYŃSKI, Z., Aims and values of structural optimization problems and modern trends in their development (in Polish), *Konf. "Metody optymalizacji ustrojów odkształcalnych", Jabłonna*, cz. I, 343–378, Ossolineum, 1968.

WASIUTYŃSKI, Z., Problem formulation in the optimum design of deformable structures (in Polish), *Konf. "Metody optymalizacji ustrojów odkształcalnych", Jabłonna*, cz. **I**, 379–456, Ossolineum, 1968.

WASIUTYŃSKI, Z., On the properties of deformation states and shapes of structures designed according to extremum criteria (in Polish), *Konf. "Metody optymalizacji ustrojów odkształcalnych", Jabłonna*, cz. **I**, 457–465, Ossolineum, 1968.

WOODS, W. J., and SAMS III, J. H., Geometric optimization in the theory of structural synthesis, *AIAA Paper* 68–330, *ASME 9th Struct. Dyn. and Mat. Conf. Palm Springs, Cal.* 1968; *RŻM*, 1969, 2B830.

WRIGHT, P. M., *Behaviour gradient matrices in the weight minimization of structures*, Doct. Diss., Univ. Colo., 1968; *RŻM*, 1970, 7B707D.

ZARGHAMEE, M. S., Optimum frequency of structures, *AIAA J.*, **6**, 749–750, 1968; *AMR*, 1968, **8**, 5655.

Życzkowski, M., Optimale Formen des dünnwandigen geschlossenen Querschnitten eines Balkens bei Berücksichtigung von Stabilitätsbedinungen, *ZAMM*, **48**, 7, 455–462, 1968; *RŻM*, 1969, 7B676.

Życzkowski, M., Optimum strength design with stability constraints (in Polish), *Konf. "Metody optymalizacji ustrojów odkształcalnych"*, *Jabłonna*, cz. **I**, 466–555, Ossolineum, 1968.

1969

Ashley, H., and McIntosh, S. C., Jr., Applications of aeroelastic constraints in structural optimization, *Proc. 12th Int. Congress of Appl. Mech.*, 101–113, Springer Verlag, Berlin 1969.

Araslanov, A. M., Design of beams and frames of minimum weight for random loading (in Russian), *Trudy KAI*, **105**, 1969.

Araslanov, A. M., On the design of uniformly reliable structures of minimum weight (in Russian), *Trudy KAI*, **105**, 1969.

Araslanov, A. M., On the design of uniformly reliable structures (systems with one degree of freedom) (in Russian), *Materialy k yubilejnoj respublikanskoj konferentsii sektsii stroitel'noj mekhaniki Tatarskogo pravleniya NTO Strojindustrii*, Kazan' 1969.

Araslanov, A. M., A discrete method for solving quasistatical design problems for uniformly reliable structures (in Russian), *Materialy k yubilejnoj respublikanskoj konferentsii sektsii stroitel'noj mekhaniki Tatarskogo pravleniya NTO Strojindustrii*, Kazan' 1969.

Arslanov, A. S., On the design of statically indeterminate trusses of minimum volume subject to stability constraints on compression members under combined loads (in Russian), *Materialy k respublikanskoj konferentsii sektsii stroitel'noj mekhaniki Tatarskogo pravleniya NTO Strojindustrii*, Kazan' 1969.

Bortniczuk, W., Some problems in the design of continuous crane beams (in Polish), *Arch. Inż. Ląd.*, **15**, 1/2, 131–150, 1969.

Brandt, A. M., Bibliographical survey of optimum structural design (in Polish), *Konf. "Metody optymalizacji ustrojów odkształcalnych" Jabłonna*, *cz.* **II**, 227–306, Ossolineum, 1969.

Brotchie, J. F., A criterion for optimal design of plates, *J. ACI*, **66**, 11, 898–906, 1969; *AMR*, 1971, **1**, 1148.

Brizgalin, G. I., On rational design of anisotropic plane bodies with a weak binder (in Russian), *Izv. AN SSSR, Mekh. Tverd. Tela*, **4**, 123–133, 1969.

Budiansky, B., Frauenthal, J. C., and Hutchinson, J. W., On optimal arches, *J. Appl. Mech.*, **36**, 1969.

Chelnokov, R. V., A circular reduced-weight beam (in Russian), *Materialy k respublikanskoj konferentsii sektsii stroitel'noj mekhaniki Tatarskogo pravleniya NTO Strojindustrii*, Kazan' 1969.

Čižas, A. P., Design of elastic-plastic rod systems with optimum prestressing (in Russian), *Lit. Mekh. Sb.*, **2** (5), 33–40, 1969.

Clemmons, T. P., Systems design for weight optimization, *Soc. Aeron. Weight Engrs*, **757**, Los Angeles 1969.

Cornell, C. A., Examples of optimization in structural design, *Study No.* 1, *SM Division, Univ. Waterloo*, Canada 1969.

Ćwik, R., Optimization of the dimensions of cross-sections of flexural elements of cranes (in Polish), *Zesz. Nauk. P. Śl.*, **262**, *Mech.*, 42, 95–112, 1969; *PBAM*, 1971, **4**, 9281.

Cypinas, I. K., On the synthesis of optimum rod systems with compression-bent members (in Russian), *Lit. Mekh. Sb.*, **2** (5), 41–48, 1969.

ČYRAS, A. A., Stochastische Modelle zur Berechnung von optimalen elas toplastischen eindmensionen Systemen, *ZAMM*, **49**, 5, 1969.

DATKA, S., and TRACZ, M., Analysis and design of optimum vertical profiles of road curves (in Polish), *Arch. Inż. Ląd.*, **15**, 3, 491–504, 1969.

DAYARATNEM, P., and PARTAIK, S., Feasibility of full stress design, *AIAA J.*, **7**, 4, 773–774, 1969.

DEMOKRITOV, V. N., On the selection of an optimum shape of crane jibs (in Russian), *Prikl. Mekh.*, **4**, 7, 97–101, 1969; *AMR*, 1970, **4**, 2788.

DE SILVA, B. M. E., The application of nonlinear programming to the automated minimum weight design of rotating discs, *Symp. Inst. Math. Appl., Univ. Kell, England*, 1968, *Optimization*, 115–150, Academic Press 1969.

DIETRYCH, M., GOLIŃSKI, J., and POGORZELSKI, W., Optimization of a pipe derrick (in Polish), *Arch. Inż. Ląd.*, **15**, 3, 491–504, 1969; *PBAM*, 1970, **2**, 8538.

DOBBS, M. W., and FELTON, L. P., Optimization of truss geometry, *Proc. ASCE*, **95**, *ST* 10, *J. Struct. Div.*, 2105–2118, 1969; *AMR*, 1970, **5**, 3589.

DZIENISZEWSKI, W., Reinforced concrete plates of uniform strength under two loading states (in Polish), *Konf. "Metody optymalizacji ustrojów odkształcalnych"*, *Jabłonna*, cz. **II**, 57–74, Ossolineum, 1969.

DZIENISZEWSKI, W., Simply supported three-dimensional rectangular plates of uniform strength (in Polish), *Konf. "Metody optymalizacji ustrojów odkształcalnych"*, *Jabłonna*, cz. **II**, 75–100, Ossolineum, 1969.

DZIENISZEWSKI, W., Certain cases of the design of simply supported three-layer plates of uniform strength (in Polish), *Rozpr. Inż.*, **17**, 4, 551–569, 1969; *AMR*, 1972, **8**, 6131.

FARKAS, J., Festigkeitseigenschaften von geschweissten auf Biegung optimal bemessen I- und Kastenträgern, *Acta Tech. Acad. Sci. Hung.*, **66**, 4, 427–439, 1969; *RŻM*, 1970, 6B764.

FOX, R. L., Mathematical methods in optimization, *Study No.* **1**, *SM Division, Univ. Waterloo, Canada*, 1969.

GAJNULLINA, S. KH., Reduced-weight three-layer frames with filler (in Russian), *Materialy k respublikanskoj konferentsii sektsii stroitel'noj mekhaniki Tatarskogo pravleniya NTO Strojindustrii*, Kazan' 1969.

GAJNULLINA, S. KH., Analysis of statically indeterminate systems of minimum volume and given reliability (in Russian), *Materialy k respublikanskoj konferentsii sektsii stroitel'noj mekhaniki Tatarskogo pravleniya NTO Strojindustrii*, Kazan' 1969.

GAJEWSKI, A., and ŻYCZKOWSKI, M., Optimal shaping of an elastic homogeneous bar compressed by a polar force, *Bull. Acad. Pol. Sc., Sér. Sc. Tech.*, **17**, 10, 479–488, 1969; *RŻM*, 1970, 6B798.

GAJEWSKI, A., and ŻYCZKOWSKI, M., Optimum design of a bar compressed by a polar force (in Polish), *Rozpr. Inż.*, **17**, 2, 299–329, 1969; *PBAM*, 1970, **2**, 8544; *RŻM*, 1970, 5B731.

GELLATLY, R. A., Structural optimization—a dream or a reality?, *Soc. Aeronaut. Weight Engrs.*, No. **814**, Los Angeles 1969.

GIRFANOV, I. S., Design of space metal trusses of optimum volume (in Russian), *Materialy k respublikanskoj konferentsii sektsii stroitel'noj mekhaniki Tatarskogo pravleniya NTO Strojindustrii, Kazan'* 1969.

GIRFANOV, I. S., Approximate methods for the design of statically indeterminate trusses economical in volume (in Russian), *Materialy k respublikanskoj konferentsii sektsii stroitel'noj mekhaniki Tatarskogo pravleniya NTO Strojindustrii*, Kazan' 1969.

GIRFANOV, I. S., On the determination of the least values of broken functions of several independent variables (in Russian), *Trudy KAI*, **113**, 1969.

GOLDSHTEIN, YU. B., and SOLOMESHCH, M. A., On the optimum design of rod systems subjected to force and temperature actions (in Russian), *Stroit. Mekh. i Raschet Sooruzh*, **2**, 4–8, 1969.

GOULD, P. L., Minimum weight design of hyperbolic cooling towers, *Proc. ASCE*, **95**, *ST* 2, *J. Struct. Div.*, 203–208, 1969.

HAUG, E. J., Two methods of optimal structural design, in: *Developments in Mechanics, Proc. 11th Midwestern Mech. Conf.*, **5**, 1, 847–860, 1969.

HEGEMIER, G. A., and PRAGER, W., On Michell trusses, *Int. J. Mech. Sc.*, **11**, 2, 209–215. 1969; *AMR*, 1970, **1**, 333.

HOLT, E. C., and HEITHECKER, G. L., Minimum weight proportions for steel girders, *Proc. ASCE*, **95**, *ST 10*, *J. Struct. Div.*, 2205–2217, 1969; *AMR*, **6**, 4472; *RŻM*, 1971, 2B1065.

HUANG, N. C., and TANG, H. T., Minimum-weight design of elastic sandwich beams with deflection constraint, *J. Opt. Theory and Appl.*, **4**, **4**, 277–298, 1969; *AMR*, 1971, **3**, 1915; *RŻM*, 1970, 6B725.

ICERMAN, L. J., Optimal structural design for given dynamic deflection, *Int. J. Solids and Struct.*, **5**, 5, 473–490, 1969; *AMR*, 1970, **1**, 330; *RŻM*, 1969, 11B178.

ISSLER, W., Membranschalen gleicher Festigkeit, *Ingen. Arch.*, **33**, 5, 330–345, 1969.

JANICZEK, R., Design of bars and beams for minimum material (in Polish), *Zesz. Nauk. P. Częst.*, **61**, *Mech.*, 3, 285–289, 1969.

JANICZEK, R., and KAPCIA, B., Axially compressed fusiform bars with filler (in Polish), *Zesz. Nauk. P. Częst.*, **64**, *Nauki Podst.* 13, 83–98, 1969.

JENDO, S., Shape optimization of two-surface prestressed cable structures for minimum auxilliary volume of the material (in Polish), *Konf. "Metody optymalizacji ustrojów odkształcalnych"*, *Jabłonna*, *cz.* **II**, 101–149, Ossolineum, 1969.

JENDO, S., An optimum design method for axi-symmetric cable structures (in Polish), *Rozpr. Inż.*, **17**, 2, 249–267, 1969; *PBAM*, 1970, **2**, 8531.

JENDO, S., Examples and Tables for the determination of optimum design parameters of axi-symmetric cable structures (in Polish), *Konf. "Metody optymalizacji ustrojów odkształcalnych"*, *Jablonna*, cz. **II**, 150–166, Ossolineum 1969.

JENDO, S., Optimum design of axi-symmetric prestressed surface structures. Application of the stress equalization criterion, *Bull. Acad. Pol. Sc., Sér. Sc. Techn.*, **17**, 4, 217–223, 1969; *AMR*, 1970, **5**, 3591.

JENDO, S., Optimum design method for suspended structures. Application of the minimum elastic energy criterion, *Bull. Acad. Pol. Sc., Sér. Sc. Techn.*, **17**, 3, 191–195, 1969.

JENDO, S., Optimum design of surface axi-symmetric suspended structures. Application of the minimum-weight criterion, *Bull. Acad. Pol. Sc., Sér. Sc. Techn.*, **17**, 3, 197–206, 1969; *AMR*, 1970, **3**, 1982.

JENDO, S., Optimum design method for suspended structures. Application of the minimum weight criterion, *Bull. Acad. Pol. Sc., Sér. Sc. Techn.*, **17**, 2, 117–126, 1969; *AMR*, 1970, **6**, 4473.

JOHNSON, D., and BROTTON, D. M., Optimum elastic design of redundant trusses, *Proc. ASCE*, **95**, *ST 12*, *J. Struct. Div.*, 2589–2610, 1969; *AMR*, 1970, **10**, 7959; *RŻM*, 1971, 1B905.

JOHNSON, W., and SOWERHY, R., Upper bounds techniques applied to plane strain extrusion minimum weight two-dimensional frames and rotationally symmetric flat plates, *Bull. Mech. Engngn. Educ.*, **8**, 3, 269–284, 1969; *RŻM*, 1970, 5B343.

KANI, G. N. J., A rational theory for the function of web reinforcement, *J. Amer. Concr. Inst.*, **66**, 3, 185–197, 1969; *RŻM*, 1969, 11B748.

KAVLIE, D., and MOE, J., Application of nonlinear programming to optimum grillage design with nonconvex sets of variables, *Int. J. for Numerical Methods in Engngn.*, **1**, 351–378, 1969.

KEITH, H. D., Optimization technique in design, *ASME*, No. **DE-14**, New York 1969; *RŻM*, 1970, 7B672.

KIM, T. S., Design of trusses of minimum volume by an integer programming method (in Russian), *Stroit. Mekh. i Raschet Sooruzh.*, **1**, 21–23, 1969.

KIM, T. S., Design of statically indeterminate trusses of minimum volume for transient loading (in Russian), *Stroit. Mekh. i Raschet Sooruzh.*, **5**, 13–17, 1969.

KHACHATURIAN, N. A., Basic concepts in structural optimization, *Study No.* **1**, *SM Division, Univ. Waterloo, Canada,* 1969.

KHUBERYAN, K. M., Optimum design methods for arch dams (in Polish), *Konf. "Metody optymalizacji ustrojów odkształcalnych", Jabłonna*, cz. **II**, 49–56, Ossolineum, 1969.

KOKORIN, N. A., Statically indeterminate trusses of minimum weight under moving loads (in Russian), *Trudy KAI*, **113**, 59–69, 1969.

KOLUPAEV, A. N., Truss design by nonlinear programming (in Russian), *Trudy KAI*, **113**, 1969.

KOLUPAEV, A. N., Truss synthesis for transient loading with application of computers (in Russian), *Trudy KAI*, **113**, 1969.

KOLUPAEV, A. N., On the design of trusses for given stiffness under transient loading (in Russian), *Materialy k respublikanskoj konferentsii sektsii stroitel'noj mekhaniki Tatarskogo pravleniya NTO Strojindustrii*, Kazan' 1969.

KORSHUNOV, A. N., Statically indeterminate trusses of minimum weight under moving loads or several alternative load variants (in Russian), *Materialy k respublikanskoj konferentsii sektsii stroitel'noj mekhaniki Tatarskogo pravleniya NTO Strojindustrii*, Kazan', 1969.

KOZŁOWSKI, W., and MRÓZ, Z., Optimal design of solid plates, *Int. J. Solids Struct.*, **5**, 8, 781–794, 1969; *AMR*, 1971, **1**, 352.

KUZNETSOV, E. N., *Introduction to the Theory of Suspended Systems* (in Russian), Iz-vo Literatury po Stroitel'stvu, Moskva 1969.

LEŚNIAK, Z. K., Methods of structural optimization with the use of computers (in Polish), *Zesz. Nauk. P. Warsz., Bud.*, **10**, Warszawa 1969.

LIND, N. C., Deterministic formats for the probabilistic design of structures, *Study No.* **1**, *SM Division, Univ. Waterloo, Canada*, 1969.

LINDISHAS, L. T., On the design of elastico-plastic hinged-rod systems of minimum weight subject to elastic stability constraints on the rods (in Russian), *Lit. Mekh. Sb.*, **1**, 4, 65–69, 1969.

ŁUKASIEWICZ, S., On the optimum design of shells loaded by concentrated forces, *Theor. thin shells*, Berlin–Heidelberg–New York, 161–175, 1969.

MACIULEVIČIUS, D. A., The problem of the synthesis of an optimum configuration of a hinged-rod structure under a fixed load and dead weight (in Russian), *Lit. Mekh. Sb.*, **2** (5), 5–15, 1969.

MACIULEVIČIUS, D. A., Synthesis of an elastic hinged-rod structure for multiple loading with dead weight consideration (in Russian), *Lit. Mekh. Sb.*, **2** (5), 17–24, 1969.

MACIULEVIČIUS, D. A., On the invariance of optimum configurations in synthesis problems for elastic hinged-rod systems (in Russian), *Lit. Mekh. Sb.*, **1** (4), 26–36, 1969.

MAJERCZYK-GOMUŁKOWA, J., and MIODUCHOWSKI, A., Optimum plastic inhomogeneity of a twisted bar with regard to limit load (in Polish), *Rozpr. Inż.* **17**, **4**, 583–599, 1969; *PBAM* 1971, **3**, 9102.

MARKS, W., Optimization of concrete elements prestressed by arbitrarily thin and arbitrarily densely distributed reinforcement (in Russian), *Conf. Nat. de Méc. Appl.*, 126–127, Bucarest, Juin 1969.

MATVEEV, S. N., Design of statically indeterminate trusses with the frequency of free vibrations not less than a prescribed value (in Russian), *Materialy k respublikanskoj konferentsii sektsii stroitel'noj mekhaniki Tatarskogo pravleniya NTO Strojindustrii*, Kazan' 1969.

MATVEEV, S. N., Design of statically indeterminate trusses of minimum volume under dynamic loads by replacing the truss by a system with one degree of freedom (in Russian), *Materialy k respublikanskoj konferentsii sektsii stroitel'noj mekhaniki Tatarskogo pravleniya NTO Strojindustrii*, Kazan' 1969.

MEDWADOWSKI, J., Optimum dimensioning of non-prismatic metal bars in simultaneous bending and torsion (in Polish), *Arch. Inż. Ląd.*, **15**, 3, 559–570, 1969; *PBAM*, 1970, **2**, 8543; *RŻM*, 1970, 3B706.

MOE, J., Optimum design of statically indeterminate frames by means of nonlinear programming, *Univ. Michigan, Rep.* **023** *and* **024**, 1969.

MOSES, F., and ONODA, S., Minimum weight design of structures with application to elastic grillages, *Int. J. of Num. Methods in Engng.* **1**, 311–331, 1969.

MOSES, F., Approaches to structural reliability and optimization, *Study* No. **1**, *SM Division, Univ. Waterloo, Canada,* 1969.

MRÓZ, Z., Optimization of structures of polyphase materials (in Polish), *Prace IPPT PAN*, **15**, 1969.

MRÓZ, Z., Optimum design of structures for maximum limit load (in Polish), *Konf. "Metody optymalizacji ustrojów odkształcalnych"*, *Jabłonna*, cz. **II**, 11–48, Ossolineum, 1969.

MOROZOV, E. P., Optimum slopes of struts in lattice masts (in Russian), *Stroit. Mekh. i Raschet Sooruzh.*, **5**, 13–17, 1969.

NEUBER, H., Der zugbeanspruchte flachstahl mit optimalen querschnittubergang, *Forsch. Ingenieurwesen*, **35**, 29–30, 1969.

NEMIROVSKI, YU. V., and REZNIKOV, B. S., Beams and plates of uniform strength in creep conditions (in Russian), *Mashinovedenye*, **58–64**, Moskva 1969.

NICHOLLS, J. I., Two examples of linear programming in optimal structural weight design, *Trend. Engng. Univ. Wash.*, **23**, 1, 14–16, 1969; *RŻM*, 1970, 1B347.

OLKOV, YA. I., and KHOLOPOV, I. S., On the optimum distribution of material in trusses (in Russian), *Stroit. Mekh. i Raschet Sooruzh.*, **1**, 24–26, 1969.

OLSZOWSKI, B., Some problems of optimum dynamic design of industrial structures (in Polish), *Czasop. Tech.*, **5**, 74, 1969; *RŻM*, 1970, 9B270.

OLSZOWSKI, B., Dynamic design of structural elements (in Polish), *Mech. Teoret. i Stos.*, **7**, 3, 299–309, 1969; *PBAM*, 1970, **2**, 8525.

OWCZAREK, S., Optimum design of columns subjected to bending moments and compressive forces (in Polish), *Konf. "Metody optymalizacji ustrojów odkształcalnych"*, *Jabłonna*, cz. **II**, 167–183, Ossolineum, 1969.

OWCZAREK, S., Determination of the shape of columns monolithically joined with beams under modified strength of materials assumptions (in Polish), *Konf. "Metody optymalizacji ustrojów odkształcalnych"*, *Jabłonna*, cz. **II**, 184–226, Ossolineum, 1969.

PALMER, A. C., Limit analysis of cylindrical shells by dynamic programming, *Int. J. of Solids and Structures*, **5**, 289–302, 1969.

PETCU, V., Generalisation of the optimum plastic design theory of reinforced concrete continuous structures, *Revue Roum. Sc. Tech., Série Méc.* Appl., 14, 2, 263–283, 1969; *AMR*, 1970, **7**, 5396.

PETCU, V., Minimum moment distribution theory of reinforced concrete continuous structures, *Revue Roum. Sc. Tech., Série Méc. Appl.*, **14**, 4, 797–817, 1969; *AMR*, 1970, **7**, 5397.

PIRAS, Z., Linear programming and optimum design of structures, Acta, Techn. *OSAV*, **15**, 6, 652–689, 1970; *RŻM*, 1971, 6B1004.

POPE, G. G., The application of linear programming techniques in the design of optimum structures, *AGARD Conf. Proc. No.* **36**, *Symp. on Struct. Opt.*, Istanbul 1969.

PRAGER, W., The deformation of rigid, workhardening optimal structures, *Nauka*, 393–396, Moskva 1969.

PRAGER, W., Optimality criteria derived from classical extremum principles, *Study No. 1, SM Division, Univ. Waterloo, Canada*, 1969.

PRAGER, W., *Optimal plastic design of rings*, Contributions to Mech., Pergamon Press, 163–169, 1969.

RADTSIG, YU. A., Topical problems in the design of continuous rod structures of minimum weight (in Russian), *Materialy k respublikanskoj konferentsii sektsii stroitel'noj mekhaniki Tatarskogo pravleniya NTO Strojindustrii*, Kazan' 1969.

RADTSIG, YU. A., Classification of synthesis problems for statically indeterminate trusses of minimum weight (in Russian), *Materialy k respublikanskoj konferentsii sektsii stroitel'noj mekhaniki Tatarskogo pravleniya NTO Strojindustrii*, Kazan' 1969.

RADTSIG, YU. A., Solution of a system of linear mirror equations with several unknowns with the use of electronic digital computers, (in Russian), *Trudy KAI*, **113**, 70–80, 1969.

REISS, R., and MEGAREFS, G. J., Minimal design of sandwich axi-symmetric cylindrical shells obeying Mises' criterion, *Acta Mech.*, **7**, 1, 72–98, 1969; *AMR*, 1970, **2**, 1116.

REJTMAN, M. I., Optimum structural design by mathematical programming methods (in Russian), *Stroit. Mekh. i Raschet Sooruzh.*, **3**, 54–62, 1969.

SACCHI, G., and SAVE, M., Le problème du poids minimum d'armature des plaques en béton armé, *Ist. Sc. Technica Costr. Politecnico di Milano*, 473, 1969.

SAFIN, R. K., On the limit state analysis of statically indeterminate steel trusses (in Russian), *Trudy KAI*, **113**, 81–86, 1969.

SAFIN, R. K., Comparative weight analysis of trusses designed in the elastic and elastic-plastic ranges (in Russian), *Materialy k respublikanskoj konferentsii sektsii stroitel'noj mekhaniki Tatarskogo pravleniya NTO Strojindustrii*, Kazan' 1969.

SCHMIT, L. A., Problem formulation, methods and solutions in the optimum design of structures, *Study No 1, SM Division, Univ. Waterloo, Canada*, 1969.

SCHMIT, L. A., Structural engineering applications of mathematical programming techniques, *AGARD Conf. Proc. No.* **36**, *Symp. on Struct. Opt.*, Istanbul 1969.

SHEU, C. Y., and PRAGER, W., Optimal design of sandwich beams for elastic deflection and load factor at plastic collapse, *ZAMP*, **10**, 289–297, 1969; *RŻM*, 1970, 3B404.

SHEU, C. Y., and PRAGER, W., Optimal plastic design of circular and annular sandwich plates with piecewise constant cross-section, *J. Mech. Phys. Solids*, **17**, 1, 11–16, 1969; *AMR*, 1970, **1**, 332.

SHINOZUKA, M., Optimum structural design based on reliability analysis, *Int. Symp. on Space Techn. and Sci.*, 245–258, Tokyo 1969.

SHROM, G. M., On the selection of a rational multispan beam design (in Russian), *Materialy k respublikanskoj konferentsii sektsii stroitel'noj mekhaniki Tatarskogo pravleniya NTO Strojindustrii*, Kazan' 1969.

SIRAZUTDINOV, YU. K., On the design of minimum-volume structures in the absence of longitudinal forces (in Russian), *Trudy KAI*, **113**, 87–97, 1969.

SIRAZUTDINOV, YU. K., Classification of rational structures with regard to design variables (in Russian), *Materialy k respublikanskoj konferentsii sektsii stroitel'noj mekhaniki Tatarskogo pravleniya NTO Strojindustrii*, Kazan' 1969.

SIRAZUTDINOV, YU. K., On the design of beams of minimum volume with strength and stiffness constraints (in Russian), *Materialy k respublikanskoj konferentsii sektsii stroitel'noj mekhaniki Tatarskogo pravleniya NTO Strojindustrii*, Kazan' 1969.

SOFRONOV, YU. D., On the design of an oscillating calibration beam (in Russian), *Trudy KAI*, **105**, 1969.

SOFRONOV, YU. D., Creeping under small number of load cycles (in Russian), *Prochnost' pri malom chisle tsiklov nagruzheniya*, M., Nauka, 1969.

SOFRONOV, YU. D., Design of minimum-weight bars for longitudinal cyclic forces (in Russian), *Stroitel'naya Mekhanika i Raschet Sooruzhenij*, **6**, 40–48, 1969.

SOFRONOV, YU. D., Design of beams of minimum weight with deformation constraint (in Russian), *Trudy KAI*, **113**, 98–108, 1969.

SOFRONOV, YU. D., Design of bars of minimum weight under impulsive loads (in Russian), *Materialy k respublikanskoj konferentsii sektsii stroitel'noj mekhaniki Tatarskogo pravleniya NTO Strojindustrii*, Kazan' 1969.

SOFRONOV, YU. D., Optimum cross-section of a beam with a maximum radius of inertia (in Russian), *Materialy k respublikanskoj konferentsii sektsii stroitel'noj mekhaniki Tatarskogo pravleniya NTO Strojindustrii*, Kazan' 1969.

SUNGATULLIN, R. Z., Design of beams of minimum volume with two-sided constraints on displacement (in Russian), *Trudy KAI*, **113**, 109–118, 1969.

SUNGATULLIN, R. Z., Design algorithms for statically indeterminate beams of minimum volume with displacement constraints (in Russian), *Materialy k respublikanskoj konferentsii sektsii stroitel'noj mekhaniki Tatarskogo pravleniya NTO Strojindustrii*, Kazan' 1969.

TAYLOR, J. E., Optimal design of structural systems: An energy formulation, *AIAA J.*, **7**, 7, 1404–1406, 1969; *AMR*, 1970, **12**, 9576.

TAYLOR, J. E., Maximum strength elastic structural design, *Proc., ASCE*, **95**, *EM3*, *J. Engng. Mech. Div.* 653–663, 1969.

TOAKLEY, A. R., The optimum plastic design of unbraced frames, Inst. Engrs., Australia. *Civ. Engng. Trans.*, *CE11*, **2**, 111–116, 1969; *AMR*, 1971, **5**, 3542.

TUJCHIEV, N. D., An algorithm for optimum design of statically indeterminate frames on computers (in Russian), *Izv. AN UzSSR, Ser. Tekh.*, **4**, 42–49, 1969.

TURNER, M. J., Optimization of structures to satisfy flutter requirements, *AIAA J.*, **7**, 945–951, 1969.

VENKAYYA, V. B., KHOT, N. S., and REDDY, V. S., Energy distribution in an optimal structural design, *Air Force Flight Dynamics Lab., Wright-Patterson AFFDL-TR*-68-156. 1969; *AMR*, 1969, **9**, 6985.

VOLKOVA, N. S., Selection of a structure-force scheme (in Russian), *Trudy KAI*, **113**, 1969.

VOLKOVA, N. S., Particular cases of the selection of a structure-force truss scheme (in Russian), *Materialy k respublikanskoj konferentsii sektsii stroitel'noj mekhaniki Tatarskogo pravleniya NTO Strojindustrii*, Kazan' 1969.

WASIUTYŃSKI, Z., A theorem on the concentration of local reinforcement effect of a structure, *Bull. Acad. Pol. Sc., Sér. Sc. Tech.*, **17**, 187–190, 1969; *PBAM*, 1970, **2**, 8532.

WOJDANOWSKA-ZAJĄC, R., and ŻYCZKOWSKI, M., Optimum design of trusses with stability constraints (in Polish), *Rozpr. Inż.*, **17**, 3, 347–367, 1969; *AMR*, 1971, **4**, 2745; *PBAM*, 1970, **314**, 8694.

WOODS, W. J., Substructure optimization in the theory of structural synthesis, *AIAA Paper No.* **69-121**, New York 1969; *RŻM*, 1969, 8B629.

Zarghamee, M. S., Minimum weight design of enclosed antennas, *Proc. ASCE*, **95**, *ST* 6, *J. Struct. Div.* 1139–1152, 1969.

Zavelani-Rossi, A., Design of sandwich plates for minimum weight (in Italian), *Ist. Lombardo, Accad. Sci. Lett., Rend. Sci., Mat. Fis. Chem. Geol., A*, **103**, 2, 323–332, 1969; *AMR*, 1972, **8**, 6288.

Zavriev, K. S., and Khuberyan, K. M., Development of the methods of optimization of deformable systems in the USSR (in Russian), *Konf. "Metody optymalizacji ustrojów odkształcalnych", Jabłonna*, cz. **II**, 9–10, Ossolineum 1969.

1970

Ajzenberg, Ya. M., and Kilimnik, L. Sh., On the optimum design criteria and limit state parameters for structures exposed to seismic effects (in Russian), *Stroit. Mekh. i Raschet Sooruzh.*, **6**, 29–34, 1970.

Akita, Y., and Kitamura, K., Studies on optimization of ship structures. Nonlinear programming and optimum design of box and I section girders, *JSNA, Japan*, **128**, Dec. 1970.

Arslanov, A. Sh., *Practical direct methods for the design of statically indeterminate steel trusses of minimum volume* (in Russian), Avtoreferat kandidatskoj dissertatsii (Thesis), Ural'skii Politekhnicheskii Institut, 1970.

Arman, J., and Vitte, W. J., Foundations of aeroelastic optimization and some applications to continuous systems, *Stanford Univ. Rep. SUDAAR*, **390**, January 1970.

Baublis, I. S., and Cypinas, I. K., Optimization of a system with centrally compressed elements working beyond the elastic limit (in Russian), *Lit. Mekh. Sb.*, **1** (6), 123–128, 1970; *AMR*, 1973, **3**, 2072.

Barnett, R. L., Optimum prestressed tubular columns, *Proc. ASCE* **96**, *ST 2, J. Struct. Div.*, 291–301, 1970; *RŻM*, 1970, 9B355.

Brandmaier, H. E., Optimum filament orientation criteria, *J. Comp. Mat.*, **4**, 422–425, 1970.

Brandt, A., Optimum design of structural sections (in Polish), *Prz. Mech.*, **29**, 21, 639–644, 1970; *PBAM*, 1972, **1**, 9390.

Chern, J. M., and Prager, W., Optimal design of rotating disk for given radial displacement of edge, *J. Opt. Theory and Appl.*, **6**, 2, 161–170, 1970.

Chern, J. M. and Prager, W., Optimal design of beams for prescribed compliance under alternative load, *J. Opt. Theory and Appl.*, **5**, 6, 424–431, 1970; *AMR*, 1971, **3**, 1913; *RŻM*, 1971, 8B826.

Cohn, M. Z., ed., An introduction to structural optimization, *Solid Mech. Div.*, **1**, Univ. Waterloo, Canada, 1970.

Corcoran, P. J., Configurational optimization of structures, *Int. J. Mech. Sc.*, **12**, 5, 459–462, 1970; *AMR*, 1971, **2**, 1147; *RŻM*, 1970, 12B859.

Cselenyi, J., Optimaldimensionierung auf Biegeschwingungen beanspruchter geschweisster I-Träger, *Hebzeuge und Fördermittel*, **10**, 3, 76–81, 1970; *RŻM*, 1970, 10B243.

Čypinas, I. K., Application of optimal control theory to the synthesis of compression bars (in Russian), *Lit. Mekh. Sb.*, **2** (7), 17–32, 1970.

Čyras, A. A., Dual optimization problems for a plastic-rigid body (in Russian), *Lit. Mekh. Sb.*, **1** (6), 5–16, 1970.

Čyras, A. A., Optimization problems in structural mechanics (in Russian), *Stroit. Mekh. i Raschet Sooruzh.*, **2**, 12–17, 1970.

Dzieniszewski, W., Optimum distribution of materials in fibrous bodies, *Bull. Acad. Polon. Sc., Sér. Sc. Techn.*, **18**, 7, 329–337, 1970.

DZIENISZEWSKI, W., Optimization of microreticular media (in Polish), *XIII Konf. Mech. Ciał Odkszt., Jaszowiec* 1970.

ESZTERGAR, E. P., and KRAUS, H., Analysis and design of ellipsoidal pressure vessel heads, *Pap. ASME*, No. **PVP-26**, 1970.

FOX, R. L., and KAPOOR, M. P., Structural optimization in the dynamic response regime: A computational approach, *AIAA J.*, **8**, 10, 1798–1804, 1970; *AMR*, 1971, **7**, 5148.

FRAJNT, M. YA., Application of the random search method to optimum design problems (in Russian), *Stroit. Mekh. i Raschet Sooruzh.*, **1**, 30–33, 1970.

GAJEWSKI, A., Some problems of the optimum design of a bar compressed by a polar force (in Polish), *Mech. Teor. i Stos.*, **8**, 2, 159–173, 1970; *PBAM*, 1971, **2**, 8996; *RŻM*, 1971, 6B621.

GAJEWSKI, A., and ŻYCZKOWSKI, M., Optimal design of elastic columns subject to general conservative behaviour of loading, *J. Appl. Math. Phys. ZAMP*, **21**, 5, 806–816, 1970; *AMR*, 1972, **10**, 7943.

GAJEWSKI, A., and ŻYCZKOWSKI, M., Influence of longitudinal non-homogeneity on the optimal shape of a bar compressed by a force pointing towards the pole, *Bull. Acad. Pol. Sci., Sér. Sci. Tech.*, **18**, 1, 19–27, **1970**; *RŻM*, **1970, 11B433.**

GAVARINI, C., and VENEZIANO, D., Minimum weight limit design under uncertainty, *Mechanics, J. of AIMETA*, **7**, 2, 1970.

GLINKIN, I. D., and KOZACHEVSKIJ, A. I., Optimum design of statically indeterminate elastic rod systems for multiple loading (in Russian), *Stroit. Mekh. i Raschet Sooruzh.*, **4**, 21–24, 1970.

GRIERSON, D. E., and COHN, M. Z., A general formulation of the optimal frame problem, *J. Appl. Mech., Trans. ASME*, **70**, 356–360, June 1970.

HEGER, F. J., Design of FRP fluid storage vessels, *Proc. ASCE* **96**, *ST* 11, *J. Struct. Div.*, 2465–2499, 1970.

HEMP, W. S., and CHAN, H. S. Y., Optimum design of pin-jointed frameworks, *Aeronaut. Res. Council, London, Rep. and Mem.*, 3672, 1970; *AMR*, 1971, **4**, 2752; *RŻM*, 1971, 6B963.

HOFMEISTER, L. D., and FELTON, L. P., Prestressing in structural synthesis, *AIAA J.*, **8**, **2**, 1970.

HOLNICKI-SZULC, J., and MARKS, W., Determination of optimal systems of prestressing cables in plane elastic media (in Polish), *XIII Konf. Mech. Ciał Odkszt., Jaszowiec* 1970.

JENDO, S., Optimization of rotationally symmetric cable lattices for uniform stress (in Polish), *Arch. Inż. Ląd.*, **16**, 1, 139–155, 1970; *AMR*, 1972, **2**, 1128; *PBAM*, 1971, **4**, 9274.

JENDO, S., Optimum design of prestressed two-surface hanging structures, *Arch. Inż. Ląd.*, **16**, 4, 585–606, 1970; *AMR*, 1972, **2**, 1100; *RŻM*, 1971, 6B965.

JENDO, S., The optimum design problem for cable structures (in Polish), *XIII Konf. Mech. Ciał Odkszt.*, Jaszowiec 1970.

KIRSCH, U., and RUBINSTEIN, M. F., Optimum prestressing by linear programming, *UCLA paper* **ENG-0670**, Dec. 1970.

KŁOSOWICZ, B., Sur la nonhomogenéité optimale d'une barre tordue, *Bull. Acad. Pol. Sci., Sér. Sc. Tech.*, **18**, 8, 611–615, 1970; *RŻM*, 1971, 3B92.

KNIZHNIK, G. G., Optimum elastic design of flexural systems (in Russian), *Prikl. Mekh.*, **6**, 6, 69–75, 1970.

KORDAS, Z., and ŻYCZKOWSKI, M., Investigation of the shape of thick-walled non-circular cylinders showing full plasticization at the collapse, *Bull. Acad. Pol. Sci., Sér. Sc. Tech.*, **18**, 10, 465–473, 1970.

KOZŁOWSKI, W., and MRÓZ, Z., Optimal design of discs subject to geometric constraints, *Int. J. Solids Structures*, **12**, 1007–1021, 1970.

KRUTINIS, A. A., and ČIŽAS, A. P., Methods for optimum distribution of loads over elastic-plastic one-dimensional systems with deformation constraints (in Russian), *Lit. Mekh. Sb.*, **2** (7), 63–74, 1970.

KUCZYŃSKA, M. and MARKS, W., Optimization of concrete elements prestressed by a dense system of wires (in Polish), *Arch. Inż. Ląd.*, **16**, 3, 1970; *PBAM*, 1971, **3**, 9128; *AMR*, 1971, **8**, 5964.

KUCZYŃSKA, M., and MARKS, W., Optimization of concrete elements prestressed with dense, thin wires, *Proc. VI Congress FIP*, Praha 1970.

KUENZI, E. W., Minimum weight structural sandwich, *US Dep. of Agr.*, *Res. Note* **FPL-086**, 1970; *AMR* 1971, **4**, 2740.

LEŚNIAK, Z. K., *Methoden der Optimierung von Konstruktionen unter Benutzung von Rechenautomaten* (translation from Polish), W. Ernst u. Sohn., Berlin 1970.

MARTIN, J. B., Optimal design of elastic structures for multipurpose loading, *J. Opt. Theory and Appl.*, **6**, 1, 22–40, 1970; *AMR*, 1971, **4**, 2743; *RŻM*, 1971, 8B825.

MASUR, E. F., Optimum stiffness and strength of elastic structures, *Proc. ASCE*, **96**, *EM* 5, *J. Engng. Mech. Div.*, 621–639, 1970.

MATVEEV, S. N., On the design of statically indeterminate trusses of minimum weight for dynamic loads and dead weight (in Russian), *Materialy tret'ej nauchno-tekhnicheskoj konferentsii Ural'skogo Politekhnicheskogo Instituta*, Sverdlovsk 1970.

MATVEEV, S. N., Design of trusses of minimum volume for the simultaneous action of the dead weight, a fixed static load and a dynamic load of pulsed type (in Russian), *Materialy respublikanskoj konferentsii Chuvashskogo pravleniya NTO Strojindustrii*, Cheboksary 1970.

MATVEEV, S. N., On the design of optimal statically determinate trusses for dynamic loads and dead weight (in Russian), *Impul'snoe nagruzhenie konstruktsii*, Cheboksary 1970.

MELCHERS, R. E., and ROZVANY, G. I. N., Optimum design of reinforced concrete tanks, *Proc. ASCE*, **96**, *EM6*, *J. Engng. Mech. Div.*, 1093–1105, 1970.

MIHAJLISHCHEV, V. YA., A practical method for the optimum design of metal structures with typical elements (in Russian), *Stroit. i Arkh.* **8**, 17–20, 1970.

MOSES, F., and STEVENSON, J. D., Reliability-based structural design, *Proc. ASCE*, **96**, *ST 2*, *J. Struct. Div.*, 1970.

MRÓZ, Z., Optimal design of structures of composite materials, *Int. J. Solids Structures*, **6**, 859–870, 1970.

MRÓZ, Z., Optimal design of elastic structures subjected to dynamic, harmonically-varying loads, *Zeitschrift für Angewandte Mathematik und Mechanik*, **50**, 5, 303–309, 1970.

MRÓZ, Z., and SHAMIEV, F. G., On optimal design of reinforced annular slabs, *Arch. Inż. Ląd.*, **16**, 4, 575–584, 1970; *AMR*, 1972, **2**, 1099; *PBAM*, 1972, **3**, 9729.

NAGYAVICHYUS, Yu. A., and ČYRAS, A. A., Optimization of rigid-plastic cylindrical shells (in Russian), *Lit. Mekh. Sb.*, **1**, 6, 25–32, 1970.

NAGASAYA, H., and HOMMA, Y., On the minimum weight of orthogonally stiffened plates under lateral pressure, *J. Soc. Naval Arch. Japan*, **127**, 1970.

NIEMIERKO, A., Optimization of prestressed bar structures for multiple loading (in Polish), *XIII Konf. Mech. Cial Odkszt.*, *Jaszowiec 1970.*

NIEMIROVSKI, YU. V., Weight consideration in structural design under creep conditions (in Russian), *Izv. AN SSSR*, *Mekh. Tverd. Tela*, **4**, 113–123, 1970.

NOWACKI, H. G., BRUSIS, F., and SWIFT, P. H., Tanker preliminary design; an optimization problem with constraints, *Trans. SNAME*, 1970.

OLHOFF, N., Optimal design of vibrating circular plates, *Int. J. Solids Structures*, **6**, 139–156, 1970.

OLSZOWSKI, B., Parametric optimization of a vibrating system with 2 degress of freedom (in Polish), *Rozpr. Inż.*, **18**, 2, 1970; *PBAM*, 1971, **3**, 9159.

OLSZOWSKI, B., Design of retaining walls by a nonlinear programming method (in Polish), *Czasop. Tech.*, **75**, 1, 1970.

OPRZĘDKIEWICZ, J., Optimization methods for structures with discrete dependences of parameters on geometric properties (in Polish), *Prace Inst. Obróbki Skrawaniem, Zesz. Nauk.*, **37**, 1–64, 1970; *PBAM*, 1970, **3**, 9136.

OWCZAREK, S., Two methods for determining an optimum profile of a junction between a beam and a column from photoelastic model studies (in Polish), *Sympozjum w zakresie badań doświadczalnych*, PTMTiS, Warszawa 1970.

OWCZAREK, S., The state of stress of a boundary deformed in the elastic range (in Polish), *XIII Konf. Mech. Ciał. Odkszt., Jaszowiec 1970.*

OWCZAREK, S., DEUX théorèmes sur l'optimization des champs des contraintes par rapport aux formes extérieures, *Revue Roumaine des Sc. Tech., Série Méc. Appl.*, **3**, 15, 1970; *RŻM*, 1971, 4B705; *AMR*, 1971, **9**.

PALMER, A. C., and SHEPPARD, D. J., Optimizing the shape of pin joined structures, *Proc. Instn. Civ. Engrs.*, **47**, 363–376, London 1970.

PEDERSEN, P., On the minimum mass layout of trusses, *AGARD Conf.* Proc. **36**, 1970.

PIRAS, Z., Linear programming and optimum design of structures, *Acta Tech. CSAV*, **15**, 6, 652–689, 1970.

PLAUT, R. H., On minimizing the response of structures to dynamic loading, *ZAMP*, **21**, 1004–1010, 1970; *RŻM*, 1971, 5B497.

POGARASI, G., The optimal form of prestressed concrete elements, *FIP Hungarian Group, FIP VI Congress, Prague 1970.*

PRAGER, W., Optimal thermoelastic design for given deflection, *Int. J. Mech. Sc.*, **12**, 8, 705–709, 1970; *AMR*, 1971, **3**, 1912; *RŻM*, 1971, 2B928.

PRAGER, W., Optimization of structural design, *J. Opt. Theory and Appl.*, **6**, 1, 1–21, 1970; *AMR*, 1971, **4**, 2742; *RŻM*, 1971, 8B398.

PRECUPANU, D., Le dimensionnement optimum des poutres métalliques prétensionnées, *Bull. Inst. Polit. Iasi*, **16**, 3, 37–46, 1970.

PROWATKE, G., Trägerkonstruktionen mit minimaler Masse unter Berücksichtigung technologischer Geschichtspunkte, *Schiffbauforschung*, **9**, 3, 100–110, 1970.

RAMASWAMY, G. S., and RAMAN, N. V., Optimum design of prestressed concrete sections for minimum cost by nonlinear programming, *Structural Engineering Research Centre Roorkee, India, FIP VI Congress*, Prague 1970.

REISS, M., and KIRSCH, U., Optimal configurations and cross-sections of structures, *Build. Sci.*, **5**, 3, 175–179, 1970.

ROSS, A. L., Optimizing the design of a product subject to shock conditions, *ASME* No. **DE-28**, New York 1970; *RŻM*, 1971, 3B1050.

ROZVANY, G. I. N., and COHN, M. Z., A lower bound approach to the optimal design of concrete frames and slabs, *Proc. ASCE*, **96**, *EM6*, *J. Eng. Mech. Div.*, 1013–1030, Dec. 1970.

ROZVANY, G. I. N., Concave programming in structural optimization, *Int. J. Mech. Sci.*, **12**, 131–142, 1970.

RUBIN, C. P., Minimum weight design of complex structures subject to a frequency constraint, *AIAA J.*, **8**, 5, 923–927, 1970; *AMR*, 1971, **3**, 1914.

SACCHI, G., and ZAVELANI-ROSSI, A., Optimal design of structures subjected to fixed and moving loads by means of linear programming (in Italian), *Ist. Lomb., Accad. Sci. Lett., Rend. Sci. Mat. Fis. Chim. Geol.*, **A 104**, 3, 485–497, 1970; *AMR* 1972, **5**, 3735; *RŻM*, 1971, 5B391.

SCHMIT, L. A., Literature review and assessment of the present position, in: *Structural design applications of mathematical programming techniques*, *AGRADograph* No. **149**, AGARD-AG-149, 1970.

SHIELD, R. T., and PRAGER, W., Optimal structural design for given deflection, *Zeit. Ang. Math. Phys.*, **21**, 2, 513–523, 1970; *AMR*, 1971, **7**, 5151; *RŻM*, 1971, 3B514.

SOFRONOV, YU. D., On the design of minimum-weight bars for longitudinal cyclic forces (in Russian), *IVUZ*, *Mashinostroenie*, **5**, 1970.

SOFRONOV, YU. D., Minimum volume beams under cyclic loads (in Russian), *Trudy KAI*, **116**, 10–17, 1970.

SPUNT, L., Optimization of redundant prestressed structures, *Proc. ASCE*, **96**, *ST* 12, *J. Stfuct. Div.*, 2589–2606, 1970; *AMR*, 1971, **5**, 3541; *RŻM*, 1971, 5B978.

SUNGATULLIN, R. Z., On the design of optimal beams with varying height (in Russian), *Trudy KAI*, **116**, 28–34, 1970.

TEMPLEMAN, A. B., Structural design for minimum cost using the method of geometric programming, *Inst. Civ. Engrs. Proc.*, **46**, 459–472, 1970; *AMR*, 1971, **4**, 2741.

TRAHAIR, N. S., and BOOKER, J. R., Optimum elastic columns, *Int. J. Mech. Sc.*, 12, 973–983, 1970; *RŻM*, 1971, 5B317.

TURKSTRA, C. J., Theory of structural design decisions, *SM Study No.* **2**, *Univ. Waterloo, Ontario*, 1970.

WALLACE, D., and SEIREG, A., Optimum design of prismatic bars subjected to longitudinal impact, *ASME* No. **DE-G**, New York 1970.

VEBRA, R. V., and ČYRAS, A. A., Optimum design of arches for limit equilibrium (in Russian), *Lit. Mekh. Sb.*, **1** (6), 85–95, 1970; *AMR*, 1973, **10**, 8098.

WANG, C. H., Optimum design of prestressed concrete members, *Proc. ASCE*, **96**, *ST* 7, *J. Struct. Div.*, 1521–1534, 1970; *AMR*, 1971, **4**, 2744; *RŻM*, 1971, 2B1023.

WEISSHAAR, T. A., An application of control theory methods to the optimization of structures having dynamic or aeroelastic constraints, *Stanford Univ. Rep. SUDAAR*, **412**, Oct. 1970.

WOJDANOWSKA-ZAJĄC, R., and ŻYCZKOWSKI, M., Optimal structural design of trusses with the conditions of elastic-plastic stability taken into account, *Bull. Acad. Polon. Sc., Sér. Sc. Techn.*, **18**, 9, 365–372, 1970; *AMR*, 1973, **5**, 3759.

YOKOO, Y., and NAKAMURA, T., The minimum weight design of Vierendeel frames, *Int. J. Solids and Struct.*, **6**. 3, 353–369, 1970; *AMR*, 1971, **7**, 5150.

ZARGHAMEE, M. S., Minimum weight design with stability constraint, *Proc. ASCE*, **96**, *ST* 8, *J. Struc. Div.*, 1697, 1970; *AMR*, 1971, **7**, 5149; *RŻM*, 1971, 3B1051.

1971

AGEJEV, A. I., and REJTMAN, M. I., Some problems of the optimum design of rod systems (in Russian), *Stroit. Mekh. i Raschet Sooruzh.*, **4**, 20–23, 1971.

ARMAND, J. L., Minimum-mass design of a plate-like structure for specified fundamental frequency, *AIAA J.*, **9**, 9, 1739–1745, 1971; *AMR*, 1972, **8**, 6285.

BABICHEV, P. E., On the design of minimum-volume trusses ensuring the allowable stiffness (in Russian), *Sopr. Mat. i Teoria Sooruzh.*, **13**, 126–137, 1971.

BARANIENKO, V. A., and POCHTMAN, YU. M., Dynamic programming in the synthesis of optimum rod systems (in Russian), *Sopr. Mat. i Teoria Sooruzh.*, **13**, 119–126, 1971.

BLOCK, D. L., Minimum weight design of axially compressed ring and stringer stiffened cylindrical shells, *AIAA Paper*, **147**, 1971.

BODACH, S., and LIEBIG, S., Optimale Gestaltung von Profilen bei einach siger Biegebeanspruchung, *Maschinenbautechnik*, **20**, 3, 117–124, 1971; *RŻM*, 1971, 8B795.

BRYZGALIN, G. I., Optimum design of locally orthotropic elastic bodies with a weak binder (in Russian), *Izv. AN SSSR, Mekh. Tverd. Tela*, **3**, 169–174, 1971.

CAMMAERT, A. B. C., Optimal design of multi-storey frames by dynamic programming, *Diss. Univ. Cambridge*, No. **7560**, 1971.

CELLA, A., and LOGCHER, R. D., Automated optimum design from discrete components, *Proc. ASCE*, **97**, *ST* 1, *J. Struct. Div.*, 1971; *RŽM*, 1971, 9B777.

CEMPEL, C., Optimization of an oscillating system subject to periodic pulsed forcing (in Polish), *Rozpr. Inż.*, **19**, 4, 657–666, 1971.

CHERN, J. M., Optimal design of beams for alternative loads and constraints on generalized compliance and stiffness, *Int. J. Mech. Sc.*, **13**, 8, 661–674; *AMR*, 1972, **8**, 6286.

CHERN, J. M., Optimal structural design for given deflection in presence of body forces, *Int. J. Solids and Structures*, **7**, 373–382, 1971.

CHERN, J. M., Optimal thermoelastic design for given deformation, *J. Appl. Mech.*, **38**, 538–540, 1971.

CHERN, J. M., and MARTIN, J. B., The multipurpose optimal design of elastic structures with a piecewise uniform cross-section, *ZAMP*, **22**, 834–855, 1971.

CHERN, J. M., and PRAGER, W., Minimum-weight design of statically determinate trusses subject to multiple constraints, *Int. J. Solids and Struct.*, **7**, 8, 931–940, 1971; *AMR*, 1972, **1**, 298.

COHN, M. Z., Optimal limit design for concrete structures, *Tech. Note* No. **2**, *Solid Mech. Div., Univ. Waterloo*, 1971.

ČYPINAS, I. K., On the optimization of elastic systems susceptible to stability loss (in Russian), *Lit. Mekh. Sb.*, **1**, 8, 35–49, 1971.

ČYPINAS, I. K., Optimization of a nonlinearly deformable rod structure under transverse-longitudinal bending (in Russian), *Lit. Mekh. Sb.*, **2**, 9, 17–26, 1971.

ČYRAS, A. A., The theory of optimization of a perfectly elastic-rigid body subjected to repeatedly varying deformation (in Russian), *Lit. Mekh. Sb.*, **1**, 8, 7–26, 1971.

ČYRAS, A. A., Proofs of the basic theorems for the optimization of a perfectly elastic-plastic body (in Russian), *Lit. Mekh. Sb.*, **1**, 8, 27–34, 1971.

DUPUIS, G., Optimal design of statically determinate beams subject to displacement and stress constraints, *AIAA J.*, **9**, 5, 981–984, 1971; *AMR*, 1973, **2**, 1210.

DZIENISZEWSKI, W., Optimization of lattice rod structures (in Polish), *Prace IPPT*, **35**, 1971.

DZIENISZEWSKI, W., Optimization of lattice rod structures, *Arch. Mech.*, **23**, 2, 223–248, 1971; *AMR*, 1973, **3**, 1645; *RŽM*, 1972, 1B949.

DZIENISZEWSKI, W., Optimization of elastic lattice structures designed on a presented surface, *Arch. Mech.*, **23**, 5, 663–680, 1971; *AMR*, 1973, **8**, 6326.

FELTON, L. P., and NELSON, R. B., Optimized components in frame synthesis, *AIAA J.*, **9**, 6, 1027–1031, 1971; *AMR*, 1972, **8**, 6287.

FOX, R. L., *Optimization methods for engineering design*, Addison-Wesley Publ. Co., Reading, Mass. 1971.

FRAJMAN, F. I., Optimum design of a thin-walled beam with box section under pure unsymmetrical bending (in Russian), *Prikl. Mekh.*, **7**, 11, 96–104, 1971.

GAJEWSKI, A., Optimum design of an elastic-plastic column subject to the general conservative behaviour of loading (in Polish), *Rozpr. Inż.*, **19**, 1, 65–83, 1971.

GAJEWSKI, A., and ŻYCZKOWSKI, M., Optimal forming of a bar compressed with subtangential force in elastic-plastic range, *Arch. Mech. Stos.*, **23**, 2, 147–165, 1971; *AMR*, 1973, **6**, 4607; *PBAM*, 1972, **3**, 9712; *RŽM*, 1971, 12B511.

GANELETS, I. B., and KOGAN, L. A., On the optimum type-design of structural elements (in Russian), *Stroit. Mekh. i Raschet Sooruzh.*, **3**, 57–60, 1971.

GELLATLY, R. A., and BERKE, L., Optimal structural design, *AFFDL-TR*-**70-165**, 1971.

GEMMERLING, A. V., On the methods of structural optimization (in Russian), *Stroit. Mekh. i Raschet Sooruzh.*, **2**, 20–22, 1971.

GERASIMOV, E. N., and VOLKOVA, N. S., Optimization of the structures of statically determinate trusses (in Russian), *Trudy KAI*, **139**, 92–98, 1971.

GJELSVIK, A., Minimum weight design of continuous beams, *Int. J. Solids Struct.*, **7**, 10, 1411–1425, 1971; *AMR*, 1972, **8**, 6284.

GLINKIN, I. D., and KOZACHEVSKIJ, A. I., Determination of load combinations in the optimum design of structures (in Russian), *Stroit. Mekh. i Raschet Sooruzh.*, **1**, 4–7, 1971.

GOBLE, G. G., and MOSES, F., Automated optimum design of unstiffened girder cross sections, *Engng. J. New York*, **8**, 2, 43–47, 1971; *AMR*, 1972, **2**, 1097.

GOBLE, G. G., and LAPAY, W. S., Optimum design of prestressed concrete beams, *J. Amer. Concr. Inst.*, **68**, 712–718, 1971; *AMR*, 1972, **10**, 8130; *RŻM*, 1972, 4B1033.

GOLDSHTEJN, YU. M., and SOLOMESHCH, M. A., Optimum design of rod systems with varying member-axis scheme (in Russian), *Stroit. Mekh. i Raschet Sooruzh.*, **4**, 25–29, 1971.

GOLIŃSKI, J., and DIETRYCH, M., Statistical methods of optimization (in Polish), *Zagadn. Drgań Nielin.*, **12**, 199–206, 1971; *PBAM*, 1972, **4**, 9784.

GRESZCZUK, L. B., and MILLER, R. J., Advanced design concepts for buckling-critical composite shell structures, *J. Aircraft*, **8**, 5, 367–373, 1971.

GRINEV, V. B., and FILIPPOV, A. P., Optimum design of structures with specified natural frequencies (in Russian), *Prikl. Mekh.*, **7**, 10, 19–26, 1971.

GROUNDY, P., Optimum beam layout for large areas, *Proc. ASCE*, **97**, *ST* 8, *J. Struct. Div.*, 2085–2096, 1971; *RŻM*, 1972, 3B1095.

HEER, E., and YANG, J. N., Optimization of structures based on fracture mechanics and reliability criteria, *AIAA Journal*, **9**, 4, 621–628, 1971.

HOLNICKI-SZULC, J., Certain methods of equalization of principal stresses in plane elastic media (in Polish), *Rozpr. Inż.*, **19**, 2, 1971.

HOLNICKI-SZULC, J., Determination of the prestressing cable systems corresponding to a given field of body forces (in Polish), *Arch. Inż. Ląd.*, **17**, 3, 515–527, 1971; *PBAM*, 1972, **4**, 9924.

HOLNICKI-SZULC, J., On the optimization of plane structures prestressed by continuous cable systems (in Polish), *Prace IPPT*, **16**, 1971.

HOLNICKI-SZULC, J., Optimization of plane elastic media subjected to a temperature field (in Polish), *XIV Konf. Mech. Ciała Stałego*, Krościenko 1971.

HUANG, N. C., Effect of shear deformation on optimal design of elastic beams, *Int. J. Solids Struct.*, **7**, 321–326, 1971.

HUANG, N. C., Optimal design of elastic beams for minimum-maximum deflection, *J. Appl. Mech., Trans. ASME, Series E* **38**, 4, 1078–1081, 1971.

HUANG, N. C., On principle of stationary mutual complementary energy and its application to optimal structural design, *ZAMP*, **22**, 3, 608–620, 1971.

HYCA, M., Some problems of the optimization of metal structures (in Czech), *Strojirenstvi*, **21**, 11, 663–668, 1971.

JENDO, S., Optimum design of axially-symmetrical suspended structures, *Rev. Roum. Sc. Tech., Sér. Méc. Appl.*, **16**, 3, 565–582, 1971; *AMR*, 1973, **2**, 1211.

JOHNSON, E. W., Approximations of optimum minimum volume frameworks, *Proc. 3rd Canad. Congr. Appl. Mech.*, 331–332, Calgary 1971; *RŻM*, 1971, 12B981.

JOHNSON, W., *et al.*, The displacement field and its significance for certain minimum weight two-dimensional frames using the analogy with perfectly plastic flow in metal working, *Int. J. Mech. Sc.*, **13**, 6, 547–561, 1971; *AMR*, 1973, **6**, 4606.

KARIHALOO, B. L., and NIORDSON, F. I., Optimum design of vibrating cantilevers, *Rep. No.* **15**, Lingby 1971.

KAVLIE, D., and MOE, J., Automated design of frame structures, *Proc. ASCE,* **97**, *ST* 1, *J. Struct. Div.*, 1971.

KEMPNY, S., Optimization of an axisymmetric hanging structure suspended by the upper ring (in Polish), *Rozpr. Inż.*, **19**, 1, 111–127, 1971; *PBAM*, 1972, **2**, 9520.

KHOLOPOV, I. S., On the rational strength of the tie-bar in prestressed trusses (in Russian), *Stroit. Mekh. i Raschet Sooruzh.*, **5**, 50–54, 1971.

KIRSCH, U., and RUBINSTEIN, M. F., Optimum design of prestressed beams, *UCLA Paper* **ENG-0671**, May 1971.

KITAMURA, K., Studies on optimization of ship structures. Optimum design of grillages, *J. Soc. Naval Arch. Japan*, **130**, 1971.

LAMBLIN, D. O., and SAVE, M. A., Minimum-volume plastic design of beams for movable loads, *Meccanica, J. Italian Ass. Theoret. Appl. Mech.*, **6**, 3, 157–163, 1971; *AMR*, 1972, **10**, 8117.

LAPAY, W. S., and GOBLE, G. G., Optimum design of trusses for ultimate loads, *Proc. ASCE*, **97**, *ST* 1, *J. Struct. Div.*, 1971; *RŻM*, 1971, 8B400.

LATOS, W., and ŻYCZKOWSKI, M., Optimum design of a rotating shaft with regard to combined fatigue strength (in Polish), *Arch. Bud. Masz.*, **18**, 2, 245–259, 1971; *AMR*, 1973, **6**, 4415; *PBAM*, 1972, **2**, 9523.

LEŚNIAK, Z. K., Optimization of the building system for steel industrial houses (in Polish), *II Krajowa Konf. Zast. Inform. Zarz. Proj. Przem. Bud.*, 321–326, Krynica 1971.

LIFSHITS, V. L., Determination of the parameters of minimum-weight structures (in Russian), *Stroit. Mekh. i Raschet Sooruzh.*, **2**, 67–68, 1971.

LIPSON, S. L., and RUSSEL, A. D., Cost optimization of structural roof systems, *Proc. ASCE*, **97**, *ST* 8, *J. Struct. Div.*, 1971; *RŻM*. 1972, 3B1162.

ŁUKASIEWICZ, S., and BORAJKIEWICZ, W., Optimum design of a bar transmitting a load into a sheet, *Arch. Bud. Masz.*, **18**, 1, 21–31, 1971; *AMR*, 1973, **4**, 2728; *PBAM*, 1972, **1**, 9399; *RŻM*, 1971, 9B168.

LUND, S., Optimum design of transverse frame structures in tankers, *Europ. Shipbuilding*, **20**, 5 and 6, 3–9 and 2–11, 1971.

MAJERCZYK-GOMUŁKOWA, J., and MIODUCHOWSKI, A., Optimum plastic nonhomogeneity of a bar subject to torsion with the limit load as a criterion, *Bull. Acad. Pol. Sc., Sér. Sc. Tech.*, **19**, 5, 369–376, 1971; *AMR*, 1973, **5**, 3584; *RŻM*, 1971, 12B510.

MAJID, K. J., and ELLIOTT, D. W., Optimum design of frames with deflection constraints by nonlinear programming, *Struct. Engr.*, **49**, 4, 179–188, 1971; *AMR*, 1972, **1**, 299.

MARKS, W., A certain criterion of optimization for prestressed structures under multiple loads (in Polish), *XIV Konf. Mech. Ciała Stałego. Krościenko 1971.*

MARTIN, J. B., The optimal design of beams and frames with compliance constraints, *J. Solids Struct.* **7**, 63–81, 1971.

MCNEILL, W. A., Structural weight minimization using necessary and sufficient conditions, *J. Opt. Theory and Appl.*, **8**, 6, 454–466, 1971; *AMR*, 1972, **8**, 6283.

MELOSH, R. J., The optimum approach to analysis of elastic continua, *Computers and Structures*, **1**, 1/2, 241–263, 1971.

MIODUCHOWSKI, A., Optimum plastic nonhomogeneity of a bar under torsion as a variational problem, *Bull. Acad. Pol., Sér. Sci. Tech.*, **19**, 6, 443–449, 1971; *RŻM*, 1972, 1B448.

MIODUCHOWSKI, A., Optimum plastic nonhomogeneity of a torsion bar as a numerical problem (in Polish), *Rozpr. Inż.*, **19**, 3, 501–512, 1971; *PBAM*, 1972, **4**, 9880.

MIODUCHOWSKI, A., Optimum plastic nonhomogeneity of a torus sector in torsion (in Polish), *Rozpr. Inż.*, **19**, 3, 523–531, 1971; *PBAM*, 1972, **3**, 9683.

MUKHOPANDHYAY, S., and RAO, J. K., Optimum design of bridge trusses by dynamic programming, *Ind. Inst. Techn., Rep.* **CE 2-71**, June 1971.

NAKAMURA, H., and MURAI, S., Optimal design and its methods (in Japanese), *J. Jap. Soc. Mech. Engrs.*, **74**, 629, 725–731, 1971; *RŻM*, 1972, 3B1057.

NEUMANN, A., LEŚNIAK, Z. K., *et al.*, Programierte Bemessung und Optimierung, *Technisch-Wissenschaftliche Abhandlung des ZIS Halle*, **86**, Halle-Saale 1971.

PIERSON, B. L., A survey of optimal structural design under dynamic constraints, *Int. J. for Numer. Methods in Engng.*, **4**, 4, 491–499, 1971.

PLAUT, R. H., On the optimal structural design for a nonconservative, elastic stability problem, *J. Opt. Theory and Appl.*, **7**, 1, 52–59, 1971.

PLAUT, R. H., Optimal structural design for given deflection under periodic loading, *Quarterly of Appl. Math.*, 315–318, 1971.

POCHTMAN, YU. M., Dynamic programming in optimization problems for structures susceptible to creep (in Russian), *D. AN SSSR*, **196**, 3, 553–555, 1971.

POCHTMAN, YU. M., and FILATOV, G. V., Optimization of the shape of the cross-sections of structural elements by the random search method (in Russian), *Stroit. Mekh. i Raschet Sooruzh.*, **4**, 23–25, 1971.

POPE, G. G., and SCHMIT, L. A., Structural design applications of mathematical programming techniques, *AGARD Report No.* **149**, 1971.

PRAGER, W., Foulkes mechanism in optimal plastic design for alternative loads, *Int. J. Mech. Sc.*, **13**, 11, 971–973, 1971, *AMR*, 1973, **4**, 2958.

PRAGER, W., Optimal design of statically determinate beams for given deflection, *Int. J. Mech. Sci.*, **13**, 10, 893–895, 1971; *RŻM*, 1972, 6B727.

RAUTU, S., and CHIROIU, V., Calcul des structures statiquement non déterminées de poids minimum soumises aux moments fléchissants et aux forces axiales, *Revue Roum. Sci. Tech. Sér. Méc. Appl.*, **16**, 2, 317–327, 1971; *AMR*, 1972, **10**, 8118.

REISS, R., and MEGAREFS, G. J., Minimal design of sandwich axisymmetric plates obeying Mises criterion, *Int. J. Solids Struct.*, **7**, 603–623, 1971.

REINSCHMIDT, K. F., Discrete structural optimization, *Proc. ASCE*, **97**, *ST* 1, *J. Struct. Div.*, 133–156, 1971.

REJTMAN, M. I., Optimum design of shells by means of the maximum principle (in Russian), *Izv. AN SSSR, Mekh. Tverd. Tela*, **3**, 175–179, 1971; *AMR*, 1973, **6**, 4428.

ROSENBLUETH, E., and Mendoza, E., Reliability optimization in isostatic structures, *Proc. ASCE*, **97**, *EM* 6, *J. Engng. Mech. Div.*, 1625–1642, 1971; *AMR*, 1972, **10**, 8116.

ROZVANY, G. I. N., Concave programming and piece-wise linear programming, *Int. J. Numer. Methods in Engng.*, **3**, 131–144, 1971.

ROZVANY, G. I. N., and ADIDAM, S. R., Optimal design of anisotropic cylindrical shells, *Monash University CE Reports*, **8**, 1971.

ROZVANY, G. I. N., and ADIDAM, S. R., Dual formulation of variational problems in optimal design, *Proc. Int. Conf. Design Automation*, Toronto 1971; *RŻM*, 1972, 5B348.

RUDISILL, C. S., and BHATIA, K. G., Optimization of complex structures to satisfy flutter requirements, *AIAA J.* **9**, 1487–1491, 1971.

RUDOLF, F., Auswahl optimaler Dachbinder unter Anwendung der dynamischen Optimierung, *Wiss. Z. Hochschule Bau. Leipzig.*, **3**, 159–163, 1971; *RŻM*, 1972, 8B676.

SACCHI, G., and ZAVELANI-ROSSI, A., Sul Progetto ottimale a shakedown di travatura continua soggetta a un generico programma di carico, *Costr. Metal.*, **23**, 4, 301–306, 1971; *RŻM*, 1972, 3B1369.

SACCHI, G., Formulation variationelle du poids minimal des structures, *Ist. Sci. Costr. Politech.* No. **510**, Milano 1971.

SHAMA, M. A., On the optimization of shear carrying material of large tankers, *J. Ship. Res.*, **15**, 1, 74–96, 1971; *RŻM*, 1971, 9B824.

SHEU, C. Y., Automatic minimum weight design of elastic redundant trusses under multiple static loading conditions, *AIAA Paper* No. **71-362**, New York 1971.

DE SILVA, B. M. E., Application of optimal control theory to some structural optimization problems, *ASME Reprint* **71-Vibr-66**, *Sept.* 1971; *RŻM*, 1972, 5B698.

SOLOV'EV, E. G., and SUCHKOV, V. N., Design of statically indeterminate thin-walled frameworks of minimum volume (in Russian), *Trudy KISI*, **13**, 264–276, 1971.

SPILLERS, W. R., A note on the design of beams, *J. Appl. Mech.*, **38**, 1073–1074, 1971.

SPUNT, L., *Optimum structural design*, N. J. Prentice-Hall, 1971; *AMR*, 1972, **7**, 5433.

STROUD, W. J., Minimum-mass isotropic shells of revolution subjected to uniform pressure and axial load, *National Aeronautics and Space Adm.*, *Tech. Note* **D-6121**, 1971; *AMR*, 1971, **4**, 2583.

SWISZCZOWSKI, S., Optimum design of continuous beams by dynamic programming (in Polish), *Czasop. Tech.*, **75**, 9, 9–12, 1971.

SZEFER, G., Deformable material continuum as a control system with spatially distributed parameters, *Arch. Mech.*, **23**, 6, 927–952, 1971.

TAYLOR, J. E., Optimal prestress against buckling: an energy approach, *Int. J. Num. Methods in Engng.*, **3**, 2, 249–259, 1971; *AMR*, 1973, **1**, 375; *RŻM*, 1971, 10B580.

TEMPLEMAN, A. B., The application of geometric programming to the optimum design of bridge structures, *Proc. PTRC Symp.*, 125–130, London 1971.

TOCHER, J. L., and KARNES, R. N., The impact of automated structural optimization on actual design, *AIAA Pap.* **71-361**, New York 1971.

TROFIMOVICH, V. V., PERMYAKOV, V. A., and MOSHKIN, L. S., On problem formulation for the optimum design of metal prestressed suspended systems (in Russian), *Sopr. Mat. i Teoria Sooruzh.*, **14**, 115–124, 1971.

VENKAYYA, V. B., Design of optimum structures, *J. Comp. Struct.*, **1**, 1, 265–309, 1971.

VINOGRADOV, A. I., Basic systems in optimum design theory (in Russian), *Prikl. Mekh.*, **7**, 10, 13–18, 1971.

VINOGRADOV, A. I., On the convergence of strength analysis in optimization problems (in Russian), *Stroit. Mekh. i Raschet Sooruzh.*, **3**, 11–13, 1971.

VINSON, J. R., and SHORE, S., Minimum-weight web-core sandwich panels subjected to uniaxial compression, *J. Aircraft*, **8**, 11, 843–847, 1971; *AMR*, 1973, 1, 374; *RŻM*, 1972, 6B274.

WANGDAHL, G., The external penalty function optimization technique and its application to ship design, *Rep. No.* **129**, *Dept. Naval Arch. Marine Engng.*, *Univ. Michigan*, 1971.

WARDASZKO, E., Optimum distribution of nodes in a lattice structure (in Polish), *Inż. i Bud.*, **28**, 10, 397–400, 1971; *PBAM*, 1972, **3**, 9698.

WASIUTYŃSKI, Z., BRANDT, A. M., DZIENISZEWSKI, W., JENDO, S., MARKS, W., and OWCZAREK, S., Foundations of optimization in the design of building structures (in Polish), Typescript elaborated by order of the ETOB, Warszawa 1971.

VOLKOVA, N. S., *Development of the Kazan' line of research in optimum structural design* (in Russian), Kazan. Aviats. Inst., Kazan' 1971.

WRIGHT, P. M., Optimized structural design—where to now?, "*Applications of Solid Mechanics*", *Study* No. **7**, *Univ. Waterloo, SM Division*, Waterloo 1971.

WRIGHT, P. M., and FENG, C. C., Optimum design of plane frames using a multi-mode scheme, *Trans. Engng. Inst. Canada*, **14**, A-5, I–IV, 1971; *AMR*, 1972, **2**, 1098.

ŻYCZKOWSKI, M., Optimal structural design in rheology, *J. Appl. Mech., Trans. ASME, Series* E **38**, 1, 39–46, 1971; *AMR* 1972, **7**, 5434.

ŻYCZKOWSKI, M., and GAJEWSKI, A., Optimal structural design in non-conservative problems of elastic stability, *IUTAM Symposium "Instability of Continuous Systems", Herrenalb 1969*, Springer-Verlag, 293–301, 1971.

1972

AKITA, Y., and KITAMURA, K., Studies of nonlinear programming optimization for ship structures and optimum design of box and I-section girders, *Jap. Shipbuild. Mar. Engng.*, **6**, 5, 15–27, 1972; *RŻM*, 1973, 6B715.

ANNAMALAI, N., LEWIS, A. D. M., and GOLDBERG, J. E., Cost optimization of welded plate girders, *Proc. ASCE* **98**, *ST* 10, *J. Struct. Div.*, 2235–2246, 1972; *RŻM*, 1973, 4B902.

ARMAND, J. L., Applications of the theory of optimal control of distributed-parameter systems to structural optimization, *NASA*, **TR-2044**, June 1972.

BALDUR, R., Structural optimization by inscribed hyperspheres, *Proc. ASCE*, **98**, *EM* 3, *J. Engng. Mech. Div.*, 503–518, 1972; *AMR*, 1973, **5**, 3757.

BATTERMAN, S. C., and FELTON, L. P., Optimal plastic design of doubly symmetric closed structures, *Int. J. Solids Struct.*, **8**, 6, 733–750, 1972.

BLACKEBY, D. R., and MANSFIELD, E. H., Optimum thickness variation in a curved strip under pure bending, *Aeronaut. Quart.*, **23**, 1, 77–86, 1972; *AMR*, 1973, **6**, 4605.

BOYKIN, W. H., and SIERAKOWSKI, R. L., Remarks on Pontryagin's maximum principle applied to a structural optimization problem, Aeronaut., **76**, 735, 175–176, 1972; *AMR*, 1972, **10**, 8120; *RŻM*, 1972, 10B893.

BRYZGALIN, G. I., On some criteria of the optimum design of non-homogeneous anisotropic bodies (in Russian), *Prikl. Mat. i Mekh.*, **36**, 4, 753–760, 1972.

CELLA, A., Properties of discrete optima in structural optimization, *Proc. ASCE* **98**, *ST* 3, *J. Struct. Div.*, 787–792, 1972; *RŻM*, 1972, 8B714.

CHARRETT, D. E., and ROZVANY, G. I. N., Extensions of the Prager–Shield theory of optimal plastic design, *Int. J. Nonlinear Mech.*, **7**, 1, 51–64, 1972; *RŻM*, 1972, 9B390.

CHERN, J. M., and PRAGER, W., Optimal design of trusses for alternative loads, *Ing.-Arch.*, **41**, 4, 225–231, 1972; *AMR*, 1975, **6**, 5102; *RŻM*, 1972, 11B852.

COHN, M. Z., and PARIMI, S. R., Optimal design of plastic structures for fixed and shakedown loadings, *Trans. ASME, J. Appl. Mech.*, pap. **72-WA/APM-9**, 1972.

COHN, M. Z., Optimal limit design for reinforced concrete structures. Inelasticity and nonlinearity of structural concrete, *SM Study*, **8**, *Univ. Waterloo Press*, 357–388, 1972.

COX, H. L., and GRAYLEY, M. E., The influence of production imperfections on design of optimum structures, *Contrib. Theory Aircraft Struct.*, 261–271, Delft 1972.

ČYRAS, A. A., Optimization theory of elastic-rigid bodies under repeated variable deformation, *Arch. Mech.*, **24**, 5, 1057–1071, 1972.

ČYRAS, A. A., Optimization theory of perfectly locking bodies, *Arch. Mech.*, **24**, 2, 203–211, 1972; *RŻM*, 1972, 11B449.

DAFALIAS, Y. F., and DUPUIS, G., Minimum-weight design of continuous beams under displacement and stress constraints, *J. Opt. Theory and Appl.*, **9**, 2, 137–154, 1972; *AMR*, 1973, **3**, 2071; *RŻM*, 1973, 1B392.

DANELOV, E. R., Design of rod systems of minimum weight from standard elements (in Russian), *Stroit. Mekh. i Raschet Sooruzh.*, **4**, 22–24, 1972.

DERMOTT, J. F., ABRAMS, J. I., and COHN, M. Z., Optimization of building systems, Inelasticity and nonlinearity of structural concrete, *AM Study* **8**, *Univ. Waterloo Press*, 493–512, 1972.

DISTEFANO, N., Dynamic programming and the optimum design of rotating disks, *J. Opt. Theory and Appl.*, **10**, 2, 109–128, 1972.

DISTEFANO, N., Dynamic programming and a max-min problem in the theory of structures, *J. Franklin Institute*, **294**, 5, 339–350, Nov. 1972.

DISTEFANO, N., and TODESCHINI, R., Invariant imbedding and optimum beam design with displacement constraints, *Int. J. Solids Struct.*, **8**, 8, 1073–1087, 1972; *RŻM*, 1973, 1B700.

DOROSHENKO, O. P., On zero-order methods in the theory of optimal systems (in Russian), *Soprot. Mat. i Teoria Sooruzh.*, **16**, 267–269, 1972.

DOROSZKIEWICZ, R. S., LIETZ, J., and OWCZAREK, S., Application of photoelastic model studies to the optimum design of plane structures (in Polish), *Mech. Teoret. i Stos.*, **4**, 10, 525–544, 1972; *RŻM*, 1973, 8B742.

DUPUIS, G., An iterative approach of structural optimization, *Int. J. Num. Meth. in Engng.*, **4**, 3, 331–336, May–June 1972; *AMR*, 1973, **8**, 6324; *RŻM*, 1972, 12B914.

DZIENISZEWSKI, W., JENDO, S., MARKS, W., OWCZAREK, S., and WASIUTYŃSKI Z., On mathematical methods of structural optimization (in Polish), *Prace IPPT PAN*, **50**, Warszawa 1972.

EIMER, Cz., Design and optimization of prestressed shells, *Arch. Inż. Ląd.*, **18**, nr 3/4, 1972.

EPPLER, R., Cylindrical shells of optimal torsional stiffness, Acta Mechanica, **14**, 4, 269–282, 1972; *AMR*, 1973, **7**, 5246.

ERMAK, E. M., Optimization of cross-sectional parameters of prestressed reinforced concrete elements (in Russian), *Soprot. Mat. i Teoria Sooruzh.*, **18**, 102–107, 1972.

FARKAS, J., Optimalbemessung und Vergleich von biegebeanspruchten dünnwandigen Trägern mit Kasten-, Kreisrohr- und Ovalquerschnitt, *Acta Tech. Acad. Sci. Hung.*, **72**, 3, 377–388, 1972.

FINIFTER, D., Structural optimization of a web frame, *S. M. and Naval Arch. Thesis, Dept. Ocean Engng., MIT*, July 1972.

FRAJMAN, F. I., A particular case of the optimum design of a thin-walled box beam in unsymmetrical bending (in Russian), *Prikl. Mekh.*, **8**, 2, 135–138, 1972.

FRANCIS, R., Optimal design of prestressed concrete continuous beams. Inelasticity and nonlinearity in structural concrete, *SM Study* No. **8**, 441–470, *Univ. Waterloo Press*, 1972.

FRAUENTHAL, J. C., Constrained optimal design of column against buckling, *J. Struct. Mech.*, **1**, 1, 79–89, 1972; *RŻM*, 1973, 8B282.

FU, K. C., The optimization of truss configuration, *Engng. J. Montreal*, **55**, 4, 1–6,1972; *AMR*, 1973, 9, 7267.

GAJEWSKI, A., Selected problems of the optimum design of bars (in Polish), *Czasop. Tech.*, **76**, 4, 25–32, 1972; *RŻM*, 1973, 4B946.

GAVARINI, C., and VENEZIANO, D., Minimum weight limit design under uncertainty, *Meccanica*, **7**, 2, 98–104, 1972.

GISVOLD, K. M., and MOE, J., A method for nonlinear mixed-integer programming and its application to design problems, *J. Engng. for Industry*, *Trans. ASME*, 353–364, May 1972.

GOLMAN, S. D., and RZHANITSYN, A. R., On the optimum distribution of reinforcement in reinforced concrete wall-beams (in Russian), *Stroit. Mekh. i Raschet Sooruzh.*, **2**, 52–53, 1972.

GUNNLAUGSSON, G. A., and MARTIN, J. B., A note on optimality conditions for trusses with a zero minimum cross-section, *Int. J. Mech. Sc.*, **14**, 10, 643–650, 1972; *AMR*, 1973, **10**, 8094; *RŻM*, 1973, 4B572.

HAUG, E. J., PAN, K. C., and STREETER, T. D., A computational method for optimal structural design. I Piece-wise uniform structures, *Int. J. for Numerical Methods in Engng.*, **5**, 2, 171–184, 1972; *RŻM*, 1973, 4B923.

HIRUTA, Y., *et al.*, Optimum design of space trusses, *9ème Congrès de l'Ass. Int. des Ponts et Charp., Amsterdam* 1972; *RŻM*, 1973, 7B815.

HOLNICKI–SZULC, J., Optimum prestressing of constant-thickness media in the plane state of stress (in Polish), *Rozpr. Inż.*, **20**, 2, 217–231, 1972.

HUANG, T., Minimum steel design of doubly reinforced sections, *J. ACI*, **69**, 510–512, 1972; *RŻM*, 1973, 1B869.

HUBER, J., Die Optimierung von Betontragwerken, *Österr. Ing. Zeit.*, **15**, 2, 40–48, 1972; *RŻM*, 1973, 6B824.

ISHIKAWA, H., and GRIERSON, D. E., Iterative optimal limit design of reinforced concrete frames. Inelasticity and nonlinearity of structural concrete, *SM Study* **8**, *Univ. Waterloo Press*, 389–412, 1972.

JENDO, S., Determination of shape of axi-symmetric hanging structures, *Proc. 1971 IASS Pacific Symp. Part II on Tension Struct. and Space Frames, Tokyo and Kyoto, Oct. 1971, Arch. Inst. Jap.*, ed. by Y. YOKOO, Tokyo 1972.

JUNG, R., Zur Frage deckengleicher Trag-Konstruktionen nach 2 Achsen, *Bauing.*, **47**, 12, 449–450, 1972; *RŻM*, 1973, 5B870.

KARIHALOO, B. L., and NIORDSON, F. I., Optimum design of vibrating beams under axial compression, *Arch. Mech.*, **24**, 5, 1029–1039, 1972; *RŻM*, 1973, 7B216.

KATO, B., NAKAMURA, Y., and ANRAKU, H., Optimum earthquake design of shear building, *Proc. ASCE*, **98**, *EM* 4, *J. Engng. Mech. Div.*, 891–910, 1972.

KIRSCH, U., Optimum design of prestressed beams, *Comp. Struct.*, **2**, 4, 573–583, 1972; *AMR*, 1973, **10**, 8095; *RŻM*, 1973, 2B767.

KIRSCH, U., REISS, M., and SHAMIR, U., Optimum design by partitioning into substructures, *Proc. ASCE*, **98**, *ST* 1, *J. Struct. Div.*, 249–267, 1972; *AMR*, 1972, **9**, 7200; *RŻM*, 1972, 6B759.

KŁOSOWICZ, B., and LURIE, K. A., On the optimal nonhomogeneity of a torsional bar, *Arch. Mech.*, **24**, 2, 239–250, 1972.

KIUSALAAS, J., Minimum weight design of structures via optimality criteria, *NASA TN* **D-7115**, 1975.

KOLUPAEV, A. N., Design of complex statically indeterminate trusses of minimum weight (in Russian), *Informatsionnyj listok* No. **239-72** *TatTsNTI*, 1972.

FU, KUAN-CHEN, The optimization of truss configutarions, *Trans. Civ. Engng. J., Engng. Inst. of Canada*, April 1972.

KUZMANOVIC, B. O., and WILLEMS, N., Optimum plastic design of steel frames, *Proc. ASCE* **98**, *ST* 8, *J. Struct. Div.* 1697–1723, 1972: *AMR*, 1973, **7**, 5447.

KWIECIŃSKI, M., and KLEIBER, M., Minimum-weight design of ideally plastic rectangular dense grid, *Arch. Inż. Ląd.*, **18**, 2, 175–183, 1972; *AMR*, 1973, **10**, 8097; *RŻM*, 1972, 10B491.

LEPIK, U., Minimum weight design of circular plates with limited thickness, *Int. J. Nonlinear Mech.*, **7**, 4, 353–360, 1972; *AMR*, 1973, **4**, 2962.

LIBRESEN, L., and BEINER, L., On the weight optimization problem for supersonic rectangular flat panels with specified flutter speed, *Rev. Roum. des Sci. Tech., Sér. de Méc. Appl.*, **17**, 5, 1087–1102, 1972.

LIU, J. S., Constrained optimization by external penalty function method and application to structural design, *Univ. Michigan, Dept. Naval Arch. and Marine Engng., Rep.* NA **591**, 1972.

LURJE, K. A., Optimum design of nonhomogeneous continuous bodies (in Russian), *XIII Mezhd. Kongres po Teoret. i Prikl. Mekh.*, Nauka, Moskva 1972.

MAIER, G., ZAVELANI-ROSSI, A., and BENEDETTI, D. A., A finite element approach to optimal design of plastic structures in plane stress, *Int. J. Num. Meth. in Engng.*, **4**, 4, 455–473, 1972; *AMR*, 1973, **6**, 4603.

MAJID, K. J., and ANDERSON, D., Optimum design of hyperstatic structures, *Int. J. Num. Meth. in Engng.*, **4**, 4, 561–578, 1972; *AMR*, 1973, **4**, 2963; *RŻM*, 1973, 2B781.

MARTIN, J. B., and PONTER, A. R. S., The optimal design of a class of beam structures for a non-convex cost function, *J. de Méc.*, **11**, 2, 341–360, 1972; *RŻM*, 1973, 3B912.

MAU, S., and SEXSMITH, G. G., Minimum expected cost optimization, *Proc. ASCE* **98**, *ST* 9, *J. Struct. Div.*, 2043–2058, 1972; *AMR*, 1973, **6**, 4610.

MIKHAJLISHCHEV, V. YA., On the theory of optimum design of statically determinate parallel-rib trusses (in Russian), *Stroit. Mekh. i Raschet Sooruzh.*, **3**, 10–15, 1972.

MORRIS, A. J., Structural optimization by geometric programming, *Int. J. Solids Struct.*, **8**, 7, 847–864, 1972; *RŻM*, 1972, 12B886.

MULLIN, J. V., and MAZZIO, V. F., Optimizing composite properties, *SAMPE Quart.*, **3**, 2, 22–27, 1972.

MUNRO, J., KRISHNAMOORTHY, C. S., and YU, C. W. Optimal design of reinforced concrete frames, *The Structural Engineer.*, **50**, 7, 259–264, 1972.

MURZEWSKI, J., Structural safety optimization for extremal distributions of load and load-carrying capacity (in Polish), *Arch. Inż. Ląd.*, **18**, 3/4, 573–583, 1972.

NAGTEGAAL, J. C., On optimal design of prestressed elastic structures, *Int. J. for Mech. Sci.*, **14**, 11, 1972.

NIEMIERKO, A., On optimization of statically indeterminate hinged rod structures for multiple loading conditions (in Polish), *Arch. Inż. Ląd.*, **18**, 1, 75–86, 1972; *PBAM*, 1973, **1**, 10099; *RŻM*, 1972, 9B642.

NIEMIERKO, A., Truss design as a linear programming problem (in Polish), *Arch. Inż. Ląd.*, **18**, 3/4, 531–541, 1972.

OLHOFF, N., Optimal design of vibrating rectangular plates, *DCAMM Rep.* **37**, *The Technical University of Denmark, Lyngby*, Dec. 1972.

OL'KOV, YA. I., and KHOLOPOV, I. S., Determination of member stability parameters in the optimization of statically indeterminate trusses (in Russian), *IVUZ, Stroitel'stvo i i Arkhitektura*, **7**, 1972.

OPARIN, A. A., KUZMINA, L. Z., and PLAKHOTNIK, A. I., On a certain method of the optimum design of building structures (in Russian), *Stroit. i Arkh.*, **9**, 27–30, 1972.

PALMER, A. C., and SHEPPARD, D. J., Optimal design of transmission towers by dynamic programming, *Comp. and Struct.*, **2**, 455–468, 1972.

PEDERSEN, P., On the optimal layout of multi-purpose trusses, *Int. J. Comp. and Struct.*, **2**, 5/6, 1972; *RŻM*, 1973, 4B970.

PIERSON, B. L., A survey of optimal structural design under dynamic constraints, *Int. J. Num. Meth. in Engng.*, **4**, 4, 491–499, 1972; *AMR*, 1973, **10**, 8096; *RŻM*, 1973, 2B200.

PLAUT, R. H., Sufficient optimality conditions for some structural design problems, *J. Appl. Math. Physics* (*ZAMP*), **23**, 257–264, 1972; *RŻM*, 1973, 2B321.

PLAUT, R. H., On the stability and optimal design of elastic structures, *Stability: Special Lectures*, **547-577**, Univ. Waterloo, Canada, 1972.

POCHTMAN, YU. M., Optimization of structures with constraints on dynamics and frequency characteristics (in Russian), *Dokl. AN SSSR*, **203**, 2, 307–309, 1972.

POCHTMAN, YU. M., and BARANENKO, A. A., Application of dynamic programming to the design of beams of minimum weight subject to stress and deformation constraints (in Russian), *Issl. Teorii Sooruzh.*, **19**, 108–113, Stroizdat, 1972.

POCHTMAN, YU. M., and FILATOV, G. V., Optimum design of beams under dynamic loads by the random search method (in Russian), *Sopr. Mat. i Teoria Sooruzh.*, **18**, 72–78, 1972.

PORTMAN, K. D., Darstellung der Abhängigkeiten und Verknupungen von Kriterien für eine optimale Entwicklung und Beurteilung von Elementen für Bausysteme, *Diss. Hannover* 1972; *RŻM*, 1973, 9B703D.

PRAGER, W., Conditions for structural optimality, *Comp. and Struct.*, **2**, 5, 833–840, 1972.

RADTSIG, YU. A., Methods of synthetic design of trusses based on the application of mirror functions (in Russian), *Arch. Inż. Ląd.*, **18**, 3/4, 479–488, 1972.

REJTMAN, M, I., An approximate design method for elastic-plastic structures of minimum weight under dynamic loads (in Russian), *Issl. po Teorii Sooruzh.*, **19**, 41–46, 1972.

ROZVANY, G. I. N., and ADIDAM, S. R., Dual formulation of variational problems in optimal design, *J. Engng. for Industry, Trans. ASME, B94*, **2**, 409–418, May 1972; *AMR*, 1973, **2**, 1208.

ROZVANY, G. I. N., and ADIDAM, S. R., Structural optimization with piece-wise concave cost functionals, *Int. J. Solids Struct.*, **8**, 661–677, 1972.

SAVE, M. A., and MASSONNET, C. E., *Plastic analysis and design of plates, shells and disks*, North-Holland Publ. Co., Amsterdam–London, 15, 478, 1972.

SCHUMANN, W., and WUTHRICH, W., Über Schalen gleicher Festigkeit, *Acta Mech.*, **14**, 189–197, 1972.

SAYIR, M., and SCHUMANN, W., Zu den anisotropen Membranschalen mit gegebenfalls gleicher Festigkeit *ZAMP*, **23**, 815–827, 1972.

SEIREG, A., A survey of optimization of mechanical design, *J. Engng. for Industry, Trans. ASME*, **B 94**, 2, 495–499, 1972; *AMR*, 1973, **6**, 4604.

SHAJKEVICH, V. D., Synthesis of optimal systems by the geometric programming method (in Russian), *Stroit. Mekh. i Raschet Sooruzh.*, **4**, 15–20, 1972.

SHEU, C. Y., and SCHMIT, L. A., Minimum weight design of elastic redundant trusses under multiple static loading conditions, *AIAA J.*, **10**, 2, 155–162, 1972; *AMR*, 1973, **3**, 2073.

SHEPPARD, D. J., and PALMER, A. C., Optimal design of transmission towers by dynamic programming, *Comp. and Struct.*, **2**, 4, 455–468, Sept. 1972; *AMR* 1973, **9**, 7265.

SHU-T'IEN LI, and RAMAKRISHNAN, V., Optimum design of prestressed concrete combined bearing and sheet piles, *J. Prestr. Concr. Inst.*, **17**, 5, 39–48, 1972; *RŻM*, 1973, 5B1015.

SIEGEL, S., A flutter optimization program for aircraft structural design, *AIAA Paper No.* **72–795**, 1972.

SIMITSES, G. J., KAMAT, M. P., and SMITH, C. V., The strongest column by the finite element displacement method, *AIAA Paper* No. **72–141**, 1972.

SOFRONOV, YU. D., Determination of the critical load for stability loss in the plane bending of a minimum-weight beam (in Russian), *Materialy respublikanskoj konferentsii po optimal'nomu proektirovaniyu*, Kazan' 1972.

SOLOVEV, E. G., and SUCHKOV, V. N., Design of thin-walled space frameworks of minimum volume (in Russian), *Sopr. Mater. i Teoria Sooruzh.*, **16**, 273–277, 1972.

SPILLERS, W. R., and FRIEDLAND, L., On adaptive structural design, *Proc. ASCE*, **98**, *ST* 10, *J. Struct. Div.*, 2155–2163, 1972; *RŻM*, 1973, 4B1052.

TEMPLEMAN, A. B., Geometric programming with examples of the optimum design of floors and roof systems, *Proc. Symp. Comp. Aided Struct. Design., Warwick Univ., 1972.*

THAKKAR, M. C. and BULSARI, B. S., Optimal design of prestressed concrete poles, *Proc. ASCE*, **98**, *ST* 1, *J. Struct. Div.*, 61–74, 1972; *AMR*, 1972, **10**, 8115; *RŻM*, 1972, 6B875.

THOMPSON, J. M. T., and LEWIS, G. M., On the optimum design of thin-walled compression members, *J. Mech. Phys. Solids*, **20**, 101–109, 1972.

VANDERPLAATS, G. N., and MOSES, F., Automated design of trusses for optimum geometry, *Proc. ASCE*, **98**, *ST* 3, 671–689, 1972; *RŻM*, 1972, 12B888.

VERMA, M. K., and KRISHNA MURTY, A. V., Effect of geometrical nonlinearity on the minimum mass design of fully stressed beams, *J. Appl. Mech., Trans. ASME*, **E39**, 2, 627–628, June 1972; *AMR*, 1973, **1**, 167.

VENKATESWARA, R. S., and DAYRATNEM, P., Minimum weight design of frameworks, *9ème Congrès de l'Ass. Int. des Ponts et Charp., Amsterdam 1972*; *RŻM*, 1973, 7B704.

WALVEKAR, A. G., MEHTA, K. C., and TESKE, C. E., Optimal design of indeterminate truss using geometric programming, *J. Aeronaut. Soc. India*, **24**, 2, 308–310, May 1972; *AMR*, 1973, **9**, 7266.

WARKENTHIN, W., Kostenoptimale Querschnittsformen für Biegeträger, Hebzeuge und Fördermittel, **12**, 5, 130–134, 1972; *RŻM*, 1972, 10B1094.

WIŚNIEWSKI, Z., Analogue methods of strength optimization of structures (in Polish), *Przegl. Mech.*, **31**, 18, 560–563, 1972; *RŻM*, 1973, 3B989.

VLADIMIRSKIJ, V. A., TROFIMOVICH, V. R., and PERMYAKOV, V. A., On the design of metal prestressed combined systems of minimum cost (in Russian), *Sopr. Mat. i Teoria Sooruzh.*, **17**, 197–214, 1972.

VOLKOVA, N. S., On trusses subjected to optimum loads (in Russian), *Arch. Inż. Ląd.*, **18**, 3/4, 489–497, 1972; *AMR*, 1975, **1**.

WOJDANOWSKA, R., and ŻYCZKOWSKI, M., Optimum design of trusses under creep conditions in connection with the Kempner–Hoff buckling theory (in Polish), *Arch.. Inż. Ląd.*, **18**, 3/4, 511–530, 1972; *RŻM*, 1973, 6B298.

WRIGHT, P. M., Optimized structural design—where to now? *Applications of Solid Mechanics, SM Study* **7**, 253–265, Univ. Waterloo, 1972.

YOKOO, Y., NAKAMURA, T., and KEII, M., The minimum weight design chart of a portal frame based upon the axial force-bending moment interaction yield condition, *Ann. Meet. AIJ*, 701–702, Oct. 1972.

YOKOO, Y., NAKAMURA, T., SAITO, K., and KEII, M., The minimum weight design of frames based upon the axial force-bending moment interaction yield condition, *Ann. Meet. AIJ*, 699–700, Oct. 1972.

ŻYCZKOWSKI, M., and WOJDANOWSKA-ZAJĄC, R., Optimal design with respect to creep buckling, *IUTAM Symposium "Creep in Structures" Gothenburg 1970*, 369–387, Springer Verlag, 1972.

1973

AGUILAR, R. J., *et al.*, Computerized optimization of bridge structures, *Comp. and Struct.*, 3, 3, 429–442, 1973; *RŻM*, 1973, 10B697.

ANG, A. H. S., Cost effectiveness of research in design optimization under uncertainty, *IUTAM Symp. "Optimization in Structural Design"*, Warszawa 1973.

ARMAND, J. L., Applications of optimal control theory to structural optimization: analytical and numerical approach, *IUTAM Symp. "Optimization in Structural Design"*, Warszawa 1973.

BANICHUK, N. V., Games problems in theory of optimal design, *IUTAM Symp. "Optimization in Structural Design"*, Warszawa 1973.

BANICHUK, N. V., Some problems of the optimum design of elastic beams for various classes of forces (in Russian), *Izv. AN SSSR, Mekh. Tverd. Tela*, **5**. 102–110, 1973.

BANNA, Al, S., A unified approach to the problem of optimization in the design of structures, *AIAA Paper No.* **73-337**, New York 1973.

BARANENKO, V. A., POCHTMAN, YU. M., and FILATOV, G. V., On simultaneous application of the methods of dynamic programming and random search in optimum design problems (in Russian), *Stroit. Mekh. i Raschet Sooruzh.*, **1**, 3–6, 1973.

BORKAUSKAS, A., and ATKOČIUNAS, J., Optimal design for cyclic loading, *IUTAM Symp. "Optimization in Structural Design"*, Warszawa 1973.

BORODACHEV, N. A., On the optimum design of beams and frames (in Russian), *Stroit. Mekh. i Raschet Sooruzh.*, **2**, 20–23, 1973.

BRACH, R. M., Optimized design: characteristic vibration shapes and resonators, *J. Acoust. Soc. Amer.*, **53**, 1, 113–119, 1973, *RŻM*, 1973, 9B257.

BRACH, R. M., On optimal design of vibrating structures, *J. Opt. Theory and Appl.*, **11**, 6, 662–667, 1973.

BRANDT, A. M., Optimization of building structures (in Polish), General lecture on *II Kongr. Nauki Polskiej, Sek. VIII, Podsek. Konstr. Inż. Most. Hydrotech.*, Politechnika Krakowska, 1973.

CHAN, H. S. Y., Symmetric plane frameworks of least weight, *IUTAM Symp. "Optimization in Structural Design"*, Warszawa 1973.

ČIŽAS, A. P., On optimal strain-hardening structures design, *IUTAM Symp. "Optimization in Structural Design"*, Warszawa 1973.

ČYRAS, A. A., On optimization theory in the mechanics of solids, *IUTAM Symp. "Optimization in Structural Design"*, Warszawa 1973.

DÖRSING, H., I-Träger mit optimalen Querschnitten, *Hebezeuge und Fördermittel*, **13**, 1, 10–11, 1973.

DOUTY, R., Optimization of light gage cold-formed steel shapes by parametric bandwith construction, *Comp. and Struct.*, **3**, 2, 299–313, 1973.

DUTTA, T. K., and RATNALIKAR, N. V., Optimal design of reinforced concrete slabs, *J. Inst. Engrs. India, C. E. Div.*, **54**, 47–52, 1973.

DWYER, W. J., BALDERES, T., and EMERTON, R. K., Optimization techniques in the structural design process, *AIAA Paper No.* **73-345**, New York 1973.

DZIENISZEWSKI, W., Spatial dense and regular network structures with equal axial deformability, *Symp. "On industrialized spatial and shell structures"*, *IASS*, 561–582, Kielce 1973.

DZIEWOLSKI, R., Computer-aided design and optimization of structures, *Symp. "On industrialized spatial and shell structures"*, *IASS*, 3–24, Kielce 1973.

EIMER, C., Design and optimization of prestressed shells, *Symp. "On industrialized spatial and shell structures"*, *IASS*, 191–200, Kielce 1973.

FARKAS, J., Minimum volume design of welded stiffened plates subjected to normal load and nonuniform temperature distribution, *Proc. IV Conf. on Dimensioning and Strength Calculation*, Budapest 1973.

FRAUENTHAL, J. C., Initial postbuckling behaviour of optimally designed columns and plates, *Int. J. Solids and Struct.*, **9**, 1, 115–127, 1973; *RŻM*, 1973, 2B302.

GALLAGHER, R. H., Fully stressed design, in: *Optimum Structural Design*, eds. GALLAGHER, R. H., and ZIENKIEWICZ, O. C., JOHN Wiley, London 1973.

GALLAGHER, R. H., and ZIENKIEWICZ, O. C., eds., *Optimum Structural Design, Theory and Application*, Wiley, New York 1973.

GANELETS, I. B., Refined approach to the optimization of the parameters of flexural reinforced concrete elements (in Russian), *Beton i Żelazobeton*, **2**, 43–45, 1973.

GELLATLY, R. A., and BERKE, L., Optimality-criterion-based algorithms, in: *Optimum Structural Design*, eds. R. H. GALLAGHER and O. C. ZIENKIEWICZ, John Wiley, London 1973.

GIERLIŃSKI, J., Some cases of rectangular uniform-strength three-layer plates supported by beam systems (in Polish), *Arch. Inż. Ląd.*, **1**, 1973.

GIERLIŃSKI, J., On the necessary and sufficient conditions of optimization of elastic lattice plates for minimum deformability, *Rozpr. Inż.*, **4**, 1973.

GLINKIN, I. D., KOZACHEVSKIJ, A. I., and PEKARSKIJ, A. L., Optimum reinforcing of reinforced concrete structures as "base-structure" systems (in Russian), *Stroit. Mekh. i Raschet Sooruzh.*, **3**, 52–55, 1973.

GWIN, L. B., and TAYLOR, R. F., A general method for flutter optimization, *AIAA J.*, **11**, 1613–1617, 1973.

HAMADA, M., On the optimum shapes of some axisymmetric shells, *IUTAM Symp. "Optimization in Structural Design"*, Warszawa 1973.

HEGEMIER, G. A., and TANG, H. T., A variation principle, the finite element method, and optimal structural design for given deflection, *IUTAM Symp. "Optimization in Structural Design"*, Warszawa 1973.

HEMP, W. S., *Optimum Structures*, Clarendon Press, Oxford 1973.

JENDO, S., and STACHOWICZ, A., Optimum design of suspension systems, *IUTAM Symp. "Optimization in Structural Design"*, Warszawa 1973.

JĘDRZEJCZYK, J., and KUBIK, J., Selection of the material parameters of a viscoelastic cable (in Polish), *Arch. Inż. Ląd.*, **19**, 4, 703–711, 1973.

KAMAT, M. P., and SMITSES, G. J., Optimal beam frequencies by the finite element displacement method, *Int. J. Solids and Struct.*, **9**, 3, 415–430, 1973.

KAPOOR, M. P., and HARIHARAN, M., Optimal design of reinforced concrete chimneys, *IUTAM Symp. "Optimization in Structural Design"*, Warszawa 1973.

KARIHALOO, B. L., and NIORDSON, F. I., Optimum design of vibrating cantilevers, *J. Opt. Theory and Appl.*, **11**, 638–654, 1973.

KEMPNY, S., Shapes of surface hanging structures from cables of minimum volume (in Polish), *Arch. Inż. Ląd.*, **19**, 4, 691–701, 1973; *AMR*, 1976, **2**, 1358.

KEMPNY, S., Determination of the shape of a surface hanging structure with one-way cables (in Polish), *Arch. Inż. Ląd.*, **19**, 4, 1973.

KESHAVA RAO, M. N., Optimal structural design of frames by nonlinear programming, *J. Struct. Engng. India*, **1**, 3, 126–137, 1973.

KHACHATURIAN, N., and TWISDALE, L. A. JR., Application of dynamic programming to optimization of structures, *IUTAM Symp. "Optimization in Structural Design"*, Warszawa 1973.

KIRSCH, U., Optimized prestressing by linear programming, *Int. J. Num. Meth. in Engng.*, **7**, 2, 125–136, 1973.

KIUSALAAS, J., Minimum weight design of structures via optimality criteria, *NASA Technical Note* **TND-7115**, 1973; *AMR*, 1973, **6**, 4609.

KIUSALAAS, J., Optimal design of structures with buckling constraints, *Int. J. Solids and Struct.*, **9**, **7**, 863–878, 1973.

KOSTEM, C. N., Optimum shaped pneumatic roofs, *Symp. "On industrialized spatial and shell structures"*, *IASS*, 77-88, Kielce 1973.

KÖNIG, J. A., On optimum shakedown design, *IUTAM Symp. "Optimization in Structural Design"*, Warszawa 1973.

KUAN-CHEN, FU, An application of search technique in truss configurational optimization, *Comp. and Struct.*, **3**, 2, 315–328, 1973.

KRISHNAMOORTHY, C. S., and MUNRO, J., Linear program for optimal design of reinforced concrete frames, *IABSE*, **33**, 1, 119–141, 1973.

KUESTER, J. L., and MIZE, J. H., *Optimization techniques with Fortran*, McGraw-Hill Book Co. Inc., 331–343, New York 1973.

KRAKOVSKIJ, M. B., On the optimum design of structures by the steep ascent method (in Russian), *Stroit. Mekh. i Raschet Sooruzh.*, **1**, 8–11, 1973.

LAI, Y. S., and ACHENBACH, J. D., Optimal design of layered structures under dynamic loading, *Comp. and Struct.*, **3**, 3, 559–572, 1973.

MAJID, K. I., and ELLIOT, D. W. C., Topological design of pin jointed structures by nonlinear programming, *Proc. Inst. Civ. Engrs.*, **55**, 2, 129–149, 1973.

MASUR, E. F., Some additional principles of optimization, *IUTAM Symp. "Optimization in Structural Design"*, Warszawa 1973.

MOLE, R. H., The minimum weight structural configuration of pin-jointed truss cantilevers of given external shape, *Int. J. Mech. Sci.*, **15**, 1, 49–63, 1973; *RŻM*, 1973, 9B669.

MIURA, H., KAVLIE, D., and MOE, J., Interactive and automated design of ship structures, *IUTAM Symp. "Optimization in Structural Design"*, Warszawa 1973.

MOSES, F., *Design for reliability. Concepts and applications. Optimum structural design*, J. Wiley and Son Inc., London 1973.

MOSES, F., Recent developments in the Case optimization program, *AGARD Conf. Proc. No. 123, 2nd Symp. on Struct. Opt.*, Milan 1973.

MOSKIN, L. S., Determination of optimum prestressing forces for combined suspension systems (in Russian), *Metallicheskie i plastmassovye konstruktsii*, Budivelnik, Kiev 1973.

MOSKIN, L. S., Optimum strengthening of prestressed suspension structures (in Russian), *Metallicheskie i plastmassovye konstruktsii*, Budivelnik, Kiev 1973.

MRÓZ, Z., and GAWĘCKI, A., Post-yield behaviour of optimal plastic structures, *IUTAM Symp. "Optimization in Structural Design"*, Warszawa 1973.

NAGTEGAAL, J. C., A superposition principle in optimal plastic design for alternative loads, *Int. J. Solids and Struct.*, **9**, 12, 1465–1471, 1973.

NAGTEGAAL, J. C., A new approach to optimal design of elastic structures, *Comp. Meth. in Appl. Mech. and Engng.*, **2**, 3, 255–264, 1973.

NAGTEGAAL, J. C., and PRAGER, W., Optimal layout of a truss for alternative loads, *Int. J. Mech. Sci.*, **5**, 7, 583–592, 1973.

NIEMIROVSKY, J. V., Account of transversal shear in optimum design of beams, plates and shells, *IUTAM Symp. "Optimization in Structural Design"*, Warszawa 1973.

NIORDSON, F. I., and KARIHALOO, B. L., Optimum design of vibrating shafts, *IUTAM Symp. "Optimization in Structural Design"*, Warszawa 1973.

NIORDSON, F. I., and PEDERSEN, P., A review of optimal structural design, *Proc. 13th Int. Congress of Appl. Mech., Moscow 1972*, eds. BECKER, E., and MIKHAILOV, G. K., 269–278, Springer Verlag, Berlin 1973.

OLHOFF, N., Formation of stiffeners by optimal design of plates, *IUTAM Symp. "Optimization in Structural Design"*, Warszawa 1973.

ONISIN, S. S., Optimum design of prestressed combined suspension systems with allowance for their deformability. Prospects of development and application to metalwork engineering (in Russian), *Ukr. resp. nauch.-tekhn. konf. 1973*, Dnepropetrovsk, Tezisy Dokladov, Kiev 1973.

PALMER, A. C., Dynamic programming and structural optimization, in: *Optimum Structural Design*, ed. by ZIENKIEWICZ, O. C., and GALLAGHER, R. H., John Wiley and Sons Inc., New York 1973.

PARIMI, S. R., and COHN, M. Z., Optimal criteria in probabilistic structural design, *IUTAM Symp. "Optimization in Structural Design"*, Warszawa 1973.

PARZNIEWSKI, Z., and STACHOWICZ, A., Rational design of plane cable girders (in Polish), *Arch. Inż. Ląd.*, **19**, **4**, 653–666, 1973.

PAVICIC, N. J., and BATTERMAN, S. C., Optimum design of fiber reinforced shells of revolution, *IUTAM Symp. "Optimization in Structural Design"*, Warszawa 1973.

PINES, S., and NEWMAN, M., Structural optimization for aeroelastic requirements, *AIAA Paper* **73-389**, New York 1973.

PLAUT, R. H., Approximate solution to some static and dynamic optimal structural design problems, *Quart. Appl. Math.*, **30**, 4, 535–539, 1973.

PLAUT, R. H., Optimal design for stability under dissipative, gyroscopic or circulatory loads, *IUTAM Symp. "Optimization in Structural Design"*, Warszawa 1973.

POCHTMAN, YU. M., and PJATIGORSKIJ, Z. I., On the optimum design of some adaptive structures (in Russian), *Stroit. Mekh. i Raschet Sooruzh.* 2, 23–26, 1973.

POPE, G. G., Optimum design of stressed skin structures using a sequence of linear programs method, *AIAA Paper* **73-342**, New York 1973.

PRAGER, W., Necessary and sufficient conditions for global structural optimality, *AGARD Conf. Proc. No. 123, 2nd Symp. on Struct. Opt.*, Milan 1973.

PRAGER, W., Minimum-weight design of a statically determinate truss subject to constraints on compliance, stress and cross-sectional area, *J. Appl. Mech.*, **40**, 1, 313–314, 1973.

RAO, S. S., Minimum cost design of concrete beams with a reliability based constraint, *Build. Sci.*, **8**, 1, 33–38, 1973; *RŻM*, 1973, 6B822.

RAVINDRA, M. K., and LIND, N. C., Theory of structural code optimization, *Proc. ASCE* **99**, *ST* 7, *J. Struct. Div.*, 1541–1553, 1973.

ROZVANY, G. I. N., Optimal force transmission by flexure—the present state of knowledge, *IUTAM Symp. "Optimization in Structural Design"*, Warszawa 1973.

ROZVANY, G. I. N., and ADIDAM, S. R., Recent advances in optimal plastic design, *Symposium on Foundations of Plasticity*, 201–217, Warsaw 1972, Noordhoff International Publishing, 1973.

ROZVANY, G. I. N., Optimal plastic design for partially pre-assigned strength distribution, *J. Opt. Theory and Appl.*, **11**, 4, 421–436, 1973.

ROZVANY, G. I. N., Basic geometrical properties of optimal flexural force transmission fields, *J. Struct. Mech.*, **2**, 4, 259–264, 1973; *AMR* 1976, **3**, 2188.

SACCHI, G., Comparison between the linear programming and the variational formulation of the optimal plastic design problem, *IUTAM Symp. "Optimization in Structural Design"*, Warszawa 1973.

SAFIN, R. K., Optimal design of arched roofs (in Russian), *Issl. po Teorii Plastin i Obolochek*, **10**, 407–411, 1973; *AMR* 1976, **3**, 2187.

SAVE, M., SACCHI, G., and MAIER, G., Optimal plastic design for movable loads, *IUTAM Symp. "Optimization in Structural Design"*, Warszawa 1973.

SCHMIT, L. A., and FARSHI, B., Some approximation concepts for structural synthesis, *AIAA paper* No. **73-341**, *AIAA(ASME)SAE 14th Struct. Dyn. Mat. Conf. Williamsburg*, Virginia, March 1973.

SHERMAN, Z., Weight minimization of axisymmetric clamped plates subject to constraints, *Int. Journ. Solids Struct.* **9**, 2, 279–290, 1973; *AMR*, 1973, **9**, 7046.

DE SILVA, M. E., and GRANT, G. N. C., Comparison of some penalty function based optimization procedures for the synthesis of a planar truss, *Int. J. Num. Methods in Engng.*, **7**, 2, 125–136, 1973.

SIMODYNES, E. E., Gradient optimization of structural weight for specified flutter speed, *AIAA Paper* **73-390**, New York 1973.

SPILLERS, W. R., A graph problem of structural design, *IUTAM Symp. "Optimization in Structural Design"*, Warszawa 1973.

SPILLERS, W. R., and AL-BANNA, S., Optimization using iterative design techniques, *Comp. and Struct.*, **3**, 1263–1271, 1973.

SPUNT, L., A dimensionless programming approach to optimal structural design, *AIAA Paper* **73-344**, New York 1973.

STACHOWICZ, A., Some problems of the optimum design of cable lattice shells (in Polish), *Arch. Inż. Ląd.*, **19**, 4, 667–677, 1973.

SWITZKY, H., Minimum weight design of finite element structures, *AIAA Paper* **73-343**, New York 1973.

SZEFER, G., Optimal control of the consolidation process, *IUTAM Symp. "Optimization in Structural Design"*, Warszawa 1973.

SWISZCZOWSKI, S., Two cases of the optimum design of beam structures by dynamic programming (in Polish), *Arch. Inż. Ląd.*, **19**, 1, 83–98, 1973; *RŻM*, 1973, 7B703.

TADJBAKHSH, I., and FARSHAD, M., On conservatively loaded funicular arches and their optimal design, *IUTAM Symp. "Optimization in Structural Design"*, Warszawa 1973.

TAIG, I. C., and KERR, R. I., Optimization of aircraft structures with multiple stiffness requirements, *AGARD 2nd Symposium on Structural Optimization*, Milan 1973, AGARD, CP-123, 1973.

TAYLOR, J. E., On generalized variational formulations for structures design problems, *IUTAM Symp. "Optimization in Structural Design"*, Warszawa 1973.

TEMPLEMAN, A. B., Optimum truss design using approximating functions, *IUTAM Symp. "Optimization in Structural Design"*, Warstawa 1973.

TEMPLEMAN, A. B., and WINTERBOTTOM, S. K., Structural design applications of geometric programming, *AGARD Conf. Proc. No.* **123**, *2nd Symp. on Struct. Opt.*, Milan 1973.

THERMANN, K., Optimal design criteria of dynamically loaded elastic structures, *IUTAM Symp. "Optimization in Structural Design"*, Warszawa 1973.

TOAKLEY, A. R., BATTEN, D. F., and WILSON, B. G., Optimum plastic design of unbraced frameworks, *IUTAM Symp. "Optimization in Structural Design"*, Warszawa 1973.

TOUMA, A., and WILSON, J. F., Design optimization of prestressed concrete spans for high speed ground transportation, *Comp. and Struct.*, **3**, 2, 265–279, 1973; *RŻM*, 1973. 8B712.

TWISDALE, L. A., and KHACHATURIAN, N., Absolute minimum weight structures by dynamic programming, *Proc. ASCE* **99**, *ST* 11, *J. Struct. Div.*, 2339–2344, 1973.

TVERGAARD, V., On the optimum shape of a filler in a flat bar with restrictions, *IUTAM Symp. "Optimization in Structural Design"*, Warszawa 1973.

VANDERPLAATS, G. N., and MOSES, F., Structural optimization of feasible directions, *Comp. and Struct.*, **3**, 4, 739–755, 1973.

VENKAYYA, V. B., KHOT, N. S., and BERKE, L., Application of optimality criteria approaches to automated design of large practical structures, *AGARD 2nd Symp. on Struct. Opt.*, *Milan*, April 1973.

VEPA, K., On the existence of solutions to optimization problems with eigenvalue constraints, *Quart. Appl. Math.*, **31**, 3, 329–341, 1973.

YING-SAN LAI and ACHENBACH, J. D., Direct search optimization method, *Proc. ASCE* **99**, *ST* 1, *J. Struct. Div.*, 19–31, 1973.

YOKOO, Y., NAKAMURA, T., and KEII, M., The minimum weight design of multi-storey building frames based upon the axial force-bending moment interaction yield condition, *IUTAM Symp. "Optimization in Structural Design"*, Warszawa 1973.

ZAVELANI, A., A compact linear programming procedure for optimal design in plane stress *J. Struct. Mech.*, **2**, 4, 301–324, 1973; *AMR*, 1976, **3**, 2189.

ZAVELANI, A., MAIER, G., and BINDA, L., On the optimal profile of plane structures, *IUTAM Symp. "Optimization in Structural Design"*, Warszawa 1973.

ZIENKIEWICZ, O. C., and CAMPBELL, J. S., Shape optimization and sequential linear programming, in; *Optimum Structural Design*, eds.: GALLAGHER, R, H. and ZIENKIEWICZ, O. C., 109–126, Wiley, N. Y. 1973.

ŻYCZKOWSKI, M., and KRUŻELECKI, J., Optimal design of shells with respect to their stability, *IUTAM Symp. "Optimization in Structural Design"*, Warszawa 1973.

1974

ALI, M. M., and GRIERSON, D. E., Design of reinforced concrete structures for strength and deformability, *Solid Mech. Div.*, **124**, *Univ. Waterloo, Canada*, 1974.

ALSPAUGH, D. W., and KUNOO, K., Optimum configurational and dimensional design of truss structures, *Comp. and Struct.*, **4**, 4, 755–770, 1974; *AMR*, 1975, **12**, 10772.

ARASLANOV, A. M., A discrete method for solving quasistatic problems of statistical dynamics of elastic systems (in Russian), *Trudy KAI*, **168**, 23–28, 1974.

ARASLANOV, A. M., On the design of plates and shells for given reliability on the basis of random variable theory (in Russian), *Trudy KAI*, **172**, 4–8, 1974.

ARORA, J. S., Inverse problem of structural optimization, *Proc. ASCE*, **100**, *ST* 11, *J. Struct. Div.*, 1974.

BERKE, L., and KHOT, N. S., Use of optimality criteria methods for large scale systems, *Lect. Ser. on Struct. Opt.*, AGARD-LS-70, 1974.

BIRKELAND, H. W., How to design prestressed concrete beams of minimum cross section, *J. ACI*, **71**, 12, 634–641, 1974; *AMR*, 1975, **9**, 7997.

BORYSEVICH, A., Design of thin-walled curved beam of minimum volume (in Russian), *Tr. Mosk. Avtomob.-Dor. Inst.*, 1974.

BORYSEVICH, A., Design of optimum continuous beams for fixed and transient loads (in Russian), *Stroit. Konstr. i Teoria Sooruzh.*, Minsk 1974.

BRADLEY, J., BROWN, L. H., and FEENEY, M., Cost optimization in relation to factory structures, *Eng. Opt.*, **1**, **2**, 125–138, 1974.

BRANDT, A. M., and MARKS, W., Optimization in structural design (in Polish), *Inż. i Bud.*, **5**, 197–200, 1974.

BROCK, J. E., Analytic treatment of minimum weight design of cantilevers, *J. Appl. Mech. Trans. ASME*, **E 41**, **2**, 512–515, 1974; *AMR*, 1975, **7**, 6055.

BRONDUM-NIELSEN, T., Optimum design of reinforced concrete shells and slabs, *Struct. Res. Lab.*, *Univ. of Denmark*, *Rep.* No. **R44**, 1974.

CARDOU, A., and WARNER, W. H., Minimum-mass design of sandwich structures with frequency and section constraints, *J. Opt. Theory and Appl.*, **14**, 6, 633–647, 1974; *AMR*, 1976, **1**, 404.

CARTER, W. J., and RAGSDELL, K. M., The optimal columns, *J. Engng. Materials and Techn. ASME* **H96**, 71–76, 1974.

ČYRAS, A., and KALANTA, S., Optimal design of cylindrical shells by the finite element technique, *Mech. Res. Comm.*, **1**, 125–130, Pergamon Press, 1974.

ĆWIK, R., Optimization of the dimensions of tubular, circular and square cross-sections of compression bars (in Polish), *Symp. "Optymalizacja w mechanice"*, *PTMTS*, 45–54, Gliwice–Wisła 1974.

DOUTY, R. T., and CROOKER, J. O., Optimalization of long-span cold-formed truss purlins, *Proc. ASCE*, **100**, *ST* 11, *J. Struct. Div.*, 2275–2288, 1974; *AMR*, 1976, **6**, 5107.

DRAG, B., and GAJEWSKI, A., Optimization of the form of a slender rod placed in a fluid stream (in Polish), *Rozpr. Inż.*, **22**, 1, 55–68, 1974.

DUTTA, T. K., DIXIT, V. D, and BHUSARI, U. B., Optimal design of two-way reinforced concrete slabs with banded reinforcement, *J. Struct. Engng.*, *India*, **2**, 3, 97–104, 1974.

DYM, C. L., On some recent approaches to structural optimization, *J. Sound and Vibr.*, **32**, 1, 49–70, 1974.

DZIENISZEWSKI, W., Derivation of physical equations for an arbitrary segment of the bar lattice of a spatial structure (in Polish), *Prace IPPT*, **7**, 1974.

DZIKIEWICZ-GOLKA, A., Optimum design of reinforced concrete frames with application of ETO (in Polish), *Inż. i Bud.*, **31**, 11, 499–502, 1974; *PBAM*, 1975, **3**, 11441.

FARSHAD, M., Optimum shape of continuous columns, *Int. J. Mech. Sci.*, **16**, 597–601, 1974.

FARSHI, B., and SCHMIT, L. A., Minimum weight design of stress limited trusses, *Proc. ASCE*, **100**, *ST* 1, *J. Struct. Div.*, 1974.

FINIFTER, D., and MANSOUR, A., Finite element analysis and optimization of a web frame of a tanker with isolated ballast systems, *J. Ship. Res.*, **18**, 2, 85–95, 1974.

FRIEL, L. L., Optimum singly reinforced concrete sections, *Proc. ACI*, **71**, 11, 556–558, 1974.

GAJEWSKI, A., and PALEJ, R., Stability analysis and shape optimization of an elastically fixed tension bar (in Polish), *Rozpr. Inż.*, **22**, 2, 265–279, 1974.

GALLAGHER, R. H., and ZIENKIEWICZ, O. C., eds., *Optimum structural design: theory and application*, John Wiley, London 1974.

GIERLIŃSKI, J., Some problems of grillage optimization (in Polish), *Symp. "Optymalizacja w mechanice"*, *PTMTS*, 105–116, Gliwice–Wisła 1974.

GIERLIŃSKI, J., and HOLNICKI-SZULC, J., Control of the deformation state in plane bar structural roofs (in Polish), *XVI Konf. Mech. IPPT PAN*, Krynica 1974.

GIERLIŃSKI, J., Optimal distribution of materials in polar gridworks (in Polish), *Rozpr. Inż.*, **22**, 1, 89–100, 1974; *AMR* 1977, **6**, 5015.

GIRARD, R., An optimization method applied to composite structures, (in French), *ONERA, Note Tech.*, **228**, 1974; *AMR* 1976, **7**, 6045.

GINZBURG, I. N., and KANTOR, B. Y., Optimal weight design of reinforced cylindrical shells in axial compression, *Soviet Aeronautics*, **17**, 1, 41–44, 1974; *AMR*, 1976, **3**, 2186.

GOLIŃSKI, J., Adaptive system of nonlinear optimization (in Polish), *Arch. Inż. Ląd.*, **20**, 3, 1974.

GRINEV, V., and FILIPOV, A., Some problems of optimization of deformable structural elements (in Russian), *Materialy Vses. Konf. "Probl. Optimiz. i Mekh. Tverd. Deform. Tela"*, Vilnius 1974.

HOLNICKI-SZULC, J., and MARKS, W., Optimization problem for structures with dense prestressing wires (in Polish), *Prace IPPT*, **1**, 1974.

HOLST, O., Automated design of plane frames, *Struct. Res. Lab., Univ. Denmark, Report No.* **R5**, 1974.

HUBER, J., Economical concrete structures by optimization, *FIP VII Congress*, New York 1974.

IJEGI, E., Mathematical model of the optimum design problem for discrete systems (in Russian), *Materialy Vses. Konf. "Probl. Optimiz. i Mekh. Tverd. Deform. Tela"*, Vilnius 1974.

JANICZEK, R., and KAPCIA, B., Optimization of a compression bar with a nonhomogeneous core (in Polish), *Symp. "Optymalizacja w mechanice"*, *PTMTS*, 117–130, Gliwice–Wisła 1974.

JENDO, S., NOWACKI, H., and YI-SUANG TEIN, Optimum design of tanker transverse web frames, *Rep. Univ. Mich.*, Ann Arbor 1974.

JENDO, S., and STACHOWICZ, A. Suspension roofs. Statical analysis and optimization (in Polish), *Seria Mechanika Konstrukcji nr.* **31**, Arkady, Warszawa 1974.

KEMPNY, S., Optimum design of a suspension roof with one-way cables (in Polish), *Rozpr. Inż.*, **22**, 3, 387–410, 1974.

KOLUPAEV, A. N., On the synthesis of trusses of minimum volume (in Russian), *Trudy KAI*, **168**, 29–33, 1974.

KOLUPAEV, A. N., Design algorithm for a minimum-volume truss (in Russian), *Trudy KAI*, **172**, 28–32, 1974.

KRAVCHENKO, V., and BURMENKO, J., Selection of optimum spans for a multisupport continuous beam (in Russian), *Vyp. Lvov. Politekhn. Inst.*, **85**, 1974.

KRZYŚ, W., and LATOS, W., Strength design of a thin-walled tube with regard to fatigue under complex loading (in Polish), *Arch. Bud. Masz.*, **21**, 4, 571–591, 1974; *PBAM*, 1975, **3**, 11430.

KUCZYŃSKA-MARKS, M., Certain cases of application of constrained media theory to the analysis and optimization of structures (in Polish), *XVI Konf. Mech. IPPT PAN*, Krynica 1974.

KUNOO, K., Optimum configuration of truss structures, *Tech. Rep. Nat. Aerospace Lab. NAL TR-388T*, Tokyo 1974; *AMR* 1976, **7**, 6044.

LEE, B., and KAPTON, J., Optimum cost design of a steel-framed building, *Eng. Optim.*, **1**, 3, 1974.

LIPSON, S. L., and AGRAWAL, K. M., Weight optimization of plane trusses, *Proc. ASCE*, **100**, *ST* 5, *J. Struct. Div.*, 1974.

LUND, S., Application of optimization methods within structural design, *Comp. and Struct.*, **4**, 1, 1974.

MAJID, K. I., *Optimum Design of Structures*, Newnes-Butterworths, London 1974.

MACIULEVIČIUS, D., Algorithm of search hypotheses in problems of structural optimization (in Russian), *Viln. Inzh. Stroit. Inst.*, *Vilnius* 1974.

MACIULEVIČIUS, D., and CHYUCHELIS, A., Application of the decomposition principle to the problem of elastic hinged-modular system synthesis (in Russian), *Viln. Inzh. Stroit. Inst.*, Vilnius 1974.

MARKS, W., Application of nonlinear programming to the optimization of beam structures (in Polish), *XVI Konf. Mech. IPPT PAN*, Krynica 1974.

MASLENNIKOV, V., and NAZARENKO, V., Optimum design of frames with regard to geometrical and physical nonlinearity (in Russian), *Stroit. Mekh. i Raschet Sooruzh.*, **6**, 1974.

MASUR, E., Optimal structural design for a discrete set of available structural members, *Comp. Method in Appl. Mech. Engng.*, **3**, 2, 1974.

MCCONNEL, R. E., Least-weight frameworks for loads across span, *Proc. ASCE*, **100**, *EM* 5, *J. Eng. Mech. Div.*, 895–901, 1974; *AMR* 1976, **3**, 2185.

MCKEOWN, J. J., A note on the maximum number and density of distribution of minimum weight under multiple loading conditions, *Int. J. Solids and Struct.*, **10**, 3, 309–312, 1974.

MIKHAJLISHCHEV, V., and MOSHINSKI, S., Practical synthesis of optimal metal trusses (in Russian), *Izv. Vyzhsh. Ucheb. Zavedenij, Stroit. i Arch.*, 1974.

MIKHAJLISHCHEV, V., Structural synthesis by means of dynamic programming (in Russian), *Gidromelior. Gidrotekhn. Bud. Resp. Mizhvid. Nauk.-Tekhn.*, 1974.

MORRIS, A. J., A primal-dual method for minimum weight design of statically determinate structures with several systems of loads, *J. Solids and Struct.*, **16**, 11, 801–867, 1974; *AMR*, 1975, **2**, 1274.

NEDOVODEJEV, V., and DEMOKRITOV, V., Optimization of bars with immovable supports, taking up a vertical load (in Russian), *Avtomatiz. Optimaln. Projekt. Inzh. Obektov i Tekhn. Protsessov*, **T 2**, Gorkij 1974.

NIEMIERKO, A., On truss optimization by nonlinear programming (in Polish), *Symp. "Optymalizacja w Mechanice"*, *PTMTS*, 195–208, Gliwice–Wisła 1974.

NOOR, A. K., Multiple configuration analysis via mixed method, *Proc. ASCE*, **100**, *ST* 9, *J. Struct. Div.*, 1974.

NURMUKHAMEDOVA, R., Investigation of the range of the objective function in the optimum design of contours (in Russian), *Vopr. Vychisl. i Prikl. Mat.*, Tashkent 1974.

OKUMURA, T., and OHKUBO, S., Optimum design of steel continuous girders using suboptimization of girder elements, *Trans. Jap. Soc. Civ. Eng.*, **5**, 1974.

OLHOFF, N., Optimal design of vibrating rectangular plates, *Int. J. Solids. Struct.*, **10**, 93–109, 1974.

OLIN, A., The optimum design problem for elastic hinged-rod systems with structural rods (in Russian), *Nauch. Tr. Zhil. i Projekt. In-t Tip. i Eksperim. Projekt. Zhil. i Obshch. Zdanij*, Kiev 1974.

PHAN DINH THANH, Optimum design of reinforced concrete beam, *Arch. Inż. Ląd.*, **20**, 1, 147–160, 1974; *AMR* 1976, **11**, 9849.

PINES, S., and NEWMAN, M., Constrained structural optimization for aeroelastic requirements, *J. Aircraft*, **11**, 6, 313–320, 1974; *AMR*, 1976, **8**, 7058.

POCHTMAN, YU. M., and PYATIGORSKIJ, Z. I., Optimum design of adaptive structures as a method of determining their limit states (in Russian), *Dokl. AN SSSR*, **216**, 6, 1237–1239, 1974.

POLIZZOTTO, C., Optimum plastic design for multiple sets of loads, *Meccanica*, **9**, 3, 206–213, 1974; *AMR*, 1976, **8**, 7057.

POLIZZOTTO, C., Optimum plastic design of structures under combined stresses, *Int. J. Solids and Struct.*, **11**, 5, 539–553, 1974; *AMR*, 1975, **8**, 7013.

PRAGER, W., *Introduction to structural optimization*, Int. Centre for Mech. Sciences (CISM), *Udine*, Springer Verlag, 1974.

RYABCHENKO, V., Optimum design of complex thin-walled load-carrying structures in terms of hierarchic algorithms (in Russian), *Avtomat. Optimal. Projekt. Inzh. Obektov i Tekhnol. Protsessov*, *T 2*, Gorkij 1974.

RADTSIG, YU. A., Application of the theory of mirror functions to the solution of ordinary differential equations (in Russian), *Trudy KAI*, **172**, 18–21, 1974.

RADTSIG, YU. A., Classification of truss synthesis problems and a synthetic method for the design of minimum-volume trusses (in Russian), *Trudy KAI*, **168**, 3–10, 1974.

RAMMERSTORFER, F. G., On the optimal distribution of the Young's modulus of a vibrating prestressed beam, *Int. J. Sound and Vibration* **37**, 140–145, 1974.

REINSCHMIDT, K. F., and RUSSEL, A. D., Applications of linear programming in structural layout and optimization, *Comp. and Struct.*, **4**, 4, 855–869, 1974.

RAO, S. S., Optimization of complex structures to satisfy dynamic and aeroelastic requirements, *Int. J. for Num. Meth. in Engng.*, **8**, 2, 249–269, 1974; *AMR* 1976, **4**, 3059.

REISS, R., Minimal weight design for conical shells, *J. Appl. Mech. Trans. ASME*, E **41**, 3, 599–603, 1974; *AMR*, 1975, 5, 4114.

ROZVANY, G. I. N., Optimal plastic design with discontinuous cost functions, *J. Appl. Mech., Trans. ASME*, E **41**, 1, 309–310, 1974; *AMR*, 1975, **2**, 1275.

ROZVANY, G. I. N., Absolute optima in plastic design for preassigned shape, *J. Appl. Mech. Trans. ASME*, E **41**, 3, 813–814, 1974; *AMR*, 1975, **8**, 7011.

ROZVANY, G. I. N., Optimization of unspecified generalized forces in structural design, *J. Appl. Mech., Trans. ASME*, **E 41**, 4, 1143–1146, 1974; *AMR*, 1975, **11**, 9848.

ROZVANY, G. I. N., Optimal flexure fields for corners, *J. Eng. Mech. Div., Proc. ASCE*, **100, EM 4**, 828–833, 1974; *AMR*, 1975, **6**, 5100.

RZHANITSYN, A. R., On a general principle of optimum structural design (in Russian), *Stroit. Mekh. i Raschet Sooruzh.*, **3**, 6–8, 1974.

SAMBURA, A., Remarks on the use of Rosenbrock's method in structural optimization (in Polish), *Arch. Inż. Ląd.*, **20**, 3, 1974.

SCHILLING, C. D., Optimum properties for I-shaped beams, *Proc. ASCE*, **100**, *ST* 12, *J. Struct. Div.*, 2385–2401, 1974; *AMR*, 1976, **2**, 1337.

SMITH, G. K., and WOODHEAD, R. G., An optimal design scheme with application to tanker transverse structure, *Engng. Opt.*, **1**, 2, 79–98, 1974; *AMR*, 1976, **11**, 9850.

SCHMIT, L. A. JR., (ed.), *Structural Optimization Symposium*, ASME, New York 1974; *AMR*, 1976, **1**, 401.

SCHMIT, L. A., and FARSHI, B., Some approximation concept for structural synthesis, *AIAA J.*, **12**, 5, 692–699, 1974.

SIRAZUTDINOV, YU. K., On a uniform-strength cross-section of a beam (in Russian), *Trudy KAI*, **168**, 11–18, 1974.

SIRAZUTDINOV, YU. K., Design of structures of minimum volume by the gradient descent method (in Russian), *Trudy KAI*, **172**, 33–39, 1974.

SOFRONOV, YU. D., On the optimum distribution of supports in constant section beams under transverse vibration (in Russian), *Trudy KAI*, **172**, 22–27, 1974.

SOFRONOV, YU. D., Design of beams of minimum weight with plane bending stability constraint (in Russian), *Trudy KAI*, **168**, 34–43, 1974.

SOLOV'EV, E. G., and ARISTOVA, N. S., Some problems concerning optimum trusses with regard to principal stresses (in Russian), *Trudy KAI*, **172**, 9–13, 1974.

SPUNT, L., A programming approach to optimal structural design using structural indices, *AIAA J.*, **12**, 6, 865–868, 1974.

SWISZCZOWSKI, S., Two cases of the optimum design of beam structures by dynamic programming (in Polish), *Arch. Inż. Ląd.*, **20**, 1, 83–98, 1974; *PBAM* 1975, **4**, 11575.

SWISZCZOWSKI, S., *Optimum design of elastic bar structures with application of the modification method in the class of structures with fixed internal forces* (in Polish), Doct. Thesis, Politechnika Krakowska, Wydz. Bud. Ląd., Kraków 1974.

SWISZCZOWSKI, S., Modification methods in the design of elastic structures (in Polish), *Arch. Inż. Ląd.*, **20**, 3, 1974.

THAKKAR, M. C., and SRIDHAR RAO, J. K., Optimal design of prestressed concrete pipes using linear programming, *Comp. and Struct.*, **4**, 2, 1974.

THOMPSON, J. M. T., and HUNT, G. W., Dangers of structural optimization, *Engng. Opt.*, **1**, 2, 99–110, 1974; *AMR* 1976, **12**, 10863.

TITAEV, B., and KUDRYAVTSEVA, V., Application of the method of local variations with directed search to the optimization of thin-walled structures (in Russian), *Materialy Vses. Konf. Probl. Optimiz. Mekh. Tverd. Deform. Tela*, Vilnius 1974.

VINOGRADOV, A., Optimization algorithms in the theory of basic systems (in Russian), *Prikl. Mekh.*, **10**, 11, 1974.

VOLKOVA, N. S., Optimum load for steel trusses (in Russian), *Vses. Konf. Probl. Optimiz. Tverd. Deform. Tela*, Vilnius 1974.

VOLKOVA, N. S., On the optimization of steel trusses (in Russian), *Trudy KAI*, **168**, 19–22, 1974.

WOJDANOWSKA, R., Optimal design of truss structures in creep conditions with reference to the Rabotnov–Shesterikov theory of buckling (in Polish), *Mech. Teoret. Stos.*, **12**, 3, 245–263, 1974; *AMR*, 1978, 1, 386.

VOLKOVA, N. S., Optimal statically determinate trusses (in Russian), *Trudy KAI*, **172**, 14–17, 1974.

YAU, W., Optimal design of simply supported beams for minimum upper bound for dynamic response, *J. Appl. Mech., Trans. ASME*, E **41**, 1, 301–302, 1974; *AMR*, 1975, **1**, 407.

ŻYCZKOWSKI, M., Optimization of shell structures (in Polish), *Symp. "Konstrukcje powłokowe, teoria i zastosowanie"*, Kraków 1974.

1975

ADALI, S., Optimum design of a Timoshenko beam and related dual extremum principles, *CANCAM 75, Proc. 5th Can. Congr. Appl. Mech.*, Fredricton 1975.

ADELMAN, H. M., WALSH, J. L., and NARAYANASWAMI, R., An improved method for optimum design of mechanically and thermally loaded structures, *NASA TN* **D-7965**; *AMR*, 1977, **5**, 4118.

ANDERSON, D., and SALTER, J., Design of structural frames to deflection limitations, *Struct. Engng.*, **53**, 8, 1975.

ARISTOV, M. V., and TROITSKII, V. A., Minimal-weight design of elastic annular plate (in Russian), *Mekh. Tverd. Tela*, **10**, 3, 172–176, 1975 (trans, in *Mech. of Solids*, **10**, 3, 153–156, 1975); *AMR*, 1977, **2**, 1343.

ARORA, J. S., HAUG, JR., E. J., and RIM, K., Optimal design of plane frames, *Proc. ASCE* **101**, *ST* 10, *J. Struct. Div.*, 2063–2078, 1975; *AMR*, 1977, **3**, 2233.

BANICHUK, N. V., On a variational problem with unknown boundaries and the determination of optimal shapes of elastic bodies, *PMM*, **39**, 6, 1037–1047, 1975.

BANICHUK, N. V., Optimal elastic plate shapes in bending problems, *Mech. of Solids*, **10**, 5, 151–158, 1975; *AMR* 1977, **8**, 6820.

BANICHUK, N. V., Determining the optimal forms of curved elastic bars, *Mech. of Solids*, **10**, 6, 107–115, 1975; *AMR* 1978, **1**, 195.

BARTON, F., and PERRY, V., Optimum design of thermally loaded structures, *CANCAM 75 Proc. 5th Can. Congr. Appl. Mech.*, Fredericton 1975.

BORISOV, V., Investigation of the properties of statically indeterminate trusses of minimum volume (in Russian), *Stroit. Mekh.*, 1975.

BIRYULEV, V., On the optimum control of stresses in continuous metal trusses (in Russian), *Izv. Vyssh. Ucheb. Zavedenij Stroit. Arkhit.*, **2**, 1975.

BRONOWICKI, A. J., *et al.*, Optimization of ring stiffened cylindrical shells, *AIAA J.*, **13**, 10, 1319–1325, 1975; *AMR*, 1976, **11**, 9845.

BRONOWICKI, A. J., and FELTON, L. P., Optimum design of continuous thin-walled beams, *Int. J. Num. Meths. in Eng.*, **9** (3), 711–720, 1975; *AMR*, 1976, **12**, 10860.

BROWN, R. H., Minimum cost selection of one-way slab thickness, *Proc. ASCE* **101**, *ST* 12, *J. Struct. Div.*, 2585–2590, 1975; *AMR*, 1977, **1**, 392.

BURAKIEWICZ, A., OWCZAREK, S., and WASIUTYŃSKI, Z., On solving optimum shape problems by investigating deformations of rubber membranes (in Polish), *Arch. Inż. Ląd.*, **20**, 1, 1975.

CHEN, T., and PLECNIK, J., Optimization algorithm with engineering applications, *CANCAM 75 Proc. 5th Can. Congr. Appl. Mech.*, Fredericton 1975.

CLAUDON, J. L., Characteristic curves and optimum design of two structures subjected to circulatory loads, *J. Méc.*, **5**, 14, 531–543, 1975.

CONTRO, R., MAIER, G., and ZAVELANI, A., Inelastic analysis of suspension structures by nonlinear programming, *Computer Methods in Appl. Mech. and Engng.*, **5**, 1975.

ĆWIK, R., Estimation of the influence of the tangential stresses of free bending on the optimal cross-sectional parameters of the bent elements (in Polish), *Zesz. Nauk. PSL*, **439**, 1975.

DISTEFANO, N., Dynamic programming and the optimization of two-point boundary value system, *J. Math. Anal. and Appl.*, **52**, 142–150, 1975.

DISTEFANO, N., and RATH, A., A dynamic programming approach to the optimization of elastic trusses, *J. Opt. Theory and Appl.*, **15**, 1, 13–26, 1975; *AMR*, 1977, **4**, 3203.

DISTEFANO, N., and SAMARTIN, A., A dynamic programming approach to the formulation and solution of finite element equations, *Comp. Methods in Appl. Mech. and Engng.*, **5**, 37–52, 1975.

EDWARDS, L. S., Optimum limit state design of highway bridge superstructures using geometric programming, *Eng. Opt.*, **1**, 4, 211–212, 1975; *AMR*, 1977, **2**, 1341.

ERBATUR, F., Optimal design of a circular plate for given deflection, *Middle East Tech. Univ. J. Pure and Appl. Sci.*, **8**, 2, 281–286, 1975; *AMR*, 1976, **9**, 8063.

FEDOROV, I., Problems of optimum structural design and application of computers in structural mechanics studies (in Russian), *Mater. Nauch.-Tekhn. Konf. Sektsii Stroit. Konstr.*, Krivirozh 1975.

FEDOROV, I., On the minimum weight of singly statically indeterminate frames (in Russian), *Mater. Nauch. Tekhn. Konf. Sektsii Stroit. Konstr.*, Krivorozh 1975.

FRANCAVILLA, A., RAMAKRISHNAN, C. V., and ZIENKIEWICZ, O. C., Optimization of shape to minimize stress concentration, *J. Strain Anal.*, **10**, 63–70, 1975.

FRIND, E., and WRIGHT, P., Gradient methods in optimum structural design, *Proc. ASCE* **101**, *ST* 4, *J. Struct. Div.*, 939–956, 1975; *AMR*, 1976, **11**, 9847.

GAJEWSKI, A., Optimum strength design for materials with physical nonlinearity (in Polish), *Zesz. Nauk. Pol. Krak.*, **12**, Kraków 1975.

GALAMBOS, A. R., HOSAIN, M. U., and SPEIRS, W. G., Optimum expansion ratio of castellated steel beam, *Engng. Opt.*, **1**, 3, 213–225, 1975; *AMR*, 1977, **2**, 1344.

GOBLE, G., and MOSES, F., Practical applications of structural optimization, *Proc. ASCE* **101**, *ST* 4, *J. Struct. Div.*, 635–648, 1975; *AMR*, 1976, **10**, 8995.

GOLDSHTEIN, J., Optimum design of elastic bar systems by the displacement method (in Russian), *Uch. Zap. Petrozavodsk. Inst.*, **20**, 5, 1975.

GORZYNSKI, J. W., and THORNTON, W. A., Variable energy ratio method for structural design, *Proc. ASCE* **101**, *ST* 4, *J. Struct. Div.*, 975–990, 1975; *AMR*, 1976, **8**, 7056.

GRINEV, V. B., and FILIPPOV, A. P., On optimal bars in stability problems (in Russian), *Stroit. Mekh. i Raschet Sooruzh.*, **2**, 21–27, Moskva 1975.

GRINEV, V. B., and FILIPPOV, A. P., Optimal bars in a problem of stability under distributed load (in Russian), *Stroit. Mekh. i Raschet Sooruzh.*, **6**, 23–27, 1975; *AMR*, 1977, **10**, 8580.

GRINEV, V. B., and FILIPPOV, A. P., Optimization of structural elements with regard to mechanical characteristics (in Russian), *Nauk. Dumka*, 1975.

HAFTKA, R. T., STARNES, J. H., JR., BARTON, F. W., and DIXON, S. C., Comparison of two types of structural optimization procedures for flutter requirements, *AIAA J.*, **13**, 1333–1339, 1975.

HALBRITTER, A., Optimization estructural con variation de geometria, *COBEM 75, An. III Congr. Bras. Eng. Mec.*, Rio de Janeiro 1975.

HAUG, E. J., PAN, K. C., and STREETER, T. D., A computational method for optimal structural design II: Continuous problems, *Int. J. Num. Meths. in Eng.*, **9** (3), 649–667, 1975.

HO, J. K., Optimal design of multi-stage structures: a neated decomposition approach, *Comp. and Struct.*, **5**, 4, 249–255, 1975.

HORNBUCKLE, J. C., NEVILLE, G. E., and BOYKIN, W. H., Structural optimization using the finite element method applied to a beam, *Int. J. Num. Meths. in Eng.*, **9**, 1, 101–107, 1975; *AMR*, 1976, **12**, 10861.

HUANG, N. C., Minimum weight design of elastic cables, *J. Opt. Theory and Appl.*, **15**, 1, 37–49, 1975; *AMR*, 1976, **5**, 4075.

HUANG, N. C., Minimum-volume design of elastic trusses with deflection constraints, *ZAMP*, **26**, 4, 437–452, 1975; *AMR*, 1976, **8**, 7055.

Hyca, M., Optimization of prismatic thin-walled beams in bending (in Czech), *Strojnicky Čas.*, **26**, 3, 1975.

Jendo, S., and Stachowicz, A., Optimum design of prestressed double-layer suspended system, *Proc. IASS Symp. on Cable Struct.*, *SVTS*, Bratislava, 1975.

Jendo, S., and Stachowicz, A., Optimum design of prestressed cable nets, *Proc. CANCAM*, 1975.

Kamat, M., Optimum Timoshenko beams, *AIAA Pap.*, **140**, 1975.

Kholopov, I., Some problems of cost optimization of metal bar structures (in Russian), *Raschet Prostranstv. Stroit. Konstr.*, Kujbyshev 1975.

Kirsch, U., Multilevel approach to optimum structural design, *Proc. ASCE* **101**, *ST* 4, *J. Struct. Div.*, 957–974, 1975; *AMR*, 1976, **7**, 6043.

Kolupaev, A., Formalization and design of optimal statically indeterminate trusses (in Russian), *Izv. Vyssh. Ucheb. Zavedenij, Stroit. i Arkh.*, **4**, 1975.

Korshunov, A., On the design of statically indeterminate trusses of minimum weight (in Russian), *Stroit. Mekh.*, 1975.

Kosolap, N. D., Isoparametric problem of structural mechanics for a truss system (in Russian), *Prikl. Mekh.*, **11**, 8, 124–127, 1975; *AMR*, 1976, **7**, 6041.

Kurshin, L. M., On the problem of determining the shape of a bar section of maximum torsional stiffness (in Russian), *Dokl. AN SSSR*, **223**, 3, 1975.

Lacey, G. C., and Breen, J. E., The design and optimization of segmentally precast prestressed box girder bridges, *Summary report* **121-3**(s), Center for Highway Research, Univ. of Texas at Austin, Aug. 1975.

Lazarev, I., Design of optimal structures by the method of representations (in Russian), *Tr. Novosib. Inst. Inzh. Zh-d. Transp.*, 1975.

Lee, B. S., and Knapton, J., Optimum cost design of a steel framed building, *Eng. Opt.*, **1**, 3, 139–153, 1975.

Maier, G., On the optimization of the shape of plastic structures (in Russian), *AN USSR, Inst. Probl. Mekh.*, 1975.

Masur, E. F., Optimal placement of available sections in structural eigenvalue problems, *J. Opt. Theory and Appl.*, **15**, 1, 69–84, 1975; *AMR*, 1976, **6**, 5106.

Mikhajlishchev, V. Ya., Synthesis of an optimally prestressed system (in Russian), *Prikl. Mekh.*, **11**, 1, 106–111, 1975; *AMR*, 1975, **12**, 10771.

Morris, A. J., A transformation for geometric programming applied to the minimum weight design of statically determinate structures, *Int. J. Mech. Sci.*, **17**, 6, 395–396, 1975; *AMR*, 1976, **3**, 2184.

Mróz, Z., and Rozvany, G. I. N., Optimal design of structures with variable support conditions, *J. Opt. Theory and Appl.*, **15**, 1, 85–101, 1975; *AMR*, 1976, **4**, 3057.

Ng, S. F., and Kulkarni, G. G., Optimum design of longitudinally stiffened simply supported orthotropic bridge decks, *J. of Sound and Vinr.*, **40**, 2, 273–284, 1975; *AMR*, 1976, **1**, 403.

Olszowski, B., Some problems of optimum design problems of vibrating systems, *Arch. Mech.*, **27**, 4, 605–616, 1975; *AMR*, 1976, **12**, 10859.

Parkes, E. W., Joints in optimum frameworks, *Int. J. Solids and Struct.* **11** (9), 1017–1022, 1975.

Nemirovskij, J. J., *Optimum Structural Design: Bibliographical Guide* (in Russian), AN SSSR, Sibirskoe Otdelenie, Novosybirsk 1975.

Pochtman, Yu. M., and Baranenko, V., *Dynamic Programming in the Problems of Structural Mechanics* (in Russian), Stroizdat, Moskva 1975.

Pochtman, Yu. M., and Pyatigorskij, Z. I., Residual stress field statically possible in optimum-adaptable structures (in Russian), *Dokl. AN SSSR*, **225**, 3, 265–268, 1975.

RAJNUS, G. E., Basis of the analysis and synthesis of instant-stiff hinged-rod systems (in Russian), *Proc. IASS Symposium on Cable Structures, SVTS, Edične Stredisko*, Bratislava 1975.

RAMAKRISHNAN, C. V., and FRANCAVILLA, A., Structural shape optimization using penalty function, *J. Struct. Mech.*, **3**, 4, 403–432, 1975.

RAO, S. S., Optimum design of structures under shock and vibration environment, *Shock and Vibration Digest*, **7**, 12, 61–70, 1975.

REINSCHMIDT, K. F., and NORABHOOMPIPAT, T., Structural optimization by equilibrium linear programming, *Proc. ASCE* **101**, *ST* 4, *J. Struct. Div.*, 921–938, 1975; *AMR*, 1976, **12**, 10857.

REPYAKH, V. I., On some problems of optimum structural design (in Russian), *Stroit. Mekh. i Raschet Sooruzh.*, **6**, 44–48, 1975; *AMR*, 1977, **6**, 5014.

ROZVANY, G. I. N., Analytical treatment of some extended problems in structural optimization: Parts 1 and 2, *J. Struct. Mech.*, **3**, 4, 359–402, 1974/1975; *AMR*, 1976, **10**, 8996 and 8997.

ROZVANY, G. I. N., A unified theory of optimal moment fields, *J. Struct. Mech.*, **3**, 2, 179–195, 1974/1975; *AMR*, 1976, **12**, 10862.

ROZVANY, G. I. N., GANGADHARAIAH, C., and HILL, R. D., Optimal slabs and grillages of constrained geometry, *Proc. ASCE* **101**, *EM* 6, *J. Engng. Mech. Div.*, 755–770, 1975; *AMR*, 1977, **5**, 4113.

ROZVANY, G. I. N., and MRÓZ, Z., Optimal design taking cost of joints into account, *Proc. ASCE* **101**, *EM* 6, *Eng. Mech. Div.*, 917–921, 1975; *AMR*, 1977, **5**, 4114.

RUDERMAN, S., SOLOMESHCH, I., and YASIN, E., Selection of optimum foundations for linearly extended flexible structures (in Russian), *Tr. Ufim. Aviats. Inst.* 1975.

SAVE, M., A general criterion for optimal structural design, *J. Opt. Theory and Appl.*, **15**, 1, 119–129, 1975; *AMR*, 1976, **11**, 9848.

SCHAMIE, J., and SCHMIT, L. A., JR., Frame optimization including frequency constraints, *Proc. ASCE* **101**, *ST* 1, *J. Struct. Div.*, 283–293, 1975.

SEYRANYAN, A. P., Optimum design of beams with deflection constraints (in Russian), *Izv. AN Arm. SSR*, **28**, 6, 24–33, 1975.

SEGENREICH, S. A., and MCINTOSH, S. C., Weight minimization of structures for fixed flutter speed via an optimality criterion, *AIAA Paper* **75-779**, 1975.

SHAMIEV, F. G., Optimal design of plates loaded by two sets of lateral loads, *Arch. Mech.*, **27**, 2, 317–324, 1975; *AMR* 1976, **9**, 8065.

SHEU, C. Y., Optimal elastic design of trusses by feasible direction methods, *J. Opt. Theory and Appl.*, **15**, 1, 131–143, 1975.

SPILLERS, W. R., On the relationship between buckling and optimal structural design, *J. Franklin Inst.*, **299**, 6, 463–466, 1975; *AMR*, 1976, **2**, 1335.

SIMITSES, G. J., and KOTRA, T., The optimal Euler–Bernoulli cantilever, *Proc. ASCE*, **101**, *EM* 6, *J. Engng. Mech. Div.*, 922–929, 1975; *AMR* 1977, 8, 6818.

SIMITSES, G. J., and UNGBHAKORN, V., Weight optimization of stiffened cylinders under axial compression, *Comp. and Struct.*, **5**, 5/6, 305–314, 1975; *AMR* 1977, **11**, 9385.

SIMITSES, G. J., and UNGBHAKORN, V., Minimum-weight design of stiffened cylinders under axial compression, *AIAA J.*, **13**, 6, 750–755, 1975; *AMR*, 1976, **7**, 6042.

SINGARAJ, N. M., and SRIDHAR RAO, J. K., Optimization in trusses using optimal control theory, *Proc. ASCE* **101**, *ST* 5, *J. Struct. Div.*, 1037–1051, 1975; *AMR*, 1977, **8**, 6819.

SOFRONOV, J., On the optimum form of the cross-section of a beam under transverse bending (in Russian), *Trudy KAI*, 1975.

SPILLERS, W. R., *Iterative structural design*, North-Holland Publ., Amsterdam 1975; *AMR*, 1976, **12**, 10853.

SPILLERS, W. R., Iterative design for optimal geometry, *Proc. ASCE* **101**, *ST* 7, *J. Struct. Div.*, 1435–1442, 1975; *AMR*, 1977, **5**, 4116.

SPILLERS, W. R., and FUNARO, J., Iterative design with deflection constraints, *Int. J. Solids and Structures*, **11** (7/8), 793–802, 1975.

SWISZCZOWSKI, S., Application of a scaling objective function in the algorithms of optimum design (in Polish), *II Konf. "Metody Komputerowe w Mechanice Konstrukcji"*, Gdańsk 1975.

TAYLOR, J. E., On the prediction of structural layout for minimum stiffness, *J. Opt. Theory and Appl.*, **15**, 1, 145–155, 1975; *AMR*, 1976, **4**, 3058.

TEMNOV, V., General mathematical model and optimization of large rod systems (in Russian), *Raschet i Projekt. Prostranstv. Konstr. Grazh. Zdanij i Sooruzh.*, 1975.

TVERGAARD, V., On the optimum shape of a fillet in a flat bar with restrictions, in: *Optimization in Structural Design*, Ed. A. SAWCZUK and Z. MRÓZ, 181–195, Springer-Verlag, New York 1975.

TWISDALE, L. A., and KHACHATURIAN, N., Multistage optimization of structures, *Proc. ASCE, J. Struct. Div.*, *ST* 5, 1005–1020, 1975, *AMR*, 1976, **9**, 8064.

VANDERPLAATS, G. N., Design of structures for optimum geometry, *NASA TM* **X-62462**, 1975; *AMR*, 1977, **4**, 3204.

VENKAYYA, V. B., and KHOT, N. S., Design of optimum structures to impulse type of loading, *AIIA J.*, **13**, 8, 989–994, 1975; *AMR*, 1977, **9**, 7725.

VINOGRADOV, A. I., Normed parameters in optimization problems of structural mechanics (in Russian), *Prikl. Mekh.*, **11**, 5, 8–13, 1975; *AMR*, 1976, **10**, 8994.

VOLYNSKII, M. I., PALCHEVSKII, A. S., and POCHTMAN, YU. M., Optimal design of ribbed cylindrical shells with large cuts under axial compression (in Russian), *Prikl. Mekh.*, **11**, 5, 118–121, 1975; *AMR*, 1977, **5**, 4110.

WEISSHAAR, T. A., Approximate solutions to idealized structural dynamic optimization problems, *J. Opt. Theory and Appl.*, **16**, 1/2, 119–133, 1975; *AMR*, 1977, **11**, 9386.

YANG, J., Application of optimal control theory to civil engineering structures, *Proc. ASCE*, **101**, *EM* 6, *J. Eng. Mech. Div.*, 819–838, 1975.

ŻYCZKOWSKI, M., and KRUŻELECKI, J., Optimal design of shells with respect to their stability, *IUTAM Symp. "Optimization in Structural Design"*, *Warsaw* 1973, Springer-Verlag, Berlin–Heidelberg–New York, 229–247, 1975.

1976

ADAMOVICH, I. S., and RIKARDS, R. B., Optimization of compressed cylindrical shells with elastic properties that vary along the length, *Polym. Mech.*, **11**, 5, 699–705, 1976; *AMR* 1977, **9**, 7521.

ALSPAUGH, D. W., and HUANG, S. N., Minimum weight design of axisymmetric sandwich plates, *AIAA J.*, **14**, 12, 1683–1689, 1976; *AMR* 1977, **8**, 6809.

ARORA, J. S., and HAUG, E. J., JR., Efficient optimal design of structures by generalized steepest descent programming, *Int. J. Num. Meths. in Eng.*, **10** (4), 747–766, 1976; *AMR*, 1977, **10**, 8577.

BAIER, H., Zur Optimierung elastischer Strukturen durch Lösung eines dualen Optimierungproblems, *ZAMM*, **56**, T 98-T, 101, 1976.

BANICHUK, N. V., Minimax approach to structural optimization problems, *J. Opt. Theory and Appl.*, **20**, 1, 111–127, 1976; *AMR*, 1977, **1**, 389.

BANICHUK, N. V., On a two-dimensional optimization problem in elastic bar torsion theory, *Mech. Solids*, **11**, 5, 38–44, 1976; *AMR*, 1977, **12**, 10210.

BANICHUK, N. V., Optimization of elastic bars in torsion, *Int. J. Solids and Struct.*, **12**, 4, 275–286, 1976; *AMR*, 1976, **10**, 8993.

BANDYOPADHYAY, N., and KAPOOR, M. P., Optimal design of prestressed concrete bridge-girder section, *J. Struct. Engng.*, **3**, 4, 179–191, 1976; *AMR*, 1977, **11**, 9383.

BANICHUK, N. V., and KARIHALOO, B. L., Minimum-weight design of multi-purpose cylindrical bars, *Int. J. Solids and Struct.*, **12**, 267–273, 1976; *AMR*, 1976, **7**, 6039.

BANICHUK, N. V., and MIRONOV, A. A., Optimization problems for plates that vibrate in an ideal fluid, *PMM*, **40**, 3, 1976.

BARTHOLOMEW, P., and MORRIS, A. J., A unified approach to fully stressed design, *Eng. Opt.*, **2**, 1, 3–15, 1976; *AMR*, 1977, **3**, 2231.

BELL, L. C., and BROWN, D. M., Guyed tower optimization, *Comp. Struct.*, **6**, 6, 447–450, 1976; *AMR*, 1978, **8**, 6733.

BORISOV, V. M., and MIKHAILOV, I. E., On the optimal distribution of material in bars under torsion (in Russian), *Zhurn. Vykh. Mat. i Mat. Fiz.*, **16**, 6, 1595–1597, 1976; *AMR*, 1978, **7**, 5746.

BROWN, M. L., Optimization of a ribbed table for minimum weight and minimum deflection, *Trans. ASME J. Engng. for Industry*, **B 98**, 1, 30–33, 1976; *AMR*, 1976, **7**, 6038.

BUDRIN, S., Optimum parameters of a thin-walled box section strengthened with longitudinal stiffening ribs (in Russian), *Podemnotransp. Mashiny*, Tula 1976.

CAPELO, A., Optimization of rigid-perfectly plastic hyperstatic structure (in Italian), *Boll. della Unione Mat. It.*, *V* **13-B**, 1, 55–91, 1976; *AMR*, 1978, **5**, 3929.

CASIS, J. H., and SCHMIT, L. A., JR., Optimum structural design with dynamic constraints, *Proc. ASCE*, **102**, *ST* 10, *J. Struct. Div.*, 2053–2071, 1976; *AMR*, 1977, **8**, 6811.

CHENG, F. Y., and BOTKIN, M. E., Nonlinear optimum design of dynamic damped frames, *Proc. ASCE* **102**, *J. Struct. Div.*, *ST* 3, 609–627, 1976; *AMR*, 1978, **4**, 2948.

CHONG, K. P., Optimization of unstiffened hybrid beams, *Proc. ASCE* **102**, *J. Struct. Div.*, *ST* 2, 401–409, 1976.

CHOU, T., Optimum elastic design of rectangular reinforced concrete beam sections (in Japanese), *Proc. Jap. Soc. Civ. Engrs.*, **250**, 99–109, 1976.

DESAYI, P., and ABID ALI, S., Optimum design of prestressed concrete girders, *J. Struct. Engng.*, **3**, 4, 192–200, 1976.

DOBBS, M. W., and NELSON, R. B., Application of optimality criteria to automated structural design, *AIAA J.*, **14**, 10, 1436–1443, 1976; *AMR*, 1977, **8**, 6810.

EIMER, C., and MĄCZYŃSKI, J., On optimal shell prestressing, *J. Struct. Mech.*, **4**, 3, 289–305, 1976; *AMR*, 1977, **8**, 6816.

EPSTEIN, E., and ALSAIGH, J., Minimum weights for circular tube cantilever beams, *Proc. ASCE*, **102**, *ST* 1, *J. Struct. Div.*, 1976.

FELTON, L. P., On optimum design of prestressed beam structures, *AIAA J.*, **14**, 3, 392–394, 1976; *AMR*, 1976, **12**, 10854.

FLEURY, C., Optimization des structures par la méthode des éléments finis, *Collect. Publs. Univ. Liège, Fac. Sci. Appl.*, **59**, 1976.

GAJEWSKI, A., Minimum weight design of a physically nonlinear rotating rod (in Polish), *Mech. Teoret. i Stos.*, **14**, 2, 261–271, 1976; *AMR*, 1978, **4**, 2949.

GANICHEV, V., and IJEGI, E., General optimum design problem for multicontour frames (in Russian), *Tr. Tallin. Politekhn. Inst.*, **394**, 1976.

GERASIMOV, E., and REPKO, V., Models and methods of vectorial optimization and their application to the structural mechanics of rod systems (in Russian), *Izv. Vys. Ucheb. Zav., Stroit. i Arkhit.*, **6**, 1976.

GIBCZYŃSKA, T., Determination of the optimum dimensions of a beam in bending (in Polish), *Przegl. Mech.*, 127–138, Feb. 1976.

HAERI, H., LEMAIRE, M., and CUBAUD, J. C., Application of the finite element method and linear programming to the optimization of a plane structure (in French), *Ann. de l'ITBTP*, **342**, 121–134, 1976; *AMR*, 1978, **4**, 2947.

HEMP, W. S., Michell framework for a force in any definite direction at the mid-point between two supports, *Eng. Opt.*, **2**, 3, 183–187, 1976; *AMR*, 1978, **7**, 5745.

HUANG, H. C., Minimum-weight design of vibrating elastic structures with dynamic deflection constraint, *J. Appl. Mech., Trans. ASME, Series E* 43, **1**, 178–180, 1976; *AMR* 1977, **6**, 5011.

IJEGI, E., and NURMUKHAMEDOVA, R., Analysis and optimum design of statically indeterminate frames by the star method (in Russian), *Tr. Tallin. Politekhn. Inst.* **394**, 1976.

JACKIEWICZ, M., Optimum design of a space truss (in Polish), *22 Konf. PZITB*, Krynica 1976.

KALININ, A., On the optimization of composite rod systems of structural type (in Russian), *Nadezhn. i Dolgovechn. Stroit. Konstr.*, Volgograd 1976.

KALININ, I. N., Design of a shell of revolution with minimum weight (in Russian), *Prikl. Mekh.*, **12**, 1, 44–50, 1976; *AMR*, 1977, **3**, 2230.

KESHAVA RAO, M. N., Optimal layout for grids by fully stressed design method, *J. Struct. Eng. (Roorkee, India)*, **3**, 4, 212–218, Jan. 1976.

KHOLOPOV, I., Choice of an optimum stress sequence for composite statically indeterminate trusses by the method of dynamic programming (in Russian), *Raschet Prostranstv. Stroit. Konstr.*, Kujbyshev 1976.

KHOT, N. S., VENKAYYA, V. B., and BERKE, L., Optimum design of composite structures with stress and displacement constraints, *AIAA J.*, **14**, 2, 131–132, 1976; *AMR*, 1977. **2**, 1340.

KHOT, N. S., VENKAYYA, V. B., and BERKE, L., Optimum structural design with stability constraints, *Int. J. Numer. Meths. in Eng.*, **10**, 5, 1097–1114, 1976; *AMR*, 1977, **7**, 5946.

KHOT, N. S., *et al.*, Optimum design of composite wing structures with twist constraint for aeroelastic tailoring, *AFFDL-TR* **76-117**, 1976; *AMR*, 1977, **11**, 9387.

KIRSCH, U., Synthesis of elastic structures with controlled forces, *Int. J. Comp. Struct.*, **6**, 2, 111–116, 1976; *AMR* 1977, **8**, 6814.

KIRSCH, U., and MOSES, F., Formulation of optimal design in the behavior variable spaces, *J. Struct. Mech.*, **4**, 4, 437–452, 1976; *AMR*, 1977, **8**, 6815.

KORSHUN, L., A practical optimization method for trusses with given configuration of axes (in Russian), *Gidromelior. i Gidrotekhn. Str. Resp. Mezhved. Nauch.-Tekhn.*, 1976.

KRISTENSEN, E. S., and MADSEN, N. F., On the optimum shape of fillets in plates subjected to multiple in-plate loading cases, *Int. J. Num. Meths. in Eng.*, **10**, 5, 1007–1019, 1976; *AMR*, 1977, **7**, 5945.

KUNAR, R., and CHAN, A., A method for the configurational optimization of structures, *Comp. Meth. in Appl. Mech. Engng.*, **13**, 3, 1976; *AMR*, 1976, **9**, 8062.

KURSHIN, L. M., and ONOPRENKO, P. M., Determination of the shapes of doubly-connected bar sections of maximum torsional stiffness, *PMM*, **40**, 6, 1020–1026, 1976.

LAMBLIN, D., and GUERLEMENT, G., Plastic minimum volume design for sandwich shells subjected to fixed or movable loads (in French), *J. de Méc.*, **15**, 1, 55–84, 1976; *AMR*, 1977, **10**, 8579.

LARICHEV, A., Some optimization problems for rods lying on elastic supports (in Russian), *Izv. AN SSSR, Mekh. Tverd. Tela*, **5**, 1976.

LIPP, W., Ein Verfaren zur optimalen Dimensionierung allgemeiner Fachwerkkonstruktionen und ebener Rahmentragwerke, *Tech.-Wiss. Mit. Inst. Konst. Ingenierbau Ruhr-Univ. Bochum*, **12**, 1976.

Lure, K. A., and Cherkaev, A. V., Prager theorem application to optimal design of thin plates, *Mechanics of Solids*, **11**, 6, 139–141, 1976; *AMR*, 1978, **3**, 2056.

Maeda, Y., Examples of computer-aided structural design, *IABSE, 10th Congress*, Tokyo 1976.

Marks, W., Optimization of bent bar sections (in Polish), *Arch. Inż. Ląd.*, **2**, 22, 251–263, 1976.

Marks, W., Optimization of prestressed beam sections (in Polish), *Prace IPPT*, **24**, 1976.

Massonnet, C., and Rondal, J., Optimal design of structures: Possibilities and limitations (in French), *Ann. des Trav. Publ. de Belgique*, **6**, 447–454, 1976; *AMR*, 1978, **5**, 3926.

Matveev, S., Application of nonlinear transformations to the solution of some optimization problems (in Russian), *Tr. Mosk. Avtomob.-Dor. Inst.*, 1976.

Mikhajlishchev, V., Synthesis of optimal metal rod systems (in Russian), *Gidromelior. i Gidrotekhn. Str. Resp. Mezhved. Nauch.-Tekhn.*, 1976.

Moshinskij, S., On the design of optimal statically indeterminate metal trusses from thin-walled elements (in Russian), *Gidromelior. i Gidrotekhn. Str. Resp. Mezhved. Nauch.-Tekhn.*, 1976.

Motro, R., Optimization of space structures with an application to double layer grid framework (in French), *Constr. Met.*, **18**, 2, 24–36, 1976; *AMR*, 1977, **12**, 10211.

Moussouros, M., Finite element optimal design of transverse web frames, *Comp. Struct.*, **6**, 3, 253–266, 1976.

Mróz, Z., and Garstecki, A., Optimal design of structures with unspecified loading distribution, *J. Opt. Theory and Appl.*, **20**, 3, 359–380, 1976; *AMR*, 1978, **4**, 2946.

Murotsu, Y., *et al.*, Optimum structural design based on reliability analysis, *Proc. 19th Jap. Congr. Mater. Res.*, Kyoto 1976.

Naaman, A. E., Minimum cost versus minimum weight of prestressed slabs, *Proc. ASCE* **102**, *J. Struct. Div.*, *ST* 7, 1493–1505, 1976; *AMR*, 1978, **5**, 3928.

Nakamura, T., and Nagase, T., Minimum weight design of multistorey, multispan plane frames subject to reaction constraints, *ASME, J. Struct. Mech.*, **4**, 3, 257–287, 1976; *AMR*, 1977, **5**, 4111.

Niemierko, A., On the possibility of the optimization of elastic trusses by discrete programming methods (in Polish), *XV Symp. PTMTS "Optymalizacja w mechanice"*, Wisła 1976.

Ohkubo, S., and Okumura, T., Graphical optimization of steel girder based on suboptimization of girder element (in Japanese), *Proc. Japan. Soc. Civ. Engrs.*, **252**, 23–34, 1976; *AMR*, 1977, **11**, 9381.

Olhoff, N., A survey of the optimal design of vibrating structural elements, Part I: Theory, Part II: Application, *Shock and Vibr. Digest*, **8**, 8 and 9, 1976; *AMR*, 1978, **8**, 6729 and 6730.

Olhoff, N., Maximizing higher order eigenfrequencies of beams with constraints on the design geometry, *DCAMM, Rep.* **108**, 1976.

Paczkowski, W., Optimization of a simply supported reinforced concrete beam, *Arch. Inż. Ląd.*, **22**, 4, 559–569, 1976; *AMR*, 1978, **2**, 1216.

Patnaik, S. N., and Srivastava, N. K., On automated optimum design of trusses, *J. Struct. Eng. (Roorkee, India)*, **3** (4), 164–178, Jan. 1976; *AMR*, 1977, **9**, 7724,

Pochtman, Yu. M., and Khariton, L. E., Optimal design for a structure considering reliability (in Russian), *Stroit. Mekh. i Raschet Sooruzh.*, **6**, 8–15, 1976.

Patnaik, S. N., and Maiti, M., Optimum design of stiffened structures with constraint on the frequency in the presence of initial stresses, *Comp. Meth. in Appl. Mech. and Engng.*, **7**, 3, 303–322, 1976; *AMR*, 1976, **10**, 8992.

PATNAIK, S. N., and SANKARAN, G. V., Optimum design of stiffened cylindrical panels with constraint on the frequency in the presence of initial stresses, *Int. J. Num. Meths. in Engng.*, **10**, 2, 283–299, 1976; *AMR*, 1978, **3**, 2055.

POPELAR, C. H., Optimal design of beams against buckling: a potential energy approach, *J. Struct. Mech.*, **4**, 2, 181–196, 1976; *AMR*, 1977, **8**, 6817.

PRAGER, W., Geometric discussion of the optimal design of a simple truss, *J. Struct. Mech.*, **4**, 1, 57–63, 1976; *AMR*, 1977, **5**, 4112.

PRAGER, W., and ROZVANY, G. I. N., Optimal layout of grillages, *Theor. and Appl. Mech., 14th IUTAM Congr. Delft*, 1976.

RAGSDELL, K., and PHILLIPS, D., Optimal design of a class of welded structures using geometric programming, *Trans. ASME*, **3**, 1976.

RAJ, P. P., and DURRANT, S. O., Optimum structural design by dynamic programming, *Proc. ASCE.* **102**, *ST* 8, *J. Struct. Div.*, 1575–1589, 1976; *AMR*, 1977, **8**, 6812.

RAJ, P. P., and PACKIE, D., Optimum structural design by dynamic programming, *Proc. ASCE* **102**, *ST* 8, *J. Struct. Div.*, 1976.

RAMESH, C. K., and KARVE, S. R., Optimization for stiffened plates—some studies, *J. Struct. Eng.* (Roorkee, India), **3**, 4, 201–211, Jan, 1976.

RAO, S. S., Structural optimization under combined blast and accoustic loading, *AIAA J.*, **14**, 2, 276–278, 1976; *AMR*, 1977, **4**, 3200.

REISS, R., The method of stress variation applied to minimal design for multiple loading, *Int. J. Solids and Struct.*, **12**, 2, 135–146, 1976; *AMR*, 1976, **7**, 6040.

REJTMAN, M., Optimization of rod structures with material blank nesting (in Russian), *Izv. Vyssh. Ucheb. Zaved. Str. i Arkh.*, **11**, 1976.

RIKARDS, D. M., Optimum design of stiffened shear webs with supplementary skin stabilization, *Int. J. Solids and Struct.*, **12**, 11, 791–802, 1976; *AMR*, 1977, **10**, 8578.

RIZZI, P., Optimization of multi-constrained structures based on optimality criteria, *Proc. AIAA-ASME-SAE 17th Struct. Dynamics and Materials Conf.*, 448–462, New York 1976.

ROMANOW, F., and ZABŁOCKI, W., Selection of optimum cross-sections for thin-walled profiles (in Polish), *Przegl. Mech.*, **35**, 19, 1976.

ROZVANY, G., Optimal design of multiload multispan systems, *Proc. ASCE* **102**, *EM*6, *J. Eng. Mech. Div.*, 1085–1087, 1976; *AMR*, 1978, **1**, 385.

ROZVANY, G. I. N., *Optimal design of flexural systems*: *Beams, grillages slabs, plates and shells*, Pergamon Press 1976; *AMR*, 1977, **3**, 2229.

ROZVANY, G. I. N., and PRAGER, W., Optimal design of partially discretized grillages, *J. Mech. and Physics of Solids*, **24**, 2/3, 125–136, 1976; *AMR*, 1977, **8**, 6813.

SEIRANYAN, A. P., Optimal beam design with limitations on natural vibration frequency and buckling load, *Mechanics of Solids*, **11**, 11, 133–138, 1976; *AMR*, 1978, **3**, 2057.

SUN, P. F., ARORA, J. S., and HAUG, JR., E. J., Fail-safe optimal design of structures, *Eng. Optim.*, **2**, 1, 1976; *AMR*, 1977, **3**, 2232.

TALEB-AGHA, G., and NELSON, R. B., Method for the optimum design of truss-type structures, *AIAA Journal*, **14**, 4, 436–445, 1976; *AMR*, 1977, **1**, 391.

TEMPLEMAN, A. B., Optimum truss design by sequential geometric programming, *J. Struct. Engng.*, **3**, 4, 155–163, 1976; *AMR*, 1977, **11**, 9382.

TEMPLEMAN, A. B., A dual approach to optimum truss design, *J. Struct. Mech.*, **4**, 3, 235–255, 1976; *AMR*, 1977, **6**, 5013.

TEMPLEMAN, A. B., Optimality criteria and dual methods in truss design, *IABSE 10th Congress, Tokyo*, 115–122, Sept. 1976.

TEPLY, B., Optimization of longitudinal reinforcement of floor and foundation RC beams and grids, *Stavebnicky Casopis*, **24**, 5, 407–414, 1976; *AMR*, 1977, **10**, 8591.

TOKARZ, B., A direct method for optimum design of flat slabs (in German), *Bautechnik*, **53**, 11, 374–379, 1976; *AMR* 1978, **4**, 2945.

TROITSKII, V. A., Optimization of elastic bars in the presence of free vibrations, *Mechanics of Solids*, **11**, 3, 139–146, 1976; *AMR*, 1978, **3**, 2058.

TURVITCH, E. L., On isoperimetric problems for domains with partly known boundaries, *J. Opt. Theory and Appl.*, **20**, 1, 65–79, 1976.

VANDERPLAATS, G. N., Structural optimization via a design space hierarchy, *Int. J. Num. Meths. in Eng.*, **10**, 3, 713–717, 1976; *AMR*, 1977, **3**, 2234.

VENKOV, L., Optimum design of prestressed flexural elements in the elastic range using computers (in Russian), *Stroitel'stvo*, **23**, 4–5, 1976.

WHEELER, L., On the role of constant-stress surfaces in the problem of minimizing elastic stress concentration, *Int. J. Solids and Struct.*, **12**, 779–789, 1976.

VINOGRADOV, A., Homogeneous functions in the gradient methods of optimization of rod systems (in Russian), *Tr. TsNII Stroit. Konstr.*, 1976.

YAMAKAWA, H., and OKUMURA, A., Optimum design of structures with regard to their vibrational characteristic: A general method of optimum design, *Bull. ISME*, **19**, 138, 1458–1466, 1976; *AMR*, 1978, **3**, 2053.

1977

ADAMOVICH, I. S., and RIKARDS, R. B., Discrete models for continuous-type problems in design optimization of structures, *Polymer Mech.*, **12**, 5, 752–758, 1977 (trans. of *Mekh. Polym.*, **5**, 852–859, 1976); *AMR*, 1978, **6**, 4796.

ADAMOVICH, I. S., and RIKARDS, R. B., Weight optimization of an orthotropic cylindrical shell with variable properties and a constraint on the vibration frequency, *Mech. of Solids*, **12**, 2, 102–107, 1977; *AMR*, 1979, 1, 352.

AGARWAL, B. L., and SOBEL, L. H., Weight comparison of optimized stiffened, unstiffened and sandwich cylindrical shells, *J. Aircraft*, **14**, 10, 1000–1008, 1977; *AMR*, 1979, **1**, 350.

ALBLAS, J. B., Optimal strength of compound column, *Int. J. Solids and Struct.*, **13**, 4, 307–320, 1977; *AMR*, 1978, 1, 384.

BAIER, H., Über Algorithmen zur Eritlung und Charaktierisierung Parteo-optimaler Lösungen bei Entwurtsaufgaben elastischer Tragwerke, *ZAMM* **57**, 5, 1977.

BANICHUK, N. V., Optimality conditions in the problem of seeking the hole shapes in elastic bodies, *PMM J. Appl. Math. Mech.*, **41**, 5, 946–951, 1977; *AMR*, 1979, **9**, 6979.

BANICHUK, N. V., Optimizing hole shape in plates working in bending, *Mech. of Solids*, **12**, 3, 72–78, 1977; *AMR*, 1979, **4**, 2790.

BANICHUK, N. V., and KARIHALOO, B. L., On the solution of optimization problems with singularities, *Int. J. Solids and Struct.*, **13**, 8, 724–733, 1977; *AMR*, 1978, **5**, 3924.

BANICHUK, N. V., KARVELISHVILI, V. M., and MIRONOV, A. A., Numerical solution of two-dimensional optimization problem for elastic plates, *Mech. of Solids*, **12**, 1, 65–74, 1977; *AMR*, 1979, **1**, 354.

BHAVIKATTI, S. S., and RAMAKRISHNAN, C. V., Optimum design of fillets in flat and round tension bars, *ASME Paper*, **77-DET-45**, 1977.

BELOV, V., Application of gradient methods to the optimum design of the elements of load-bearing structures (in Russian), *Uch. Zap. Tsentr. Aero-Gidrodinam. Inst.*, **8**, 1, 1977.

BŁACHUT, J., Optimum design of a flexible bar by means of dynamic programming (in Polish), *Mech. Teoret. i Stos.*, **15**, 1, 125–130, 1977; *AMR*, 1978, **10**, 8586.

BORKOWSKI, A., Optimization of slab reinforcement by linear programming, *Comp. Meth. in Appl. Mech. and Engng.*, **12**, **1**, 1–17, 1977; *AMR*, 1979, **2**, 1199.

BRANDT, G. D., Direct feasible and optimal design of laterally unsupported beams, *Engng. J.*, **14**, 2, 78–84, 1977; *AMR*, 1979, **2**, 1200.

CALAFELL, D. O., and WILLMERT, K. D., Automated resizing optimization of generally loaded planar frames via linear programming, *Proc. Symp. on Applications of computer meths. in Eng., School of Eng.*, Univ. of Southern California, Los Angeles, August 1977.

CARMICHAEL, D., and GOH, B. S., Optimal vibrating plates and a distributed parameter singular control problem, *Int. J. of Control*, **26**, 1, 19–31, 1977; *AMR*, 1979, **2**, 1202.

CARPENTER, W. C., and SMITH, E. A., Computational efficiency of nonlinear programming methods on a class of structural problems, *Int. J. Num. Meths. in Eng.*, **11**, 8, 1203–1223, 1977.

CHOU, T., Optimum reinforced concrete T-beam section, *Proc. ASCE* **103**, *J. Struct. Div.*, *ST* 8, 1605–1617, 1977.

CHANG, D. C., and BARONE, M. R., Structural optimization in panel design, *SAE Trans.*, **86**, 3, 2273–2281, 1977; *AMR*, 1979, **5**, 3822.

CINQUINI, C., LAMBLIN, D., and GUERLEMENT, G., Variational formulation of the optimal plastic design of circular plates, *Comp. Meth. in Appl. Mech. and Engng.*, **11**, 1, 19–30, 1977; *AMR*, 1977, **12**, 9981.

DAVIDSON, J. W., FELTON, L. P., and HART, G. C., Reliability-based optimization for dynamic loads, *Proc. ASCE* **103**, *J. Struct. Div.*, *ST* 10, 2021-2035, 1977; *AMR*, 1979, **3**, 2114.

DAVIDSON, J. W., FELTON, L. P., and HART, G. C., Optimum design of structures with random parameters, *Comp. and Struct.*, **7**, 3, 481–486, 1977; *AMR*, 1978, **10**, 8582.

DEKHTYAR, A. S., Optimization of rigidly plastic shells of revolution, *Soviet Appl. Mech.*, **13**, 5, 473–477, 1977.

DZYUBA, A., On a certain method of solving optimum design problems (in Russian), *Soprotivl. Materialov i Teoria Sooruzh. Resp. Mezhved. Nauch.-Tekhn.*, 1977.

DREWNOWSKI, S., and MARKS, W., Optimization of prestressing and post-compression of beams (in Polish), *Arch. Inż. Ląd.*, **23**, 4, 593–608, 1977.

DISTEFANO, N., and BELLOWS, G., Application of coupled dynamic programming to two-point boundary-value systems, *J. Opt. Theory and Appl.*, **23**, 1, 1977.

DUDA, G., Optimization of space lattice domes (in Polish), *23 Konf. PZITB*, Krynica 1977.

ERBATUR, F., and MENGI, 1., Optimal design of plates under the influence of dead weight, and surface loading, *J. Struct. Mech.*, **5**, 4, 345–356; *AMR*, 1978, **10**, 8585.

ERBATUR, F., and MENGI, Y., On the optimal design of plates for a given deflection, *J. Opt. Theory and Appl.*, **21**, 1, 103-110, 1977; *AMR*, 1977, **9**, 8574.

FARKAS, J., Optimum design of compressed column of constant welded square box cross-section considering the effect of residual welding stresses, *Acta Tech.*, **84**, 3/4, 335–348, 1977; *AMR*, 1979, **2**, 1046.

FELTON, L. P., and DOBBS, M. W., On optimized prestressed trusses, *AIAA J.*, **15**, 7, 1037–1039, 1977; *AMR*, 1979, **3**, 2133.

FENG, T. T., ARORA, J. S., and HAUG, E. J., JR., Optimal structural design under dynamic loads, *Int. J. for Num. Meth. in Engng.*, **11**, 1, 39–52, 1977; *AMR*, 1978, **3**, 2052,

FOLEY, M., and CITRON, S. J., A simple technique for the minimum mass design of continuous structural members, *J. Appl. Mech. Trans. ASME*, E **44**, 2, 285–290, 1977; *AMR*, 1977, **12**, 10206.

FRANGOPOL, D., and RONDAL, J., Optimum probability-based design of plastic structures, *Engng. Opt.*, **3**, 1, 17–25, 1977; *AMR*, 1978, **2**, 1210.

GAWECKI, A., GAWECKI, M., GARSTECKI, A., and RAKOWSKI, J., Engineering optimization of elastic frames with varying cross-sections (in Polish), *III Konf. "Metody komputerowe w mechanice konstrukcji"*, **1**, Opole 1977.

GOLDSHTEIN, J., Design of optimal structures with statically admissible forces due to multiple loading (in Russian), *Izv. Vyssh. Ucheb. Zaved. Stroit. i Arkhit.*, **6**, 1977.

GORDEEV, V. N., and GRINBERG, M. L., Selection of optimal parameters for lattice roofs (in Russian), *Stroit. Mekh. i Raschet Sooruzh.*, **6**, 12–18, 1977; *AMR*, 1979, **6**, 4644.

GRIERSON, D. E., Second-order optimal plastic design of steel frames, *Proc. Conf. on Mech. in Engng.* 1976, Univ. Waterloo Press, 223–249, 1977; *AMR*, 1979, **9**, 7305.

GRINEV, V. B., and FILIPPOV, A. P., Optimum circular plates, *Mech. of Solids*, **12**, **1**, 122–128, 1977; *AMR*, 1979, **2**, 1203.

GRINEV, V. B., and FILIPPOV, A. P., Optimization of circular plates in stability problems (in Russian), *Stroit, Mekh. i Raschet Sooruzh.*, **2**, 16–21, Moskva 1977.

GURA, N. M., and SEIRANYAN, A. P., Optimum circular plate with constraints on the rigidity and frequency of natural oscillations, *Mech. of Solids*, **12**, 1, 129–136, 1977; *AMR*, 1978, **5**, 3925.

GUREVICH, I. B., *et al.*, Weight optimization of eccentrically reinforced cylindrical shells (in Russian), *Prikl. Mekh.*, **13**, **7**, 113–116, 1977 (also in *Sov. Appl. Mech.*, **13**, 7, 724–726, 1978); *AMR*, 1978, 11, 9470.

HAUG, E. J., ARORA, J. S., and FENG, T. T., Sensitivity analysis and optimization of structures for dynamic response, *J. Mech. Design, Trans. ASME*, **100**, 2, 311–318, 1977; *AMR* 1979, **2**, 1197.

HARTMAN, D., Optimization of shallow hyperbolic paraboloidal shells (in German), *Beton und Stahlbet.*, 72, 9, 216–222, 1977; *AMR*, 1979, **11**, 9589.

HILL, R. D., and ROZVANY, G. I. N., Optimal beam layouts: the free edge paradox, *J. Appl. Mech.*, **44**, 4, 696–700, 1977; *AMR* 1978, **7**, 5742.

JENDO, S., On certain problems of optimum design of cable nets, *Euromech 69, Int. Coll. on Large Elastic Deformations of Discrete Systems*, Balatonszemes, Hungary, 1977.

KHALIFA, M. M. K., and MERWIN, J. E., Optimum plastic design of space frames, *Instn. Civ. Engrs., Proc.*, **63**, 769–783, 1977; *AMR*, 1979, **5**, 3821.

KHOT, N. S., Computer program (OPTCOMP) for optimization of composite structures for minimum weight design, *Tech. Rep.* **AFFDL-TR-76-149**, 58 pp., 1977; *AMR*, 1977, **9**, 7726.

KHOT, N. S., VENKAYYA, V. B., *et al.*, Experiences with minimum weight design of structures using optimality criteria methods, *SAE Trans.*, **86**, **3**, 2244–2254, 1977; *AMR*, 1979, **4**, 2975.

KIRSCH, U., and MOSES, F., The relationship between plastic and prestressed elastic optimal design, *Univ. Waterloo Press*, 207–222, 1977; *AMR*, 1979, **6**, 4627.

KIRSCH, U., and MOSES, F., Optimization of structures with control forces and displacements, *Engng. Opt.*, **3**, 1, 37–44, 1977 *AMR*, 1978, **6**, 4795.

KISELEV, V., Optimization of a thin-walled rod subjected to forces and moments applied to its ends (in Russian), *Mater. II Nauch. Konf. Molodykh Uchen. Mekh.-Mat. Fak.*, Gorkov 1977.

KORDAS, Z., and SKRABA, W., Shaping of thick-walled cylinders subject to internal pressure and bending under the condition of full plasticization at the state of collapse (in Polish), *Rozpr. Inż.*, **25**, 1, 37–52, 1977; *AMR*, 1979, **10**, 8125.

LEIMBACH, K. R., and LIEKWEG, H., Simple weight-optimization scheme for steel frames, *Proc. ASCE* **103**, *ST* 12, *J. Struct. Div.*, 1977.

LELLEP, YA. A., Optimal design of beams under conditions of steady-state creep, *Mech. of Solids*, **12**, 1, 193–197, 1977; *AMR*, 1979, **3**, 2115.

LEPIK, U., and MRÓZ, Z., Optimal design of plastic structures under impulsive and dynamic pressure loading, *Int. J. Solids and Struct.*, **13**, 7, 657–674, 1977; *AMR*, 1979, **1**, 351.

LEŚNIAK, Z., System optimization by the decomposition method (in Polish), *III Konf. "Metody komputerowe w mechanice konstrukcji"*, **2**, Opole 1977.

LEŚNIAK, Z. K., *OSY—Optimization system for steel hall structures* (in Polish), Arkady, Warszawa 1977.

LEV, O. E., A structural optimization solution to a branch-and-bound problem, *Quart. Appl. Math.*, **34**, 4, 365–371; *AMR*, 1977, **12**, 10208.

LEV, O. E., Optimum choice of determinate trusses under multiple loads, *Proc. ASCE*, **103**, *ST* 5, *J. Struct. Div.*, 1977; *AMR*, 1978, **2**, 1212.

LIFSHITS, V., Optimization of composite statically indeterminate systems (in Russian), *Rukopis den. TsNIIT Strojmash*, 3 marta, 79, 1977.

LIPSON, S. L., and GWIN, L. B., The Complex Method applied to optimal truss configuration, *Comp. Struct.*, **7**, 3, 1977.

LIPSON, S. L., and GWIN, L. B., Discrete sizing of trusses for optimal geometry, *Proc. ASCE*, **103**, *ST* 5, *J. Struct. Div.*, 1031–1046, 1977; *AMR*, 1978, **7**, 5743.

LUKYANTSEVA, V., Optimization of statically indeterminate beam systems by the method of uniform search (in Russian), *Issl. po Stroit. Konstr. i Stroit. Mekh.*, Tomsk 1977.

MAJID, K. I., and SAKA, M. P., Optimum shape design of rigidly jointed frames, *Proc. Symp. on Applications of computer meths. in Eng., School of Eng.*, Univ. of Southern California, Los Angeles, August 1977.

MAKOWSKI, M., and SZEFER, G., Optimum design of a beam on elastic foundation with constraints on normal stresses (in Polish), *Mech. Teor. i Stos.*, **15**, 3, 1977.

MAQUOI, R., and RONDAL, J., Optimal layout of cables in prestressed indeterminate bridges, *Rev. Roum. des Sci. Appl.*, **22**, 3, 447–463, 1977; *AMR*, 1979, **9**, 7307.

MARKS, W., Prestressed and prestressed post-compressed beams of minimal cross-section (in Polish), *Prace IPPT*, 75, 1977.

MARKS, N., Optimization of the cross-sections of prestressed and post-compressed beams (in Polish), *Prace IPPT*, *41*, 1977.

MAXWELL, R., and JOHNSON, C., Optimization of structural design (analysis) testing, *Symposium Philadelphia* 1977.

MCKEOWN, J. J., Optimal composite structures by deflection-variable programming, *Comp. Mech. in Appl. Mech. and Engng.*, **12**, 2, 155–199, 1977; *AMR*, 1978, **6**, 4794.

MIKHAJLISHCHEV, V., Optimization of metal trusses (in Russian), *Soprot. Mater. i Teoria Sooruzh., Resp. Medvezhd. Nauch.-Tekhn.*, 1977.

MORLEY, C. T. and GULVANESSIAN, H., Optimum reinforcement of concrete slab elements, *Instn. Civ. Engrs. Proc.*, **63**, 441–454, 1977; *AMR*, 1979, **6**, 4628.

MOSES, F., Structural system reliability and optimization, *Comp. and Struct.*, **7**, 2, 283–290, 1977; *AMR*, 1978, 1, 388,

NA, T. Y., and KUAJIAN, G. M., On optimal arch design, *J. Engng. for Industry, Trans. ASME*, **B 99**, 1, 37–40, 1977; *AMR*, 1977, **8**, 6808.

NARUSBERG, V. L., RIKARDS, R. B., and TETERS, G. A., Optimization of reinforced cylindrical shells with nonuniform thickness, Polymer Mech., **12**, 2, 257–262, 1977; *AMR*, 1978, **4**, 2943.

ODA, J., and YAMAZAKI, K., On a technique to obtain an optimum strength shape of an axisymmetric body by the finite element method, *Bull. JSME*, **20**, 150, 1524–1532, 1977; *AMR*, 1979, **1**, 348.

OLHOFF, N., and RASMUSSEN, S. H., On single and bimodal optimum buckling loads of clamped columns, *Int. J. Solids and Struct.*, *13*, *7*, 605–614, 1977; *AMR*, 1978, *5*, 3759.

ORANGUN, C. O., and AKEJU, T. A. I., A study of some characteristics of optimum design of composite bridge deck, *Materials and Structures*, **10**, 55, 39–44, 1977; *AMR*, 1979, **2**, 1214.

OSIŃSKI, Z., and WRÓBEL, J., *Theory of Structures* (in Polish), Wyd. Polit. Warszawskiej, Warszawa 1977.

PARBERY, R. D. and KARIHALOO, B. L., Minimum-weight design of hollow cylinders for given lower bounds on torsional and flexural rigidities, *Int. J. Solids and Struct.*, **13**, 1271–1280, 1977; *AMR*, 1978, 8, 6728.

PIERSON, B. L., An optimal control approach to minimum-weight vibrating beam design, *J. Struct. Mech.*, **5**, 2, 147–178, 1977; *AMR*, 1978, **8**, 6596.

POPELAR, C. H., Optimal design of structures against buckling: A complementary energy approach, *J. Struct. Mech.*, **5**, 1, 45–66, 1977; *AMR*, 1977, **11**, 9380.

POWELL, G. H., and RAMAKRISHNA, L. V., Practical automated design of continuous prestressed concrete bridges, *Proc. Symp. on Applications of computer methods in Eng.*, *School of Eng.*, Univ. of Southern California, Los Angeles, August 1977.

PRAGER, W., and ROZVANY, G. I. N., Optimal layout of grillages, *J. of Struct. Mech.*, **5**, 1, 1–18, 1977; *AMR*, 1979, **2**, 1201.

RAMMERSTORFER, F. G., Increase of the first natural frequency and buckling load of plates by optimal fields in initial stresses, *Acta Mechanica*, **27**, 217–238, Springer-Verlag 1977.

ROZVANY, G. I. N., Optimal plastic design: Allowance for self-weight, *Proc. ASCE* **103**, *J. Eng. Mech. Div. EM* **6**, 1165–1170, 1977; *AMR*, 1979, 6, 4626.

ROZVANY, G. I. N., and MRÓZ, Z., Column design: optimization of support conditions and segmentation, *J. Struct. Mech.*, **5**, 3, 279–290, 1977; *AMR*, 1978, 7, 5744.

SCHMIT, L. A., JR., and FARSHI, B., Optimum design of laminated fibre composite plates, *Int. J. for Num. Meth. in Eng.*, **11**, 4, 623–640, 1977; *AMR*, 1978, **3**, 2051.

SCHULDT, S., Application of a new penalty function to design optimization, *J. Eng. Industry Trans. of the ASME Series* **B 99**, 1, 1977.

SEGENREICH, S. A., ZOUAIN, N. A., and HERSKOVITS, J., An optimality criteria method based on slack variables concept for large scale structural optimization, *Symp. Appl. Comp. Meths. in Engm. Univ. of Southern Calif.*, Los Angeles 1977.

SEIRANYAN, A. P., Quasioptimum solutions of optimum design problems with various constraints (in Russian), *Prikl. Mekh.*, **13**, 6, 18–26, 1977.

SELLERI, F., and SPADACCINI, O., Optimal design of prestressed plane cable structures, *J. Struct. Mech.*, **5**, 2, 179–205, 1977; *AMR*, 1978, **2**, 1213.

SHAMIEV, F. G., and GASANOVA, D. D., On the optimal design of plates supported by an incompressible fluid (in Russian), *Izv. AN Azerb. SSR*, **3**, 25–32, 1977.

SHAPOSHNIKOV, N., and KUZIN, K., Some problems of the optimization of statically indeterminate trusses (in Russian), Tr. Mosk. Inst. Inzh. Zh-d. Transp., 1977.

SIMITSES, G. J., and ASWANI, M., Minimum-weight design of stiffened cylinders under hydrostatic pressure, *J. Ship Res.*, **21**, 4, 217–224, 1977; *AMR*, 1979, **1**, 349.

SIMITSES, G. J., and GIRI, J., Minimum weight design of stiffened cylinders subjected to pure torsion, *Comp. and Struct.*, **7**, 5, 667–677, 1977; *AMR*, 1979, **3**, 2113.

SIMITSES, G. J., and SHEINMAN, I., Minimum-weight design of stiffened cylindrical panels under combined loads, *J. Aircraft*, **14**, 5, 419–420, 1977.

SPILLERS, W. R., Iterative structural design, *Solid Mech. Arch.*, **2**, 4, 369–401, 1977; *AMR* 1979, **2**, 1198.

STADLER, W., Uniform shallow arches of minimum weight and minimum-maximum deflection, *J. Opt. Theory and Appl.*, **23**, 1, 137–165, 1977; *AMR*, 1979, **7**, 5551.

SURTEES, J. D., and TORDOFF, D., Optimum design of composite box girder bridge structures, *Inst. Civ. Engrs., Proc., Part 2, Res. and Theory*, **63**, 181–194, 1977; *AMR*, 1978, **10**, 8584.

TAYLOR, J. E., Optimal truss design based on an algorithm using optimality criteria, *Int. J. Solids and Struct.*, **13**, 10, 1977.

TAYLOR, J. E., and ROSSOW, M. P., Optimal truss design based on an algorithm using optimality criteria, *Int. J. Solids and Struct.*, **13**, 10, 913–923, 1977; *AMR*, 1978, **4**, 2944.

THOMAS, H. R., JR., and BROWN, D. M., Optimum least-cost design of a truss roof system, *Comp. and Struct.*, **7**, 1, 13–22, 1977; *AMR*, 1979, **1**, 353.

TOAKLEY, A. R., and WILLIAMS, D. G., The optimum design of stiffened panels subject to compression loading, *Eng. Opt.*, **2**, 4, 239–250, 1977; *AMR*, 1978, 8, 6727.

VAN KEUREN, G. M., JR., and EASTEP, F. E., Use of Galerkin's method for minimum-weight panels with dynamic constraints, *J. Spacecraft and Rockets*, **14**, 7, 414–418, 1977; *AMR*, 1978, **3**, 2050.

VITIELLO, E., Standardization and optimum structural design by dynamic programming, *J. Opt. Theory and Appl.*, **23**, 1, 183–191, 1977; *AMR*, 1978, **8**, 6726.

WILBY, C. A., Optimization of design of circular tanks, *Instn. Civ. Engrs., Proc.* **63**, 921–924, 1977; *AMR*, 1979, **7**, 5550.

YAMAKAWA, H., and OKUMURA, A., Optimum design of structures with regard to their vibrational characteristics, *2nd and 3rd Report, Bull. JSME*, **20**, 141, 292–306, 1977; *AMR*, 1979, **9**, 7303 and 7304.

ZHURAJEV, T., and FROLOV, V., On a method of solving multi-criteria optimization problems for load-bearing structures (in Russian), *Uch. Zap. Tsentr. Aerogidrodinam. Inst.*, **8**, 2, 1977.

1978

ADAMOVICH, I. S., and RIKARDS, R. B., Optimization of the mass of rotation shells with a variable geometry and reinforcing structure, *Polymer Mech.*, **13**, 3, 421–428 and 4, 566–571, 1978.

ADIDAM, S. R. S., *et al.*, A utility function model for optimal material choice for multifunctional performance, *Comp. and Struct.*, **8**, 5, 583–587, 1978; *AMR*, 1979, **8**, 6360.

AMAZIGO, J. C., Optimal shape of shallow circular arches against snap-buckling, *J. Appl. Mech.*, **45**, 3, 591–594, 1978; *AMR*, 1979, **10**, 8144.

ARMAND, J. L., and LODIER, B., Optimal design of bending elements, *Int. J. Num. Meth. Engng.*, **13**, 373–384, 1978.

AZAD, A. K., Economic design of homogeneous I-beams, *Proc. ASCE* 104, *J. Struct. Div. ST* 4, 637–648, 1978; *AMR* 1979, **5**, 3824.

BENEDETTI, D., and VITIELLO, E., Optimal aseismic structural standards for the replacement of existing buildings, *Engng. Optimization*, **3**, 221–227, 1978.

BOISSERIE, J. M., and GŁOWINSKI, R., Optimization of the thickness law for thin axisymmetric shells, *Comp. and Struct.*, **8**, 3/4, 331–343, 1978; *AMR*, 1979, **8**, 6357.

BURY, K. V., Reliability-constrained optimum static design for random load sequences, *Engng. Opt.*, **3**, 4, 215–220, 1978; *AMR*, 1979, **7**, 5554.

CARMICHAEL, D. G., Probabilistic design of structures in state equation form, *Engng. Opt.*, **3**, 83–92, 1978.

CHAN, A.S.L., and TURLES, E., An approximate method for structural optimization, *Comp. and Struct.*, **8**, 3/4, 357–363, 1978; *AMR*, 1979, **9**, 7300.

CHERKAEV, A. V., On the question of formulating the problem of optimal design of freely oscillating structures, *PMM* **42**, 1, 185–188, 1978.

CHUN, Y. W., and HAUG, E. J., Two-dimensional shape optimal design, *Int. J. Num. Meth. Engng.*, **13**, 311–366, 1978.

COURBON, J., Optimization of hyperstatic structures (in French), *Ann. de l'ITBTP*, 362, 1–32, 1978; *AMR*, 1979, **9**, 7299,

FLEURY, C., and GERADIN, M., Optimality criteria and mathematical programming in structural weight optimization, *Comp. and Struct.*, 8, 1, 7–17, 1978; *AMR*, 1979, **7**, 5549.

FRANGOPOL, D., and RONDAL, *J.*, Le dimensionnement probabiliste optimal des structures, *Ann. de l'ITBTP*, **363**, 21–30, 1978; *AMR*, 1979, **10**, 8366.

FU, K. C., and LEVEY, G. E., Discrete frame optimization by complex-simplex procedure, *Comp. and Struct.*, **9**, 2, 207–217, 1978; *AMR*, 1979, **10**, 8360.

GARSTECKI, A., and GAWĘCKI, A., Experimental study on optimal plastic rings in the range of large displacements, *Int. J. Mech. Sci.*, **20**, 12, 823–832, 1978; *AMR*, 1979, **10**, 8143.

HARTMANN, D., A generally applicable structural optimization method and its application to the design of cylindrical shells, *Engng. Opt.* **3**, 4, 201–213, 1978; *AMR*, 1979, **11**, 9584.

HAUG, E. J., JR., and FENG, T. T., Optimal design of dynamically loaded continuous structures, *Int. J. Num Meth. Engng.*, **12**, 2, 299–317, 1978; *AMR*, 1979, **11**, 9587.

GIBCZYŃSKA, T., General analysis of the optimum design of a box beam subject to bending (in Polish), *Arch. Bud. Masz.*, **25**, 2, 325–339, 1978; *AMR*, 1979, **11**, 9591.

HORNBUCKLE, J. C., and BOYKIN, W. H., JR., Equivalence of a constrained minimum weight and maximum column buckling load problem with solution, J. *Appl. Mech.*, **45**, 1, 159–165, 1978; *AMR*, 1978, **10**, 8580.

HOWELL, G. C., and DOYLE, W. S., Dynamic programming and direct iteration for the optimum design of skeletal towers, *Comp. and Struct.*, **9**, 6, 621–627, 1978, *AMR* 1979, **11**, 9583.

JACKIEWICZ, M., *Multiparameter optimization of a selected class of bar structures—a particular case of application of network theory* (in Polish), Doct. Thesis, *Inst. Bud. Pol. Wrocł.*, I-2/K-89/78, Wrocław 1978.

KESHAVA RAO, M. N., Optimization of variable thickness plates under multiloads, *J. Struct. Engng.*, **6**, 2, 87–92, 1978; *AMR*, 1979, **8**, 6356.

KHAN, A. R., Optimization of trusses, *J. Struct. Engng.*, 6, 2, 78–86, 1978; *AMR*, 1979, **5**, 3823.

KRUŻELECKI, J., Problems of optimal design of axially symmetrical shells in membrane state, *Bulg. Acad. Sci., Theor. and Appl. Mech.*, **IX**, 2, 91–95, 1978.

KUMAR, P., Optimal force transmission in reinforced concrete deep beams, *Comp. and Struct.*, **8**, 2, 223–229, 1978; *AMR*, 1979, **7**, 5552.

LAMBLIN, D., *et al.*, Application of linear programming to the optimal plastic design of circular plates subject to technological constraints, *Comp. Meth. in Appl. Mech. and Engng.*, **13**, 2, 223–243, 1978; *AMR*, 1978, **11**, 9468.

LEPIK, U., Optimal design of beams with minimum compliance, *Int. J. Non-Lin. Mech.*, **13**, 1, 33–42, 1978; *AMR*, 1979, **3**, 2112,

LEPIK, U., and MRÓZ, Z., Optimal design of impulsively loaded plastic beams for asymmetric mode motions, *Int. J. Solids and Struct.*, **14**, 10, 841–850, 1978; *AMR* 1979, **9**, 7302.

MARKS, W., Optimization of prestressed flexural elements (in Polish), *Prace IPPT*, **52**, 1978.

MASUR, E. F., Optimal design of symmetric structures against postbuckling collapse, *Int. J. Solids and Struct.*, **14**, 4, 319–326, 1978; *AMR*, 1979, **1**, 173.

MCINTOSH, S. C., JR., On the optimization of discrete structures with aeroelastic constraints, *Comp. and Struct.*, 8, 411–419, 1978.

MORRIS, A. J., On condensed geometric programming in structural optimization, *Comp. Meth. Appl. Mech. Engng.*, **15**, 2, 139–148, 1978; *AMR*, 1979, **7**, 5548.

NARAYANAN, S. and NIGAM, N. C., Optimum structural design in random vibration environment, *Engng. Opt.*, **3**, 2, 97–108, 1978; *AMR*, 1979, **8**, 6358.

PARIMI, S. R., and COHN, M. Z., Optimal solutions in probabilistic structural design, *J. de Méc. Appl.*, **2**, 1, 47–92, 1978; *AMR*, 1979, **11**, 9594 and 9595.

NIEMIERKO, A., Optimization of trusses by mathematical programming methods (in Polish), *Prace IPPT*, **2**, 1978.

POCHTMAN, YU. M. and PYATIGORSKY, Z. I., The design of bending plates of minimal weight, optimal in shakedown (in Russian), *Izv. AN Arm. SSR*, **3**, 1978.

RAO, S. S., Optimum design of bridge girders for electric overhead traveling cranes, *J. Engng. for Industry, Trans. ASME* **100**, 3, 375–382, 1978; *AMR*, 1979, **7**, 5609.

RAO, S. S., *Optimization, Theory and Applications*, Wiley Eastern Ltd., New Delhi 1978.

ROZVANY, G. I. N., Optimal elastic design for stress constraints, *Comp. and Struct.*, **8**, 3/4, 455–463, 1978; *AMR*, 1979, **11**, 9586.

ROZVANY, G. I. N., and HILD, R. D., Optimal plastic design: superposition principles and bounds on the minimum cost, *Comp. Meth. in Appl. Mech. and Engng.*, **13**, 2, 151–173, 1978; *AMR*, 1978, **11**, 9467.

SCHMIT, L. A., JR., and RAMANATHAN, R. K., Multilevel approach to minimum weight design including buckling constraints, *AIAA J.*, **16**, 2, 97–104, 1978; *AMR*, 1979, **1**, 347.

THIERAUF, G., A method for optimal limit design of structures with alternative loads, *Comp. Meth. in Appl. Mech. and Engng.*, **16**, 135–149, 1978.

TRIKHA, D. N., and MURTHY, N. S., Semi-automatic optimization approach for reinforced concrete frame spacing by limit analysis, *J. Struct. Engng.*, **6**, 2, 96–101, 1978; *AMR*, 1979, **8**, 6359.

UMETANI, Y., and HIRAI, S., Shape optimization for beams subject to displacement restrictions on the basis of the growing-reforming procedure, *Bull. JSME* **21**, 1113–1119, 1978.

VACHAJITPAN, P., and ROCKEY, K. C., Design method for optimum unstiffened girders, *Proc. ASCE* **104**, *J. Struct. Div. ST* 1, 141–155, 1978; *AMR*, 1979, **4**, 2973.

VAVRICK, D. J., and WARNER, W. H., Minimum mass design with torsional frequency and thickness constraints, *J. Struct. Mech.*, **6**, 2, 211–213, 1978; *AMR*, 1979, **11**, 9588.

VAVRICK, D. J., and WARNER, W. H., Duality among optimal design problems for torsional vibration, *J. Struct. Mech.*, **6**, 2, 233–241, 1978; *AMR*, 1979, **10**, 8364.

YANG, W. H., On a class of optimization problems for framed structures, *Comp. Meth. in Appl. Mech. and Engng.*, **15**, 1, 85–97, 1978; *AMR*, 1979, **8**, 6355.

YASSERI, S. F., Optimal design of slabs on a plastic foundation, *Int. J. Mech. Sci.*, **20**, 6, 327–333, 1978; *AMR*, 1979, **9**, 7301.

WOJDANOWSKA, R., Optimal design of weakly curved compressed bars with Maxwell-type creep effects, *Arch. Mech.*, **30**, 845–851, 1978.

ŻYCZKOWSKI, M., and RYSZ, M. Optimal design of a thin-walled pipe-line cross-section in creep conditions, *Proc. Symp. on Mech. in Inelastic Struct.*, Warszawa 1978.

1979

DURELLI, A. J., and RAJAIAH, K., Optimum hole shapes in finite plates under uniaxial load, *J. Appl. Mech.*, **46**, 691–695, 1979.

FLEURY, C., A unified approach to structural weight minimization, *Comp. Meths. in Appl. Mech. and Engng.*, **20**, 17–38, 1979

FUCHS, M. G., and BRULL, M. A., A new strain energy theorem and its use in the optimum design of continuous beams, *Comp. and Struct.*, **10**, 4, 647–657, 1979; *AMR*, 1979, 9, 7298.

HAUG, E. J., and ARORA, J. S., *Applied optimal design*, Wiley-Interscience, New York 1979.

HILL, R. D., ROZVANY, G. I. N., WANG, C. M., and LEONG, K. H., Optimization, spanning capacity and cost sensitivity of fully stressed arches, *J. Struct. Mech.*, **7**, 375–410, 1979.

HIRANO, Y., Optimum design of laminated plates under axial compression, *AIAA Journal*, **17**, 9, 1017–1019, 1979.

JOURON, C., On some structural design problems, *Bull. Soc. Math. France, Mém.* **60**, 87–93, 1979.

KARIHALOO, B. L., and WOOD, G. L., Optimal design of multipurpose sandwich tie-columns, *Proc. ASCE*, **105**, *EM* 3, *J. Engng. Mech. Div.*, 465–469, 1979.

KHOT, N. S., BERKE, L., *et al.*, Comparison of optimality criteria algorithms for minimum weight design of structures, *AIAA J.*, **10**, 2, 182–190, 1979; *AMR.* 1979, **10**, 8357.

KRISHNAMOORTHY, C. S., and MOSI, D. R., A survey on optimal design of engineering structural systems, *Engng. Opt.*, **4**, 1979.

MARKS, W., Optimization of prestressed and postcompressed beams (in Polish), *Rozpr. Inż.*, **27**, 2, 245–262, 1979.

MIDDLETON, J., Optimal design of torispherical pressure vessel end closures, *Eng. Opt.*, 4, 3, 129–138, 1979.

ROZVANY, G. I. N., and PRAGER, W., A new class of optimization problems: optimal archgrids, *Comp. Meth. Appl. Mech. Engng.*, **19**, 49–58, 1979.

YAMAKAWA, H., Optimum design of structures with regard to their vibrational characteristics, *Bull. JSME*, **19**, 1976; **20**, 1977; **21**, 1978; **22**, 1979.

YOO, C. H., Optimization of triangular laced truss columns with tubular compressed members for space application, *AIAA J.* **17**, 8, 921–924, 1979.

1980

BANICHUK, N. V., *Optimization of Shape of Elastic Bodies*, Nauka, Moscow 1980.

BANICHUK, N. V., Perturbation methods for structural optimization, *Conf. "Optymalizacja wytrzymałościowa konstrukcji", PAN, Jabłonna, 17–22 Nov. 1980.*

BŁACHUT, J., and GAJEWSKI, A., Unimodal and bimodal optimization of vibrating bars and arches liable to stability loss (in Polish), *Conf. "Optymalizacja wytrzymałościowa konstrukcji", PAN, Jabłonna, 19–80 Nov. 1980.*

BORKOWSKI, A., Analysis and optimization of structures within the framework of limit load theory (in Polish), *Conf. "Optymalizacja wytrzymałościowa konstrukcji", PAN, Jabłonna 17–22 Nov. 1980.*

CHENG, K. T., and OLHOFF, N., An investigation concerning optimal design of solid plates, DCAMM Report 174, 1980.

DÖPKER, B., and THIERAUF, G., Applications of optimization techniques in computer aided design, *Conf. "Optymalizacja wytrzymałościowa konstrukcji", PAN, Jabłonna, 17–22 Nov. 1980.*

FUCHS, M. B., Linearized homogeneous constraints in structural design, *Int. J. Mech. Sci.* **22**, 33–40, 1980.

GIBCZYŃSKA, T., Problems of the optimum design of thin-walled box girder sections (in Polish), *Zesz. Nauk. Pol. Krak., Mechanika*, **4**, 1980.

GIBCZYŃSKA, T., Optimum design of a box bar section in uniform bending and torsion (in Polish), *Rozpr. Inż.*, **28**, 2, 1980.

GIBCZYŃSKA, T., Optimum design of thin-walled box girder sections for combined loads (in Polish), *Conf. "Optymalizacja wytrzymałościowa konstrukcji", PAN, Jabłonna, 17–22 Nov. 1980.*

HAUG, E. J., Shape optimal design of elastic structural elements, *Conf. "Optymalizacja wytrzymałościowa konstrukcji", PAN, Jabłonna, 17–22 Nov. 1980.*

JACKIEWICZ, M., and SIECZKOWSKI, J. M., Optimization of hinged bar structures (in Polish), *Conf. "Optymalizacja wytrzymałościowa konstrukcji", PAN, Jabłonna, 17–22 Nov. 1980.*

JENDO, S., Introduction to mathematical methods of optimization (in Polish), *Conf. "Optymalizacja wytrzymałościowa konstrukcji", PAN, Jabłonna, 17–22 Nov. 1980.*

JENDO, S., A review of iterative methods of nonlinear programming (in Polish), *Conf. "Optymalizacja wytrzymałościowa konstrukcji", PAN, Jabłonna, 17–22 Nov. 1980.*

JENDO, S., and MARKS, W., Dynamic programming (in Polish), *Conf. "Optymalizacja wytrzymałościowa konstrukcji", PAN, Jabłonna, 17–22 Nov. 1980.*

JENDO, S., and MARKS, W., Optimization of structures by the methods of stochastic programming (in Polish), *Conf. "Optymalizacja wytrzymałościowa konstrukcji", PAN, Jabłonna, 17–22 Nov. 1980.*

KRUŻELECKI, J., and ŻYCZKOWSKI, M., Optimum design of cylindrical shells under combined loads based on the concept of uniform stability (in Polish), *Conf. "Optymalizacja wytrzymałościowa konstrukcji", PAN, Jabłonna, 17–22 Nov. 1980.*

NOWACKI, H., Optimization in a computer environment, *Conf. "Optymalizacja wytrzymałościowa konstrukcji", PAN, Jabłonna, 17–22 Nov. 1980.*

OLSZOWSKI, B., and TOMANA, A., Optimum design of vibrating rod systems (in Polish), *Conf. "Optymalizacja wytrzymałościowa konstrukcji", PAN, Jabłonna, 17–22 Nov. 1980.*

PACZKOWSKI, W., Optimization of prestressed concrete span structures of highway bridges (in Polish), *Conf. "Optymalizacja wytrzymałościowa konstrukcji", PAN, Jabłonna, 17–22 Nov. 1980.*

PRAGER, W., and ROZVANY, G. I. N., Optimal coupola of uniform strength, *Ing. Arch.*, **40**, 1980.

ROZVANY, G. I. N., Layout optimization and the effect of design dependent loads, *Conf. "Optymalizacja wytrzymałościowa konstrukcji", PAN, Jabłonna, 17–22 Nov. 1980.*

ROZVANY, G. I. N., NAKAMURA, H., and KUHNELL, B. T., Optimal archgrids: allowance for selfweight, *Comp. Meth. Appl. Mech. Engng.*, **24**, 1980.

RZEGOCIŃSKA-PEŁECH, K., and WASZCZYSZYN, Z., Numerical optimization of elastic bars and annular plates for buckling (in Polish), *Conf. "Optymalizacja wytrzymałościowa konstrukcji", PAN, Jabłonna, 17–22 Nov. 1980.*

THERMANN, K., Application of Pontryagin's maximum principle to a class of optimization problems in structural design, *Conf. "Optymalizacja wytrzymałościowa konstrukcji", PAN, Jabłonna, 17–22 Nov. 1980.*

THIERAUF, G., A quadratic approximation of the structural design problem, *Conf. "Optymalizacja wytrzymałościowa konstrukcji", PAN, Jabłonna, 17–22 Nov. 1980.*

ŻYCZKOWSKI, M., and SWISTERSKI, W., Optimal structural design of flexible beams with respect to creep rupture time, in "*Structural Control*", H. H. E. LEIPHOLZ, ed., 795–810, North-Holland Publ. and SM Publ., IUTAM 1980.

Author Index

Subject Index